ENCYCLOPEDIA OF
Geography

EDITORIAL BOARD

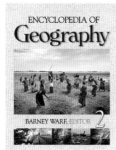
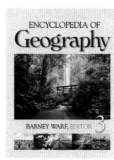
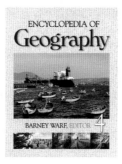
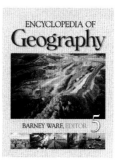
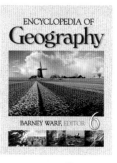

ENCYCLOPEDIA OF
Geography

BARNEY WARF, EDITOR

University of Kansas

1

Los Angeles | London | New Delhi
Singapore | Washington DC

For information:

 SAGE Publications, Inc.
2455 Teller Road
Thousand Oaks, California 91320
E-mail: order@sagepub.com

SAGE Publications Ltd.
1 Oliver's Yard
55 City Road
London, EC1Y 1SP
United Kingdom

SAGE Publications India Pvt. Ltd.
B 1/I 1 Mohan Cooperative Industrial Area
Mathura Road, New Delhi 110 044
India

SAGE Publications Asia-Pacific Pte. Ltd.
33 Pekin Street #02-01
Far East Square
Singapore 048763

Printed in Singapore.

Library of Congress Cataloging-in-Publication Data

Encyclopedia of geography, volume I - VI / edited by Barney Warf.
     6 v., p. cm.
Includes bibliographical references and index.
ISBN 978-1-4129-5697-0 (cloth)
   1. Geography—Encyclopedias. I. Warf, Barney, 1956-

G63.E554 2010
910.3—dc22   2010009453

This book is printed on acid-free paper.

10 11 12 13 14 10 9 8 7 6 5 4 3 2 1

Publisher:	Rolf A. Janke
Assistant to the Publisher:	Michele Thompson
Acquisitions Editor:	Robert Rojek
Developmental Editor:	Diana E. Axelsen
Reference Systems Manager:	Leticia Gutierrez
Reference Systems Coordinator:	Laura Notton
Production Editor:	Tracy Buyan
Copy Editors:	QuADS Prepress (P) Ltd.
Typesetter:	C&M Digitals (P) Ltd.
Proofreaders:	Penelope Sippel, Rae-Ann Goodwin, Annette Van Deusen
Indexer:	Julie Sherman Grayson
Cover Designer:	Ravi Balasuriya
Marketing Manager:	Amberlyn McKay

Volume 1 cover photo: The fjords of Sogn og Fjordane dominate this image of the western coastline of Norway. Carved out of mountains by glaciers, the flooded U-shaped valleys are surrounded by steep cliffs, verdant slopes, and snowcapped mountain tops. Courtesy of the U.S. Geological Survey.

CONTENTS

LIST OF ENTRIES

READER'S GUIDE

The Reader's Guide is designed to assist readers in finding articles on related topics. Headwords are organized into six major categories: Physical Geography; Human Geography; Nature and Society; Methods, Models, and GIS; History of Geography; and People, Organizations, and Movements. Note, however, that many topics defy easy categorization and belong to more than one grouping.

Physical Geography

Biogeography

Adaptive Radiation
Animal Geographies
Biodiversity
Biogeochemical Cycles
Biogeography
Biome: Boreal Forest
Biome: Desert
Biome: Midlatitude Deciduous
    Forest
Biome: Midlatitude Grassland
Biome: Tropical Deciduous
    Forest
Biome: Tropical Rain Forest
Biome: Tropical Savanna
Biome: Tropical Scrub
Biome: Tundra
Bioregionalism
Biosphere Reserves
Biota and Climate
Biota and Soils
Biota and Topography
Biota Migration and Dispersal
Coral Reef
Ecoregions
Ecoshed
Ecosystem Decay

Ecosystems
Ecotone
Exotic Species
Extinctions
Gaia Theory
Great American Exchange
Human-Induced Invasion of
    Species
Invasion and Succession
Island Biogeography
Keystone Species
Nutrient Cycles
Organophosphates
Peat
Permaculture
Phosphorus Cycle
Prairies
Pyrogeography
Resilience
Single Large or Several Small
    (SLOSS) Debate
Species-Area Relationship
Taphonomy
Wetlands

Climatology

Adiabatic Temperature
    Changes
Air Masses

Albedo
Anthropogenic Climate
    Change
Atmospheric Circulation
Atmospheric Composition
    and Structure
Atmospheric Energy
    Transfer
Atmospheric Moisture
Atmospheric Particulates
    Across Scales
Atmospheric Pressure
Atmospheric Variations in
    Energy
Carbonation
Carbon Cycle
Carbon Trading and Carbon
    Offsets
Chinooks/Foehns
Chlorinated Hydrocarbons
Chlorofluorocarbons
Climate: Dry
Climate: Midlatitude, Mild
Climate: Midlatitude, Severe
Climate: Mountain
Climate: Polar
Climate: Tropical Humid
Climate Change
Climate Policy

Map Visualization
Multimedia Mapping
Multivariate Mapping
Participatory Mapping
Poles, North and South
Portolan Charts
Resource Mapping
Self-Organizing Maps
Surveying
Trap Streets
Typography in Map Design
Virtual Globes
Voronoi Diagrams

GIS

Analytical Operations in GIS
Business Models for Geographic
  Information Systems
CAD Systems
Cellular Automata
Client-Server Architecture
Collaborative GIS
Conflation
Critical GIS
Database Management Systems
Database Versioning
Data Compression Methods
Data Editing
Data Format Conversion
Data Indexing
Data Querying in GIS
Digital Terrain Model
Digitizing
Distributed Computing
Dynamic and Interactive
  Displays
Enterprise GIS
Error Propagation
Exploratory Spatial Data
  Analysis
Geocoding
Geocollaboration
Geocomputation
Geographic Information Systems
Geospatial Semantic Web
GIS, Environmental Model
  Integration and
GIScience
GIS Design

GIS Implementation
GIS in Archaeology
GIS in Disaster Response
GIS in Environmental
  Management
GIS in Health Research and
  Health Care
GIS in Land Use Management
GIS in Local Government
GIS in Public Policy
GIS in Transportation
GIS in Urban Planning
GIS in Utilities
GIS in Water Management
GIS Software
GIS Web Services
Global Positioning System
Google Earth
Ground Reference Data
Heuristic Methods in Spatial
  Analysis
High-Performance Computing
Humanistic GIScience
Internet GIS
Interoperability and Spatial
  Data Standards
Legal Aspects of Geospatial
  Information
Linear Referencing and
  Dynamic Segmentation
Location-Based Services
Metadata
Mobile GIS
Neogeography
Network Analysis
Network Data Model
Ontological Foundations of
  Geographical Data
Open Geodata Standards
Open Source GIS
Privacy and Security of
  Geospatial Information
Public Participation GIS
Scale in GIS
Semantic Interoperability
Semantic Reference Systems
Spatial Data Infrastructures
Spatial Data Integration
Spatial Data Mining
Spatial Data Models

Spatial Data Structures
Spatial Decision Support
  Systems
Spatialization
Supervised Classification
Temporal GIS
Temporal Resolution
Terrain Analysis
Three-Dimensional Data
  Models
Topological Relationships
Triangulated Irregular
  Network (TIN) Data
  Model
Unsupervised Classification
Usability of Geospatial
  Information
Vagueness in Spatial Data
Vectorization
Viewshed Analysis
Virtual and Immersive
  Environments
Web Geoprocessing
  Workflows
Web Service Architectures
  for GIS

Qualitative Techniques

Fieldwork in Human
  Geography
Interviewing
Participant Observation
Participatory Learning and
  Action
Participatory Rural Appraisal
Positionality
Qualitative Methods
Writing

Quantitative Models

Agent-Based Models
Bayesian Statistics in Spatial
  Analysis
Ecological Fallacy
Fieldwork in Physical
  Geography
Geographically Weighted
  Regression

Herodotus
Hettner, Alfred
Hipparchus
Hou Renzhi
Hoyt, Homer
Humboldt, Alexander von
Huntington, Ellsworth
Ibn Battuta
Ibn Khaldūn
Isard, Walter
Jackson, John Brinckerhoff
Jefferson, Thomas
Johnston, R. J.
Kant, Immanuel
Kates, Robert
Köppen, Wladimir
Kropotkin, Peter
Kuhn, Werner
Kwan, Mei-Po
Lefebvre, Henri
Lewis, Peirce
Ley, David
Lillesand, Thomas
Livingstone, David
Lösch, August
Lynch, William
MacEachren, Alan
Mackinder, Sir Halford
Magellan, Ferdinand
Mahan, Alfred Thayer
Marcus, Melvin G.
Mark, David M.
Marsh, George Perkins
Massey, Doreen
Mather, John Russell
Maury, Matthew Fontaine
McKnight, Tom L.
Meinig, Donald
Mercator, Gerardus
Miller, Harvey J.
Mitchell, Don
Monmonier, Mark
Morrill, Richard
Morse, Jedediah
Nyerges, Timothy
Okabe, Atsuyuki
Olsson, Gunnar
Ortelius
Parsons, James
Peet, Richard

Penck, Walther
Peuquet, Donna
Pickles, John
Powell, John Wesley
Pred, Allan
Ptolemy
Raisz, Erwin
Ratzel, Friedrich
Reclus, Élisée
Relph, Edward
Ritter, Carl
Rose, Gillian
Sauer, Carl
Schaefer, Fred
Scott, Allen
Semple, Ellen Churchill
Smith, Neil
Soja, Edward
Storper, Michael
Strabo
Strahler, Arthur
Sui, Daniel
Taylor, Griffith
Taylor, Peter
Thales
Thornthwaite, C. Warren
Thrift, Nigel
Tobler, Waldo
Tomlinson, Roger
Trewartha, Glenn
Troll, Carl
Tuan, Yi-Fu
Turner, Billie Lee, II
Unwin, David
Vance, James
Varenius
Vidal de la Blache, Paul
Virilio, Paul
Waldseemüller, Martin
Walker, Richard
Watts, Michael
Weber, Alfred
White, Gilbert
Whittlesey, Derwent
Wilson, John
Wittfogel, Karl
Wood, Denis
Wright, Dawn
Wright, John Kirtland
Zelinsky, Wilbur

Geographical Organizations

American Geographical Society
Association of American
    Geographers
Association of Geographic
    Information Laboratories
    for Europe
Canadian Association of
    Geographers
Conference of Latin
    Americanist Geographers
Institute of British Geographers
International Geographical
    Union
National Center for
    Geographic Information
    and Analysis
National Council for
    Geographic Education
National Geographic Society
Open Geospatial Consortium
    (OGC)
Open Source Geospatial
    Foundation
Regional Science Association
    International (RSAI)
Royal Geographical Society
Russian Geographical Society
University Consortium for
    Geographic Information
    Science

*Political and Economic
Organizations*

African Union
Asia-Pacific Economic
    Cooperation
Association of Southeast
    Asian Nations
Commonwealth of
    Independent States
Council for Mutual Economic
    Assistance (COMECON)
European Union
Food and Agriculture
    Organization (FAO)
International Criminal Court
International Monetary Fund

ABOUT THE EDITOR

Barney Warf is a professor of geography at the University of Kansas. His research and teaching interests lie within the broad domain of human geography. Much of his career effort concerns economic geography, with an emphasis on producer services and telecommunications. His work straddles contemporary political economy and social theory on the one hand and traditional quantitative, empirical approaches on the other. He has studied, among other things, New York as a global city, telecommunications, offshore banking, international networks of financial and producer services, and the geographies of the Internet. He has also written on military spending, voting technologies, the U.S. Electoral College, and religious diversity in different national contexts. He has coauthored or coedited 6 books, including 2 textbooks, 35 book chapters, and 90 refereed journal articles. He has served or serves on the editorial boards of *Geographical Review*, *Annals of the Association of American Geographers*, *Growth and Change*, *The Professional Geographer*, *Urban Geography*, *International Regional Science Review*, and *Geografiska Annaler*. His articles have appeared in the *Annals*, *The Professional Geographer*, *Economic Geography*, *Urban Geography*, *Political Geography*, *Geographical Review*, *Regional Studies*, *Urban Studies*, *Tijdschrift voor Economische en Sociale Geografie*, *International Regional Science Review*, *Transactions of the Institute of British Geographers*, *Area*, *The Southeastern Geographer*, *Environment and Planning* A and D, and *Journal of Geography*. His teaching interests include urban and economic geography, the history of geography, globalization, social theory, and East Asia. This is the second encyclopedia he has edited.

CONTRIBUTORS

Scott R. Abella
*University of Nevada,
    Las Vegas*

Yohannes Aberra
Addis Ababa University

Damian Absalon
University of Silesia

Michael Adams
University of Wollongong

Fenda Aminata Akiwumi
University of South Florida

Todd H. Albert
Bowling Green State University

Jochen Albrecht
Hunter College CUNY

Josep Antoni Alcover
*Institut Mediterrani d'Estudis
    Avançats (CSIC-UIB)*

Derek H. Alderman
East Carolina University

Jared Aldstadt
*State University of New York
    at Buffalo*

David E. Alexander
University of Florence

Melinda Alexander
Arizona State University

Saleem H. Ali
University of Vermont

John All
*Human Environment Linkages
    Program*

John Logan Allen
University of Wyoming

Scott Anderson
*State University of New York
    at Cortland*

Sharolyn Anderson
University of Denver

Gennady Andrienko
Fraunhofer Institute IAIS

Natalia Andrienko
Fraunhofer Institute IAIS

Yuko Aoyama
Clark University

Seth Appiah-Opoku
University of Alabama

J. Clark Archer
University of Nebraska–Lincoln

Kevin Archer
University of South Florida

Santa Arias
University of Kansas

Richard Peter Armitage
University of Salford

Michael Christopher
    Armstrong
DePaul University

David L. Arnold
Frostburg State University

Jennifer S. Arrigo
East Carolina University

R. Rochelle Arrington
Florida State University

Yasushi Asami
University of Tokyo

Bjørn Asheim
Lund University

Walker S. Ashley
Northern Illinois University

Genevieve Atwood
University of Utah

Ivonne Audirac
Florida State University

Louis Awanyo
*Luther College, University of
    Regina*

Rikke Baastrup
Danish Cancer Society

Andrew J. Bach
*Western Washington
    University*

Sharmistha Bagchi-Sen
*State University of New York
    at Buffalo*

John E. Bailey
University of Alaska

Keiron Bailey
University of Arizona

Mark Ian Bailey
University of Newcastle upon Tyne

Robert G. Bailey
USDA Forest Service

Robert Bailis
Yale University

Earl J. Baker
Florida State University

John Richard Bale
Keele University

Shivanand Balram
Simon Fraser University

Greg J. Bamber
Griffith University

Abhijit Banerjee
Northeastern Illinois University

Aniruddha Banerjee
Indiana University–Purdue University Indianapolis

Kezia Barker
Birkbeck University of London

Justin R. Barnes
North Carolina Solar Center

Trevor J. Barnes
University of British Columbia

Clive Barnett
Open University

Keith Barney
York University

Stewart Barr
University of Exeter

Roger G. Barry
University of Colorado

Bipasha Baruah
California State University, Long Beach

S. Willis Bassett
Florida State University

Pratyusha Basu
University of South Florida

Ranu Basu
York University

Dean Bavington
Nipissing University

Jamie Baxter
University of Western Ontario

Ryan E. Baxter
Pennsylvania State University

Kate Beard-Tisdale
University of Maine

Hartmut Behr
Newcastle University

Matthew F. Bekker
Brigham Young University

Scott Bell
University of Saskatchewan

Thomas L. Bell
University of Tennessee

Karina Benessaiah
McGill University

Lawrence D. Berg
University of British Columbia

Susan Bergeron
West Virginia University

Richelle M. Bernazzoli
University of Illinois at Urbana-Champaign

Ruggero Bertani
Enel

Nicholas L. Betts
Queen's University Belfast

Kirsten M. M. Beyer
University of Iowa

William Beyers
University of Washington

Keshav Bhattarai
University of Central Missouri

Ling Bian
University at Buffalo

Jess Bier
City University of New York, Graduate Center

Phil Birge-Liberman
Syracuse University

Mark D. Bjelland
Gustavus Adolphus College

Chris Blackden
University of Kentucky

Stefano Bloch
University of Minnesota

Andrew Blohm
University of Maryland

Talja Blokland
Institut für Sozialwissenschaften

Darcy Boellstorff
Bridgewater State College

Hans-Georg Bohle
University of Bonn, Germany

Gregory S. Bohr
California Polytechnic State University, San Luis Obispo

Lowe Börjeson
Stockholm University

Saturnino M. Borras Jr.
Erasmus University Rotterdam

Keith Bosak
University of Montana

Fernando Javier Bosco
San Diego State University

Pablo Bose
University of Vermont

Brian H. Bossak
Georgia Southern University

Pere Bover
Institut Mediterrani d'Estudis Avançats (CSIC-UIB)

Dawn S. Bowen
University of Mary Washington

John Bowen
Central Washington University

Mark W. Bowen
University of Kansas

Ben Bradshaw
University of Guelph

Richard R. Brandt
Salem State College

Anthony J. Brazel
Arizona State University

Leah Bremer
San Diego State University/ University of California, Santa Barbara

Amanda DeeAnn Briney
California State University, East Bay

Ray Bromley
University at Albany, SUNY

John O. Browder
Virginia Polytechnic Institute and State University

Dwight A. Brown
University of Minnesota

Marilyn A. Brown
Georgia Institute of Technology

Michael Brown
University of Washington

Alec Brownlow
DePaul University

Joe Bryan
University of Colorado, Boulder

Karin Bryan
University of Waikato

William R. Buckingham
University of Wisconsin at Madison

Michaela Buenemann
New Mexico State University

Rebecca Burnett
University of Washington

Amy C. Burnicki
University of Wisconsin

Jeffrey Bury
University of California, Santa Cruz

Jason Byrne
Griffith University

John Byrne
University of Delaware

Maria Caffrey
University of Tennessee

David A. Call
Ball State University

Angus Cameron
University of Leicester

Harrison S. Campbell Jr.
University of North Carolina at Charlotte

Étienne Cantin
Université Laval

Huhua Cao
University of Ottawa

Andrew M. Carleton
Pennsylvania State University

David L. Carr
University of California, Santa Barbara

John I. Carruthers
U.S. Department of Housing and Urban Development

Eric D. Carter
Grinnell College

Jennifer L. Carter
University of the Sunshine Coast

Norman Carter
San Diego State University

Guillermo Castilla
University of Calgary

Noel Castree
Manchester University

Brian Ceh
Ryerson University

Ryan Centner
Tufts University

Jayajit Chakraborty
University of South Florida

Charlotte N. L. Chambers
Otago University

Kang-Tsung (Karl) Chang
Kainan University

Thomas Chapman
Old Dominion University

Igal Charney
University of Haifa

Omair Chaudhry
Ordnance Survey

Deborah Che
Independent Scholar

DongMei Chen
Queen's University

Jin Chen
Pennsylvania State University

Nalini Chhetri
Arizona State University

Netra B. Chhetri
Arizona State University

Timothy M. Chowns
University of West Georgia

Nicholas Chrisman
Université Laval

George Christakos
San Diego State University

Torben R. Christensen
Lund University

Brett Christophers
Uppsala University

Robert W. Christopherson
*American River College/
Geosystems*

Julie Cidell
University of Illinois

Eric Clark
Lund University

Jennifer Clark
*Georgia Institute of
Technology*

Abigail R. Clarke-Sather
*Michigan Technological
University*

Sebastian Cobarrubias
University of North Carolina

Sharon C. Cobb
University of North Florida

Craig A. Coburn
University of Lethbridge

David M. Cochran Jr.
*University of Southern
Mississippi*

Alisa W. Coffin
U.S. Geological Survey

David Cohen
*Indiana University of
Pennsylvania*

Kim M. Cohen
Utrecht University

Jill S. M. Coleman
Ball State University

Ashley R. Coles
University of Arizona

Elizabeth R. Congdon
Georgia Southern University

Heather Congdon Fors
University of Gothenburg

Jamison Conley
Pennsylvania State University

Dennis Conway
Indiana University

Michael P. Conzen
University of Chicago

Karen S. Cook
University of Kansas

Jon Corbett
*University of British
Columbia–Okanagan*

Dov Corenblit
King's College London

Marci Cottingham
*Indiana University of
Pennsylvania*

Lloyd L. Coulter
San Diego State University

Rosie Cox
*Birkbeck, University of
London*

James Craine
*California State University,
Northridge*

Valorie A. Crooks
Simon Fraser University

John A. Cross
*University of
Wisconsin–Oshkosh*

Sean M. Crotty
San Diego State University

Jeff R. Crump
University of Minnesota

Karen Culcasi
West Virginia University

Julie Cupples
University of Canterbury

Judd Michael Curran
Grossmont College

Giorgio Hadi Curti
San Diego State University

Kevin M. Curtin
George Mason University

Scott Curtis
East Carolina University

Donald C. Dahmann
*George Washington
University*

Peter H. Dana
University of Texas at Austin

Raju J. Das
York University

Conny Davidsen
University of Calgary

Fiona M. Davidson
University of Arkansas

Joyce Davidson
Queen's University

Jason Davis
*University of California,
Santa Barbara*

Alastair Dawson
*Aberdeen Institute for
Coastal Science and
Management*

Shari Daya
University of Cape Town

Anthony P. D'Costa
Asia Research Centre

Mark de Socio
Salisbury University

Keith G. Debbage
*University of North
Carolina–Greensboro*

Eric Delmelle
*University of Charlotte, North
Carolina*

Urška Demšar
*National University of Ireland,
Maynooth*

Hannes Dempewolf
University of British Columbia

Mary Dengler
*Royal Holloway, University
of London*

Elizabeth R. DeSombre
Wellesley College

Pierre Desrochers
University of Toronto Mississauga

Pauline Deutz
University of Hull

Boris Dev
San Diego State University

Raymond J. Dezzani
University of Idaho

Jose R. Diaz-Garayua
Kent State University

Jacqui Dibden
Monash University

Jason Dittmer
University College London

Jerome E. Dobson
University of Kansas

Katarina Doctor
George Mason University

Klaus Dodds
Royal Holloway

William Donner
Indiana University of Pennsylvania

Richard Donohue
University of Wisconsin–Madison

Ronald I. Dorn
Arizona State University

Percy H. Dougherty
Lehigh County/Kutztown University

Ian Douglas
University of Manchester

Victoria S. Downey
University of Minnesota

Joni A. Downs
University of South Florida

Suzana Dragicevic
Simon Fraser University

Christine Drennon
Trinity University

Jiunn-Der Duh
Portland State University

Robert A. Dull
University of Texas at Austin

Michael Dunford
University of Sussex

Cheryl Morse Dunkley
University of Vermont

Jeffrey Dunn
University of Connecticut

Leslie A. Duram
Southern Illinois University

Joshua D. Durkee
Western Kentucky University

Chris S. Duvall
University of New Mexico

Mark J. Dwyer
University of Cambridge

Robert Edsall
University of Minnesota

Stephen L. Egbert
University of Kansas

Patricia Ehrkamp
University of Kentucky

Kajsa Ellegård
Linköping University

Deborah L. Elliott-Fisk
University of California, Davis

Jean T. Ellis
University of Southern California

James B. Elsner
Florida State University

Michael Emch
University of North Carolina at Chapel Hill

Ewald Engelen
University of Amsterdam

Dirk Enters
Universität Bremen

Stefan Erasmi
University of Goettingen

Christine Eriksen
University of Wollongong

Tahire Erman
Bilkent University

Veronica Escamilla
University of North Carolina at Chapel Hill

Jamey Essex
University of Windsor

Jürgen Essletzbichler
University College, London

Lawrence E. Estaville
Texas State University

Andres Etter
Javeriana University

Frances Fahy
National University of Ireland, Galway

Juliet J. Fall
Université de Genève

Carol Farbotko
University of Tasmania

Kathleen A. Farley
San Diego State University

David Featherstone
University of Glasgow

Tobias Federwisch
Friedrich-Schiller–Universität Jena

Robert Feick
University of Waterloo

Michael P. Ferber
King's University College

Dinali Nelun Fernando
Rutgers University

Carlos Ferras
*University of Santiago de
Compostela*

Peter F. Ffolliott
University of Arizona

Richard Field
University of Nottingham

Brian L. Finlayson
University of Melbourne

Manfred M. Fischer
*Vienna University of
Economics and Business*

Peter Fisher
University of Leicester

Robin Flowerdew
University of St. Andrews

Steven Flusty
York University

Emily A. Fogarty
*Suffolk County Community
College*

Kenneth Foote
*University of Colorado at
Boulder*

Timothy W. Foresman
*Johnson, Mirmiran &
Thompson, Inc.*

Benjamin Forest
McGill University

Jennifer Forkes
University of Toronto

A. Stewart Fotheringham
*National University of Ireland,
Maynooth*

Witold Fraczek
ESRI

Grant Fraley
San Diego State University

Richard V. Francaviglia
*University of Texas at
Arlington*

Benjamin Fraser
*Christopher Newport
University*

James Freeman
Concordia University

Lisa M. Freeman
University of Toronto

Hugh M. French
University of Ottawa

Kenneth French
*University of
Wisconsin–Parkside*

Karen E. Frey
Clark University

Anders Friis-Christensen
*Joint Research Center, Aalborg
University*

Joy A. Fritschle
West Chester University

Patricia Frontiera
San Francisco Estuary Institute

Christopher M. Fuhrmann
University of North Carolina

Jayson J. Funke
Clark University

Maia Gachechiladze
Central European University

Karsten Gaebler
*Friedrich-Schiller–University
Jena*

Stefan Gaertner
*Institute for Work and
Technology*

Jennifer L. Gagnon
Virginia Tech

J. C. Gaillard
University of Grenoble

Francis A. Galgano
Villanova University

Nikolaos Galiatsatos
Durham University

Nick Gallent
University College London

Ryan E. Galt
University of California, Davis

Julia A. Gamas
*U.S. Environmental Protection
Agency*

Peng Gao
Syracuse University

Isabel-María García-Sánchez
University of Salamanca

Rita Gardner
*Royal Geographical Society
and Institute of British
Geographers*

Gregg M. Garfin
University of Arizona

Rebecca R. Gasper
University of Maryland

Roland Gehrels
University of Plymouth

Jonathan Henry Geisler
Georgia Southern University

Christian Geißler
Universität Tübingen

Paul Gepts
University of California, Davis

Peter Gething
University of Oxford

Gennady Gienko
*University of the South
Pacific, Fiji*

Alan Gilbert
University College London

Aaron Gilbreath
University of Kansas

Robert Welch Gillespie
Population Communication

Thomas Gillespie
*University of California,
Los Angeles*

Jessey Gilley
University of Kansas

Mark Giordano
*International Water
Management Institute*

François Girard
CEMAGREF

Sonya Glavac
*University of New England,
Australia*

Brian J. Godfrey
Vassar College

Joseph Godlewski
*University of California,
Berkeley*

Christopher Malcolm Gold
University of Glamorgan

Daniel W. Goldberg
*University of Southern
California*

Dominic Golding
Worcester Polytechnic Institute

Michael F. Goodchild
*University of California,
Santa Barbara*

Hugh S. Gorman
*Michigan Technological
University*

Alan Grainger
University of Leeds

Dennis Grammenos
*Northeastern Illinois
University*

Nicholas Gray
University College London

Michael R. Greenberg
Rutgers University

Richard P. Greene
Northern Illinois University

Nicholas Grenier
*University of Nevada,
Las Vegas*

William J. Gribb
University of Wyoming

Carl J. Griffin
Queen's University, Belfast

Liza Griffin
University of Westminster

Daniel A. Griffith
University of Texas at Dallas

John R. Grimes
Eastern Kentucky University

Karl Grossner
*University of California,
Santa Barbara*

Diansheng Guo
University of South Carolina

Barry Haack
George Mason University

Karl R. Haapala
*Michigan Technological
University*

Joshua Hagen
Marshall University

Chris Hagerman
Portland State University

Euan Hague
DePaul University

Martin J. Haigh
University of Chicago

David Hall
University of Greenwich

Sarah J. Halvorson
University of Montana

Jeffrey D. Hamerlinck
University of Wyoming

Dean M. Hanink
University of Connecticut

Kathy Hansen Crawford
Montana State University

Paul E. Hardwick
*San Diego State
University Research
Foundation*

James W. Harrington Jr.
University of Washington

Richard Harris
McMaster University

Trevor M. Harris
West Virginia University

John Harrison
Loughborough University

Francis J. Harvey
University of Minnesota

David Havlick
*University of
Colorado–Colorado
Springs*

Kingsley E. Haynes
George Mason University

John W. Hearne
*Royal Melbourne Institute of
Technology*

Brent Hecht
Northwestern University

C. Patrick Heidkamp
*Southern Connecticut State
University*

Michael K. Heiman
Dickinson College

Carola Hein
Bryn Mawr College

Eileen H. Helmer
USDA Forest Service

Sabine Henning
*United Nations, Population
Division*

Jeffrey W. Henquinet
*Michigan Technological
University*

John Heppen
University of Wisconsin–River Falls

Martin Herold
Global Observation for Forest and Land Cover Dynamics

Darrel Hess
City College of San Francisco

Amy E. Hessl
West Virginia University

Nina Hewitt
York University

Miriam Helen Hill
Jacksonville State University

Tobin Hindle
Florida Atlantic University

William Hipwell
Victoria University of Wellington

Kyle R. Hodder
University of Regina

Briavel Holcomb
Rutgers University

Meg Holden
Simon Fraser University

William N. Holden
University of Calgary

Peter J. Holmes
University of the Free State

Mark W. Horner
Florida State University

Peter Hossler
University of Georgia

Chris Houser
Texas A&M University

Serin D. Houston
Syracuse University

Thomas Frederick Howard
Armstrong Atlantic State University

Richard Howitt
Macquarie University

Shunfu Hu
Southern Illinois University, Edwardsville

Matthew T. Huber
Clark University

Matthew B. Hufford
University of California, Davis

Christopher Hugenholtz
University of Lethbridge

Peter J. Hugill
Texas A&M University

Hilary Hungerford
University of Kansas

Thomas A. Hutton
Centre for Human Settlements, Vancouver

Niem Tu Huynh
Texas State University–San Marcos

Sungsoon Hwang
DePaul University

Kevin Hyde
University of Montana

Jennifer Hyndman
Syracuse University

Jonathan Iliffe
University College London

Jungho Im
SUNY College of Environmental Science and Forestry

Moshe Inbar
University of Haifa

Jack D. Ives
Carleton University

Andrés D. Izeta
Consejo Nacional de Investigaciones Científicas y Técnicas

Mark Jackson
University of Bristol

Trish Jackson
University of Kansas

R. Daniel Jacobson
University of Calgary

Donald G. Janelle
University of California, Santa Barbara

Marta Jankowska
San Diego State University

Piotr Jankowski
San Diego State University

Krzysztof Janowicz
University of Muenster

Helen Jarvis
Newcastle University

Timothy L. Jenkins
Michigan Technological University

Ryan R. Jensen
Brigham Young University

Erika S. Jermé
University of Toronto

Jason Bryan Jindrich
Brown University

Scott Jiusto
Worcester Polytechnic Institute

Pascale Joassart-Marcelli
San Diego State University

Ola Johansson
University of Pittsburgh at Johnstown

Jennifer Johns
University of Liverpool

Corey M. Johnson
University of North Carolina–Greensboro

Diana N. Johnson
Geosciences Consultants

Donald L. Johnson
University of Illinois

Douglas L. Johnson
Clark University

Jay T. Johnson
University of Kansas

Merrill Johnson
University of New Orleans

Nicholas H. Johnson
Michigan Technological University

Timothy Lawrence Johnson
U.S. Environmental Protection Agency

William C. Johnson
University of Kansas

Zachary Forest Johnson
University of Texas

Ron Johnston
University of Bristol

Andrew E. G. Jonas
University of Hull

David Jordhus-Lier
University of Manchester

Gabriel Judkins
Arizona State University

Patricia Julio-Miranda
Universidad Autónoma de San Luis Potosí

Jaan-Henrik Kain
Chalmers University of Technology

Ronald V. Kalafsky
University of Tennessee

Ezekiel Kalipeni
University of Illinois at Urbana–Champaign

Nagaraj Kanala
Ohio State University

Pavlos Kanaroglou
McMaster University

Kristin Kane
Global Footprint Network

David H. Kaplan
Kent State University

Hassan A. Karimi
University of Pittsburgh

Aharon Kellerman
University of Haifa

John Kelly
University of Kansas

Ilan Kelman
Center for International Climate and Environmental Research, Oslo

Thomas Kemeny
University of California, Los Angeles

Paul S. Kench
University of Auckland

Dylan Keon
Oregon State University

Thembela Kepe
University of Toronto

Carsten Keßler
University of Münster

Lisa Keys-Mathews
University of North Alabama

Lawrence Morara Kiage
Georgia State University

Changjoo Kim
University of Cincinnati

Daehyun Kim
Texas A&M University

Richard Kingston
University of Manchester

Scott Kirsch
University of North Carolina

Danielson R. Kisanga
Miami University Middletown

Erik Kjems
Aalborg University

Dorothea Kleine
Royal Holloway, University of London

Alexander Klippel
Pennsylvania State University

Dan Klooster
University of Redlands

Gerrit Jan Knaap
University of Maryland

Lawrence Knopp
University of Washington, Tacoma

Audrey Kobayashi
Queen's University

Julia Koschinsky
Arizona State University

John Kostelnick
Illinois State University

Ashish Kothari
Kalpavriksh

Kraig H. Kraft
University of California, Davis

Barry Kronenfeld
George Mason University

Rob J. Krueger
Worcester Polytechnic Institute

Richard Kujawa
St. Michael's College

Lado Kurdgelashvili
University of Delaware

Hiroyuki Kusaka
University of Tsukuba

Kristopher Kuzera
San Diego State University

Reginald Yin-Wang Kwok
University of Hawaii at Manoa

Derrick Yuk Fo Lai
McGill University

Jennifer S. Lalley
University of the Witwatersrand

Marcus B. Lane
Commonwealth Scientific and Industrial Research Organisation, Australia

Ruth Lane
Royal Melbourne Institute of Technology

Andrei G. Lapenas
University at Albany

Soren C. Larsen
University of Missouri

Stephen S. Y. Lau
University of Hong Kong

Ian T. Lawson
University of Leeds

Yves-François Le Lay
National Center for Scientific Research, University of Lyon

James L. LeBeau
Southern Illinois University Carbondale

Valerie Ledwith
National University of Ireland, Galway

Jonathan Leib
Old Dominion University

Robin M. Leichenko
Rutgers University

Walter Leimgruber
University of Fribourg

Michael Leitner
Louisiana State University

Dave Lemberg
Western Michigan University

Jonathan D. Lepofsky
Independent Scholar

Samuli Leppälä
Turku School of Economics

Deborah Leslie
University of Toronto

Agnieszka Leszczynski
University of Washington

Shawn Lewers
Florida State University

Laura R. Lewis
University of Maryland, Baltimore County

Martin W. Lewis
Stanford University

Selma Lewis
University of Maryland

Mei-Hui Li
National Taiwan University

Wei Li
Arizona State University

Xingong Li
University of Kansas

Zhilin Li
Hong Kong Polytechnic University

Johan Liebens
University of West Florida

Joshua Lieberman
Traverse Technologies

Arika Ligmann-Zielinska
Michigan State University

Yangrong Ling
National Oceanic and Atmospheric Administration

Christopher D. Lippitt
San Diego State University

Lisa Jordan
Florida State University

Emanuele Lobina
University of Greenwich

Tatiana Loboda
University of Maryland

Gary Lock
Oxford University

Mahtab A. Lodhi
University of New Orleans

Alex Loftus
Royal Holloway, University of London

Paul A. Longley
University College London

Anna Carla Lopez
San Diego State University

Andrew Lothian
Scenic Solutions

Alexander Malcolm Lovell
Queen's University

Chris Lukinbeal
University of Arizona

Jan Lundqvist
Stockholm University, Sweden

Jennifer K. Lynes
University of Waterloo

William S. Lynn
Williams College

Kin M. Ma
Grand Valley State University

Juliana A. Maantay
Lehman College, City University of New York

Neil Macdonald
University of Liverpool

William Mackaness
University of Edinburgh

Virginia W. MacLaren
University of Toronto

Phil Macnaghten
Durham University

Alan D. MacPherson
University at Buffalo

Yahia Mahmoud
Lund University

Minelle Mahtani
University of Toronto

Philippe Maillard
Universidade Federal de Minas Gerais

Jacek Malczewski
University of Western Ontario

Melissa L. Malin
Northland College

Julia Mambo
*University of the
Witwatersrand*

Charles Manyara
Radford University

Everisto Mapedza
*International Water
Management Institute*

Jennifer Mapes
*University of Southern
California*

Hug March Corbella
*Universitat Autònoma de
Barcelona*

W. Andrew Marcus
University of Oregon

Bryan G. Mark
Ohio State University

Philip M. Marren
University of Melbourne

Stuart E. Marsh
University of Arizona

Lisa Marshall
*University of North
Carolina–Chapel Hill*

Richard A. Marston
Kansas State University

Robert J. Mason
Temple University

Aron Massey
Kent State University

Joy Nystrom Mast
Carthage College

Christine W. Mathenge
Indiana University

Anne E. Mather
University of Plymouth

Ian J. Mauro
University of Victoria

Jeff May
University of Toronto

Chris Mayda
Eastern Michigan University

Clive McAlpine
University of Queensland

James McCarthy
Pennsylvania State University

Linda McCarthy
*University of
Wisconsin–Milwaukee*

George F. McCleary Jr.
University of Kansas

Sarah McCormack
University of Kentucky

Daniel McGowin
Florida State University

Molly McGraw
*Southeastern Louisiana
University*

Sara McLafferty
*University of Illinois at
Urbana–Champaign*

Brenden E. McNeil
West Virginia University

Thomas G. Measham
CSIRO Sustainable Ecosystems

David B. Mechem
University of Kansas

Kimberley E. Medley
Miami University

Lars Meier
*Institute for Employment
Research (IAB)*

Mariselle Meléndez
University of Illinois

Assefa M. Melesse
*Florida International
University*

Jaymie R. Meliker
Stony Brook University

Tom Mels
University of Gotland, Sweden

V. Ernesto Méndez
University of Vermont

Qingmin Meng
*University of North Carolina
at Charlotte*

Timothy Mennel
American Planning Association

Jessica Mercer
Macquarie University

Christopher D. Merrett
Western Illinois University

Peter Merriman
Aberystwyth University

Peter Meserve
Fresno City College

Victor Mesev
Florida State University

Sara Metcalf
*State University of New York
at Buffalo*

Sandra Metoyer
Texas A&M University

Michelle Marie Metro-Roland
Indiana University

Peter Meusburger
Heidelberg University

Burghard C. Meyer
*Technische Universität
Dortmund*

Bryon D. Middlekauff
Plymouth State University

Byron Miller
University of Calgary

Jennifer Miller
University of Texas at Austin

Matthew Miller
University of Georgia

Hugh Millward
Saint Mary's University

Phillipa Marlis Mitchell
University of Auckland

Ivan Mitin
*Russian Institute for Cultural
& Natural Heritage*

Ines M. Miyares
Hunter College

Sami Moisio
*University of Turku,
Finland*

Mariana Mondini
*Consejo Nacional de
Investigaciones Científicas y
Tecnológicas*

Burrell Montz
East Carolina University

Debnath Mookherjee
*Western Washington
University*

Toby Moore
RTI International

Bruno Moriset
*University of Lyon–Jean
Moulin*

Penny L. Morrill
*Memorial University of
Newfoundland*

Richard Morrill
University of Washington

William G. Moseley
Macalester College

Alison Mountz
Syracuse University

Aditi Mukherji
*International Water
Management Institute*

Joshua Muldavin
Sarah Lawrence College

Samantha L. Muller
Macquarie University

Tiffany Muller Myrdahl
University of Minnesota

Beverley Mullings
Queen's University

Dustin R. Mulvaney
*University of California,
Santa Cruz*

A. Sebastián Muñoz
*Consejo Nacional de
Investigaciones
Científicas y Tecnológicas*

Yuji Murayama
University of Tsukuba

James T. Murphy
Clark University

Alan T. Murray
Arizona State University

Brian Murton
University of Hawaii

Christine Mango Mutiti
Miami University

Garth Myers
University of Kansas

Soe Win Myint
Arizona State University

Harini Nagendra
Indiana University

David Nally
University of Cambridge

Nazanin Naraghi
Simon Fraser University

Rizwan Nawaz
University of Leeds

Elisabeth S. Nelson
*University of North
Carolina at Greensboro*

David J. Nemeth
University of Toledo

Markus Neteler
Fondazione Edmund Mach

Benjamin Newton
Southern Illinois University

Andrea Joslyn Nightingale
University of Edinburgh

Silvia Nittel
University of Maine

Bram F. Noble
University of Saskatchewan

Emma Norman
University of British Columbia

Fungisai Nota
Wartburg College

Samuel Nunn
*Indiana University–Purdue
University Indianapolis*

Benjamin Kofi Nyarko
University of Cape Coast

Timothy Nyerges
University of Washington

Ann M. Oberhauser
West Virginia University

Nancy J. Obermeyer
Indiana State University

Karen L. O'Brien
University of Oslo

Segun Ogunjemiyo
*California State University,
Fresno*

Oladele Ogunseitan
University of California, Irvine

Shannon O'Lear
University of Kansas

Thomas O'Loughlin
University of Wales, Lampeter

John A. Olson
Syracuse University

Ola Olsson
University of Gothenburg

Giok-Ling Ooi
*Nanyang Technological
University*

Kathleen O'Reilly
Texas A&M University

Ibrahim M. Oroud
Mu'tah University

William Y. Osei
Algoma University

Francis Owusu
Iowa State University

Tonny J. Oyana
Southern Illinois University

Rachel Pain
Durham University

Joni M. Palmer
University of Colorado at Boulder

Anupam Pandey
St. Mary's University, Halifax

Maria Paradiso
University of Sannio

Mark Paterson
University of Exeter

Marianna Pavlovskaya
Hunter College

Serge Payette
Centre d'études nordiques

Margaret Wickens Pearce
Ohio University

Mauri S. Pelto
Nichols College

Jim Penn
Grand Valley State University

Clifford Pereira
Royal Geographical Society

Reed Perkins
Queens University of Charlotte

George L. W. Perry
University of Auckland

Evelyn J. Peters
University of Saskatchewan

Frederick E. Petry
Naval Research Laboratory

Martin Phillips
University of Leicester

Alan G. Phipps
University of Windsor

James B. Pick
University of Redlands

Jenny Pickerill
University of Leicester

Michael Pidwirny
University of British Columbia–Okanagan

Sonja K. Pieck
Bates College

Bruce Pietrykowski
University of Michigan–Dearborn

Stephanie Pincetl
U.S. Forest Service/University of California, Los Angeles

John S. Pipkin
State University of New York

Tara M. Plewa
University of South Carolina

Thomas Poiker
Simon Fraser University

George Pomeroy
Shippensburg University

Robert Gilmore Pontius Jr.
Clark University

Jessie P. H. Poon
State University of New York at Buffalo

Barbara S. Poore
U.S. Geological Survey

Gregory A. Pope
Montclair State University

Gabriel Popescu
Indiana University South Bend

Deborah E. Popper
College of Staten Island/ City University of New York

William Todd Powell
University of North Carolina at Greensboro

José-Manuel Prado-Lorenzo
University of Salamanca

Valerie Preston
York University

Patricia L. Price
Florida International University

Narcisa Gabriela Pricope
University of Florida

Jesse Proudfoot
Simon Fraser University

Ruiliang Pu
University of South Florida

Hardy Pundt
University of Applied Sciences Harz

Virginia Pungo
Independent Scholar

Darren Purcell
University of Oklahoma

Heather R. Putnam
University of Kansas

Xiaomin Qiu
Missouri State University

Steven Quiring
Texas A&M University

Grzegorz Rachlewicz
Adam Mickiewicz University

Elspeth Rae
University of New South Wales

Atiqur Rahman
Jamia Millia Islamia University

Steven James Rainey
McNeese State University

Kevin Ramsey
University of Washington

Mahesh Rao
Humboldt State University

Samuel Ratick
Clark University

Martin Raubal
*University of California,
Santa Barbara*

Julia Rauchfuss
University of Minnesota

Lisa Rausch
University of Kansas

Steven Reader
University of South Florida

S. M. Reaney
Durham University

Rajyashree Reddy
University of Minnesota

Maureen G. Reed
University of Saskatchewan

Tim Reiffenstein
Mount Allison University

Wolfgang Peter Reinhardt
UniBw Muenchen

Dennis Reinhartz
*University of Texas at
Arlington*

Robert N. Renner
*U.S. Department of Housing
and Urban Development*

Lynn M. Resler
*Virginia Polytechnic Institute
and State University*

Kevon Christopher Rhiney
University of the West Indies

Pedro Ribeiro de Andrade
*National Institute for Space
Research*

Edward Rice
City University of New York

Robert A. Rice
Smithsonian Institution

Douglas Richardson
*Association of American
Geographers*

Mathieu Richaud
*California State University,
Fresno*

Jeffrey P. Richetto
University of Alabama

Laurie S. Richmond
University of Minnesota

Claus Rinner
Ryerson University

George F. Roberson
University of Massachusetts

David J. Roberts
University of Toronto

David J. Robinson
Syracuse University

George W. Robinson
*Michigan Technological
University*

Pamela Robinson
*School of Urban and Regional
Planning*

Peter J. Robinson
University of North Carolina

Wilder Robles
University of Manitoba

Agustin Robles-Morua
*Michigan Technological
University*

Dianne E. Rocheleau
Clark University

Gilbert L. Rochon
*Purdue University, Purdue
Terrestrial Observatory*

Christine M. Rodrigue
*California State University,
Long Beach*

Jean-Paul Rodrigue
Hofstra University

Paul Rollinson
Missouri State University

José Luis Romanillos
University of Bristol

Adam Rose
University of Southern California

Deborah Bird Rose
Macquarie University

William I. Rose
*Michigan Technological
University*

Mark W. Rosenberg
Queen's University

Amy Ross
University of Georgia

David A. Rossiter
*Huxley College, Western
Washington University*

Robert E. Roth
Pennsylvania State University

Robin J. Roth
York University

Graham D. Rowles
University of Kentucky

Shouraseni Sen Roy
University of Miami

James M. Rubenstein
Miami University

Simón Ruiz
*Mediterranean Institute for
Advanced Studies*

Bradley C. Rundquist
University of North Dakota

Matthias Ruth
University of Maryland

Stephanie Rutherford
Macalester College

Alan Saalfeld
Ohio State University

Dorothy Sack
Ohio University

Ahmad Safi
University of Nevada, Las Vegas

Robin Saha
University of Montana

Dipti Saletore
University at Buffalo

David S. Salisbury
University of Richmond

Luis Sánchez
Universidad de los Andes

Heather Sander
University of Minnesota

Mary V. Santelmann
Oregon State University

Antoinette W. Satterfield
Kansas State University

David Saurí
*Universitat Autònoma de
Barcelona*

Nathan F. Sayre
*University of California,
Berkeley*

Bastian Schaeffer
University of Muenster

Michael E. Schaepman
University of Zurich

Simon Scheider
University of Muenster

Kolson Schlosser
Clarkson University

Ginger L. Schmid
Minnesota State University

Thomas W. Schmidlin
Kent State University

Dietrich Schmidt-Vogt
*Asian Institute of Technology,
Thailand*

Laura C. Schneider
Rutgers University

Justin T. Schoof
Southern Illinois University

Yda Schreuder
University of Delaware

Zachary N. Schulman
George Washington University

Nadine Schuurman
Simon Fraser University

Joan M. Schwartz
Queen's University

Leonard J. Scinto
*Florida International
University*

Christopher Scott
University of Arizona

Leonie Seabrook
University of Queensland

David Seamon
Kansas State University

Anne-Marie Séguin
*Insitut National de la
Recherche Scientifique*

Christian Sellar
University of Mississippi

Jason C. Senkbeil
University of Alabama

Bongman Seo
Hitotsubashi University

Susanna Servello
University of Parma

Gary W. Shannon
University of Kentucky

Martha B. Sharma
National Cathedral School

Ian Graham Ronald Shaw
University of Arizona

Shih-Lung Shaw
*University of Tennessee,
Knoxville*

Marshall Shepherd
University of Georgia

Douglas J. Sherman
Texas A&M

Kathleen Sherman-Morris
Mississippi State University

Xun Shi
Dartmouth College

Michael Shin
*University of California,
Los Angeles*

Takeshi Shirabe
*Vienna University of
Technology*

John F. Shroder
*University of Nebraska at
Omaha*

J. Matthew Shumway
Brigham Young University

Laura Kathryn Siebeneck
University of Utah

Jennifer J. Silver
Simon Fraser University

Sunhui Sim
Florida State University

Daniel Simberloff
University of Tennessee

Nicole Simons
San Diego State University

Alex David Singleton
University College London

Todd Sink
Indiana State University

Henry Sivak
*University of California, Los
Angeles*

Emily Skop
*University of Colorado at
Colorado Springs*

André Skupin
San Diego State University

Adrian Smith
*Queen Mary, University of
London*

Janet L. Smith
University of Illinois at Chicago

Jeffrey S. Smith
Kansas State University

Jonathan M. Smith
Texas A&M University

William James Smith Jr.
University of Nevada, Las Vegas

Thomas A. Smucker
Ohio University

Michael Solem
*Association of American
Geographers*

Barry D. Solomon
*Michigan Technological
University*

Kean Huat Soon
Pennsylvania State University

André Sorensen
University of Toronto

John Sorensen
*Oak Ridge National
Laboratory*

Catherine Souch
Royal Geographical Society

Jane Southworth
University of Florida

Dale Splinter
*University of Wisconsin–
Whitewater*

Dana Sprunk
*Friedrich-Schiller–University
Jena*

Anna Stanley
*National University of Ireland,
Galway*

David Stea
Center for Global Justice

Johannes Steiger
University Blaise Pascal Geolab

Philip E. Steinberg
Florida State University

Stefan Steiniger
University of Calgary

Susanne Stenbacka
Uppsala University

Rolf Sternberg
Montclair State University

Kristin L. Stewart
Florida State University

Olof Stjernström
Umeå University

Kristin Stock
University of Nottingham

Glenn Davis Stone
Washington University

Roger R. Stough
George Mason University

Douglas Alan Stow
San Diego State University

Jeppe Strandsbjerg
Copenhagen Business School

Elaine Stratford
University of Tasmania

Daniel Z. Sui
Ohio State University

Selima Sultana
*University of North Carolina
at Greensboro*

Björn Surborg
University of British Columbia

Laurel Suter
*University of California, Santa
Barbara*

John W. Sutherland
*Michigan Technological
University*

Paul C. Sutton
University of Denver

Brendan Sweeney
Queen's University

Martin Swobodzinski
San Diego State University

Mattias Tagseth
*Norwegian University of
Science and Technology*

Tyson Taylor
Boise State University

Yvette Taylor
Newcastle University

Philippe M. Teillet
University of Lethbridge

Rajiv Thakur
University of South Alabama

Rajesh Bahadur Thapa
University of Tsukuba

Lisa Theo
*University of Wisconsin–
Stevens Point*

Colin Edward Thorn
*University of Illinois at
Urbana-Champaign*

Jane R. Thorngren
San Diego State University

Jocelyn Thorpe
*University of British
Columbia*

Grant Thrall
University of Florida

Graham A. Tobin
University of South Florida

Marina A. Tolmacheva
American University of Kuwait

Brian Tomaszewski
Pennsylvania State University

R. Tomlinson
Queen's University Belfast

Daoqin Tong
University of Arizona

Nancy K. Torrieri
*American Community Survey
Office*

Ramzi Touchan
University of Arizona

Kostas A. Triantis
University of Oxford

Ming-Hsiang Tsou
San Diego State University

Lysandros Tsoulos
National Technical University of Athens

Steven Tufts
York University

Benjamin Tuttle
University of Denver

James Tyner
Kent State University

Elizabeth Underwood-Bultmann
University of Washington

Jon Unruh
McGill University

Michael A. Urban
University of Missouri

Julie Urbanik
University of Missouri–Kansas City

Philip Edward van Beynen
University of South Florida

Hein-Anton van der Heijden
University of Amsterdam

Mark van der Meijde
International Institute for Geo-Information Science and Earth Observation (ITC)

Harald van der Werff
International Institute for Geo-Information Science and Earth Observation (ITC)

Thom van Dooren
University of Technology, Sydney

Corné P. J. M. van Elzakker
International Institute for Geo-Information Science and Earth Observation (ITC)

Scott Van Keuren
University of Vermont

William M. Van Lopik
College of Menominee Nation

Dalia Varanka
U.S. Geological Survey

Maria Vasardani
University of Maine

Stefano Vassere
Archivio dei nomi di luogo, Bellinzona, Switzerland

Geir Vatne
Norwegian University of Science and Technology

Elizabeth Vaughan
Ball State University

Gregory L. Vert
University of Nevada

Paolo Vignolo
Universidad Nacional de Colombia, Bogotá

Robert Voeks
California State University, Fullerton

Igor Vojnovic
Michigan State University

Tim Vorley
University of Cambridge

Monica Wachowicz
Wageningen University and Research Centre

Mathis Wackernagel
Global Footprint Network

Roland M. Wagner
University of Muenster

Jayme Walenta
University of British Columbia

Johnathan Walker
James Madison University

Alan Walks
University of Toronto

Candice Wallace
University of Kentucky

Julie Wallace
McMaster University

Andy Walter
University of West Georgia

Sarah M. Wandersee
San Diego State University

Enru Wang
University of North Dakota

Jinfei Wang
University of Western Ontario

Shaowen Wang
University of Illinois at Urbana-Champaign

Wenfei Winnie Wang
University of Bristol

Yi-Chen Wang
National University of Singapore

Young-Doo Wang
University of Delaware

Barney Warf
University of Kansas

Timothy A. Warner
West Virginia University

John F. Watkins
University of Kentucky

Peter Waylen
University of Florida

Joe Weber
University of Alabama

John R. Weeks
San Diego State University

Robert Weibel
University of Zurich

Tony Weis
University of Western Ontario

Sally A. Weller
Victoria University

Jeremy Wells
Clemson University

Qihao Weng
Indiana State University

Elizabeth A. Wentz
Arizona State University

Benno Werlen
Friedrich-Schiller–University of Jena

George W. White
Frostburg State University

Mark Whitehead
Aberystwyth University

Robert J. Whittaker
University of Oxford

Thomas J. Wilbanks
Oak Ridge National Laboratory

Wolfgang Wilcke
Johannes Gutenberg University

Margaret Wilder
University of Arizona

Forrest D. Wilkerson
Minnesota State University

Richard Wilkie
University of Massachusetts, Amherst

Michael Williams
Oxford University

Stewart Williams
University of Tasmania

Cort J. Willmott
University of Delaware

Carl Wilmsen
University of California, Berkeley

David Wilson
University of Illinois at Urbana-Champaign

Matthew W. Wilson
University of Washington

Gordon M. Winder
Ludwig-Maximilians-Universität

Jamie Winders
Syracuse University

Antoinette M. G. A. WinklerPrins
Michigan State University

Morton D. Winsberg
Florida State University

Ben Wisner
Oberlin College

Frank Witlox
Ghent University

Eric B. Wolf
University of Colorado

Poh Poh Wong
National University of Singapore

Joseph S. Wood
University of Baltimore

Jared Wouters
City Colleges of Chicago

Pamela Wridt
University of Colorado Denver

Dawn J. Wright
Oregon State University

Ningchuan Xiao
Ohio State University

Zhen Xiong
University of New Brunswick

Bisheng Yang
Wuhan University

Lakshman Yapa
Pennsylvania State University

Robert Yarbrough
Georgia Southern University

Stephen R. Yool
University of Arizona

Kenneth R. Young
University of Texas at Austin

Donald M. Yow
Eastern Kentucky University

Hongbo Yu
Oklahoma State University

May Yuan
University of Oklahoma

Kathryn Yusoff
University of Exeter

Alessandro Zanazzi
Georgia Southern University

Peyman Zawar-Reza
University of Canterbury

Donald J. Zeigler
Old Dominion University

Chuanrong Zhang
University of Connecticut

Tong Zhang
Wuhan University of Technology

Yun Zhang
University of New Brunswick

Dennis Zhao
Chinese Academy of Sciences

Tingting Zhao
Florida State University

Wolfgang Zierhofer
University of Basel

Kaj Zimmerbauer
University of Oulu

Matija Zorn
Scientific Research Centre of SASA

Ashley B. Zung
University of Kansas

Alex Zvoleff
San Diego State University

INTRODUCTION

Long misunderstood by the general public as an infantile field obsessed with drawing boundaries and memorizing capital cities, contemporary geography exhibits, in contrast, a fecund, robust, and increasingly sophisticated understanding of the world. Previously confined to the margins of academia and earning little respect, geography in the past four decades has been decisively transformed into an innovative and dynamic field that has earned the respect, even the admiration, of scholars in the sciences, humanities, and social sciences.

Literally "earth description" (*geo-graph*), geography may broadly be defined as the art and science of understanding the space of Earth's surface. At its simplest, geography studies the locations of things and the explanations that underlie spatial distributions of different types. Fundamentally, geographers ask "Why are things where they are?" Now, such a wide definition obscures the enormous diversity of the discipline and the multiple meanings of space itself. Space may mean, for example, the biophysical surface of Earth, an arena of class confrontation, or an intangible set of meanings. Geography is notorious, even infamous, for its eclecticism and for the very wide range of topics that it addresses. Not only do geographers tackle a great span of issues, but they do so from a multitude of theoretical and conceptual angles. Some fear that this diversity robs the discipline of a central narrative: At first glance, quantitative geomorphologists and feminist deconstructivists might seem to share little in common. Moreover, geographers approach their issues from an astonishingly diverse set of conceptual perspectives, ranging from complex mathematical models and geographic information systems (GIS) to poststructuralist discourse analysis. Geography speaks with more than one voice, which is part of the key to its growing contemporary popularity and vitality. The various ways in which space can be understood will be discussed in more detail below.

The Rising Popularity of Geography Today

The surge in geography's popularity today reflects much more than its own internal intellectual metamorphosis. Profound forces at work throughout the world have made geographical knowledge increasingly important to understanding numerous human dilemmas and our capacity to address them. Four of these are emphasized here: (1) globalization, (2) environmental destruction, (3) new geospatial technologies, and (4) cyberspace.

Globalization

Contrary to much received opinion, globalization—that complex set of often contradictory processes that operate at a transnational scale—has made geographical location more, not less, relevant. As neoliberal capital operates ever more effortlessly on a worldwide stage, small differences among regions become increasingly important. The persistent inequality between the global North and South, that is, the world's economically developed and underdeveloped regions, as well as continuing war and famine have drawn many people's attention to the role of space in perpetuating, and at times challenging, such problems. Moreover, far from simplistic and naive assertions about the "death of distance" or the "end of geography," place has become increasingly central to the behavior of transnational corporations, trade, immigration, tourism, and global financial flows. Discourses of national competitiveness in a global economy thrive as place-centered ways of coping with increasingly mobile capital, which operates on a worldwide scale. Opposition to globalization, including terrorism, has also increased the importance of learning about other cultures. At a moment when more than half the human race lives in cities, questions of how urban areas are interconnected (or not) loom large, as does the need for an understanding of deeply geographical issues such as rural-to-urban migration. Rising inequality worldwide speaks to processes that have disproportionately favored the elites at the expense of the working classes and the poor. For many people, matters of social justice are thus inescapably spatial ones. Increasingly, what happens in one part of the world has important ramifications in other parts—a fundamentally geographic phenomenon. Moreover, globalization itself plays out in different ways in different places, so that the global and the local are deeply entangled with one another.

Contemporary globalization has also undermined commonly held notions of absolute, Euclidean space by forming linkages among disparate producers and consumers intimately connected over vast distances through flows of capital, information, and goods. In generating new landscapes twisted like origami and filled with wormholes, in giving rise to ever-changing patterns of centrality and peripherality, the tsunami of globalized time-space compression has forced many people to recognize that geographies are always and everywhere dynamic, incomplete, forever coming into being, and perpetually in flux. The period of rapid globalization in the late 20th and early 21st centuries was accompanied, not coincidentally, by an explosion of theoretical work in human geography, in which the discipline firmly came to view space as a social construction, embedded in its historical context and filled with political and cultural meanings. The relational perspective on the world's landscapes today, which depicts them in terms of flows, rhizomes, and networks, is a far cry from the archaic study of places as discrete, bounded, and isolated and has played a large role in electrifying human geography.

Environmental Destruction

Another patently obvious role that geography plays today concerns the rapidly rising seriousness of global ecological and environmental problems. Issues that once could be understood and contained within relatively localized contexts, such as air or water pollution, are increasingly approachable only on a worldwide basis. Acid rain crosses national borders with ease. The enormous crisis of biodiversity unfolding across the world is inseparable from global patterns of population growth, habitat destruction, deforestation, desertification, soil erosion, and resource consumption. The oceans have been confronted with systematic overfishing and coastal pollution, including agricultural runoff and chemical spills. As protection and preservation of the environment have risen on the political agenda, geographers' work in these domains has gained attention accordingly. Hazards, such as the devastating Indian Ocean tsunami of 2004, have drawn the attention of worldwide audiences. The 2010 earthquake in Haiti triggered a response from millions around the globe. Combating global transmissions of diseases, such as AIDS, entails a geographical understanding of their genesis and diffusion. Limited global supplies of resources, including petroleum, have helped elevate the notion of the limits to growth in popular consciousness.

And of course, looming over all these issues is the dark shadow of global warming: As evidence mounts daily of rapidly melting ice caps and glaciers, of rising sea levels and more extreme weather events, of ecosystems turned upside down overnight and mass extinctions, the origins, impacts, and control of greenhouse gases have become increasingly problematic issues. Because environmental issues are unevenly distributed across space and because geography as a discipline has a long history of investigating human-environment interactions, space and spatiality have become crucial dimensions in understanding and tackling these problems.

Geospatial Technologies

Geography and cartography have always been close cousins, and for many people, mapping is their first entry into the world of spatial analysis. As technologies for surveying, mapping, and monitoring Earth's surface have proliferated in number, accuracy, and sophistication, geographical awareness about Earth's surface, both popular and academic, has grown correspondingly. Geospatial technologies, including GIS, global positioning systems (GPS), remote sensing, and location-based services, have infiltrated many corners of everyday life. Many cars and cell phones today have them. The enormous flood of data that such techniques have generated have allowed geographers to study the world's social and environmental issues in novel ways, testing models and hypotheses, enhancing forecasting abilities, and attracting widespread attention and enthusiasm. Increasingly, mapping has become not the preserve of an isolated elite group of experts but a popular activity, in which users create their own displays uniquely suited to their own contexts and problems. More people today create, disseminate, and use larger volumes of geographic information than ever before. Novel quantitative approaches (e.g., spatial autocorrelation) have supplemented the ability to understand spatial patterns in new and fecund ways. Indeed, many people would argue that such techniques form the engine that drives the growth of much academic geography. The engagement between human geographers and practitioners of GIS has opened fertile new fields to explore, including neogeography, humanistic GIS, and Web 2.0 mashups.

Cyberspace

Finally, the rise of cyberspace and the Internet has also elevated issues of spatiality in several ways. The Internet, an unregulated electronic network connecting an estimated 1.7 billion people (or 26% of the planet) in 2009, allows users to transcend distance virtually instantaneously. Telecommunications systems have become the central technology of postmodern capitalism, vital not only to corporations but also to individual consumption, communication, entertainment, education, politics, and numerous other domains of social life. As Internet usage becomes more popular and widespread in nature, including diverse applications such as e-mail, online shopping, banking, airline and hotel reservations, multiplayer video games, electronic job searches, instant messaging, e-marketing, chat rooms,

VOIP telephony, distance education, downloading of music and television shows, digital pornography, blogs, YouTube and MySpace, and simply "Googling" of information, cyberspace has profound effects on social relations, everyday lives, culture, politics, and many other spheres of social activity. Indeed, for many people who spend a great deal of time in the digital world, cyberspace has become such an important part of everyday life that the once solid boundary between the real and the virtual has essentially dissolved: It is difficult today to tell where one ends and the other begins. In allowing people and firms to "jump scale," to connect effortlessly with others around the world at the click of a mouse, cyberspace has been instrumental in the production of complex, fragmented, jumbled spaces of postmodernity.

There are, of course, many other reasons why geography remains important. Shortages of resources including water as well as a dwindling oil supply have intensified competition for the remaining stocks. The world's oceans are being emptied of their food reserves and are suffering ecosystem collapse at dizzying rates. Massive, ongoing technological changes have led to a proliferation of new industries, occupations, and products, even as old ones disappear. Knowing where one is located is vital to knowing one's identity, and as spaces have been reconfigured, so, too, have individual and collective notions of self and the other. Poverty-stricken populations the world over have spoken out against the causes of their oppression and made their voices heard. Concerns over distant strangers have multiplied as geographical imaginations have expanded in the face of rising long-distance social and personal relations and media coverage. In short, in a world in which the future arrives all too soon, whenever larger numbers of people must daily contend with questions of difference, when nature repeatedly inserts itself into the news, matters of geography have become not a luxury but a necessity.

The Multiple Meanings of Space

One reason for the diversity of geographical knowledges is that space has many dimensions and can be understood in widely different ways.

Space as a Biophysical Landscape

Space as the biophysical surface of the Earth is one such dimension. For physical geographers, Earth's surface has traditionally been seen as a set of atmospheric, hydrologic, biological, and geomorphologic processes. Biogeographers focus on the spatial distribution of life forms, as evolutionary forces have generated webs of ecological relations that vary enormously over time and space. Many physical geographers traditionally paid little attention to human dynamics and often felt that they had more in common with their intellectual brethren in geology, biology, or meteorology than they did with fellow geographers. Conversely, for many human geographers, the natural environment has amounted, at best, to a set of resources. However, the schism between physical and human geography is gradually closing, and increasingly, physical geographers have come to recognize that social activities must be taken seriously, just as many human geographers have slowly come to see that social relations do not simply unfold over a blank surface but are intensely wrapped up with a complicated natural terrain that exhibits a complex logic in its own right. Just as human geographers have grudgingly come to accept that the biophysical world does play an important role in shaping the distributions of human activity, so too have many physical geographers gradually acknowledged that nature does not simply exist outside society but is shaped by human beings and given meaning only within a variety of social contexts.

Space as Places That People Create and Inhabit

A second way in which space may be approached is as a set of places that people create and inhabit. Human geography—the analysis of how societies are stretched over Earth's surface—is thus the study of people in space in much the same way as history views people in time. Indeed, because space and time are inseparable, historical and geographical analyses are necessarily fused. Thus, human geography includes the study of how societies construct places, how different types of activities (cultural, economic, and political) are differentially located and connected, and how all these constantly change, with important ramifications for their inhabitants' quality of life. Regional geographers have long been interested in the specific conjunctions of processes in individual locales, and the creative tension between understanding the unique and the general continues to reverberate throughout the discipline, as it has for centuries. The generation and transformation of human geographies occur at a variety of spatial scales, and scale has become important not only to cartographers but also to those seeking to understand how different social processes play out at different scales, ranging from the local to the global. Indeed, scale itself has widely become viewed as a human product. Some geographers focus on the most personal and intimate of scales, the body: While bodies appear as "natural," or given, human geographers view them as social constructions, filled with gendered, racialized, and other meanings. Increasingly, as globalization has linked the planet's diverse societies into an increasingly integrated totality, it has become impossible to separate the local from the global. Integral to this perspective are questions regarding how people use the surface of the Earth, modify it, and in turn are affected by it. The interrelations between people and nature are a long-standing and hugely important part of the discipline. As the ties between human and physical geography have grown, many have rejected the long-standing notion that nature lies "outside" social relations; rather, everywhere, nature has been changed by people and been given meaning by them, even as it changes societies, a notion well encapsulated by relatively recent approaches such as political ecology.

Space as Representation of Earth's Surface

Third, space exists as a set of representations of Earth's surface, the most literal meaning of *geo-graph*. Geography is concerned with much more than the tangible worlds of mountains or factories, rivers or farms; it also includes the study of how people bring the world into consciousness, make sense of it, and give it meaning. Cartography, for example, is perhaps the most common way in which geographical knowledge has been constructed throughout its history. Maps of one form or another date back deep into human prehistory, when hunters and gatherers formed their own symbolic geographies. Voluminous, detailed, often encyclopedic descriptions of the locations of places are another form of geographical representation.

Shortly after World War II, many geographers adopted the scientific method; using quantitative techniques, they initiated a tradition of modeling space that continues to this day. Geographical models take a diversity of forms and serve many different purposes, ranging from traffic planning to watershed analysis. Models are central to many academic and nonacademic applications of geography, as they simplify the dynamics of the world to reveal its causal properties. This tradition was greatly assisted by the rise of GIS, which allow for the integration and analysis of diverse forms of spatial information. Often integrated with remotely sensed high-altitude or satellite photography, GIS has become a hugely popular, powerful, and important part of the discipline.

In stark contrast, other geographers approached the question of spatial representation in a dramatically different way. The behavioral tradition explored mental maps and cognitive models of space. Humanistic geographers, drawing on a rich philosophical tradition of existentialism and phenomenology, are concerned with the textures of human experience, the role of language and discourse, and how individual understandings are wrapped up with social configurations of power and knowledge. Discourse analysis, which lies at the core of contemporary poststructuralist approaches, insists that geographic knowledges are always social in origin and social in consequence and thus inevitably linked to power interests.

A *Very* Brief Historical Sketch

Geography has a long, rich, and fascinating history. For any society to exist, it must create a landscape, and its geography is central to how it is organized and changes. Thus, as long as there have been human beings, there have been human geographies.

The history of geography is important because all ideas, including geographical ones, are embedded within and reflect their historical contexts. For example, classical Greek geography, which included both encyclopedic descriptions and mathematical treatises, reflected the location of city-states at the intersections of Africa, Asia, and Europe and the trade relations that sutured them

together. In the medieval era, European geography labored under the yoke of theological suppression, and its intellectual development suffered according. Conversely, Arab geographers were among the world's best, skilled in mapping and providing rich portrayals of the places strewn throughout the Muslim ecumene.

During the Renaissance and Enlightenment, periods in which European empires extended their tentacles across the planet and the "West" as we know it began to come into being, cartography grew explosively—a testimony to geography's long history of service to imperial interests. Exploration and geography have long been tied together, but geography was vital in mapping the world's resources and understanding the diverse peoples falling under the tsunami of colonialism.

In the late 19th century, infected by a racist and unscientific form of Darwinism, geography in the form of environmental determinism was widely used to justify and naturalize colonial inequalities. Geopolitics became a central part of colonial rivalries, and political geography made its first appearance. Finally, new strains such as anarchist geography made their appearance. Simultaneously, geography became institutionalized within academia, with a variety of professional organizations and university departments.

In the early 20th century, retreating from the disaster of environmental determinism, geographers entombed themselves within empty empiricist descriptions—an era that continues to haunt the discipline's public image today. In this paradigm, the region, untheorized and taken for granted, ruled as the core of the geographer's craft. However, cultural geographers, such as Carl Sauer, paved the way for new understandings of landscapes as social creations, infusing the discipline with historical sensitivity.

Following World War II, scientific cartography was born. Armed with statistics and an enthusiasm for the scientific method, the quantitative revolution decisively transformed the field, offering a concern for methodological rigor, models, and conceptual abstraction. Physical geography, drawing on 19th-century roots, expanded into quantitative climatology, geomorphology, and biogeography.

The late 20th century and the current era gradually witnessed an intellectual renewal, which took various forms. Diverse schools of thought rejuvenated human geography, such as Marxism, feminism, humanistic thought, and, more recently, poststructuralism and postcolonialism. The discipline exhibited mounting methodological sophistication, including GIS, remote sensing, and quantitative modeling. Work concerned with the social construction of nature and human-environment relations began to heal the long-standing schism between human and physical geographers. Understandings of natural hazards, anthropogenic climate change, complexity theory, contingency and path dependency, gender, everyday life, race and ethnicity, Eurocentrism, sexuality, class, power, and uneven development proliferated. Long a borrower of ideas from other fields, geography today has experienced a renaissance that has made it increasingly interesting to nongeographers, as exemplified by the "spatial turn" widespread throughout the social sciences and humanities.

A Note on This Project

This encyclopedia sets for itself an ambitious task: to offer a reasonably comprehensive and useful summary of the state of the discipline in the early 21st century. As geographical knowledge has exploded in quantity, diversity, and sophistication, many people, including students and the informed lay public, find it difficult to know where to turn. This project attempts to fill that gap. No single project, of course, can do justice to the gargantuan amount of information that characterizes any discipline today; at best, it can hope only to trace the broad contours. Thus, readers should not expect the entries to attempt a comprehensive portrait of their topics; space limitations have forced the authors to focus only on the most important aspects of their respective essays. With 1,224 entries, the encyclopedia encapsulates an enormous diversity of topics in all subdisciplines. Cross-references at the end of each entry point to the innumerable connections among them, which, in conjunction with the suggestions for further readings, allow all readers to pursue topics as deeply as they wish. Because a picture is worth a thousand words, many entries are accompanied by art and graphics that lend depth and realism to the text. In addition to the more than 900 pieces of art in the A-to-Z entries, the encyclopedia includes an atlas with maps, found in the last volume.

To assist readers in locating relevant topics, the Reader's Guide at the beginning of Volume 1 decomposes geography into the six broad subject areas listed below. While the ones used here are relatively conventional, it should be noted that there are as many ways to organize geographical knowledge as there are geographers. (Indeed, geography is such a notoriously diverse and broad discipline that any categorization of it is bound to offend someone.) Note, too, that many topics defy easy categorization and belong to more than one grouping.

1. Physical Geography
   a. Biogeography
   b. Climatology
   c. Geomorphology

2. Human Geography
   a. Economic Geography
   b. Geographical Theory
   c. Medical Geography
   d. Political Geography
   e. Regional Geography
   f. Social and Cultural Geography
   g. Urban Geography

3. Nature and Society
   a. Agriculture
   b. Environment and People
   c. Hazards and Disasters
   d. Pollution and Waste
   e. Resources and Conservation
   f. Water

4. Methods, Models, and GIS
   a. Cartography
   b. GIS
   c. Qualitative Techniques
   d. Quantitative Models
   e. Remote Sensing

5. History of Geography
   a. Cartography
   b. GIS
   c. Human Geography
   d. Physical Geography

6. People, Organizations, and Movements
   a. Biographies
   b. Geographical Organizations
   c. Political and Economic Organizations

d. Scientific Organizations
e. Social Movements

Acknowledgments

*This work is dedicated to all
geographers, wherever they may be,
past, present, and future.*

A multivolume encyclopedia is such a large and complex undertaking that it demands the time and energy of a large number of people. So many people have contributed to this project that it is impossible to name them all. All I can do here is give my heartfelt thanks to all of those responsible for bringing these six volumes into being.

First, the bulk of the credit lies with the authors—all 942 of them, from dozens of countries—of the 1,224 entries, who agreed to write on their respective topics, often under short deadlines, revised their contributions as necessary, corrected me when I was wrong, and often came up with a wonderful array of graphics. Their works reveal the richness, diversity, and sophistication of contemporary geography, and I learned more from reading their contributions—1.78 million words—than I can put into words. I thank you all.

Second, the associate editors of this venture played an enormously valuable strategic role. Piotr Jankowski handled the GIS and cartography entries with aplomb; Barry Solomon rounded up a large number of skilled authors to write great essays on human-environmental relations; Mark Welford collected excellent physical geography entries with admirable grace; and Jonathan Leib, working quietly behind the scenes, served as an effective managing editor.

Third, the staff at SAGE demonstrated the utmost professionalism, working far beyond the call of duty. Rolf Janke and Robert Rojek served as acquisitions editors and made this encyclopedia possible. Diana Axelsen deserves special merit for her extraordinarily thorough review of each and every entry and helpful feedback. Among numerous other coordinators, assistants, copy editors, and graphics specialists who deserve a round of applause are Laura Notton, Yvette Pollastrini, Tracy Buyan, Eileen Gallaher, Leticia Gutierrez, Rebecca Johnson, Sheri Gilbert, and Shamila Swamy. Their meticulous assessment of all the contributions, identification of errors, suggestions for improvement, and patient corrections greatly improved the appearance and quality of the entire encyclopedia. There are many others who also lent a hand, and their work is much appreciated.

All these parties, with thousands of e-mails and often enduring drudgery, have worked tirelessly to produce a product of which they, and I, can be justly proud.

Barney Warf

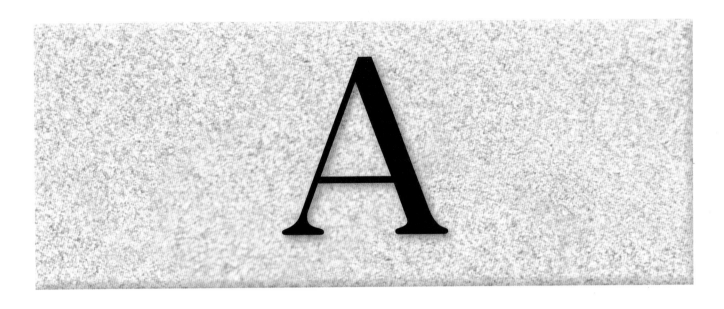

ABLER, RONALD (1939–)

Ronald F. Abler, president of the International Geographical Union (IGU) during 2008–2012, former president of the Association of American Geographers (AAG) (1985–1986), and executive director of the AAG (1989–2002), contributed to the geographical study of communication while fostering collaborative geographical research through communication among geographers and between geographers and the wider academy.

Abler received his geography degrees from the University of Minnesota (PhD, 1968) and served on the faculty (1967–1995) and as head of the geography department (1976–1982) at the Pennsylvania State University, where he is professor emeritus. His research explored how societies use communication technologies, most recently with respect to telecommunications and cyberspace, to shape space and place. In addition to numerous research publications, Abler coauthored a groundbreaking 1971 textbook on spatial organization and led two atlas projects significant for their scale and scholarly importance, one on America's major metropolitan regions and the other a state atlas on Pennsylvania.

As director of the Geography and Regional Science Program at the National Science Foundation (1984–1988), Abler helped build important research capacity in universities, expanded funding for physical geography, and championed the National Center for Geographic Information and Analysis (NCGIA). As president and executive director of the AAG, he led a substantial increase in the scale of the organization's activities and outreach, bolstered the development of geographic information systems (GIS), and was scientific administrator (1994–2002) for the AAG's Global Change and Local Places project.

Abler has worked to ensure geography's visibility by serving on the boards of the American Council of Learned Societies, the Consortium of Social Science Associations, the National Humanities Alliance, the International Geographic Information Foundation, the NCGIA, the Renewable Natural Resources Foundation, and the University Consortium for Geographic Information Sciences, among others. He was a member of the IGU's Commission on Geography of Communications and Telecommunications (1984–1992), and he organized the IGU's Columbian Quincentenary meeting in Washington, D.C. (1992), for which he also coedited a critical overview of contemporary American geography. After serving as IGU vice president (1996–2000 and 2006–2008) and IGU secretary general (2000–2006), he was elected IGU president in 2008 for a 4-year term.

Among Abler's many awards and honors are Fellow of the American Association for the Advancement of Science (1985), Centenary Medal of the Royal Scottish Geographical Society (1990), Association of American Geographers Honors (1995), Victoria Medal of the Royal Geographical

Society/Institute of British Geographers (1996), and the American Geographical Society's Samuel F. B. Morse Medal (2004) for encouragement of geographical research. The AAG also named its Distinguished Service Honors Award in honor of Abler's service to the profession.

Joseph S. Wood

See also Association of American Geographers; Communications Geography; International Geographical Union; Telecommunications and Geography

Further Readings

Abler, R. (Ed.). (1976). *A comparative atlas of America's great cities: Twenty metropolitan regions*. Minneapolis: University of Minnesota Press.

Abler, R., Marcus, M., & Olson, J. (Eds.). (1992). *Geography's inner worlds: Pervasive themes in contemporary American geography*. New Brunswick, NJ: Rutgers University Press.

Adams, J., Abler, R., & Gould, P. (1971). *Spatial organization: The geographer's view of the world*. Englewood Cliffs, NJ: Prentice Hall.

ABSOLUTE SPACE

Geographical space may be viewed from a variety of conceptual vantage points, all of which reflect and sustain various ideologies and interests in society at large. The division between absolute and relative space is one of the most long-standing and important of these contending theoretical frameworks. Absolute space is typically represented as fixed, mathematized, geometrified, asocial, and atemporal.

Absolute space has a long lineage in Western history that can be traced back to classical Greece. Plato, for example, equated light with the good, in which the cave is a metaphor for the *kosmos*, the world of human ignorance. For Plato, time was a moving image of eternity, an imperfect and untrustworthy mirror of eternal forms embedded in space that transcend time. Time is derived from change in space, which preexisted time. Space exists in its own right by virtue of its visibility, whereas time is derivative of changes in the spatial order. Time, then, was a facade, and space as the realm of changeless, permanent forms took precedence in his thought. Intellectual progress consisted of attempting to discover transcendental forms located outside of time. For Plato, the process of becoming belonged to the domain of time and illusion, in contrast to the state of being, which was empirically observable in space and thus more "real."

Similarly, Greek geometry reflected the mathematization of terrestrial and celestial space. Aristotle reasoned that since Earth's shadow on the moon was an orb, and only a sphere could throw such a shadow, the Earth must therefore be spherical. Many Greek intellectuals subscribed to the dichotomy between *chora*, or space as an empty container, and *topos*, space that is bound, inhabited, and given meaning. Hipparchus came up with the system of locating every place in a coordinate grid, dividing Earth's circumference into 360°, inventing the mathematical vocabulary of cartography still in use today. Euclidean geometry, grounded in the assumption of one uniform, continuous space, dominated the mathematics of spatial representation for the next two millennia. Ó Tuathail (1996) asserts that "Greek geography, geometry, and cartography are all suffused with the teleological dream of displaying space as a simultaneous, synchronic totality" (p. 70). The tradition of Greek geometry initiated by Euclid, Aristotle, Plato, and Pythagoras powerfully informed Western metaphysics for the next two millennia and initiated a long history of Western ocularcentrism.

The explicit division between absolute and relative space can be traced back to the famous debate in the 17th century between two geniuses, Sir Isaac Newton and Gottfried Leibniz. Newton, greatly influenced by the recent popularity of the clock, viewed time and space as abstract, absolute entities that existed independently of their measurement; that is, their existence was absolute, for their reality remained real regardless of what they contained or how they were measured. Leibniz, in contrast, held that time and space were relational—that is, comprehensible only with reference to specific frames of interpretation: Distance, for example, could only be understood through appeal to the space between two or more objects situated in space. Space and time, therefore, had no independent existence but were derivative of how we measured them. Eventually, for reasons having little to do with inherent intellectual merit and much to do with the emergence of early capitalist modernity, Newton's view triumphed.

Absolute space was greatly advanced in popularity, utility, and sophistication by Renee Descartes, a central figure of the Renaissance, who may be regarded as the founder of the doctrine that equates perspective with the abstract subject's mapping of space. Descartes proposed a mechanical view of the world centered on what has come to be known as the Cartesian cogito: a disembodied, rational mind without distinct social or spatial roots or location (but implicitly male and white). Cartesian rationalism was predicated on the distinction between the inner reality of the mind and the outer reality of objects; the latter could only be brought into the former, rationally at least, through a neutral, disembodied gaze situated above space and time. Such a perspective presumes that each person is an undivided, autonomous, rational subject with clear boundaries between "inside" and "outside," that is, between self and other, body and mind. With Descartes' cogito, vision and thought became funneled into a spectator's view of the world, one that rendered the emerging surfaces of modernity visible and measurable and rendered the viewer body-less and placeless. Illumination was conceived to be a process of rationalization, of bringing the environment into one's consciousness through the modality of vision, which is but one of several competing forms of gaining understanding. Cartesianism was thus simultaneously a model of knowledge and of the "individual." Ideologically, this process led to the mathematization of the sciences, the search for a single set of universal laws, and an enormously powerful scientific worldview that greatly expedited Europe's technological progress.

Co-catalytic with the Cartesian model of the human subject was the geometric view of space that it suggested; the ascendance of vision as a criterion for truth merged Euclidean geometry with the notion of a detached observer. This worldview had powerful social and material consequences. In Renaissance Europe, Cartesian views

of space were instrumental in the development of geometry, cartography, land surveying, civil engineering, and architecture. The Cartesian/Euclidean notion of space—infinite, absolute, and homogeneous—was replicated throughout the Renaissance and the Enlightenment, forming the basis for Newtonian physics and the theory of gravity. For European navigators, it was instrumental in smoothing space by reducing it to distance, rendering the oceans navigable, and ordering the multitude of the world's places in a comprehensible schema. The projection of Western power across the globe necessitated a Cartesian conceptualization of space as one that could be easily crossed, a function well performed by the cartographic graticule. Inserting various places in all their unique complexity into a global graticule of meridians and longitudes positioned countries, and locales within them, into a single, unified, coherent, and panopticonic understanding of the world designed by Europeans for Europeans, allowing places to be compared and normalized within an affirmation of a godlike view over Cartesian space at the global level. In all these cases, space was robbed of substantive social context to become an ordered, uniform system of abstract linear coordinates. By presenting space as fixed, asocial, and unchanging, Cartesian space accentuated the long-standing Western practice of emphasizing the temporal over the spatial. Some observers maintain that contemporary geographic information systems continue to uphold the old Euclidean/Cartesian view of absolute space, relying as they do on geometric coordinate systems.

Barney Warf

See also Historicism; Logical Positivism; Relative/Relational Space; Time-Space Compression

Further Readings

Cosgrove, D. (1988). The geometry of landscape: Practical and speculative arts in sixteenth-century Venetian land territories. In D. Cosgrove & S. Daniels (Eds.), *The iconography of landscape* (pp. 254–276). Cambridge, UK: Cambridge University Press.

Gregory, D. (1994). *Geographical imaginations.*
    Cambridge, MA: Blackwell.
Harvey, D. (1990). Between space and time:
    Reflections on the geographical imagination.
    *Annals of the Association of American
    Geographers, 80,* 418–434.
Ó Tuathail, G. (1996). *Critical geopolitics.*
    Minneapolis: University of Minnesota Press.

ACCESSIBILITY

Accessibility refers to the ability of people to overcome distance or other constraints and reach locations, activities, or services that are important to them. A number of different accessibility measures and concepts exist, though these can be classified into two distinct categories: (1) proximity (or zonal or aggregate) and (2) individual (or space-time or disaggregate) measures. Proximity measures show the nearness of one or more origin locations to a set of destinations, whether measured using distance, travel time, or other impedances. The closer an origin is to more destinations, the greater the accessibility of that origin. This notion is commonly calculated for points (such as cities) or zones (such as tracts) representing areas within a city. Origins might represent home locations, while destinations include jobs, recreation facilities, and health care or other services. In contrast, space-time measures are based on the mobility and time constraints of individuals during the course of their daily activities. The greater an individual's mobility and the fewer the time constraints, the greater his or her accessibility will be.

Aggregate Accessibility Measures

The topological or Shimbel measure is based on graph theory and the representation of a network as nodes (origins and destinations) and links (connections between nodes). This measure was introduced in geography in 1960 by William Garrison. Row sums in a filled impedance matrix represent the sum of shortest paths from one origin to all destinations and the accessibility of that origin.

The market or population potential measure of accessibility incorporates the attractiveness of destinations (such as the size or square footage of retail space), with higher accessibility resulting from being closer to more and larger destinations. As in the related gravity model of spatial interaction, distance is usually weighted to reflect the declining importance or utility of destinations farther away. There is no consensus about the specification of the weights attached to distance or the problem of self-potential, in which the distance of an origin to itself is zero and so does not allow the population or attractiveness of the origin to be included.

Simpler measures count the number of destinations within a certain radius of origin locations. These may use a single radius, such as a quarter mile, although more elaborate conceptualizations of distance are possible. In some cases, the counts of destinations may simply refer only to those within the origin zone.

Accessibility Studies

Accessibility studies became common in the 1960s during the quantitative revolution in geography. These studies provided a measure of the spatial structure of economies and cities, and they related socioeconomic indicators to accessibility to jobs or services at the intraurban level or economic variables at the interurban level. Before-and-after studies were (and remain) common. In this type of study, accessibility within a city or region is compared at multiple points in time before and after a major transport project, such as a freeway or railroad system, is completed. Accessibility improvements are then related to changes in economic development. Causality in these studies remains uncertain, and there is also the possibility that in developed countries the provision of transport infrastructure in the past half-century has lessened the importance of accessibility.

In recent decades, accessibility is often used as a variable in other studies, including those on commuting, shopping, recreation, voting, or health care provision or outcomes. The measurement of accessibility has also been expanded to incorporate the presence of competition for jobs or services or to serve as a measure of network survivability. In all such cases, low levels of accessibility would be treated as a problem.

Time-Space Accessibility

In 1970, Torsten Hagerstrand introduced the concept of time-geography, which examined how the presence of authority, coupling, and capability constraints affected human mobility. Each person has a daily space-time path that reflects these constraints and the need to move around to take part in activities. Within a three-dimensional space, time-geography maps out a space-time prism that can be reduced to a two-dimensional potential path area. Time-geography provides a map of all areas that an individual could have reached during the course of his or her daily activities. This space and its contents have been used as accessibility indicators, with a larger space containing a larger set of potential destinations that can be visited for longer periods indicating higher accessibility.

Accessibility in Geographic Information Systems

Calculating aggregate measures is relatively simple, though if more than a few origins or destinations are present, it quickly becomes labor intensive. The use of computers was therefore essential for early studies. The availability of geographic information systems (GIS) street network data sets with the topological data model has allowed accessibility to be carried out quite easily and has facilitated the use of travel time or cost instead of straight-line distances. GIS has also allowed much richer representations of origins and destinations at increasingly finer spatial scales.

For individual measures, GIS has made calculation within network space feasible since the early 1990s. Studies have explored the accessibility of individuals within differing urban environments and household types and found that individuals living at the same location can possess considerably different accessibility levels, which will also vary by time of day. These measures are currently undergoing considerable development.

Aggregate and individual measures remain distinct and produce different spatial patterns when mapped at the urban level. Recently, an entirely different notion of accessibility has emerged, based on laws such as the Americans with Disabilities Act, which ensure that public facilities have suitable access for the disabled. While this approach has so far rarely appeared in geographical studies, it suggests a future focus for the concept beyond infrastructure and intra-urban mobility.

Joe Weber

See also GIS in Transportation; Gravity Model; Network Analysis; Quantitative Revolution; Time-Geography; Transportation Geography

Further Readings

Garrison, W. (1960). Connectivity of the interstate highway system. *Papers and Proceedings of the Regional Science Association, 6,* 121–137.
Hagerstrand, T. (1970). What about people in regional science? *Papers of the Regional Science Association, 24,* 7–21.
Kwan, M., & Weber, J. (2003). Individual accessibility revisited: The implications of changing urban form and human spatial behavior. *Geographical Analysis, 35,* 341–353.

ACID RAIN

Acid rain refers to a mixture of wet and dry deposition of chemical compounds from the atmosphere containing higher than normal amounts of nitric acid and sulfuric acid (HNO_3 and H_2SO_4, respectively). Rain with a pH of 5.0 or stronger is considered acidic, which is slightly more acidic than clean or unpolluted rainwater (around 5.2–5.6). Extreme cases have been measured at a pH of between 3.0 and 4.5. Pure, unpolluted water has a pH of around 7.0. The phenomenon of acid rain and deposition has resulted in geographically and regionally pronounced effects, leading in some cases to dead lakes, fish, and trees, and damaged soils, crops, and national monuments in many countries. Nitric acid and sulfuric acid also damage human health. Concern with this problem began in the late 1960s and 1970s and,

Acid-rain-damaged trees at dusk, petrochemical plant in the distance.

Source: David Woodfall/Getty Images.

following substantial research efforts, culminated in government action in Europe, North America, and East Asia in the 1980s and 1990s.

The precursors of acid rain formation come from both natural sources, for example, decaying vegetation and volcanoes, and anthropogenic sources, especially emissions of sulfur dioxide (SO_2) and nitrogen oxides (NO_x) from coal and oil combustion. High-sulfur coal-fired power plants and copper smelters are particularly important sources, and in some areas, motor vehicles. Acid rain occurs when these gases react in the atmosphere with rain, snow, or fog and oxygen and other chemicals to form various acidic compounds (acid aerosols). If the weather is dry, the acid chemicals can mix with dust particles or smoke. Either way, the result is a mild mix of HNO_3 and H_2SO_4. When SO_2 and NO_x are released from fossil-fueled power plants and other

sources, prevailing winds can blow the compounds across state and national borders, sometimes over hundreds of miles from the combustion sources. The widespread use of tall chimneys in coal-fired power plants and other industrial facilities in the 1970s and 1980s in North America and Europe led to an increase in this problem.

In the United States, there are two national networks (both supported by the U.S. Environmental Protection Agency) that monitor acid rain and dry-acid deposition: the National Atmospheric Deposition Program and the Clean Air Status and Trends Network. Similar monitoring networks exist in Europe and East Asia. While some areas have naturally acidic lakes and soils, the regional patterns of the problem have been fairly clear for decades. Areas suffering from especially strong levels of acid rain and deposition include the Adirondack and Catskill Mountains (New York), the rest of the northeastern United States, and some mountainous areas of the southeastern United States; southeastern Canada; Poland, Czech Republic, Slovakia, Germany, Greece, Russia, northeastern England, Southern Sweden, Norway, and Finland in Europe; and China, India, Japan, and Malaysia in Asia.

The most serious adverse effects of acid rain are those affecting freshwater aquatic environments (lakes, streams, creeks, marshes, and fish), forests, and human health; serious effects can also occur to buildings, other materials, automotive coatings, and visibility. The lower pH and higher aluminum levels in acid rain can damage a variety of ecosystems. This is especially problematic in watersheds whose soils have a limited ability to neutralize acidic compounds (called "buffering capacity"). At extreme levels, fish will develop tumors, and their eggs will not hatch; mortality rates increase; and biodiversity decreases. In the case of forests, acid rain does not usually kill trees but instead weakens them by damaging their leaves, limits the nutrients available to them, or exposes them to the slow release of toxic substances in soils. Human health effects from acid rain result from the important role of sulfate and nitrate aerosols as fine particles, as well as the role of NO_x in tropospheric ozone formation. These pollutants can result in a variety of morbidity and mortality risks associated with lung inflammation, including asthma, chronic bronchitis, emphysema, vascular inflammation, and atherosclerosis. Particulate pollution can also result in lower levels of visibility. Building or monument damage can occur when H_2SO_4 rain reacts with calcium deposits in the stones to create gypsum, which will eventually flake off.

Lakes and streams suffering the ill effects of acid rain have often been limed to increase the alkalinity and pH level, clearly a short-term fix. Unfortunately, liming can have adverse effects on wetland plant species. Long-term control of acid rain requires a significant reduction in the precursor pollutants, SO_2 and NO_x. After researching the problem for many years, acid rain control programs have been enacted and implemented in Europe (Convention on Long-Range Transboundary Air Pollution of 1979, or LRTAP); North America (U.S. Clean Air Act Amendments of 1990—Title IV and the U.S.–Canada Air

Quality Agreement of 1991); and China (which approved its so-called total Amount Control Program for Eastern China in 1998). The first European Country to take acid rain seriously, which ultimately led to LRTAP, was Norway, followed by Sweden. Norway was also the first to instigate legal, cross-border action against any perpetrator—in this case, the United Kingdom. It has also seen the greatest post-1979 improvement in its lakes and freshwater fisheries. The U.S. Acid Rain Program is most well-known for its groundbreaking SO_2 emissions trading program, which was implemented among fossil-fueled power plants and is widely seen as successful, setting the stage for greenhouse gas trading.

Given the great attention paid to acid rain in the 1980s and 1990s, many people consider the problem under control. While great strides have been made in the reduction of emissions, the recovery of acid-damaged ecosystems is a long-term process that will take decades. Thus, the acid rain problem will be around for a while, and emission control programs need to be assessed for their interaction with other serious environmental problems, most famously global climate change.

Barry D. Solomon

See also Ambient Air Quality; Atmospheric Pollution; Market-Based Environmental Regulation; Point Sources of Pollution

Further Readings

Burtraw, D., Krupnick, A., Mansur, E., Austin, D., & Farrell, A. (1998). Costs and benefits of reducing air pollutants related to acid rain. *Contemporary Economic Policy, 16,* 379–400.

Ellerman, A. D., Joskow, P. L., Montero, J. P., Schmalensee, R., & Bailey, E. M. (2000). *Markets for clean air: The U.S. Acid Rain Program.* New York: Cambridge University Press.

Jenkins, J. C., Roy, K., Driscoll, C., & Buerkett, C. (2007). *Acid rain in the Adirondacks.* Ithaca, NY: Cornell University Press.

Menz, F., & Seip, H. (2004). Acid rain in Europe and the United States: An update. *Environmental Science & Policy, 7,* 253–265.

ACTOR-NETWORK THEORY

Actor-network theory is an approach that attempts to capture the complexity of the social world by uncovering the relations among human and nonhuman actors. Supporters of this approach argue that it is only through the tracing of all types of connections between actors that one can come to understand how different social processes and phenomena arise. Because of its emphasis on relations, actor-network theory is also called the sociology of associations: The focus is always on the connections among disparate things rather than on the similarities or regularities that may appear to be grouping actors together.

Actor-network theory shares some antiessentialist sentiments (an opposition to fixed entities and essences) with other strands of postmodern and poststructural thinking. As a result, the approach is suspicious of explanations that draw on common social theory constructs or conceptual aggregates such as social ties, structure, context, and so on. Actor-network theory also adopts a unique view of agency. Rather than being limited to intentional human action, actor-network theory locates agency in any "actant" (any human or nonhuman entity) that can be indentified as the source of action. However, those engaging in this approach should keep in mind that the ultimate aim of actor-network theory is to reassemble the social through detailed studies of all the elements and connections that come together to produce whatever phenomena are being studied. In this sense, actor-network theory moves away from mere critique or deconstruction and instead offers explanations that are based on analysis of the effects of the relations among different types of entities.

Actor-network theory had its origins in studies of the sociology of science and technology carried out by some of the theory's main proponents, Bruno Latour, John Law, and Michael Callon. Today, the approach has transcended its original disciplinary boundaries and goals. Actor-network theory is one of several relational perspectives adopted in the past decade by critical human geographers. In the process, geographers have further refined the vocabulary of actor-network theory and built on some of its foundations in the development of other conceptual approaches such as nonrepresentational theory and hybrid geographies. Geographers also have developed more complex accounts of the geographic implications of adopting the actor-network perspective. For example, despite being a theory identified with networks, the approach does not lend itself to the traditional representations of networks based on Euclidean space common in some strands of spatial and network analysis. Instead, actor-network theory adopts a relational view of spatiality where society and space are mutually constituted, where the positions and roles of actors in a network lead to different configurations of power relations through space-time, and where places are not territorially circumscribed but rather are always in the process of being formed through relations. Rather than imagining space as a container, actor-network theory imagines multiple spatialities operating simultaneoulsy, performed through networks and constantly becoming. Geographers have fruitfully adopted this imagery and vocabulary to conduct studies that range from analyses of urban and environmental planning to tracings of global wildlife protection networks.

Fernando Javier Bosco

See also Critical Human Geography; Hybrid Geographies; Network Analysis; Nonrepresentational Theory; Poststructuralism; Relative/Relational Space; Science and Technology Studies

Further Readings

Latour, B. (2005). *Reassembling the social: An introduction to actor-network theory.* New York: Oxford University Press.

Law, J., & Hassard, J. (1999). *Actor-network theory and after.* New York: Wiley.

Murdoch, J. (2006). *Postructuralist geography: A guide to relational space.* Thousand Oaks, CA: Sage.

Whatmore, S. (2002). *Hybrid geographies: Natures, cultures, spaces.* Thousand Oaks, CA: Sage.

ADAPTATION TO CLIMATE CHANGE

International efforts to seriously curtail rising atmospheric greenhouse gas levels seem unlikely to be effective in the next decade, and inertia in the climate system means that some climate change will occur no matter what mitigation strategies are undertaken. Thus, many policymakers and researchers are focusing on the more realistic and politically palatable policy of "adaptation to climate change," which is becoming increasingly vital. This approach assumes that environmental systems will become less stable as greenhouse gas levels increase, and thus humans must undertake actions that increase social and ecological resilience to perturbations. Such a policy has the additional benefit of offering "no regrets" or "win-win" outcomes—that is, even if global climate change does not unfold as expected, there are no regrets for having made these adaptations because they address other current issues of societal concern. Current adaptation targets include improved irrigation for agriculture, coastal protection from severe storms, preparation for extreme events, a strong public health infrastructure, and biodiversity preservation. The recent Bali Action Plan was created under the United Nations Framework Convention on Climate Change to focus on stronger climate change policies, and it urges far more aggressive action on adaptation than do past agreements. Those most vulnerable to climate change impacts are also those least able to adapt, so capacity building in all sectors will lead to the best adaptation strategies.

Regional Case Studies

Because specific climate change impacts cannot be perfectly modeled, case studies that examine social and ecological responses to climate variability are a valuable tool for climate change adaptation. Past responses to drought or extreme events in sensitive areas can be extrapolated to other regions that may experience these stresses in a climatologically altered world. Some examples are cost-benefit analysis that indicates the need for a planned retreat from hurricane-prone coastlines and vulnerability studies in areas that are currently climatologically marginal that examine how producers respond to current droughts or other weather anomalies. These studies are occurring in less developed areas, such as the African Sahel region (grazing impacts), Southern Africa (agriculture, wildlife, and flooding), and Southern Asia (agriculture and extreme events), and also in more developed regions, such as London (heat stress), Arizona (water availability), and Greenland (resource extraction).

Environmental Management Goals

Specific management goals for climate change adaptation differ depending on the environmental systems examined, but efforts have generally focused on the most climatologically sensitive systems—human health, agriculture, water resources, forestry, coastal areas, and natural ecosystems. Climate change impacts will depend on system sensitivity as well as ability to adapt. For example, rain-fed cotton production in West Texas is already sensitive to subtle shifts in precipitation, and even a small change will have a large impact. Small changes in more humid parts of the world will require less adaptation. Humans have adapted to the environment throughout history. The climate change adaptation process will merely be a more rapid and better thought out process instead of an ad hoc one—proactive rather than reactive—so that the overall costs to society can be minimized, at least in the wealthier nations. Current and historic losses from climate variability (droughts, hurricanes, etc.) without any change in greenhouse gas levels have been extreme at times, and thus no-regrets policies are relatively easy for society and policymakers to support. However, poor or low-lying countries such as Bangladesh and the Maldives have little or no ability to adapt to climate change effects such as a rise in the sea level.

Autonomous planning—where individuals respond to changing conditions—can be skewed by unavailability of information, lack of financial and capital resources, and cultural preferences.

Planned adaptation at the policymaker level will be far more effective because incentives to individuals can be changed wholesale through insurance, capital availability, land tenure, and subsidies to encourage society to maximize resilience-seeking activities. Individual cost-benefit decisions will have to be influenced by government because individuals do not have sufficient information to plan effectively in a rapidly changing climate and environment. Unfortunately, this type of societal decision making is difficult and requires cross-sectoral support for success.

Adaptation Targets

Climate change is expected to lead to an intensification of extreme events. Society has adapted to a range of climate and weather variability based on past events. However, extreme events will increasingly exceed society's ability to withstand them and lead to major societal impacts. Increases in heavy precipitation events already have been documented by numerous authorities in the midlatitude regions. Natural variability masks trends in tropical storm frequency and intensity, and this, compounded by inadequate lengths of records, means that while larger and more frequent storms are modeled to occur, they have not yet been documented unequivocally. The most important adaptation measure for extreme events is to encourage people to remove themselves from harm's way—that is, not building and living on hurricane-prone shorelines, in floodplains, and in drought-prone agricultural areas. This is difficult to do in many parts of the world where population densities are high and space is limited. Resilience to storms can also be developed through improved building codes and shoreline protection. Shoreline protection includes government infrastructure improvements, removing subsidies on insurance, removing development from inappropriate locations, and strong enforcement of building ordinances through inspections. In the long term, none of these measures will likely suffice, and more stringent development limits on vulnerable shorelines will probably be needed.

Public health infrastructure, especially disease early-warning systems based on climate conditions, can enable public health officials to engage in timely interventions including immunizations

and focused vector habitat removal. Education of populations regarding long-term risks will yield major dividends, and monitoring of current fringe populations will enable rapid response. These are classic "no-regrets" policies because they make sense immediately and their benefits will grow through time no matter what happens with climate change.

Agricultural adaptation includes major farm-level responses such as changes in crop types and more minor changes such as shifts in crop-planting dates. Climate changes will not necessarily be negative for agriculture, and adaptation measures must both reduce risk and help capture opportunities. Major challenges will be tackling new pest species, coping with more frequent drought, and weathering extreme events. Many necessary agricultural adaptations are identical to proposals within the sustainable agriculture literature, and links between the two will no doubt grow.

There is a major risk of new pest species that are resistant to variable climate moving into croplands during the expected poleward movement of pest species into new habitats. Controlling new pest insects and diseases can be accomplished through a variety of techniques, including increased surveillance and response capabilities, elimination of breeding and overwintering sites, and targeted pesticide use. However, recent dramatic increases in insect resistance to individual pesticides and environmental concerns limit the efficacy of these chemicals in many cases.

Climate variables pay such a critical role in agricultural success from year to year that small-scale adaptations are likely to have marginal effects. Agricultural irrigation and other major structural adjustments are more likely to provide long-term benefits—if adequate water supplies are available. Additionally, crops better suited to the new climate conditions will be the most successful in the long run. Even so, the uncertainties in modeled future climates are large enough so that specific crop replacements will likely be an autonomous adaptation, while policy-level adaptation is best focused on removing the structural impediments to capital shifts—likely poleward—while providing alternative options for areas experiencing drier or less favorable growing conditions. Changing tillage practices to reduce water loss is a short-term solution that is providing immediate benefits. Unfortunately, this can also increase insect or weed population pressures and compound the problem of new pests. Drought early-warning systems also can provide current and long-term benefits.

The effect of extreme events will have the greatest impact on agriculture—both frequent intense drought and storms that will physically damage crop plants—and these effects are the most difficult to model and to create policy adaptations to help mitigate. While drought measures are noted above, severe storm damage is typically addressed only through crop insurance. Yet crop insurance may slow agricultural adaptation because it subsidizes past behavior that may no longer be appropriate and could discourage reallocation of capital to more climatologically logical uses. Finally, the major no-regrets global policies will be to reduce food security risk and protect genetic resources so that new climate-adapted crops can potentially be discovered.

Biological systems will be the hardest hit by climate change because they require the greatest amount of time to adapt. The recent United Nations Millennium Ecosystem Assessment projected that climate change will be the dominant driver of biodiversity loss in the coming century. Biodiversity adaptations will also be difficult because of lower economic investments in this sector. Adaptation must include both new research and policy measures. Necessary research includes establishing current environmental condition baselines, increasing the availability of species data, and selecting the indicators of change. Once this data process has begun, policy goals include maintaining native ecosystems and ecosystem services, controlling alien species, developing and maintaining wildlife corridors to facilitate species migration, increasing management outside of protected nature reserves, and potentially translocating species into new habitat—"assisted colonization." Many authorities posit that adaptation at the species level will be insufficient and may lead to even more widespread extinction than is already occurring. Species may be able to thrive in new environments if they are able to reach them. However, the fragmented nature of land use indicates that the major adaptation goals must be to promote the dispersal of species and reduce the non-climate-change pressures on habitats.

Ecosystem services such as soil conservation, clean water, and other benefits are no-regrets measures that assist society in multiple ways during climate change adaptation.

Water is already a critical global environmental issue because of increasing populations and degrading water quality. Climate change will worsen this situation dramatically. Required adaptations are already under way to address current crises in many parts of the world, and climate change adaptation will hasten this process. Important measures include developing the infrastructure for water resources, managing societal risk, and addressing uncertainties about water quantity and quality. In spite of rising human populations, it will be necessary to reduce the demand for water through increased conservation and water use efficiency, especially in agricultural use (which is the number one global use of water). Also, populations will have to be protected from extreme events—both drought and flood—as noted above. Water use adaptation can be facilitated through community education and outreach programs, including those for more efficient water use.

Confounding Issues

Human population growth is the major factor confounding climate change adaptation. Increasing numbers of people require an increasing number of resources, and these may not be available in a transitioning world. Human barriers to migration, such as a country's borders, will be an ever-increasing problem, as more people seek to flee climatologically sensitive regions. Many governments are beginning to explore the potential national security issues involved. Additionally, unanticipated potential nonlinear responses to climate forcing cannot be modeled or addressed currently in adaptation planning. Finally, while case studies provide analogies, there still are major knowledge gaps regarding the specific adaptations required for a given region. Overall, adaptation becomes far more difficult depending on how much climate change occurs before mitigation has reduced the risks. A significant reduction in greenhouse gas concentrations is the only long-term solution.

John All

See also Anthropogenic Climate Change; Climate
Policy; Environmental Management; Greenhouse Gases;
Human Dimensions of Global Environmental Change;
Resilience; Social and Economic Impacts of Climate
Change; Symptoms and Effects of Climate Change

Further Readings

Easterling, W., Hurd, B., & Smith, J. (2004). *Coping
with global climate change: The role of adaptation
in the United States*. Arlington, VA: Pew Center
on Global Climate Change.
Food and Agriculture Organization of the United
Nations. (2007). *Adaptation to climate change in
agriculture, forestry and fisheries*. Rome, Italy:
FAO Press.
Intergovernmental Panel on Climate Change. (2007).
*Climate change 2007: Impacts, adaptation and
vulnerability* (Working Group II contribution to
the fourth assessment report of the IPCC).
Cambridge, UK: Cambridge University Press.
National Research Council. (2001). *Climate change
science*. Washington, DC: National Academies
Press.

ADAPTIVE HARVEST MANAGEMENT

Adaptive harvest management is a methodology used in natural resources and environmental management. It is based on the concept of adaptive management, which evolved in the 1970s because geographers and other scientists realized that conventional resource management could not guarantee that species were used sustainably. Emphasizing that human activities ultimately affect resource sustainability, adaptive management evaluates the additional social factors that affect how humans use their resources rather than simply managing the nonhuman resources.

Adaptive approaches conceptualize management as a learning process that is either inherent in data analysis or part of a less formal but, nevertheless, rational and structured decision process. The approach is a relatively recent methodological innovation and, despite its limitations, can facilitate faster and less expensive problem solving.

The Need for a New Approach

Managing commercially harvested resources such as game and timber species, water, and other renewable resources traditionally relied on quantitative data analysis and predicting the optimal harvest level. These analyses necessitated overharvesting a species before managers could determine an optimal take. Reducing harvest effort after that optimal take had been regulated was difficult to implement because the authorities had to enforce a harvest reduction or buy back licenses issued for harvest.

Even after calculating and enforcing an optimal harvest level, managers noticed dramatic fluctuations in biological populations and needed objective decisions about their exploitation, especially where species crossed multiple jurisdictional boundaries. Continuous species monitoring showed that population demographics remained dependent on factors such as stochastic events, unknown habitat features, or underlying landscape processes. Managing commercially harvested resources was more complicated than estimating the populations of nonharvested species because the behaviors of people, political decisions, market fluctuations, and the values of stakeholders, including those allocated some of the resource, affected management efficacy.

Adaptive Management as a Learning Experiment

Adaptive management forefronts the complexity and interdependence of natural and social systems, their constant adaptation to change, and the inherent risk and uncertainty characteristic of management processes. In complex systems, a positive feedback loop perpetuates change, and a negative feedback loop retards change. Some responses to change reduce the natural variability and diversity of a system, and thus its resilience and capacity to create alternative responses.

Adaptive management relies on a feedback learning loop after managers implement alternative policies as field "experiments" that inform progressive decisions made at regular intervals. Information about the state of a system is collected and interpreted; alternative management scenarios are considered; decisions are implemented, evaluated, and reinterpreted; and responses are altered accordingly. The cycle starts again as the manager or researcher observes the state of the new system.

Harvested Populations

Adaptive harvest management incorporates ecological data about the abundance and reproductive parameters of a species, as well as information about the impacts of various harvest policies (e.g., the costs and benefits of annual quotas, of daily bag limits or sizes and equipment limits in fish and waterfowl hunting, or of retaining brood stock to increase populations).

Dominant approaches to adaptive harvest management apply a recursive computer-generated algorithm to estimate the expected utility of future harvests based on present information. Over a sequence of time intervals, alternative harvest policies are implemented, and their probability as the optimal harvest strategy for the current system is estimated. These estimates are combined with updated ecological data (e.g., the new population size after restricting hunting to a particular area or season) to inform a subsequent iteration of the algorithm. Each iteration recursively adjusts the model probabilities using the impacts of the previous decision and the calculated change in population size. The optimal harvest strategy maximizes the expected cumulative harvest value at any particular point in time.

Types of Adaptive Harvest Management

Adaptive management is sometimes categorized as either active or passive, depending on whether uncertainty is made explicit. Active adaptive management refers to more systematic and rigorous experimentation and decision-making processes in which feedback learning about the harvest policy and the state of the resource are incorporated into various mathematical models that have weighted parameters based on previous analyses. Passive adaptive management is sometimes used

to refer to nonexperimental analyses that rely on trial-and-error learning, which can be more pragmatic and attractive to practitioners.

Humans have always adapted to their environments, and many indigenous societies, small-scale farmers, and people with local knowledge of resources use locally adaptive harvest management. Sometimes comanagement systems develop where community representatives establish and enforce agreed-on harvest rules, and public agencies benefit because the local people observe resource trends at a finer scale than broadscale commercial management with its faster and more cost-effective responses. Local strategies cannot be applied, however, to control harvesters from outside a community without government devolution of power to local groups to enforce the rules. Furthermore, local knowledge may not reflect broader regional trends.

Criticisms and Challenges

Although natural and human-induced changes occur in socionatural systems, adaptive harvest management is criticized because change may be attributed to one, human intervention that is inferred as causal. Furthermore, causal factors are themselves subject to change and uncertainty.

Other criticisms of adaptive management arise from difficulties with measuring social processes (e.g., when political pressures override agreement on how to value and allocate harvest policies) or suggest that the adaptive approach may be used to justify taking a particular policy stance.

Adaptive management cannot take the place of rigorous and systematic scientific experimentation, but adaptive management is faster and less expensive. The traditional approaches frequently cannot produce environmental solutions in the time required by managers, which is critical in the face of global climate change and other rising uncertainties about the sustainability of ecosystems and their components.

Jennifer L. Carter

See also Renewable Resources; Sustainability Science; Sustainable Agriculture; Sustainable Fisheries; Sustainable Forestry; Sustainable Production

Further Readings

Berkes, F., Colding, J., & Folke, C. (2000). Rediscovery of traditional ecological knowledge as adaptive management. *Ecological Applications, 10*, 1251–1262.

Holling, C. S. (Ed.). (1978). *Adaptive environmental assessment and management.* London: Wiley.

Walters, C. (1986). *Adaptive management of renewable resources.* New York: Macmillan.

Walters, C., & Holling, C. (1990). Large-scale management experiments and learning by doing. *Ecology, 71*, 2060–2068.

 # ADAPTIVE RADIATION

Adaptive radiation is a term used to describe a period in evolutionary history when a single or a few species diversified and underwent relatively rapid speciation. The event is distinguished from the background pattern of evolution in two ways: in pace and in the extent of diversification. These deviations from the slow and gradual flow of evolution are "adaptive" in that they are driven by natural selection as new species expand into vacant ecological niches. They are "radiation" events in that they result in expansion of the taxonomic group from one or a few to several or many species.

This rapid speciation is often accompanied by an increase in extinction rates as well, because not all the initial emergent forms are successful. For example, if one were to consider an incomplete assemblage of the fossil record, a seemingly linear history could be traced to today's genus of horses (*Equus*) from its ancestor *Hyracotherium* some 55 mya (million years ago). However, the true lineage of this group of mammals actually involves several branching events with subsequent extinctions. Rather than a straight line from one species evolving into the next, its history resembles a bush with many branches cut short and a rather crooked path leading from *Hyracotherium* to *Equus*. Adaptive radiations can be considered times of evolutionary experimentation, when many new traits are tested but only a fraction survive the process of natural selection. The Cambrian explosion was a rapid diversification in animal body forms that occurred between 545 and 525 mya. Over the past 100 years, studies of the fossil record of one particular area, the Burgess Shale in the Canadian Rocky Mountains, have provided unique insights into this process. Of the 60,000 unique fossils found at the location, most represent species now extinct, with some of them not clearly related to any animals existing today.

Adaptive radiations typically follow one of three events: (1) the emergence of an evolutionary innovation, (2) colonization of a new habitat, or (3) a mass extinction. In each case, relatively rapid speciation is driven by natural selection favoring individuals that have escaped their competitors and/or reached unoccupied ecological niches. An ecological niche is defined as the totality of an individual's role in its environment, including its habitat requirements and competitive constraints.

First, an adaptive radiation can follow the evolution of an innovative characteristic that gives a taxon access to a novel niche. These radiations are typically widespread, even worldwide, rather than regional. Examples include the radiations following the colonization of land by plants, insects, and tetrapods, each involving the evolution of innovative characters facilitating success on land. A more specific case would be the evolution of flight in an ancestral dinosaur that led to the radiation of birds, which are now the most diverse terrestrial vertebrates, with approximately 10,000 species. Another example is the radiation of angiosperms (flowering plants) into more than 250,000 species after divergence from gymnosperms between 200 and 140 mya.

Second, a population may reach a novel habitat by colonizing a relatively empty system free from traditional competitors, such as an island. Adaptive radiation can be seen in the divergence following the initial colonization of these relatively isolated ecosystems. Examples of these regional events are found in archipelagoes and island-like systems around the world. The most famous example are the finches of the Galapagos, which served as a partial muse for Charles Darwin's theory of natural selection and descent with modification. Current evidence suggests that an ancestral population of approximately 30 individuals colonized the islands some 2.3 mya, having been carried on ocean wind currents, and then evolved to include

15

the 14 species known today. An example of an island-like system that displays the same principle can be found in the cichlid fishes in Lakes Malawi, Victoria, and Tanganyika in East Africa. The genetic and geographic origin of these fishes appears to be Lake Tanganyika, with subsequent speciation leading to 2,500 species in East Africa in as little as 100,000 years.

Third, large-scale adaptive radiations have followed all the five mass extinctions of the past, when the survivors expanded into the many vacated niches of the victims. The best-known example of this type of expansion is the radiation of the mammals approximately 65 mya following the extinction of the dinosaurs. Prior to this extinction event, most mammals were small and probably nocturnal. Once released from competition and predation pressure from dinosaurs, mammals underwent rapid speciation to reach today's 5,300 species with diverse morphologies and ecologies. While this view of mammal evolution is common and a clear example of an adaptive radiation, some recent studies have suggested that the timing of the diversification may not have coincided with the extinction of the dinosaurs. More research is needed on this topic.

Elizabeth R. Congdon

See also Archipelago; Biogeography; Island Biogeography

Further Readings

Grant, P. (Ed.). (1998). *Evolution on islands.* Oxford, UK: Oxford University Press.
Schluter, D. (2000). *The ecology of adaptive radiation.* Oxford, UK: Oxford University Press.

 # ADIABATIC TEMPERATURE CHANGES

When air rises or subsides in the atmosphere, its temperature changes as a result of the change in pressure. This is called adiabatic temperature change, and understanding what happens in this process is essential for understanding condensation and evaporation in the atmosphere, which in turn explain cloud formation and precipitation. The rates of temperature change of rising or subsiding air are known, but they vary according to several factors, such as whether there is condensation or evaporation taking place and the temperature of the air.

Rising and Subsiding Air

Rising air experiences a drop in temperature, *even though no heat is lost to the surrounding environment.* Air is a poor conductor of heat, so a rising air parcel will tend to remain discrete from the surrounding air, not mixing rapidly. The temperature drops because there is a decrease in atmospheric pressure at higher altitudes. As the pressure of the surrounding air is reduced, the rising air parcel will expand. This results in cooling, since there is an inverse relationship between the volume of an air parcel and its temperature. During either expansion or compression, the total amount of energy in a discrete parcel of air remains the same (none is added or lost). The energy can be used either to do the work of expansion or to maintain the temperature of the parcel, but not both. If no heat is added or lost to the surroundings, then when an air parcel rises and expands, its temperature drops. Conversely, when the parcel is compressed, its temperature rises. So if the parcel of air descends into altitudes where the pressure is greater, it would be compressed and would warm up again without taking in heat from the outside.

This is the process of adiabatic heating and cooling. The term *adiabatic* implies a *change in temperature* of the air parcel *without gain or loss of heat* from outside the air parcel. Air parcels may rise or subside in the atmosphere as the result of variations in temperature and of the dynamics of the atmosphere, and this in turn affects their pressure and temperature. Adiabatic processes are very important in the atmosphere, and the cooling of rising air is necessary for cloud formation.

Adiabatic Lapse Rates

Dry Adiabatic Lapse Rate

In the atmosphere, the decrease in temperature of rising, unsaturated air is about 10 °C per

1,000 m (meters) altitude. For example, if a parcel of air is at 24 °C at sea level and it rises to 1,000 m, its temperature will go down to 14 °C. If unsaturated air subsides, it *warms up*, also at 10 °C per 1,000 m. So if a parcel of air at 4,000 m altitude has a temperature of −10 °C and it subsides to 3,000 m, its temperature will warm up to 0 °C. If it continues to subside, then at sea level it would have a temperature of 30 °C. This rate of temperature change of unsaturated air that is rising or subsiding is called the *dry adiabatic lapse rate* (DALR).

Note that this applies to *moving air*, not *still air*. The change in temperature of still air (air that is not rising or subsiding) follows the environmental (or average or normal) lapse rate, which varies considerably but averages about 6.5 °C per 1,000 m. Thus, the rate of temperature change in still air is not as great as the rate of change in rising air; that is, the air parcel does not cool off as fast.

Saturated Adiabatic Lapse Rate

When the moving air parcel is saturated, the change in temperature is more complicated. Consider a parcel of air that is rising—and getting colder. One of the basic rules in meteorology is that that there is an inverse relationship between the temperature of a parcel of air and its relative humidity. As a parcel of air cools, its relative humidity increases. As a parcel of air warms, its relative humidity decreases. (Note that the relative humidity changes, even though the total moisture content of the air parcel remains the same in both cases.)

So if we look at a parcel of air that is rising and cooling at the rate of 10 °C per 1,000 m, eventually it's going to cool off enough for the relative humidity to reach 100%, and at that point condensation can take place. The dew point is the temperature at which air becomes saturated and condensation begins. The lifting condensation level is the altitude at which air reaches the dew point.

This condensation of moisture influences the temperature. When water changes its physical state (evaporating to form water vapor or condensing to form liquid water in this case), energy is involved. The energy gained or released when

water changes state is called latent heat. When water evaporates, latent heat is absorbed (by the water molecule as it evaporates). When water condenses, latent heat is released (by the water molecule to the surrounding air). The amount of latent heat released by condensation is almost 600 calories per gram of water vapor. So when condensation takes place, latent heat is released to the surrounding air. When the rising air parcel reaches 100% relative humidity, there are then two opposing trends going on at the same time within the parcel. The parcel is rising and cooling, but there is also condensation, which warms up the air. The question is, will the air get colder, or will it get warmer?

What happens is that the air will still cool off as it rises, but more slowly. If water vapor in rising air is condensing, the rate of cooling is *less* than the DALR. This is the *saturated adiabatic lapse rate* (SALR), also called the wet adiabatic lapse rate or moist adiabatic lapse rate. The SALR is not constant, like the DALR, but varies inversely with the original temperature of the air parcel. A typical value for the SALR would be about 5 to 6 °C per 1,000 m. If the original temperature of the air is relatively high, the SALR of rising air will be lower, while if the temperature of the air is low to begin with, the SALR will be higher—and closer to the DALR. The reason for this is that a parcel of air with a high temperature will be able to contain a greater mass of water vapor than a parcel with a lower temperature. So when this larger mass of water condenses, a lot of latent heat is released. A parcel of air with a lower temperature to begin with will have less water vapor. So when this smaller amount of water condenses, less latent heat is released (Figure 1).

For example, if wind runs into a mountain and starts moving up the mountainside, it will likely start cooling at the DALR as it rises. However, as it cools, the relative humidity increases. When the air cools enough for the relative humidity to reach 100%, the rising air will continue to cool, but more slowly, following the SALR. Using a reasonable value of 6 °C per 1,000 m for the SALR, if an air parcel reaches saturation at a temperature of 5 °C and an altitude of 2,000 m, then as it continues rising to 3,000 m, it will cool 6 °C, to −1 °C. If the air subsequently descends on the leeward side of the mountain, it will warm up—at

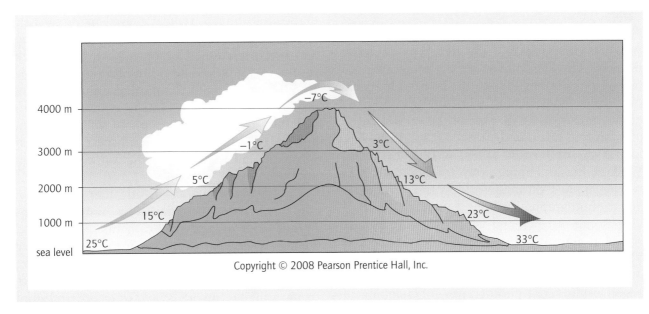

Figure 1   Temperature changes in a hypothetical parcel of air passing over a 4,000-meter-high mountain

Source: McKnight, T., & Hess, D. (2008). *Physical geography: A landscape appreciation* (9th ed., p. 155). Copyright © 2008. Reproduced by permission of Pearson Education, Inc., Upper Saddle River, New Jersey.

the SALR if there is evaporation taking place or at the DALR if there is no evaporation.

In general, the windward side of a mountain range, with rising air, is likely to have more precipitation than the leeward side, with subsiding air. The leeward side of a mountain range is often said to be in the "rain shadow" of the mountains. The rising and adiabatic cooling of air are necessary for cloud formation and the development of precipitation.

Jane R. Thorngren

See also Atmospheric Pressure; Clouds; Humidity; Lapse Rate; Latent Heat; Precipitation Formation

Further Readings

Aguado, E., & Burt, J. (2006). *Understanding weather and climate.* Upper Saddle River, NJ: Prentice Hall.

Barry, R. G. (2003). *Atmosphere, weather, and climate.* New York: Routledge.

McKnight, T., & Hess, D. (2008). *Physical geography: A landscape appreciation* (9th ed.). Upper Saddle River, NJ: Prentice Hall.

National Weather Service. (2009, January). Synoptic weather. *JetStream: Online School for Weather.* Retrieved January 19, 2008, from www.srh .weather.gov/jetstream/synoptic/clouds.htm

Thorngren, J. (2002, August). Adiabatic processes and lapse rates. *Humidity* [Electronic version]. Retrieved January 19, 2009, from http://daphne .palomar.edu/jthorngren/humidity.htm

 AERIAL IMAGERY: DATA

As one of the main sources of high-resolution images, aerial imagery is the primary data of airborne remote sensing, which involves taking photographs and images using an airborne sensor. Thus, the term *aerial imagery* refers to aerial photographs or aerial images. Airborne sensors, mounted on airplanes, helicopters, balloons, rockets, or other platforms, automatically collect aerial images of Earth's surface. A photograph or an image taken by a handheld camera is also a type of aerial imagery; that is, the camera should not be placed on a ground structure.

Development

Aerial photographs are the first type of aerial imagery. The first known aerial photograph was taken from a hot balloon by the French photographer Felix Tournachon, who used the nom de plume Nadar, in 1858. One of the earliest aerial photographs taken in the United States is the Balloon View of Boston, taken in 1860 by James Wallace Black. Photography from aircraft was widely applied during World War I. The early aerial photographs are typical nondigital photographs and in black and white. Data are usually stored in films or prints. The American Society of Photogrammetry was founded in 1934 to advance aerial photography. The U.S. Geological Survey began using aerial photographs for mapping in the 1930s.

Aerial photography evolved to a high level of sophistication during World War II and the Korean War. Since then, color infrared photography has become important in identifying vegetation types, detecting disease, and quantifying biomass. Multispectral aerial imagery, taken at the same time with different portions of the electromagnetic spectrum, was developed and tested with the data of true-color and false-color aerial photographs. Data are saved in magnetic, optical, solid-state media with digital images of the matrix of numbers, using the technology of digital photography. Now, most aerial imagery is stored in digital image format.

Innovation

The most important innovations in the development of aerial imagery are black-and-white film, color film, conventional cameras, multiband cameras, digital cameras, aerial hyperspectral imagery, side-looking airborne radar (SLAR), and aerial light detection and ranging (LIDAR). The black-and-white film traditionally used in aerial imagery is panchromatic black-and-white film, which is often used for a wide range of resolutions, gradations, and sensitivities. Compared with color film, an advantage of black-and-white film is that it can be combined with appropriate filters for bad weather, such as a misty day.

Color film includes color diapositive film, color negative film, and color infrared film. Color diapositive film is typically used in large-scale and medium-scale photogrammetry for mapping and interpretation. Color negative film, which results in a high geometric resolution, is suitable for all types of applications. Color infrared film, also called false-color film, is typically used to generate a color photograph of yellow, green, and the red of the near-infrared spectrum, which offers the most useful information in vegetation analysis.

Traditional cameras include mapping frame cameras, frame reconnaissance cameras, panoramic cameras, and strip cameras. A multiband camera is used for capturing images in a variety of wavelength bands and usually consists of a set of three or more cameras to suit a specific spectrum. A digital camera is used to record and store photographic images in digital format. Images are usually created through a photosensor with a charge-coupled device (CCD). The captured images can be stored in the camera or directly uploaded on a computer. Images also can be archived on a photographic disc or external hard drive. Currently, digital cameras often have a liquid crystal display (LCD) for viewing both captured and archived images. With the help of Web-based instruments, aerial imagery can be proofed by clients immediately.

Aerial hyperspectral remote sensing, also called aerial imaging spectroscopy, is a relatively new technology employed to detect and identify minerals, terrestrial vegetation, man-made landscapes, and other ground materials and background. Hyperspectral imagery generally is composed of more than 100 spectral bands with a narrow bandwidth of 5 to 10 nanometers; therefore, a detailed pixel spectrum indicates much more information about a surface than multispectral remote sensing, and hyperspectral data can be seen as continuous points in an n-dimensional data space. A very advanced aerial hyperspectral imaging system is the Airborne Real-Time Cueing Hyperspectral Enhanced Reconnaissance (ARCHER), with an attached global positioning system (GPS) and an inertial navigation system. ARCHER is the most sophisticated hyperspectral imaging technology and generates one square meter per pixel of hyperspectral images and 8 cm (centimeters) × 8 cm per pixel panchromatic imagery.

SLAR images are acquired by transmitting microwave energy to the ground and recording the signals reflected and scattered from the ground. The signals are recorded in digital values that also can be represented on photographic film. SLAR imagery can be collected and produced in all kinds of weather and at any time of the day.

Aerial LIDAR is a remote sensing technology applying light detection and ranging systems to produce accurate and detailed aerial imagery or maps of ground objects. It uses an active sensor with a laser and is attached with airborne GPS to measure distances to objects and the ground. Compared with the conventional methods of topographic mapping using photogrammetry, aerial LIDAR technology is a fast, economical, accurate, and reliable airborne method for creating a digital terrain model with the three-dimensional (3D) data obtained and for creating digital orthophotomaps (orthoimages) with intensity values. However, LIDAR-intensity imagery could be noisy because of dynamic survey geometry and complex laser interactions among ground features.

Data Types

Oblique and vertical images are the two types of aerial imagery data. Vertical imagery is taken directly above the ground, so that there are relatively small errors of scale and azimuth in it. It is relatively easy to complete measurements with moderate accuracy. Scale also is relatively constant across the whole captured area. But vertical imagery provides a view of the subject from above Earth's surface, which is an unfamiliar view of Earth objects. Oblique imagery is usually taken from the side of an airplane. In this case, the aircraft need not fly directly over the area to be imaged. The horizon can be seen in a high oblique image but not in a low oblique one. Disadvantages of oblique images are that scale decreases from foreground to horizon and measurements are difficult to take even by way of trigonometric computation.

Orthophotography is probably the most widely used aerial imagery in analysis. An orthophotograph is a maplike aerial image with the spatial attributes of a map and the spectral characteristics of imagery. Directions, dimensions, and plan positions can all be scaled from orthophotographs, which makes them very useful, although an orthophotograph looks like an aerial photograph. An orthophotograph contains all visible information without the simplified symbols of a map. Orthophotographs can be stored both in hardcopy and as digital imagery.

According to the aerial sensor and media characteristics, aerial imagery can be classed into black-and-white aerial images, natural-color aerial images, color infrared aerial images (false-color aerial images), SLAR imagery, aerial hyperspectral images, and LIDAR-intensity imagery.

Aerial LIDAR-intensity imagery is highly accurate black-and-white digital imagery, which is generated directly from the intensity of the LIDAR return signals.

Data Sources

Data sources of aerial imagery are usually of three types: paper, film, and digital. Earlier, captured aerial imagery was usually saved in paper or film format. Digital cameras are now widely used, and most of the aerial imagery is saved in MrSID, GeoTiff, or TIFF format.

Digital imagery has many advantages over traditional film. For instance, (a) the photographer can immediately review the picture and correct the problem or take another picture if necessary; (b) numerous shots of the same scene with slightly different settings can be taken, and only the best one needs to be printed; and (c) computer- and Web-based approaches make it much easier to store, distribute, review, process, analyze, and print aerial imagery. However, digital imaging needs a powerful battery. Sometimes, there is apparent multicolored image noise in digital pictures.

Many countries have specific agencies to collect and distribute aerial imagery, in addition to numerous private companies. The National Aerial Photography Program (NAPP), formerly the National High Altitude Photography (NHAP) program, has been an invaluable source of high-quality, cloud-free, quad-based photography for the conterminous United States since 1980. NAPP supplies 1-m (meter) aerial imagery. The National Agriculture Imagery Program (NAIP), established

Three different typhoons were spinning over the western Pacific Ocean on August 7, 2006, when the Moderate Resolution Imaging Spectroradiometer (MODIS) on NASA's Aqua satellite acquired this image. The strongest of the three, Typhoon Saomai (lower right), formed in the western Pacific on August 4, 2006, as a tropical depression. Within a day, it had become organized enough to be classified as a tropical storm. While Saomai was strengthening into a storm, another tropical depression formed a few hundred kilometers to the north, and by August 6, it became tropical storm Maria (upper right). Typhoon Bopha (left) formed just as Maria reached storm status and became a storm itself on August 7. As of August 7, the University of Hawaii's Tropical Storm Information Center predicted that Bopha and Saomai would continue on tracks that would take each into China, while Maria would move northwest across the southern end of Japan. Saomai was predicted to gather strength, while Maria and Bopha were projected to remain near their current strengths. This photo-like image was acquired at 12:35 p.m. local time (04:35 UTC) on August 7. It is unusual, but certainly not unprecedented, to have three storm systems all in the same general area at one time. Bopha, the youngest at just a few hours old, shows only the most basic round shape of a tropical storm. Maria, a day older, shows more distinct spiral structure with arms and an apparent central eye. Despite their differences in appearance, both storms were around the same size and strength, with peak sustained winds of around 90 and 100 km/hr. (58 and 63 mph), respectively. A day older than Maria is the much more powerful Typhoon Saomai. At the time of this image, the typhoon had sustained winds of around 140 km/hr. (85 mph), and forecasters predicted that it would continue to gather strength before coming ashore in China, according to the University of Hawaii's Tropical Storm Information Center. The typhoon's well-developed structure (including a distinct, closed eye in the center) in comparison to Maria is clear in this image. The slanting diagonal feature through the image is sunlight bouncing off the ocean into the MODIS instrument, a phenomenon called sunglint. The very bright patch is where the reflection is strongest.

Source: Jeff Schmaltz.

In the south-central reaches of Northern Africa's Sahara Desert, a dust storm is whipping up the sand (top center) and moving it southward over an already-struggling Lake Chad (left center edge). Regional water demands, and perhaps climate change, have shrunk the surface area of the lake dramatically, and sand dunes (tan striations) can be seen encroaching into the dense vegetation surrounding the small lake. Scattered fires (red dots) were detected as well in this true-color MODIS image captured on October 31, 2002, by the Terra satellite.

Source: Jeff Schmaltz, MODIS Land Rapid Response Team, NASA/GSFC.

in 2001 and 2002, acquires leaf-on imagery and delivers aerial imagery to the U.S. Department of Agriculture County Service Centers to maintain land boundaries and to assist in crop management. NAIP supplies two types of aerial images: (1) 1-m ground sample distance image with a horizontal accuracy of within 5 m of a reference orthoimage and (2) 2-m ground sample distance image with a horizontal accuracy of within 10 m of a reference orthoimage.

Applications

Aerial imagery is used by scientists, administrators, developers, planners, engineers, and surveyors to explore and analyze geographic phenomena with accurate spatial information of the environment, natural resources, agriculture, urban infrastructure, telecommunications, and corridor arrangement. Aerial imagery is usually processed by geographic information systems (GIS) through data integration with scanned or digitized analog images, digital images, and other digital geographical data on population, air pollution, and transportation. Aerial imagery is extensively applied in the following fields: cartography (especially topographic mapping), 3D modeling and geovisualization, land cover/land use planning, environmental studies, natural resources assessment, urban planning, transportation planning, surveillance, and archaeology.

Social Impacts

From a simple landscape shot to natural resources management and assessment to environmental and hazard analysis to urban and infrastructure geovisualization and planning, aerial imagery, as the data source of aerial remote sensing, has become a part of our daily lives. Privacy laws usually do not apply to what one can clearly view from a public space. When people are in aircrafts or balloons, they are in a public space. Therefore, taking aerial photographs or images does not break the law of privacy. From high above Earth's surface, one can legally image everything, including private events on private lands, which can lead to distressing situations.

Challenges

Although the advances in optics and camera technologies have been rapid, there are still some challenges. Aerial imagery is often taken at a certain angle to the object on the ground. This may result in an incorrect perspective, so that near objects look too large when compared with far-away targets. The process of perspective correction distorts the image, with the result that equal-sized features on the ground could have equal size in aerial imagery. When we need aerial imagery with high spatial resolution for a large area, many images will have to be taken and then compiled together, which seems like a simple process but may take hours or days of preparation.

There are some special challenges in using aerial hyperspectral imagery, although it is fascinating. Hyperspectral imagery often needs accurate atmospheric corrections before further analysis. Many hyperspectral analyses also require the use

of known material spectra for spectral classification or segmentation. Additionally, aerial hyperspectral imagery is not as often available as other types of remotely sensed data. The airborne visible/infrared imaging spectrometer (AVIRIS), developed by the National Aeronautics and Space Administration, is one of the main airborne hyperspectral sensors.

There are also some challenges in the use of aerial LIDAR. Constraints are typically associated with the collection and delivery of LIDAR imagery: Flights can be delayed due to the weather; postprocessing of a huge number of raw data points could be time-consuming; currently, there is no standard file format, although raw point data often are preprocessed and delivered in ASCII format; dense vegetation makes it difficult to map Earth's surface; and certain ground surfaces, such as water, asphalt, tar, clouds, and fog, absorb near-infrared radiation (NIR) and result in null or poor returns, since LIDAR lasers usually use NIR radiation.

Qingmin Meng

See also Aerial Imagery: Interpretation; Biophysical Remote Sensing; Image Interpretation; Image Processing; Image Registration; Image Texture; Imaging Spectroscopy; LiDAR and Airborne Laser Scanning; Multispectral Imagery; Multitemporal Imaging; Remote Sensing

Further Readings

Colwell, R. (Ed.). (1983). *Manual of remote sensing* (2nd ed.). Falls Church, VA: American Society of Photogrammetry.
Jensen, J. (Ed.). (2007). *Remote sensing of the environment: An Earth resource perspective* (2nd ed.). Upper Saddle River, NJ: Prentice Hall.

AERIAL IMAGERY: INTERPRETATION

Aerial images are photographic or digital images captured from a position above the ground. These images provide visual information over a wide area that is difficult to obtain at the ground level. Interpretation is the act of examining these images for the purpose of identifying, characterizing, and measuring objects and phenomena and judging their significance. Interpretation is not considered to be an exact science, and the interpretations are often based on the probability that an object is identified accurately. Successful interpretations depend on practical training and experience, disciplined approaches using knowledge of aerial imagery and application of that knowledge, as well as the inherent talents of the interpreter. Computer technology can aid in interpretation, but final identification by a knowledgeable human is often necessary. Interpretation is key for linking remotely sensed information to ground information on Earth's surface. This entry discusses the fundamental approaches to aerial imagery interpretation.

Interpretation Tasks

There are four tasks in image interpretation, listed in order of increasing sophistication: *detection, identification, measurement*, and *problem solving*. Detection is the lowest order and determines the presence or absence of an object or phenomenon on the ground. An example is the detection of buildings or roads in an urban environment. Identification is more advanced and involves the labeling of detected objects, such as schools or highways. Measurement allows the quantification of ground objects measured directly from the imagery, such as the number and area of buildings or the length of a highway. Problem solving is the most complex task and uses information from the first three tasks for higher-level identification, such as commercial building density.

Types of Aerial Imagery

The type of aerial imagery used can determine the level of interpretation. Aerial imagery is captured at varying spectral wavelengths, from panchromatic imagery across visible light wavelengths, to color infrared imagery, which captures energy in the visible and near-infrared wavelengths (see Figure 1), to hyperspectral imagery, which divides captured energy into many narrow

bands of information. Figure 1 shows a typical color infrared image, where green wavelengths are represented by blue in the RGB spectrum, red wavelengths are green, and near-infrared wavelengths are red. The wavelengths at which imagery is captured and viewed can affect successful interpretation. For instance, in Figure 1, healthy vegetation can be easily identified as red, because high amounts of energy are reflected by plants in the near-infrared wavelengths, shown here as the red band. Imagery is also captured at a number of angles, from directly above to oblique. Stereoscopic imagery, with multiple oblique views of the same landscape, can aid interpretation by giving the image a three-dimensional perspective. The spatial resolution of imagery can also vary based on the type of camera used, as well as the height of the camera from the ground. The spatial resolution refers to the smallest possible feature that can be distinctly detected from its surroundings. Imagery captured at a higher altitude will have coarser spatial resolution than the same imagery from a lower altitude.

Interpretive Elements

There are several elements that can be interpreted from aerial imagery. The most basic elements can be easily interpreted using a computer. However, computers have difficulty with more complex elements, which require expert knowledge from a human for interpretation, where cognition can vary from person to person.

The primary and most basic element is *tone* or *color*. Tone represents the level of grayness in panchromatic (black-and-white) imagery, where black represents low energy return and white is high energy return. In color imagery, differences in color are based on energy returned at specific wavelengths. Tone can vary in three dimensions: intensity, hue, and saturation. Intensity is the brightness of the energy. Hue is the dominant wavelength that controls the color appearance in the image. Saturation is the purity of color relative to the gray level.

The secondary level of image interpretation includes *size, shape*, and *texture*. The size of an object within an image can be a clue to the identification of that object. The size can be measured relative to other objects in an image, or it can be

Figure 1   False color digital ortho quarter quad (DOQQ) of Del Mar, California, on June 15, 2002, at 3-feet spatial resolution

Source: Center for Earth Systems Analysis Research (CESAR), Department of Geography, San Diego State University.

absolute, based on the known scale of the imagery. Certain features can have characteristic shapes that aid in their identification. Cultural features tend to be geometric with distinct boundaries, such as buildings and roads. Natural features tend to be less regular, such as rivers and forests. Texture represents the subtle changes in tone based on nearby features within the image. Texture gives the impression of "smoothness" or "roughness" of an object within an image. Texture tends to be scale dependent, where more discrete patterns can be revealed with higher resolution.

The tertiary level of interpretation includes *pattern, height,* and *shadow.* Pattern represents the spatial arrangement of objects in an image. Repeated pattern characteristics can aid in the identification of features in an image, such as

houses in a neighborhood. Shadow indicates the absence of direct illumination. The length, size, and shape of the shadow can determine an object's height as well as its shape. Shadow can also be a hindrance within an image by obscuring information with low illumination. Height can be determined in the absolute and the relative sense, and it can be better identified with oblique or stereographic imagery.

The most complex level of interpretation includes *context,* in terms of *site* and *association.* Context is the highest level of image cognition and integrates all other elements of interpretation. Site is the location of an object in relation to its environment. Association is its location in proximity to other objects being studied, often related to these features or activities.

Humans generally use the higher-order elements of interpretation more successfully than a computer due to their qualitative nature, while computers are more successful at characterizing the lower-order, more quantitative elements. Computers are less biased than human interpreters, but they may also be less accurate in higher-order interpretation, due to the difficulty of understanding context in interpretation. Most successful interpretations use a combination of both human and computer interpretive approaches.

Kristopher Kuzera

See also Aerial Imagery: Data; Panchromatic Imagery; Remote Sensing; Remote Sensing: Platform and Sensors; Spatial Resolution; Stereoscopy and Orthoimagery

Further Readings

Jensen, J. R. (2007). *Remote sensing of the environment: An Earth resource perspective* (2nd ed.). Upper Saddle River, NJ: Pearson Prentice Hall.

Paine, D. P., & Kiser, J. D. (2003). *Aerial photography and image interpretation* (2nd ed.). Hoboken, NJ: Wiley.

 # AFRICAN UNION

The African Union (AU) is a regional organization that aims to strengthen continental solidarity, territorial integrity and sovereignty, and development. The AU officially began in 2002, emerging out of what had been the Organization of African Unity (OAU). It has 53 member states, comprising all of Africa's continental and island independent states except Morocco and including the nonindependent Sahrawi Arab Republic (Western Sahara). The AU is particularly pertinent to political geographers as an example of the 21st-century phenomenon of growing significance for supranational and international organizations amid contemporary globalization.

In 1999, the OAU decided to relaunch itself, but it took three more years to officially become the AU. The AU can be distinguished from its predecessor mostly through its commitment to invoking African unity to further the development interests of member states. The AU forthrightly prioritizes economic development through integration of economic policies and solidarity on global trade issues. Four of its eight "portfolios" are directly related to economic development (Economic Affairs, Rural Economy and Agriculture, Trade and Industry, and Infrastructure and Energy). Fostering the idea of a smaller group of member countries' leaders for a New Partnership for Africa's Development (NEPAD) has been chief among the AU's achievements, following the 2001 adoption of NEPAD as an AU program. The AU seeks to implement NEPAD's economic agenda for what is characterized as African solutions to African problems.

The AU has retained the strategic agenda of the OAU, with portfolios dedicated to Political Affairs and Peace & Security. The AU supports NEPAD's peer review mechanism, an institutionalized means for strengthening democracy and governance on the continent. Another key area of operation is peacekeeping—notably, in the AU's early years, in Sudan's Darfur region. In both peer review and peacekeeping, the AU strongly advocates for pan-Africanism, but with a practical orientation. In 2008, the introduction to its Web site contained the key pan-Africanist phrase "Africa Must Unite," but with this caption: "An Efficient and Effective African Union for a New Africa." The AU also operates in social and cultural spheres, but its role in these areas is muted in comparison with its political and economic priorities.

The AU has many supporters but also a wide array of critics. One common vein of criticism is that the AU remains captive to development policies imposed from the Global North. Critics from the Global North in turn often cite the failure of the AU to condemn violations of its own standards for human rights and social justice in nondemocratic member states. Others have seen its peacekeeping operations in Darfur as inadequate. Although its record thus far is mixed, the AU can legitimately claim to have improved on the OAU's performance in both the political and the economic spheres.

Garth Myers

See also Decolonization; Developing World; Geopolitics; Political Geography; Supranational Integration

Further Readings

Makinda, S. M., & Okumu, F. W. (Eds.). (2008). *The African Union: Challenges of globalization, security, and governance.* New York: Routledge.
Murithi, T. (2005). *The African Union: Pan-Africanism, peacebuilding and development.* Burlington, VT: Ashgate.

 # AGAMBEN, GIORGIO (1941–)

Giorgio Agamben is one of the most prominent Italian philosophers of the late 20th and early 21st centuries. His oeuvre has been particularly influential for those dealing with questions of law, modern citizenship, subjectivity, power, and community. While Agamben has conducted a wide range of philosophical inquiries, his writing generally hones in on contemporary issues of ethics and biopower (a form of power through which political entities articulate their subjects as populations and manage them through control over life). Within these issues, Agamben portrays the modern world as strongly shaped by the legacy of the Nazi Holocaust as well as characterized by the large number of people who exist in ambiguous relationships to states, such as Europe's large number of international *sans-papiers* (people without legal documents), those across the world living in refugee camps, or those excluded from civic life in their countries of origin. Such a vision reveals a fundamental disjuncture between existing as what Agamben labels *homo sacer* ("bare life") versus living life as a fully endowed political subject. For Agamben, and the geographers influenced by his work, the key questions of politics center on controlling the relations of power that coalesce to uphold or break down this distinction, as well as how those relations of power gain spatial expression in the world.

Agamben often identifies the camp (as in the Nazi concentration camps) as the space most evident of the contemporary world's manifestation of ethics and biopower. He uses the camp as a metaphor for and an example of an actually existing space within the modern world to show how spaces of exclusion from the law create political subjects who exist merely as bare life and who can then become the legitimate outlets for state-sponsored violence—a space beyond the reach of ethics. Agamben repeatedly argues that the production of this space outside the law is a key component of the state's capacity to uphold its power and employ tactics of biopolitics. Agamben posits that such spaces are historically generated during periods of crisis and emergency and represent a movement away from the normal functions of the political body. However, Agamben shows how the modern state normalizes this state of emergency so that such spaces become indistinguishable from spaces of the law. He further argues that the basis of sovereign power in the modern world depends on the capacity to produce and maintain an indistinguishable relationship between spaces inside the law (inhabited by those granted political life) and those outside the law (inhabited by those banned from political life and thrust into bare life). This indistinctness structures the relations between those who can no longer clearly recognize their position as being within or outside the camp. From this, ethics breaks down into relations of violence as biological, not political, creatures encounter each other. In short, Agamben portrays a world in which the camp is the norm, not the exception, and a space inherent to and inside the space of the law, not outside it.

Agamben's work is of significance to geographers in three primary ways. First, Agamben's investigations of *homo sacer* as a condition of modern political subjectivity have allowed geographers to investigate the spaces in which bare life is cultivated, maintained, and resisted through biopower. This approach has allowed geographers to locate places such as the U.S. military prison in Guantánamo Bay, Cuba, or international refugee camps within a map of modern political life and explain them not as spaces outside the normal functions of the law but as central to legitimated state-space. The geographies of *homo sacer* extend political geographers' capacity to explain the ways in which law and space come together to produce political subjects in the world, political subjects who seem to exist outside the law because of their exposure to seemingly illegitimate and arbitrary violence. Geographers, guided by Agamben, have been able to give a logic to these spaces.

Second, Agamben's investigations of the hybridity of human beings as political and biological forms focuses attention on the animality of human beings. This offers an expansion of the idea of bare life to explain more clearly how bare life gets produced and how this expression of bare life can become rearticulated as the basis for ethical relationships. In Agamben's work, and opening up possibilities for geographers to follow suit, it identifies the spaces between the human world of the *polis* (the forerunner of the nation-state and its guarantees of citizenship) and the wilderness outside the boundaries of the polis as spaces of ethical encounter (rearticulating the content of those ethics as something more than what is codified into law). Geographers have extended this to explore the ways in which modern political subjects reclaim their biopolitical power against the machinations of the state and produce spaces of biopolitical resistance.

Third, building from Agamben's contributions to rethinking the idea of community as a way to envision the ethics that might emerge from the normalization, and therefore collectivization, of *homo sacer*, geographers have been able to extend his theory of the "coming community" into a spatialized articulation of a poststructuralist collective subject. This subject is recognized by certain attributes: (a) that of becoming (over being), (b) that of nonessentialism (over classification), and (c) that of situated ethics (over universal morality). Agamben argues that the politics of community should be postrepresentational; that is, it should not strive for recognition as a representational organ within the body politic of the state but rather recognize its ontological condition of commonality. As such, geographers have deployed Agamben's theory to find a radical politics of potentiality that transforms spaces of exclusion from a repressive apparatus of state power into a space of pure (and ethical) being together, imbuing bare life with human being.

Jonathan D. Lepofsky

See also Human Rights, Geography and; Poststructuralism

Further Readings

Agamben, G. (1998). *Homo Sacer: Sovereign power and bare life* (D. Heller-Roazen, Trans.). Stanford, CA: Stanford University Press.

Gregory, D., & Pred, A. (Eds.). (2006). *Violent geographies: Fear, terror and political violence.* New York: Routledge.

Lepofsky, J. (2003). Towards community without unity: Thinking through dis-positions and the meaning of community. *disClosure, 12,* 49–75.

Minca, C. (2007). Agamben's geographies of modernity. *Political Geography, 26*(1), 78–97.

Mitchell, K. (2006). Geographies of identity: The new exceptionalism. *Progress in Human Geography, 30*(1), 95–106.

Special issue on Agamben. (2006). *Geografiska Annaler, Series B: Human Geography, 88*(4), 363–485.

 # AGENT-BASED MODELS

Agent-based models (ABMs) are digital representations of systems, composed of a community of heterogeneous and interacting individuals distributed within a shared environment, which they transform to achieve their objectives. Agents, implemented as computer programs, are viewed as autonomous entities that decide for themselves what needs to be done to meet their goals. Equipped with atomic decision rules that drive their behavior, agents are the basic units of action in the model. They can epitomize humans, animals, and higher-order entities such as companies, tribes, agencies, or herds. Within the geographic domain, the critical element of agent-based modeling is the spatially differentiated environment that represents economies, ecosystems, and societies. ABMs are characterized by a variety of complex system facets that are relevant to modeling dynamic processes within human-environment systems. For that reason, ABMs are often referred to as computational laboratories that allow experimentation with changes within artificial societies. Due to their modular nature, ABMs are implemented using object-oriented programming. They are applied to address various spatially explicit problems, from urban growth to deforestation, to migration, to traffic congestion. The most important challenges of ABMs are the construction of proper cognitive models to represent decision making, the scarcity of social data, and the extensive parameterization that leads to a plethora of outcomes.

Rationale for Agent-Based Modeling

Complex systems involve dynamic and nonlinear linkages between the constituting parts that lead to unexpected and often perpetuating changes at the aggregate level, which are called emergent phenomena. As such, they are difficult to study in vivo, and hence, we must turn to computational environments to explore complex geographic realities.

ABMs offer a solid framework for the direct inclusion of many aspects of complex systems, such as heterogeneities of actors and landscapes, dynamic feedbacks, path dependence, and bottom-up emergence. As an example, consider the process of slum formation, which occurs throughout urban outskirts in developing countries. Due to rural overpopulation, people are forced to move to cities looking for employment. Yet local labor markets cannot accommodate such a rapidly growing urban population, which leads to poverty of the migrants. Consequently, they settle on the urban-rural fringe in makeshift houses that constitute the shantytowns. Since existing slums are the breeding ground for additional victims of poverty, the process leads to a self-reinforcing inner feedback of constantly growing slums.

Consider an ABM that emulates this phenomenon. The basic components of the model would involve migrants as the core decision makers and, in addition, villages, urban areas, and local labor markets situated within a particular region. As is the case in the real-world situation, the migrating agents would have incomplete information about the available opportunities. Restricted in their capabilities and disparate in sociodemographic characteristics such as age, family size, or health,

the migrants would be programmed to move around the landscape in search of work and a new place to live.

The complex system of local market interdependencies could be implemented as a nested hierarchy of socioeconomic institutions with built-in feedback loops. Positive feedbacks lead to path dependence, in which the outcomes of a system are very sensitive to specific initial conditions. Following the example, the existence of shantytowns in a particular locality could be induced by illegal access to electricity, growing with each new dwelling tapped to the grid.

Within the spatial context, two types of changes pertaining to an agent can occur during model execution: (1) *attribute value* change (e.g., from an unemployed state to employment) and (2) *relocation* (such as a move from a village to a city). ABMs also enable us to observe two types of system-level changes: (1) quantitative changes (such as an areal increase of slums) and (2) qualitative, often abrupt changes (such as an outbreak of a disease due to poor sanitation). In short, by offering a better symbolic representation of societies, ABMs facilitate system comprehension by professionals and novices alike.

Components of Agent-Based Models

An ABM is a dynamic simulation model that integrates the system structure with its processes. Agents and environments are two key elements of ABMs. An agent is equipped with *attributes* describing its characteristic states, *rules* guiding its decision making, and *actions* performed after the decision is made. Unlike the rational "averaged" individuals present in traditional economic models, ABMs offer a nonuniform representation of system actors. Multiformity can manifest itself as parametric heterogeneity or functional diversity. The former can be exemplified by different levels of economic status and the latter by variable decision rules.

Agents can be animate (farmers, residents) or inanimate (firms). They can be also grouped into fixed entities (building footprints, cities, and road networks) and movable objects (pedestrians, vehicles, relocating firms, and migrating households). Based on agent architecture, we deal with weak and strong agents. Whereas weak agents have a simplified internal structure and primitive behavior, strong agents, with their roots in artificial intelligence, are capable of learning, problem solving, and planning.

Direct representation of individual behavior lies at the heart of ABMs. Various cognitive frameworks have been employed in agents, ranging from *reactive* models, based on a simple stimulus-response mechanism that excludes past experiences, to *proactive* ones, with attentive agents that can take the initiative. More elaborate agents are able to learn and adapt to the exogenous changes based on "memory." Finally, agents can be deliberative and socially aware, or they can be equipped with "personality" in the form of values, beliefs, attitudes, and desires.

Agent decision making is mostly implemented using two approaches: rule based and mathematical. Rule-based systems operate on the basis of a series of *if-then* conditional statements ("if the agent is unemployed, it moves to a city"). Mathematical models use various forms of utility functions that attach a weight to a particular choice. Traditional models deal with homogeneous optimizers that make rational decisions based on complete information about their environment. An alternative approach is based on bounded rationality, which assumes that an agent has limited access to information about options. As a result, the perception of option utility varies from individual to individual. Within the bounded-rationality realm, decision-making heuristics can be grouped into random choice, satisfier choice, and ordered choice. In random choice, the agent randomly picks one of the opportunities with equal probability of selection. The satisfier choice is characterized by picking any alternative that is "good enough" with respect to a threshold value of a selected criterion. With ordered-choice heuristics, the agent orders options based on their utilities. It then picks the opportunity with the highest score. Hybrid-choice heuristics have also been proposed.

Unlike cellular automata, where interaction is constrained to geometrically fixed neighborhoods, ABMs permit "action at a distance," in which the agent's decisions influence and are influenced by drivers physically situated all over the landscape and, possibly, comprising social or political networks.

The spatial environment may differ in its extent, from buildings and neighborhoods, through villages, cities, and local habitats, to large-scale regions. An important aspect of agent-based simulations is their spatiotemporal resolution, which defines the smallest spatial unit that builds the landscape, the duration of model execution (the number of iterations), and the characteristic time step. More complex models, containing diverse agent types with varying dynamics, cause a challenge for model design. An ABM that emulates a city could include buildings that change their structure over years, as well as pedestrians, who change their location over minutes. Depending on the modeling objectives, such dilemmas are resolved by fixing state values of some objects while updating others.

As noted above, the agent framework is particularly suitable to model different interactions, such as cognitive, social, and biophysical processes. Interactions occur among agents or between agents and their environment. In particular, agent-agent interaction is expressed as cooperation among nonantagonistic agents, negotiation among self-interested agents, and competition where agents work against each other. In addition, the endogenous interrelationships in spatial ABMs include topological and nontopological relations such as adjacency, accessibility, or containment, which affect the behavior of the model over geographic space.

Agent-Based Modeling Applications

Two types of spatially explicit ABMs have been recognized: abstract and empirical. Abstract models, which dominated during the 1980s and the early 1990s, are often composed of designed agents and/or environments. They are aimed at building theories and testing hypotheses about macrostructures arising from a limited set of local behavioral rules.

More recently, agent-based modeling has been applied to many real-world geographic problems, such as resource management, the evolution of settlements and societal collapse, and the effects of policy regulations on land development, or for simulating travel behavior, to name but a few examples. Empirically grounded ABMs comprise detailed, multiscale, and multisource data, coupled with multiple modeling techniques. Such models are extensively parameterized, calibrated, validated, and used for scenario analysis, policy testing, and forecasting. However, predictive applications of agent-based modeling, which are based on trend extrapolation, are often criticized due to the high level of uncertainty associated with the processes built into the model.

Constructing Agent-Based Models

Implementation of agent-based modeling starts from a conceptual model that reflects the objectives of the research. With a conceptual model at hand, we can move to the design phase. At this step, the key challenge is to decide which components, driving mechanisms, and interactions are crucial to the problem studied. Since ABMs tend to be highly dimensional, it is important to construct a sufficiently simple model that, nevertheless, accurately mimics the system. In addition, choices should be made about the necessary data (such as elevation, roads, or preservation areas) and data representation (discrete objects vs. continuous fields). What follows is the coding stage. Implementation of agent-based modeling is carried out using object-oriented programming languages such as Java, C++, and Python or specialized agent toolkits such as REPAST (REcursive Porous Agent Simulation). In addition to two broad modules that stand for the agents and the environment, an ABM is typically equipped with a scheduler that manages the sequence of model execution.

Once the program has been built, it undergoes a process of verification that focuses on checking whether the implemented model represents the conceptual ABM. Sensitivity analysis is then performed, with the objective of measuring the effect of alterations in the initial conditions on the results of model execution. The changes in input values are usually drawn from boundary conditions that constrain the parameter space. The final phase of building an ABM is the most demanding. It involves model validation, which tests whether the model really emulates the system under study. Comprehensive validation enhances model credibility and defensibility, which is especially necessary for subsequent applications.

Challenges

Over the past decades, ABMs have experienced momentous development. However, there are still some fundamental obstacles that pose a challenge to the agent-based modeling community. Among other things, empirical ABMs are limited by expensive social-behavioral data. If such data are available, their quality may be inadequate due to the subjectivity of qualitative judgments, inappropriate scale of representation, or lack of longitudinal observations. A closely related problem is the accurate representation of choice behavior. In this respect, the disaggregated decision making of ABMs constitutes both their advantage and their shortcoming.

Arika Ligmann-Zielinska

See also Cellular Automata; Complexity Theory; Complex Systems Models; Land Use and Land Cover Mapping; Models and Modeling; Path Dependency

Further Readings

Benenson, I., & Torrens, P. M. (2004). *Geosimulation automata-based modeling of urban phenomena.* New York: Wiley.

Gimblett, R. (Ed.). (2002). *Integrating geographic information systems and agent-based modeling techniques for simulating social and ecological processes* (Santa Fe Institute Studies in the Sciences of Complexity). New York: Oxford University Press.

Matthews, R., Gilbert, N. G., Roach, A., Polhill, J. G., & Gotts, N. M. (2007). Agent-based land-use models: A review of applications. *Landscape Ecology, 22,* 1447–1459.

Parker, D., Berger, T., & Manson, S. (Eds.). (2002). *Agent-based models of land-use and land-cover change: Report and review of an international workshop, October 4–7, 2001, Irvine, California, USA.* Bloomington: LUCC Focus 1 Office, Indiana University. Retrieved May 20, 2008, from www .globallandproject.org/Documents/LUCC_No_6.pdf

Parker, D., Manson, S., Janssen, M., Hoffman, M., & Deadman, P. (2003). Multi-agent systems for the simulation of land use and land cover change:

A review. *Annals of the Association of American Geographers, 93,* 314–337.

REPAST. (2008). *Agent simulation toolkit.* Retrieved May 20, 2008, from http://repast.sourceforge.net

Wooldridge, M. (1999). Intelligent agents. In G. Weiss (Ed.), *Multiagent systems: A modern approach to distributed artificial intelligence* (pp. 27–42). Cambridge: MIT Press.

AGGLOMERATION ECONOMIES

By clustering in close proximity to one another, firms can lower their production costs and raise profits. This process is enormously important to many forms of production and is titled agglomeration economies—that is, the benefits derived from grouping together. By forming dense webs of production and embedding themselves within them, firms usually can produce more efficiently and profitably.

Agglomeration economies take several forms. Production linkages accrue to firms locating near producers that manufacture their inputs. By clustering, transportation and assembly costs are reduced. Service linkages occur when enough firms locate in one area to avail themselves of specialized support services. For example, the advertising industry in New York is concentrated within a short distance of Madison Avenue, investment banks form a dense wad in southern Manhattan, and film companies agglomerate in Hollywood. By locating near one another, firms can acquire up-to-date information on the latest trends in their sector, technological changes, shifts in policies and markets, clients, hires, and new products and processes, and they can keep an eye on the competition. Marketing linkages occur when a cluster of similar firms is large enough to attract specialized distribution services. The small firms of the garment industry in New York City have collectively attracted advertising agencies, showrooms, buyer listings, and other aspects of product distribution that deal exclusively with the garment trade. Firms within the cluster have a cost advantage over isolated firms, which must acquire these benefits for themselves.

Agglomeration economies may be temporary, are found to different extents in different industries, and may be offset through various forms of economic, technological, and geographic change. Typically, agglomeration economies reflect firms' need for close interaction with clients and suppliers, often on a face-to-face basis. Thus, they are most pronounced in vertically disintegrated types of production, in which firms have many linkages "upstream" and "downstream" in the production process. (In contrast, vertically integrated firms, with relatively few external linkages, are less dependent on agglomeration.) Firms in markets with low degrees of uncertainty (usually due to slow rates of technical change, the market structure, or the regulatory environment), in contrast, are less dependent on agglomeration to minimize costs and maximize profits. As firms grow, they often become more vertically integrated and more capital intensive, have fewer external linkages, and come to substitute economies of scale for agglomeration economies.

Because agglomeration economies provide powerful incentives for firms to locate in close proximity to one another, they are most heavily manifested in large metropolitan areas. The prime motivation behind the agglomeration of firms in metropolitan regions is the ready access they offer to clients, suppliers, and ancillary services, most of which is accomplished through face-to-face interaction. Often, personal relationships of trust and reputation are of paramount significance. Agglomeration thus maximizes access to information, much of which is tacit, irregular, and nonstandardized, and helps firms minimize uncertainty. Firms in these locations have an advantage, within limits, over similar firms in rural areas. Cities provide the markets, specialized labor forces and services, utilities, and transportation connections required by manufacturing. *Urbanization economies,* therefore, are a combination of production, service, and marketing linkages concentrated at a particular location. Agglomeration forms the basis for the comparative advantage of cities in forms of production, which typically consist of relatively labor-intensive, vertically disintegrated firms in highly competitive markets with high degrees of uncertainty and change.

Agglomeration economies have been manifested in numerous industries throughout the historical geography of capitalism. They were critical during the early Industrial Revolution, when many small firms in industries such as watch making or gun manufacturing clustered in the cores of British cities. Since the emergence of post-Fordist "flexible production" in the late 20th century, the competitiveness of regions such as California's Silicon Valley, Italy's Emilia-Romagna, or Germany's Badden-Wurtenburg has relied heavily on agglomeration. Finally, producer services (business and financial services that cater primarily to other firms) rely heavily on agglomeration economies, often in "global cities," forming complexes of service firms comparable to other types of highly concentrated production.

Barney Warf

See also Economic Geography; Flexible Production; Industrial Districts; Knowledge, Geography of; Knowledge Spillovers; Learning Regions

Further Readings

Stutz, F. P., & Warf, B. (2005). *The world economy: Resources, location, trade, and development* (4th ed.). Upper Saddle River, NJ: Prentice Hall.

 # AGING AND THE AGED, GEOGRAPHY OF

See Elderly, Geography and the

 # AGNEW, JOHN (1949–)

One of the leading human geographers of the late 20th and early 21st centuries, John Agnew has made major contributions to the fields of political geography, international relations, and international political economy. Born in England in 1949, Agnew received his BA from the University of Exeter and his MA and PhD from Ohio State University, receiving his doctorate in 1976. Since 1995, Agnew has been a professor of geography at the University of California, Los Angeles. From 1975 to 1995, he taught at Syracuse University. In 2008, he was elected president of the Association of American Geographers.

The author or editor of two dozen books and approximately 150 journal articles and book chapters, Agnew has been at the forefront of new research directions in several different areas of human geography. Starting in the late 1980s and continuing through today, Agnew has been the leading figure in the effort to create a theoretically informed, place-based electoral geography. In his work on politics in Italy and the United States, Agnew has moved the subfield beyond its former rampant empiricism and lack of concern for social theory. As part of this effort, Agnew has played an important role in theorizing and problematizing the concept of place.

Agnew has been a key figure in the formation of the field of critical geopolitics. Critical geopolitics seeks to make apparent the geographical assumptions and understandings, and implicit and explicit meanings that foreign policymakers and others give to places that serve to justify their geopolitical actions. In his work, Agnew traces the historical foundations of the modern geopolitical imagination through the 19th and 20th centuries.

Agnew has also played an important role in reconceptualizing the main concepts of international relations, most famously through his critique of the "territorial trap," problematizing the state-centrism that dominates much of geopolitical and international relations thought. In addition, he has been a leading writer in the areas of hegemony, sovereignty, and globalization.

In studies of international political economy, Agnew has demonstrated the important role that regional, national, and global economies and economic policies play in (geo)politics; detailed the uneven spatial impacts of globalization within countries; and recognized the role of core-periphery impacts and relations within countries, as well as across countries, within the world economy.

Through his many publications, John Agnew has played a critical role in political geography's renaissance in the late 20th century and is recognized today as one of human geography's leading scholars.

Jonathan Leib

See also Critical Geopolitics; Political Geography

Further Readings

Agnew, J. (1987). *Place and politics: The geographical mediation of state and society.* Boston: Allen & Unwin.

Agnew, J. (1987). *The United States in the world economy.* Cambridge, UK: Cambridge University Press.

Agnew, J. (2002). *Place and politics in modern Italy.* Chicago: University of Chicago Press.

Agnew, J. (2003). *Geopolitics: Re-visioning world politics* (2nd ed.). London: Routledge.

Agnew, J. (2005). *Hegemony: The new shape of global power.* Philadelphia: Temple University Press.

Angew, J. (2009). *Globalization and sovereignty.* Lanham, MD: Rowman & Littlefield.

Agnew, J., & Corbridge, S. (1995). *Mastering space: Hegemony, territory and international political economy.* London: Routledge.

AGRICULTURAL BIOTECHNOLOGY

Agricultural biotechnology has been a topic of considerable social and environmental controversy since it was first pursued by researchers and later adopted by industrial agriculture. The term usually refers to living agricultural inputs, such as seeds that use genetic engineering, though some institutions define it much more broadly. According to the Organisation for Economic Co-operation and Development, *biotechnology* means the application of science and technology to living organisms and their parts, products, and models thereof and to alter materials for the production of knowledge, goods, and services. Such a broad definition applied to food and agriculture can describe everything from beer brewing and bread baking to all of plant breeding since the start of sedentary civilization. The international Convention on Biological Diversity defines biotechnology in similar terms in its biosafety protocol and revenue-sharing agreements for genetic resources. The use of broader meanings normalizes genetic engineering techniques and lumps them together with traditional and conventional plant breeding.

The first pronouncements of the promise of agricultural biotechnology in the early 20th century suggested that bioreactors could produce single-cell proteins that could feed the poor. Similarly, today agricultural biotechnology proponents cite the promise of genetic engineering solutions to mitigate hunger, chemical pollution, fertilizer use, and soil erosion. The use of genetic engineering techniques allows scientists to move and manipulate an organism's DNA (deoxyribonucleic acid), which gives them an unprecedented range of novel genetic combinations. The rapid adoption of genetically engineered corn, soy, canola, and cotton on over 100 million hectares by 2008 has led some to call this seed technology the most rapidly adopted in history. These seeds predominantly (>99.9%) exhibit the traits of either insect resistance (using genes from *Bacillus thuringiensis*, Bt) or herbicide tolerance, though many other kinds of crop traits are in development.

The first product of agricultural biotechnology to receive considerable public attention was the deliberate release of the "ice minus" bacterium developed by a University of California biologist. The concern with this genetically engineered organism (GEO) aroused local legal reactions in the California Bay Area cities of San Francisco, Berkeley, and Oakland. The project was to spray potatoes in Northern California with an "ice-nucleation-active" bacterium in order to inhibit the formation of frost on the plants. The National Institute of Health's Recombinant DNA Advisory Committee eventually approved these field tests. But the activist Jeremy Rifkin, of the Foundation on Economic Trends, obtained a court injunction to stop the release, arguing that the experiment posed an environmental hazard and that there were no adequate containment protocols in place. The fallout from the controversy led to a new round of discussions about how GEOs should be regulated, ultimately leading to a decision by the U.S. Congress in 1985 to regulate them through the existing regulatory system. Under what is known as the "Coordinated Framework," the U.S. Food and Drug Administration evaluates food safety concerns, the U.S. Environmental Protection Agency oversees concerns about toxicity from mobile plant tissues such as pollen and root exudates, and the U.S. Department of Agriculture's Animal and Plant Health Inspection Service oversees problems related to increased weediness and biological invasion. Many activists and ecologists simply saw the Coordinated Framework as an effort to manage GEO introduction instead of regulating it.

Since the first plans to release them in the early 1980s, products of agricultural biotechnology have received criticism from the Ecological Society of America, which subsequently released a statement noting the potential ecological and environmental hazards associated with releasing GEOs. They noted that the products of genetic engineering posed no new classes of ecological hazards. However, the novelty of the new technology warranted regulatory oversight, because there is potential for more extreme and uncertain ecological hazards. They strongly advocated a robust system of containment to prevent the escape of the genes into the environment, by either seed dispersal or cross-pollination. They also argued for early ecological risk assessments of GEOs to ensure that they did not pose significant risk to native biota. The early planning for ecological risks poses challenges, given the importance of trade secrets in the development of these commercial products.

Ecologists' opinions on the risks from GEOs vary widely. Some ecologists suggest that some organisms pose threats to the environment, while others argue that GEOs will suffer greater fitness consequences from having their physical expression altered in ways that make them more dependent on human intervention to live in areas outside of agricultural production. They urge an approach to evaluating the risks of biotechnology, recognizing that uncertainty, complexity, and incomplete knowledge must be factored into any regulatory approach, from postrelease monitoring to more comprehensive ecological risk assessments.

The specific categories of risks discussed regarding agricultural biotechnologies include those related to food safety—for example, potential toxics and allergies from the novel combinations or antibiotic resistance from the marker genes used in the new genetic constructs. They include a range of ecological risks, from those associated with invasive plants or weeds to those related to loss of genetic diversity. Additionally, there may be nontarget impacts, as was discovered by one scientist who investigated the impact of Bt corn on the monarch butterfly.

Of the more controversial products of agricultural biotechnology is Posilac (originally developed by Monsanto and now controlled by Eli Lilly and Company), a recombinant bovine growth hormone given to cows to gain considerable rises in milk production per cow. It receives far less oversight in the United States than in Europe, where the growth hormone is banned. While there were few ecological risks presented by Posilac, its role in fostering utter infections is partly responsible for the high levels of antibiotics in milk. This also raises ethical questions about the use of these hormones in the context of animal health. For cows using Posilac, some argue, there are significant productivity gains that are painful for the animal. It also raised questions about the role of agricultural biotechnology in raising productivity, and forcing down commodity prices, a situation that could foster concentration in the food system. Several other Monsanto products have received negative publicity, including their RoundUp Ready seed lines for corn, cotton, canola, and soy. While makers of herbicide-tolerant seeds argue there are benefits for preventing soil erosion and simplifying crop management, others argue that overuse of the herbicides tied to these seeds will foster the evolution of weeds that are resistant to the chemical, eventually requiring a transition to more toxic chemicals. Monsanto has also received negative publicity for its terminator technology that renders seeds sterile as a sort of patent protection.

Dustin R. Mulvaney

See also Biotechnology and Ecological Risk; Biotechnology Industry; Ecological Risk Analysis; Factors Affecting Location of Firms; Genetically Modified Organisms (GMOs)

Further Readings

Kloppenburg, J. (2005). *First the seed: The political economy of plant biotechnology* (2nd ed.). Madison: University of Wisconsin Press.

Mills, L. (2003). *Science and social context: The regulation of recombinant bovine growth hormone in North America*. Montreal, Quebec, Canada: McGill-Queen's University Press.

National Research Council. (2000). *Genetically modified pest-protected plants*. Washington, DC: National Academies Press.

AGRICULTURAL INTENSIFICATION

There are a variety of definitions of agricultural intensification, but they all essentially refer to a process whereby inputs of capital and/or labor are increased to raise the productivity or yield (output) of a fixed land area. The academic interest in agricultural intensification has been concerned with the technical, social, and ecological details of local intensification processes as well as more broadly with agricultural intensification as key in the rise of complex societies and development in general. Agricultural intensification is a process that involves both societal (social, political, and economic) and environmental processes of change, and it has been an important research topic in geography, particularly during the second half of the 20th century. The challenge of meeting future global food demands, as well as framing policies for more sustainable food production and land use systems, means that agricultural intensification is likely to remain an important concept during the 21st century.

It is common to distinguish between two ways of intensifying agricultural production by increasing inputs of capital or labor. Capital-intensive agriculture depends on high inputs of capital, such as machinery, energy, and biotechnology, while labor-intensive agriculture primarily depends on high inputs of manual labor. If high inputs of land are used in an agricultural system, while capital and labor inputs are kept at a minimum, it is referred to as *extensive* agriculture, indicating its character of being the opposite strategy to that of *intensive* agriculture. Hence, a farming system with long fallow periods (e.g., swidden systems, slash-and-burn, or shifting cultivation) that in its totality encompasses a relatively large land area is more extensive than a system based on permanent tillage (i.e., with no or minimum fallow periods) of the same amount of land. Levels of agricultural intensity are thus commonly graded on a scale from "extensive" to "intensive" systems. A labor-intensive farming system is thus characterized by relatively high inputs of manual labor on rather small arable fields, while a capital-intensive system requires high inputs of capital investments per land area. This grading is part of the standard evolutionary model of agricultural and societal development, where societies are understood to progress from a primitive state—that is, with low technical and social complexity, low population density, and extensive production systems—to more advanced, complex, densely populated, and socially stratified societies with increasing levels of agricultural intensity. But, as detailed studies of agricultural intensification have shown, this historical scenario does not fit in all places. In some areas, intensive practices were also part of early agricultural strategies.

Studies of agricultural intensification have been closely linked to a broad range of issues related to the development of human societies and their use of natural resources, such as the evolution of political centralization, urbanization, land degradation, and population dynamics. Rapid population growth combined with limits on expanding cultivated land typically strengthens the imperative to intensify agricultural production, and from a policy perspective, concerns with agricultural intensification are generally grounded in the fundamental question of how to produce enough food for future consumers. To what extent processes of agricultural intensification can indeed satisfy the three critical development goals of (1) agricultural growth, (2) poverty alleviation, and (3) sustainable resource use is a critical issue in contemporary development research and policy debates. The Green Revolution, by which agricultural productivity was greatly enhanced in many countries during the 1950s to 1980s, is a classical and much debated case of the positive and negative effects of agricultural intensification.

A major contribution to the study of agricultural intensification as a historical and geographical process was the work by the Danish economist Ester Boserup (1910–1999). In her 1965 book *The Conditions of Agricultural Growth*, she argued against the Malthusian doctrine that population growth would necessarily lead to problems of insufficient food production and environmental and social stress. What Boserup convincingly showed was how farming societies in different parts of the world and in different historical contexts instead demonstrated a remarkable capacity to intensify agricultural production as population numbers increased and in many cases also succeeded in meeting the demands of a growing population. Boserup's theory of population-pressure-induced agricultural intensification has been extensively discussed, criticized, and modified, but it still remains a fundamental reference for contemporary studies of agricultural intensification.

By definition, agricultural intensification is a process that takes place in a specific geographical and historical context, and the academic study of agricultural intensification has consequently mainly relied on individual case studies of agricultural change, over both long and shorter time periods. But comparisons between different areas, regions, and historical periods have also been an important analytical method. Because agricultural intensification is a process that is always situated in a local context, it feeds on a complex set of interrelating factors (societal and environmental). In contemporary research, it is thus common to stress that the processes of agricultural intensification are often multicausal and dynamic. This means that there is great variety in how and why agricultural production has been intensified in different places. Nonetheless, most studies of agricultural intensification focus on a few key driving forces. Besides population pressure, the roles of markets, infrastructure, political pressures, and biophysical conditions (e.g., soils and climate) are the most important.

Lowe Börjeson

See also Agricultural Land Use; Agriculture, Preindustrial; Environmental Impacts of Agriculture; Malthusianism; Neo-Malthusianism; Population and Land Use; Population, Environment, and Development; Shifting Cultivation; Soil Conservation; Sustainable Agriculture

Further Readings

Boserup, E. (1965). *The conditions of agricultural growth*. London: George Allen & Unwin.
Brookfield, H. (1984). Intensification revisited. *Pacific Viewpoint, 25,* 15–44.
Lee, D., & Barret, C. (Eds.). (2001). *Tradeoffs or synergies? Agricultural intensification, economic development and the environment*. Wallingford, UK: CABI.
Turner, B., II, Hyden, G., & Kates, R. (Eds.). (1993). *Population growth and agricultural change in Africa*. Gainesville: University Press of Florida.

 AGRICULTURAL LAND USE

Agricultural land is land fit for producing crops and livestock. Land used for (a) cultivation of food products; fibers such as cotton, wool, and silk; and tobacco and medicinal plants and (b) livestock production, such as aquaculture, cattle and beef, dairy products, hogs and pork, and sheep, lamb, and mutton can be called agricultural land. Agriculture is the largest economic sector in the world, and it employs more than 30% of the world's population and is the economic backbone of many countries.

Cropland and pasture, vineyards, nurseries, orchards, horticultural areas, and groves are the general categories used in agricultural land use classifications. Land that is confined to feeding operations; breeding facilities in farms; and ditches, canals, and small farm ponds are also categorized under agricultural land use. Cropland and pasture, the most basic category of agricultural land use, include harvested lands, fallow lands, annual or seasonal croplands, and croplands used for improving soils and legumes. Land used for permanent crops such as fruit plantations and nuts and land used for cultivation of vines are classified as orchards, groves, and vineyards. Specialized livestock production facilities and large enterprises with confined feeding, poultry farms, and built-up or open facilities with a large animal population can be categorized under agricultural land with confined feeding operations. Horse farms and storage facilities cannot be included in this category. Breeding facilities, farm lanes, canals, small farm ponds, and horse farms cannot be classified under any of the above-specified categories and are termed as other agricultural lands. Arable land, land under permanent crops, and permanent pastures are the basic classifications used by the Food and Agriculture Organization of the United Nations. Distinguishing boundaries for agricultural land use is difficult in places where there is a mixture of land uses, such as wetlands. Cropland, grassland, pasture and range land, and forests account for 76% of land in the world, and land classified as agriculture constitutes more than 46% of land in the United States (Figure 1).

Agricultural land use often gets converted to other incompatible, nonagricultural land uses, such as residential, commercial, recreational, and industrial land uses, which can be broadly classified as urban land use. Understanding the factors driving agricultural land use conversions is critically important because policy initiatives need to be formulated to confront this dangerous trajectory.

Agricultural land value frequently determines agricultural land use and is often considered in calculating farm sector economic indicators. Because of its economic value as a source of crop production, as a habitat for wildlife, and as a source of open space, agricultural land plays a vital role in the agrarian economy. Socioeconomic factors, policy reforms, demographic changes, technological developments, and urban growth are some of the major factors leading to agricultural land use conversion. Over the past

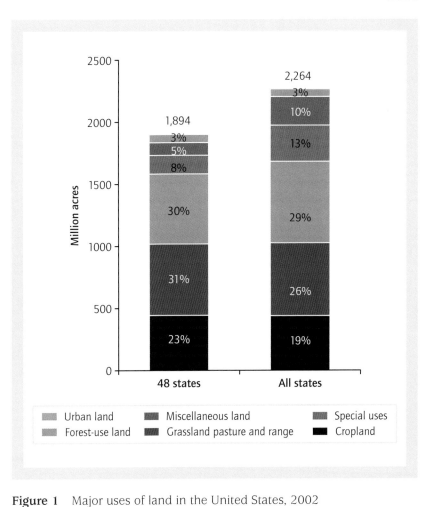

Figure 1   Major uses of land in the United States, 2002

Source: U.S. Department of Agriculture, Economic Research Service. (2002). *Land use, value and management: Major uses of land*. Retrieved December 10, 2008, from www.ers.usda.gov/Briefing/LandUse/majorlandusechapter.htm.

few years, the majority of agricultural land has been converted to forest land, urban land, or built-up land. Conversion of agricultural land to forest land is attributed to reclassification and policy reforms. Net shifts between cropland and forest land are illustrated in Figure 2. Conversion of agricultural land to urban or built-up land is mostly due to urban sprawl and the higher values placed on land that is developed within suburban and peri-urban areas.

The majority of the lands converted from nonagricultural uses to agricultural use between 1982 and 1997 were forest lands (55%). During the same period, most of the agricultural lands that were converted to nonagricultural lands became some sort of "built-up" land. Net land use shifts are demonstrated in Table 1.

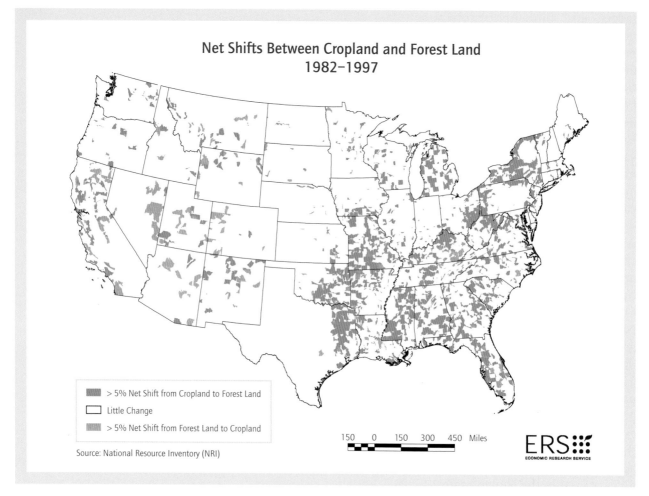

Figure 2 Net shifts between cropland and forestland, 1982–1997

Source: U.S. Department of Agriculture, Economic Research Service. (2002). *Land use, value and management: Major uses of land.* Retrieved December 10, 2008, from www.ers.usda.gov/Briefing/LandUse/agchangechapter.htm.

Land Use	Nonagricultural Shifts to Agriculture (1,000 acres)	Agricultural Shifts to Nonagricultural Uses (1,000 acres)	Net Land Use Shifts (1,000 acres)
Forest land	8,304	22,720	−14,416
Minor land	3,120	6,481	−3,361
Urban and built-up land	3	13,924	−13,921
Rural transportation	383	685	−302
Water areas	654	1,500	−846
Federal land[a]	2,685	4,101	−1,416
Total	15,148	49,410	−34,262

Table 1 Sources of land use shifting into and out of agriculture, 1982–1997

Source: U.S. Department of Agriculture, Economic Research Service. (2002). *Land use, value and management: Major uses of land.* Retrieved December 10, 2008, from www.ers.usda.gov/Briefing/LandUse/agchangechapter.htm.

a. "Federal land" is not a land use. While the Natural Resources Inventory (NRI) shows federal land, it does not account for land use changes on federal land.

Cropland (million acres)	1992	1993	1994	1995	1996	1997	1998	1999	2000	2001	2002	2003[a]	2004[a]
Cropland used for crops[b]	337	330	339	332	346	349	345	344	345	340	340	341	336
Cropland harvested[c]	305	297	310	302	314	321	315	316	314	311	307	315	312
Crop failure	8	11	7	8	10	7	10	8	11	10	17	10	9
Cultivated summer fallow	24	22	22	22	22	21	20	20	20	19	16	16	15
Cropland idled by all													
Federal programs[b]	55	60	49	55	34	33	30	30	31	34	34	34	35
Annual programs	19	23	13	18	0	0	0	0	0	0	0	0	0
Conservation Reserve Program	35	36	36	36	34	33	30	30	31	34	34	34	35
Total, specified uses[b,d]	392	390	388	388	380	382	375	374	376	374	374	376	372

Table 2 Major uses of cropland, United States, 1992–2004. Includes the contiguous 48 states. Fewer than 200,000 acres were used for crops in Alaska and Hawaii.

Sources: Adapted from Economic Research Service (ERS), U.S. Department of Agriculture. (2006). *Land use*. Washington, DC: Economic Research Service; National Agricultural Statistics Service (NASS), 1999; and unpublished data from the USDA Farm Service Agency, ERS, and NASS. Retrieved December 10, 2008, from www.ers.usda.gov/publications/arei/eib16/chapter1/1.1.

a. Preliminary, subject to revision.

b. Breakdown may not add to totals due to rounding.

c. A double-cropped acre is counted as 1 acre.

d. Does not include cropland pasture or idle land not in federal programs that is normally included in the total cropland base.

Decreases in total cropland constitute the largest decrease in agricultural land use during the same period. Except for conservation reserves and croplands used only for pasture, all other forms of agricultural land, such as cropland, idle cropland, and grassland pastures, have decreased during this period (Table 2).

Trends in land use shifts were different among the regions. Urban sprawl has negatively affected croplands in the northeastern part of the country for a considerable period of time. At the same time, grassland and pastureland conversions have led to an increase in the number of agricultural lands in the mountain regions. Figures 3, 4, and 5 depict a series of images that reflect land use shifts by region: major land covers, 1987 (Figure 3); shares of land in major uses, 2002 (Figure 4); and increase in developed land, 1992–1997 (Figure 5).

Factors Driving Agricultural Land Use Change

Land use change is often based on the opportunities and constraints involved in socioeconomic and demographic factors and policy reforms. Population growth or increase in population density, housing growth, and residential development

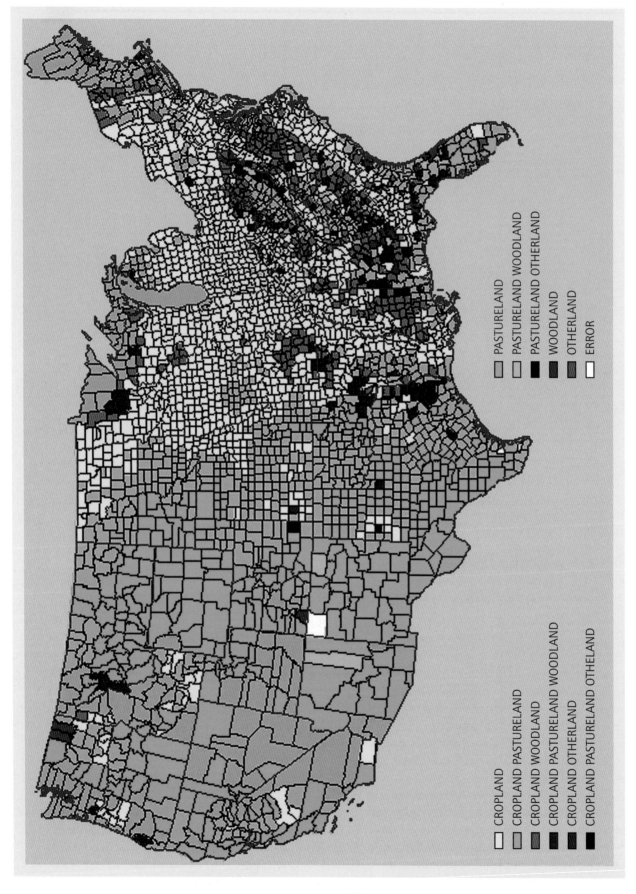

CROPLAND
CROPLAND PASTURELAND
CROPLAND WOODLAND
CROPLAND PASTURELAND WOODLAND
CROPLAND OTHERLAND
CROPLAND PASTURELAND OTHELAND

PASTURELAND
PASTURELAND WOODLAND
PASTURELAND OTHERLAND
WOODLAND
OTHERLAND
ERROR

Figure 3 Major land covers from the 1987 census of agriculture

Source: Gisiger, A., Cooper, E. S., Riggle, S., Limp, W. F., & Hodgson, T. W. (1997, August). *RCA III cultural resources in America and agricultural land use: An initial national profile* (Working Paper No. 17, Figure 20, p. 21). Fayetteville: University of Arkansas, Center for Advanced Spatial Technologies.

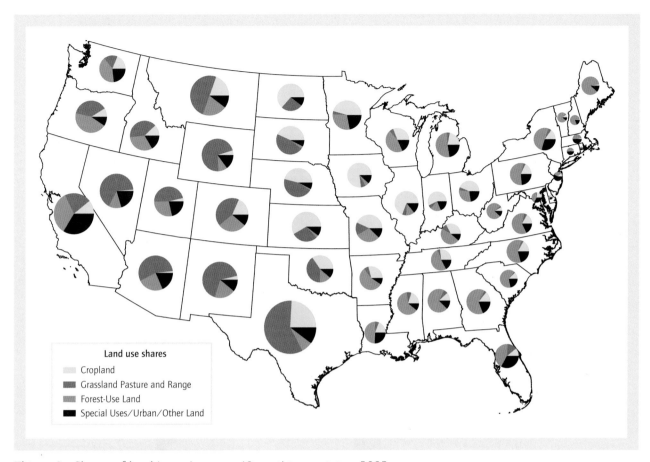

Figure 4 Shares of land in major uses, 48 contiguous states, 2002

Source: U.S. Department of Agriculture, Economic Research Service. (2007). Data sets—Major land uses: Map. Retrieved March 17, 2010, from http://www.ers.usda.gov/data/majorlanduses/map.htm.

are some of the demographic factors that lead to decisions to convert agricultural land use to other uses. Household incomes, type of employment, accessibility to markets, income earnings on crop production, and distance from urban areas are some of the socioeconomic factors leading to land use conversions. Most of these factors are attributed to urbanization, population growth, migration, rise in land values, changes in employment, income growth, increasing demand for residential land, rise in house rents, and increasing housing and road densities.

Increase in urban and built-up areas forces the urban fringe to develop rapidly, and often agricultural land is the most economical and accessible land located at the urban-rural fringe. Increase in housing density leads to construction of new roads and infrastructure facilities, which

leads to more land use conversions. Typically, suburban development consists of small parcels for single-family and multifamily housing developments extending well beyond the urban core. Eventually, as the urban area expands, these areas develop into major activity centers, driving the urban area to more rapid growth.

High growth rates associated with increased land values help developers lure farmers to make more money by selling their land. Increased marketing opportunities lead to shifts in agricultural operations that result in a decrease in traditional farming communities. Increased access to labor and higher land values further help these conversions. Employment patterns change over time, and so do land values. Over a period of time, this becomes a cyclic process, and the newly developed urban fringe starts looking for expansion

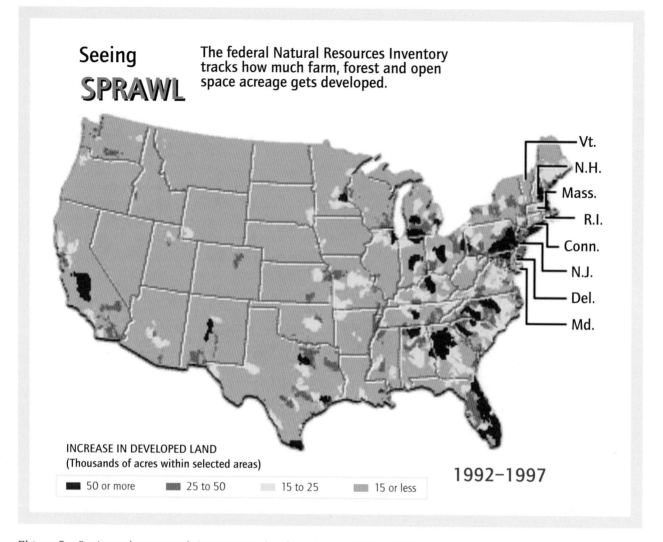

Seeing SPRAWL

The federal Natural Resources Inventory tracks how much farm, forest and open space acreage gets developed.

Vt.
N.H.
Mass.
R.I.
Conn.
N.J.
Del.
Md.

INCREASE IN DEVELOPED LAND
(Thousands of acres within selected areas)

■ 50 or more   ▨ 25 to 50   ░ 15 to 25   ▧ 15 or less

1992–1997

Figure 5   Seeing urban sprawl: Increase in developed land, 1992–1997

Source: U.S. Department of Agriculture. (2002). *Land use*. Delaware, OH: Northeastern Research Station. Retrieved December 10, 2008, from www.fs.fed.us/ne/delaware/biotrends/trends_humanpop.html.

again. Local governments tend to ignore or overlook the land use policies targeted at land management and protection of agricultural lands. Policies such as purchases of development rights and those targeted at residential and infrastructure development are also factors leading to conversions. Changes in land value calculation and tax assessment policies may result in a decrease in the annual benefits accruing from land that is still under agricultural production. Lack of stringent policies for protection of open spaces, soil conservation, and biodiversity also helps developers hunt for suitable land beyond the permissible limits.

Protecting Agricultural Land

Throughout the world, the demand for food products has resulted in programs and policies for protecting agricultural lands. In the United States, several farmland protection measures at various levels (right-to-farm laws, transfer of development rights, agricultural zoning, agricultural districts, smart growth and growth management strategies) have been developed to strengthen local agrarian communities and provide for rural amenities and open space. The U.S. Department of Agriculture coordinates with other federal, state, and local agencies and nongovernmental organizations to protect farmland through various programs.

Nagaraj Kanala

See also Agricultural Intensification; Agriculture, Industrialized; Agrofoods; Food, Geography of; Land Reform; Land Use Analysis; Land Use and Cover Change (LUCC); Rural Geography; Sustainable Agriculture

Further Readings

Fischel, W. A. (1982). The urbanization of agricultural land: A review of the National Agricultural Lands Study. *Land Economics, 58*(2), 236–259.

Lubowski, R., Claassen, R., Bucholtz, S., Cooper, J., Gueorguieva, A., Johansson, R., et al. (2006). *Environmental effects of agricultural land-use change: The role of economics and policy.* Washington, DC: U.S. Department of Agriculture. Retrieved December 10, 2008, from www.ers.usda.gov/Publications/ERR25

Lubowski, R., Vesterby, M., Bucholtz, S., Baez, A., & Roberts, A. (2006). *Major uses of land in the United States.* Washington, DC: U.S. Department of Agriculture. Retrieved December 10, 2008, from www.ers.usda.gov/publications/EIB14

Wiebe, K., Tegene, A., & Kuhn, B. (1996, November). *Partial interests in land, policy tools for resource use and conservation* (AER-744). Washington, DC: U.S. Department of Agriculture, Economic Research Service.

AGRICULTURE, INDUSTRIALIZED

The industrialization of agriculture is the foundation of modern societies in dominant theories about economic development. Embedded in this vision of progress is the assumption that increasingly mechanized agriculture will constitute a declining share of the employment structure over time. This shift has been most dramatic in the temperate heartlands of industrial agriculture in North America, Europe, Australia, and the southern cone of South America. On a global scale, agriculture has also constituted a declining share of the total workforce, a phenomenon sometimes described as "de-peasantization." Though the pace of global de-peasantization has been highly uneven, the net magnitude is indicated by the fact that while the world's population was still predominantly rural and agrarian at the outset of the 20th century, only a century later half of humanity was living in urban areas.

The positive association with this trajectory is best encapsulated in the imagery of industrialized agriculture releasing or freeing people from the drudgery of farmwork and rural life. Immeasurably less attention has been given to its precarious biophysical foundations, intractable externalities, and subsidized competitiveness (both directly and indirectly) and the social upheavals associated with fast-paced and job-scarce de-peasantization.

Yield Gains Versus Ecological Problems

One of the basic systemic tendencies in the industrialization of agriculture is to simplify farm environments—reducing biological diversity on a range of scales, from extensive monoculture fields and massive animal factories to soil micro-biota and plant and animal genetics—in order to replace labor with technology and standardize methods of production. When coupled with the heavy use of inputs and scientific innovations in crop breeding ("enhanced seeds"), this transformation was associated with phenomenal yield gains, particularly in the 1960s and 1970s. But this should not obscure the fact that nature is an uncooperative partner in the construction of scale economies in agriculture and that mechanization and homogenization of farm environments lead to a host of inexorable biophysical problems.

Monocultures and mechanized plowing, planting, and spraying increase vulnerability to soil erosion and nutrient loss, such that industrial farms are often described as mining the soil. To address the chronic degradation of soil fertility, industrial farms require regular application of external inputs, most crucially nitrogen and secondarily phosphorus and potassium, which come overwhelmingly from inorganic chemical sources.

The yield gains of industrial agriculture throughout the 20th century are inconceivable apart from the soaring production of inorganic fertilizer, in particular the natural-gas-dependent manufacture of synthetic nitrogen fertilizer (the Haber-Bosch process), which accounts for almost 60% of the world's total fertilizer consumption. Inorganic fertilizers also have the advantage of getting absorbed more quickly by plants than do most organic sources, which enhances the consistency of nutrient uptake, especially when coupled with irrigation. However, inorganic fertilizers ultimately rest on a nonrenewable resource base and are hence only a temporary fix for the soil problems of industrial agriculture.

Biological homogenization also increases the problems posed by pests, weeds, fungus, and disease, which entail the use of a range of petrochemical-based pesticides, herbicides, and fungicides. This dependence is compounded by the notorious treadmill effect, which describes the need for protracted chemical usage as natural predators and controls are eliminated and for new chemicals to be introduced as resistance develops over time. High-yielding, high-input crop varieties have necessitated massive expansions in irrigation infrastructure and water usage, as well as the unsustainable drawdown of underwater aquifers, most famously the vast Ogallala Aquifer, which underpins the productivity of the arid Midwestern United States.

The Livestock Revolution

The yield gains of industrial monocultures are tightly linked to what has been called the "livestock revolution," the rising place of animal flesh and derivatives in agricultural systems and ultimately in human diets. The growth of livestock populations since 1950 has dwarfed that of humankind, reflected in a momentous dietary shift on a global scale: Human beings now eat roughly twice as much meat as they did half a century ago, with industrialized countries consuming more than two and a half times as much meat per capita as developing countries. Pioneered in the United States, the integration of industrial grain and livestock production is now occurring very quickly and dramatically in China and other rising economies.

The increasing cycling of grains (led by maize) and oilseeds (primarily soybeans) through livestock has allowed farm animal populations to increase far beyond rangeland stocking capacities. This has gone hand in hand with the proliferation of concentrated animal-feeding operations (CAFOs), or "factory farms," and animal-breeding innovations that have accelerated weight gain and egg and milk production. Because large percentages of usable plant protein, carbohydrates, and fiber are lost in the process of cycling grains and oilseeds through livestock, as the level of meat production expands, so too does the area of land needed for agriculture and the volume of industrial inputs required.

As with monoculture farming, the application of industrial techniques and the drive for economies of scale in livestock rearing face serious biophysical constraints. Crowded animal populations are highly susceptible to disease outbreaks, induce deviant behaviors that are detrimental to production, and generate levels of fecal waste and byproducts that cannot be absorbed in nearby landscapes. The biophysical limits to scale in industrial livestock rearing are overridden through a combination of technological fixes, such as mass applications of antibiotics, disinfectants, and hormones; large fecal cesspools and lagoons; and the assembly-line debeaking of chicks.

Role of Multinational Corporations

One of the most important dimensions of the industrialization of agriculture is the rising control over decision making and value possessed by transnational corporations. The agro-input industry is a complex and ever more oligopolistic web in which corporations control large shares of the global markets in chemicals, fertilizers, seeds, and animal pharmaceuticals and weave together input usage. Corporate control over agricultural processing, distribution, and retailing is also intensifying, with vertical and horizontal integration occurring at a swift pace. The concentration of value and power on both the input and the output sides of industrial agriculture is a significant factor in the long-term increase in input costs and decline in farm-gate prices, which has pushed farmers to grow in scale or be driven out of business.

According to the U.S. Department of Agriculture, in the beginning of July 2009, there were 11.6 million head of cattle being fattened for slaughter on feedlots in the United States. About 84% of these (9.8 million) were on large feedlots, with a capacity of 1,000 or more head. Shown above is the former Montfort feedlot near Greeley, Colorado. Established in 1930, it was the first U.S. feedlot with a capacity of 100,000 head. It was sold to ConAgra in 1987.

Source: Glowimages/Getty Images.

In sum, industrial agriculture has radically reconfigured the biophysical nature of farming from relatively closed nutrient and energetic systems into a through-flow process of corporate-dominated inputs and outputs. This transformation hinges on fossil energy and petrochemical derivatives to a tremendous and often underappreciated extent, from heavy machinery to the manufacture, application, and transport of fertilizer and agrochemicals, to the long-distance movement of food durables. In addition to being thus implicated in the rising carbon emissions, industrial agriculture, and the rising livestock populations in particular, is also a large source of methane emissions. In 2008, the Intergovernmental Panel on Climate Change identified global livestock production as a major force in the destabilization of Earth's atmosphere.

External Costs and Subsidies

The environmental costs of industrial agriculture extend much further than this atmospheric burden. The collective impact of fertilizer and agrochemical runoff, CAFOs, and large-scale abattoirs makes industrial agriculture the principal cause of water contamination wherever it is practiced, with diffuse and difficult-to-measure impacts on human and ecosystem health. The cost of soil degradation is even harder to measure, given the longer temporal horizon needed. The intensive confinement of suffering animals in battery cages

Large operation in northeastern Colorado with many silage choppers and trucks harvesting corn silage (ensilage) in a circular field, with a beef feedlot in the distance

Source: Dunbar Arnold/AgStock Images/Corbis.

and gestation crates might also be seen to constitute another cost beyond measure. Yet whatever challenges there are in defining and measuring the environmental costs of industrial agriculture, it should be clear that they are overwhelmingly "externalized"; that is, they go unmeasured in conventional accounting systems and hence have no bearing on the competitiveness of industrial foods. Viewed another way, these unaccounted environmental costs might also be seen as an implicit subsidy to industrial agriculture.

Government subsidies have also enhanced the competitiveness of industrial agriculture, most notably in the United States and the European Union (EU). The agrosubsidy regimes of the United States and the EU were partly a response to the market problems facing farmers as the booming productivity of mechanized, input-intensive agriculture outpaced domestic demand, and these subsidies have helped keep these systems operational amid the "cost-price squeeze" noted earlier. However, they have done so by exacerbating the polarizing tendencies at work: At the same time as low unit margins have driven strong pressures for farmers to grow in scale,

agricultural subsidies have been concentrated very disproportionately on the largest farmers.

Given that one of the primary motives for these subsidy regimes has been to manage a state of chronic surplus, they are linked to an export imperative that was fostered over time via aid, trade promotion, and various forms of "dumping" (i.e., dispensing exports at prices below those in domestic markets). The flipside is that many of the world's poorest countries have come to depend on the relatively cheap, industrial food imports and run persistent net food trade deficits. These imports—in which various implicit and explicit subsidies are embedded—have also presented competitive pressures for small farmers in many nations.

Current Social Issues

The precarious position of small farms in the face of market liberalization and subsidized industrial competition and the projected impacts of climate change pose vexing social challenges. In spite of the magnitude of global de-peasantization in the 20th century, small farmers still constitute more than two fifths of humanity, and in most of the world's poorest nations (especially in sub-Saharan Africa and South Asia), large shares of the workforce are still engaged in farming and face limited prospects of absorption into industry or the services.

At the same time, cracks in the biophysical foundation of industrial agriculture have begun to appear, manifested in the sharply rising global food prices since 2006. As global petroleum reserves decline and bring inevitable cost increases, a crucial implicit subsidy underpinning cheap industrial food correspondingly wanes. This is augmented by the fact that industrial agriculture has become a crucial source of liquid biofuels in response to declining oil reserves, especially in the United States and Brazil. The global volume of grains and oilseeds devoted to biofuel production is on a sharp incline, a trend that is again led by the temperate heartlands of industrial agriculture and with subsidies an important part of the process. The volume of industrial grains and oilseeds devoted to motor vehicles and to the still fast-rising livestock populations is projected to continue increasing in the coming decade.

The sudden price spikes in industrial foods after decades of steady decline in real terms has intersected profoundly with the food import dependencies in the world's poorest countries. Prior to the price increases, there were already an estimated 854 million people who were chronically malnourished, at the same time as an estimated 1.2 billion people were obese. With the primary agents directing industrial agriculture principally motivated by effective demand (i.e., purchasing power) rather than real human needs, the food-related social tremors witnessed in 2007–2008 threaten to intensify.

Conclusion

Ultimately, in place of the assumption that agrarian livelihoods will be progressively substituted by technology in modern societies, we see tremendous ecological and social instability and uncertainty. Machines and inputs that hinge on fossil energy and derivatives cannot continue to be substituted for human labor indefinitely, and agriculture that works with nature to enhance functional diversity, rather than to override and standardize it, is necessarily knowledge and labor intensive. The challenge of resubstituting scientific and local ecological knowledge, skilled labor, more diverse cropping patterns for machines and inputs, and biological simplification may appear daunting, even counterhistorical, but it does not mean going back in time. Rather, this may be one of humanity's most urgent challenges in the 21st century.

Tony Weis

See also **Agricultural Biotechnology; Agricultural Intensification; Agrochemical Pollution; Agrofoods; Biofuels; Food, Geography of; Land Use Analysis; Land Use and Cover Change (LUCC); Rural Geography**

Further Readings

Kimbrell, A. (Ed.). (2002). *The fatal harvest reader.* Washington, DC: Island Press.

Mazoyer, M., & Roudart, L. (2006). *A history of world agriculture.* New York: Monthly Review Press.

Weis, T. (2007). *The global food economy: The battle for the future of farming.* London: Zed Books.

AGRICULTURE, PREINDUSTRIAL

Throughout much of the world's history (indeed, dating back to the Neolithic Revolution 8,000 to 10,000 yrs. [years] ago), societies fed themselves through an assortment of preindustrial agricultural systems. Preindustrial or nonindustrial agricultural systems differ from industrialized ones in a variety of respects. Most important, preindustrial systems do not use the inanimate sources of energy that are vital to industrialized agricultural systems (e.g., fossil fuels) and, therefore, are markedly less energy intensive in nature. Rather, work in preindustrial farming systems is accomplished entirely through human or animal labor power. Thus, these types of farming are much more labor intensive. In societies fed predominantly through preindustrial agriculture, the vast bulk of people are engaged as farmers or peasants. Second, because preindustrial societies are often not fully commodified, that is, capitalist social relations have not come to dominate every aspect of production, preindustrial agricultural systems are

Farmer working in paddy field in Sekinchan Malaysia
Source: © Can Stock Photo Inc.

generally organized around production for subsistence rather than production for profit. Food, in other words, is mostly grown for local consumption rather than for sale in a market.

Preindustrial agricultural systems played an enormous role in the slave-based and feudal social systems that existed for millennia in most of the world. Roman *latifundia*—large estates worked by slaves—for example, formed the backbone of agricultural production in the Roman Empire. The expansion of medieval agriculture in the dense soils of Northern Europe was made possible by the introduction of the heavy plow and, later, the three-field system. The manorial system that formed the social and economic basis of feudal Europe involved peasants and serfs who rented land from large landowners, paying rent with a fraction of their output. Variations of peasant-based production continue to be important in many contexts.

Today, various forms of preindustrial agricultural systems are still found throughout the world, with large variations in the types of crops grown,

the methods used to plant and harvest them, their productivity, labor relations, and their relative vulnerabilities to drought or other hazards. While it is not technically a form of agriculture, many observers classify nomadic herding in this category, although it involves only the domestication of animals, not crops. Typically, nomadic herders measure their wealth in terms of livestock (generally cattle, goats, or reindeer) and often follow their herds in annual migratory cycles, such as transhumance, the movement between summer pastures in higher elevations and winter pastures in lower ones. Nomadic herding has been slowly vanishing throughout the world over the past two centuries, but contemporary examples include the Masai of East Africa, the Mongols in Mongolia and Northern China, the Tuareg of Northern Africa, and the Lapps of Northern Finland.

The best-known example of preindustrial agriculture is slash-and-burn, also known as swidden or shifting cultivation. This form is only found in tropical areas such as parts of Central America, the Amazon rain forest, West and Central Africa, and southeast Asia, to which it is ideally suited. Roughly 50 million people continue to be fed this way in these regions. Due to heavy rainfall and the subsequent leaching of nutrients, tropical soils are generally quite poor, and most nutrients are stored in the biomass. The first step in slash-and-burn, therefore, is to cut down existing trees and bushes in a given plot of land and burn them, releasing nutrients into the soil through the ash. Crops are then planted for several years. However, because the rate of nutrient extraction exceeds the rate of replenishment, the site can only be used for a brief period—generally 2 to 6 yrs.—and then farmers must move on to a new site. Abandoned sites may gradually recover with a sufficient fallow period. If there is rapid population growth and fallow periods are reduced, the soil may permanently decline in fertility. This form of farming was widely practiced in the Mayan kingdoms prior to the Spanish conquest, and declining soil fertility may have played a role in the collapse of the Mayan states.

A third form of preindustrial agriculture is the method of Asian rice paddy cultivation, which is widely practiced throughout a region stretching from Japan, Korea, and Southern China through southeast Asia into Eastern India. Rice is the staple crop for billions of people in Asia, and its cultivation in this form goes back millennia. Young rice plants require standing pools of water, and to create spaces in which this can occur, Asian societies carved countless terraces out of hillsides along elevation contour lines to control the flow of water with vast networks of dikes and small levees. Furrows may be dug using water buffalos. Often, small fish are grown in these pools of water as a source of protein. The planting of rice is exceedingly laborious, giving rise to stereotypes of Asian peasants bent over their paddies. The supply of water may depend on monsoon rainfall.

Preindustrial agricultural systems have functioned effectively for thousands of years and continue to do so in many parts of the developing world. Boserup showed that the rising populations in such places often stimulate growth in productivity. In most places, the preindustrial systems are marginalized or threatened by the expansion of globalized, capitalist industrialized farming systems, including imports of subsidized grains from Europe or North America. However, preindustrial systems enjoy advantages of their own, including a diversity of crops (in contrast to industrialized monocultures) and freedom from dependence on pesticides and petroleum. Thus, it may be helpful to view these systems not as backward remnant forms of food production but as historical adaptations to particular social and environmental contexts—that is, as nonindustrial rather than preindustrial.

Barney Warf

See also Agricultural Intensification; Agriculture, Industrialized; Bush Fallow Farming; Food, Geography of; Nomadic Herding; Peasants, Peasantry; Population and Land Use; Population, Environment, and Development; Shifting Cultivation

Further Readings

Boserup, E. (1965). *The conditions of agricultural growth: The economics of agrarian change under population pressure.* Chicago: Aldine.

Food and Agriculture Association, United Nations. (1984). *Changes in shifting cultivation in Africa.* Rome, Italy: Author.

Peters, W. (1988). *Slash and burn: Farming in the Third World forest.* Moscow: University of Idaho Press.

AGROBIODIVERSITY

The Convention on Biological Diversity (CBD) defines biodiversity as the variety of life on Earth and the natural patterns it forms. Our world's wide variety of plants, animals, and microorganisms and the genetic diversities within each species; the varieties of patterns they form in ecosystems; and the valuable products, services, and opportunities they provide are increasingly threatened by human exploitation. The CBD calls on all countries to conserve and sustainably use biodiversity in collaboration with local actors, such as farmers, because they take most of the actions that affect biodiversity.

Agriculture greatly affects biodiversity. Much of the world's farmlands are cultivated by small-landholding farmers in the developing countries, and the practices of these farmers affect biodiversity in a large part of the world. On the one hand, the practices of these farmers, such as the slash-and-burn agriculture of shifting cultivators in forest ecologies, have been derided as the primary cause of biodiversity loss. The conservation community has consequently called for biodiversity rehabilitation in such regions. On the other hand, there is an influential body of primary-research-based literature that is challenging this discourse on biodiversity endangerment by farmers in the developing world and instead emphasizes what can be learned from these farmers about in situ biodiversity conservation and sustainable use at both farm and landscape levels. This literature indicates that the farms and farming landscapes of hundreds of millions of farmers in developing countries are characterized by great diversity compared with the large uniform fields of farmers in the more developed world. Researchers and practitioners have noted the considerable extent to which small-landholding agriculture is related to the details of the landscape in which it occurs. Such agricultural strategies are the outcome of an accumulated body of local knowledge derived from experience and experimentation. This leads many small-landholding farmers to use the diversities of plant life on their fields to meet agronomic and livelihood goals; make optimal use of variations in agroecological conditions on their fields (such as differences in microclimate, soil properties, and vegetation); and modify or transform their landscapes to produce a great variety of crops and genetic diversities of particular crops and also create local crop varieties (known as landraces). The varieties of crops cultivated are not static. They are conditioned by ecological, cultural, and socioeconomic factors and thus change through time.

This diversity of crops and other plants used by or useful to people in actively managed agricultural ecosystems is referred to as agrobiodiversity. The diversity of crops and other plants found in the fields of many small-landholding farmers is the result of the interaction of three factors: (1) biophysical diversity in the landscape; (2) the diversity, flexibility, and adaptability of agricultural management practices; and (3) the diversity of socioeconomic processes that condition the organization of agriculture. Biophysical diversity in landscapes refers to diversities of the biotic environment (such as plant genetic resources) and the abiotic environment. The diversity, flexibility, and adaptability of agricultural practices for managing this biophysical diversity are based on indigenous ecological knowledge and include selection and management of crop and plant varieties; combinations and rotations to suit particular ecological niches; management of successional vegetation during fallows; management of slopes, water, and soil fertility; and management of weeds, pests, and diseases. The agricultural factors and processes that affect biodiversity include the different types and modes of access to and control over land, labor, and manufactured capital—all of which vary by the income, gender, generation, and citizenship of farmers and may facilitate or undermine biodiversity. Processes relating to government policies and the uncertainties and risks of markets are also important aspects of agricultural organization.

Agrobiodiversity occurs in various kinds of agricultural fields. In sub-Saharan Africa, some of these are permanent intensively cultivated fields in close proximity to the residences of farmers or extensively cultivated fields farther away from residences, temporary bush fallow fields (which are alternately cropped and fallowed), planted and managed agroforests, and mixed home gardens. In West Africa, some innovative research has shown how the following practices have conserved and allowed sustainable use of biodiversity: (a) soil fertility and soil property enhancement, weed control practices such as mounding and mulching of soils, and intercropping of specific crop combinations and (b) tree-biodiversity-friendly practices such as minimum tillage to protect the soil tree seed bank and seedlings; protection of saplings; coppicing; multiplying trees from suckers; selective weeding; deliberate fallow management, including planting and transplanting of trees; pruning and thinning of trees, giving priority to the preservation of tree species over crop yields; nurturing a variety of tree species that interact favorably with crops; and the extinction of fires. Several studies have documented the expansion of forests and their biodiversity as a result of such practices.

Agrobiodiversity also occurs in a variety of ecological niches where farmers use different crop plant combinations of immense genetic diversity and crop rotations over a range of land types and altitudes. In the Central Andes of Peru, Bolivia, and Ecuador, researchers have analyzed agroecosystems containing diverse world-renowned food plants, such as Andean potatoes, maize, quinoa, ulluco, barley, wheat, fava beans, and tarwi. They describe tiered agrobiodiversity zones of different elevations, although other researchers working in the same area suggest that the flexibility in practices has contributed to the current considerable overlap of land uses at different elevations and a patchwork of different crop types in adjacent fields.

Government policies, such as those that emphasize the Green Revolution and the gene revolution, food and labor market processes, and labor migration are among the factors that are severely undermining the capacities of small-landholding farmers to sustain their cultivation of biodiversity. The decline in the varieties of rice produced in India from 30,000 to only 10 is stark evidence of the magnitude of the threat to agrobiodiversity.

Louis Awanyo

See also **Agricultural Biotechnology; Agricultural Land Use; Agriculture, Preindustrial; Biodiversity; Biotechnology and Ecological Risk; Bush Fallow Farming; Population and Land Use; Population, Environment, and Development; Shifting Cultivation**

Further Readings

Brookfield, H. (2001). *Exploring agrodiversity*. New York: Columbia University Press.
Brookfield, H., Padoch, C., Parsons, H., & Stocking, M. (Eds.). (2002). *Cultivating biodiversity*. London: ITDG.

AGROCHEMICAL POLLUTION

An optimum nutrient level is essential for productive agriculture. Nutrients can be supplied either by natural sources or in the form of synthetic chemical fertilizers. However, the use of both natural and chemical fertilizers may result in an excess of nutrients, which can cause problems in water bodies and to health. This excess use of various fertilizers may pollute food crops and groundwater; this process is commonly called agrochemical pollution. Nitrates are highly soluble and, therefore, may quickly reach water bodies. Phosphates tend to be fixed to soil particles and can reach watercourses when soil is eroded. Phosphate-saturated soils and high phosphate levels in groundwater are found in many countries. This entry describes the main types of agrochemicals, the reasons for their use, and their effects.

Types of Agrochemicals

Agrichemical (or *agrochemical*) is a generic term for various chemical products used in agriculture. In most cases, agrichemical refers to the broad range of insecticides, herbicides, and fungicides, but it may also include synthetic fertilizers, hormones and other chemical growth agents, and concentrated stores of raw animal manure. Most agrochemicals are toxic, and all agrochemicals in bulk storage pose significant environmental and/or health risks, particularly in the event of accidental spills. Pollutants from agrochemical sources include fertilizers, manure, and pesticides. To these, we may add accidental spills of hydrocarbons used as fuel for agricultural machinery. The main effect of this is to introduce heavy metals and their compounds in soils—that is,

cadmium, arsenic, magnesium, zinc, and so on. Various harmful agrochemicals and their effects include the following:

Cadmium (Cd): Cd mainly comes from impurity in zinc-containing fertilizers. Its accumulation in the living system leads to growth retardation, diarrhea, kidney damage, liver injury, hypertension, and so on. Cd is soluble in water and is a major pollutant.

Chromium (Cr): Cr mostly comes from the fertilizer industry. It can lead to gastrointestinal ulceration, CNS (central nervous system) disease, cancer, and so on.

Copper (Cu): Cu also comes from chemical fertilizers. When Cu gets into living systems, it can lead to sporadic fever, hypertension, uremia, and even coma.

Mercury (Hg): Hg is a component of certain fungicides. It is a highly toxic chemical and acts as a cumulative poison, causing abdominal pain, headache, diarrhea, hemolysis, and so on.

Selenium (Se): Se is an ingredient of various fungicides and insecticides. Its effects can include damage to the liver, kidney, and spleen, fever, nervousness, vomiting, low blood pressure, blindness, and so on.

Zinc (Zn): Zn is an essential component of most fertilizers. It is one of the most common soil pollutants and can cause vomiting, cramps, and renal damage.

Effects of Agrochemicals

The use of pesticides in agriculture has been a major cause of pollution in soil ecosystems. Toxic, nondegradable chemicals from various pesticides accumulate in the soil and build up in concentrations that can threaten animals, crops, and plants. Agrochemicals pollute the soil directly by affecting the organisms that reside in it. Soil, however, acts as a vector for the pollution of surface water and groundwater. Persistent agrochemicals accumulate in soil and aquatic ecosystems, causing their biomagnifications in living organisms.

Agrochemicals may contaminate food crops and drinking water and endanger human health. They can lead to destruction of the soil's

The air-curtain orchard sprayer driven by the technician Andrew Doklovic uses multiple cross-flow fans to disperse pesticide to apple trees. Under some conditions, the smoother, gentler flow of air can reduce spray drift by half.

Source: Keith Weller/ARS/USDA.

microflora and fauna, leading to both physical and chemical deterioration and a severe reduction in crop yield. Agrochemical pollution has many ill effects on our agricultural systems. The most prevalent effect of agrochemicals is that they accumulate in the bodies of animals without killing them. If an animal receives small quantities of pesticides or other persistent pollutants in its food and is unable to eliminate them, the concentration within the animal's body increases over time. This process of accumulating higher and higher amounts of material within the body of an animal is called bioaccumulation. This phenomenon of increasing levels of a substance in the bodies of

higher-trophic-level organisms is known as biological magnification. Another problem associated with insecticides is the ability of insect populations to become resistant to pesticides, thus producing resistant species population. A third problem associated with pesticides is that they prove harmful for nontarget organisms, thus killing certain beneficial species as well.

Another dimension of agrochemical pollution was associated with the introduction of the Green Revolution in the 1970s for high production of food grains and multiple cropping. One side effect was the greatly increased use of agrochemical fertilizers in developing countries such as India and Mexico. Most high-yielding varieties of food crops can be sustained only with the help of an adequate freshwater supply, modern farming techniques, massive input of synthetic manures, and the use of chemical pesticides to ward off a variety of pests. Pests vary from insects and nematodes to fungi, herbs, and rodents. Generally, pesticides are used against a target group of organisms, usually insect pests, which are aimed to be controlled or exterminated. The important pesticides introduced were dichlorodiphenyl trichlororethane (DDT), BHC (benzene hexachloride), aldrin, and endrin, among others, and many of these are now banned by international treaty. It is unfortunate that when an insecticide is used with a particular insect pest as a target, many benevolent organisms are also exposed. These nontarget organisms are also affected. Any pesticide used in agriculture or horticulture and even for environmental sanitation in public health programs is a poison designed to kill a particular form of life and should not have negative effects on other forms of life. The effect that pesticides have on target as well nontarget organisms needs to be well studied and documented.

In India, pesticides were first introduced primarily to control mosquitoes and subsequently to control agricultural pests, starting with a modest use of a few tons. It has now reached a stage where there are approximately 120,000 metric tons of these poisonous chemicals used per year. Currently, India has 131 pesticides registered for use, some of which are under restricted use for public health purposes only. Yet some of these banned chemicals are easily available in the market. In addition, some of the chemical pesticides, such as DDT, have a long residency period, about 20 years in the environment. The question that should be addressed is how much of the total amount of pesticide used in an agricultural area reaches the target organism. Unfortunately, some useful microorganisms such as nitrogen-fixing bacteria, which are responsible for the fertility of soil; insects such as the honeybee; and certain organisms such as frogs, birds, and snakes, which are natural allies in our war against insects, rodents, and pests, are also affected.

Soil organisms acquire pesticides by coming in direct contact with them in soil or plant roots. The longer a pesticide remains in the environment, the greater are the chances that it will spread or likely be transported to another area from the one where it was applied. Drift and run-off also result in sufficiently high quantities of pesticides getting into water bodies. Thus, it is well documented that pesticides on application in various ecosystems for control of a particular target pest also find their way into almost all the parts of the ecosystem. Once they reach different systems, including biotic ones, they are bound to affect the apex-level organisms in food chains. Biomagnification has been common with persistent organochlorines and particularly with those that are not readily metabolized by vertebrates (e.g., DDT, aldrin). Ultimately, this leads to death of the organisms at the top of the food chain or indirectly to reproductive failure. Younger animals and those in the reproductive phase are more susceptible to pesticides than others. Of course, the nontarget organism we are most concerned with is humans. Although pesticides have never been used to control the human population, almost all of us carry a heavy load of persistent pesticides in our bodies. Recent studies indicate that expectant mothers are transmitting these to unborn fetuses through the placenta and, after delivery, through milk to the newborn.

Other commonly used agrochemicals include herbicides, which are the group of chemicals used to kill undesirable vegetation. They can be divided into two categories—selective and nonselective herbicides. The first one kills certain types of vegetation and is used depending on the need, for example, a pesticide designed to control or kill plants, weeds, or grasses. Almost 70% of all pesticides used by farmers and ranchers are herbicides. These chemicals have wide-ranging effects on nontarget species. Another important type of agrochemical is fungicides, which are used to control fungal diseases in agricultural crops. Common fungicides are sulfur and organic mercuric compounds such as methyl mercury and copper compounds. A third type is rodenticides, which are used against rodents, that is, rats and mice, which destroy agricultural crops. Important rodenticides are norbormide and strychnine.

Conclusion

Humans have been continuously increasing the quantity of pesticides in the environment. Also, some of the crop insect pests have developed immunity and resistance to these chemicals. On the other hand, most of the nontarget species that were taking care of these pests by devouring them may be killed or may have their reproductive systems impaired. Even humans, who have been targeting insects and other pests, may not be spared from the adverse effects of these silent killers, the pesticides. According to the United Nations Environment Programme, every year more than 10,000 people are killed from pesticide poisoning from agrochemical pollution in developing countries.

Atiqur Rahman

See also Agriculture, Industrialized; Agroecology; Agrofoods; Chemical Spills, Environment, and Society; Chlorinated Hydrocarbons; Heavy Metals as Pollutants; Nonpoint Sources of Pollution; Water Pollution

Further Readings

Merrington, G. (2002). *Agricultural pollution: Problems and practical solutions.* London: Taylor & Francis.

Prabhakar, V. (2001). *Environment and agricultural pollution.* New Delhi, India: Anmol.

Scheierling, S. (1995). *Overcoming agricultural pollution of water.* Washington, DC: World Bank.

⊕ AGROECOLOGY

Agroecology is one of several fields that emerged as a response to the environmental and social impacts of industrialized agriculture (also termed *conventional agriculture*). Industrial agriculture's focus on maximizing yields and profits has resulted in negative impacts on social and ecological systems around the world. Many of the practices associated with industrial agriculture (e.g., soil tillage; excessive use of water, pesticides, and fertilizers) have led to increasing degradation of the long-term productivity and health of agricultural land. The term *agroecology* began appearing more frequently in the sustainable agriculture literature in the 1970s through contributions of academics based mostly in the United States and Latin America. Some of the most influential among these people were Stephen Gliessman, from the University of California, Santa Cruz; Miguel Altieri from the University of California, Berkeley; John Vandermeer, from the University of Michigan; and Charles Francis, from the University of Nebraska. However, in what is widely recognized as the first textbook in the field, Gliessman traces the first forms of agroecology to the German geographer K. Klages, who published an article on crop ecology in 1928. This was followed by the first actual use of the term *agroecology* by the Czechoslovakian agronomist Basil Bensin in 1930, as part of a proposal to the then International Institute of Agriculture in Rome for an agroecologically based research agenda for agriculture. However, it was not until the past 25 years that agroecology has become a vibrant field of research and practice, with increasing importance in policy, academic, and field applications.

Early definitions of agroecology focused on the application of ecological concepts and principles to the design and management of sustainable agroecosystems. This initial conception of agroecology remains the most widely known, but a recent key publication by a group of renowned agroecologists has redefined and expanded the term as an interdisciplinary field that explicitly addresses social, economic, and ecological factors associated with food systems. This new definition by Francis and colleagues (2003) defines agroecology as "the integrative study of the ecology of the entire food system, encompassing ecological, economic and social dimensions" (p. 100). In Francis and colleagues' article and in subsequent contributions on the theoretical basis of the field, agroecologists have strongly shifted the focus from a farm-based approach to addressing the entirety of the food system, including production, processing, transportation, financial intermediation, marketing, and consumption.

Ecological Basis

Agroecology argues for a whole-systems approach to analyze agroecosystems and their surrounding environments, which examines components and their arrangements (structure), as well as their interactions and their impact on ecological processes (function). The ecological characteristics of the agroecosystem are linked to its plant, animal, and management components (Figure 1).

Natural Ecosystem and Traditional Agricultural Models

Agroecologists have usually relied on two main sources of inspiration to guide the design and management of agroecosystems and agricultural landscapes. The first one seeks to understand ecological processes in natural ecosystems, as they have proven to be resilient over time. The theory is that most of the ecological processes found in natural ecosystems can be replicated in agricultural fields, albeit in modified forms. An understanding of natural ecological processes is used to provide insight on how best to replicate these in an agricultural setting, with the goal of minimizing external synthetic inputs (e.g., fertilizers and pesticides) and maintaining important agroecosystem conditions, such as soil fertility.

The second source of inspiration for agroecological models is traditional agriculture, which is sometimes synonymously termed *local* or *indigenous*. Again, the idea is that agricultural systems that have persisted through time can provide lessons to design more sustainable modern agroecosystems. Many of the farming systems used as models by agroecologists come from the tropical regions, where agriculture has been practiced for thousands of years and where traditional systems

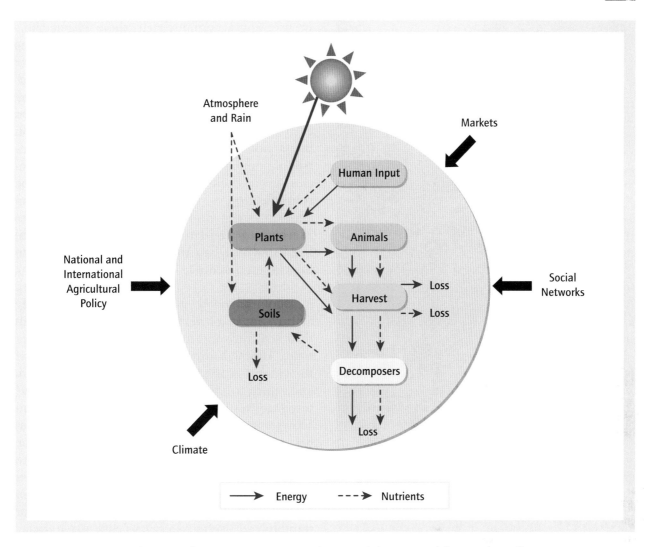

Figure 1 Schematic diagram of an agroecosystem and some of the external factors that affect it

Source: Adapted from Gliessman, S. (2007). *Agroecology: The ecology of food systems* (2nd ed.). Boca Raton, FL: Taylor & Francis.

have not yet been displaced by industrial systems. An example of traditional agriculture, with valuable lessons for sustainability, is agroforestry, a practice of integrating and managing trees and agricultural crops in the same land area. Many agroforestry systems illustrate important agroecological principles, such as replicating the structure and some of the ecological processes of local natural ecosystems (tropical forests). They have also survived for long periods of time through processes of cultural and ecological adaptation. Specific examples of tropical agroforestry with desirable agroecological characteristics include shade coffee and cacao plantations, mixtures of

trees and annual crops, and agroforestry home gardens (see photo).

A focus on supporting and maintaining key ecological processes on the farm is at the core of agroecological management. Gliessman has identified the following four broad ecological processes to be managed and monitored.

Energy Flow. In natural ecosystems, the primary source of energy is the sun, and this light energy gets captured and stored in plant biomass. In agroecosystems, additional energy is derived from human inputs, and plant biomass is removed from the system with each harvest.

An example of an agroforestry home garden in Cupilco, Tabasco, Mexico. Researchers at the University of California, Santa Cruz, and the Colegio de Postgraduados in Tabasco have documented that this home garden has maintained a very similar structure and crop composition for at least several decades. Some of the components of the home garden include fruit and timber trees, cacao, vanilla, medicinal and cooking herbs, and bananas.

Source: Author.

Nutrient Cycling. Nutrients are the chemical compounds organisms use to maintain their basic life functions; they include different forms of nitrogen, phosphorus, potassium, carbon, and many others. In natural ecosystems, nutrients get recycled in a relatively closed loop, while in agroecosystems there are more losses through leaching and harvest.

Population-Regulating Mechanisms. A population is a group of organisms of the same species.

In ecological communities, these populations interact with each other in different ways, such as predation (one organism feeding on the whole or part of the other), competition (two species competing for the same limited resources), or facilitation (interactions that do not harm or can benefit each other). In natural ecosystems, population interactions usually result in a self-regulating control of populations that prevents a particular species from excluding others. In contrast, the simplification of the ecological community in agroecosystems results in loss of self-regulation. This usually translates into large population outbreaks, which can turn into pests and require human interference to control them.

Resilience/Stability. Natural ecosystems usually attain a state of relative stability, maintaining their structure and function over long periods of time. Although disturbances continue to occur, the natural ecosystem is able to respond quickly and return to the state it was in prior to the disturbance (resilience). The simplification of agroecosystems, in terms of structure and function, has resulted in loss of resilience and stability. The high levels of disturbance that agroecosystems are subjected to and the complete removal of biomass through harvesting require that humans invest labor and inputs to maintain a certain level of stability.

The four ecological processes broadly outlined above can be used to compare natural ecosystems and agroecosystems. Agroecology has used these types of comparisons to try to design agricultural systems that "mimic" natural ecosystems as much as possible, in order to increase the efficiency and resilience of agroecosystems.

Social, Economic, and Cultural Basis

In agroecology, the agricultural and ecological characteristics of farming communities are seen as part of a broader social and ecological landscape in which farmers and other rural actors interact and negotiate to pursue their particular goals. The livelihoods of farming families and communities are central to agroecological research and practice. Livelihood refers broadly to the multiple resources and capacities that people use to make a living. Understanding the social and ecological processes in local or traditional agriculture has been of particular importance to agroecological research. For example, agroecological work on agroforestry home gardens (see photo) in Mexico and Central America has shown that the ecological structure of these agroecosystems, in terms of plant arrangements and horizontal allocation of space, was mainly influenced by each family's livelihood strategies.

A notable contribution to the socioeconomic and cultural basis of agroecology has been provided by a group led by the rural sociologist Eduardo Sevilla-Guzmán at the University of Cordoba, Spain. The group's publications (most of them in Spanish) have critically examined agroecology as an option to support small-scale and peasant agriculture in the context of accelerating globalization. Through an analysis of the epistemological roots and characteristics of agroecology, these researchers have concluded that the field is an ideal foundation for "participatory and endogenous" rural development, with a focus on small-scale agriculture. These characteristics include (a) agroecology's value, use, and integration of different forms of knowledge, including scientific, indigenous, local, and experiential; (b) the ecological, not economic, basis of agroecological management; and (c) agroecology's concern for the social and cultural factors that interact with and affect agriculture.

Participatory Approaches and Direct Applications

Agroecology's concern for small-scale agriculture has led to interaction with and support for farmer organizations and social movements. Examples of this can be found in Spain (ecological olive grove cooperatives), Latin America (the Agroecological Movement of Latin America and the Caribbean [MAELA]), Brazil (the Landless Rural Workers' Movement [MST]), and Central America and Mexico (the Campesino a Campesino Movement [MCAC]).

Direct support for farmers has also taken the form of participatory approaches in agroecological research and implementation that seek to directly benefit farming communities. Most of these initiatives have used long-term participatory action research, a process that combines research with direct action to directly benefit participating stakeholders (i.e., farmers, researchers, and rural communities). Examples of this work can be found among the shade coffee communities of Mesoamerica, olive growers in Spain, and strawberry farmers in California.

An increased focus on the food system has resulted in many agroecologists supporting alternative food network initiatives, such as Fair Trade, direct marketing, and eco-friendly farming. This has been done through participatory research on the role that alternative food networks (e.g., certifications, direct marketing) might play in increasing the sustainability of food systems. A recent example of this work addressed shade coffee production in Mexico and Central America and included an examination of "sustainable coffee" and the certifications associated with it (i.e., Fair Trade, organic, and eco-friendly). Other direct applications of agroecology exploring alternative food networks have been accomplished through researcher-NGO partnerships, such as the Community Agroecology Network (CAN, www.communityagroecology.net), which works on marketing innovations for small-scale coffee farmers in Mexico and Central America.

Agroecology and Sustainability

Although there is an ongoing debate on the meaning of sustainability and how to measure it, most would agree that it relates to maintaining or enhancing social and ecological conditions indefinitely. In agriculture, this refers to a way of farming that will not degrade the land in a way that hinders agricultural production in the future. It also includes factors related to social equity and economic viability for future generations. Agricultural sustainability has been one of the core issues addressed by agroecology. A large portion of current agroecological research is devoted to analyzing and developing agricultural management that is more sustainable. This is a daunting task and rife with debate and controversy. However, agroecology's integration of ecology, agriculture, and social and economic concerns renders it one of the most promising fields to help us better understand and resolve the challenge of sustainable food systems.

Critiques, Challenges, and Future Trends

Since the publications and interest in agroecology first appeared in the 1970s, the field has been challenged from supporters of industrialized agriculture and narrow disciplinary scientists. Some of the criticism against agroecology has been directed at its potential for major reduction in agricultural production and farmer profit, which would result from its widespread adoption. In addition, as an emergent, interdisciplinary field, agroecology has been questioned by the disciplines that it sought to integrate, mostly from purists in the fields of agronomy and ecology. Many of these critiques have dissipated because of the general consensus that industrialized agriculture has had negative social and ecological impacts; the need to keep sustainability in mind; evidence that organic agriculture, as a form of agroecological management, can produce comparable yields to industrialized agriculture; and a wider acceptance of interdisciplinary approaches as part of scientific inquiry. Nonetheless, there is still a strong financial and ideological sector that continues to support industrialized agriculture and that will continue to defy agroecology.

One of the greatest tasks faced by agroecologists is to be able to develop a fully integrated interdisciplinary approach. Agroecology has strong ecological roots, and early work in the 1970s and 1980s focused on analyzing and improving agronomic and ecological management. In-depth analyses of social, economic, and cultural aspects were incorporated later. One of agroecology's current challenges is to develop a flexible and adaptive interdisciplinary framework that fully integrates the ecological, social, economic, and cultural dimensions.

V. Ernesto Méndez

See also Agricultural Biotechnology; Agricultural Intensification; Agricultural Land Use; Agriculture, Industrialized; Agrobiodiversity; Agrochemical Pollution; Agrofoods; Agroforestry; Ecosystems; Nonpoint Sources of Pollution; Population, Environment and Development; Sustainable Agriculture

Further Readings

Altieri, M. (1995). *Agroecology: The science of sustainable agriculture*. Boulder, CO: Westview Press.

Bacon, C., Méndez, V., Gliessman, S., Goodman, D., & Fox, J. (Eds.). (2008). *Confronting the coffee crisis: Fair trade, sustainable livelihoods and ecosystems in Mexico and Central America*. Cambridge: MIT Press.

Clements, D., & Shrestha, A. (Eds.). (2004). *New dimensions in agroecology for developing a biological approach to crop production*. Binghamton, NY: Food Products Press.

Francis, C., Lieblein, G., Gliessman, S., Breland, T., Creamer, N., Harwood, R., et al. (2003). Agroecology: The ecology of food systems. *Journal of Sustainable Agriculture, 22*(3), 99–118.

Gliessman, S. (2007). *Agroecology: The ecology of sustainable food systems* (2nd ed.). Boca Raton, FL: CRC Press.

Uphoff, N. (Ed.). (2002). *Agroecological innovations: Increasing food production with participatory development*. London: Earthscan.

 # AGROFOODS

Sometimes spelled *agri-foods*, this term captures the increasingly long and complicated path food takes to our table. While we may like to think that the food we eat comes from a farm, it is only one place among many. Most farming is only possible with industrial inputs such as tractors and combines and chemical inputs such as fertilizers and pesticides. Farmers often require loans each season to buy what is needed to produce their crop. Farming also depends on energy to run the machines, pump water, produce fertilizers, and transport the finished product to the consumer. Farmers need expert information on what and when to plant, how to diagnose and treat plant diseases and destroy pests, the weather, and when and at what price to sell their crop. When one considers what goes into farming, it becomes apparent that farms are linked to and have become dependent on many other places, such as places of industrial production, places of petrochemical and fuel production, banking centers, and universities and government institutions. What is done with farm outputs is equally complicated.

Farm output can remain in its original form and simply be graded, washed, and shipped to the consumer. However, most of the food that we consume in the developed world is not in an unprocessed or "raw" form. Most of the food consumed has been substantially modified and processed to be made durable through canning, freezing, or other methods. This series of changes is an important stage in the agrofood system because durable foods allow the spatial separation of production from consumption and make long-distance trade possible. In fact, the distinction between agriculture and industry has become so blurred that many foods have become known by the industrial process that have transformed them, for example, *homogenized* milk, *pasteurized* cheese, *refined* sugar, and so on. Agricultural products can be further industrialized by breaking them down into their constituent parts. For example, a starch, a sweetener, oil, and protein can be extracted from a grain. Processors attempt to break the product of the farm into as many parts as possible and then find profitable uses for all of them. These different "fractions" of whole farm products are used as inputs for manufactured foods or in other industrial processes.

The producers of manufactured foods have an advantage over farmers because they buy the farm output and have flexibility over what ingredients to use and where to source them. For example, the manufactured food requires a sweetener but not necessarily sugar derived from the sugarcane plant. It requires oil yet not necessarily oil from corn. It requires a starch, but that could be derived from a potato or wheat or a number of other grains. The production of potato chips provides a good example of this substitution effect: Producers can fry the chips in whatever oil is cheapest at the moment of production. This illustrates why farmers are often at a disadvantaged position within the agrofood system.

More toward the consumer end of the agrofood system is food distribution. Food reaches consumers via food wholesalers, food retailers, and the restaurant and catering industry. Powerful economic entities in food distribution can shape the agrofood system with their purchasing power, such as when a fast-food restaurant chain decides to fry its french fries in healthier oil or add salad to its menu or when large grocery chains decide to carry some food products and not others.

At the end of the agrofood system is the final consumer. Food is unlike other commodities because we must eat daily to survive. Food is taken into the body and metabolized or used by our cells to provide energy. Our food choices affect our own bodies, but they also reverberate back and reshape the agrofood system. What we eat reflects demographic characteristics, such as the size and growth of the population and purchasing power, and social relations, such as the structure of the family. Obviously, advertising influences our food choices. But more subtly, the ever-quickening pace of the economy has led to the proliferation of "fast" foods (which can be consumed without the use of utensils) and other convenience foods meant to be consumed "on the go," in the car or at the desk.

The backlash against foods high in carbohydrates has reduced the consumption of potatoes, rice, bread, donuts, and orange juice (to name just a few) and has had consequences for the places that produce these foods. But the changing attitudes also provide opportunities. For example, a food high in fat such as fried chicken, which was criticized during the time when a healthy diet was believed to be one low in fat, can now present itself as a healthier food choice because it is low in carbohydrates.

The geographies of the agrofood system are continuing to change as food technologists attempt to bypass the farm altogether by creating "nonfood foods," or foods that are consumed but not metabolized by the body. These substances are made in the lab, not grown on the farm, and allow food producers to avoid the risks inherent in farming while providing greater control over the production process. Another important agrofood trend is "functional foods" (also called *nutraceuticals*), which attempt to marry foods and pharmaceuticals to create a substance consumed to create a desired effect in the body.

Examples are grape juice with added antioxidants intended to fight cancer and so-called smart drinks that have ginseng, caffeine, and vitamins added to them. Functional foods blur the line between drugs and food, and their producers know that foods that make health claims often have an advantage in a competitive marketplace. The changing agrofood system not only affects our bodies, but its changing technology and consumer choices have significant impacts in reshaping our geography.

John R. Grimes

See also Agricultural Biotechnology; Agriculture, Industrialized; Agrochemical Pollution; Agroecology; Consumption, Geographies of; Food, Geography of

Further Readings

Bonanno, A., Busch, L., Freidland, W., Gouveia, L., & Mingione, E. (Eds.). (1994). *From Columbus to ConAgra: The globalization of agriculture and food*. Lawrence: University Press of Kansas.

Friedmann, H. (1993). The political economy of food. *New Left Review, 197,* 29–57.

Goodmann, D., & Redclift, M. (1990). *Refashioning nature*. London: Routledge.

 AGROFORESTRY

Agroforestry is a term related to land use whose exact meaning has changed as different approaches to studying land use have arisen. The oldest use of the term, as the adjective *agroforestal*, was to name a category of land use. This meaning was introduced in the 1930s and was firmly established by the 1950s, a period when the areal differentiation approach in geography was prominent. Originally, *agroforestry* was used when an observer considered agriculture and forestry as distinct land uses but wished to combine these categories in mapping or analysis. This concept reflected the traditional view that agriculture did not include the cultivation or management of trees and that forestry did not pertain to herbaceous plants. Although this narrow meaning has not been entirely superseded, in the 1950s and 1960s many geographers recognized that the distinction between agriculture and forestry is neither absolute nor clear in many landscapes, especially from the perspective of rural land managers in traditional societies. Thus, *agroforestry* came to designate land use in which farming, often including livestock husbandry, and tree cultivation and/or management are practiced, whether simultaneously or sequentially. In situations where livestock is prominent, the adjective *agro-silvo-pastoral* has been used.

The second broad meaning of *agroforestry* is a system of land use in which trees and crops, and often livestock, are deliberately associated in space and time. In other words, this meaning refers to land and resource management practices in areas where land use can be classified as agroforestry, according to the second sense described above. This meaning also has two distinct senses. First, "scientific agroforestry" is a field of study that uses agronomic methods to analyze productivity in settings where trees and crops are grown together. The concept and pursuit of scientific agroforestry became prominent beginning in the 1960s, although there were earlier "scientific" efforts to combine agriculture and forestry. For example, during the 1880s in colonial Burma, British foresters developed the *taungya* system, now widely practiced in southeast Asia, in which forests are cleared for timber plantations and crops are grown among saplings for several years. Scientific agroforestry represents a positivist approach to agricultural development that many international research and development institutions have promoted. While scientific agroforestry has been more important in the applied natural sciences than in geography, the social and environmental effects of its promotion interest many geographers. Second, agroforestry has been used to describe the resource management practices of rural land managers in traditional societies where trees are an integral part of farming systems. In this sense, the term is often qualified as "indigenous agroforestry," although many nonindigenous, traditional peoples—such as the Brazilian *caboclos*—also rely on trees and crops for food and other resources. Many indigenous/traditional agroforestry practices serve to manage

Planting peanuts in Mali, West Africa

Source: Author.

soil fertility, as in the Southern African *citimene* system, where tree branches are collected from a large area and burned on a field to release nutrients. Many geographers have studied indigenous/ traditional agroforestry, mostly taking the perspective of either cultural ecology or political ecology. For instance, a cultural ecological study might examine how agroforestry practices relate to local soil and vegetation conditions, while a political ecological study might examine how multiscale political economic processes affect agroforestry practices.

Many cultural, social, demographic, and environmental processes are embedded in agroforestry systems. Indigenous/traditional agroforestry systems generally reflect culturally specific values and histories, and their persistence depends on the persistence of associated knowledge cultures. Ancient agroforestry systems unlike any known today can be deduced from archaeological evidence,

such as the raised beds found widely in the Amazon Basin. Indigenous/traditional agroforestry systems reflect the constraints of specific biophysical environments, and many systems are sustainable if certain environmental conditions remain relatively constant. Conversely, alteration of indigenous/traditional agroforestry practices may lead to significant environmental changes, such as erosion or changes in the composition of vegetation. Indeed, the characteristic vegetation of many cultural landscapes is the product of long, co-adaptive histories of human-environment interaction within agroforestry systems. For example, although farmers clear new fields every 5 to 10 years in West Africa's woodlands, this practice does not cause permanent deforestation because trees that can reproduce vegetatively dominate the vegetation (see photo). Farmers fell trees to reduce their competition with crops but strictly avoid damaging the tree roots, which will sprout and become trees again, reconstituting the woodland vegetation.

Indigenous/traditional agroforestry systems continuously change as people develop new practices in response to demographic, environmental, and social change. Although Esther Boserup did not use the term *agroforestry*, she argued that farming systems including trees are expected in situations of intermediate population density. Agroforestry systems generally represent moderately intensive land use; agroforestry often supplants forests, but in highly intensive farming, most trees are removed to allow greater crop-plant density. Environmental changes can force people to develop new practices, but agroforestry systems can show significant resilience because trees are less susceptible than crops to short-term events, such as drought.

Social processes are more important than environmental adaptation in shaping indigenous/traditional agroforestry practices over the short term. Tree and crop management practices can vary considerably among individuals based on gender, age, wealth, and other social factors, reflecting the varying rights, roles, or authority people have within societies. Thus, knowledge of plant use is strongly gendered because in many societies women have less control over land and labor resources and must rely on noncrop plants to meet subsistence needs during difficult times, whereas men can more easily travel and seek wage labor. Struggles over agroforestry practice may express struggles over social rights, roles, or authority. Agroforestry systems also vary among societies. In particular, the practices and principles of scientific agroforestry often differ from those of indigenous/traditional agroforestry systems. Historically, promoters of scientific agroforestry portrayed indigenous/traditional agroforestry systems as unproductive, environmentally damaging, or even atavistic. For instance, African colonial regimes forcefully imposed scientific agroforestry practices because European scientists believed that indigenous practices created permanent deforestation. Negative perceptions of indigenous/traditional agroforestry linger in some institutions. While scientific agroforestry has widely contributed to agricultural productivity and sustainability, it has also been socially disruptive and has served to extend state authority. Indeed, promotion of the *taungya* system in colonial Burma helped end anticolonial resistance.

Chris S. Duvall

See also Deforestation; Forest Fragmentation; Forest Land Use; Indigenous Agriculture; Indigenous Forestry

Further Readings

Brookfield, H. (2001). *Exploring agrodiversity*. New York: Columbia University Press.
Buck, L., Lassoie, J., & Fernandes, E. (Eds.). (1999). *Agroforestry in sustainable agricultural systems*. Boca Raton, FL: CRC Press.
Carne, R. (2004). Agroforestry land use: The concept and practice. *Australian Geographical Studies*, *31*(1), 79–90.
Dove, M. (1983). Theories of swidden agriculture, and the political economy of ignorance. *Agroforestry Systems*, *1*(2), 85–99.

 # AIDS, GEOGRAPHY OF

See Disease, Geography of; HIV/AIDS, Geography of

 # AIR MASSES

The most common definition of an air mass is a large body of air, relatively homogeneous with respect to temperature, humidity, and vertical lapse rates. While this definition is generally valid, it is true only for air masses that remain within their source regions. Once in motion, air masses begin to modify, in some cases quickly, especially if they move significant latitudinal distances. Eventually, modification renders the air mass unrecognizable with respect to its source region.

Source Regions

Air masses form in regions of generally uniform surfaces where winds are typically light for extended periods of time, usually days to weeks. Over periods of time, the overlying air takes on the properties of the underlying surface through radiative and convective processes. These regions tend to be located near the centers of semipermanent high-pressure centers due to the weak pressure gradients found in these regions. Climatologically, these semipermanent highs are located in the polar and subtropical latitudes.

In the Northern Hemisphere there is little land in the polar latitudes, so in most cases air mass source regions are located over pack ice. The thermal character of pack ice is much different from that of snow-covered land, with the underlying Arctic Ocean exchanging heat with the overlying atmosphere. As a result, winter polar air masses, with the Siberian exception, are milder than in the Southern Hemisphere. For this reason, air masses that form in the "High Arctic" are commonly referred to as polar, while only Antarctic and Siberian air masses are considered truly Arctic in nature.

There is also much variability among subtropical source regions due to land cover type. In general, land surfaces within subtropical highs are desert-like, with very little water available. As a result, more of the incoming solar radiation that reaches the surface is available to heat the air, so dry bulb temperatures are quite high and dew point temperatures quite low. The result is very low relative humidity and great drying power of the air. Ocean surfaces, on the other hand, convert a significant proportion of solar irradiance to latent heat, which leaves much less sensible heat to increase the air temperature.

Air Mass Types

Based on common reference to air mass source regions as polar, subtropical, land covered (continental), or maritime, four primary air mass types are recognized: (1) continental polar (cP), (2) maritime polar (mP), (3) continental tropical (cT), and (4) maritime tropical (mT). Continental arctic (cA) air is usually considered a special case of cP air, while maritime arctic air (mA) is an air

mass rarely recognized because for air to be considered truly arctic, it must develop over an ice-covered surface.

This classification scheme has become popular among geographers because of its simplicity, which has been considered an important consideration in introductory courses. For example, there are cold air masses originating in the higher latitudes that are either moist (mP) or dry (cP and cA), depending on whether the surface type over which they formed was ocean or land, and warm air masses that develop in the lower latitudes that are either moist (mT) or dry (cT).

There are two other air masses referred to on occasion: equatorial and superior air. The term *equatorial (E) air* is sometimes used to separate the slighter, warmer, and less humid air of the subtropics from the cooler, more humid air of the tropics, although attempting to distinguish between the two at the surface is difficult at best. Superior (S) air is a special form of cT air in the Southern Plains of the United States with a source region over the Mexican Plateau. As this already hot dry air advects northeastward, it's mixed down to the surface, and with compression, it further heats and dries. The boundary that separates this air at the surface from the mT air to the east is often referred to as the "dryline" or "Marfa front."

The vertical character of the primary air mass types is typically considered to be consistent for each air mass. For example, cP air and cA air typically exhibit deep inversions from the surface up to the midlevel of the troposphere, with rapid cooling with height; that is, both air masses are considered to be absolutely stable. mT air, on the other hand, is assumed to be conditionally unstable, while cT air is absolutely unstable, although typically with little more than high-based cumulus humilus clouds due to the lack of water vapor. More often than not, however, multiple air masses exist at any midlatitude point location, which adds complexity to the situation. For example, in the Northern Hemisphere it is not uncommon to observe cP air advecting southward at the surface in response to a surface high to the north, while a southwesterly mid- and upper-level flow brings in warmer and sometimes moister air from the lower latitudes up and over the top of the cP air. When multiple air masses are moving through a region at different heights, the situation is often referred to as "differential advection." Differential advection can lead to air mass stabilization or destabilization depending on the resulting vertical thermal stratification.

Air Mass Modification

As air masses move out of their source regions, they continue to take on the characteristics of the underlying surfaces but at a rate proportional to the speed of migration. Generally speaking, polar air masses will become warmer as they advect toward lower latitudes due to increased solar angles of incidence, while subtropical air masses cool as they move toward lower latitudes not only due to the smaller angles of incidence but, in winter, also due to the existence of snow cover. It should also be understood that air masses either lose (precipitation) or gain (evaporation) water vapor as they migrate out of their source regions. This not only changes the static stability of the air mass but can result in a completely different air mass type as time progresses.

In general, there are five factors involved in understanding air mass modification: surface temperature, surface moisture, topography, trajectory, and migration rate. Each of these plays an important role depending on the time and location of modification.

The difference in temperature between the air mass and the surface results in a vertical exchange that commonly affects air mass stability. Since gradients are always directed from an area of surplus to an area of deficit, air masses that are colder than the surface over which they move experience a net gain in heat from that surface. In time, this decreases the vertical thermal lapse rate, which has a stabilizing effect on the air mass; however, this vertical exchange can be vigorous while the difference is still large, with intense convective cells developing in response. These intense convective cells are a condition of deep moist convection (DMC), which tends to produce thunderstorms and generally brief periods of heavy rainfall. When air masses are warmer than the surface over which they are moving, then the exchange of heat is from the air mass to the ground surface. Over time, this reduces the condition of absolute static stability to that of

neutrality; however, as long as a gradient exists, a strong capping inversion will work to keep the air stagnant.

The static stability of air masses is so important to the resulting weather that an additional letter, w or k, is added to the standard air mass acronym. The letter w is used when the air mass is warmer than the surface over which it lies, implying a statically stable environment, and the letter k refers to a condition where the air mass is colder than the surface and the air is conditionally or absolutely unstable. For example, an mTk air mass is unstable, promoting upward vertical exchange between the surface and the air mass, and mTw represents a stable condition favoring downward vertical exchange. The only air mass for which this convention is not used is S air.

Air masses may also be modified by a vertical exchange of water (vapor, liquid droplets and drops, and ice). Water is removed from air masses via precipitation and gained through evaporation. These exchanges can be in response to either dynamic (forced ascent) or thermodynamic (free convection) processes. In general, the greater the cumulative precipitation, the more water the air mass loses; however, moisture advection and convergence on meso- and local scales can resupply limited areas within larger air masses with enough water vapor to support heavy precipitation at the same time, so the air mass as a whole loses no water at all.

In terms of evaporation, larger water bodies tend to supply a greater amount of water to an air mass, but since this supply is a function of evaporation, the warmer and windier the surface, the more water vapor will be available. The addition of water is greatest when an upward vertical exchange occurs under unstable conditions, such as when a cold air mass moves over a warm body (e.g., cPk). In these situations, however, the large addition of water vapor into an unstable air mass will usually result in convective precipitation and, ultimately, loss of the water gained.

Topography plays a role in air mass modification through mechanical lift that in some cases produces precipitation, thus removing water from the air. Once the air crosses the crest and its density is greater than that of the surrounding air, it will subside down the slopes of the leeward side, resulting in considerable compressional heating. If water vapor is not added to the air in this process, then the relative humidity decreases to very low levels, perhaps contributing to desert-like environments commonly referred to as a "rain shadow."

The trajectory an air mass follows—that is, in terms of its curvature, which is the result of a force either pushing (anticyclonic) or pulling (cyclonic) the air mass—has an effect on the conditional stability or instability of that air. For anticyclonic flow, air masses tend to stabilize due to subsidence, while cyclonic flow tends to support convective instability in response to lift. It should be noted here that convective instability and conditional instability refer to two different situations. Convective instability occurs when the upper parts of an air layer cool at more rapid rates than its base when lifted. Conditional instability, on the other hand, exists when the air is stable if unsaturated but unstable if saturated.

The speed of air mass migration affects air mass modification rates as faster-moving air is exposed to the underlying surface for shorter periods of time than slower-moving air. In other words, an inverse relationship exists between the rate of change of air mass character and the speed of air mass migration. This means that for arctic air to reach Central North America and Europe, it must travel very quickly from Siberia over the pole into North America, what is commonly referred to as a cross-polar flow.

David L. Arnold

See also Adiabatic Temperature Changes; Anthropogenic Climate Change; Atmospheric Circulation; Atmospheric Composition and Structure; Atmospheric Energy Transfer; Atmospheric Moisture; Atmospheric Pressure; Atmospheric Variations in Energy; Climate Change; Climate Types; Climatology; Temperature Patterns

Further Readings

Barry, R., & Chorley, R. J. (2003). *Atmosphere, weather, and climate* (8th ed.). London: Routledge.

Blair, T. (1946). *Weather elements*. New York: Prentice Hall.

Carlson, T. (1998). *Midlatitude weather systems*. Boston: American Meteorological Society.

Rauber, R., Walsh, J., & Charlevoix, D. (2002). *Severe and hazardous weather*. Dubuque, IA: Kendall.

Wallace, J., & Hobbs, P. V. (2006). *Atmospheric science* (2nd ed.). St. Louis, MO: Elsevier.

 AIR POLLUTION

See Atmospheric Pollution

 ALBEDO

Albedo refers to the amount of solar radiation reflected by an object. Specifically, it is the ratio of reflected to incident solar radiation. *Albedo* is derived from the Latin *albus*, which means "to be white." Albedo is a unitless measure of reflectivity that varies from 0 (*no reflectance*) to 1 (*complete reflectance*), although it is also commonly shown as a percentage (0% to 100%). The albedo of an object is a function of the surface properties of the object (i.e., color, roughness, transparency), the zenith angle of the sun, and the wavelength of the incident radiation. Albedo is usually determined based on the visible portion of the spectrum (0.4–0.7 μm [micrometer]).

Planetary albedo is the fractional amount of solar radiation reflected by Earth (both the atmosphere and Earth's surface). Based on measurements from satellites and modeling experiments, it has been determined that Earth's albedo is ~0.31 (31%). The albedo of a surface is extremely important because it determines how much of the incident solar radiation is reflected and how much is absorbed. Small changes in albedo can have a large influence on the surface energy budget of an object. For example, if Earth's albedo decreased by 0.01, it would increase Earth's temperature by 1 °C.

Geographic and Seasonal Variations

Because Earth's surface is varied with regard to its composition, the amount of solar radiation that is reflected and absorbed varies by location and by season. Albedo for natural surfaces ranges from ~0.03 to 0.95 (Table 1). Generally, dark-colored surfaces (e.g., dark soils) have low albedos, and light-colored surfaces (e.g., snow and white sand) have high albedos. Albedo is also influenced by the solar zenith angle. Water generally has a relatively low albedo (<0.10), except when the sun is near the horizon. The albedo of snow also varies greatly. Freshly fallen snow initially has a high albedo, but as the snow ages and dust and soot are deposited on top of the snow surface, its albedo decreases. The presence of snow can significantly lower the albedo of the land surface, which in turn leads to colder temperatures that support further expansion/accumulation of snow (e.g., a positive feedback).

Many of the changes in albedo take place naturally due to the changing of the seasons, which can lead to changes in vegetation or snow cover, or variations in the weather, which can influence the density and health of vegetation and soil saturation. However, human activities can also have a major influence on albedo. Deforestation and urbanization are two examples of how human activities can change the albedo of the land surface, which then modifies the local and regional climate.

Steven Quiring

See also Anthropogenic Climate Change; Climate Change; Land Use and Cover Change (LUCC); Radiation: Solar and Terrestrial; Weather and Climate Controls

Further Readings

Kiehl, J. T., & Trenberth, K. E. (1997). Earth's annual global mean energy budget. *Bulletin of the American Meteorological Society, 78,* 197–208.

Oke, T. R. (1987). *Boundary layer climates.* New York: Routledge.

AL-IDRISI (CA. AD 1100–1165)

al-Idrisi was arguably the greatest medieval geographer. A descendant of the prophet Muhammad, this Arab scholar was titled al-Sharif al-Idrisi but known in the West as *Geographus Nubiensis,* the Nubian geographer. Born in Morocco and educated in Cordoba, he worked in Sicily at the Palermo court of the Norman king Roger II, after whom his book of world geography was called *Kitab Rujjar (Book of Roger)* in 1154; its full Arabic title is *Nuzhat al-mushtaq fi 'khtiraq al-afaq* (Entertainment for One Desiring to Travel Far). A later, shorter world geography, sometimes called "the Little Idrisi," is known by various Arabic titles. Both are extensively illustrated with maps; the maps in *Nuzhat al-mushtaq* are full or half-page in size and oriented to the south, while the maps in the later book are smaller and often oriented to the east.

al-Idrisi presided over a collective project, and his *Geography* is a synthesis of information and cartographic traditions from both Islamic and European cultures. He had traveled in Asia Minor, Europe, and North Africa, and in Sicily,

Surface	Details	Albedo
Bare soil	Dark and wet	0.05–
	Light and dry	0.40
Sand		0.15–0.45
Grass		0.16–0.26
Agricultural crops		0.18–0.25
Forests	Deciduous	0.15–0.20
	Coniferous	0.05–0.15
Water	Low zenith angle	0.03–0.10
	High zenith angle	0.10–1.0
Snow	Old	0.40–
	Fresh	0.95
Ice	Sea	0.30–0.45
	Glacier	0.20–0.40
Clouds	Thick	0.60–0.90
	Thin	0.30–0.50

Table 1 Representative albedos of natural surfaces

Source: Created using data from Oke (1987).

he was able to consult both European and Islamic sources. Because the data were procured from earlier books and travel reports (only 10 authors are named), they are sometimes out-of-date, but the book and accompanying maps are unsurpassed in medieval geography.

The text is a detailed description of the map, supposedly based on the Arabic version of Marinus's map, created under Caliph al-Ma'mun (AD 813–833) and engraved on a silver disk. The silver prototype was lost, but the text is accompanied by a round, schematic map of the world and 70 rectangular maps of the 70 parts into which al-Idrisi had divided the world. The projection used remains unexplained, although al-Idrisi followed the Ptolemaic foundation, adopted by the early Islamic scholars, whereby the round Earth is divided into quarters and only the Inhabited Quarter is described. It is astronomically divided into seven latitudinal belts, "climates," leaving off the extreme north and the equatorial south. Although familiar with coordinates, al-Idrisi did not use them. His innovation was in using, instead of meridians, 10 longitudinal divisions of the parallel latitudinal climates. The climates are numbered from south to north, as in the scheme developed by Eratosthenes, and the numbers of sections go west to east. After a brief general introduction, the text follows the map's division into 70 sections. al-Idrisi names the most place names since Ptolemy, expands and updates the medieval Arabic geographical inventory, and describes for the first time many identifiable locations, which he connects by itineraries.

Earth is shown surrounded by Encircling Sea al-Bahr al-Muhit (the Greek Ocean). The western limit is the prime meridian drawn through the westernmost part of Africa (the Fortunate Isles in the Atlantic are also included). The easternmost country is Sila (Korea), supposedly at 180°. The southern portion of the round world map is filled with the African landmass, not shown on sectional maps. Africa is extended eastward to form the southern coast of the Indian Ocean, which remains open in the Far East. The northern limit is 64°, practically coinciding with the Polar Circle. The maps are color coded and distinctive and, together with the special features of the text, make it possible to speak of the "Idrisi school," which influenced both Arabic and European cosmographers.

The book *Nuzhat al-mushtaq* became the first secular Arabic work printed in Europe (in Rome in 1592); a Latin translation was published in 1619 in Paris. These editions did not include the maps, but a pieced-together Latin version of the sectional map of the world was produced in Paris by Petrus Bertius in the 1620s. In honor of the Arab scholar, the name IDRISI has been given to a grid-based geographic analysis system and related GIS (geographic information system) and remote sensing software and to an Arabic electronic search engine with bilingual (Arabic/English) indexing and publishing capabilities.

Marina A. Tolmacheva

See also Biruni; Ibn Battuta; Ibn Khald n; Travel Writing, Geography and

Further Readings

Ahmad, S. (1992). Cartography of al-Sharīf al-Idrīsī. In J. Harley & D. Woodward (Eds.), *The history of cartography: Vol. 2, Book 1. Cartography in the traditional Islamic and South Asian societies* (pp. 156–174). Chicago: University of Chicago Press.

Tolmacheva, M. (1995). The medieval Arabic geographers and the beginnings of modern Orientalism. *International Journal of Middle East Studies, 27*(2), 141–156.

Tolmacheva, M. (1996). Bertius and al-Idrīsī: An experiment in Orientalist cartography. *Terrae Incognitae, 28*, 49–59.

 # ALTITUDE

Altitude is the distance between a position and a vertical reference surface. Distance might be measured in angular or length units. The position might be the point location of an object or a specified point along a vehicle track or satellite orbit. The surface might be one of many vertical datums, such as the center of the Earth, the surface of the ocean, the topographic surface of the Earth, the top of the built environment, or a constant barometric pressure surface.

The precise distance between a surface and a position depends on the definition of the line between them. For example, the line might be perpendicular to a plane tangent to the reference surface, or it might extend from the position toward the center of the mass of the Earth.

The terms *altitude, elevation*, and *height* are sometimes used interchangeably. In different contexts, these words take on different meanings, and the modifiers attached to them can sometimes clarify their usage. *Elevation* is often associated with the distance from a defined surface, such as the geoid, the theoretical equipotential gravity surface of the Earth, or a physically defined gravity surface model, such as a specific mean-sea-level datum, or with respect to the actual local-level plane as measured at the position. *Height* is sometimes reserved for the distance between a reference ellipsoid and a position or for the distance from the bottom to the top of an entity, such as a building or a mountain peak. The height of an aircraft might be the distance above the topographic surface of the Earth, while the height of a geodetic survey monument might be its vertical distance from a reference ellipsoid.

This sentence appears in a text on surveying principles: "Therefore, the altitude at which the plane must fly is calculated by adding the elevation of the mean datum to the flying height." The *altitude* of the aircraft is above mean sea level, the flying *height* is the distance between the aircraft and the ground, and the ground (the mean datum) has an *elevation* with respect to mean sea level.

Altitude is modified by words that further specify the meaning. *Absolute altitude* refers to the distance above the physical surface of the sea or land. Angular *altitude* is the vertical angle between some plane (such as local level) and a line from the observation point to an object such as a mark on a surveyor's rod or a star. *Barometric altitude* is the distance from one constant pressure surface (an isobaric surface) to another. *Meridian altitude* is the vertical angle to an object measured along a line of longitude.

Peter H. Dana

See also Datums; Geodesy; Latitude; Longitude; Surveying

Further Readings

American Society of Civil Engineers, American Congress on Surveying and Mapping, & American Society for Photogrammetry and Remote Sensing. (1994). *The glossary of mapping sciences*. Bethesda, MD: Author.

Kavanagh, B. F., & Glenn Bird, S. J. (2000). *Surveying: Principles and applications*. Upper Saddle River, NJ: Prentice Hall.

National Geodetic Survey. (1986). *Geodetic glossary*. Rockville, MD: National Geodetic Information Center, Charting and Geodetic Services, National Ocean Service, National Oceanic and Atmospheric Administration.

AMBIENT AIR QUALITY

Ambient air quality refers to the concentration of gases, particles, or other elements in the outdoor atmosphere—from either natural or anthropogenic sources. The term *quality* is objective and gives an indication of what the concentration of a constituent of air is in relation to certain thresholds or standards. If a concentration is above the threshold, air quality is considered poor or, in some cases, even hazardous to human health. An applicable geographic area is often termed a "nonattainment" area. In these instances, the air is considered "polluted," and the elevated constituent is a pollutant. Thresholds for different pollutants are determined with reference to human health. Standards are set by the World Health Organization (WHO), though in some countries thresholds have been modified by national or local authorities.

As an example, carbon dioxide (CO_2) is a naturally occurring gas in the atmosphere, which in very small concentrations is necessary for life. If its concentration builds over certain thresholds, however, it can become detrimental and so will be termed a hazardous air pollutant. This process was tragically exemplified in 1986 by the Lake Nyos disaster in Cameroon, West Africa, when an upwelling of CO_2 from the lakebed had lethal consequences for the local human population and animals. In this case, both the pollutant and the emission source were naturally occurring. Pollutants such as sulfur oxides, nitrogen oxides, and heavy metals, however, can only be caused by human activities.

Air quality issues came to the forefront of scientific and societal attention during the 1950s. In 1952, London's "killer fog" was linked to a large increase in human fatalities (4,000 excess deaths), especially among people with a history of cardiopulmonary problems. The episode resulted in the implementation of air pollution mitigation measures by the London authorities. The air pollution in question was caused by smoke emissions from the burning of coal and other raw materials into a winter atmosphere already dense with mist and fog from the River Thames. This type of pollution is now termed London-type smog (smoke plus fog). In the same decade, another type of photochemical smog formed by chemical reactions in the atmosphere gained notoriety and became known as Los Angeles smog. Later, during the 1970s, acid rain emerged as a top environmental concern.

Most urbanized regions in the developed world have monitoring stations dedicated to assessing air quality on an hourly basis. Some, like Melbourne, Australia, even provide daily forecasts for ambient air quality, much like a typical weather forecast. This is made possible by the strong link between ambient air quality and weather.

Air quality is influenced by two major factors. The first is the emission of polluting substances into the air. Polluting emissions can be from urban, industrial, agricultural, or rural sources, or they can be totally natural. The second of these factors is the pollution potential of the atmosphere, or its ability to transport, diffuse, chemically transform, and remove pollutants. This involves three important processes: (1) *dispersion*, which is the horizontal and vertical spread and movement of pollutants; (2) *transformation*, which involves chemical reactions between pollutants or in pollutants under certain temperature and sunlight conditions; and (3) the *removal* of pollutants through mechanisms such as dry and wet deposition.

These three processes are controlled in turn by synoptic conditions, the intermediate scale of atmospheric activity between the global and

Don Bock, senior air pollution specialist with the Minnesota Pollution Control Agency, explains the functions of an air particle monitor located at the Anoka County Airport in Blaine, Minnesota, on June 2, 2003. Clean Air Minnesota officials announced their plans to decrease harmful ozone in the twin cities of Minneapolis–St. Paul.

Source: AP Photo/Jim Mone.

regional scales. Synoptic conditions include circulation systems such as anticyclones and cyclones. Atmospheric circulation at this scale is an important control on air pollution, influencing clouds (amount, thickness, height, and type), the temperature and relative humidity of air, the type and amount of precipitation, and wind speed and direction. These parameters, in turn, influence the vertical temperature structure of the atmosphere, which—by controlling the vertical extent pollutants can disperse—determines atmospheric stability. The layer through which the mixing of pollutants occurs is called the *mixed layer*, and the height to which pollutants mix is termed the *mixing depth* or *mixing height*. Poor dispersion occurs when a region is under the influence of anticyclonic synoptic-scale circulation. Here, vertical mixing is suppressed by an elevated inversion; combined with light winds, this suppression of vertical mixing can cause air quality to deteriorate rapidly. The clear skies created by the anticyclone can also allow the formation of nocturnal, surface-based inversion layers, which contribute to poor air quality at night.

Pollutants often undergo chemical transformations in the air, producing what are called secondary pollutants. The damaging London smog of 1952 ultimately killed 12,000 people with sulfuric acid fogs, which remained stagnant for 4 days. Such fogs occur when sulfur dioxide (SO_2)—which is generated by fuel combustion, particularly coal with high sulfur content—is oxidized in the air to sulfur trioxide (SO_3). This, in the presence of catalysts, then reacts with water vapor (H_2O) to form sulfuric acid mist. More common in recent decades is the formation of photochemical smog, where elevated levels of ozone have been detected. Ozone is a secondary pollutant formed by photochemical reaction of precursors such as NO_x ($NO + NO_2$), reactive organic gases, and many other gases.

Air quality can be improved by pollutant removal through processes such as gravitational settling and dry and wet deposition. Gravitational settling removes larger particulates, with the rate of removal related to the size and density of the particles and the strength of the wind. Dry deposition is a turbulent process, in which there is a downward flux of pollutants to the underlying surface. The amount of deposition depends on the characteristics of the turbulence, with increased rates under stronger turbulence.

Wet deposition involves the removal of air pollutants through absorption by precipitation elements (water droplets, ice particles, and snowflakes) and consequent deposition to the Earth's surface during precipitation. This removal process includes the attachment of pollutants to cloud droplets during cloud formation, in-cloud scavenging (also known as *washout*), and coalescence of raindrops with material below the cloud (*rainout*).

Peyman Zawar-Reza

See also Acid Rain; Atmospheric Circulation; Atmospheric Particulates Across Scales; Atmospheric Pollution; Photochemical Smog

Further Readings

Arya, S. P. (1998). *Air pollution meteorology and dispersion*. New York: Oxford University Press.
Seinfeld, J. H., & Pandis, S. N. (2006). *Atmospheric chemistry and physics* (2nd ed.). New York: Wiley-Interscience.

AMERICAN GEOGRAPHICAL SOCIETY

Founded in 1851, the American Geographical Society (AGS) is the oldest professional geographical organization in the United States. It is known globally as a pioneer in geographical research and education, whose mission is to link the business, government, professional, and scholarly worlds in the creation and application of geographical knowledge, methods, and techniques to address economic, social, and environmental problems.

For 158 years, the AGS has informed public policy. Specific priorities and programs evolve, but the commitment of service to government, the business community, and the world at large continues unchanged. Throughout World War I, President Woodrow Wilson commissioned the AGS to lead "The Inquiry," a massive analysis of foreign intelligence to support the peace negotiations that would follow. As part of that effort, the AGS was responsible for drafting his famous "14 Points" and for supporting the American delegation at the Paris Peace Conference. Earlier, the AGS played an instrumental role in siting the Panama Canal, the Transatlantic Telegraph Cable, and the Transcontinental Railway.

From its earliest days, the AGS was a leading proponent and sponsor of Arctic, Antarctic, and Andean exploration. A notable artifact from that era is the AGS Fliers' and Explorers' Globe, which has been signed by 80 of the world's most renowned explorers of the past century.

Through its journals, the *Geographical Review* and *FOCUS on Geography*, and through consulting, lectures, and educational travel services, the AGS presents clear, concise, and relevant geographical information that can be understood by policymakers and the public as well as by professional geographers.

The AGS is best known for its pioneering work in exploration and cartography, its invaluable geographical research library, and its research, mapping, and consulting services for government and business at the highest levels. In 1978, its library, then valued at $15,000,000, was given to the University of Wisconsin at Milwaukee. Then and now, the American Geographical Society Library contains the largest collection of geographical literature, maps, charts, and globes in the Western Hemisphere and perhaps the world.

From 1925 to 1945, the AGS mapped all of Latin America at 1:1,000,000 scale, the most authoritative maps of that area until well after World War II. Currently, the AGS is promulgating a new world standard for cartographic representation of land mines, minefields, and mine actions.

In 1912, AGS sponsored the Transcontinental Excursion, a 13,000-mile lecture and study tour on which more than 40 of the world's most prominent foreign geographers and 100 American geographers traversed the country by rail from New York to the Pacific Coast and back. A *New York Times* correspondent traveled along and sent news dispatches from every stop. Today, the AGS continues to sponsor educational trips around the globe that include lectures by leading geographers.

The AGS is led by a council of 27 professional geographers and other devotees of geography from academia, business, and government. Most AGS fellows are Americans, but 40% of its journal subscriptions are from outside the United States. Individuals, corporations, and businesses wishing to support the AGS can join its Galileo Circle or Humboldt Club. Its office is located at 120 Wall Street in New York City.

Jerome E. Dobson

See also **Association of American Geographers; Bowman, Isaiah; Human Geography, History of**

Further Readings

American Geographical Society: www.amergeog.org

ANALYTICAL OPERATIONS IN GIS

Geographic information systems (GIS) have grown to be big and somewhat unwieldy software packages with hundreds, sometimes thousands, of functions. The majority of these are, however, devoted to facilitating data input/output and general data management. Only a small fraction is devoted to what arguably distinguishes a GIS from other software packages, the set of analytical operations. A practical definition of "analytic" in the context of this entry is an operation that fully supports the "I" in GIS. In other words, an analytical GIS operation creates new information beyond what has explicitly been stored in the database.

Based on this definition, a database query is not an analytical GIS operation because it merely retrieves what has been stored before, though possibly in a different organizational structure. The boundary between analytical and other GIS operations is not always clear-cut because complex query operations can be expanded to be analytical operations proper. The litmus test in the end is whether the operation results in new data that did not exist before.

A classic example is the *overlay* operation. As opposed to conditional queries or recoding operations, the overlay results in the creation of new geometries not previously stored in the database, and these new geometries inherit their attributes from their parent geometries according to user- or system-specified rules (Figure 1).

The logic of this overlay operation is typically based on Boolean operands such as AND, OR, NOT, or XOR. Together with its cousin, the *buffer* operation, this analytical operation is responsible for more than half of the generally used analytical capabilities of GIS. Following an often cited characterization by Mike Goodchild, we can use the data model underlying the input

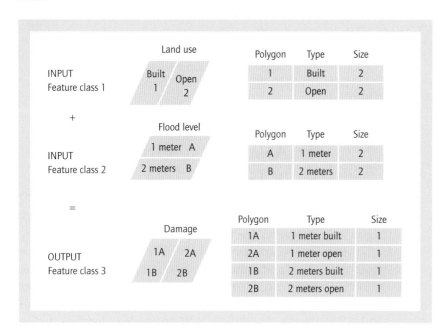

Figure 1 Schematic overlay operation illustrating the combination of new area definitions and attribute value calculations for a flood damage scenario

Source: Author.

and/or output to develop a hierarchy of ever more complex analytical GIS operations. On the object-centered side of analytical operations, one may (a) just describe point patterns, (b) create new geometries and attributes as in our overlay example, (c) perform some rather involved operations on transportation or hydrological networks, or (d) model the interaction between two or more objects. If we follow the field perspective of geospatial data, then we have the whole range of Map Algebra operations that truly get us into the modeling of geographic phenomena.

Jochen Albrecht generated a list of 20 so-called universal GIS operations from a user interface perspective that encompass a wide range of GIS tasks. Although this list seems surprisingly short and already contains a few exotic or rarely used examples, it has been shown that it covers most of what a wide range of GIS users actually want to see in a GIS. The question then arises, What makes analytical GIS operations so important? And the answer lies not in their quantity but in their overwhelming usefulness once they become part of a larger geoprocessing workflow.

While originally developed for desktop GIS, this set of data model-independent analytical GIS operations is now the focus of numerous endeavors to standardize the analytical component of GIS Web services. In both cases, there is a need for well-specified operations that can be concatenated to workflows and larger models in support of spatial decision making. The description of analytical GIS operations is hence not aimed anymore at improving GIS user interfaces but at the automatic generation of scripts that allow clients to create mash-ups for Web-based GIS applications that range from object-based image analysis and environmental model integration to geocollaboration and multicriteria decision support.

Well-defined analytical GIS operations also form the building blocks for libraries, not of data but of models that help us analyze data and that can be shared or sold. These model libraries will eventually capture geographic knowledge and serve the electronic central role that the discipline of geography plays in the orchestra of sciences. But it all starts with a small number of analytical GIS operations.

Jochen Albrecht

See also Agent-Based Models; Cellular Automata; Geocomputation; GIScience; Spatial Analysis; Spatial Optimization Methods

Further Readings

Albrecht, J. (2007). *Key concepts and techniques in GIS*. London: Sage.

Goodchild, M. (1991). Spatial analysis with GIS: Problems and prospects. In *Proceedings GIS/ LIS'91* (pp. 41–48). Falls Church, VA: ASPRS/ ACSM.

ANARCHISM AND GEOGRAPHY

Anarchism is a political philosophy named from the Greek word *anarchos*, which means "without a ruler." The key tenets of anarchist thought are antiauthoritarianism and the formation of a new social order based on mutual collaboration and decision making. Advocates of anarchism see the elimination of the state and other forms of authority as part of a necessary progression toward the formation of a voluntary and self-sustaining society. Anarchists may emphasize individualism, as an extension of liberalism, or socialism, which rejects private property and espouses cooperative ownership. Anarchists often argue for ecological preservation, advocating that societies should live in harmony or equilibrium with the natural world. Two well-known anarchists writing during the late 19th century, Peter Kropotkin and Élisée Reclus, were also influential geographers. During the 1960s and 1970s, geographers engaged with anarchist thought as part of reinvigorating approaches to theory, research, and teaching. As radical geographies emerged, scholarly focus on anarchism dwindled; instead, Marxism, feminism, queer theory, and postcolonialism received further attention. While there has been little theoretical development of anarchist ideas in geography, recent work by a small number of scholars explores anarchist ideas as expressed and practiced in alter-globalization and environmental direct-action movements.

The intersection of anarchism and geography remains underexplored. The main contributions of anarchism include the principle of mutual aid as explained by Peter Kropotkin, environmental preservation, theories and practices that attempt new forms of social order, and uses of space that disrupt and decentralize authority.

Early Anarchist Geographers

Élisée Reclus (1830–1905) was a French geographer who participated in the Paris Commune of 1871, when Parisian workers seized control of the city government. He was imprisoned and then exiled to Switzerland, where he wrote several volumes of geographic works that were thoroughly interlaced with anarchist thought. Reclus proposed an equitable distribution of the world's resources, arguing that decisions about production and distribution should be made by local communities. This was a direct response to imperialist claims on the resources of the global South; he argued that existing social inequalities were the result of authoritarian social and political organization rather than productive capabilities or overpopulation. Reclus was also an early advocator for environmental preservation.

Peter Kropotkin (1842–1921) was a Russian prince, high-ranking military officer, and physical geographer, who credited fieldwork in Siberia with readying him to become an anarchist. His theory of mutual aid analyzed cooperation among animals, using examples in the natural world as a scientific basis for collaborative human organization. This was a response to the Social Darwinists of the time, who were proponents of competition. Kropotkin's essay "What Geography Ought to Be" proposed educational reform and emphasized shared humanity across international borders, an important argument that weakens the legitimacy of the state.

Their background in geographic scholarship influenced both men, engendering a distinctive understanding of humanity in different cultural and economic contexts and a broader comprehension of possible social and economic structures.

Anarchism and Radical Geography: The 1960s and 1970s

During the late 1960s and 1970s, a number of geographers drew from anarchist thought in attempts to reinvigorate and rethink aspects of the discipline, especially to inject practical and socially useful practices into research and teaching methodologies. In a special issue of the journal *Antipode*, meant to stimulate new interest in anarchist ideas, Richard Peet criticized geographical scholarship and challenged his colleagues to use anarchist principles as transformative tools. He wrote that geography, as a descriptive discipline, had become an instrument for the scientific justification of existing inequalities rather than an active challenger of the status quo.

The Situationist International (SI) was a political and artistic movement mostly active from the mid 1950s through the mid 1970s in Western

Europe. The Situationists practiced drifts (or *dérives*) and psychogeography to disrupt the authority of capitalist spectacle. The anarchist scholar and sociologist Richard Day drew on Situationist practices to discuss global social movements.

Current Scholarship

Despite spurring a brief moment of interest in the 1960s and 1970s, anarchism fell to the wayside as radical scholars instead turned to Marxism, feminism, queer theory, and postcolonialism as modes of inquiry. More recent activist scholarship, drawing on the principles of mutual aid and decentralization of authority, is more broadly linked to autonomous, alter-globalization, and other social movements. Anarchist practice and activism outside the academy have survived and taken myriad forms; geography and anarchism would benefit from a renewed encounter.

Melinda Alexander

See also Human Geography, History of; Kropotkin, Peter; Marxism, Geography and; Radical Geography; Reclus, Élisée; Social Geography; Social Justice

Further Readings

Blunt, A., & Wills, J. (Eds.). (2000). *Dissident geographies: An introduction to radical ideas and practice*. New York: Prentice Hall.

Peet, R. (1979). The geography of human liberation. *Antipode, 10*(3), 119–134.

Pickerill, J., & Chatterton, P. (2006). Notes towards autonomous geographies: Creation, resistance, and self-management as survival tactics. *Progress in Human Geography, 30*(6), 1–17.

 # ANAXIMANDER (CA. 610–546 BC)

Anaximander (born in Miletus ca. 610 BC, died in Miletus ca. 546 BC) was a Grecian pre-Socratic philosopher who belonged to the Milesian

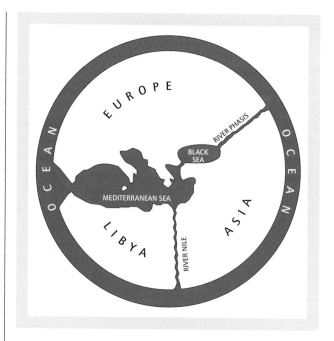

Figure 1 Re-creation of Anaximander's map

Source: Public domain.

school, whose thinkers observed natural phenomena to study and explain their rules independently of mythological belief. Only a few of his life anecdotes and works are known today; nevertheless, his contributions to geography, cartography, cosmology, and astronomy are extraordinarily important. He was the first to draw a map of the ecumene (inhabited land), and he is also said to have invented the gnomon (the part of a sundial that casts the shadow) and discovered the equinoxes, the solstices, and the inclination of the ecliptic (the plane on which the orbit of the Earth around the sun lies). He supposedly authored a work on nature that was the first philosophical text written in prose. Other works attributed to him are known nowadays under the titles *Around the Earth*, *The Sphere*, and *On Fixed Bodies*.

As a result of his observations of nature, Anaximander understood the hydrologic cycle by realizing how water on the Earth's surface evaporates and consequently comes down again in the form of precipitation. He also elaborated a theory that explained how lightning and thunder are generated by the collision of clouds and how earthquakes occur when high temperatures and heavy rains cause land to split. According to his view of the universe and the Earth, the first living beings were aquatic and later adapted to life on the land. The seas are what still remains of primordial humidity, and they will disappear from Earth's surface when all the water evaporates because of increased solar energy. On the other hand, steam created the winds and caused the rotation of the sun and of the moon, which are both much bigger in size than our planet.

Anaximander conceived of the universe as a system of concentric blazing wheels with flames coming out through some holes and explained eclipses as being caused by the occlusion of those openings. The sun circles the nearest circle, while the stars revolve around the farthest. He considered Earth to be a cylindrical entity that hovers motionless in the universe without any support and is equidistant from other celestial bodies. Life has developed on its flat circular plane, which in the beginning was completely covered by water.

As far as cartography is concerned, Anaximander created the first map of the world by drawing three continents (Europe, Asia, and "Libya") separated by the Black Sea, the Phasis River, the Nile, and the Mediterranean Sea. He imagined waters to surround all those lands.

Susanna Servello

See also Cartography, History of; Human Geography, History of

Further Readings

Couprie, D., Hahn, R., & Naddaf, G. (2002). *Anaximander in context: New studies in the origins of Greek philosophy.* Albany: State University of New York Press.

De Santillana, G. (2000). *The origins of scientific thought: From Anaximander to Proclus 600 B.C. to 300 A.D.* Chicago: University of Chicago Press.

Kahn, C. (1994). *Anaximander and the origins of Greek cosmology.* Indianapolis, IN: Hackett.

 # ANIMAL GEOGRAPHIES

The study of animal geographies is a burgeoning subfield of cultural geography that examines the interplay between culture, society, and animals. Animal geographers examine a broad range of human-animal concerns, including, for example, habitat loss and species endangerment, domestication, animal entertainment and display, wildlife conservation, and more. Essentially, animal geographies explore nonhuman animals and their *place* in society—place meaning in both physical boundaries (material practices that shape the spaces where some animals are welcomed and others are not) and conceptual margins that call up matters of human identity and animal subjectivity. While the study of contemporary animal geographies is varied and diverse, we can think loosely in terms of three organizational themes: (1) animals and the making of place, (2) human identity and animal subjectivity, and (3) the role of ethics and how humans *ought to* treat animals. These categories often overlap and dovetail with concepts such as animal instrumentalism, anthropocentrism, and the human-animal continua. Moreover, animal geographers recognize the fluidity of boundaries, emphasizing not only the distinctions but also the connections, overlaps, and similitudes between human and animal worlds.

"Old" Animal Geography

While geography has always been concerned with the interface between human culture and the natural world, until recently, nonhuman animals were largely overlooked. Historically, animals in geography were considered no more than biological pieces of a larger ecological system, instruments for human use, or forms of symbolic natural capital. It is not that geographers showed no interest in animals; indeed, there was a field called *animal geography* as early as 1913, consisting of studies of animal populations and examination of floral and fauna regions. In the early days of animal geography, two approaches emerged, mirroring the widening gap between physical and human geography. Zoogeography, which considered mainly animal distributions, was rooted in zoology and physical geography; the other approach aligned with human geography and social sciences and focused on animal domestications. By the 1960s, however, owing to the low status of cultural geography (due partly to the Berkeley School's treatment of culture–economy relations),

Franz Marc, *The Dream* (1912)

Source: © Museo Thyssen-Bornemisza, Madrid. Photography: José Loren. Used with permission.

questions about human-environment relations receded from view. By the last quarter of the 20th century, the term *animal geography* had disappeared from the discipline altogether.

Today's animal geographies differ substantially from the "old" animal geography. The interplay between geography and social theory, cultural studies, and environmental ethics in the 1990s led to a rebirth of interest in nonhuman animals. The increased focus on animals, culture, and society came on the heels of growing public and academic concern about environmental degradation, habitat loss, species endangerment, and the plight of animals relegated to a dismal life (and untimely death) in shelters, labs, and factory farms. The 1970s and 1980s witnessed hundreds of new organizations created to lead social movements involving animals and the environment. Animal rights groups (especially the more radical organizations such as People for the Ethical Treatment of Animals and the Animal Liberation Front) challenged people to reconsider their relationships with animals by suggesting, for example, that speciesism is equivalent to racism and sexism; that animal captivity is as heinous as human slavery; and that factory farms, fur farms, and research labs are tantamount to genocide. Alongside this tumultuous public activity, the legacies of modernity and modernist ways of thinking came under attack as critics argued (and still argue) that the achievements of modernity rested on race, class, and gender domination, as well as colonialism and imperialism, anthropocentrism, and the destruction of nature. Given this, scholars in social theory and cultural studies began to rethink culture, and geographers (along with other social and natural scientists) began thinking about "the animal question" and the need to unpack the black box of nature.

Edited books by Jennifer Wolch and Jody Emel, and by Chris Philo and Chris Wilbert, brought human-animal interactions to the foreground in the past decade. Reviving and remaking the face of animal geography, today, animal geographers recognize animals both as foundational to our ontology and epistemology *and* as beings with inner lives and intentionality, worthy of consideration in their own right.

Material Boundaries: Animals and the Making of Place

Discussions in human geography about the social construction of landscapes have led to the exploration of how animals and their networks leave their imprint on places, regions, and landscapes over time. Animal geographers consider tangible places such as zoos, farms, experimental laboratories, and wildlife reserves as well as economic, social, and political spaces, such as the worldwide trade of captive wild animals. Even a relatively new space through which animals are woven into human culture—the "electronic zoo"—has been explored as an emerging form of animal display trading in digital images, rather than animal bodies such as traditional zoos and aquariums.

Animal geographers also consider places characterized by the presence or absence of animals and how human-animal interactions create distinctive landscapes. Researchers have studied the impact of land use practices on wildlife survival in the Peruvian Amazon, in boundary-making policy conflicts between urban and rural New Yorkers over the proper place of wolves, and in the changing relationships between people and mountain lions in California. Animal geographers also foreground links between humans and other animals—those consumed as meat, medicine, clothing, and beauty products, for example—that go mostly unseen in contemporary society, given the distance created by globalized commodity chains. Such borderland communities, where humans and animals share public and/or private space, where some are loved while others are despised, and where so many are unconsciously consumed, reveal the contingent and often contradictory ways in which humans and animals interact with one another.

Borderland communities can span various places and spaces. Investigating human-dolphin encounter spaces, for example, requires a look at the material and well-defined (if often contested) boundaries of zoos and aquariums, as well as more amorphous boundaries. For example, borders are in flux in the open seas where humans and dolphins meet: While a growing number of tourism operators seek out dolphins for paying customers to closely interact and even swim along with, government officials and some animal advocates want to keep people and dolphins at least some distance apart from one another. Animal geographers interrogate such material, social, and political boundaries; what's more, they acknowledge both human and nonhuman roles in the making, maintaining, and changing of animal geographies. In this case, to understand human-dolphin encounter spaces, scholars would strive to explicate not just human actions and viewpoints but also how the dolphins encourage or defy the human ordering of contested border waters. Examining these material places—from the zoo and the open ocean to the economic and policy arenas considered by those investigating human-dolphin encounter spaces, in this example—helps animal geographers illuminate the complex relationships between human and nonhuman worlds.

Conceptual Boundaries: Human Identity and Animal Subjectivity

Breaking from the traditional geographic approach to animals, contemporary animal geographers think about nonhuman animals as more than biotic elements of ecological systems. Not only are animals appreciated as foundational to countless cultural norms and practices, they are also valued as individuals with mental and emotional lives. Thus, animal geographers call for a more theoretically inclusive approach to thinking about humans and animals; both are considered to be embedded in social relations and networks with others on whom their social welfare depends.

Franz Marc, *Elephant* (1907)

Source: Public domain.

Such thinking suggests a reconceptualization of the "human-animal divide," in which humans are vastly different from (and superior to) animals, and instead points toward a continuum that allows for a kinship of (although still acknowledging the differences between) humans and other animals.

Stressing the importance of *relations* in their work, some animal geographers conceptualize humans and animals in ways that resist typical characterizations. For example, some scholars highlight the many "inappropriate/d others" of our society that resist being represented within the conventional taxonomy—*cyborgs*, which are part human, part organic, and part mechanical. As such, any claims to hard-line distinctions are ontologically unstable, even those seemingly constant beings we label as human or nonhuman. To illustrate, animal geographies scholars who examined human-elephant relations have shown that *becoming an elephant* is a contingent process. Depending on whether the life (social and otherwise) of an elephant is experienced in the openness of a savanna or in the closed spaces of a zoo, those extremely varied life experiences will shape the elephant as an individual in entirely different

ways. Which, we might ask, should be called a "real" elephant (see illustration)?

Ethics, Humans, and Other Animals

Human relationships with animals have always been diverse and deeply complex, ranging from magnificent to malignant. In every case, humans remain the regulators of whether animals are conceived of as either "in place" or "out of place," and it is moral sensibility that defines such orderings, with significant ethical implications. In many cases, animal geographers attribute instances of instrumentalism, exclusion, and exploitation of the nonhuman world to a history of anthropocentric, or human-centered, thinking. Critical of such activities, much of the animal geographies literature is concerned with the ethical task of advancing the well-being of animals. This is where animal geography largely departs from the theoretical positioning in the contemporary nature/culture debates in geography, which remain largely anthropocentric.

One way to advance this apparently normative project is to explicate societal values, which certainly determine human treatment of animals. For example, some animal geographers have considered how an animal's position in the scientific community's hierarchy of value (as determined by the rarity of species) can have significant influence on its fate. A crocodile that belongs to a species that is included in a global conservation policy, for instance, is "protected" and therefore privileged over animals that are not included in such a policy. With a change in conservation policy, or the "down-listing" of a particular species from the rank of endangered species, the same crocodile once protected and perhaps flourishing in its natural habitat could likely be removed for human use to a shortened and no doubt impoverished life as a factory farm animal.

Animal geographies also encourage thinking about animal agency and subjectivity, recognizing that animals have intentions and are communicative subjects with potential viewpoints, desires, and projects of their own. Many animal geographers suggest that nonhuman animals are best seen as "strange persons" or as marginalized, socially excluded people. But because animals cannot organize and challenge human activities for themselves, animal geographers recognize that they require human representatives to speak and act in their interests.

Seeking to advance the well-being of both humans and animals, many animal geographers explicitly locate animals in the moral landscape, recognizing that ethical questions are present in all human and animal geographies. In these instances, animal geographers argue for the inclusion of animals in the moral community, valuing animals as ends in themselves rather than as simply the means to human ends. The practical consequences of such inclusion are considerable: For example, how are we to decide what is most important in environmental policy making? And who, exactly, gets to decide? Especially when human-animal needs clash in a world of finite space, a framework of normative principles suggested by animal geographies—principles inclusive of animal interests and desires—can guide human-animal relations and resolve the moral dilemmas that relate to conflicting wants and needs of both humans and animals. Thus, the study of animal geographies is an academic discipline as well as a moral and political project that seeks, at its basic level, to foreground nonhuman animals so that their needs and desires are not unthinkingly ignored or automatically secondary to our own but are thoughtfully considered as part of a more than human geography.

Kristin L. Stewart

See also Berkeley School; Biogeography; Coupled Human and Animal Systems; Cultural Geography; Domestication of Animals; Extinctions; Hybrid Geographies; Landscape and Wildlife Conservation

Further Readings

Fox, R. (2006). Animal behaviours, post-human lives: Everyday negotiations of the animal-human divide in pet-keeping. *Social & Cultural Geography, 7,* 525–537.

Lynn, W. S. (1998). Contested moralities: Animals and moral value in the Dear/Symanski debate. *Ethics, Place and Environment, 1,* 223–242.

Midgley, M. (1983). *Animals and why they matter.* Athens: University of Georgia Press.

Philo, C., & Wilbert, C. (Eds.). (2000). *Animal spaces, beastly places: New geographies of human-animal relations.* London: Routledge.

Plumwood, V. (2002). *Environmental culture: The ecological crisis of reason.* London: Routledge.

Risan, L. C. (2005). The boundary of animality. *Environment and Planning D: Society and Space, 23,* 773–787.

Whatmore, S. (2001). *Hybrid geographies: Natures cultures spaces.* London: Sage.

Wolch, J., & Emel, J. (Eds.). (1998). *Animal geographies.* London: Verso.

 ## ANNALES SCHOOL

The founders of the French Annales School of history, Lucien Febvre (1878–1956) and Marc Bloch (1886–1944), as well as its most prominent late-20th-century member, Fernand Braudel (1902–1985), were all heavily influenced by geography, both in their formative education and in their subsequent historical research. Febvre and Bloch were students of the French geographer Vidal de la Blache at the École Normale Supérieur, and Braudel fully absorbed the founders' enthusiasm for *la tradition vidalienne* (Vidalian tradition).

Specifically, Annalistes took from Vidal the notion that geographical landscapes are the result of human and natural processes in mutual adaptation, both historically and in the present. The constraints and opportunities provided by the environing natural environment thus play a crucial role in terms of the manner in which humans materially and culturally reproduce themselves and their societies. The human past thus involves the historical-geographic creation of "social natures" (*genres de vie*) that influence the day-to-day activities of individuals in society. In short, variations in the way nature becomes humanized provide the very foundation for the variations in the historical and cultural trajectories of societies.

The key to this Vidalian vision is that nature and society play equal roles in this process. This is known as *possibilisme*, a term popularized by Lucien Febvre in his book *La terre et l'évolution humaine*, in which he strongly promoted this geographical approach to his fellow historians. Vidal's notion of *genres de vie* thus provided the very basis for what Braudel calls the "geohistorical" approach of the Annales School. Specifically, *genres de vie* are constructed and reproduced only very slowly in time. This necessitates a consideration of what Annalistes call the "longue durée" of historical time compared with the medium time frame of institutions and the day-to-day time frame of specific historical events. Social natures are long in the making and reproduce and change only slowly. Institutions—economic, social, and political—are less long in the making and less slow to change. Finally, historical events such as wars, political contests, and economic depressions occur and end relatively rapidly. To write history means to consider occurrences within each of these time frames in interrelation or otherwise as mutually influencing.

This layered view of historical time necessitates a holistic, interdisciplinary approach. To this end, Febvre and Bloch founded the journal *Annales: Histoire, Sciences sociales* in 1929. The format of this journal was certainly influenced by a journal published earlier, *La synthèse historique*, which also advocated interdisciplinarity. But significantly, the holistic format of the *Annales de Géographie*, first published in 1891 by Vidal de la Blache and Marcel Dubois, was also a crucial influence. This is not often acknowledged because, after Vidal, French geographers mostly lost his holistic, historical vision, particularly as Febvre in his book rendered *la tradition vidalienne* in terms that largely relegated geography to the description of physical landscape to make disciplinary room for the Annales kind of history. In the end, however, this new history is really merely a more truly Vidalian study of the human past.

Kevin Archer

See also Cultural Geography; Febvre, Lucien; Historical Geography; Human Geography, History of; Vidal de la Blache, Paul

Further Readings

Braudel, F. (1992). *Civilization and capitalism, 15th–18th century: Vol. 1. The structure of everyday life*. Berkeley: University of California Press.
Clavel, P. (2003). Fernand Braudel. *Geographers Biobibliography, 22*, 28–42.
Clavel, P. (2004). Lucien Febvre. *Geographers Biobibliography, 23*, 35–49.

 # ANSELIN, LUC (1953–)

The year 2008 marks the 20th anniversary of the book *Spatial Econometrics: Methods and Models*, for which Luc Anselin is best known and which has been cited more than 2,600 times. One of Anselin's principal academic achievements has been his contributions to moving the discipline of spatial econometrics from the margins in 1988 to current acceptance in mainstream econometrics, thereby advancing the econometric foundations of geographic information science (GIScience). His publications include several hundred articles and seminal edited books (including *New Directions in Spatial Econometrics* in 1995 and *Advances in Spatial Econometrics* in 2004) in the fields of quantitative geography, regional science, GIScience, econometrics, economics, and computer science.

His development of spatial software further facilitated the establishment of spatial econometrics. Prominent software tools include *SpaceStat* (spatial econometrics), *GeoDa* (exploratory spatial data analysis and spatial regression modeling), and collaborative efforts such as *PySAL*, an open-source library of spatial analytic functions based on the Python programming language. The increased popularity of spatial analysis methods is illustrated by *GeoDa*'s worldwide adoption by more than 36,000 users within 5 years.

A native of Belgium, Luc Anselin graduated magna cum laude with a BS in economics in 1975 and summa cum laude with an MS in statistics, econometrics, and operations research in 1976, both from the Vrije Universiteit Brussel. Around this time, the origins of spatial econometrics began to take shape in economics departments in the Netherlands and geography/regional science departments in the United Kingdom. In 1977, Anselin moved from Belgium to the United States to enroll in Cornell University's interdisciplinary doctoral program in regional science. This change provided the opportunity to work with Walter Isard, one of the founders of regional science in the United States, and William Greene, author of the standard textbook in econometrics. Luc Anselin earned his doctorate in regional science in 1980.

Anselin is currently Foundation Professor of Geographical Sciences and director of the School of Geographical Sciences and Planning at Arizona State University (ASU), where he attracted some of the leading spatial econometrics scholars of the next two generations. He also founded and directs the *GeoDa Center for Geospatial Analysis and Computation* at ASU to develop, implement, apply, and disseminate spatial analysis methods. He held prior appointments at the University of Illinois, Urbana-Champaign; University of Texas at Dallas; West Virginia University; University of California, Santa Barbara; and the Ohio State University. His (joint) appointments included a range of disciplines, including geography, urban and regional planning, economics, agricultural and consumer economics, political economy, and political science.

In recent years, several national and international awards recognized Luc Anselin's lifetime achievements, including his development of new spatial methodologies (e.g., local indicators of statistical association) and his widely adopted spatial software tools. The Regional Science Association International elected him as a fellow in 2004 and awarded him their Walter Isard Prize in 2005 and their William Alonso Memorial Prize in 2006. In 2008, Luc Anselin was awarded one of the nation's highest academic honors by being elected to the National Academy of Sciences.

Julia Koschinsky

See also Automobility; Quantitative Methods; Regional Science; Regional Science Association International (RSAI); Spatial Analysis; Spatial Autocorrelation; Spatial Econometrics

Further Readings

Anselin, L. (1988). *Spatial econometrics: Methods and models*. Berlin, Germany: Springer.

 # ANTEVS, ERNST (1888–1974)

Ernst Antevs was a Swedish American geologist who specialized in glacial geology and did pioneering work in geomorphology.

Antevs was born Ernst Valdemar Eriksson in Vartofta-Åsaka, Southern Sweden, and adopted the name Antevs after finishing school in 1909. In 1917, he received his PhD at Stockholm University with a thesis on shell banks in southwestern Sweden. As a paleobotanist, Antevs joined expeditions within Scandinavia as well as to Bjørnøya in 1916 and Spitsbergen in 1918. In 1920, he came to North America on an expedition with his professor, Gerard De Geer, inventor of the clay-varve dating method. He returned to America and obtained U.S. citizenship in 1939. Supported by Swedish and American foundations and societies, he worked at the University of Arizona, where he was appointed to a research assistantship in 1957. Antevs's work there earned him an honorary doctorate of science in 1965. The Antevs Library at the University of Arizona is named after him.

Antevs studied glacial geology in New England and Canada, applying the clay-varve method to trace the recession of the Laurentide ice sheet. Although partly revised, parts of the work still hold true in detail. As a specialist in glacial geology, he was entrusted with the compilation of maps of Pleistocene glaciers and ice sheets, which he presented in 1928 at the symposium The Centenary of the Glacial Theory in New York.

Antevs was fascinated by the possibility of identifying climate variations by means of tree rings. In 1922, his lifelong work on North American archaeology and climate variations began with studies of pluvial lakes and arroyos in the Western United States, particularly in the Great Basin. One of his main interests was human-environment interaction. Together with his glacial-geological work, this made Antevs a pioneer in geomorphology. He proposed a subdivision of the Holocene into Meditthermal (0–4500 BP [before present]), Altithermal (4500–7000 BP), and Anathermal (7000–10500 BP). The scheme was criticized and partly revised, but the term *Altithermal* has remained in use. Antevs's pollen-analytical work (pollen analysis is the reconstruction of former vegetation and climate by means of pollen grains and spores in sediment layers) in the Great Basin led him into discussions resulting in the creation of the science of palynology (the science of pollen and spores).

Antevs's later works focus on climate change and dating problems and on the relationship between radiocarbon and other dating. They are published mainly in *American Antiquity* and *Journal of Geology*.

Jan Lundqvist

See also Geomorphology; Glaciers: Continental; Glaciers: Mountain; Physical Geography, History of

Further Readings

Antevs, E. (1925). On the Pleistocene history of the Great Basin. In J. C. Jones, E. Antevs, & E. Huntington (Eds.), *Quaternary climates* (Monograph Series, No. 352, pp. 51–114). Washington, DC: Carnegie Institution.

Antevs, E. (1925). Retreat of the last ice sheet in Eastern Canada. *Canada Department of Mines, Geological Survey, Memoir, 146*(126), 1–142.

Antevs, E. (1929). Maps of the Pleistocene glaciations. *Bulletin of the Geological Society of America, 40*, 631–720.

Smiley, T. L. (1977). Memorial to Ernst Valdemar Antevs 1888–1974. *Memorials: Geological Society of America, 6*, 1–7.

 # ANTHROPOGENIC ATMOSPHERIC CHANGE

See Anthropogenic Climate Change; Stratospheric Ozone Depletion

 # ANTHROPOGENIC CLIMATE CHANGE

In 2008, climate scientists celebrated the 50th anniversary of the Keeling Curve, one of the iconic images of science, rivaling the double helix or Darwin's sketches of finches. March 1958 marked the inception of the historical recording of atmospheric carbon dioxide levels at a small observatory on top of Hawaii's Mauna Loa. As the measurements progressed over the years, Dr. Keeling noted a steady increase of about 1.5 ppm (parts per million) per year (Figure 1). His plot provides clear evidence that carbon dioxide accumulates in the atmosphere as the result of mankind's use of fossil fuels, turning speculations about increasing CO_2 from a hypothesis into a fact.

The World Meteorological Organization and the United Nations Environment Program established the Intergovernmental Panel on Climate Change (IPCC) in 1988, assigning it to assess the scientific, technical, and socioeconomic information relevant to human-induced climate change and its potential risks. The IPCC does not support research directly or monitor climate-related data, but its assessment reports have often inspired scientific research in many climate-oriented fields, leading to new findings. The IPCC has published a total of four assessment reports, becoming the de facto accurate and relevant standard regarding scientific facts on global climate change and its impacts.

The IPCC has three Working Groups and one Task Force:

Working Group I (WGI) assesses the scientific aspects of the climate system and climate change.

Working Group II (WGII) measures the vulnerability and adaptation of socioeconomic and natural systems to climate change.

Working Group III (WGIII) evaluates the mitigation options for limiting greenhouse gas emissions.

The Task Force is responsible for the IPCC National Greenhouse Gas Inventories Program.

The WGI for the First Assessment Report (FAR) was completed in May 1990. Its nonquantitative findings for an anthropogenic interference

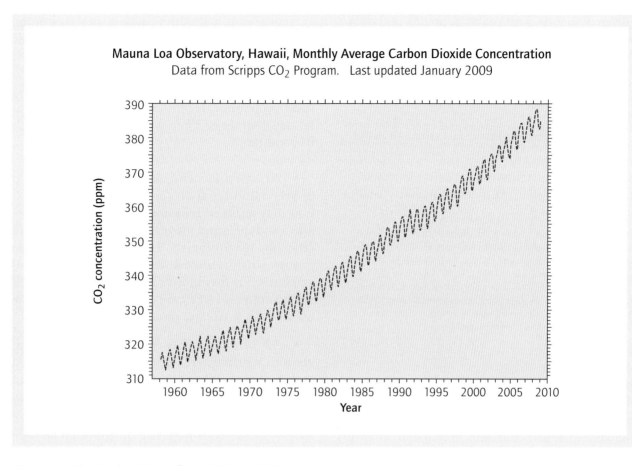

Mauna Loa Observatory, Hawaii, Monthly Average Carbon Dioxide Concentration
Data from Scripps CO_2 Program.   Last updated January 2009

Figure 1   The Keeling Curve from 1958 to 2009

Source: Data from Scripps CO_2 Program. Retrieved December 10, 2009, from http://scrippsco2.ucsd.edu/graphics_gallery/mauna_loa_record/mauna_loa_record.html. Last updated January 2009.

Notes: Monthly average atmospheric carbon dioxide concentration versus time at Mauna Loa Observatory, Hawaii (20° N, 156° W), where CO_2 concentration is in parts per million (ppm) in the mole fraction. The curve is a fit to the data based on a stiff spline plus a 4-harmonic fit to the seasonal cycle with a linear gain factor.

with the climate system remain valid today. It correctly concluded that "emissions resulting from human activities are substantially increasing the atmospheric concentrations of the greenhouse gases: CO_2, CH_4, CFCs, N_2O." The FAR introduced varying levels of confidence, ranging from "certainty" to expert "judgment" (Table 1).

The Second Assessment Report (SAR) was released in November 1995. The report reaffirmed that "the balance of evidence suggests a discernible human influence on global climate." SAR provided key input to the negotiations that led to the adoption in 1997 of the Kyoto Protocol.

The Third Assessment Report (TAR) was approved at the government plenary in January 2001. The main conclusion from TAR strengthened the finding from SAR: "The Earth's climate system has demonstrably changed on both global and regional scales since the pre-industrial era, with some of these changes attributable to human activities."

From observations of the increased global average air and ocean temperatures, the widespread melting of snow and ice, and the rising global average sea level, per the IPCC words in the latest assessment report released in November 2007, "there is very high confidence that the net effect of human activities since 1750 has been one of warming" over each continent (including Antarctica). Most of the observed increase in global average temperatures since

Confidence Terminology	Degree of Confidence in Being Correct
Very high confidence	At least 9 out of 10 chance
High confidence	About 8 out of 10 chance
Medium confidence	About 5 out of 10 chance
Low confidence	About 2 out of 10 chance
Very low confidence	Less than 1 out of 10 chance

Table 1 Standard terms used to define levels of confidence as given in the IPCC uncertainty guidance note

Source: Adapted from Le Treut, H., Somerville, R., Cubasch, U., Ding, Y., Mauritzen, C., Mokssit, A., et al. (2007). Historical overview of climate change. In S. Solomon, D. Quin, M. Manning, A. Chen, M. Marquis, K. B. Averyt, et al. (Eds.), *Climate change 2007: The physical science bases* (Contribution of Working Group I to the Fourth Assessment Report of the Intergovernmental Panel on Climate Change; p. 120, Box 1.1). Cambridge, UK: Cambridge University Press.

the mid 20th century is very likely due to the observed increase in anthropogenic greenhouse gas (GHG) concentrations.

Potential Future Climate Change and Its Impacts

Between 1970 and 2004, GHG emissions have increased 70%. The *IPCC Special Report on Emissions Scenarios* (SRES) projects a further increase of GHG emissions by 25% to 90% CO_2-eq (carbon dioxide equivalent) between 2000 and 2030, as fossil fuels maintain their dominant position as energy source. As the emissions continue, it is very likely that the warming will induce many changes in the global climate system during the 21st century that will be larger than those observed during the 20th century (Figure 2). Furthermore, warming would reduce terrestrial and ocean uptake of atmospheric CO_2, increasing the fraction of anthropogenic emissions remaining in the atmosphere.

The SRES includes six marker scenarios for GHG emissions. These emissions are driven by demographic development, socioeconomic development, and technological change, and though their future evolution is highly uncertain, they assist in climate change analysis, assessment of impacts, adaptation, and mitigation. This exercise helps climate scientists determine the scope of consequences for a variety of possible fuel-use scenarios.

Four different narrative storylines were developed, each representing different demographic, social, economic, technological, and environmental developments. They were designated A1, A2, B1, and B2.

- *A1 storyline:* This describes a future world of very rapid economic growth, a global population that peaks in midcentury and declines thereafter, and the rapid introduction of new and more efficient technologies. It develops into three groups distinguished by their technological emphasis: (1) fossil intensive (A1FI), (2) nonfossil energy sources (A1T), or (3) a balance across the use of both fossil and nonfossil fuels (A1B).

- *A2 storyline:* This describes a very heterogeneous world where national identities and local and regional solutions to environmental and social equity issues predominate. Global fertility patterns result in continuously increasing world population. Economic development, per capita economic growth, and technological change are slower than in other storylines.

- *B1 storyline:* This describes a world with a peaking population midcentury and declining thereafter, as in the A1 storyline. The economy changes to a service and information economy and the introduction of clean and resource-efficient technologies. Environmental, economical, and social solutions are found globally but without additional climate initiatives.

- *B2 storyline:* This describes a world in which the emphasis is on local solutions to economic, social, and environmental sustainability. It is a world with a continuously increasing global population at a rate lower than in the A2 storyline, intermediate levels of economic development, and less rapid and more diverse technological change than in the B1 and A1 storylines. While the scenario is also oriented toward environmental protection and social equity, it focuses on local and regional levels.

Future emissions differ tremendously between the different scenarios (Figure 3). The largest

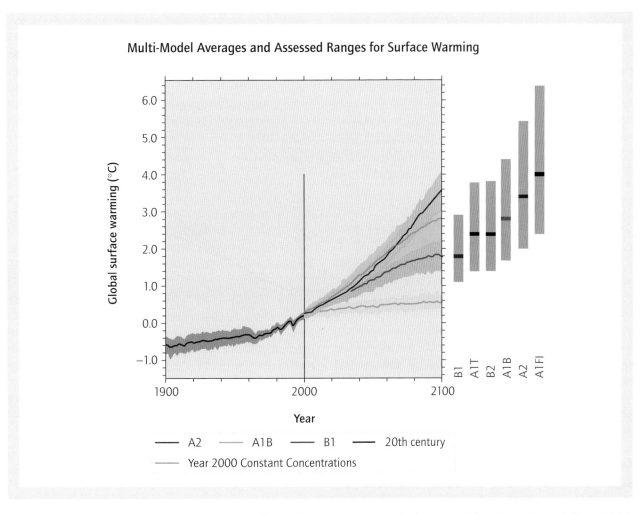

Figure 2 Total global annual CO_2 emissions from all sources (energy, industry, and land use change) from 1900 to 2100 (in GtC/yr., gigatons of carbon per year) for the six scenario groups, presented by the four storylines (A1, A2, B1, and B2)

Source: © Intergovernmental Panel on Climate Change (IPCC), 2007.

Notes: A1FI is the fossil-intensive fuel (comprising the high-coal and high-oil-and-gas scenarios); A1T is the predominantly nonfossil fuel; A1B is the balanced scenario. A2, B1, and B2 are presented in Figures 4b, 3c, and 3d, respectively. Each colored emission band shows the approximate range of carbon dioxide concentrations.

growth and cumulative release of CO_2 is associated with the A1FI fossil-fuel-intensive scenarios (1,550 ppm in 2100, or about 30 GtC/yr. [gigatons of carbon per year] emitted; Figure 4), while the smallest is associated with the B1 scenario (600 ppm in 2100, or less than 10 GtC/yr. emitted; Figure 4). The IPCC scientists made no attempt to estimate the likelihoods of any of these possible scenarios actually occurring; the uncertainties are simply too large.

The findings of these scenarios reinforce our understanding that the main driving forces of future greenhouse gas trajectories will continue to be demographic change, social and economic development, and the rate and direction of technological change.

Climate Models

Climate scenarios and predictions for the future rely on the use of numerical models. Their evolution over the past three decades has enabled scientists to run ever more computationally demanding and complex models that incorporate more and more components and processes of the climate system (Figure 5).

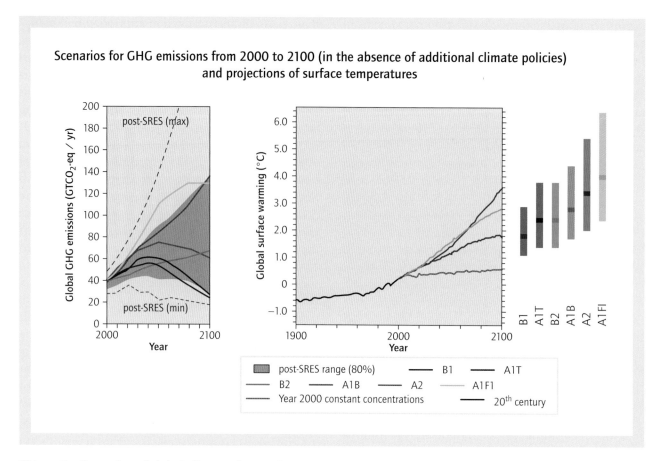

Scenarios for GHG emissions from 2000 to 2100 (in the absence of additional climate policies) and projections of surface temperatures

Figure 3   Scenarios of global climate change (Intergovernmental Panel on Climate Change, IPCC)

Source: © 2007 Intergovernmental Panel on Climate Change.

Notes: *Left panel*: Global greenhouse gas emissions in GtCO$_2$-eq (gigatons of carbon dioxide equivalent) in the absence of climate policies: six illustrative scenarios (colored lines) and the 80th percentile range of recent scenarios published since the release of the *IPCC Special Report on Emissions Scenarios* (SRES) in 2000 (gray-shaded area). *Right panel*: Solid lines are global averages of surface warming from climate models for scenarios A2, A1B, and B1 (see text). The bars at the right of the figure indicate the best estimate (solid line within each bar) and the likely range assessed for the six SRES marker scenarios from 2090 to 2099. All temperatures are relative to the period 1980 to 1999.

Climate models are computer codes expressing mathematical representations of the physical, chemical, and biological processes of the climate system, as well as their interactions. They are built to simulate the climate system as it exists today and are judged by how well they do. If their simulation of the modern-day climate (called the *control* case) is good, models are further tested against the very different conditions found in the past. The models that succeed in reproducing past climates are then used to simulate and predict future climate change.

The confidence in models arises from the fact that model fundamentals are based on established physical laws, such as conservation of mass, energy, and momentum, along with a wealth of observations. The confidence factor is higher for some climate variables (e.g., temperature) than for others (e.g., precipitation). A second source of confidence comes from models being routinely and extensively assessed by comparing their simulations with observations of the atmosphere, ocean, cryosphere, and land surface. The Paleoclimate Modelling Intercomparison Project (PMIP) has been instrumental in bringing unprecedented levels of evaluation over the past decade in the form of organized multimodel "intercomparisons."

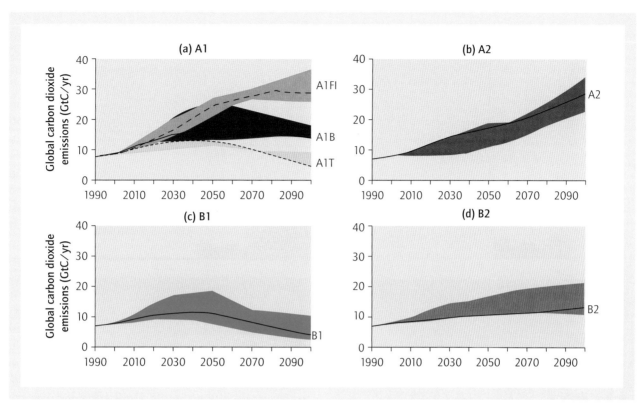

Figure 4   Projections of future global carbon dioxide emissions (GtC/yr., gigatons of carbon per year) for the A2, A1B, and B1 scenarios

Source: © 2007 Intergovernmental Panel on Climate Change.

Different Models, Different Uses

There are many different types of climate models—some very simple, some very complex. Simplicity can be achieved by a reduced number of equations (e.g., a single equation for the global surface temperature); by the reduced dimensionality of the model (one-dimensional [1D], vertical or latitudinal; two-dimensional [2D], latitude-altitude); or by restriction to a few processes (e.g., a midlatitude quasi-geostrophic atmosphere with or without the inclusion of moist processes).

One-dimensional climate models focus on the balance between incoming solar radiation and outgoing terrestrial energy ("heat"). Though simple, these models account for the greenhouse effect and feedback loops to determine temperature changes on Earth. Three-dimensional (3D) "general circulation models" (GCMs) provide the most complete numerical representations and simulations of the climate system. They take into account the 3D structure of the atmosphere and oceans, the shape and arrangement of the continents, the bathymetry of the oceans, and the topography of the land (Figure 6). Many climatic variables are simulated; land surface temperatures, precipitation, atmospheric pressure, wind direction and strength at the surface and upper levels, ocean currents, temperatures, and salinity (Figure 5). There is considerable confidence that these models provide credible quantitative estimates of future climate change, particularly at continental and larger scales. Global circulation models can be complemented by regional models for a higher resolution over a given area or by process-oriented models, resolving clouds or large oceanic eddies. Earth models of intermediate complexity are used to investigate long timescales, such as a glacial to interglacial oscillation.

Although models have largely improved since their integration in the FAR of the IPCC, there is awareness among the climate research community

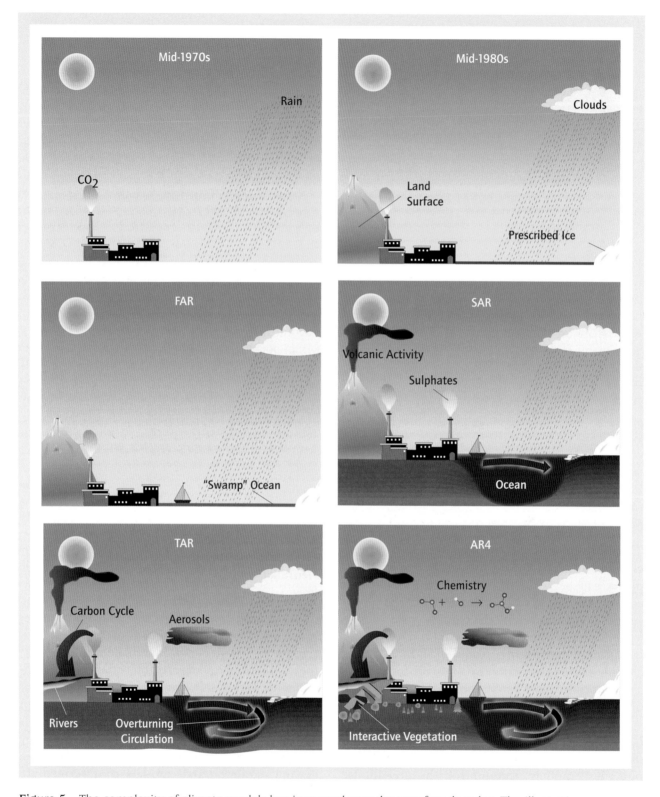

Figure 5   The complexity of climate models has increased over the past few decades. The illustrations represent the additional physical features incorporated in the models since the mid 1970s.

Sources: © Intergovernmental Panel on Climate Change (IPCC). FAR (First Assessment Report, IPCC, 1990), SAR (Second Assessment Report, IPCC, 1996), TAR (Third Assessment Report, IPCC, 2001), and AR4 (Assessment Report 4, 2007).

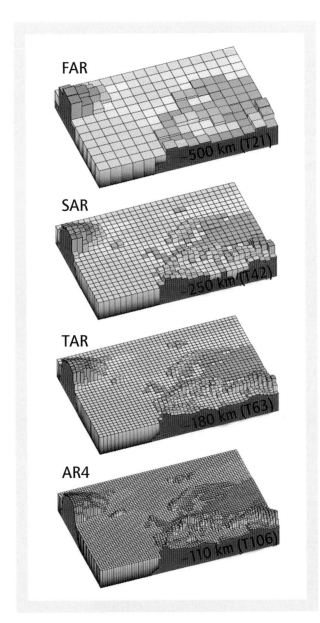

Figure 6 Geographic resolution characteristic of the generations of climate models used in the Intergovernmental Panel on Climate Change (IPCC) assessment reports

Sources: © 2007 Intergovernmental Panel on Climate Change. First Assessment Report (FAR, IPCC, 1990), Second Assessment Report (SAR, IPCC, 1996), Third Assessment Report (TAR, IPCC, 2001), and Assessment Report 4 (AR4, 2007).

Notes: Successive generations of general circulation models (GCMs) increasingly resolved Northern Europe. These illustrations represent the most detailed horizontal resolution used for short-term climate simulations. Century-long simulations cited in IPCC assessment reports after the FAR were typically run with the previous generation's resolution. Nowadays, the vertical resolution in both atmosphere and ocean models is about 30 levels in both atmosphere and ocean.

that models do not provide a perfect simulation of reality. Indeed, resolving processes at all important spatial or timescales remains far beyond current computing capabilities; clouds, vegetation, and oceanic convection cannot be represented in full detail in GCMs. In this case, physical parameters are said to be parameterized.

Analysis of model results showed a substantial model dependency over the simulated cloud processes and feedbacks. Models have produced changes in global average surface temperatures ranging from 1.9 to 5.4 °C (in a doubled atmospheric CO_2 concentration) by simply altering the way clouds would radiate heat and solar energy.

Furthermore, the behavior of any complex nonlinear system, such as climate, may be chaotic. As first described by H. A. Lorentz in the 1960s, climate models are intrinsically affected by uncertainty. Unpredictability, bifurcations (the equilibrium climate point leads to either a stable or an unstable equilibrium climate), and transitions (model goes from a stable to an unstable equilibrium climate state, or vice versa) to a chaotic mode happen even in simple models of ocean-atmosphere interactions, climate-biosphere interactions, or climate-economy interactions.

Feedback Mechanisms

Feedbacks are internal processes to Earth's climate system that either amplify changes in climate (positive feedback) or dampen/moderate them (negative feedback).

Positive Feedback

- An important positive feedback is the "water vapor feedback loop." An initial increase in CO_2 in the atmosphere causes the atmosphere to warm. This warming translates as a greater evaporation of surface water, which adds water vapor in the atmosphere, thereby warming it even more. Furthermore, if this warming takes place at high latitudes, snow and ice can melt. Exposed ground or ocean will absorb more solar radiation because either surface is less reflective than snow (lower albedo index), warming the Earth still further.
- This "ice-albedo feedback" constitutes another very important positive feedback loop. Although

simple, a quantitative understanding of the effect is far from complete. For instance, it is difficult to ascribe high-latitude amplification of the warming signal to retreated ice.

- When scientists realized that global warming signals are amplified in high latitudes, research on the "permafrost-climate feedback" started. As permafrost thaws due to warmer climate, GHGs trapped in it are released to the atmosphere. The subsequent increase in atmospheric temperature results, then, in more thawing.
- A warmer ocean has less ability to absorb carbon dioxide.

Negative Feedback

- The "water vapor feedback loop" has also a negative feedback component to it. As water vapor condenses, clouds form. Models and observations point toward a net cooling effect of the clouds as they reflect solar radiation back to space.
- Another possibility lies with increased precipitation patterns that would enhance the growth of ice sheets and glaciers, both regionally and globally, which would result in regions with a higher albedo and lower thermal conductivity.
- The acidification of the surface oceans, by reducing the production of corals and phytoplankton, will slightly increase the ocean's ability to take up carbon dioxide. When these organisms die, the $CaCO_3$ skeletons break down, and CO_2 is released to the water, lowering the ocean's ability to take up atmospheric CO_2.

Nevertheless, positive feedback outweighs negative feedback in the climate system on all but multimillennial timescales.

How Will Climate Change in the Next Century?

In its latest assessment report published in 2007, the IPCC writes that "observational evidence from all continents and most oceans shows that many natural systems are being affected by regional climate changes, particularly temperature increases." Using different models and scenarios, scientists have come up with a range of possible trajectories for the future climate. The

predicted increase in global average from 2000 to 2100 is between 1 and 3 °C (1.8–5.4 °F) for the storyline that is the most aggressive toward cutting GHG emissions (i.e., B1), 1.5 and 4.5 °C (2.7–8.1 °F) for the "middle-of-the-road" storyline (A1B), and 2.5 and 6.5 °C (4.5–11.7 °F) for the least aggressive emission scenario (A1FI).

Precipitation Change

If the average global temperature increases, so does the overall rate of evaporation. More water vapor in the atmosphere will drive higher overall rates of precipitation. Thus, one might expect a faster water cycle, as more water will be "sucked up" into the atmosphere and then proceed to fall back out as rain and snow. There is a projected increase in winter precipitation in polar, subpolar, and equatorial regions, while the subtropics will see a decrease in rain precipitation (Figure 7).

However, the timing and regional shift in precipitation will not be distributed evenly over all regions. Some locations will get more snow, others will see less rain. Some places will have wetter winters and drier summers, and so on. Models indicate more frequent extreme precipitation, with drenching downpours more prevalent. The regions predicted to experience overall declines in average annual precipitation will actually be hit with short periods of more intense rainfall followed by longer periods between rain. The mid-continental areas will dry more during summer, indicating a greater risk of droughts in those regions.

Melting Ice and Rising Sea Level

A warmer climate leads to higher sea levels via two mechanisms: (1) the melting glaciers and ice sheets add water to the oceans, raising the sea level, and (2) the thermal expansion of the ocean displaces a greater volume, thus also raising sea levels.

During the 20th century, sea levels rose about 10 to 20 cm (centimeters; 4–8 in. [inches]). Thermal expansion and melting ice each contributed probably about half of the rise. Sea level is predicted to rise with global warming by 0.1 to 0.4 m (meters; 4 to 16 in.) by 2100.

The various IPCC emission scenarios predict that by 2100, sea level will rise between about

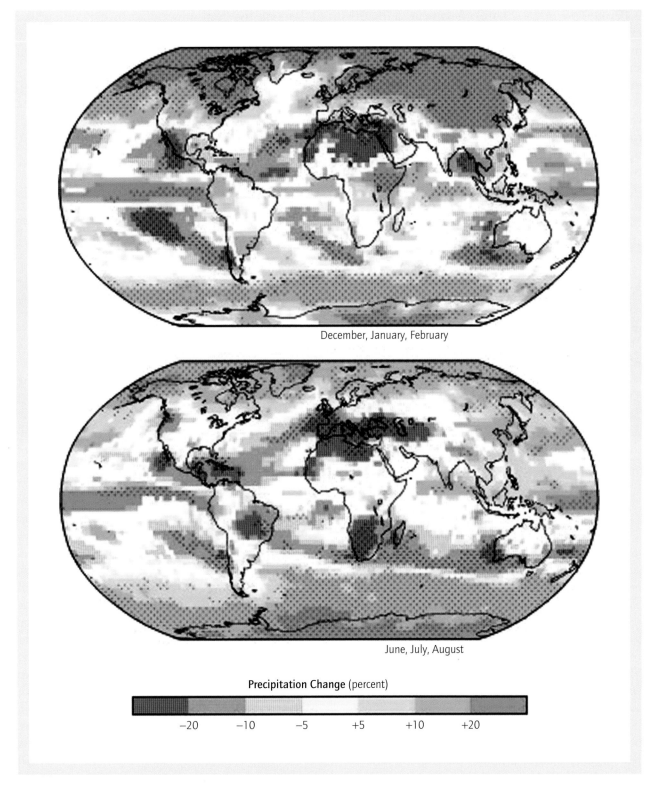

December, January, February

June, July, August

Precipitation Change (percent)

−20   −10   −5   +5   +10   +20

Figure 7 Modeled precipitation map for the winter months *(upper panel)* and the summer months *(lower panel)*

Source: © 2007 Intergovernmental Panel on Climate Change.

Note: With some exceptions, the tropics will likely receive less rain (orange) as the planet warms, while the polar regions will receive more precipitation (green). White areas indicate that fewer than two thirds of the climate models agreed on how precipitation will change. Stippled areas indicate where more than 90% of the models agreed.

0.19 to 0.59 m (8–23 in.) above the late-20th-century levels. However, these numbers do not include how the Greenland and Antarctic ice sheets will react to rising temperatures because modeling the response is very complicated.

However, in August 2008, a team of researchers reported in *Nature Geoscience* that sea level rise from the Greenland ice sheet could be double or triple the current estimates over the next century. According to this new study, and if the trend continues, sea level might rise 1 m (3.3 ft. [feet]) or more by 2100.

Changes in Extreme Weather

Since the TAR, there is a higher level of confidence in the projected increase in droughts, heat waves, floods, and frosts, as well as their adverse impacts. Changes are predicted to vary regionally.

The Unknowns in Predicting Climate Change

There are several "known unknowns" and "unknown unknowns" regarding the possible pattern of climate change ahead of us. Among the "known unknowns," one should consider the following:

- Since 2000, the growth rate of global emissions has been about 3%. In the IPCC's storylines, these emissions range between 1.4% and 3.4% between 2000 and 2010. This has attracted scientific and policy attention and could be evidence that these scenarios are too conservative. However, it should be clear that these emissions scenarios are long-range predictions, so it is also possible that the present trend is a short-term trend that will not result in a long-term deviation from the possibilities described by these emissions scenarios.
- What will be the steps taken by societies to mitigate GHG build up in the atmosphere, and how successful will these efforts be?
- As the major ice sheets are melting, just how rapidly will they melt, and how much will the sea level rise? These are still subjects of study.
- Could the release of freshwater from ice sheets and glaciers trigger a weakening of the thermohaline circulation ("conveyor belt"), and if so, when?

Among the "unknown unknowns," one of the most important ones concerns whether the response of the climate to increased GHG will take an unpredicted path.

Mathieu Richaud

See also Atmospheric Composition and Structure; Carbon Trading and Carbon Offsets; Climate Change; Climate Policy; Climate Types; Greenhouse Gases

Further Readings

Keeling, C. D., Piper, S. C., Bacastow, R. B., Wahlen, M., Whorf, T. P., Heimann, M., et al. (2005). Atmospheric CO_2 and $13CO_2$ exchange with the terrestrial biosphere and oceans from 1978 to 2000: Observations and carbon cycle implications. In J. R. Ehleringer, T. E. Cerling, & M. D. Dearing (Eds.), *A history of atmospheric CO_2 and its effects on plants, animals, and ecosystems* (pp. 83–113). New York: Springer.

Le Treut, H., Somerville, R., Cubasch, U., Ding, Y., Mauritzen, C., Mokssit, A., et al. (2007). Historical overview of climate change. In S. Solomon, D. Qin, M. Manning, Z. Chen, M. Marquis, K. B. Averyt, et al. (Eds.), *Climate change 2007: The physical science basis* (Contribution of Working Group I to the Fourth Assessment Report of the Intergovernmental Panel on Climate Change). Cambridge, UK: Cambridge University Press.

Mann, M., & Kump, L. (2009). *Dire predictions: Understanding global warming*. New York: DK Publishing.

Nakicenovic, N., Davidson, O., Davis, G., Grübler, A., Kram, T., Lebre La Rovere, E., et al. (2000). *Special report on emissions scenarios: A special report of Working Group III of the Intergovernmental Panel on Climate Change*. Cambridge, UK: Cambridge University Press. Retrieved December 3, 2009, from www.grida.no/climate/ipcc/emission/index.htm

Paleoclimate Modelling Intercomparison Project: http://pmip2.lsce.ipsl.fr

Van Vuuren, D., & Riahi, K. (2008). Do recent emission trends imply higher emissions forever? *Climatic Change, 91*, 237–248.

ANTHROPOGEOGRAPHY

Anthropogeography is a term used predominantly in the late 19th and early 20th centuries that means roughly "the geography of humans." The term comes from *Anthropogeographie*, the title of a two-volume work published in 1882 and 1891 by the German geographer Friedrich Ratzel, who is well known for his influence in early human geography, particularly his thought in political geography. Ratzel strongly influenced the thought of the well-known American geographer Ellen Churchill Semple, who presented Ratzel's anthropogeography to North America through her 1911 work *Influences of Geographic Environment: On the Basis of Ratzel's System of Anthropo-Geography*. Some thinkers, like the geographer J. K. Wright, criticized Semple for not clearly stating which ideas were hers and which originated from Ratzel.

The British geographer Halford Mackinder had a positive view of anthropogeography and believed that it required the ability to understand and use many subjects, including physics, biology, and geology. In Mackinder's opinion, this knowledge of various disciplines made anthropogeographers well-rounded scholars. Generally speaking, anthropogeography was an attempt to study and organize the relationships between people and their environments. More specifically, its goal was to systematically and scientifically describe the economic, historical, and political characteristics of people and the places they inhabited.

Unlike much geographical thought today that explores how humans affect their environments, anthropogeography focused on how humans were affected or "influenced" by their environments. In her essay "The Anglo-Saxons of the Kentucky Mountains: A Study in Anthropogeography," Semple begins with a physical description of the Appalachian Mountains in Kentucky. She goes on to describe how the isolation caused by the rough terrain influenced its people to be inbred, violent, backward, and lawless. Since the publication of this essay in 1901, similar images of Appalachia have persisted to this day. The scientific appearance of Semple's descriptions made them seem all the more believable. Semple described Appalachia and its people as cut off from civilization and progress. At the same time, however, she praised the residents of the Kentucky Appalachians as people of good stock. Many considered Appalachians to be the purest Americans: They were white and were not corrupted by the negative aspects of progress like the rest of the country. They were thought to be frozen in time. This view was widespread and was not limited to Semple or anthropogeographers, but it does highlight the fact that anthropogeography was influenced by prevailing cultural discourses. It is also likely that it helped reinforce those discourses.

The term *anthropogeography* is rarely used today because of its overtones of social Darwinism and environmental determinism. Occasionally, it is used as a synonym for human geography or to refer to the aspects of human geography that intersect with anthropology. In its heyday, anthropogeography focused predominantly on the environment's influence on humans. Despite its flaws, however, anthropogeography of the late 19th and early 20th centuries helped open the field of geography and its goal of exploring the relationships between humans and nature.

Jessey Gilley

See also Environmental Determinism; Human Geography, History of; Mackinder, Sir Halford; Ratzel, Friedrich; Semple, Ellen Churchill; Social Darwinism

Further Readings

Hartshorne, R. (1939). The nature of geography: A critical survey of current thought in light of the past. *Annals of the Association of American Geographers, 39*, 173–645.

Mackinder, H. (1895). Modern geography, German and English. *Geographical Journal, 6*, 367–379.

Semple, E. (1901). The Anglo-Saxons of the Kentucky mountains: A study in anthropogeography. *Geographical Journal, 17*, 588–623.

Semple, E. (1911). *Influences of geographic environment*. New York: Henry Holt.

ANTIGLOBALIZATION

Globalization—the diverse, complex set of processes that transcend national borders—is not a "natural" or inevitable phenomenon but a historical product and therefore contingent and malleable. Because globalization does not benefit everyone, contrary to popular neoclassical economic assertions that it forms a tide that lifts all boats, globalization often breeds resentment among those who bear its costs but enjoy relatively few of its benefits. For this reason, it may be said that globalization inevitably breeds its own opposition.

Resistance to globalization is as old as globalization itself. Thus, from the beginning, European colonial empires were met with heated opposition, often violent and generally unsuccessful. Examples include the Incan uprisings against the Spanish in the early 16th century, Zulu attacks on Dutch and British pioneers in Southern Africa, the great Sepoy Rebellion of India in 1857, and the long series of anticolonial and guerrilla struggles in Vietnam, Algeria, and much of sub-Saharan Africa, some of which persisted into the 1970s. For many contemporary opponents of globalization, current engagements are part of a long history of opposition that reaches back centuries.

Contemporary antiglobalization movements take a variety of forms. For many, the movements are relatively peaceful in nature, including protests, boycotts, demonstrations, and working through nongovernmental organizations (NGOs). For others, opposition takes on a decidedly more active, even violent form. Benjamin Barber's famous book *Jihad vs. McWorld* noted this diversity of forms, using *jihad*, the Arabic expression for holy war, as a metaphor for the vast umbrella of groups opposed to contemporary globalization and "McWorld" as a metaphor for the American-led, information-intensive penetration of various societies by crass commercialism, including fast food, entertainment and media (particularly Hollywood cinema), and fashion, all of which are the most visible faces of Western hegemony and are seen by many as cultural imperialism.

Nonviolent Antiglobalization

Because of the dominant role of the United States in the contemporary world system, globalization in the minds of many people is synonymous with Americanization. Much of the world has a love-hate relationship with the United States, often adoring its popular culture but abhorring the foreign policies of the American government, which has earned enmity in part for its long support of unsavory dictatorships in many regions, particularly during the Cold War. Globalization, in this reading, consists of the one-way export of American culture to the rest of the world, a process that threatens cultural diversity through the imposition of a monoculture across the planet. Opposition to the United States, for example, may take the form of attacks on Ronald McDonald, the clown statue that serves as a mascot for that famous fast-food chain, as a symbol of American commercialism.

American-style globalization is most powerful when it seduces the young, for whom it promises fun, status, hope, sexual appeal, and the appeal of wealth and power. In this sense, McWorld infantilizes everyone, turning adults and children alike into teenagers, wearing the same clothes, listening to the same music, and watching the same movies. The young may rapidly adopt Western customs at the expense of time-honored traditions. For the elderly, however, globalization can present a bewildering mix of new customs, leading to a generation gap. Perhaps the biggest contest between globalization and its opponents is in the minds of youth.

A different form of antipathy to globalization is found in Western Europe. Many residents of France or neighboring countries, for example, are disgusted with the crassest aspects of American culture: its obsession with money, commodities, and status and its neglect of tradition and leisure. (Vacation times in Europe tend to be considerably longer than in the United States.) For example, when the French farmer José Bové drove his truck into a McDonald's restaurant in 1999 to demonstrate his hostility to American fast food, he became a national hero. Other Europeans strenuously object to the import of American genetically modified foods. Given the state of social

Filipino protesters burn a sign during a rally in front of the main gate of the U.S. embassy in Manila, November 9, 2001, to protest the World Trade Organization's (WTO) ministerial meeting.

Source: Reuters/Corbis.

democracies in Europe, many Europeans abhor the intensity of American individualism, its denial of the social origins of people, and the correspondingly conservative lack of empathy for the poor and unfortunate that this ideology often produces. The United States, in this view, is overly tolerant of inequality and social injustice. Moreover, in Europe, an increasingly secular continent, there is widespread dislike of the prominent role that organized religion plays in American public life and of the profound religiosity of the American people, manifested, for example, in the rise of the "religious right" and attempts to limit the teaching of evolution in schools. Such phenomena are seen as symptomatic of a generalized anti-intellectualism in U.S. culture, which is stereotyped (accurately or not) as a culture in which

ideas and the life of the mind are held in low regard. Finally, many Europeans view the United States as a cowboy culture, given the widespread ownership of firearms and rates of violent crime that greatly exceed those of almost all industrialized countries. This celebration of violence extends to the American use of the death penalty, which is absent in all other economically developed states.

Grassroots Antiglobalization

Another variant of opposition to globalization emanates from a vast assortment of organizations of varying sizes and purposes that seek to resist neoliberal initiatives in various ways. Many such groups view globalization as something of a cabal or conspiracy fomented by transnational corporations, producing policies that are secretive and antidemocratic, that work against the interests of the poor and disadvantaged, and that are environmentally destructive. There are tens of thousands of nongovernmental organizations (NGOs) worldwide that work to soften the bluntest edges of globalization, including charities, philanthropies, community organizers, human and animal rights groups, research institutes, environmental activists, and watchdog organizations. Such groups are part of the civil society of the world's peoples (i.e., belonging neither to the state nor to the market). Their numbers exploded in the late 20th century in the wake of neoliberal assaults on various parts of the world, and many use the Internet to form alliances and to work cooperatively.

Environmental groups, for example, may be concerned with sustainable development, desertification, soil erosion, the preservation of wilderness and biodiversity, endangered species, the treatment of domesticated and wild animals, global warming, rivers and wetlands, and environmental justice. If, in their opinion, transnational corporations view the world as a set of resources to be harvested, their goal is to protect the commons shared by all. Examples include Greenpeace, the World Wildlife Fund, the Global Institute for Sustainable Forestry, the Rainforest Alliance, Friends of the Earth, and more locally based groups such as the Philippine Association for Conservation and Development or the Mountain Trust in Nepal.

Human rights and labor movements form another dimension of grassroots globalization, including those concerned with protecting the interests and rights of women, children, refugees, the handicapped and retarded, and workers. Topics embraced by such groups range from poverty alleviation and human trafficking and slavery to the protection of endangered ethnic cultures. Some target the International Monetary Fund (IMF) and its Structural Adjustment Programs, as well as the World Bank, and demand debt relief for impoverished countries. Some engage in attacks on meetings of the IMF or World Trade Organization (WTO), such as the "Battle in Seattle" in 1995, the assault on the G-8 in Genoa in 2001, and the demonstrations against the WTO in the Yucatan in 2005, which included strenuous opposition to subsidized American agricultural exports. Others are concerned with the preservation of tribal cultures and indigenous peoples, such as Australian aborigines, Native Americans, and the Ogoni of Nigeria. Global organizations in this vein include Amnesty International, Human Rights Watch, Doctors Without Borders, Handicap International, Academics for Justice, and the Union of Concerned Scientists. More local examples include the Ethiopian Women's Organization, the Pachamama Alliance, and Indian groups mobilizing on behalf of the *dalits*, or *untouchables*. The World Social Forum forms an umbrella for many such groups, with the slogan "Another World Is Possible."

Violent Antiglobalization

Although most opposition to globalization is peaceful, one variant engages in armed resistance. The origins of such groups, commonly labeled terrorists, lie in the complex transformations of many societies in the face of intense globalization. As a long tradition of Weberian social science has noted, modernity generates numerous secularizing changes in identity and behavior associated with the rise of markets, individualism, and commodity-based norms. Essentially, it may be held that Western, modern forms of life reduce identity to that of a buyer or seller of commodities, obliterating many time-honored, noncapitalist forms of life grounded in tradition. Particularly for people who experience severe disorientation through rural-to-urban migration, torn from local support systems, and the annihilation of systems of meaning that provided ontological security (honor, family, ancestors, God, etc.), modernity can be viewed as a sinister, morally offensive force. In the view of such victims of globalization, all that is holy is rendered profane. David Harvey (1990) notes that the disorienting changes in time and space that have accompanied the latest round of globalization often provoke a retreat into the local:

The more global interrelations become, the more internationalised our dinner ingredients and our money flows, and the more spatial barriers disintegrate, so more rather than less of the world's population clings to place and neighbourhood or to national, region, ethnic grouping, or religious belief as specific marks of identity. Such a quest for visible and tangible marks of identity is readily understandable in the midst of fierce time-space compression. (p. 427)

Given that globalization is often viewed as a secularizing force, it is not surprising that some of the most heated violent opposition has emanated from religious groups. Indeed, in the wake of the end of the Cold War, religious fundamentalism has erupted around the world, often coupled with antiglobalization sentiments. In the United States, for example, the upsurge in the religious right, in conjunction with the Republican Party, has ignited conservative political activists. On the extreme fringes of this ideology are militias that engage in xenophobic violence against immigrants; others view the federal government in terms of paranoid fantasies, including Timothy McVeigh, who destroyed a federal office building in Oklahoma City. In Europe, such movements include British skinheads, German attacks on guest workers, and French xenophobes led by LePen. In India, Hindu fundamentalists include the Bharatiya Janata Party, which instigated the destruction of the Ayodhya mosque and attacks on Indian Muslims. In the world of Judaism, the most rapid growth has been among ultraorthodox Hasidim, who support conservative policies vis-à-vis the Palestinians; the Gush Emunim movement to establish Eretz Israel; and the settlers on the West Bank.

Of course, the upsurge in religious fundamentalism also includes Islam, the world's second largest religion, with 1.5 billion followers. While the Western media often portray the Muslim world in negative terms, the fact remains that the vast majority of Muslims are peaceful and law-abiding. Nonetheless, a tiny minority, fueled by indifferent, corrupt governments in the Arab world, blame their culture's relative powerlessness in the world, and particularly against Israel, on an ostensible departure from the teachings of the Holy Koran. Radical Islamists toppled the Shah of Iran in 1979 and installed the Ayatollah Khomeini, turning Iran into a medieval theocracy. Others include the Muslim Brotherhood, which

assassinated Anwar Sadat in 1982. In Afghanistan, the Taliban drove out the Soviets in 1989, reorganized the country along strictly fundamentalist lines, hosted al-Qaeda in 2001, and has fought the United States since the American invasion later that year. Among Palestinian nationalists, hitherto a relatively secular movement, as exemplified by the Palestine Liberation Organization, fundamentalist groups such as Hamas and Hezbollah have been associated with suicide attacks against Israel. And obviously, al-Qaeda (Arabic for "the base") exemplifies the most pernicious aspects of this trend: Led by the infamous Osama bin Laden, member of a Yemeni family that grew rich in Saudi Arabia, al-Qaeda has become the most visible face of Muslim opposition to globalization and the United States, as spectacularly exemplified by the attacks on the World Trade Center and the Pentagon on September 11, 2001. To understand such movements, it is imperative not to view them in simplistic terms as acts of insanity or cowardice but to understand the social origins that drive individuals and groups to seek such goals, the poverty, frustration, humiliation, and sense of powerlessness that globalization often generates.

Conclusion

Antiglobalization, like globalization, is a diverse set of groups and ideologies, some of which find common ground. At their best, such efforts help protect the defenseless and preserve endangered cultures and environments. At its worst, antiglobalization can commit sins worse than the global forces it opposes. Because globalization has increasingly come to mean Western cultural and economic neocolonialism and neoliberalism, antiglobalization will inevitably rise not simply as its opponent but as its birth child.

Barney Warf

See also Antisystemic Movements; Colonialism; Globalization; Human Rights, Geography and; International Monetary Fund; Neocolonialism; Neoliberalism; Nongovernmental Organizations (NGOs); Resistance, Geographies of; Social Movements; Structural Adjustment; Terrorism, Geography of; World Bank

Further Readings

Barber, B. (1995). *Jihad vs. McWorld: How globalism and tribalism are reshaping the world.* New York: Ballantine.

Hawken, P. (2007). *Blessed unrest: How the largest social movement in history is restoring grace, justice, and beauty to the world.* New York: Penguin.

Stump, R. (2000). *Boundaries of faith: Geographical perspectives on religious fundamentalism.* Lanham, MD: Rowman & Littlefield.

Warf, B., & Grimes, J. (1997). Counterhegemonic discourses and the Internet. *Geographical Review,* 87, 259–274.

 # ANTIPODES

The term *antipodes* refers both to an upside-down world at the other side of the planet and to its inhabitants, who supposedly live head down and feet up. The term, derived from the Greek *anti* ("against" or "opposed") and *podus* ("foot"), appeared for the first time in Plato (ca. 427–347 BC), who used it to indicate a place diametrically opposite the city-state of Athens on the surface of the globe. Since then, the term became a crucial corollary of the theory of zones, the milestone of geographic speculation for almost 2,000 years. According to this theory, there are two temperate zones suited to human life, respectively in the boreal and in the austral hemisphere. The latter zone, called *antichtone* or antipodes, is unknowable because of the impossibility of crossing the Torrid Zone. Its scientific importance developed in parallel with its literary fortune: The antipodes challenged common sense and subverted the established order, becoming a great source of inspiration for artistic and social criticism.

A major problem arose when early Christian thinkers endeavored to reconcile hegemonic Greek-Roman geographical knowledge with Biblical topography. The existence of the antipodes contradicted both the common origins of humanity from Adam and the ecumenical aspiration of the evangelical message, jeopardizing the entire Christian vision of the world. Augustine of Ippona successfully proposed considering the antipodes as monsters. The presence of fabulous beings in faraway lands, attested to by ancient authorities, confirmed the great variety of products from the Creation and did not contradict any religious dogma. From then on and throughout the feudal era, the antipodes were represented as anthropomorphic freaks with deformed feet, turned backward.

Starting in the 11th century, revolutionary changes in the topography of the Christian afterlife, related to the invention of Purgatory, challenged the traditional geographies of the faraway. The rising merchant class needed a more sophisticated image of otherness to accompany its commercial expansion and found in the antipodes a transitional paradigm to shift from the late-medieval cosmology to a modern vision of geographical space. Antipodes become a question of puzzling political actuality in the second half of the 15th century, when Portuguese expeditions in Western Africa crossed the equator, demonstrating the fallacy of the theory of zones and opening the way to their conquests.

During the Renaissance, the antipodes acted as a powerful performative and rhetorical device that allowed imagining new possible worlds through a sophisticated system of inversions and a rich *imaginary* built throughout the centuries. At least four great myths nourished it: (1) the Universal Empire, revitalized by the rising national states; (2) the Golden Age that humanism borrowed from ancient thought; (3) the Jewish-Christian quest for Eden; and (4) the carnivalesque land of Cocagne of medieval popular culture.

Practices and representations related to the antipodal *imaginary* reshaped naturalistic taxonomies and transformed the frontiers among animals, monsters, and human beings, contributing decidedly to the building of a new colonial order and of a modern subjectivity. As the material and symbolic appropriation of the world by European powers progressed, the antipodes progressively lost their importance. Nevertheless, it is precisely at the antipodes that the most powerful of modern myths originated—the myth of utopia.

Paolo Vignolo

See also Biblical Mapping; Human Geography, History of

Further Readings

Fausett, D. (1995). *Images of the antipodes in the eighteenth century: A study in stereotyping.* Amsterdam: Rodopi.

ANTISYSTEMIC MOVEMENTS

Antisystemic movements (ASMs) may be defined as political groupings that oppose and resist the prevailing productive forces and relations in a given historical era. Thus, antisystemic movements can be said to have existed throughout human history. Antisystemic movements achieved profound economic and political success in the post-1945 period in overthrowing structures of formal colonialism in the global political economy and establishing the acceptance of social democratic norms of limited wealth redistribution and the state provision of social welfare in rich, developed states. However, they simultaneously failed to achieve their principal objective of transforming the unequal relations of exchange among the different zones of the global political economy. Present-day antisystemic movements have their origins in the "new social movements" of the 1960s and 1970s, many of which were identity based, oppositional, and exclusively concerned with single-issue politics, for example, women's rights, racial issues, antiwar and antinuclear movements, and environmentalist movements.

Themes of Protest

There are five principal themes of protest within contemporary anticapitalist, antisystemic movements: (1) trade regulation, (2) environmental degradation, (3) labor conditions, (4) militarism (especially in the guise of the "War on Terror"), and (5) human rights. Implicit within each campaign theme is the recognition that as each of these areas is becoming increasingly globalized, so too must counterhegemonic protest. The overlying theme is therefore the delivery of social, political, and economic justice to those who have been or may be threatened by the globalization of a particularly fundamentalist form of free market capitalism.

Patterns and Aspects of Protest

ASMs campaign on a huge variety of issues, from trade unions to organizations of indigenous peoples, whose very existence is threatened by the process of capitalist globalization. A strong commitment is maintained to a nonhierarchical "politics of difference" within and between antisystemic strategies and movements. Of particular importance is the emphasis that anticapitalist groups have placed on the *global* scope of resistance, while acknowledging that small-scale actions in groups' own locales contribute to a greater political-economic struggle.

Because of the failures of previous ASMs to enact systemic transformation through the capture of state power, current antisystemic movements emphasize the need to develop participatory forms of democratic practice "from the ground up." Political power is thus concentrated at the lowest possible level, flowing upward only through the establishment of consensus through a loose network of affiliations and coalitions. Contemporary ASMs have also concentrated on constructing transnational connections between developed and developing world movements. The declared objective is the construction of forms of global civil society to combat exploitation of the global poor through the transnationalization of production and its concomitant creation of a class-conscious, transnational capitalist elite, especially in the post–Cold War era.

Most ASMs aim at political empowerment through the promotion of an agenda of constructive change in the global political economy. By encouraging participation in mass protests and other campaigns, there is an educative agenda that encourages activists to learn more about their world, the workings of the transnationalized networks of finance and production, and those anticapitalist movements and campaigns in operation in other areas of the world, thereby building consciousness of the global nature of the ASM agenda.

Methods and Tactics of Protest

ASM protests take a variety of forms. At the smallest scale, protest may constitute the boycotting of particular brands or acts of "culture jamming," the defacement of advertisements. At the other end of the scale are mass protests, such as the 1999 Seattle anti-WTO protests and the 3-day anti-G8 protests in Genoa in July 2001. Such protests may not be limited to one location—the June 18, 1999, "Ambush of Global Capitalism" saw mass demonstrations take place across the world.

Participants in mass protests are activists and members of organizations from across the world. The Seattle demonstrations, for example, comprised groups from Europe, the Americas, Africa, and Asia and included environmentalists, community self-help groups, trade unions, debt and poverty relief campaigners, and organizations of indigenous peoples. Mass demonstrations have been specifically intended to disrupt, or even halt completely, major multilateral summits through direct action (as in Seattle). They also act as an affirmation of solidarity for activists involved within the ASMs, providing opportunities to meet and discuss future strategies for action or to attend teach-ins on the effects of globalization around the world. Third, mass protests attract the attention of the global media to movements' antisystemic agendas, theoretically promulgating them to a global audience. Serious differences remain, however, between different movements over the use of violent tactics attracting unfavorable media attention at the expense of more substantive debate.

Paramount is the use of humor and absurdism within a tactical context of civilly disobedient direct action. The use of humor in civil disobedience can trace its origins to the "situationist" school that grew out of the student dissident movement in Paris in the late 1960s. In the present era, humor and the celebration of life without commercial intrusion have been the underlying themes behind the spontaneous (if immaculately organized) hijacking of major road junctions by organizations such as Reclaim the Streets to hold street parties/street theater. Subversive humor was also evident in the tactics of "Guerrilla Gardening," employed during the A16 demonstrations in Washington, D.C., in April 2000 and in the "Mayday Monopoly" theme of the 2001 protests in London.

Other forms of direct mass action include organizations such as San Francisco's Critical Mass, which has organized mass bicycle rides to block traffic in urban centers around the world. In each case, the intention is to generate a global debate concerning a capitalist mode of social relations that regards material consumption as the key to happiness.

Mark Ian Bailey

See also Antiglobalization; Countermapping; Globalization; Human Rights, Geography and; Justice, Geography of; Neoliberalism; Nongovernmental Organizations (NGOs); Resistance, Geographies of; Social Justice; Social Movements

Further Readings

Barlow, M., & Clarke, T. (2002). *Global showdown: How the new activists are fighting corporate rule.* Toronto, Ontario, Canada: Stoddart.

Burbach, R., & Danaher, K. (Eds.). (2000). *Globalize this: The battle against the World Trade Organization and corporate rule.* Monroe, ME: Common Courage Press.

Klein, N. (2000). *No logo: Taking aim at the brand bullies.* London: Flamingo.

Wallerstein, I., Hopkins, T., & Arrighi, G. (1989). *Antisystemic movements.* London: Verso.

 # APPLIED GEOGRAPHY

First mentioned in the late 1800s, applied geography is a relatively recent discipline that has enjoyed controversy, acclaim, and change in its short life. Beginning as a merger of natural sciences and social sciences, applied geography has faced critics from both sides of science; however, it has also been hailed by both as having the ability to help humanity.

The term first appeared during a time when educational programs at the high school and college level were reevaluating the curriculum being taught then. Until this time, the discipline of geography had included only the natural sciences, such as geology and meteorology. John Scott Keltie (1890) was influential in suggesting that it is possible for the gap between natural and social sciences to be bridged through the application of geographic science to human behaviors. Most of what was written about applied geography during this time emerged from Europe. The first college to develop an applied geography program in the United States was the University of California, Berkeley, and even then it was only included as a part of an economics program. Applied geography uses geographical theory and methodology to solve problems on many topics as long as a problem has a geographical component, and therefore, the field has found a home in disciplines outside of geography.

Some 20 years after the first academic program was created, applied geography classes and research emerged throughout the country. This was partly due to a U.S. federal requirement passed in 1914. The requirement mandated that all land grant universities must disseminate research and findings on agriculture and home economics to the public and surrounding communities in a way that was helpful and easily understood. Here, a distinct shift in the importance placed on theory versus working geography was shown. For the universities to maintain their government-granted land, geography, among other disciplines, had to prove its usefulness to society.

Although the field continued to grow in popularity over time, it was not until 1965 that an official definition for the term *applied geography* was created. A committee of professional geographers interested in applied geography was assembled under the name Committee on Applied Geography, and the definition they created reads as follows: "The application of geographical knowledge, methods, techniques, and ways of thinking to the solution of practical problems" (*Association Affairs*, 1996, p. 168).

Under the impetus of job cutbacks and program cuts at universities during the 1980s, the need for a clear application of programs traditionally centered in theory became apparent yet again. Professors who once simply taught geography began redefining their curriculum in applied geography terms, such as environmentalism, to make their jobs seem less dispensable. It was during the 1980s that much of the current controversy emerged within the field, perhaps as a result of the shortage of jobs in academia during this period.

Past arguments show that applied geography was treated with some suspicion and uncertainty by academia. There was concern within the field that teaching a skills-based curriculum in addition to theory would cause a loss of prestige, a devaluation of theory, brain drain within the field, and a shortage of planning jobs. In a skills-based curriculum, students would be taught marketable skills and quantitative methodology that are not unique to geography and can be learned outside of the discipline, which would cause geography to become useless. Along with this was the fear that if students were being taught solely skills-based courses, the most successful students would try to find employment outside of academia in the hope of earning a higher salary. This trend would cause some of the best researchers, who could theoretically make great contributions to the field, to leave, thus creating a brain-drain effect.

For proponents of applied geography, a shift in focus from theory to skills allowed for more options for students, professors, and planners. For students who are not academia-bound, the chance to take courses that teach particular skills would increase the number of employment options they have after graduation.

However, setting the debate aside, geography has always been applied, long before applied geography became an identified academic discipline. Since the discipline of geography was established, individuals have used their knowledge in a variety of contexts and for various types of clients. Outside the universities, some of those trained as geographers have applied their skills to solve a range of problems, including, but not limited to, the following areas: locational analysis for retail or industrial site location; selecting sites for public services; identifying environmental problems and creating public policy to deal with such problems; selecting appropriate sites for solid waste disposal facilities; modeling the least-cost paths for transportation; identifying natural and environmental hazards; making military decisions; addressing issues in climatology; and allocation of resources. Understanding and resolving these types of problems are at the core of applied geography.

Applied geographers have careers in areas such as resource management; land use planning; remote sensing; geodemographics; retail analysis; site planning; town and city planning; international trade; the military; local, federal, and state governments; and geographic information systems (GIS) applications, just to name a few. It is this last field that has caused a resurgence in the popularity of applied geography. GIS turn geographic data into maps and decision-making tools by creating large databases of geographic information that enable geographers to store, manipulate, map, and analyze spatial data to solve a multitude of environmental, economic, urban, recreational, and social problems. Although this field is seen as a new frontier in the geographic discipline, it can also be seen as a modern type of applied geography. Although there is debate over the appropriateness and place of applied geography within the strata of university and occupational systems, most authors within the field agree that the future looks hopeful, especially with the popularity of GIS in solving real-world problems.

Sonya Glavac

See also Business Geography; Geographic Information Systems; Geospatial Industry; GIS in Environmental Management; GIS in Land Use Management; GIS in Public Policy; GIS in Transportation; GIS in Urban Planning; GIS in Utilities; Location Theory; Models and Modeling

Further Readings

Bennett, D., & Patton, J. (2007). *Applied human geography*. New York: Kendall/Hunt.

Ford, L. (1982). Beware of new geographies. *The Professional Geographer, 34*(2), 131–135.

Keltie, J. S. (1908). *Applied geography: A preliminary sketch* (Rev. ed.). London: G. Philip.

Ludwig, G. (1984). Extension geography: Applied geography in action. *The Professional Geographer, 36*, 236–238.

Moorlag, S. (1983). Applied geography and new geographies. *The Professional Geographer, 35*(1), 88–89.

Pacione, M. (1999). *Applied geography*. London: Routledge.

APPROPRIATE TECHNOLOGY

See Sustainable Development; Sustainable Development Alternatives; Sustainable Production

AQUACULTURE

Aquaculture is the controlled and/or targeted cultivation of specific freshwater or marine species by humans. An activity often likened to agriculture, aquaculture involves selecting a species for cultivation either because of market demand or subsistence requirements and because it displays traits amenable to highly successful reproduction or growth under regulated environmental conditions. Intensification in aquaculture has generally increased over time and currently varies depending on the aquaculturists' desired level of capital investment and scale of production. Aquaculture should be considered as an umbrella term that encompasses a variety of marine and freshwater culture activities such as captive grow-outs, fish farming (including genetic manipulation and embryo rearing/hatching), alga farming, mariculture, aquaponics, and so on.

Alongside resource extraction and food production more generally, seafood production has experienced rapid, and often state-supported, industrialization during the late 20th and early 21st centuries. Market demand for seafood products has also steadily increased during this time. Consequently, there is unprecedented interest in aquaculture, and according to the Food and Agriculture Organization (FAO) of the United Nations, it is currently one of the fastest growing food production sectors in the world. Thus, aquaculture is receiving much regulatory, academic, and corporate attention, and it is a growing research area in both the social and the physical sciences.

Development and Evolution

There seems to be general agreement that the earliest practice of aquaculture began in China

Fisheries in Chios Island, Greece

Source: Iraklis Klampanos/iStockphoto.

between 2000 and 2500 BC. At this time, methods involved holding, feeding, and growing fish in ponds or coastal entrapments until they reached the desired size or other food sources became scarce. However, there are many references to aquaculture activities scattered across ancient and modern history. A small selection of these include marine and freshwater habitat modification by indigenous peoples, capture and experimentation by the Egyptians and the Romans, and the hatching and release of numerous fish species in efforts to repopulate the dwindling recreational fish stocks in North America and Western Europe.

At present, aquaculture for subsistence purposes remains widely practiced, particularly in Asia and Africa. Precise data regarding volumes harvested and species cultured are limited; however, the significance of subsistence-driven aquaculture in these regions is substantial. Concurrently, industrial aquaculture is increasingly lucrative. By

incorporating scientific, technical, and marketing expertise, selected species are bred, reared from embryonic stages, harvested, and then fed into national and international networks of markets. Products commonly cultured currently include salmon, shrimp, trout, carp, tilapia, bivalves, kelp and other aquatic plants, and pearls. Tensions between these two modes of aquaculture stand to increase, especially in cases where they operate in close proximity to one another.

Impacts

In recent decades, aquaculture has been dubbed the "blue revolution," a direct reference to the "green" revolution of the agricultural sciences in the 1960s. Indeed, it is frequently stated that industrial aquaculture has the potential to increase the economic productivity of coastal or freshwater social-ecological systems, supply protein, and contribute to national gross domestic products. In addition, it offers an alternative to exhausted wild-harvest fisheries; and, for several reasons, many believe that cultured seafood products are more amenable to high-end seafood markets than is wild seafood. First, because it is reared so carefully, cultured seafood tends to have a more uniform or "appealing" appearance. Second, because of the controlled inputs and known growth and harvest rates, industrial aquaculture makes for a more predictable and steady supply. Third, and related to predictable supply, it is often better positioned than wild fisheries to achieve efficient economies of scale and pass cost savings on to consumers.

Yet concerns about the negative impacts linked to industrial aquaculture abound, particularly in European and North American public policy, academic, and activist circles. Harmful environmental impacts have thus far been the most frequently alleged by such groups, though consensus on their true implications has yet to be reached. Concerns include the introduction of exotic species, pollution/debris, and concentrated levels of feed and excrement into the surrounding environment; interactions between cultured and wild fish, including disease transfer and genetic weakening in wild stocks; and the harvest of smaller fish and crustaceans to manufacture feed for carnivorous fish.

Less frequently cited, though equally important, are a variety of aquaculture's potential socio-economic and cultural impacts. These include its interplay with long-standing systems of coastal tenure and potential to alter traditional arrangements for resource management; its contribution to, or conflict with, local food security; its ability to draw high levels of investment into rural communities; its frequent necessitation of new training and job skills, often among workers historically involved in wild fisheries or harvests; and its potential to contribute to resource user–stakeholder tension.

Frontiers in Aquaculture Research

Industry-driven research is focused foremost on improving production efficiencies and market-ability and secondarily on reducing aquaculture's ecological footprint. Promising avenues for increased efficiencies include new species and product development, including genetic modification; integrated multitrophic aquaculture; and open-ocean/deep-water mariculture. The development of closed-containment and filtration systems, reductions in feed wasting, and plant-based fish feed all provide promising potential to diminish the intensity of industrial aquaculture's environmental impacts. Interest from academics and development practitioners in aquaculture's role in rural food security, particularly for less industrialized regions, also continues to grow.

Currently, most geographical research regarding aquaculture focuses on GIS or remote sensing technology for aquaculture siting and/or impact assessment; the social-environmental impacts and appropriateness of industrial aquaculture; and the politics and discourse surrounding aquaculture development, expansion, and siting. However, much space remains for political ecologists, economic and cultural geographers, environmental historians, hydrologists, and biogeographers to contribute to dialogue regarding aquaculture's range of impacts and future development. Given the weakened state of many key fish stocks, the fluidity of industrial investment and capital, and growing concerns regarding food security, it seems unlikely that interest in either subsistence or industrial aquaculture will diminish.

Jennifer J. Silver

See also Agrofoods; Environment and Development; Fish Farming; Food and Agriculture Organization (FAO); Industrialization; Marine Aquaculture; Sustainable Fisheries

Further Readings

Food and Agriculture Organization of the United Nations. (2007). *World fisheries and aquaculture atlas* (4th ed.) [CD-ROM]. Rome, Italy: Author.
Taylor, W. W., Schechter, M. G., & Wolfson, L. G. (Eds.). (2007). *Globalization: Effects on fisheries resources*. Cambridge, UK: Cambridge University Press.

◉ ARCHIPELAGO

The term *archipelago* is most commonly used to refer to a group or chain of islands, although it can also refer to a sea containing a large number of islands. These islands are typically tectonically formed island arcs resulting from volcanoes (e.g., the Galapagos Islands) or hot spots in the lithosphere (e.g., the Hawaiian Islands) but can also form from erosion, deposition, or land elevation. They are usually found in the open ocean, although occasionally archipelagoes are near large landmasses.

The word *archipelago* is ultimately derived from the Greek *arkhon*, meaning chief or ruler, and *pelagos*, meaning sea. The ancient Greek name for the Aegean Sea, which is dotted with small islands, was *Aigaion Pelagos*, which later became simply *Archipelago*. Today, the word is used more broadly for any group of islands or sea containing many islands.

The largest archipelago in the world by number of islands is the Canadian Arctic Archipelago, with 94 major islands and 36,469 minor islands covering more than 1.4 million km^2 (square kilometers). The largest archipelago by area is the Malay Archipelago, which includes several nations spanning the distance between southeast Asia and Australia, including approximately 20,000 islands (depending on season and tide) and covering more than 2 million km^2. The Malay Archipelago encompasses the nations of Indonesia, the Philippines, Singapore, East Timor, Brunei, Malaysia, and Papua New Guinea. Some of these could be considered separate archipelagoes in their own right, particularly the Philippines. Other nations that are primarily archipelagoes include Japan, New Zealand, and the United Kingdom.

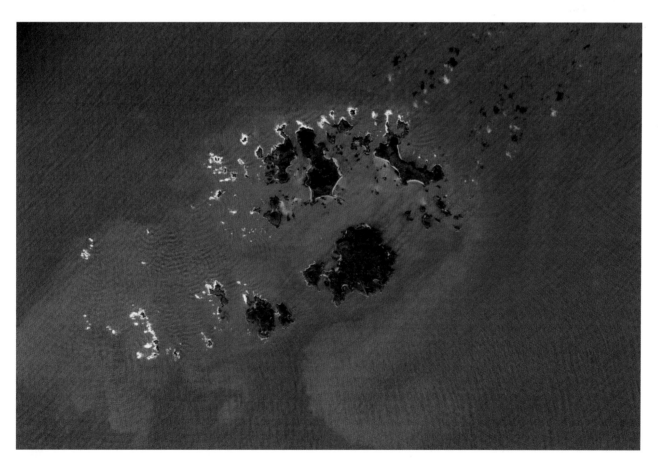

The Isles of Scilly, an archipelago of approximately 150 islands, is located some 44 km southwest of the westernmost point of England.

Source: NASA.

Archipelagoes have a unique relevance and usefulness among land formations for the study of evolution. According to island biogeography theory, the size of islands and the distance between them will interact to affect species immigration rates, colonization abilities, and local extinction rates. Their isolation from other systems increases the endemism of resident species (i.e., the likelihood that they are found nowhere else on Earth), and they are often the site of adaptive radiations. In fact, both Charles Darwin and Alfred Russel Wallace developed many features of their groundbreaking hypotheses with regard to natural selection and biogeography by studying archipelagoes—Darwin in the Galapagos, and Wallace in the Malay Archipelago.

Elizabeth R. Congdon

See also Adaptive Radiation; Island Biogeography; Islands, Small

Further Readings

Grant, P. R. (Ed.). (1998). *Evolution on islands: Originating from contributions to a discussion meeting of the Royal Society of London.* New York: Oxford University Press.
Wallace, A. R. (2000). *The Malay archipelago*: Vols. 1 & 2. The land of the orang-utan, and the bird of paradise. A narrative of travel, with sketches of man and nature. Berkeley, CA: Periplus Editions. (Original work published 1890)

 # ARCHITECTURE AND GEOGRAPHY

Geographers have had a long-standing interest in architecture, extending from early studies of the distribution and diffusion of folk-architectural styles in the 1930s to more recent analyses of the cultural and political symbolism, meanings, and uses of buildings and architectural environments.

Folk Architecture and the Vernacular Landscape

Architecture emerged as an important concern of scholars associated with the Berkeley School of cultural geography, most notably Fred Kniffen, in his studies of the distribution and diffusion of folk housing types from the 1930s to the 1960s. Kniffen approached folk houses as important cultural artifacts that could reveal a great deal about settlement and migration patterns in the United States, and he mapped the geographies of particular categories of house and barn in an attempt to identify regional differences in architectural styles and reconstruct historical changes in settlement patterns. Folk housing was approached as one component of the "vernacular" or "everyday" landscape of North America, but this work was not limited to the study of rural buildings or the rural landscape. The prolific landscape writer and editor J. B. Jackson documented diverse, everyday landscapes, including suburban, urban, and roadside architectures and landscapes, and his magazine *Landscape* (first published in 1951) contained contributions from historians, architects, landscape architects, planners, geographers, and others who had a significant influence on geographical writings on architecture and landscape in North America and farther afield.

Architecture, Power, and Cultural Symbolism

Architecture and the built environment have emerged as topics of concern for a broad array of geographers since the 1960s. At times, the planning and political decision making behind the production of the built environment have been of primary interest, but geographers have also shown how architectural environments reflect and refract political aspirations, contexts, and symbolisms in a range of ways. David Harvey famously showed how the Basilica de Sacré-Coeur emerged as a highly contested building project, dividing the Catholic Church, Republicans, politicians, and the Communards in late-19th-century Paris, while in later writings, he examined how shifts from modernism to postmodernism were reflected in changing attitudes and aesthetics in architectural

A luxurious hotel lobby interior. Geographers have examined the complex meanings and cultural-political symbolism of buildings and architectural environments.

Source: iStockphoto/Nikolai Okhitin.

and urban design. In these and other studies, architectural geographies are much more than the geographical distribution of building types, and in a wide array of studies influenced by the cultural turn of the 1980s and 1990s, geographers have examined the complex meanings and cultural-political symbolism of buildings and architectural environments.

Important studies have emerged of the cultural and political symbolism of colonial architecture and imperial cities, skyscrapers, shopping malls, zoos, war memorials, and other structures. Studies have focused on the broader discourses surrounding particular architectural aesthetics and movements, as well as the geographies of individual buildings such as St. Paul's Cathedral in London, the Red Road flats in Glasgow, the Mall of America in Minnesota, the New York World Building, and the Vancouver Public Library.

Geographers have often focused their attention on monumental and spectacular architectural structures, and less attention has been paid to the architectural geographies of more modest, mundane, or "ordinary" buildings, or the work of landscape architects and designers who knit structures into their broader surroundings. Architectural geographies are clearly closely entwined with the geographies of planning, engineering, and landscaping structures, while the start of the 21st century has seen an increasing number of academics calling for a focus on the inhabitation and use of architectural structures.

Inhabiting Architectural Spaces

In 2001, Loretta Lees drew on the emerging body of geographical writings on nonrepresentational theory and cultures of consumption to suggest that geographers who examined the cultural and political symbolism of buildings and the production of architectural environments frequently overlooked or discounted the multiple and often mundane ways in which users inhabit and consume architectural environments. Earlier studies of consumption in shopping centers by Daniel Miller, Peter Jackson, and others had already shown how consumers inhabited the spaces of the mall or shopping center in ways not foreseen or desired by designers and managers, and Lees was keen to emphasize the role of building users in actively "producing" architectural environments through their consumption practices. Architectural environments are performed and encountered through the embodied movements of a diverse range of human actors, and since Lees's call for geographers to turn their attention to the affective, performative, and nonrepresentational dimensions of buildings, a number of geographers have begun exploring the embodied geographies of architectural inhabitation.

Mark Llewelyn advanced a polyvocal methodology that could be used to explore the lived experience of a modernist housing estate in London—Kensal House—between the mid 1930s and 1950s. Combining archival research on the visions of architects and planners with oral histories of pioneer residents, Llewelyn constructed a critical geography of the diverse experiences of a utopian modernist housing project, showing how it was not merely "produced" by architects and "consumed" by residents but was continually "reproduced" by users on an ongoing, daily basis. Buildings affect their diverse users in all manner of ways. As Rob Imrie has shown, architects often design buildings to be inhabited by people with "normal" or "standard" bodies, with many architects (unless legislation requires it) neglecting to take account of the embodied capacities and spatialities of different users (including the disabled) who may be affected by a building in different ways.

Architectural geographies have come a long way since studies of the distribution and diffusion of building types in the 1930s. Geographers have engaged with the writings of architectural historians and architectural and urban theorists, tracing the spectacular *and* mundane, symbolic *and* affective geographies of buildings and the people and things inhabiting them.

Peter Merriman

See also Art and Geography; Berkeley School; Cultural Landscape; Jackson, John Brinckerhoff; Landscape Architecture; Symbolism and Place; Urban Planning and Geography; Zoning

Further Readings

Goss, J. (1988). The built environment and social theory: Towards an architectural geography. *The Professional Geographer, 40,* 392–403.

Jacobs, J. M. (2006). A geography of big things. *Cultural Geographies, 13,* 1–27.

Kraftl, P., & Adey, P. (2008). Architecture/affect/inhabitation: Geographies of being-in buildings. *Annals of the Association of American Geographers, 98,* 213–231.

Lees, L. (2001). Towards a critical geography of architecture: The case of an ersatz colosseum. *Ecumene, 8,* 51–86.

ARGUMENTATION MAPS

Argumentation maps were proposed to facilitate participatory planning using Internet geographic information systems (GIS) technology. Argumentation mapping combines online discussion forums with online mapping to enable participants in discussions to provide explicit geographic references for their contributions. The concept draws on principles of geospatial data modeling and participatory GIS and poses new challenges for geographic information science. This entry describes the basic idea and conceptual model of argumentation maps, their multidisciplinary origin, and their potential applications.

The argumentation map concept combines elements from geographic data modeling and argumentation theory. Geographic data modeling defines various approaches for representing spatial phenomena, including the vector and the raster data models. Argumentation theory is concerned with formal modeling of human discourse.

Combining these two foundations, argumentation maps describe relations between geographic objects and argumentation elements. On the map side, geographic objects are being used as reference objects for contributions to discourses. On the argumentation side, elements of text messages are being linked to those geographic references.

Argumentation maps respond to an observation from participatory planning processes, where participants can be found to refer to different places in their expressed views. Whether in oral statements, written submissions, or online comments, people regularly refer to their home location, neighborhood, and place of work, as well as to the location of planned development projects. The argumentation map model aims at making these implicit linkages more explicit and available for further computer-based processing.

Conversely, argumentation between humans usually involves interconnected expressions of personal opinions, which include direct responses to existing statements as well as references to other contributions. More elaborate argumentation models include issue-based information systems, which use typed argumentation elements. The available types include issues, positions, and arguments that can be placed in predefined relations with each other. The older Toulmin model uses claims, data, warrants, and relationships between these elements to structure human argumentation. The linking of this argumentation logic to geographic objects adds a layer of complexity to the spatial relations between geographic objects. In this respect, argumentation mapping goes beyond map annotation, which adds comments to places on maps but does not connect these annotations themselves.

Related concepts in cartography and geographic information science include hypermaps and geotagging. The common principle is the use of standard geographic references (geographic features, coordinates, geospatial tags) to localize text and multimedia documents. GIS databases are usually not able to handle complex data of this nature. Argumentation tends to be relatively simple (e.g., the question-reply model), but through its above-mentioned internal logical structure, it poses additional demands on the data modeling and analytical functions of participatory GIS tools.

More fundamentally, argumentation mapping benefits from a paradigm shift in computer science that expanded the view of the computer as a data processor to include its function as a medium for human-to-human communication. An argumentation map provides a cartographic representation of a discussion and can be used to explore, query, analyze, and participate in this discussion. Participatory GIS and, more recently, geocollaboration examine the theoretical and practical implications of using computer networks to support collaborative decision making in the geospatial domain.

Within planning, the need for a combined analytic-deliberative approach has been recognized where spatial analysis can be used to model and forecast decision outcomes, while actual decision making is often based on deliberation in groups of stakeholders. Although GIS are powerful tools for the analytic approach, argumentation maps are among the few computer-based, structured approaches to support deliberation.

A number of case studies of using argumentation maps have been conducted by academic researchers. The applications ranged from university campus planning to sustainable neighborhood development and social housing debates. Technologies used for implementation of argumentation maps include the University of Minnesota MapServer, the MySQL database, the Apache Web server with the tomcat Servlet engine, and, more recently, the Google Maps API, as well as custom Java applets and PHP scripts.

Claus Rinner

See also GIS in Urban Planning; Internet GIS; Multimedia Mapping; Participatory Planning

Further Readings

Goodchild, M. F. (1997). Towards a geography of geographic information in a digital world. *Computers, Environment and Urban Systems, 21*(6), 377–391.

Horn, R. E. (2003). Infrastructure for navigating interdisciplinary debates: Critical decisions for representing argumentation. In P. A. Kirschner, S. J. Buckingham Shum, & C. S. Carr (Eds.), *Visualizing argumentation: Software tools for collaborative and educational sense-making* (pp. 165–184). London: Springer.

Jankowski, P., & Nyerges, T. (2001). GIS-supported collaborative decision-making: Results of an experiment. *Annals of the Association of American Geographers, 91*(1), 48–70.

MacEachren, A. M. (2001). Cartography and GIS: Extending collaborative tools to support virtual teams. *Progress in Human Geography, 25*(3), 431–444.

Rinner, C. (2001). Argumentation maps: GIS-based discussion support for online planning. *Environment and Planning B: Planning and Design, 28*(6), 847–863.

Sieber, R. (2003). Public participation geographic information systems across borders. *The Canadian Geographer, 47*(1), 50–61.

ARID TOPOGRAPHY

One third of Earth's land surface is classified as arid or semiarid, and approximately 15% of the world's population lives in these regions. As population and the demand for resources continue to grow, these sensitive landscapes face intensifying pressures and the threat of greater environmental disturbance. Understanding arid topography—the principal landforms found in arid-region structural settings—and the major geomorphic (landforming) processes operating in arid lands is fundamental to limiting disturbance and minimizing hazards in these areas. Weathered rock, crusts and pavements, slope failures, desert streams, desert piedmonts, desert lakes, and landforms made by the wind are all influenced by the limited moisture availability in deserts.

Arid-Region Physical Geography

The terrain in arid climate regions has a distinctive appearance that can be attributed to the nature of the precipitation that falls in those landscapes. In addition to being low in average annual amount, precipitation in desert regions tends be sporadic in time and place. When storms do occur, they are often intense, delivering a substantial amount of precipitation over a short time. Although arid lands are typically vegetated, desert plants are specialized to cope with the moisture stress and do not form a continuous cover across the ground surface. With barren ground exposed between individual plants, loose rock and soil particles are easily moved downslope during sudden cloudbursts. As a result, unlike humid regions that have plenty of available moisture, a continuous coverage of vegetation, and a thick surface mantle of fine-grained (small particle size) soil covering the underlying rock, arid landscapes are more sparsely vegetated and are often visually dominated by outcrops of solid rock or loose, coarse-grained (large particle size) sediments. Far from being a wasteland, however, arid environments provide an austere and beautiful, yet fragile, home for a wide variety of specially adapted flora and fauna. In addition, because landforms and rock formations are not hidden under thick vegetation and soils, arid lands offer some of the most spectacular scenery on Earth.

Structural Setting

Volcanic and tectonic processes originate inside the Earth. Over geologic time, these processes have subjected Earth's surface to local and regional accumulations of volcanic rock and to various stresses, leading to the formation of mountains, basins, ridges, valleys, and plateaus. In many arid regions, these elements of geologic structure are well exposed and easy to identify. They form the large-scale relief features that the geomorphic processes originating on Earth's surface modify by breaking down rock (weathering), by eroding weathered rock materials from highlands, and by depositing those materials in lowlands.

Two generalized contrasting styles of arid-region topography are recognized, although they are best considered end members of a range of possible types. Mountain and basin deserts are those that retain considerable relief in the form of tectonically generated alternating uplands and depressions (see photo). This topography is created by a succession of uplifted and down-dropped fault blocks or large-scale upfolds and downfolds.

Upland, piedmont, and basin floor in a mountain and basin desert of active tectonism, Western Utah

Source: Author.

The second general topographic style is marked by wide expanses of low relief, which is typical of large structural plateaus or tectonically stable continental shield locations. The Basin and Range physiographic province of Western North America exemplifies the mountain and basin desert setting; the Sahara in North Africa represents the shield and platform category. Most desert landforms are found in each of these structural settings, but some geomorphic features are especially well developed, or occur in distinctive associations, in mountain and basin or shield and platform deserts. Other differences exist within various mountain and basin deserts depending on whether or not the relief-building tectonic processes are still active.

Geologically, desert areas of tectonically active mountain and basin topography continue to undergo uplift of the highlands and depression of the lowlands. Streams transport eroded rock material from the mountains and deliver it to the adjacent basin. However, because the ranges continue to be uplifted and the basins lowered through tectonic activity, the mountain ranges stay high despite erosion, and the depressions remain low despite considerable sedimentation. In an arid climate in this tectonic setting, the amount of surface water that accumulates on the floor of the topographic basin is not sufficient to spill over the drainage divide. Most tectonically active mountain and basin deserts, therefore, consist of closed basins of interior drainage.

If tectonism in mountain and basin deserts has become inactive, erosion of the uplands and sedimentation in the intervening basin will eventually decrease the overall relief by lowering the highlands and raising the basin floor. If sedimentation raises the level of the basin floor to the lowest point on the drainage divide, surface water will flow over the divide, creating a stream channel

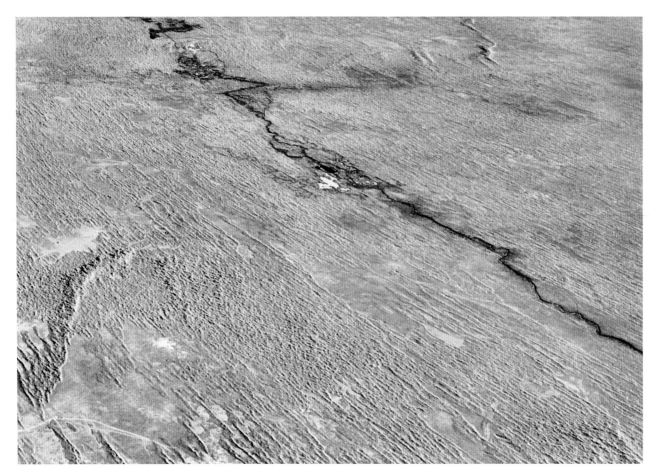

Extensive arid region of low relief, dominated by deflated areas and linear sand dunes in Turkmenistan

Source: Author.

that will begin to convey water and an accompanying load of sediment out of the basin. If the outflowing stream connects to a drainage system that ultimately reaches the ocean, it will have achieved exterior drainage.

Dominated by broad expanses of low relief, shield and platform deserts have the most extensive systems of wind-created landforms on Earth (see photo). Other features in this setting include isolated knobs and ridges of resistant rock, uplifted plateaus, lava flows, active and relict stream channels, shallow lakes, and extensive gravel-strewn plains.

Weathering, Varnish, and Crusts

In arid lands, the same chemical and physical weathering processes that break down rocks into ions and sediments in other climate regions operate.

Because of the high removal rates of weathered products from arid-land slopes, exposed desert rocks commonly display visible evidence of chemical and physical weathering, including cracks, fractures, flaking, exfoliating layers, microchannels (rillen), loss of mineral grains, holes and recesses (tafoni), pedestals, and iron oxide deposits. Alternatively, some rocks that remain exposed and stable for long periods in the alkaline desert environment acquire dark coatings of rock varnish, which are atmospherically derived iron- and manganese-rich substances concentrated on the rock surface by microorganisms. Likewise, some stable areas of sediment and soil acquire substantial deposits of chemical precipitates in the void spaces between particles. These deposits may reach the surface and form hard surface crusts, or duricrusts. These substances accumulate when soil water or ground water evaporates through the surface to the atmosphere, leaving behind salts that were dissolved in the water. Crusts develop from calcium carbonate (calcrete), calcium sulfate (gypcrete), iron compounds (ferrecrete), or silica compounds (silcrete).

Desert Pavement

Desert pavement consists of a surface layer of close-fitting stones that overlies a zone of much finer sediment, which in many cases is completely stone free. Varnish often forms on the exposed stone surfaces. Several explanations for desert pavement have been suggested. Some argue that the concentration of surface stones proceeds by wind and/or water removing surface fines from a deposit that originally consisted of a mix of gravel and fine-grained sediment. In these cases, the pavement represents a lag of clasts that were too heavy to be moved by the wind or water. Another explanation involves heaving processes in soils with expansive clays. The small cracks that form in the soil when the clays contract on drying allow only fine-grained sediments to fall to lower positions, not the gravel clasts, which over time become concentrated at the surface. Others have suggested that weathering rates may be higher within the soil than at the ground surface, leading to an accumulation of fine-grained products of weathering beneath the surface. Evidence from the American Southwest has demonstrated that desert pavement can also form from deposition of fines brought in as wind-blown dust, which sifts beneath larger stones. Thus, desert pavement represents an example of equifinality, with different processes working alone or in combinations to produce the same end product.

Mass Wasting

Desert hillslopes commonly have an angular appearance, with gentler ramps composed of weaker rocks and steeper cliffs representing rocks that are more resistant to weathering and erosion. When the force of gravity pulling down on cliff or slope material exceeds the forces binding that material to the upland, the rock matter moves downslope in a process called mass wasting. Mass wasting can occur in any environment and in many ways, ranging from a single rock falling off a cliff to the catastrophic failure of an entire hillside. Common types of mass wasting affecting arid-land slopes are rockfall, rockslide, and debris flows. Rockfall involves a small or large mass of rock that breaks off from a cliff and falls freely through air before crashing down onto a flatter surface below. Evidence of rockfall consists of a steep slope of broken, angular rock pieces piled up at the base of the cliff. Rockslides occur when tilted rock strata fail and move downslope in continuous contact with stationary rock strata below. Debris flows often begin as rockfalls or rockslides triggered by intense precipitation that moves the chaotic jumble of poorly sorted broken rock fragments into and down canyon stream channels toward the mountain front. Tongue-shaped (lobate) masses of debris-flow sediments are commonly deposited on the more gentle slopes beyond the mountain front, where the flowing mass spills out of the channel.

Surface Runoff and Desert Streams

When precipitation occurs, some water enters the voids between soil particles at the ground surface. With little vegetation to delay delivery of the precipitation to the surface, the voids in arid-land soils fill with water quickly, rapidly instigating overland flow as unconcentrated sheet wash. Because of gravity, sheet wash preferentially flows toward low points, and it soon starts to concentrate in small channels. Small channels direct the flow to larger channels in a stream network.

Most streams in arid climates are ephemeral, flowing only during and shortly after precipitation events. When precipitation falls, these ephemeral channels, also known as washes or wadis, can conduct sudden flows of significant amounts of water. In the American Southwest, steep-walled, ephemeral gullies known as arroyos are created when there is disequilibrium in the natural system. Arroyos commonly develop because of changes in vegetation and runoff resulting from various human-induced alterations of the landscape, such as overgrazing.

Unlike most humid-region stream channels, which receive water directly into the channel sides and bottom from the subsurface, most arid-region streams lose water to the subsurface by seepage into the gravelly channel bed as well as to the atmosphere by evaporation. Channel systems in some arid regions, such as those with hyperarid climates or active relief-building tectonics, do not reach the ocean, like most humid-region stream systems. Instead of having that exterior drainage to the sea, desert drainage systems of interior drainage terminate in a regional basin low point, which often contains a terminal lake (without outflow), such as Great Salt Lake in Utah or Lake Eyre in Australia. A few major rivers, such as the Nile and Colorado rivers, have enough water to flow all year across the desert and reach the ocean. These externally drained, exotic streams arise in distant regions of humid climate that provide sufficient water for the rivers to survive to the ocean despite high infiltration and evaporation losses when crossing the desert.

Badlands

Badlands develop in areas of very easily eroded rocks, such as shale, or geologically recent sediments, including ancient lakebeds. Rapid erosion rates make it difficult for vegetation, with its stabilizing influence, to become established, or to become reestablished once the plant cover is breached. The signature characteristic of badlands is that they are intensely channelized, leaving the terrain difficult to traverse. Areas of badlands have the highest drainage densities (total length of all channel segments per unit area) in the world. Although surface runoff dominates in most badlands, some badlands also exhibit considerable evidence of diversion of flowing water into shallow subsurface conduits, a process known as piping.

Desert Piedmonts

Adjacent basins and ranges comprise the principal geologic structural components of mountain and basin deserts, but not the only landforms. Geomorphic processes originating at Earth's surface modify the internally driven structural elements by weathering, erosion, transportation, and deposition. A major landscape element formed by surface processes in these desert settings is the piedmont, a transition zone between the steeply sloping mountain front of the upland and the near-horizontal surface of the basin.

Piedmonts constitute a zone of intermediate slope that links the upland and basin subsystems. The nature of this transition zone differs between mountain and basin deserts of active versus those of stable tectonism.

Alluvial Fans

In tectonically active mountain and basin deserts, continued uplift means that upland precipitation remains comparatively high and slopes steep. Under these conditions, mass wasting and surface runoff persist in delivering considerable sediment to upland stream channels. When stream flows or debris flows occur, they mobilize this sediment and transport it down-canyon. Just beyond the mountain-front canyon mouth, where stream channels are no longer as constrained by side slopes, channel widening leads to a decrease in flow velocity and therefore sediment deposition. Coarsest sediment is deposited first nearest the mountain front, with finer particles deposited farther out into the basin. During subsequent flow events, the stream changes course beyond the mountain front due to erosion of unstable channel banks or because of blockage of the channel by previous deposits. Debris flows overtop the channel, spread out, and form a lobate deposit. After many flow events, a fan-shaped deposit of stream and/or debris-flow sediments, called an alluvial fan, is constructed. Alluvial fans have their point of origin, the apex, at the canyon mouth and spread out toward the basin. The alluvial-fan surface is intermediate in slope angle between the mountain front and the basin, but its slope gradually decreases with grain size toward the basin floor. Where alluvial fans constructed at the mouths of several adjacent canyons coalesce into a laterally continuous deposit, the resulting landform is known as a bajada.

Pediments

In mountain and basin deserts of stable tectonism, external drainage transports products of upland erosion out of the local area. Little sediment accumulating in the piedmont zone and transportation of sediment across it create the pediment, an erosional piedmont, instead of a depositional one. Like alluvial fans and bajadas, pediments slope from the uplands to the lowlands

The Eastern Utah desert. Although arid lands are typically vegetated, desert plants are specialized to cope with the moisture stress and do not form a continuous cover across the ground surface.

Source: John J. Mosesso/NBII.Gov.

at an intermediate angle. Pediments have only a thin layer of surface sediment overlying coherent bedrock. Knobs of resistant rock that project above the relatively planar pediment surface are inselbergs.

Desert Lakes

Basins of topographic closure, in which surface water may accumulate as lakes, ponds, or marshes, form in many ways. Particularly common in arid regions are structural basins created by tectonics and generally smaller depressions formed by wind erosion. The former are common in mountain and basin deserts, the latter in shield and platform deserts. Unlike most humid-region lakes that have outflowing as well as inflowing streams, most desert lakes occupy terminal points in surface drainage; that is, they lack outflowing streams. Water can enter a terminal lake from streams and sheet wash, through direct precipitation onto the lake, and by seepage from groundwater. Water

leaves a terminal lake by evaporation to the atmosphere and by seepage to the subsurface. The Dead Sea in Israel and the Caspian Sea in Central Asia are examples of large terminal lakes that are perennial; that is, they contain water all year.

Terminal lakebeds that contain a shallow water body on an ephemeral or intermittent basis, rather than perennially, are playas. Many arid-region structural basins contain a playa, which occupies the lowest part of the depression. Playas dominated by the inflow of surface water consist of fine-grained sediment (clays and silts) and are very flat and hard. Those dominated by groundwater discharge have a high salt content and often considerable microrelief. Numerous playas likewise exist on extensive wind-blown plains in arid and semiarid settings, such as the U.S. Southern High Plains of Western Texas and Eastern New Mexico, a region that has thousands of small playas.

Many desert basins that are topographically and hydrologically closed today contained permanent lakes during times of greater effective wetness in the recent geologic past. Evidence of these paleolakes consists of abandoned coastal landforms and lake sediments perched high above the present playa or perennial desert lake. When dated, these relict landforms and sediments provide excellent evidence for reconstructing past climate changes.

Wind Erosion

Wind is the least effective agent of erosion at Earth's surface, but it is capable of landscape modifications, particularly in arid environments with their limited vegetation and available loose sediment. Localized removal of sediment by the wind leaves deflation hollows—depressions in a surface of unconsolidated sediments. Once the wind acquires sediment, it has abrasive tools that can sculpt even solid rock into aerodynamic forms or carve out pits, flutes, and grooves. Wind-abraded stones with planar surfaces that meet at sharp edges are called ventifacts; larger wind-sculpted, aerodynamic ridges of rock or consolidated sediments are yardangs.

Sand Dunes

Sand dunes constitute some of the most beautiful scenery in arid regions. Although sheets of wind-deposited sand exist, the wind typically deposits sand in distinctive mounds or hills. Most sand dunes have one side, known as the slip face, which is steeper than its other sides. Dune shape and the location of the slip face reflect details of the sediment supply and wind regime. Major categories of dune types include those formed in association with obstacles, those with slip faces oriented perpendicular to the sand-transporting wind, and those elongated parallel to the sand-transporting wind. Where multiple individual dunes cluster together, they comprise a sand dune field. Huge sand seas, or ergs, of repeating dunes are best developed in shield and platform deserts, particularly those (such as the Sahara in North Africa and the deserts of interior Australia) that are influenced by the subtropical high-pressure zone along the Tropic of Cancer and the Tropic of Capricorn.

Dorothy Sack

See also Basin and Range Topography; Climate: Dry; Desert Varnish; Dunes; Environmental Management: Drylands; Mass Wasting; Playas; Rock Weathering; Wind Erosion; Xeriscaping

Further Readings

Cooke, R., Warren, A., & Goudie, A. (1993). *Desert geomorphology*. London: UCL Press.

Laity, J. (2008). *Deserts and desert environments.* Chichester, UK: Wiley-Blackwell.

Lancaster, N. (1995). *Geomorphology of desert dunes*. New York: Routledge.

Mabbutt, J. A. (1977). *Desert landforms*. Cambridge: MIT Press.

Parsons, A. J., & Abrahams, A. D. (Eds.). (2009). *Geomorphology of desert environments.* Guildford, UK: Springer.

Thomas, D. (Ed.). (1998). *Arid zone geomorphology.* Chichester, UK: Wiley.

ARISTOTLE (384–322 BC)

Many of the writings of the enormously influential Greek philosopher and scientist Aristotle (born in Stageira, 384 BC, died in Chalcis, 322 BC) played a fundamental role in the development of the discipline of geography. His geocentric model, which was perfected by Hipparchus and Ptolemy, considered Earth to be immobile at the center of the universe and was held by medieval Christianity as an incontestable truth. It had a huge influence on Western cosmology until the 16th century, when it was replaced by the heliocentric theory supported by Copernicus.

Aristotle believed the universe to be unique and finite and thought that nothing existed outside it. From his point of view, four elements (air, water, earth, and fire) combined to form the planet, while ether was considered to be the fifth element, which does not belong to this world: It is in the cosmos, which makes it immutable and thus differentiates it from Earth, where things continuously change.

Regarding the Earth, Aristotle asserted that its circumference was 400,000 stadia (1 stadion = about 185 meters) and presupposed the existence of an enormous southern continent that provided a counterweight to the Northern Hemisphere. He called this land *Terra Australis Incognita*, and its discovery was the aim of many explorative missions of the following centuries, such as that of Captain James Cook in 1768. Aristotle had a critical opinion about representing the ecumene (the habitable zone of the world) as a circle, but he deduced the roundness of Earth and the other celestial bodies by observing that other planets cast round shadows during eclipses. To confirm the validity of his theory, he reported that at different latitudes, stars were not seen at the same place in the sky. He was not sure about the existence of what is now called the Dead Sea, he considered the origin of the Danube River to be in the Pyrenees, and he believed the Caucasus to be the highest mountains in the East.

In his *Meteorologica* (*Meteorology*) text, Aristotle observed the distribution of animals and vegetation on Earth's surface and drew the conclusion that it was related to latitude, realizing how the poles and the equator could be considered the least favorable areas for living beings. He also made observations of atmospheric phenomena such as wind, hail, frost, and rainbows, trying to analyze the reason why they occur in certain moments or places. He also explained his ideas about topics such as the origins of the rivers and the salinity of the seas, floods, earthquakes, and astronomical questions such as comets, shooting stars, and the Milky Way.

Susanna Servello

See also Antipodes; Equator; Hipparchus; Human Geography, History of; Latitude; Poles; Ptolemy

Further Readings

Martin, G. (2005). *All possible worlds. A history of geographical ideas* (4th ed.). New York: Wiley.

ARMSTRONG, MARC (1952–)

Marc P. Armstrong is a geographer who specializes in geographic information science (GIScience) and has done pioneering work in computational geography.

Armstrong earned degrees in geography from the State University of New York College at Plattsburgh (baccalaureate), the University of North Carolina at Charlotte (master's), and the University of Illinois at Urbana-Champaign (doctorate). He has been a faculty member at the University of Iowa since 1984, rising to the rank of professor in 1998. He has held several academic leadership positions at the university, including continuous service as departmental executive officer of the department of geography since 2000 as well as other appointments, such as administrative fellow in the College of Liberal Arts and Sciences in 2006–2009, interim dean for research in the college in 2006, and interim director of the School of Journalism and Mass Communication, 2007–2009.

Dr. Armstrong's research is in the interdisciplinary area of GIScience. He has published more than 100 academic papers and is also coauthor or coeditor of several monographs. He has served as North American Editor of the *International Journal of Geographical Information Science*, and he now serves as an editorial board member of several journals in the broad area of GIScience.

Dr. Armstrong is known as a pioneer in computational geography, specifically for his contributions to the development of parallel processing methods for geographic problem solving. His approaches span novel problem representations, new methods and algorithms, benchmark evaluations of solution qualities, and a variety of geographic applications. For example, with his collaborators, Armstrong has developed a variety of evolutionary programming approaches to multicriteria spatial decision-making problems that often require evaluation by various stakeholders. Other recent work has investigated how geographic information technologies may be used to compromise personal privacy and how mobile geographic information technologies can improve geographic education.

Shaowen Wang

See also GIScience; Spatial Multicriteria Evaluation

Further Readings

Armstrong, M., & Ruggles, A. (2005). Geographic information technologies and personal privacy. *Cartographica, 40*(4), 63–73.

Armstrong, M., Xiao, N., & Bennett, D. (2003). Using genetic algorithms to create multicriteria class intervals for choropleth maps. *Annals of the Association of American Geographers, 93*(3), 595–623.

 # ART AND GEOGRAPHY

Geography, visual images, and creative visual representation have long been intertwined. These links may be most long standing and evident in cartography and in the conception and production of maps, which the International Cartographic Association defines as "symbolized images of geographic reality, representing selected features or characteristics, resulting from the creative effort of their authors' execution of choices, and are designed for use when spatial relationships are of primary relevance." But the interconnection between geography and art extends well beyond cartography and map design. This relationship can be discussed in terms of landscapes and representation, geography and the production of art, and the role of art in remaking places.

Landscapes and Representation

Geographers have long been interested in studying the visual arts. In particular, geographers have looked at art such as landscape paintings as texts that reveal the influence of particular places on the artists' work as well as broader themes of climate change, attitudes toward and interactions with nature, and regional and national consciousness. In the 1980s, Stephen Daniels politically interpreted 19th-century English paintings to raise issues of how land is represented, changing social and power relations, and the sense of belonging. Likewise, Cosgrove noted that geographers needed to view landscape as more than an areal concept and to take on the view of artists (i.e., painters, literary writers, poets, and novelists) who explored links between landscape, perspective, and visual ideology. Recently, geographers influenced by Cosgrove and Daniels have studied disparate art and themes, such as the discourse of antiurbanism in Edward Hopper's paintings and murals in postapartheid South Africa, which provide insights into changing cultures and places and increasingly present local struggles over place, environment, identity, and history.

Geography and the Production of Art

Cultural geographers such as Donald Meinig and Yi-Fu Tuan have called on geographers to produce works of art. In his article "Geography as an Art," Meinig noted that although geography has been described as both an art and a science, geography has tenuous links to the humanities. He felt that even much of humanistic geography remained tied to the sciences, with its analytical frameworks used to examine behavior and meaning through art and literature. Consequently, Meinig suggested that geographers become creators of literature and pointed to the few examples where regional geographic interpretations moved into the world of creative literature (i.e., Estryn Evans's *Mourne Country: Landscape and Life in South Down*, William Bunge's *Fitzgerald: Geography of a Revolution*, and Henry Glassie's *Passing the Time in Ballymenone*). Meinig suggested that other geographers make the shift from science to emotional, personal statements of art by following the example of sociologist Raymond Williams and his novel *The Country and the City*.

Perhaps in response to Meinig's charge, geographers have increasingly been involved in the production of creative works. New outlets for publishing geography as an art have surfaced, including *Aether: The Journal of Media Geography*, which explores links between cultural politics, cultural industries, lived experiences, and the imagination, and *You are Here: The Journal of Creative Geography*, which delves into the concept of place through creative works including articles, fiction, poetry, essays, maps, photographs, and artwork.

Remaking Places Through Art

In addition to being produced and used to examine landscapes, art also can be a means for economically reshaping places. Portrayal of a city through art (e.g., "booster" U.S. Post Office murals supported by the New Deal era's Works Progress Administration) has been seen as a way to shape its economic future—whether or not those images fit reality. More recently, given the dominance of name brands in shaping consumer behavior, art can help brand cities or regions attract tourists, capital investment, and new residents. Given the substitutability of places, art events and attractions that are a good strategic and cultural fit can help create destination brand winners. According to Nigel Morgan, Annette Pritchard, and Roger Pride, brand winners are places that are rich in emotional meaning, have great conversational value, and create high anticipation for tourists. For art tourism, such destination winners include Santa Fe, New Mexico, with its Native American and U.S. western and southwestern regional art as well as its designation as the first UNESCO Creative City of the United States, and reinvented industrial cities such as Glasgow, Scotland, which, in its branding as "Scotland With Style," prominently highlights its museums and galleries.

Art as a creative industry can also reshape geography. Recent economic geography literature has linked creative industries such as art with economic development. Richard Florida has argued that urban space should be transformed to draw footloose workers such as artists, engineers, writers, and entertainers. These workers' talent, or human capital, can help cities build new-economy, high-technology, high-income industries. In addition to being a creative industry, art can attract needed talent to cities. Florida contended that cities that rank high in terms of factors such as coolness (measured by the availability of nightlife and cultural amenities such as art) and diversity, or openness to all people (i.e., racial, ethnic, and sexual minorities, immigrants), will attract the creative workforce they need to

The Dotty Wotty House on Heidelberg Street in Detroit, Michigan. The different dots connote diversity as well as equality.

Source: Permission to use kindly granted by The Heidelberg Project.

thrive in a postindustrial economy. Ann Markusen has stressed the need to disaggregate the fuzzy, creative class notion into its occupational components, as they have different spatial distributions in metropolitan areas, based on political leanings, levels of self-employment, and degrees of being footloose. Relatively footloose artists who gravitate toward central cities can boost regional growth through import-substituting consumption activities for residents, direct export activity, and contribution toward the design, production, and marketing of other products and services. While Florida's ideas have been simultaneously lauded and critiqued, his emphasis on the need for "cool

The OJ (Obstruction of Justice) House and The Oval Room (depicting the blight the city of Detroit is experiencing and how to clean up the town and its government)

Source: Permission to use kindly granted by The Heidelberg Project.

cities" to attract and keep the creative class has influenced many involved in city, regional, and state planning and economic development.

Art can also literally reshape specific places in a direct, hands-on manner. Through the use of abandoned objects, polka dots, and bright orange paint, urban environmental art projects such as Detroit's Heidelberg Project and "Detroit. Demolition. Disneyland" (Project Orange) have visually called attention to blight and the flight of people, capital, and jobs. These projects have issued a direct challenge to the status quo. While people might like safe, comfortable art, controversial art may best effect change. Geographers are working with artists and with children, asylum seekers, racial minorities, and so on to develop explicitly political, collaborative art practices involving critical spatial practices to reshape places.

Deborah Che

See also Cosgrove, Denis; Cultural Geography; Cultural Landscape; Popular Culture, Geography and; Representations of Space; Symbolism and Place; Vision and Geography

Further Readings

Cosgrove, D. (1985). Prospect, perspective and the evolution of the landscape idea. *Transactions of the Institute of British Geographers, 10*(1), 45–62.

Daniels, S. (1984). Human geography and the art of David Cox. *Landscape Research, 9*(3), 14–19.

Florida, R. (2002). *The rise of the creative class.* New York: Basic Books.

International Cartographic Association. (2008, April 21). *ICA mission.* Retrieved June 5, 2008, from http://cartography.tuwien.ac.at/ica/index.php/TheAssociation/Mission

Markusen, A. (2006). Urban development and the politics of a creative class: Evidence from a study of artists. *Environment and Planning A, 38,* 1921–1940.

Meinig, D. W. (1983). Geography as an art. *Transactions of the Institute of British Geographers, 8*(3), 314–328.

Morgan, N., Pritchard, A., & Pride, R. (2002). Introduction. In N. Morgan, A. Pritchard, & R. Pride (Eds.), *Destination branding: Creating the unique destination proposition* (pp. 3–10). Oxford, UK: Butterworth Heinemann.

ASIA-PACIFIC ECONOMIC COOPERATION

The Asia-Pacific Economic Cooperation (APEC) forum is a supranational association of Pacific Rim economies whose leaders and other representatives meet with the purpose of improving economic linkages within the region. Australia played a pivotal role in establishing APEC in 1989 at the end of the Cold War to promote economic growth and to strengthen the economies of the region.

APEC was formed against the background of increasingly protectionist sentiment around the world, particularly in the United States and the European Union (EU). APEC was conceived partly as the result of concern that the world could divide into three trading blocs: Asia, the European Union (EU), and the Americas. APEC exemplifies the growing interdependence among Asia-Pacific economies. The heads of government of all APEC members meet annually in a summit that rotates between APEC's member economies.

APEC is not a trade bloc, but it includes other trade blocs such as the Association of South East Asian Nations + 3 (ASEAN's 10 members plus China, Japan, and South Korea); North American Free Trade Agreement signatories (Canada, Mexico, and the United States); and Australia and New Zealand, which have a long-standing Closer Economic Relations (CER) Free Trade Agreement (FTA).

APEC is the largest loosely coupled economic community of its type in the world. It currently has 21 members: Australia, Brunei, Canada, Chile, China, Hong Kong, Indonesia, Japan, South Korea, Malaysia, Mexico, New Zealand, Papua New Guinea, Peru, Philippines, Russia, Singapore, Taiwan, Thailand, the United States, and Vietnam (Figure 1).

APEC began as a relatively informal group of government officials, mainly from departments of trade or the equivalent. It claims to be the only intergovernmental group in the world operating on the basis of nonbinding commitments, open dialogue, and equal respect for the views of all participants. APEC's decision making by consensus and the diversity of its membership are sources of strength and weakness.

In contrast with the EU, the World Trade Organization (WTO), and other multilateral trade bodies, APEC does not require treaty obligations of its participants; members make commitments only on a consensual and voluntary basis. Thus, many of APEC's goals are expressed in vague terms that are more practicable for national governments' diplomats to agree on. The vague terms provide diplomats with space to negotiate trade agreements that cannot be covered in WTO treaties. There are benefits of using APEC as a channel for signing such bilateral trade agreements, which otherwise might not be practicable in other forums.

APEC has become a vehicle for promoting the liberalization of trade and investment around the Pacific Rim. This agenda includes promoting FTAs and other forms of practical economic cooperation such as facilitating economic development, the growth of small- and medium-sized enterprises, the mobility of businesspeople, and more equitable participation by women and young people in the labor market.

While APEC countries include approximately 40% of the world's population, they represent approximately 49% of world trade and 55% of world gross domestic product (GDP). APEC represents a dynamic economic region in the world, which has been fuelled especially by the rapid growth rates of China and some other Asia-Pacific economies.

In light of the diversity of its membership, it will not be easy for APEC to achieve an Asia-Pacific

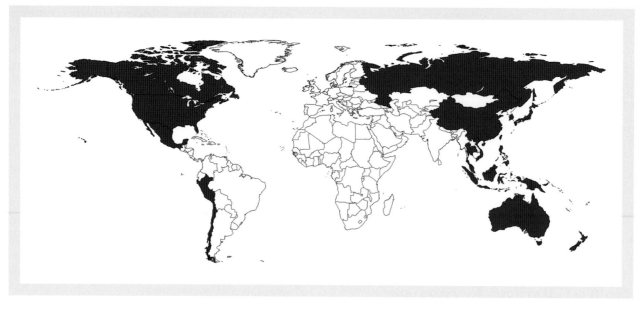

Figure 1 APEC members. The Asia-Pacific Economic Cooperation Forum comprises 21 member states circling the Pacific Ocean.

Source: Map created by Barney Warf.

FTA among all of its members. APEC has been subject to criticism such that it lacks much of a social agenda. The criticisms are especially pointed in comparison with the EU. Another criticism is that at APEC meetings, for members' leaders, a significant amount of the focus appears to be on bilateral side meetings between other members rather than on the APEC agenda itself. Nevertheless, APEC is an important forum and has much potential.

Greg J. Bamber

See also Association of Southeast Asian Nations; European Union; Globalization; Newly Industrializing Countries; Trade; Transnationalism; World Trade Organization (WTO)

Further Readings

Asia-Pacific Economic Cooperation: www.apec.org
Bamber, G. (2005). How is the Asia-Pacific Economic Cooperation (APEC) forum developing? Comparative comments on APEC and employment relations. *Comparative Labor Law & Policy Journal, 26*(3), 423–444.

 # ASSOCIATION OF AMERICAN GEOGRAPHERS

The Association of American Geographers (AAG) is a professional and scholarly association representing educators, researchers, and practitioners in geography. Founded in 1904 as a nonprofit scientific and educational society, its 10,000-plus members today share interests in the theory, methods, and practice of geography. The association is governed by an elected council, which sets the overall policy and appoints an executive director, who manages the affairs of the association, supported by numerous committees and a professional staff. The AAG headquarters is located in Washington, D.C.

The association hosts approximately 7,000 attendees each year at its annual meetings, sponsors international workshops and specialty conferences, and conducts research programs on a wide range of geographical topics. The AAG also publishes some of the world's most distinguished and influential scholarly journals, newsletters, books, and research reports. Publications include its flagship journals, *The Annals of the Association of American Geographers* and *The*

Professional Geographer, and the monthly *AAG Newsletter.*

Representing a discipline with a strong international perspective, the AAG also supports and recognizes geographers around the globe through its many professional development programs and prestigious grants and awards programs. The AAG outreach network consists of more than 30,000 colleagues worldwide. Scientific research and educational programs are conducted internationally, and scientific exchange and interaction occurs through the AAG's annual meetings, its leading scholarly journals, its newsletters, an extensive network of online interactive media and discussion groups, reciprocal membership programs, and other collaborative activities.

The AAG also sponsors 60 specialty groups, which promote geographic research and discussion on specific regional or topical geographic research areas. AAG specialty groups address a variety of geographic research topics, ranging across human and physical geography, geographic education, and geographical information sciences and technologies. AAG specialty groups are focused on Africa, China, Remote Sensing, Geographic Information Systems and Science, Spatial Analysis and Modeling, Regional Development and Planning, Economic, Urban, and Political Geography, as well as a host of other thematic topics crucial for sustainable development, such as Human Dimensions of Global Change, Climate, Hazards, Biogeography, and many others.

Fostering and conducting geographic research is also a key focus of the association's activities and mission. The AAG designs, develops, manages, and implements special research, education, and outreach projects that advance geography, both within the United States and internationally. The AAG manages several dozen special projects and research programs with approximately $6 million in annual funding. Many of these initiatives are supported by grants from U.S. government agencies, such as the U.S. Agency for International Development, the National Science Foundation, the Department of Agriculture, the Department of Education, NASA, and others. Research projects often are conducted in partnership with universities or other organizations, including private sector and nonprofit entities as well as international agencies.

As the scientific, societal, and university context within which the discipline of geography continues to evolve, the AAG and the discipline of geography continue to interactively lead, adapt, and evolve in order to sustain and strengthen geography's intellectual and practical contributions to the needs of a rapidly changing world.

Douglas Richardson

See also American Geographical Society; Institute of British Geographers; Regional Science Association International (RSAI)

Further Readings

Association of American Geographers: www.aag.org

Association of American Geographers. (2009). *Guide to geography programs in the Americas and AAG handbook and directory of geographers* (2008–2009 ed.). Washington, DC: Author.

James, P., & Martin, G. (1979). *The Association of American Geographers: The first seventy-five years*. Washington, DC: Association of American Geographers.

Richardson, D., & Solis, P. (2004). Confronted by insurmountable opportunities: Geography in society at the AAG's centennial. *The Professional Geographer, 56*(1), 4–11.

 ## ASSOCIATION OF GEOGRAPHIC INFORMATION LABORATORIES FOR EUROPE

The Association of Geographic Information Laboratories for Europe (AGILE) was established in early 1998 to ensure the continuation of the networking activities that have emerged as a result of the EGIS (Exhibition on Geographical Information Systems) Conferences and the European Science Foundation GISDATA Scientific Programmes. AGILE represents a community committed to promoting academic teaching and state-of-the-art research on advancing GI (geographic information) science. It consists of 96 research laboratories across Europe, counting hundreds of GI researchers, students, and academics (see Table 1).

Country	AGILE Member
Austria	Salzburg Research Forschungsgesellschaft Salzburg University Vienna University of Economics and Business Administration Technical University of Vienna Carinthian Technical Institute
Belgium	Brussels Free University Catholic University of Leuven University of Liege
Bosnia-Herzegovina	Energoinvest
Czech Republic	Masaryk University VSB Technical University
Denmark	Aalborg University University of Copenhagen Danish Forest and Landscape Research Institute Technical University of Denmark National Environmental Research Institute
Finland	Technical Research Centre of Finland Finnish Environment Institute SYKE Finnish Geodetic Institute University of Turku
France	University of Avignon University of Franche-Comté Lab. d'informatique, Université de Marne-la-vallée U.M.R. CEMAGREF-ENGREF Institut Géographique National (IGN-France) Louis Pasteur University CNRS/INSA de Lyon
Germany	Humboldt University HafenCity University Hamburg Leibniz University of Hannover Fraunhofer Institute for Factory Operation and Automation University of Applied Sciences FH Mainz University of Muenster University of the Bundeswehr University of Applied Sciences Oldenburg (FH OOW) University of Osnabrück University of Rostock University of Applied Studies and Research, Hochschule Harz
Greece	NTUA Harokopio University Foundation for Research and Technology-Hellas (FORTH) Aristotle University of Thessaloniki
Hungary	Applied GeoInformatics Laboratory (AGIL) University of West-Hungary
Ireland	University College Cork National University of Ireland
Israel	Tel Aviv University
Italy	Università di Bologna ENEA Università degli Studi di Cagliari

Country	AGILE Member
	Università di Cagliari
	Università degli Studi di Firenze
	Joint Research Centre
	Università di Padova
	Università di Milano-Bococca
	Università di Roma la Sapienza
	Politecnico di Torino
The Netherlands	Free University of Amsterdam
	Technical University of Delft
	TNO – STB
	ITC
	Van Hall Instituut
	University of Utrecht
	Wageningen University and Research Centre
Norway	University of Bergen
Poland	Jagiellonian University
	Institute of Geodesy and Cartography (IGiK)
Portugal	University of Minho
	ISEGI/UNL
Romania	Romanian Academy
Serbia	University of Nis
Slovenia	University of Ljubljana
	Scientific Research Centre of the Slovenian Academy of Sciences and Arts (ZRC SAZU)
Spain	University of Jaume I
	University of Girona
	University of Santiago de Compostela
	Technical University of Madrid
	University de les Illes Balears
	University of Zaragoza
Sweden	University of Gävle
	Linkoping University
	Lulea University of Technology
	Royal Institute of Technology (KTH)
	Uppsala University
Switzerland	Swiss Federal Institute of Technology at Lausanne (EPFL)
	University of Zurich
Turkey	Anadolu University
U.K.	Forestry Commission
	University of Newcastle
	University of Nottingham
	University of Sheffield
	University of Edinburgh
	University of Glamorgan
	City University London
	University College London

Table 1 AGILE members

Source: Reprinted by permission of the Association of Geographic Information Laboratories for Europe (AGILE).

AGILE also maintains liaison with several organizations through memoranda of understanding, including Bentley Definiens, Environmental Systems Research Institute, Geomatics for Informed Decisions, Intergraph, Open Geospatial Consortium, Partnership for European Environmental Research, and University Consortium for Geographic Information Science.

The association is a registered nonprofit organization in the Netherlands and is coordinated by an eight-person council elected by its members every 4 years. The council meets two times a year, in addition to the association's annual general meeting and conference. Each member of the council holds a portfolio responsibility for various aspects of the organizational structure, dissemination activities, and member initiatives. The council is also supported by a number of external officers.

Mission

The aim of the association is to sustain and promote a vital GI research and education community in Europe by

- enhancing networking and new cooperation within AGILE members,
- promoting cooperation with other national educational research associations in Europe, and
- encouraging initiatives to support a leading research culture committed to open inquiry and improvement of education.

AGILE ensures that the different views and experiences of the GI community are fully represented in discussions on a solid European GI forum by focusing on promoting communication among its members and institutions in Europe, exchanging their know-how in conferences and workshops, and initiating initiatives on GI education, research, mobility, and strategic cooperation.

Annual Conferences

The AGILE conferences are hosted by a member every year. They provide a multidisciplinary forum for an increasingly varied landscape of scientific knowledge production and dissemination to scientists around the world. The call for papers includes a full-paper submission track of original, unpublished, fundamental scientific research, the results of which are published by Springer Series on Lecture Notes in GeoInformation and Cartography (acceptance rate is usually around 40%). The call for papers also includes an abstract submission track of original scientific and strategic research work for presentation at the conference and published in the AGILE proceedings volume (with ISBN). All proceedings are available at the organization's Web site.

The authors include computer scientists, geographers, geomatic engineers, planners, and GIScience practitioners, just to mention a few. The contributions come from all four corners of Europe, as well as from countries on other continents. There are specific education sessions in the annual AGILE conferences and the biannual European GI Education Seminars, as well as specialist meetings on urgent topics in GI education. The main findings identified from these activities demonstrate the increase of public–private partnerships to pursue GI education toward meeting the needs of the GI sector.

Monica Wachowicz

See also Geographic Information Systems; GIScience

Further Readings

Association of Geographic Information Laboratories for Europe: www.agile-online.org

 # ASSOCIATION OF SOUTHEAST ASIAN NATIONS

An important trend in the global economy since the early 1990s has been the proliferation of regional initiatives among countries. The North American Free Trade Agreement (NAFTA) and the European Union (EU) best illustrate such a trend; however, regionalism among developing countries has also become popular. The Association of Southeast Asian Nations (ASEAN) represents

Figure 1 ASEAN members. ASEAN comprises virtually all countries in southeast Asia.

Source: Association of Southeast Asian Nations. Retrieved June 8, 2008, from www.aseansec.org.

the most advanced regional bloc in Asia (Figure 1). Conceived and formed in 1967, it has grown from 5 to 10 members (Table 1) and developed from an informal to a more formal entity, creating the ASEAN Free Trade Area (AFTA) and ASEAN Regional Forum (ARF) along the way. However, ASEAN has often been compared with the EU or NAFTA with the general observation that it has produced a system of practices that are undergoverned and underinstitutionalized. To understand the nature of ASEAN regionalism, this entry will highlight three themes, namely, (1) the historical context of regional processes in southeast Asia, (2) the ASEAN model of regionalism, and (3) the AFTA.

Historical Context

Unlike NAFTA, the impetus for ASEAN was not rooted in a desire among countries in southeast Asia for increased intraregional trade and investment. Rather, postcolonial southeast Asia in the 1960s was preoccupied with nation building and issues of sovereignty and identity. Galvanized by the outbreak of war in Vietnam, the five original members, Indonesia, Malaysia, Philippines, Singapore, and Thailand, embarked on a model of "relational control" that was effectively driven by concerns of regional security and stability. Such control included the containment of communist expansion from Indochina and the construction

Year	Member
1967	Indonesia, Malaysia, Philippines, Singapore, Thailand
1984	Brunei
1995	Vietnam
1997	Myanmar, Laos
1999	Cambodia

Table 1 Development of the Association of Southeast Asian Nations

Source: Association of Southeast Asian Nations. Retrieved June 8, 2008, from www.aseansec.org.

of an indigenous geopolitical space in the context of ideological competition between the Soviet Union and the United States.

A few attempts to develop regional consciousness and identity had preceded the formation of ASEAN but failed to gain widespread support among the countries. Japan, for example, had pursued a vision of regional co-prosperity in the 1920s by proposing an East and southeast Asian economic bloc. However, its activities in World War II precluded any role for Japanese leadership in the proposed regional initiative. Britain, too, had tried to promote the then Filipino leader Manuel Quezon's pan-Malayan union comprising the Philippines, British Malaysia, and the Dutch East Indies in the 1940s, but with little success.

By the 1960s, however, events were favorable for the formation of ASEAN. First, with the victory of Viet Minh forces in Vietnam in 1954, the United States began to pay greater attention to the region. The seeds of regional cooperation were sown, for instance, when the Unites States agreed to contribute funds and help establish the Asian Development Bank. Second, the ascension of new leadership under Indonesian President Suharto greatly reduced tensions between two of the original ASEAN members, namely, Malaysia and Indonesia, paving the way for improved regional relations. Third, conflicts in Indochina provided an incentive for increased engagement in regional problems.

However, given the region's tremendous ethnic, religious, economic, and political diversity as well as the fragility of recently acquired sovereignty, postcolonial southeast Asia was faced with the challenge of forging a regional entity where national identities had hitherto been constructed from the outside. Indeed many of the countries' economic and political links were historically oriented toward Europe, and any decline in southeast Asian/European interactions was promptly replaced by increased economic and security interactions with the United States. Like Japan, South Korea, and Taiwan, some southeast Asian countries began to adopt export promotion policies, and many exports targeted the U.S. market. Table 2 summarizes the economic profiles of ASEAN countries. It shows a tremendous variation in economic development levels and sizes. For example, Singapore has about one-twelfth the population size of Myanmar, but its gross domestic product (GDP) is almost 13 times higher. The export-GDP ratio also confirms that countries such as Singapore and Malaysia are trade oriented. Furthermore, the United States is the top market for many countries here. Strong trans-Pacific interactions may have supported the export-promotion strategies of many southeast Asian countries, but they also pose a problem for the indigenous construction of a regional space, since the region tends to exhibit greater extraregional than intraregional tendencies. What, then, is the nature of ASEAN regionalism?

The ASEAN Model

The logic of regionalism in the West is thought generally to be underscored by several factors, of which two will be highlighted here. First, the EU experience suggests that institution building is an important mechanism for achieving greater integration between member countries. For this reason, institutionalists in the international relations literature view the lack of ASEAN institutional building in the first 30 yrs. (years) of its formation to be an impediment to the development of a regional identity. Central to the institutionalist argument is that the governance of regional space is necessarily supported by a regulatory process that lowers transaction costs. This, in turn, encourages the formation of institutional norms that ensure predictable regional practices. Such norms are further accompanied by legal rules of enforcement, so that the governance of EU space is highly formalized.

Country	Population[a] (thousands)	GDP[a] (US$ million)	GDP per Capita[a] (US$)	Exports[b] (US$ million)	Imports[b] (US$ million)	Foreign Direct Investment[b] (US$ million)
Brunei Darussalam	396	12,317	31,076	7,619	1,489	434
Cambodia	14,475	8,662	598	3,514	2,923	483
Indonesia	224,905	431,718	1,920	100,799	61,066	5,556
Laos PDR	5,608	4,128	736	403	588	187
Malaysia	27,174	186,961	6,880	157,226	128,316	6,060
Myanmar	58,605	12,631	216	3,515	2,116	143
Philippines	88,875	146,895	1,653	47,410	51,774	2,345
Singapore	4,589	161,547	35,206	271,608	238,482	24,055
Thailand	65,694	245,702	3,740	121,580	127,108	10,756
Vietnam	85,205	71,292	837	37,034	40,236	2,360

Table 2 Summary of economic statistics among ASEAN countries. Exports and imports are based on merchandise trade.

Source: Association of Southeast Asian Nations. Retrieved June 8, 2008, from www.aseansec.org.

Note: a. 2007, b. 2006.

The constructivist school, however, maintains that norms are socially constructed, and institution building the American or European way fails to account for ASEAN's variegated geography; for example, propinquity has not led to much trade between neighboring countries. Analysis of trade patterns indicates that the Philippines, Thailand, and Indonesia are much more connected to Japan than to their neighbors. Yet economic theory informs us that regional economic blocs tend to arise out of "natural" geography, because lower transportation costs and cognitive distances encourage neighboring countries to trade more with one another than with distant countries. Hence, free trade areas such as NAFTA merely consolidate and formalize the high levels of trade that exist between the neighboring countries of Canada, United States, and Mexico. Second, more than 60% of Europe's trade is intraregional. On the other hand, intraregional trade among ASEAN countries has rarely risen above 30%, and this share even decreased during the 1980s. Unlike the EU or NAFTA, extraregional trade dominates the trading relationships of most southeast Asian countries. Furthermore, a significant share of intra-ASEAN trade is dominated by bilateral trade between Singapore and

Malaysia. Both the constructivist theory of international relations and the economic reality of relatively low intra-ASEAN trade suggest that it is difficult to transfer the experience of North American or European regionalism to southeast Asia.

Furthermore, security concerns were responsible for the creation of ASEAN. ASEAN's secretary general from 1998 to 2002, R. C. Severino, observed that regional supranational governance was generally viewed with suspicion in the 1960s among developing countries. In the case of southeast Asia, the countries were plagued by territorial disputes: They also had little, if any, historical experience of regional cooperation. Indeed, territorial disputes today continue to be adjudicated outside of ASEAN, with, for example, the role of the International Court of Justice in territorial disputes between Malaysia and Indonesia regarding the islands of Sipadan and Ligitan and between Malaysia and Singapore over Pedra Branca.

The "ASEAN way" emerged in response to southeast Asia's overriding concern over sovereignty issues in supranational initiatives. Regional interactions are characterized by informal agreements and diplomacy, consensus building, and noninterference in members' internal affairs. The ASEAN way of regionalism, however, has seen a mixed record. On the one hand, it has been relatively successful in getting the original five ASEAN members to work toward a united front on the Indochina conflict, specifically by opposing the incursion of Vietnam to Cambodia. This helped pave the way for the eventual membership of countries in Indochina. On the other hand, the isolationist tendencies of Myanmar, along with its poor human rights record (e.g., the country's refusal to accept international aid despite widespread human suffering caused by a destructive cyclone in May 2008), have increased criticism regarding the ineffectiveness of ASEAN's principle of noninterference and its informal institutions to resolve regional problems. On the whole, however, constructivists consider the ASEAN model to be relatively successful and potentially a model of regionalism to emulate for other developing countries. They point to the region's engagement with large and middle powers through the ASEAN Regional Forum (ARF) as evidence of this success. The ARF was established in 1993, and it encourages security dialogues among some 26 participants ranging from countries in the EU, North America, and East Asia to Australia and New Zealand. More recently, in 2007, the 10 ASEAN members also signed the ASEAN Charter, which establishes the institutional framework for the organization, thereby allowing the organization to move toward a more formal means of governance.

ASEAN Free Trade Area

In 1992, ASEAN decided to form a regional free trade agreement, creating Asia's first such entity, AFTA. Despite relatively low levels of intraregional trade, concern about competition from China and India for foreign direct investment (FDI) increased the amenability among members for a more formal means of attracting economic and financial flows back to the region from its northern neighbors. Here again, commentators suggest that the logic of regionalism that underscores the formation of free trade areas played a less important motivation than the fear of FDI diversion from the region to China and India.

AFTA was preceded by other preferential trade arrangements (PTAs) in the 1970s. But like most PTAs among developing countries in that period, low income and development levels did not allow the realization of scale economies that the regional pooling of national markets was expected to bring about. Intraregional trade in ASEAN was around 15% in the 1970s. AFTA was conceived in a climate when both globalization and regionalism converged as multinational companies relocated their production and investment worldwide but often at the scale of the region. Hence, Singapore's then prime minister observed that

> unless ASEAN can match the other regions in attractiveness both as a base for investments as well as a market for their products, investments by multinational companies are likely to flow away from our part of the world to the SEM (Single European Market) and NAFTA (North American Free Trade Agreement). If we do not synergise our strengths, ASEAN will risk missing the boat. (Severino, 2006, p. 224)

Right from the start, however, members sought temporary exclusions in various industries from the targeted list for tariff reductions. With the exception of Singapore, most ASEAN countries were pursuing import substitution industrialization that was supported by protectionist policies. Malaysia's national car project is one such example. Furthermore, targets of tariff reductions were pegged at 0% to 5% instead of 0% even for fast-track items. Fast-track items are products whose tariffs were already less than 20% at the time of tariff implementation. Tariff reductions for these items were targeted at 0% to 5% within a period of 5 yrs. The 5% figure arises from the fact that tariffs constitute a source of revenue for some countries; hence, some level of tariff remains important for these countries. In addition, industrial policies that promote specific industries are still popular, so that less developed southeast Asian countries have also been given more time to meet the tariff reduction targets.

Clearly, the above reflects the difficulties of regional governance among developing countries; as ASEAN Secretary General Ong Keng Yong (2003–2007) wrote recently,

Furthermore, in ASEAN, there are no supranational institutions to mandate region-based action. Member countries are still very much their own sovereign power. Regional action on a particular area works only if the national and regional agendas are aligned. Many issues in the social sector are ultimately national responsibilities. (Yong, 2004)

ASEAN's approach is to build institutions only when other arrangements have been exhausted. ASEAN's early development was unique. Uncertainty and confrontation characterized the region even as the grouping was being nurtured.

Nonetheless, AFTA represents the first step toward more formal regulation and institution building, which ASEAN has eschewed for nearly 3 decades. Indeed, many of the original six ASEAN members are already meeting their targets of tariff reductions.

Overall, ASEAN regionalism is a top-down process driven by political elites' concern for national sovereignty. This has resulted in a regional process that has largely resisted institutionalization in the first 3 decades. However, the countries appear to be broadening their regional relationships to include economic affairs and environmental problems, not just security. AFTA thus represents an important expansion of interests toward a broader scope of regional interactions. While this contrasts with the wider participation of EU actors that has seen the engagement of the organization with business, labor, environmental, and civic groups, it is too early to tell just how the recent broadening of the ASEAN agenda will result in regionalism that is more inclusive and that is less narrowly defined.

Jessie P. H. Poon

See also Globalization; Newly Industrializing Countries; North American Free Trade Agreement (NAFTA); Trade

Further Readings

Acharya, A. (1997). Ideas, identity, and institution-building: From the "ASEAN way" to the "Asia Pacific way"? *Pacific Review, 10,* 319–346.

Association of Southeast Asian Nations: www.aseansec.org

Beeson, M. (2005). Rethinking regionalism: Europe and East Asia in comparative historical perspective. *Journal of European Public Policy, 12,* 69–85.

Bowles, P. (1997). ASEAN, AFTA and the "new regionalism." *Pacific Affairs, 70,* 219–233.

Eaton, S., & Stubbs, R. (2006). Is ASEAN powerful? Neo-realist versus constructivist approaches to power in southeast Asia. *Pacific Review, 19,* 135–155.

Moller, K. (1998). Cambodia and Burma: The ASEAN way ends here. *Asian Survey, 38,* 1087–1106.

Poon, J., Sajarattanachote, S., & Bagchi-Sen, S. (2006). The role of US exports in Asia Pacific regionalism. *Political Geography, 25,* 715–734.

Poon, J., Thompson, E., & Kelly, P. (2000). Myth of the triad: Geography of trade and investment blocs. *Transactions of the Institute of British Geographers, 25,* 427–444.

Severino, R. (2006). *Southeast Asia in search of an ASEAN community.* Singapore: Institute of Southeast Asian Studies.

Yong, O. K. (2004, August 18). *Community building in ASEAN's social sector.* In the ASEAN Plenary Session on Civil Society and Regional Cooperation, 31st International Conference, International Council on Social Welfare, Kuala Lumpur, Malaysia. Retrieved August 20, 2009, from www.aseansec.org/16324.htm

ATMOSPHERIC CIRCULATION

Occurring over various time periods and at various geographic scales, atmospheric circulation refers to the mass movement of air and energy through the Earth-atmosphere system. The driving force behind atmospheric circulation is the amount and intensity of solar energy, or *insolation*, which differs according to latitude, season, and hour of day. Topography, unequal land and water distribution, and land surface type also contribute to atmospheric circulation characteristics. In combination with ocean circulation, the atmosphere is responsible for moving the surplus energy of the low latitudes poleward to counterbalance the energy deficit of higher latitudes.

Atmospheric circulation systems are classified according to their spatial and temporal extent. A direct relationship exists between the size of the atmospheric phenomenon and the timescale involved, with larger (smaller) systems usually enduring for longer (shorter) time periods. There are four major divisions of atmospheric features: planetary scale, synoptic scale, mesoscale, and microscale; sometimes, the planetary and synoptic scales are combined into a single class known as macroscale. However, not all atmospheric phenomena display characteristics exclusive to a single division, as some features may vary considerably in size and duration.

Macroscale Circulation

Planetary scale features are the largest, occurring on spatial scales ranging from 10,000 to 40,000 km (kilometers) (6,000–24,000 mi. [miles]) and with life cycles on the order of weeks to months. These features include three major latitudinal wind and pressure bands encircling the globe in each hemisphere: the convective Hadley cells of the tropics; the Ferrel cells of the midlatitudes, and the polar cells of the high latitudes. The three circulation cells represent an idealized model of the dominant planetary wind motions by means of two major simplifications: (1) a uniform surface (to eliminate complications with differential surface heating) and (2) no seasonal or diurnal insolation changes (i.e., the model assumes solar noon on the equinox so that all latitudes are receiving 12 hrs. [hours] of daylight and the sun is always directly over the equator).

The low latitudes, between approximately 0° and 30° latitude, are governed by the heat-driven Hadley cells, named after the English physicist and meteorologist George Hadley (1685–1768), who first proposed their existence in 1753. Around the equatorial region, warm, humid air ascends from daytime convective heating, often condensing into towering cumulonimbus clouds. These convective thunderstorms release large amounts of latent heat energy that increases the buoyancy of the air and provides the energy to direct the Hadley cell. The ascension of air is further enhanced by surface convergence of the northeast and southeast trade winds that produce a low-pressure zone known as the *equatorial trough* or *intertropical convergence zone* (ITCZ). Consequently, locations around the ITCZ are associated with high annual precipitation, a marked daily regularity in rainfall, and a tropical wet climate with thick, broadleaf vegetation forests. Tropical rain forests are common around the Congo River basin in Africa, the Amazon River basin in South America, and the South China Sea region of southeast Asia.

Aloft, the upper-level flow of the Hadley cells is characterized by poleward-moving air that cools along the *tropopause* (the boundary between the weather-producing troposphere and the stable stratosphere). As the air propagates farther away from the ITCZ, it loses buoyancy and latent heat energy, and radiational cooling commences. The density of the air increases and begins to sink between 20° and 35° latitude. Since the Coriolis effect increases with increasing latitude, the poleward-moving air is also deflected, forcing the air into a west-east flow. The result is a convergence of air aloft that strengthens the subsidence and increases the surface pressure to produce an elongated subtropical high-pressure belt. Subtropical air is comparatively dry with a relatively low humidity, a by-product of (a) most of the moisture content being liberated near the equator and (b) the further warming of the air by adiabatic compression. Accordingly, the subtropics are often associated with dry conditions and clearer skies and are the site of the largest deserts in the world. The Sahara Desert of Northern Africa, the Atacama Desert of Western South America, the Arabian Desert of the Middle East, and the Great Sandy Desert of Australia are all positioned along this subtropical high-pressure zone.

In the middle latitudes lies the Ferrel cell, a secondary circulation feature arising from the separation of the more-defined Hadley and Polar cells. Named after the American meteorologist William Ferrel (1817–1891), the Ferrel cell is a thermally indirect cell in which cool air rises in the subpolar region around 60° latitude and warm air sinks in the subtropics around 30° latitude. Air diverges from the subtropical high-pressure belt and moves poleward, deflecting into a west-east flow. These prevailing westerlies generally strengthen with height, reaching their maximum velocities near the tropopause level (around 10–12 km, or 6–7 mi.). Known as jet streams, these high-speed upper tropospheric winds, often in excess of 125 km/hr. or 75 mi./hr., form along a region of sharp thermal contrast and are embedded within the dominant westerly flow of the midlatitudes.

Two principal jet streams exist around the margins of the Ferrel cells, the polar or midlatitude jet stream on the poleward side and the subtropical jet stream on the equatorward side. The polar jet stream occurs alongside the polar front, the major frontal boundary separating air masses of polar origin from those of tropical origin. Frequently, midlatitude cyclones form along the polar front and are propelled eastward by the jet stream. The path of the polar jet stream is typically not linear or continuous but instead meanders northward and southward and may split into separate sections of high-velocity flow. Midlatitude locations are characterized by highly variable weather conditions depending on juxtaposition to the prevailing polar jet (and front). Somewhat slower and less defined than the polar jet, the subtropical jet is characterized by strong midtropospheric temperature contrasts that are most prevalent during the cold season; hence, the subtropical jet is a wintertime phenomenon. Occasionally, the polar and subtropical jet streams will merge, providing upper-level dynamic support for surface midlatitude cyclone development and/or intensification.

In comparison with the Hadley cells, the thermal structure of the high-latitude polar cells is weaker and less understood. Surface air circulation in the high latitudes is governed by the polar easterlies, arising from the Coriolis effect deflection of equatorward-moving air from the polar high-pressure regions. Rising motion occurs along the polar front, thereby establishing a low-pressure belt around 60° latitude. The circulation cell is completed as a portion of the subpolar air aloft is further cooled and subsides near the poles. The polar highs are further sustained by limited insolation (i.e., low sun angle, limited daylight in the cold season, and highly reflective snow-covered surface) and clear nighttime skies that promote radiational cooling.

The three-cell model is a simplified representation of the general atmospheric pressure and wind patterns. In reality, the global atmospheric circulation patterns diverge from the model in accordance with insolation variability, topography, and land/water contrasts. The uniform zonal patterns of the model are replaced by discontinuous pressure belts or semipermanent cells of high and low pressure that vary seasonally. In the Northern Hemisphere, the Aleutian low in the Gulf of Alaska and the Icelandic low in the northern Atlantic are two such semipermanent pressure cells that bring much storminess to the midlatitudes.

Global atmospheric circulations also include expansive waves thousands of kilometers long encircling the Earth in the upper atmosphere, varying in amplitude to create a series of ridges and troughs (i.e., areas of higher and lower pressure regions). The waves are measured as the distance between the nadir of two successive troughs or the crest of two successive ridges. Longer (shorter) wavelengths with lower (higher) amplitudes indicate a principally zonal (meridional) flow and are associated with more (less) variable weather conditions. In the midlatitudes, these long waves are known as Rossby waves and largely propagate eastward within the prevailing westerly flow, but in rare instances, they have been known to move from east to west in a state known as retrogression. Rossby waves are responsible for transporting warm, subtropical air masses from lower latitudes toward the higher latitudes and cold, polar air masses to lower latitudes, particularly during strong meridional (north-south) flow. Between two and seven Rossby waves encircle the hemispheres at any given time, with three or four waves more typical. The number, length, and strength of Rossby waves vary seasonally, with the cold season displaying longer, fewer wavelengths in the company of strong wind speeds. Shorter-wavelength features (<10,000 km or 6,000 mi.) with comparatively diminished life spans (e.g., synoptic scale systems) often move through Rossby waves quickly in comparison with the overall net-long-wave movement.

Synoptic scale systems have life spans of days to weeks and vary considerably in size, from 100 to 10,000 km wide (60–6,000 mi.), with 2,000 to 3,000 km (1,200–1,800 mi.) being more common. Larger scale synoptic circulation features include anticyclones and cyclones that preside over regions of hundreds to thousands of square kilometers. Midlatitude cyclones are low-pressure systems that develop along the polar front and have winds converging at their centers, rotating counterclockwise (clockwise) for the Northern (Southern) Hemisphere. The configuration of upper-level Rossby waves is also important in the development and intensification of surface cyclones and anticyclones. Midlatitude cyclones are associated with one or more additional synoptic scale features known as fronts, boundaries separating air masses with different temperature and moisture characteristics. At fronts, warmer, more buoyant air rises over cooler, denser air and initiates widespread cloud cover and likely precipitation development.

The smaller end of the synoptic scale includes tropical storms and hurricanes, though these features sometimes can be rather small and short-lived, bordering on the mesoscale class. Known also as typhoons in the western Pacific and cyclones in the Indian Ocean, mature hurricanes average about 600 km (375 mi.) in diameter but can range in size from a couple of hundred kilometers to around 1,500 km (or 930 mi.). In comparison with midlatitude cyclones, hurricanes are much smaller and more intense, with a steeper pressure gradient, lower central pressure, and stronger surface winds in excess of 33 m/s (meters per second, or 74 mph [miles per hour]).

Tropical cyclones generally form between 5° and 20° latitude during the late warm season when ocean temperatures are 27 °C (80 °F) and the vertical wind shear (changing wind speeds with height) is minimal. Hurricanes will not develop within 5° latitude of the equator, since the Coriolis effect is too weak to initialize surface convergence and rotation. Warm, moist air converges into the low center or eye and is forced upward, producing towering cumulonimbus clouds of intense convective precipitation and very strong winds around the storm center in a region known as the eye wall. Spiral rain bands of clouds and relatively weaker convective activity are located successively outward from the eye. Near the top of the tropical cyclone, the air diverges and transports air away from the storm center, sustaining the inward flow at the surface. Hurricane dissipation occurs when the system (a) travels along cooler sea surface temperatures and loses latent heat energy, (b) encounters the stronger upper-level westerlies of the middle latitudes, and vertical wind increases, and/or (c) travels across land and friction slows wind speeds, and the system is removed from the primary moisture (and energy) source.

Mesoscale Circulation

Mesoscale circulations occur at scales ranging from approximately 2 to 2,000 km (1–1,200 mi.). Although markedly smaller than macroscale circulations, mesoscale circulations can cause equally intense weather events, such as torrential rain, thunderstorms, and dramatic changes in temperature. Mesoscale circulations include monsoons, sea breeze circulation, and thunderstorms.

The term *monsoon* itself means "season" and refers to the seasonal reversal of periodic sustained winds that establish prolonged periods of wet or dry conditions. Monsoonal circulations occur around the world. Perhaps the best-known example is over India, but other areas that experience monsoonal circulations include Indonesia, China, and northwestern Mexico and the southwestern United States. Monsoons result from movement of macroscale circulation features on a seasonal basis. In India, for example, the sun's rays intensify in spring as the subsolar point moves into the Northern Hemisphere. This results in strong heating, causing air over the subcontinent to become very warm and buoyant. Cooler but wetter oceanic air begins moving over the subcontinent, creating clouds, showers, and thunderstorms. Orographic lifting by the Himalaya mountains also enhances the rainfall. During the winter monsoon, the process is reversed; maximum heating shifts south of India, and cold, dry air drains southward from Central Asia.

Other regional circulations, such as sea breezes, Santa Ana winds, and sandstorms, while smaller than monsoonal circulations, also can have significant effects on life and property. These circulations are all trigged by large physical features, such as bodies of water, mountains, and deserts.

Sea breezes and lake breezes are found near coastlines. On a clear, sunny day when no major weather systems—such as fronts—are nearby, land surfaces heat more quickly than adjacent bodies of water, because land has a lower specific heat capacity (i.e., the amount of energy required to raise a substance 1 °C at atmospheric sea level pressure). As the temperature difference increases, colder, more dense air moves ashore, cooling areas near the body of water. After sunset, land cools more quickly than adjacent water and the process reverses, causing a land breeze.

Areas cooled by a sea breeze are separated from warmer air by a sea breeze front, which is merely a mesoscale cold front. Like other fronts, convergence and upward motion are common along the boundary; such motions may trigger clouds, showers, and even thunderstorms. On narrow peninsulas, such as the one comprising the state of Florida, converging sea and lake breeze fronts are responsible for a large percentage of summertime rainfall.

Like bodies of water, mountains also spawn mesoscale winds. On a calm, sunny day, air near the base of a mountain warms more quickly than air above it, leading to unstable conditions, rising air, and clouds. This wind pattern is termed a *valley breeze*. Similar to sea breezes, these breezes are most pronounced during the warmest time of the day (usually late afternoon). At night, the process is reversed, and cold air spills back down into the valleys, causing a mountain breeze, also known as a nocturnal drainage wind.

Katabatic winds are strong downslope winds that are generally more intense than typical mountain breezes. At their strongest, these winds can be as fast as hurricane wind speeds (>33 m/s or 74 mph). Most katabatic winds have location-based names. For example, warm, dry winds that descend the eastern slopes of the Rockies are called chinooks, while similar winds on the eastern slopes of the European Alps and Andes are called foehn winds and zonda winds, respectively. Such winds can increase temperatures 20 °C (or 46 °F) per hour and cause extremely low relative humidities. A similar katabatic wind sometimes blows from east to west into Southern California. This "Santa Ana wind" quickly dries out vegetation and contributes to California's annual wildfire season. Not all katabatic winds are warm and dry, however. The *bora* (southeastern Europe), *mistral* (Southern France), and *coho* (northwestern United States) all result from very cold air moving downslope. Although this air warms as it descends, it is still much colder than the coastal air that it displaces.

Arid environments also create mesoscale circulations. Dust storms and sandstorms are endemic to most of the world's subtropical deserts. Perhaps the most spectacular example is the haboob, a virtual wall of dust hundreds of meters high precipitated by thunderstorm outflow. Haboobs are native to areas near the Sahara desert (especially the Sudan) but have also been observed in the southwestern United States and Australia.

A mesoscale circulation that can occur almost anywhere is a thunderstorm. Thunderstorms are deep convection composed of cumulonimbus clouds that extend high into the atmosphere. Most thunderstorms are single cells; these usually last less than an hour and rarely cause severe weather. Thunderstorms also occur in multicellular groups such as linear lines of single cells (often near fronts) or mesoconvective complexes (large nocturnal groups of storms). Unusually large and long-lived thunderstorms are called supercells. Supercells last for hours and can be hundreds of kilometers in size. Identified by their rotating updraft, supercells are usually responsible for exceptional weather reports such as large hail (>5 centimeters) or strong tornadoes.

Microscale Circulation

Small-scale circulations (<2 km or 1 mi. in size) are considered microscale. Most of these are short-lived (typically lasting a few minutes) and nonhazardous—although tornadoes are an obvious exception. Tornadoes are rotating columns of violently rising air spawned by thunderstorms (primarily supercells). Although dramatic in appearance, most are short-lived and responsible for minor damage. Dust devils appear similar to tornadoes but form in a very different environment: clear skies with sunshine. Dust devils occur when air near the ground heats rapidly, rises, and begin to rotate. Most dust devils are too weak to cause any damage whatsoever. Waterspouts are rotating, rising columns of air over water. These form when very cold air moves over a warm body of water, causing great instability in the lower atmosphere. While typically longer-lived than other rotating updrafts, waterspouts usually do not cause widespread destruction. (It is also worth noting that a tornado that moves over water is called a waterspout, although this is clearly a very different phenomenon.)

Humans are responsible for many microscale circulations as well, such as the channeling effect of wind between tall buildings, small-scale circulations between parking lots and adjacent undeveloped forests or fields, and pollution effects. Although they are often difficult to measure individually, collectively these circulations affect weather. For example, urban areas are associated with significantly warmer temperatures at night, lower net wind speeds, and downwind increases in clouds and precipitation.

Jill S. M. Coleman and David A. Call

See also Adiabatic Temperature Changes; Air Masses; Anthropogenic Climate Change; Atmospheric Composition and Structure; Atmospheric Energy Transfer; Atmospheric Moisture; Atmospheric Pressure; Atmospheric Variations in Energy; Chinooks/Foehns; Climate Types; Climatology; Coriolis Force; Fronts; Hurricanes, Physical Geography of; Monsoons; Thunderstorms; Tornadoes; Wind

Further Readings

Ahrens, C. D. (2009). *Meteorology today* (9th ed.). Belmont, CA: Brooks/Cole.

Grenci, L. M., & Nese, J. M. (2006). *A world of weather: Fundamentals of meteorology*. Dubuque, IA: Kendall-Hunt.

McIlveen, J. F. R. (1992). *Fundamentals of weather and climate*. London: Chapman & Hall.

Rohli, R. V., & Vega, A. J. (2008). *Climatology*. Sudbury, MA: Jones & Bartlett.

⬤ ATMOSPHERIC COMPOSITION AND STRUCTURE

Generalizations of the size of Earth's atmosphere compared with its lithography (or the solid part of our planet) liken the gaseous shell surrounding Earth to the skin on an apple. Although many might believe the atmosphere to be uniform in gaseous composition or linear in temperature change, this gaseous envelope is much more complex, similar to most environmental systems on Earth. In fact, Earth's atmosphere, while technically "thin," is nonuniform and consists of many distinct layers that can be classified according to individual parameters. Such parameters include gaseous composition, temperature (termed as the "thermal structure of the atmosphere"), and differences in ionic charge.

Composition of the Atmosphere

The atmosphere is composed of a mixture of gases and aerosols that, when considered in their elemental states, result in a surprisingly small number of major gases, trace gases, and a number of variable components that can be present depending on the location and conditions. The most important of these atmospheric constituents are the major gases, as their percent by volume of the atmosphere rarely changes from day to day or location to location. Given a random sample of air at sea level (therefore, representative of a 100% gaseous mixture), we can express the relative composition of individual gases and components in the atmosphere by their percentages. The three most prominent gases in the atmosphere, based on a relatively recent composition analysis are (1) nitrogen in its diatomic state (N_2), comprising approximately 78.1% of our air sample; (2) oxygen in its diatomic state (O_2; perhaps the best-known component of the atmosphere due to its respiratory function within the Kingdom Animalia), which comprises approximately 21% of the atmosphere; and (3) the noble gas argon (Ar) in its elemental state, which comprises less than 1% of the atmosphere. If one were to add up a more scientifically precise percent by volume for our sample of dry air by not rounding for simplicity's sake, these three components would comprise approximately 99.964% of all the detectible gases in the air mixture. Therefore, for each breath inhaled by human beings, the vast majority of that air consists of inert diatomic nitrogen and the relatively inert noble gas argon, with less than 21% of the air that we breathe consisting of oxygen. What gases and aerosols then comprise the remaining 0.036%?

The remaining 0.036% is changing in percent by volume at all times due to the dramatic increase in the fourth most prevalent atmospheric gas, carbon dioxide (CO_2). At the time when careful measurements of the atmosphere were being conducted to determine atmospheric composition, CO_2 comprised nearly the entire 0.036% remaining. However, in recent years, the level of CO_2 in the atmosphere has been increasing both in percent by volume and in the number of molecules of a selected gas per million molecules sampled, known as parts per million or ppm. Expressed in this way, a few years ago, the concentration of CO_2 in the atmosphere was nearly 360 ppm (or almost 0.036%). Today, the concentration of CO_2 in the atmosphere is slightly more than 380 ppm (or a little more than 0.038%), resulting in slightly less of a percent by volume of the three major gases but only by a small fraction of a percent. However, this small fraction of a percent is leading many scientists today to worry about "climate change," as there are inherent properties associated with the concentration of CO_2 in the atmosphere that affect the temperature of the atmosphere as well.

Occupying the fifth through ninth places on the list of most prevalent atmospheric gases are neon (Ne), helium (He), methane (CH_4), krypton (Kr), and hydrogen (H), respectively. The concentration of hydrogen is so low that expressed as a percent by volume, hydrogen makes up only 0.00005%, or just 0.5 ppm—this means that, on average, it would take a random sample of 2 million molecules present in air to find a single molecule of hydrogen. These figures should make it very clear as to why we refer to these gases (from CO_2 through H) as the trace gases.

Given that a sample of air will contain the gaseous constituents discussed above in approximately the same concentrations as described, it is important to note that accounting for 100% of the elemental gases in air does not equal accounting for 100% of the air mixture itself. In fact, there are other constituents of air that we must consider, and the first of these is water vapor.

Water vapor is nothing more than the gaseous phase of water (H_2O)—however, its concentration on Earth varies depending on weather conditions, the time of day, and even the individual location selected. The percent by volume of water vapor in the air can vary between 1 and 4. If one imagines a dry desert during a scorching summer afternoon, an impression of dryness and lack of moisture will often be generated. This is not far from reality, wherein high temperatures and the scientific properties of water vapor result in very low humidities reflective of low atmospheric water vapor content. It is important to note that even in such environments, the amount of water vapor in the air is not equal to zero; some water vapor will still be present in the air. On the other hand, consider a rain forest in the Amazon region of Brazil following a torrential summer downpour, and an impression of high humidity, "stickiness," and of condensation may be generated. Here, atmospheric moisture levels may approach the 4% high end of the spectrum. Similar to CO_2 in the atmosphere, water vapor plays a role in climate change because it is also an effective absorber of energy—resulting in higher temperatures at the surface. This will be discussed in more detail in the next section. It is also of interest to note that the contrails of jet aircraft are formed as water vapor is produced as a by-product of the combustion process in jet engines. This water vapor quickly cools in the cold atmosphere surrounding modern jet aircraft in flight, resulting in the formation of minute ice crystals and, hence, the human-process-generated clouds named contrails. A study of a contrail-less sky conducted in 2001 during the 3 days following the attacks of September 11 (the only lengthy period in modern times in which *all* civilian aircraft were grounded) found that the contrails produced in jet aircraft are actually playing a role in Earth's climate—resulting in a difference of 3% in temperature changes than would take place without such humanmade clouds.

The atmosphere also contains a variable amount of suspended particulates that may be of solid or liquid consistency. Together, these airborne particulates are known as aerosols. There are many sources for these aerosols as well as different types of aerosols. For example, winds of sufficient intensity moving across the surface of a desert region may pick up the loose dust or clay particles forming the surface of such desert (the heavier sand composed of quartz or other minerals may become airborne during winds of significant force but typically remain within a few meters of the planetary boundary layer). Depending on the actual material hoisted into the air by these winds (clay vs. dust), these particulates can travel for hundreds or even thousands of miles, riding the air currents that flow ribbon-like throughout Earth's atmosphere. As in this example, the concentration of actual aerosols in a given sample of air is highly dependent on many factors, including prevailing wind direction, wind speed, proximity to urban versus rural areas, and location near a coastline (e.g., salt spray) or power plant (e.g., soot from combustion). Due to the effect of gravity on these particulates, they are more concentrated at ground level and typically reduce in presence as one gains altitude.

Ozone also is present in the atmosphere but at widely differing concentrations. Ozone is the common name for triatomic oxygen, or O_3. Ozone is present in minuscule amounts in the lower atmosphere (generally less than one part per 100 million) but is found in much higher concentrations in a layer high above Earth's surface called the ozone layer. This layer is present in a thermal layer of the atmosphere called the stratosphere, which exists from about 6 to about 30 mi. (miles) above the surface of the Earth. Ozone, at ground level, acts as a respiratory irritant, and is produced as a by-product of combustion, such as the burning of fossil fuels by automobiles, power plants, and other items of urban infrastructure. Temperature plays a role in the concentration of ozone—therefore, hot summer days in large urban areas exhibit the highest risk of ozone-induced respiratory distress such as asthma attacks, and there is clear evidence of increased emergency room visits for asthma attacks on summer days in which elevated ozone levels have been recorded. Many large cities now issue ozone warnings, advising susceptible individuals to stay indoors on days in which dangerous ground-level ozone limits may be achieved. At the level of the ozone layer, however, high up in the stratosphere, ozone is absolutely vital to life. Here, ozone acts as a shield, protecting the surface of the Earth from deadly ultraviolet (UV) radiation emitted by the sun by incorporating the UV light into a complex, self-sustaining chemical reaction. If the ozone layer were not present, it is likely that most life on Earth would not be able to survive open exposure on a sunny day, including humans. Unfortunately, man-made chemicals called chlorofluorocarbons (CFCs) were used as refrigerants for much of the

20th century, and these chemicals had a tendency to reach stratospheric heights—whereby these CFCs worked to destroy ozone and allow more UV light to strike Earth's surface. The result was a depletion of the ozone layer called the ozone hole. This "hole" in Earth's ozone layer was, and is, concentrated over the South Polar regions (Antarctica) because the chemical reaction that allows CFCs to destroy ozone progresses faster as the temperature of the air drops, and cold, snowy Antarctica has some of the coldest air on the planet. Fortunately, an international agreement was signed in 1987 that called for the abolishment of the use of CFCs due to the destruction of the ozone layer. As a result, the ozone layer is expected to recover to its formerly "normal" size—but only by the middle of the 21st century (a 50-yr. [year] recovery period).

Density of the Atmosphere

Since gravity affects even small molecules of gas just as it does items with larger mass, including people, the density of atmospheric gases is much higher at sea level than it is with even slight altitude increases. In fact, if one climbs to the top of a tall mountain, the amount of oxygen in the air is dramatically decreased, such that one can have difficulty breathing with even minimal exertion! One half of the atmosphere lies below an altitude of 3.5 mi., and the remainder decreases nonlinearly as altitude increases (diagramming this decrease demonstrates a curve with rapid decline in the lowest 50% of the atmosphere, with a slowing rate of decline above approximately 10 mi. in elevation). Therefore, the pressure exerted by all the gases in the atmosphere behaves in this fashion. This rapid decrease in atmospheric density is the reason why many mountain climbers must bring supplemental oxygen canisters to breathe while scaling high peaks and, similarly, why military high-altitude parachutists must wear oxygen tanks to keep from passing out when leaping from airplanes at cruising altitudes.

Thermal Structure of the Atmosphere

The most commonly used parameter in describing the structure of the atmosphere is the thermal structure. The atmosphere is divided into four primary layers that share similar thermal characteristics. This does not mean that the temperature remains the same within each of these four layers; rather, this thermal homogeneity refers to similar temperature trends within these layers—for instance, temperatures may decrease with increasing altitude in one layer, whereas temperatures may increase with increasing altitude in another layer. The fact that temperatures may actually rise as one gets farther away from the surface of the Earth may come as some surprise for those not familiar with meteorological concepts; however, careful scientific measurements and observations have confirmed that this phenomenon is genuinely reflective of the thermal structure of the atmosphere. On the other hand, some reflection on the meaning of the word "temperature" will be required to fully comprehend the dramatic see-saws in temperature parameters that occur as altitude above the Earth increases.

The lowest thermal layer of the atmosphere is called the troposphere. It extends from the actual surface of the Earth upward to a height of approximately 7.5 mi. when averaged, although it can range from a low of near 5 mi. to a high boundary of approximately 10 mi. in altitude. The troposphere contains the majority of atmospheric gases due to the pull of gravity and a complicated concept called hydrostatic equilibrium—this, then, is the thermal layer in which almost all of Earth's weather occurs. Although temperatures generally decrease throughout the troposphere, during certain conditions, temperatures can increase slightly with height before resuming their decrease. These small increases in temperature with height in the troposphere layer are called inversions.

At some point above the Earth's surface, at the top of the troposphere, temperatures stop decreasing with altitude and remain fairly stable. This "pause" in the decrease of temperatures with increasing elevation is known as the tropopause, and it represents the boundary between the troposphere and the next thermal layer of the atmosphere—the stratosphere. Just as the height of the troposphere varies over the Earth's surface due to climate patterns, so too does the tropopause vary in height.

The stratosphere extends from approximately 7.5 mi. in height (average) to approximately 30 mi. in height—this is higher than most jet aircraft ever fly! There is very little "weather" in the stratosphere, although occasionally, strong thunderstorms can breach the tropopause and extend into the stratosphere in what are called "overshooting tops." In the stratosphere, temperatures actually do not decrease with height as in the troposphere. Rather, they remain fairly stable just above the tropopause and then begin increasing with height. In fact, at an altitude of 30 mi. above the Earth, the temperature of the outside air can be as warm as the freezing point of water (32° F), much warmer than the air outside a passenger jet at cruising altitude! This brings up two important questions: (1) Why does the stratosphere get warmer with increasing height? (2) What do we mean when we use the word "temperature"?

As to the first, remember that the ozone layer is located in the stratosphere. In fact, the layer itself is concentrated in the middle of the stratosphere, between 10 and 20 mi. in altitude. The chemical reactions produced between the incoming UV sunlight and the ozone present in the layer produce heat as a by-product. Therefore, this chemical reaction serves to heat up the stratosphere so that temperature trends are essentially opposite to those in the troposphere. Now, as to the second item, *temperature* refers to the atomic motion within an object. At higher temperatures, this atomic motion increases. Therefore, an object with high temperatures exhibits rapid atomic motion, whereas colder objects have less atomic motion. All atomic motion ceases at *absolute zero*, which is −459.67 °F—quite cold indeed. Above absolute zero, all objects contain atomic movement or motion. In the atmosphere, then, this means that as temperatures increase, the individual molecules of gas in the air have greater atomic motion and therefore move faster through the air mixture.

At the top of the stratosphere, there is another region of stable temperatures called the stratopause, which represents the boundary between the stratosphere and the next thermal layer, which is called the mesosphere. The mesosphere extends from the top of the stratosphere, approximately 30 mi. in height, to approximately 50 mi. in height, where the mesopause occurs. Temperatures decrease with altitude in the mesosphere, similar to the troposphere, and the lowest temperatures in the atmosphere occur in this layer (as low as −130 °F, on average). At the top of the mesosphere lies the mesopause, a region of stable temperatures with height.

Above the mesopause, at 50 mi. in altitude, the final thermal layer of the atmosphere exists, which is called the thermosphere. Here, the small fraction of atmospheric gas that remains exists between Earth and space, and in fact, the exact boundary between the atmosphere and what we consider to be space is not easy to define. In the thermosphere, there are few gas molecules present in a given volume; however, due to excitation by incoming solar energy, these gas molecules have extremely high temperatures. Approaching the top of the thermosphere at 90-plus mi. in altitude, temperatures can reach above and beyond 1,800 °F. However, because the gas density is so low, it would not feel warm if you were to expose your hand to this environment. In fact, despite temperatures of such magnitude, your hand would likely freeze solid on such exposure, due to the large distance between such individually energetic gas molecules.

Additional Atmospheric Structure Conventions

In addition to the definitions of the composition, density, and thermal structure of the atmosphere, there are additional topics that can address the manner in which the atmosphere is constructed. One is the division of the atmosphere into two separate "shells" depending on the consistency of the atmospheric composition. From the mesopause down to the surface of the Earth, the ratio of gaseous components is fairly consistent, and this "shell" is therefore known as the homosphere. Above the mesopause, individual molecules form stratified layers in the atmosphere, with the heaviest elements at lower elevations and the lighter elements at higher elevations. This other "shell" of the atmosphere is known as the heterosphere, due to this stratification.

The atmosphere also contains a layer of positively charged nitrogen and oxygen ions, which is called the ionosphere. This portion of the atmosphere consists of three layers (D, E, and F), classified on the basis of their ionization properties. The F layer is the only layer to be charged both day and night; as the sun's energy sets below the horizon each day, the D and E layers generally lose their charges fairly rapidly. The ionosphere's unique properties are often used in communications, such as by radio, to allow for extremely long-distance transmission and receiving by "skipping" radio waves off of the D and E layers and therefore extending transmission beyond the curve of the Earth.

Brian H. Bossak

See also Adiabatic Temperature Changes; Air Masses; Anthropogenic Climate Change; Atmospheric Circulation; Atmospheric Energy Transfer; Atmospheric Moisture; Atmospheric Pressure; Atmospheric Variations in Energy; Chinooks/Foehns; Climate Types; Climatology; Coriolis Force; Fronts; Hurricanes, Physical Geography of; Monsoons; Thunderstorms; Tornadoes; Wind

Further Readings

Lutgens, F. K., & Tarbuck, E. J. (2007). *The atmosphere* (10th ed.). Upper Saddle River, NJ: Pearson Prentice Hall.

Wallace, J. M., & Hobbs, P. V. (2006). *Atmospheric science* (2nd ed.). London: Elsevier/Academic Press.

ATMOSPHERIC ENERGY TRANSFER

There are two primary energy sources within the atmosphere: heat energy originating from the sun and angular momentum from Earth's rotation about its axis. While solar irradiance is the primary control on Earth's surface temperature, angular momentum and its variation over time have a large impact on the state of the climate system on subseasonal timescales.

Much temporal variability exists in both solar output and angular momentum, and to a certain extent the two may be related. But variability in Earth's angular momentum clearly varies on smaller timescales than solar output. There are some who believe that variance of only a few watts per square meter is negligible and therefore meaningless. However, one only needs to look at mean temperature time series plots to see just how close the association has been between solar output and temperature over the past 50 yrs. or more. This is not a well-recognized association, often hidden by the focus on global warming as a function of anthropogenic CO_2 increase. Nevertheless, it should always be remembered that solar radiation is the primary climate control.

Heat Energy

Heat energy available at the Earth's surface as a function of solar irradiance tends to converge in the warm season and diverge in the cold season. Variance in the direction of the net flux is due to seasonal and subseasonal temperature gradients.

During the warm season, heat convergence is typically large, with gradients operating both upward into the cooler overlying air and downward toward the cooler soil that lies beneath the surface. While there are many thermal characteristics to consider with respect to the material through which heat flows via conduction, the upward flux of heat (and in many cases water vapor) is usually very strong due to the superadiabatic (where temperature decreases with height at a rate of greater than 10 °C per kilometer) nature of the lower planetary boundary layer (PBL).

In general, heat energy is not particularly well transmitted through air, since the molecules are only briefly in contact with one another; however, the ground surface is in immediate and frequent contact with atmospheric molecules in the lower atmosphere. From here the heat is transferred upward or downward in response to an upward- or downward-directed temperature gradient. If the gradient is directed downward, then the heat is transferred back into the surface layer, and if it is upward, then the transfer is accomplished via conduction through the convective boundary layer (CBL). It is at the top of the CBL that the upper boundary of the PBL is found.

As long as the capping inversion at the top of the PBL is not strong (generally less than 2 °C), the convection process continues to drive heat upward until a layer of neutral static stability is found. If this does not occur until the air reaches

the upper levels of the troposphere, then deep moist convection (DMC) is the likely result.

Heat is also transferred upward via long-wave radiative flux. In this case, one must remember that Wien's law tells us that the wavelength of emission is inversely related to heat intensity. In cases where the greenhouse gasses of water vapor and ozone are highly concentrated, the upwelling long-wave radiation is rapidly absorbed and counterradiated in all directions, with some being directed downward toward Earth's surface. For this reason, humid and/or cloudy nights generally result in greater than average overnight low temperatures. This is especially true in cities, where upwelling radiation is contributed by asphalt surfaces and concrete buildings, not to mention heat input from the buildings' mechanical systems. Ozone levels also tend to be much greater in cities due to automobile traffic, which improves the ability of the air to absorb upwelling long-wave radiation.

So there is little doubt that humans are warming local environments as urban sprawl continues to encroach on rural lands. In some cases, for example, morning low temperatures in cities can be 10° to 15° warmer than in rural areas surrounding the same cities. As urbanization overruns climate stations that had been within rural environments for many years, there is a marked increase in the mean temperatures of those stations simply due to the increase in overnight low temperatures.

Heat and Temperature

Temperature and *heat* are not synonymous terms. Heat represents the total molecular motion of a substance, while temperature represents the average speed of the molecules that strike the temperature sensor. It is said that temperature is a proxy for heat; however, temperature in reality represents only the heat intensity for the volume of substance in which the temperature sensor is immersed. Using temperature to estimate heat energy is generally valid within the troposphere, stratosphere, and much of the mesosphere; however, once the thermosphere is reached, temperature is a poor estimate of heat. This is because the molecules at that altitude move very quickly, since they absorb solar radiation directly with no interference from other molecules, and as a result, the temperature is quite high. However, there are so few molecules that the heat content of the air is small.

Baroclinic Systems

In the middle latitudes, extratropical cyclones are accompanied by frontal systems, and in general, the life cycle of an extratropical cyclone involves the evolution of an occlusion process from the incipient stages of a stationary front. While there are two primary types of extratropical cyclones—the Type A frontal wave cyclone and Type B shortwave cyclone—the Type A cyclone is probably the most common. Type A cyclones begin as thermal advection strengthens along a stationary front, which tightens the pressure gradient and strengthens the wind. As thermal advection continues in response to increasing winds, the temperature gradient tightens and grows vertically into the midlevels. This, in turn, increases the pressure gradient there, and wind speeds also increase. As this process continues to develop upward, the pressure gradient strengthens due to an increasingly sloping pressure surface, and in combination with decreasing density and friction, wind speeds increase to jet streak speeds of more than 35 m/s (meters per second). The jet streaks, due to their accelerations, produce ageostrophic (where there is a vector difference between the real wind and the geostrophic wind) components that generate transverse circulations, so that uplift is found under the left front and right rear quadrants, and subsidence (the descending motion of air in the atmosphere) under the right front and left rear quadrants. While ascent in the left front and right rear quadrants work to increase the pressure gradient and cause the cycle to strengthen even further, subsidence in the right front and left rear quadrants work to weaken the thermal and pressure gradients, and the system begins to weaken, becoming quasi-barotropic (where fronts in the region exist but are weak), with a vertically stacked low-pressure center. This process involves converting potential energy stored within the initial stationary front to the kinetic energy of motion that powers cyclogenesis (i.e., the development and strengthening of a cyclone).

In the case of Type B cyclones, a preexisting shortwave passing through the flow, usually with an accompanying jet streak, initiates cyclogenesis through positive differential vorticity (e.g., clockwise or counterclockwise spin in the troposphere) advection. This increases thermal advection while the cyclone strengthens, resulting in fully developed fronts at the time of occlusion. In other words, while the Type A cyclogenesis represents a "bottom-up" process, a Type B cyclone represents top-down development.

On the other hand, moist deep convection is a thermally direct process by which potential energy, generally building within the PBL due to diabatic heating and thermal and moisture advection, is eventually "released" (i.e., converted from potential to kinetic energy), producing a vertical circulation that forces warm air rapidly upward, where it cools, and cold air rapidly downward, where it warms. This circulation works to minimize the vertical temperature gradient and return the system back to thermal equilibrium.

Global Angular Momentum

The term *angular momentum* simply refers to the speed of a rotating body such as the Earth. Earth's rotation at the equator is generally considered to exhibit a speed of about 1,040 mi. (1,674.4 km) per hour, while the mean angular velocity of that rotation equals 7.2921150×10^{-5} radians per second.

The physical definition of angular momentum (L) is $L = mvr$, where m = mass, v = speed, and r = radius. Angular momentum is a conserved quantity; that means that it will remain constant unless a torque that is external to the system acts on it.

The term *torque* represents the movement of a force about a point, or the tendency of a force to rotate an object about an axis. In many cases, a torque is considered a "twist."

The transfer of angular momentum in the lower troposphere is affected by two independent torques: mountain and frictional torque. Mountain torque, referred to as a normal stress, is associated with a difference in pressure between the eastern and western slopes of north-south orographic barriers. In the Northern Hemisphere, these include the Himalayas and the Rocky

Mountains, while the Andes Mountains are the primary barriers in the Southern Hemisphere. When the pressure gradient is oriented east-west, producing an easterly flow, the mountain torque is positive. When the gradient is oriented west-east, it is negative. Positive mountain torque has been shown to have a dominant impact on global angular momentum (GLAM) on short times scales of between 14 and 20 days. Positive mountain torque will increase the GLAM, actually shortening the length of day (LOD), while negative torque will lengthen it (timescales of ≈ 0.1 ms [milliseconds]). It has been shown that the seasonal dependence of mountain torque variance is proportional to the frequency of midlatitude extratropical cyclones near the Himalayas, Rockies, and Andes mountains, where the largest mountain torques occur. Surprisingly, the Andes Mountains torque is more than two times greater than either of the Northern Hemisphere's primary torques.

Frictional torque, on the other hand, is referred to as a tangential stress and becomes much more important than mountain torque on timescales exceeding about 3 wks. (weeks). It is interesting to note, and quite useful operationally to recognize, that global friction torque peaks about 10 days before global mountain torque. The frictional torque, usually at a maximum near 20° N and 20° S, transfers the Earth's angular momentum to the atmosphere in the presence of organized tropical convection to near 30° N and 30° S, which either increases or decreases the wind speed. Significant increases in wind speed produce the subtropical jet, which, unlike the thermally driven polar front jet, is higher in altitude (33,000–52,000 ft. [feet] above ground level) with generally weaker wind speeds.

Pacific friction torques associated with convective activity over the Indian Ocean/Western Pacific warm pool are closely tied to frictional GLAM changes. This is part of the Madden-Julian oscillation (MJO) that begins with organized convective clusters near Africa, which first migrate eastward to the Indo-Pacific warm pool, then out into the central Pacific to near or past the International Dateline. Rossby wave trains originating from the tropical convective clusters migrating eastward across the Pacific on 40- to 60-day (MJO) cycles dictate the midlatitude weather regimes and intra-annual climate variability. Thus, both heat energy originating from the sun and angular momentum from the Earth's rotation about its axis contribute to the exchange of energy from the equator to the poles that balances the Earth's energy budget.

David L. Arnold

See also Adiabatic Temperature Changes; Air Masses; Anthropogenic Climate Change; Atmospheric Circulation; Atmospheric Circulation; Atmospheric Composition and Structure; Atmospheric Moisture; Atmospheric Pressure; Atmospheric Variations in Energy; Chinooks/Foehns; Climate Types; Climatology; Coriolis Force; Fronts; Hurricanes, Physical Geography of; Monsoons; Thunderstorms; Tornadoes; Wind

Further Readings

Bluestein, H. (1992). *Synoptic-dynamic meteorology in midlatitudes, principles of kinematics and dynamics.* New York: Oxford University Press.

Horvath, K., Lin, Y., & Ivančan-Picek, B. (2008). Classification of cyclone tracks over the Apennines and the Adriatic Sea. *Monthly Weather Review, 136,* 2210–2227.

Marshall, J., & Plumb, R. A. (2008). *Atmosphere, ocean, and climate dynamics: An introductory text.* Burlington, MA: Elsevier/Academic Press.

Weickmann, K. M., & Berry, E. (2009). The tropical Madden-Julian Oscillation and the global wind oscillation. *Monthly Weather Review, 137,* 1601–1614.

Weickmann, K. M., & Sardeshmukh, P. D. (1994). The atmospheric angular momentum cycle associated with a Madden-Julian Oscillation. *Journal of the Atmospheric Sciences, 51,* 3194–3208.

 # ATMOSPHERIC MOISTURE

The atmospheric branch of the hydrologic cycle comprises moisture in three forms: (1) water vapor, (2) condensed (liquid) and sublimated (ice) water in clouds, and (3) precipitation (primarily rain and snow). Although radiation and energy budgets are the fundamental basis of physical

geographic processes, by themselves they are insufficient to generate weather and climate: Moisture is essential. Latent heat absorbed when water evaporates at Earth's surface, and released in condensation when clouds form, links the energy and moisture budgets. Moreover, atmospheric moisture modulates the incoming solar (i.e., shortwave, SW) radiation and Earth-emitted long-wave (LW) radiation streams, significantly affecting the surface net radiation (SW + LW) and energy budget. Much like CO_2 (carbon dioxide) and methane, water vapor is a greenhouse gas (GHG): More of it in the atmosphere reduces the surface diurnal temperature range (DTR) by limiting the overnight temperature drop compared with when there is little water vapor. The liquid water comprising most clouds lowers the clear-sky daytime temperature at Earth's surface and helps heat the atmosphere by absorbing SW radiation. Also, thick or multilayered clouds behave like "blackbodies" for LW radiation; they efficiently absorb and reradiate that energy back to Earth's surface, contributing to the natural greenhouse effect and, thereby, reducing the DTR compared with cloud-free skies. Clouds are integrators and tracers of atmospheric moisture and energy (Figure 1), and their large-scale patterns show associations with climatic teleconnections, particularly the El Niño Southern Oscillation (ENSO).

Most clouds exert a negative radiative forcing (i.e., the surface temperature is lower in their presence) because their reduction of the SW radiation received at Earth's surface exceeds their enhancement of the greenhouse effect. Cirrus clouds are the exception: They both transmit SW radiation and enhance the greenhouse effect, for most situations. Precipitation influences the atmospheric energy budget by cooling the air via evaporation.

Compared with Earth's surface and subsurface components of the water budget, atmospheric moisture amounts are very small (<0.05% of all freshwater), and residence times are only on the order of days; yet they are highly significant, participating directly in both weather and climate fluctuations. Considered globally, the dominant surface sources (sinks) of atmospheric moisture are the oceans, especially in the tropics and subtropics (land surfaces, especially in the subtropics and middle latitudes). Because the atmospheric storage of moisture is negligible, the water budget can be estimated as the difference between evaporation (E) and precipitation (P). Maps of the long-term, annually averaged E-P show maximum positive values (i.e., E ≫ P) over subtropical oceans, and maximum negative values (E ≪ P) in the tropics and over middle-latitude oceans and adjacent land areas. Values are weakly negative over polar regions and weakly positive over subtropical land areas and the interior midlatitude deserts. The vertical exchanges of moisture (as E, P) between land surfaces and atmosphere, or "recycling," characterize the tropics all year and middle latitudes in summer. Elsewhere and at other times (e.g., middle latitudes in winter), the quasi-horizontal movement of moisture by the winds (i.e., advection) from warmer and/or moister regions dominates.

Characteristics

Most atmospheric moisture is found within the planetary boundary layer (PBL), or lowermost 1,000 m (meters) or so of the atmosphere, with lesser amounts in the free atmosphere. This arises because of the PBL's proximity to surface moisture in oceans, lakes, rivers, vegetation, and soil and the positive net radiation (SW + LW > 0) required for evaporation. The atmospheric water vapor content is expressed by a number of humidity measures, including the relative humidity (RH, %), specific humidity (q, in grams [g] per kilogram of dry air), absolute humidity (in g/m³ [cubic meter]), vapor pressure (in hPa [hectopascals]), and column-integrated precipitable water (PW, in mm [millimeters] of centimeters). All measures except RH give actual moisture amounts; the RH depends on the air temperature, which determines the maximum amount of water vapor that the air can hold (i.e., saturation). This Clausius-Clapeyron relationship is positive and nonlinear (i.e., warm air holds much more water vapor than cold air). Thus, latitude plots of the annually averaged absolute humidity or specific humidity show a maximum in the tropics and minima at the poles, whereas the relative humidity has two maxima (tropics, middle latitudes) and two minima (subtropics, polar regions). Seasonally, the hemisphere experiencing summer has the highest water vapor

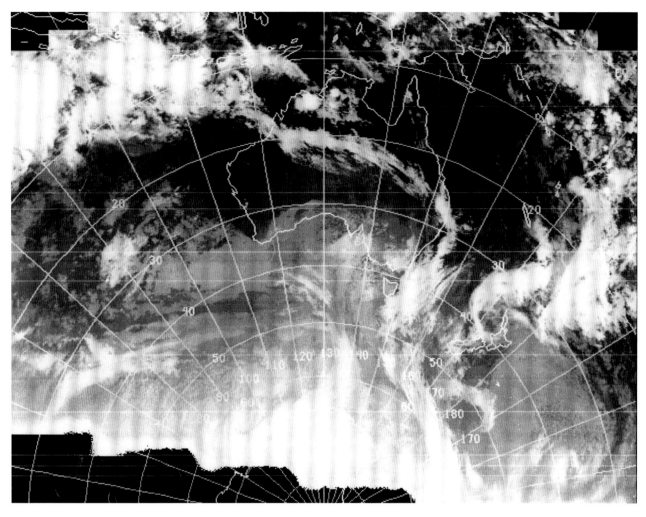

Figure 1   Sample satellite long-wave (LW) (cloud) image of the Australian sector (Geostationary Meteorological Satellite [GMS-5] of the Japanese Meteorological Agency)

Source: Australian Bureau of Meteorology, Melbourne (Peter Boemo).

amounts, because of the higher temperatures. Spatially, the area of greatest atmospheric moisture content is colocated with the "warm pool" of highest sea-surface temperatures (SSTs) in the tropical western Pacific, demonstrating the temperature-evaporation-humidity feedback. Moreover, the relationship between humidity and cloud cover is positive; high humidity in a given atmospheric layer increases the likelihood of clouds forming or persisting there. Indeed, in computer models of weather and climate that do not explicitly predict clouds, simulated high humidity values are a proxy for cloud cover.

The lapse rate of temperature with height increase ($-dT/dZ$) is combined with moisture in measures of the static stability, or the propensity for air to rise (unstable), sink (stable), or remain at the same level (neutral) after removal of a mechanism that initially displaces air upward (i.e., solar heating of Earth's surface; orographic uplift). Stability indices denote not only the dominant cloud type (cumuliform or stratiform) but also the probability of severe weather (thunderstorms, hail, tornadoes). Unstable air dominates over land in the warm season, helping move PBL moisture into the free atmosphere via the development of cumuliform (convective) clouds. The convective instability of air is greatest for high surface temperatures, large PBL moisture, and dry air aloft. Stable conditions prevail over cool surfaces (i.e., temperature inversion), the free atmosphere dries because of sinking air, and PBL clouds are stratiform with any precipitation being light (e.g., drizzle).

Horizontal contrasts in lower-atmosphere humidity occur on a range of scales, from individual fields (e.g., irrigated agriculture adjacent to dryland; forest adjacent to crops), to regional and larger scales. For example, the *dry line* of the Central United States forms when warm, dry air from the plateaus of the southwest deserts overruns moist air from the Gulf of Mexico. The dry line promotes convective instability and severe weather development in the Great Plains. Both humidity and temperature contrasts characterize air-mass fronts. The widespread lifting of warm, moist air along a front, or *slantwise convection*, greatly increases the precipitation efficiency relative to cool air. Particularly in the colder season, moisture incorporated into frontal cyclones can originate within the tropics as deep convection and be carried to subtropical and middle latitudes, and even to polar regions, by the jet streams and fronts as very long yet narrow *tropical-extratropical cloud bands* (TECBs), or *moisture bursts* (Figure 1). These features may produce widespread flooding. The climatology of TECBs varies between hemispheres and also according to teleconnections, particularly ENSO.

On synoptic scales in the free atmosphere, spatial variations in water vapor—including RH—and clouds show associations with the vertical motion of air, as follows: (a) ascent, or *uplift*, and moistening within low pressure, and (b) descent, or *subsidence*, and drying in high pressure. Similarly, high clouds accompany rising air on the equatorward side of a jet stream; clear skies or only low clouds occur on the poleward side, owing to subsidence (e.g., over Central Australia in Figure 1).

On hemispheric to global scales, the advection of moisture by the vector wind (V), or *moisture transport* ($q\mathbf{V}$), is a consequence of the atmospheric general circulation's (GC) continual attempt to balance Earth's heat and momentum budgets across latitude zones. On average, moisture is transported poleward everywhere except in the lower atmosphere within the tropics; there, the trade winds originating in the subtropical highs supply moisture to the Intertropic Convergence Zone (ITCZ) of low pressure, realized as deep convection and latent heat release in the thunderstorm *hot towers* that power the ascending limb of the Hadley cell. The total (i.e., time and space averaged) $q\mathbf{V}$ can be separated into its *mean* and *eddy* components, which vary in importance by latitude and also by season. The mean circulation dominates in the tropics all year (Hadley cell), while eddy transport by the planetary waves and traveling high- and low-pressure systems dominates the extratropics, especially in the colder season. In middle latitudes, the moisture transport occurs rapidly along elongated yet narrow *atmospheric rivers*, ranging in size from mesoscale *low-level jets* to TECBs. Calculation of the net movement of moisture into or out of an atmospheric column due to the horizontal wind determines whether moisture flux convergence (i.e., ascending air and moistening) or divergence (subsidence and drying) is occurring. In the PBL, frictional contrasts between major land cover types (e.g., forest adjacent to crops), and along air density boundaries resulting from prior convection, encourages moisture convergence and renewed convection, which may result in precipitation.

Monitoring

Atmospheric moisture is monitored using direct observation and also remotely by satellites and radar. At Earth's surface, temperature, moisture (RH, dew point temperature), and precipitation observations are made at least once daily at manned weather and climate stations; these quantities, along with clouds, are monitored more frequently at higher-order and automatic weather stations. However, site characteristics significantly influence the measurements of temperature and humidity, such as when measurements are taken at a place that differs markedly from a standard grass surface (e.g., adjacent to a parking lot) or when they are taken at a nonstandard height, such as the top of a building, as well as precipitation where gauges are inadequately exposed (e.g., when partially covered by a roof or overhang). Spurious climate changes on local scales can result from these biases. Vertical variations of moisture within and above the PBL, along with temperature, pressure, and winds, are determined twice daily from rawinsondes launched mostly from land stations. Retrieved humidity values are most reliable in the lower-to-middle troposphere; in the upper troposphere, low temperatures and small moisture amounts can result in substantial errors. Thus, detecting climate changes in moisture at high altitudes using rawinsonde data is problematic.

Although of shorter duration (since 1979), global observations of atmospheric layer-averaged RH and the PW made operationally from satellites are only limited in areas of frequent thick clouds and over ice and snow surfaces. Retrieval of the humidity in cloudfree or only partly cloudy conditions relies on the absorption and reemission of LW radiation by water vapor (thermal infrared [IR] sensing, microwave radiometry); in the midtroposphere (approximately 600- to 400-hPa pressure levels), this reemission of LW radiation peaks around 6.7 μm (micrometers). Determining water vapor amounts in layers closer to Earth's surface—important for prediction of severe weather—uses slightly longer IR wavelengths. Information on atmospheric water vapor is also acquired using global positioning system technology: The electronic signal is delayed as PW increases. Cloud detection in visible wavelength data relies on the increased reflectance of solar radiation to the satellite sensor by clouds over low reflectance Earth surfaces (ocean, snowfree land). To retrieve cloud information on climatically significant attributes such as cloud-top temperature and cloud height (the gray tones in Figure 1), the IR window region (8–14 μm) is used: The emission from a blackbody radiator decreases with decreasing temperature (Stefan-Boltzmann's law), and cloud temperature typically decreases with increased cloud height (normal lapse rate). Other climatically important cloud attributes, particularly the cloud liquid water content (CLW), rain occurrence and rain rate (mm/hour), and the presence of hail and large

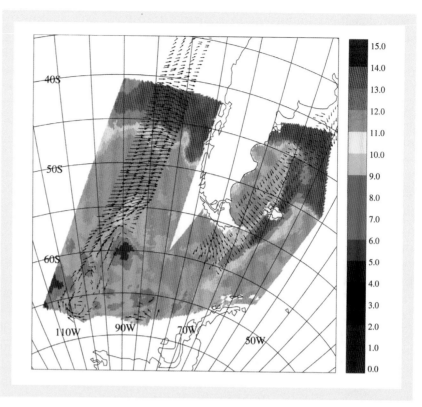

Figure 2   Sample SSM/I water vapor (in kilograms per square meter) and satellite radar-determined V, with Southern South America for reference

Source: Defense Meteorological Satellite Program (DMSP) image processed by Lynn McMurdie (University of Washington) and Andrew M. Carleton.

snow aggregates in the upper parts of deep convective clouds, use passive microwave radiometry, notably, the Special Sensor Microwave/Imager (SSM/I): lower frequencies (i.e., longer wavelengths) for CLW and rain; higher frequencies (shorter wavelengths) for ice scattering. Combining spatial information on water vapor with near-surface winds (**V**)—the latter determined remotely from the wind's effect on a water surface—can depict mesoscale PBL moisture convergence and precipitation (the area near 55° S, 90° W in Figure 2).

High-resolution vertical profiles of cloud types and cloud structure, revealed by CLW, cloud ice content, and precipitation retrievals along a narrow subsatellite track, have been acquired by the cloud profiling radar on NASA's CloudSat for most latitudes since 2006. For cirrus clouds, lidar (light detection and ranging) based at Earth's surface and also satellite platforms yields information on cloud height and thickness as well as dominant ice crystal attributes (characteristic sizes, densities) important in radiative transfer studies. Compared with passive microwave retrievals of rain, satellite-based radar—such as the tropical rainfall measuring mission (TRMM)—generates higher-resolution rain-rate estimates for land as well as ocean, along with three-dimensional images of precipitating cloud clusters and mesoscale convective systems (MCSs).

Climatology

Spatial variations in annually and seasonally averaged atmospheric moisture parameters depict broad associations with the GC. The zones of maximum PW, cloud amount (e.g., from satellite-retrieved CLW), and precipitation rate located in the tropics are associated with convergence of air and deep convection in the ITCZ. In the middle latitudes, slantwise convection associated with the extratropical cyclones gives the secondary maxima in RH, clouds, and precipitation (e.g., Figure 2). Zones of low RH and a lack of deep precipitating clouds—the subtropics and polar regions, particularly Antarctica—are associated with sinking air, in the subtropical highs, where air subsides and warms adiabatically over a considerable depth, and near the poles, where cold, dry air is located at Earth's surface, respectively. Seasonally, cloud and precipitation zonal maxima and associated circulation features either shift latitude (the ITCZ, which follows the zone of maximum heating, or *thermal equator*) or change intensity (middle-latitude storm tracks are more pronounced in winter and weaker in summer). Longitudinal variations in humidity and cloud frequency accompany the summer monsoons of West Africa, South Asia, and also Northern Australia and Indonesia and the West Pacific—the so-called maritime continent (Figure 1).

Interannual variations in atmospheric moisture are large and dominated by ENSO. Over the tropical Pacific, water vapor increases during El Niño events relative to non-ENSO and La Niña conditions. The deep convection and heavy precipitation associated with the warm pool shift eastward to the international dateline and also increase in the tropical eastern Pacific during El Niño as the subtropical high weakens. There, the normally stable atmosphere containing extensive

yet shallow marine stratus is disrupted by convection resulting from greater surface receipts of solar radiation and increased SSTs.

Climatologies of rain rate for the low, subtropical, and lower-middle latitudes yielded by satellite microwave data (SSM/I, TRMM) are especially improved for ocean areas poorly sampled by conventional data. Moreover, the TRMM climatologies both confirm and add detail to the large spatial and magnitude variations in tropical rain rates associated with ENSO. Analysis of thermal IR "enhanced" images and SSM/I data on ice scattering have permitted development of climatologies of MCSs for a number of lower-latitude regions.

Changes

Recent (past 30–50 years) changes in atmospheric moisture include, on large scales, an increase in lower-atmosphere humidity for the middle latitudes accompanying reductions of DTR, especially in the warmer season, and a moistening of the normally dry lower stratosphere. For the 20th century considered as a whole, land-based total cloud amounts increased and precipitation increased generally in middle latitudes and, to an extent, in the tropics, while it decreased in many subtropical locations. Although the role of human activities in these large-scale changes is debated, particularly their connection to GHG-induced "global warming," some regional-to-local-scale changes are more definitively tied to humans. These include increased summertime dew point temperatures in the intensively cropped Midwest of the United States and increased precipitation adjacent to areas of extensive irrigation in the High Plains (i.e., intensified recycling of moisture due to agriculture); reduced recycling in areas of tropical deforestation; increases in cirrus cloud cover in regions of frequent air traffic (from contrails); greater reflectivity of marine stratus due to pollution from ship stacks; and increasing convective precipitation downwind of large urban areas (from combinations of the urban heat island effect and greater atmospheric loading of particulates) yet suppression of frontal precipitation at locations downwind of major industrial areas. The last results from air pollution particles enhancing the formation of cloud droplets (cloud reflectance increases), but these droplets have a smaller size than is optimal for precipitation formation. Some of the observed changes in atmospheric moisture, at least those on larger scales, are consistent with the intensified hydrologic cycle predicted by climate models under global warming.

Andrew M. Carleton

See also Anthropogenic Climate Change; Atmospheric Circulation; Atmospheric Energy Transfer; Atmospheric Remote Sensing; Atmospheric Variations in Energy; Clouds; Cyclones: Occluded; El Niño; Humidity; Hurricanes, Physical Geography of; Hydrology; Latent Heat; Monsoons; Precipitation Formation

Further Readings

Carleton, A., Travis, D., Master, K., & Vezhapparambu, S. (2008). Composite atmospheric environments of jet contrail outbreaks for the United States. *Journal of Applied Meteorology and Climatology, 47,* 641–667.

Dai, A., Trenberth, K. E., & Karl, T. R. (1999). Effects of clouds, soil moisture, precipitation, and water vapor on diurnal temperature range. *Journal of Climate, 12,* 2451–2473.

Daly, C., Gibson, W., Taylor, G., Doggett, M., & Smith, J. (2007). Observer bias in daily precipitation measurements at United States cooperative network stations. *Bulletin of the American Meteorological Society, 88,* 899–912.

DeRubertis, D. (2006). Recent trends in four common stability indices derived from U.S. radiosonde observations. *Journal of Climate, 19,* 309–323.

Dirmeyer, P. A., & Brubaker, K. L. (2007). Characterization of the global hydrologic cycle from a back-trajectory analysis of atmospheric water vapor. *Journal of Hydrometeorology, 8,* 20–37.

Mo, K. C., Chelliah, M., Carrera, M. L., Higgins, R. W., & Ebisuzaki, W. (2005). Atmospheric moisture transport over the United States and Mexico as evaluated in the NCEP regional reanalysis. *Journal of Hydrometeorology, 6,* 710–728.

Mohr, K. I., & Zipser, E. J. (1996). Mesoscale convective systems defined by their 85 GHz ice scattering signature: Size and intensity comparison over tropical oceans and continents. *Monthly Weather Review, 124,* 2417–2437.

Neimann, P. J., Ralph, F. M., Wick, G. A., Lundquist, J. D., & Dettinger, M. D. (2008). Meteorological characteristics and overland precipitation impacts of atmospheric rivers affecting the west coast of North America based on eight years of SSM/I satellite observations. *Journal of Hydrometeorology, 9,* 22–47.

Rosenfeld, D. (2000). Suppression of rain and snow by urban and industrial air pollution. *Science, 287,* 1793–1796.

Sandstrom, M. A., Lauritsen, R. G., & Changnon, D. (2004). A Central-U.S. summer extreme dew-point climatology. *Physical Geography, 25,* 191–207.

Sassen, K., & Wang, Z. (2008). Classifying clouds around the globe with the CloudSat radar: 1-year of results. *Geophysical Research Letters, 35,* L04805.

ATMOSPHERIC PARTICULATES ACROSS SCALES

Atmospheric particulates, also called particulate matter (PM), are a complex mixture of extremely small solid particles and liquid droplets suspended in the air. Particles are made up of several components, including acids (i.e., nitrates and sulfates), organic chemicals, metals, and soil or dust. PM sources can be anthropogenic or natural. Natural sources include volcanoes, dust storms, vegetation fires, and sea spray. Anthropogenic sources include coil and oil combustion, fossil-fuel power plants, and other industrial activity. The anthropogenic contribution to total atmospheric PM is estimated to be approximately 10%. Some of these particles are emitted directly into the air (primary emissions), while some are emitted as gases and subsequently form particles in the air (secondary emissions).

PM is categorized by size fractions. As particles are often nonspherical, categorization depends on their aerodynamic diameter. A particle with an

Fraction	Size Range (μm)	Residency Time in Air	Travel Distance
PM$_{10}$ (thoracic fraction)	≤10	Minutes to hours	Up to tens of kilometers
PM$_{10}$–PM$_{2.5}$ (coarse fraction)	2.5–10		
PM$_{2.5}$ (respirable fraction)	≤2.5	Days to weeks	More than hundreds of kilometers
PM$_1$	≤1		
Ultrafine (UFP)	≤0.1		

Table 1   Particulate characteristics

Source: Author.

Note: PM = particulate matter.

aerodynamic diameter of 10 μm (micrometers) moves in the air like a sphere with a diameter of 10 μm. PM diameters range from less than 10 nm (nanometers) to more than 10 μm. These dimensions represent the continuum from a few molecules up to the size where particles can no longer be carried by the atmosphere and settle because of gravity or turbulent impact on the surface. A typical notation such as PM$_{10}$ is used to describe particles of 10 μm or less, which are called *coarse* particles; and PM$_{2.5}$ represents *fine* particles equal to or less than 2.5 μm in aerodynamic diameter. Particles between these two sizes are considered inhalable coarse particles. Everything below 100 nm, down to the size of individual molecules, is classified as ultrafine particles (UFP). In comparison, an average human hair is about 70 μm in diameter.

In general, the smaller and lighter a particle is, the longer it will stay in the air. Larger particles (greater than 10 μm in diameter) tend to settle to the ground in a matter of hours, whereas the smallest particles (less than 1 μm) can stay in the atmosphere for weeks and are mostly removed by precipitation (see Table 1).

The composition of particulates is determined by their source. Wind-blown mineral dust tends to be made of mineral oxides and other material, whereas sea salt consists mainly of sodium chloride originated from sea spray yet can include magnesium, sulfate, calcium, and potassium.

Secondary particles derive from the oxidation of primary gases such as sulfur and nitrogen oxides into sulfuric acid (liquid) and nitric acid (gaseous). The precursors for these acid aerosols may have an anthropogenic origin (e.g., coal combustion) and/or a biogenic origin. The chemical composition of PM affects how it interacts with solar radiation, determining how much light is scattered and absorbed. In this manner, particulates can influence the global climate. Recent evidence suggests that anthropogenic PM can control cloud development by altering the size distribution of cloud condensation nuclei.

The effects of inhaling PM have been widely studied in humans and animals and include asthma, lung cancer, cardiovascular disease, and premature death. The large number of deaths and other health problems associated with particulate pollution was first demonstrated in the early 1970s. As a result, PM is subject to strict regulatory control.

The size of the particle is a main determinant of where in the respiratory tract it comes to rest when inhaled. Larger particles are generally filtered in the nose and throat and do not cause serious health problems, but PM$_{10}$ can settle in the bronchi and lungs and damage human health. PM$_{10}$ does not represent a strict boundary between respirable and nonrespirable particles but is used as a standard for monitoring of airborne PM by most regulatory agencies. Similarly, PM$_{2.5}$ tends to penetrate deep into the gas-exchange regions of the lungs, and very small particles (<100 nm) can pass through the lungs to affect other organs. In particular, PM$_{2.5}$ leads to high plaque deposits in the arteries, causing vascular inflammation and atherosclerosis, which can lead to heart attacks and other cardiovascular problems. Research suggests that even short-term exposure at elevated concentrations could significantly contribute to heart disease.

The smallest particles, less than 100 nm (nanoparticles) in size, may be even more damaging to the cardiovascular system. There is evidence that particles smaller than 100 nm can pass through cell membranes and migrate into other organs, including the brain. It has been suggested that PM can cause similar brain damage as that found in Alzheimer patients. In addition, particulates also carry carcinogenic components such as benzopyrenes adsorbed on their surface. One particle of 10-μm diameter has approximately the same mass as 1 million particles of 100-nm diameter, but it is clearly much less hazardous, as it probably will never enter the human body—but if it does, it can be quickly removed.

Because of the human health effects of PM, maximum ambient standards have been set by governments (the World Health Organization's guideline is that daily averaged PM_{10} should not exceed 50 μg/m³ [micrograms per cubic meter] of air). Many urban areas in the developed world still frequently violate PM standards, especially in Southern California and Pittsburgh, Pennsylvania, in the United States and in Belgium, Germany, Italy, and Albania, although in many areas there has been a downward trend. Much of the developing world exceeds the standards by a wide margin. The most concentrated PM pollution tends to be in densely populated metropolitan areas, especially Cairo (Egypt), Delhi and Kolkata (India), and Tainjin, Chongqing, and Linfen (China), all with PM levels commonly above 100 μm/m³. Thus, PM pollution control programs are very important worldwide.

Peyman Zawar-Reza

See also Ambient Air Quality; Atmospheric Pollution; Clouds; World Health Organization (WHO)

Further Readings

Jacobson, M. (2002). *Atmospheric pollution: History, science, and regulation.* New York: Cambridge University Press.

Seinfeld, J. H., & Pandis, S. N. (2006). *Atmospheric chemistry and physics: From air pollution to climate change.* New York: Wiley-Interscience.

ATMOSPHERIC POLLUTION

Atmospheric pollution describes harmful gases, solid particles, and liquid droplets that are present in the atmosphere in quantities above natural ambient levels. The main gases in the atmosphere are nitrogen (78.08%), oxygen (20.95%), and argon (0.93%). Carbon dioxide, a greenhouse gas, accounts for 0.038%. Given this composition, pollutants are present in very small proportions, usually measured as the mixing ratio of the number of molecules of the pollutant per total number of air molecules, such as parts per million (ppm) or parts per billion (ppb), or as a mass concentration per unit volume of air, such as micrograms per cubic meter (μg/m³). Each year, millions of tons of pollutants are emitted into the atmosphere, and the effects on the health of the planet and its life forms can be devastating. Pollutants may be "primary" in origin, if they are emitted directly from a source, or "secondary," if they are products of chemical reactions involving primary pollutants.

Air quality is directly linked to the levels of pollutants in the atmosphere and is rated on the basis of air quality indices. The most pervasive and toxic compounds, known as *criteria air pollutants*, are used to compute such indices. Though they vary with jurisdiction, the most common or "criteria" air pollutants (pollutants for which criteria have been established for their level and length of exposure to protect human health and welfare) are carbon monoxide, oxides of nitrogen, ozone, sulfur dioxide, volatile organic compounds, and fine particles. While some of these compounds occur naturally, anthropogenic activities significantly increase the levels of pollutants, introduce many of which that do not occur naturally, and elevate concentrations in populated areas. Particularly high concentrations are usually observed in urban areas where most pollution results from industrial activities and transportation, leading to enhanced human exposure.

Criteria Air Pollutants

Carbon Monoxide

Carbon monoxide (CO) is a by-product of fuel combustion, resulting from incomplete oxidation

Mexico City, which is partially surrounded by mountains, suffers from prolonged temperature inversions and severe smog episodes.

Source: Morguefile.

of carbon in the combustion process. In North America, about 80% of CO emissions come from internal combustion engines used in transportation. Natural sources include forest fires (which may also be anthropogenically driven) and volcanoes. Indoor sources of CO include wood burning and gas stoves or furnaces.

On a global scale, CO levels average about 100 ppb. In urban areas, the levels are elevated to a few ppm, while in traffic, values may be measured in tens of ppm. CO is a precursor to greenhouse gases such as methane and carbon dioxide.

In the human body, CO is attracted to hemoglobin in the bloodstream, replacing oxygen. Exposure to low levels may cause fatigue in healthy people and is more dangerous for persons with heart disease. High levels can be fatal and because of its colorless and odorless characteristics, it is often called a "silent killer."

Oxides of Nitrogen

Harmful oxides of nitrogen include nitric oxide or nitrogen monoxide (NO) and nitrogen dioxide (NO_2), which are collectively called NO_x. High energy is required to break the strong triple bond of molecular nitrogen to generate elemental nitrogen for oxidation. Hence, the formation of nitric oxide generally results from high temperature processes such as fuel combustion in transport engines or industries, such as those involved in thermal power generation. In North America, approximately 60% of anthropogenic NO_x gases result from transportation emissions and about 35% from industry, including electric power plants. Natural sources include lightening and microbial activity in soils. Nitric oxide has a lifetime of minutes, so once emitted it quickly oxidizes to NO_2. NO_x species play pivotal roles in the complex chemistry of the atmosphere, reacting with other gases, for example, in the creation and depletion of ground-level ozone.

Both nitric oxide and NO_2 are soluble in water. Nitric oxide creates weak nitrous acid (HNO_2), and NO_2 dissolves to create strong nitric acid (HNO_3). These gases contribute to the acidity in rain, ground, and surface waters and soils. The acids may also react with other compounds, such as ammonia, to form nitrate particles that are common components of particulate matter in the atmosphere.

NO_2 is a toxic gas that has detrimental health effects on the respiratory system, can damage the lining of the lungs, and exacerbate diseases such as asthma and bronchitis. The acids will irritate eyes and skin and cause damage to the respiratory system, while the particles contribute to health effects of particulate matter. The acids will also damage plants and infrastructure.

Ozone

Ozone (O_3) is a secondary pollutant created by reactions between precursor gases such as NO_2 and volatile organic compounds. It is one of the most consistently monitored pollutants, globally. O_3 occurs both in the stratosphere, where it forms a protective layer to shield the Earth's surface from ultraviolet radiation, and in the troposphere, where it has detrimental effects on biota.

Sunlight is an essential component of O_3 production. Photolytic reactions use energy from the sun to dissociate NO_2, producing nitric oxide and an oxygen atom (R1). The oxygen atom reacts with molecular oxygen, O_2, to form ozone, O_3 (R2). Nitric oxide will also react readily with O_3 to produce NO_2 and oxygen gas (R3), thereby reducing the levels of O_3.

$$NO_2 + sunlight \rightarrow NO + O \text{ (R1)}$$
$$O + O_2 \rightarrow O_3 \text{ (R2)}$$
$$O_3 + NO \rightarrow O_2 + NO_2 \text{ (R3)}$$

The reactions often involve other gases such as volatile organic compounds. This cycle of reactions is most prolific in urban areas, where traffic and industry provide abundant precursor gases. The scavenging of O_3 by NO in high traffic areas leads to depletion in urban areas compared with rural locations.

Because of the sun's role, O_3 levels have diurnal and seasonal cycles, with highest concentrations on a summer day. O_3 can be transported for thousands of miles by prevailing winds, leading to high levels at downwind locations.

The O_3 level in unpolluted air is approximately 10 to 15 ppb. In polluted environments, this may rise to more than 100 ppb. O_3 irritates the respiratory system and aggravates diseases such as asthma and bronchitis. It is a primary component of "smog," a toxic mix of pollutants that causes poor air quality and severely impairs cardiopulmonary function. Severe smog episodes can lead to premature death.

Sulfur Dioxide

Sulfur dioxide (SO_2) results from the burning of compounds that contain sulfur. Coal is a primary source, and thus, coal-fire generating plants emit high levels, as do smelters, which extract metals from sulfur-bearing ores. Some fuels, such as those used in marine or rail transportation, also have high sulfur content. However, industry and electric power plants have traditionally been the major source.

SO_2 is soluble in water and forms sulfuric acid (H_2SO_4), creating acid rain that can devastate ecosystems. Further reactions will lead to the formation of sulfate particles in the air. This pollutant adversely affects the human respiratory system, worsening symptoms of lung diseases. In North America, emissions have been reduced significantly in the past 15 to 20 years but still remain a concern.

Volatile Organic Compounds

Volatile organic compounds (VOCs) include a large suite of carbon-based chemicals that evaporate readily under normal temperature and pressure conditions. While thousands of these compounds exist, about a hundred are considered when assessing air quality. Evaporated fuels such as those used in transportation and industry are the major contributor. Other significant sources include solvents used in commercial and industrial processes and in products such as cleaning fluids or paints. Natural sources include vegetation, such as plants that emit isoprene. VOCs are primary precursors for the creation of ground-level O_3. Studies suggest that some are cancer-causing agents and can aggravate asthma and other lung diseases.

Fine Particulate Matter

Particulate matter describes solid or liquid particles in the atmosphere. Most particles are composed of sulfates, nitrates, carbon, soil particles, and organic materials such as pollen. Particulate matter is ubiquitous, with numerous anthropogenic sources that include industry, power plants, construction sites, road dust, open fields, and diesel fuel combustion. Secondary particles result from chemical interactions, which convert gases, such as SO_2 or NO_x, to sulfate and nitrate particles.

Natural sources include sea salt, desert sand, volcanoes, and carbon particles from forest fires. The finer particles are easily transported over long distances, and dust from the Sahara Desert has been tracked by satellite to locations in the middle United States and Northern Europe.

Particulate matter exists in various shapes and sizes and is classified according to the aerodynamic diameter of the particles, measured in micrometers (μm). Particles of diameter less than 10 μm (PM_{10}) and less than 2.5 μm ($PM_{2.5}$) are of particular concern. Unpolluted $PM_{2.5}$ levels may be 0 to 5 $\mu g/m^3$, while polluted areas may have values exceeding 50 $\mu g/m^3$. Numerous epidemiological studies suggest that the finer particles penetrate and settle deep into the respiratory tract, causing inflammation. They are considered to be among the most detrimental air pollutants and have been implicated in cardiopulmonary diseases and premature death. They are also a primary component of smog.

Dispersion and Dilution

Dilution and dispersion of pollutants are key factors in reducing concentrations near the source. Meteorology and physiography can assist or inhibit dispersion rates. Winds transport pollutants vertically and horizontally away from the source and allow dilution through mixing with less polluted air. While reducing pollution near the source, transport increases concentrations downwind. If the air becomes stable, as in the case of temperature inversions, pollution levels rise quickly. Physiographic features such as basins or valleys restrict airflow and also promote the formation of temperature inversions. Locations such as Los Angeles and Mexico City, which are partially surrounded by mountains, suffer from prolonged temperature inversions and severe smog episodes.

Monitoring Air Pollution

In an effort to mitigate the health effects of air pollution, many countries establish air quality criteria for selected contaminants, which are monitored regularly. Monitoring can be accomplished in various ways. Most commonly, fixed systems comprising gas and particulate matter monitors are strategically placed in urban or rural environments to measure pollution levels continuously. This system works well to monitor local pollution levels near the monitoring unit and to record the long-term pollution patterns at that location. Mobile systems, which comprise monitors mounted on aircraft or motor vehicles, can be used for periodic measurements over selected areas. These systems offer flexibility in selecting the time, location, and extent of the area to be monitored. Personal monitors are also available to assess an individual's exposure level. On a regional or global scale, air pollution can be detected by satellites, which orbit the Earth. These data are useful for monitoring pollution levels over large urban centers, tracking long-range transport of pollutants, and assessing the effects on global climate.

Atmospheric pollution also can be modeled using algorithms developed for computer simulations. Estimates of concentrations can be determined based on information about the pollution source and meteorology. These models can be used to simulate pollution effects under different scenarios, and in climate change models.

Reducing Air Pollution

The personal costs of air pollution include illness and premature death. There are also economic costs associated with health care and lost productivity as well as environmental damage and climate change. Efforts to reduce emissions include the development of new technologies such as hybrid or fully electric engines and the hydrogen fuel cell. Alternatives such as solar, wind, wave, and tide energy, and biofuels continue to be researched and developed. Legislation in some jurisdictions mandates cleaner fuels, and many countries encourage reduced reliance on car use. While there have been ongoing efforts to reduce pollution in North America and Europe, rapidly developing countries have increased pollution output immensely in recent years, leading to greater need for global cooperation to maintain clean air for health and the total environment.

Pavlos Kanaroglou and Julie Wallace

See also Acid Rain; Ambient Air Quality; Anthropogenic Climate Change; Atmospheric Particulates Across Scales; Greenhouse Gases; Point Sources of Pollution; Urban Environmental Studies; Urban Heat Island

Further Readings

Gibilisco, S. (2006). *Alternative energy demystified.* New York: McGraw-Hill Professional.

Office of Air and Radiation, United States Environmental Protection Agency: www.epa.gov/air

World Health Organization. (2006). *Air quality guidelines global update 2005.* Retrieved November 9, 2009, from www.euro.who.int/Document/E90038.pdf

ATMOSPHERIC PRESSURE

The most basic atmospheric element is air pressure. Pressure itself is a measure of the force per unit area. The concept of force is usually best understood by considering Newton's second law, which states that the acceleration of a body is directly proportional to the force exerted on that body and inversely proportional to its mass. In other words, force (F) is equal to the product of mass (M) and acceleration (a).

Using International System (SI) units of measure, M is given in kilograms (kg) and acceleration in units of meters per second per second (m/s^2, also expressed as m s^{-2}). Given $F = Ma$, then force is described in units of kg m s^{-2}, which is a newton. A newton (N) is equal to the amount of force required to accelerate a mass of 1 kg at the rate of 1 m/s^2, or 1 m s^{-2}. So if force is given in newtons, then pressure must equal to N per unit area, or N/A or N A^{-1}. This unit of measure for pressure is known as a pascal (Pa). Given that a unit area would be provided in square meters (m^2), a pascal would be equal to kg m s^{-2} m^{-2}, or kg m^{-1} s^{-2}.

In general, the mass of the atmosphere varies slightly depending on location. When dealing with air, however, we need to consider Pascal's law, which states that fluids (gases and liquids) transmit pressure in all three dimensions. For this reason, mass of the atmosphere varies depending on the volume of space considered, so to compare pressure across space, one must standardize the unit of measure to a constant volume, a measure known as "density." Typically, the mass is given in kg and the volume in cubic meters (m^3); therefore, we consider atmospheric mass, or density, in units of kg/m^3 (or kg m^{-3}).

Gravitational acceleration, the other term in force, also varies depending on distance away from the center point of Earth. If Earth were a perfect sphere, then gravitational acceleration would be constant at the same height. However, even though Earth is best described as a *geoid*, the variation in gravitational acceleration due to a nonspherical body within all heights of the troposphere are negligible. For this reason, and to simplify the situation, we assume gravitational acceleration within the troposphere to be a constant 9.8 m/s^2. To ensure that this assumption is recognized, all heights above Earth's surface are referred to as "geopotential"; that is, with a vertical datum of mean sea level on a spheroid-shaped Earth.

Atmospheric pressure, then, is the summation of all mass within a 1-m^2 column of air that extends to outer space, where pressure is 0 Pa. On average, if we consider all sea level point locations over Earth 24 hrs. (hours) per day for 30 yrs. (years) or more, the pressure would be 101,325 Pa. Since geographers and atmospheric scientists frequently work with pressure, the SI units are simply too large to deal with on an operational basis. As a result, a new unit of measure called the millibar (mb) was developed in 1909 by Sir Napier Shaw. An mb equals 100 Pa, so average atmospheric pressure at sea level is 1,013.25 mb, or simply 1,013 mb.

Another common unit of measure for air pressure is inches (in.), used primarily for aviation purposes and included on most home barometers. A barometer is the instrument used to measure atmospheric pressure. First developed by Torricelli in the 17th century, the instrument includes a 30-in.-tall glass filled partially with mercury. The mercury rises and falls in response to changing pressure, so that the average mean sea level pressure produces a mercury column height of 29.92 in. from its base.

Pressure Gradients

Atmospheric motion is a function of imbalance in mass and/or heat across space. In other words, the Earth-atmosphere system is always working to achieve a balance based on the spatial distribution of mass and heat energy. When imbalances exist, surplus mass or heat is moved by forces from regions of surplus to regions of deficit until equilibrium, or at least some steady state, is achieved. This is the basis of quasi-geostrophic (QG) theory.

The primary force that produces atmospheric motion is commonly referred to as the pressure gradient force (PGF). A PGF exists when an imbalance in atmospheric pressure exists over space. When this occurs, air is moved from areas of high (surplus) pressure to areas of low (deficit) pressure. In other words, the pressure gradient is directed from high to low pressure, so the PGF moves the air in that direction and at a speed proportional to the strength of the gradient. This process operates on all space scales from micro to planetary and is the basis for the study of atmospheric dynamics.

Pressure Change

Atmospheric pressure is not constant over space and time but varies in response to dynamic and thermodynamic processes. If we consider the equation of state, $p = \rho RT$, where p = pressure, ρ = density, R = gas constant for dry air, and T = temperature, then pressure change is proportional to density and temperature change. If we break density down into its component parts, then change in pressure is proportional to change in mass and inversely proportional to a change in volume. In other words, there are two primary ways in which pressure can change; through increases and decreases in temperature and mass and through changes in volume. Given that volume and temperature are inversely related, then temperature and density must also be inversely related. This means that as temperature increases (decreases), volume increases (decreases) and density decreases (increases), if we hold mass constant. On the other hand, if mass decreases (increases), pressure also must decrease (increase). Considering that 99% of atmospheric mass is made up of oxygen and nitrogen with atomic weights of 15.9994 and 14.0067, respectively, then replacing any nitrogen or oxygen atoms with hydrogen atoms with an atomic weight of 1.0079, which is what occurs when humidity increases,

will decrease the mass of the air and decrease the pressure.

When we consider pressure changes at some fixed point location, then thermal, moisture, and mass advection will have an effect on pressure. If warm air advection results in thermal convergence, then the pressure will fall as the air becomes lighter (less dense). If that warm air advection also results in moisture convergence, then the pressure will fall even further. These effects are the result of thermodynamic processes.

When convergence of mass occurs within a column of air anchored at some fixed point location, then the pressure will increase. The processes that may bring an increase of mass through a mass convergence process other than moisture advection are termed "dynamic." Dynamic processes are most commonly the result of mass convergence or divergence that occurs within the middle and upper levels of the troposphere, the result of jet streaks and shortwaves. Jet streaks and shortwaves are sometimes considered together as *upper-level disturbances*. In the absence of thermal and/or moisture advection that occurs primarily within the planetary boundary layer (PBL), decreasing surface pressure over time can indicate the approach of an upper-level disturbance and an increased likelihood of precipitation.

Vertical Pressure Profile

Given the gravitational attraction that the Earth presents to air molecules, a great majority are found close to the Earth's surface. This means that mass within the atmosphere decreases with increasing height and does so in an exponential fashion. As a result, almost half the atmospheric mass is reached at an altitude of about 18,300 ft. (feet).

The relationship between pressure and height is so reliable, varying only a few tens of meters depending primarily on season, that it is commonly used as a vertical coordinate.

For example, while we consider the sea level pressure to average near 1,013 mb, by the time a geopotential height of 4,781 ft. (1,457 m) is reached, the pressure has decreased to 850 mb. A further increase in altitude results in a pressure of 700 mb at about the 9,882-ft. (3,012-m) level,

500 mb at the 18,289-ft. (5,574-m) level, 300 mb at the 30,065-ft. (9,164-m) level, and 150 mb at 44,647 ft. (13,608 m). The 150-mb level typically marks the base of the tropopause during the warm season.

These average pressure-height relationships and associated values for temperature and density, based on international standards, were first established by the U.S. Committee on Extension to the Standard Atmosphere and published by the U.S. Government Printing Office in 1958. The most recent update to the standard atmosphere was published jointly in 1976 by the National Oceanic and Atmospheric Administration (NOAA), the National Aeronautics and Space Administration (NASA), and the U.S. Air Force.

It is common for geographers and atmospheric scientists to refer to different height levels in terms of the average air pressure found there. For example, instead of analyzing the wind pattern at the 30,065-ft. (9164-m) level, we refer to it as the 300-mb level.

For purposes of many atmospheric analyses, the troposphere is divided up into three layers defined by specific pressure levels. Assuming that the surface is close to sea level (within 1,000 ft., or 305 m), the PBL would on average extend up to the 850-mb level, the midlevels to a height of 500 mb, and the upper levels to the tropopause (usually about 150 mb in the middle latitudes during summer and between 300 and 250 mb in winter).

When balloon soundings are developed from rawinsondes, measurements are required at specific altitudes as determined by their corresponding pressures. These are called the *mandatory levels*, which include measurements taken at 1,000, 850, 700, 500, 400, 300, 200, 150, 100, and 50 mb.

David L. Arnold

See also Adiabatic Temperature Changes; Air Masses; Anthropogenic Climate Change; Atmospheric Circulation; Atmospheric Composition and Structure; Atmospheric Energy Transfer; Atmospheric Moisture; Atmospheric Variations in Energy; Chinooks/Foehns; Climate Types; Climatology; Fronts; Coriolis Force; Hurricanes, Physical Geography of; Monsoons; Thunderstorms; Tornadoes; Wind

Further Readings

Barry, R. G., & Chorley, R. J. (2003). *Atmosphere, weather, and climate* (8th ed.). London: Routledge.

Gray, D. E. (1972). *American Institute of Physics handbook* (3rd ed.). New York: American Institute of Physics.

Holton, J. R. (2004). *An introduction to dynamic meteorology* (4th ed.). St. Louis, MO: Elsevier.

Marshall, J., & Plumb, R. A. (2008). *Atmosphere, ocean, and climate dynamics: An introductory text.* Burlington, MA: Elsevier/Academic Press.

Wallace, J. M., & Hobbs, P. V. (2006). *Atmospheric science* (2nd ed.). St. Louis, MO: Elsevier/Academic Press.

ATMOSPHERIC REMOTE SENSING

Remote sensing can be broadly defined as the process of collecting data through means that do not involve physical contact. For example, the process of reading this text is an example of remote sensing, since your eye is gathering information about written letters on the page. A more technical definition of remote sensing describes the process of gathering data about a physical object using radiant energy, wave fields, or particles emanating from or interacting with the physical world. The NASA Remote Sensing Tutorial offers an appropriate definition:

A technology for sampling electromagnetic radiation to acquire and interpret non-contiguous geospatial data from which to extract information about features, objects, and classes on the Earth's land surface, oceans, and atmosphere (and, where applicable, on the exteriors of other bodies in the solar system, or, in the broadest framework, celestial bodies such as stars and galaxies). (http://rst.gsfc.nasa.gov, "Introduction: The Concept of Remote Sensing," I-2)

Atmospheric remote sensing applies remote sensing techniques for data collection and analysis of physical parameters or variables associated with the atmosphere such as temperature, moisture,

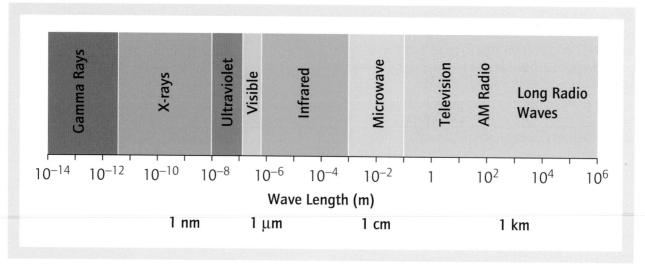

Figure 1 The electromagnetic spectrum

Source: NASA (http://rst.gsfc.nasa.gov).

clouds, precipitation, aerosols, chemical constituents, wind, or weather systems.

Characteristics of Atmospheric Remote Sensing

Active Versus Passive Instruments

Instruments used in atmospheric remote sensing can be characterized as active or passive. An active system generates and transmits an electromagnetic signal that interacts with a medium through reflection, absorption, or scattering. Examples of active systems include Doppler radar (radio detection and ranging) and optical lidar (light detection and ranging). A passive system does not transmit a signal but makes a measurement by receiving scattered, reflected, or emitted radiation. Examples of passive systems include visible cameras, infrared radiometers, and pyranometers. Figure 1 shows the electromagnetic spectrum.

Resolution

Atmospheric remote sensing highly depends on understanding spatial, temporal, spectral, and radiometric resolution. Such factors determine what component or constituent of the atmosphere can effectively be measured or quantified. Spatial resolution represents the small area or pixel resolvable by the remote sensor. Smaller areas represent higher spatial resolution. For example, a 10-megapixel digital camera has higher resolution than a 5-megapixel digital camera because more small pixels can be resolved in the image. In the same manner, a 1-km (kilometer) satellite field of view (1-km × 1-km pixel) is higher in spatial resolution than a 10-km satellite field of view (10-km × 10-km pixel). Temporal resolution represents the time interval between measurements by the remote sensor. Spectral resolution represents the wavelength or frequency intervals of the electromagnetic spectrum that can be applied to make a measurement. Atmospheric remote sensing instruments can use one wavelength interval or many intervals (e.g., multispectral). Many advanced sensors such as the Moderate Resolution Imaging Spectroradiometer (MODIS) (http://modis.gsfc.nasa.gov) are multispectral. Some instruments are even classified as hyperspectral, if they can use an extremely large number of spectral bands. The width of the interval is a measure of the relative magnitude of the spectral resolution. Radiometric resolution represents the smallest quantization of steps of the measurements or the number of bits. It can be useful for characterizing the precision of the instrument measurement.

Vantage Point

Atmospheric remote sensing can be conducted from the vantage point of the ground, air, or space.

Doppler radar systems, upward-pointing radiometers, and acoustic sounders are often operated from the ground. Active and passive sensors are commonly placed on airborne or balloon-borne sensors to improve the field of view or enable three-dimensional (3D) measurements of the atmosphere. For even greater spatiotemporal fidelity, space-borne instrumentation can be applied. Geosynchronous and polar orbits are the two most common orbit strategies that have been employed for space-based atmospheric remote sensing.

Geosynchronous or *geostationary* orbits are achieved when a spacecraft is placed in an orbit approximately 36,000 km (~22,000 mi. [miles]) above Earth. At this altitude, the spacecraft is orbiting Earth in a circular path at rates near 11,052 km/hour (6,802 mph [miles per hour]). With this orbit strategy, the satellite is "geostationary" to a fixed area on the Earth. In other words, the satellite is synchronized with Earth's rotation rate, or "geosynchronous." The United States and several other countries maintain a fleet of geostationary satellites for monitoring atmospheric and environmental conditions.

Polar orbiting satellites are typically found at altitudes ranging from 300 to 1,500 km. Their orbits are typically highly inclined (i.e., low angles with respect to longitude lines) relative to Earth's rotation and elliptical. This type of orbit causes the satellite to pass over the high-latitude polar regions. Many polar orbiting satellites are also sun-synchronous, which means that they pass over the same reference point on the Earth at the same local time. Other orbit strategies can be configured so that they cross equivalent latitudes in a sequence enabling observations associated with the diurnal or daily cycle. NASA's Tropical Rainfall Measuring Mission (TRMM) has an orbit at approximately 36° to 38° inclination so that it can measure the diurnal cycle of precipitation at tropical latitudes (http://trmm.gsfc.nasa.gov). Relatively low inclination orbits that capture the diurnal cycle may be referred to as "walking" orbits.

Atmospheric remote sensing clearly can be described in many ways, but satellite-based remote sensing is quickly emerging as an important methodology. Therefore, the forthcoming discussion will focus primarily on the satellite-based perspective.

Evolution of Space-Borne Atmospheric Remote Sensing

The heritage of atmospheric science remote sensing can be traced to early missions such as the failed Vanguard 2 mission of 1959 and the early Explorer 6 and 7 missions. Using revolutionary radiometric technology by pioneering scientist Verner Suomi, Explorer 7 measured solar and terrestrial radiative energy associated with Earth's global energy balance. The launch of TIROS I (Television and InfraRed Observation Satellite) on April 1, 1960, marked the beginning of a new era of observing Earth's weather conditions from the vantage point of space. TIROS I was launched into a 450-mi. orbit and demonstrated the basic principles of using spectral regions of the electromagnetic spectrum (EM), or wavelengths of light. Many atmospheric observations are imaged in the visible portion of the EM. But like a camera or human eye, the visible wavelengths are not very useful under night conditions or obscured view. Many components of the atmosphere such as clouds, water vapor, and other gases have distinctive absorption-emission wavelengths, or bands. Such information can be applied to thermally imaged atmosphere features or constituents. For example, the advanced very high-resolution radiometer (AVHRR) has been a standard instrument on meteorological satellites over the past several decades. Table 1 provides a typical overview of visible and infrared channels used by the AVHRR and the associated meteorological applications.

From the simplistic video camera methods of TIROS I, the instrument evolution for atmospheric remote sensing advanced at a rapid pace.

The Nimbus Series (1964–1978) significantly advanced atmospheric remote sensing by introducing the use of the microwave portion of the EM and atmospheric sounding. Atmospheric constituents such as water vapor, cloud droplets, raindrops, and ice crystals can absorb (then reemit) or scatter microwave energy depending on the wavelength (or frequency). Microwave atmospheric remote sensing enabled observations and analysis in cloudy or cloud-free environments and provided new opportunities to measure water vapor, precipitation, and clouds. Atmospheric

Channel	Wavelengths (µm)	Atmospheric/ Meteorological Application
1	0.58–0.68	Day clouds, snow, ice
2	0.725–1.10	Surface water; snow, ice
3	3.55–3.93	Fires, clouds at night
4	10.30–11.30	Day/night cloud and surface temperatures and mapping
5	11.50–12.50	Same as 4, water vapor

Table 1 Typical channels of the advanced very high-resolution radiometer (AVHRR)

Source: Adapted from NASA (http://rst.gsfc.nasa.gov).

sounders provide measurements of the vertical distribution of infrared or microwave radiation emitted by atmospheric constituents at different levels of the atmosphere. This distribution can be use to provide temperature or humidity profiles (measurements from top to bottom). This type of information is traditionally provided by radiosondes (weather balloons) launched at specific locations two times per day. Satellite-based soundings can significantly extend the spatial and temporal coverage for meteorological, climatological, and numerical modeling applications. The textbook *Satellite Meteorology: An Introduction* by Kidder and Vonder Haar is an excellent reference for more information on historical satellite series, imagers, and sounders.

Though satellite remote sensing had its origins in the United States, many nations contribute an array of atmospheric remote sensing satellites. These countries include, but are not limited to, Japan, the European Union, China, Canada, and India. The Committee on Earth Observation Satellites is a good resource on the scope of international involvement (http://idn.ceos.org).

Current and Future Atmospheric Remote Sensing Capabilities

Space-Borne Systems

Advanced satellite-based remote sensing capabilities in the post-2000 era and beyond have revolutionized atmospheric sciences. It would be difficult to imagine living in a coastal community of Japan or Mexico and not being forewarned of an approaching hurricane. However, this is exactly what happened to the people of Galveston, Texas (United States), in 1900. Extensive loss of life and property occurred due to lack of warning that is now commonplace because of satellite remote sensing. As valuable as the space-borne perspective is to research and operational communities, the atmospheric community continues to advance forward with development. Space-borne platforms now employ active remote sensing instrumentation such as lidar (light detection and ranging) and precipitation radar (e.g., CloudSat, http://cloudsat.atmos.colostate.edu, or the Tropical Rainfall Measuring Mission [TRMM] Precipitation Radar, http://trmm.gsfc.nasa.gov) to extend research and prediction capabilities related to aerosols, cloud properties, and precipitation. For example, a 3D view or "cat-scan" through Hurricane Katrina was enabled by a suite of passive and microwave instruments aboard TRMM. Missions such as the planned Global Precipitation Measurement Mission will use a constellation of satellites carrying microwave instruments to further extend coverage and temporal resolution. When the technology becomes available, active instruments will move from low Earth orbiting to geosynchronous platforms.

The next generation of operational satellites, such as GOES-R and NPOESS in the United States (www.ipo.noaa.gov), is capitalizing on advanced instruments, new technologies, and higher resolutions (spatial, spectral, temporal, radiometric) that continue to emerge from coordinated research and development programs. New satellites can detect lightning (www.goes-r.gov), atmospheric aerosols (http://envisat.esa.int), 3D profiles of temperature and moisture (http://aqua.nasa.gov), atmospheric chemical composition (http://aura.gsfc.nasa.gov), and greenhouse gases (http://oco.jpl.nasa.gov). Even the international global positioning system satellite network used for navigation can be exploited to extract information on atmospheric moisture. Additionally, innovative formation-flying or constellation orbit strategies are being implemented to leverage unique capabilities of individual platforms.

Ground-Based Systems

Satellite remote sensing of the atmosphere has seen explosive development, but there are numerous exciting developments in ground-based or suborbital remote sensing as well. Next-generation weather radars will exploit information derived from polarimetric and continuous wave measurements. Ground-based Raman lidars and acoustic sounders provide important information on the vertical structure of temperature and moisture that, along with other remotely sensed data sets, can be used in weather analysis or as input for numerical weather prediction (NWP) models. Wind profilers provide high-temporal-resolution data on the vertical structure of wind, another critical element of weather analysis and NWP.

Obsolete or Not?

The elapsed time between the writing of this encyclopedia entry and your reading of it represents a relatively long period of time in technological development supporting atmospheric remote sensing. New ground, airborne, and spaceborne remote sensing techniques and instruments are constantly in some phase of conception, development, testing, implementation, or operation. Readers interested in atmospheric remote sensing should stay abreast of dynamic sources of information to remain at the forefront of the topics. New capacities will be critical as the requirements of the weather, climate, environmental, and societal stakeholder communities evolve.

Marshall Shepherd

See also Aerial Imagery, Data; Aerial Imagery: Interpretation; Ambient Air Quality; Anthropogenic Climate Change; Atmospheric Moisture; Atmospheric Particulates Across Scales; Atmospheric Pollution; Biophysical Remote Sensing; El Niño; Floods; Greenhouse Gases; Hurricane Katrina; Precipitation, Global; Precipitation Formation; Remote Sensing; Remote Sensing: Platforms and Sensors; Weather and Climate Controls

Further Readings

Kidder, S. Q., & Vonder Haar, T. H. (1995). *Satellite meteorology: An introduction.* San Diego, CA: Academic Press.

Short, N. (2009). *Remote sensing tutorial.* Retrieved November 9, 2009, from http://rst.gsfc.nasa.gov

ATMOSPHERIC VARIATIONS IN ENERGY

When we consider the atmospheric envelope that surrounds our solid Earth, there exists an upper boundary, the edge of outer space, and a lower boundary, the Earth's surface. Given the spherical nature of Earth, energy can only enter the atmosphere from the upper and lower boundaries.

Energy that enters the atmosphere through the upper boundary originates from the sun and passes through the voids of outer space, reaching the upper atmosphere at an altitude of somewhere near 650 km (kilometers). How much energy passes through the thermosphere ($\approx$500-km depth), mesosphere ($\approx$85 km), and stratosphere ($\approx$50 km) before reaching the troposphere is a function of the mass that exists within these layers. It is important to note that while the sun emits radiation across a large spectral range, the large majority of radiation originates from the ultraviolet, visible, and near-infrared portions of the electromagnetic spectrum. The manner in which this "shortwave" radiation interacts with the atmospheric mass it encounters varies depending on the character of the mass, particularly its size relative to the wavelength of radiation.

As solar radiation passes through the atmosphere, it can be scattered, absorbed, or transmitted. That which is absorbed raises the temperature of the absorbing body, and if a thermal gradient exists away from that body, heat will be radiated (in many cases "re-radiated") in that direction. Radiation that is scattered will in many cases be scattered multiple times before being absorbed or returned back to outer space. When the scattering occurs back in the general direction from where it came, it is referred to as "backscattering," sometimes also known as reflection. Radiation that is not backscattered is "forward scattered."

On average, about 25% of the solar radiation that reaches the troposphere is absorbed and 25% is backscattered, which leaves about 50% that reaches the Earth's surface. About 4% of the radiation that reaches the Earth's surface is backscattered, so only about 47% of the roughly 1,352 W/m^{-2} (watts per square meter is a measure of radiant flux density using SI units) that reaches the top of the atmosphere (a value commonly referred to as the *solar constant*) is absorbed by the Earth's surface. If we are considering anything other than a global average, all these values would vary considerably, and the solar angle of incidence would also have to be considered.

To understand that partitioning of local-scale atmospheric energy between its sensible and latent states we would also have to consider adiabatic processes, especially those associated with cloud and precipitation development and dissipation. So the variation of atmospheric energy on anything less than the planetary scale is infinitely complex, and interactions of all the processes is still not well understood. Nevertheless, numerical models continue to be constructed in an attempt to model these processes for purposes of forecasting future states of the atmosphere on a variety of space and timescales. Some of these are used to forecast weather from a couple of days to periods of 2 wks. (weeks), while others attempt to characterize the state of our climate system decades into the future. At this point in time, with our lack of knowledge regarding the interactions of some of these important processes, the difficulty in simulating cloud and precipitation processes is a limiting factor that cannot be overlooked. In fact, for meteorological purposes, there tends to be too much reliance on numerical models in developing weather forecasts. This situation has grown to the point where it has earned the title "meteorological cancer," since too heavy a reliance on forecast models has resulted in a general deterioration of our forecasting skills.

Given the difficulty in addressing anything less than planetary scale atmospheric variations in energy, we will focus only on the general inputs and outputs of energy.

The primary inputs of energy into our atmosphere originate from the sun, the rotation of Earth about its axis, and Earth's surface. Technically, we should also consider the revolution of Earth about the sun and the motion of our solar system with the galaxy, although this motion is generally considered negligible. We also need to consider that Earth's surface is covered primarily with water, and given water's large heat capacity, more energy is stored in the oceans with potential input to the atmosphere than in any other reservoir, including the atmosphere. In addition, the land surface is both a heat source and a heat sink, and it does provide orographic (of or relating to mountains) barriers that generate mountain torque, which affects the general atmospheric circulation.

Solar Energy

The sun is a mass of gas with a core, body, and surface. Altogether, the mass of the sun is a million times greater than that of Earth.

Solar energy is generated within the sun's core primarily by the fusion of hydrogen nuclei into helium. Once that energy is produced, it must pass through the body of the sun, which is made up of thick layers of compressed gas. This takes about 200,000 yrs. before it can reach the sun's surface, where it is released into outer space. This surface of the sun from where the energy is emitted is known as the *coronosphere*. Solar radiation is emitted from the coronosphere at the speed of light, but even so, it takes about 8 minutes to reach Earth's atmosphere.

Despite use of the term *solar constant*, the sun's output is anything but constant. For example, when Earth formed, some 4.6 billion yrs. (years) ago, solar output was only 70% of what it is now. At present, the sun is in the middle ages of its life cycle. As it continues to age, it will emit more solar energy, although we can assume it will radiate roughly within the same general range for the next 5 billion yrs. or so.

Even in its middle-age stage, solar irradiance varies over time. And while this variation is only on the order of a few watts per square meter at the top of the atmosphere, even minor variation can have significant impacts on temperature, as was the case during the Little Ice Age from the mid 16th to mid 19th centuries. For this reason, the sun must be considered the primary climate control.

Solar variation appears to follow an 11-yr. cycle associated with sunspots and solar flares, which affect solar irradiance. This cycle was discovered in 1843 by Samuel Heinrich Schwabe. Rudolf Wolf reconstructed the cycle history back to the days of Galileo during the 17th century, although he began the modern solar cycle era with Solar Cycle 1 covering the period 1755 to 1766. During the year 2009, we entered Solar Cycle 24. This is the point where sunspots, solar flares, and solar output are at an 11-yr. minimum. Over the following 11 yrs., which is actually the average duration of a solar cycle (range of 10–13 yrs.), solar output will first increase for roughly 5½ yrs., then decrease over the next 5½ yrs. So holding all else constant, we would expect global temperatures to increase, reaching a peak in 2013, with another minimum in about 2019.

While cycle-to-cycle variability exists, there has been a smoothed 75-yr. cycle (ranges between 50 and 100 yrs.) where cycle-to-cycle peaks progress from large to small. When looking back at the Little Ice Age, temperatures were the coldest between 1650 and 1700, when almost no sunspots were observed, a period now known as the Maunder Minimum. A second extremely cold surge occurred from about 1780 to 1830 during the Dalton Minimum. Since about 1940, we have been within a period known as the Modern Maximum, although there are some solar scientists who believe we are heading back toward another long-term solar cycle minimum. Whether this would lead to another Little Ice Age is extremely difficult to say.

The significance of even very minor variations in solar irradiance is important and becomes obvious if annual values are plotted against U.S. annual mean temperatures during a 20th-century time series, as it corresponds much better with temperature than does exponentially rising atmospheric CO_2.

Volcanic Eruptions

Major volcanic eruptions inject large volumes of sulfate aerosols into the troposphere and stratosphere that can circle the Earth for years, reflecting incoming solar radiation back to outer space and increasing cloud cover. This will reduce incoming solar radiation, which can cause significant global cooling.

When Mt. Pinatubo erupted in the Philippines in June 1991, it became the largest volcanic eruption of the 20th century and the largest since the eruption of Mt. Krakatau in 1883. While Mt. St. Helens was significant on local scales, it was 10 times weaker than Pinatubo.

The Pinatubo blast spewed over 10 billion metric tons of lava and released between 15 and 30 million metric tons of sulfur dioxide into the atmosphere. In just 2 wks., the sulfate aerosol cloud had spread around Earth, covering the entire planet within a year. The reduction in the Northern Hemispheric average temperature was 0.5 to 0.6 °C, and for the entire globe, the reduction was 0.4 to 0.5 °C. In August 1992, the greatest reduction, of 0.73 °C, was recorded. Following the eruption, the United States recorded its third coldest and third wettest summer in 77 yrs.

When sulfur dioxide mixes with water and oxygen, it becomes sulfuric acid, which ultimately results in the destruction of ozone. During the period 1992 to 1993, after the Pinatubo eruption, the ozone hole over the South Pole reached record size.

In summary, the global impact of the Mt. Pinatubo eruption was greater than that of the El Niño the following summer, which shows how great an effect this single eruption had on the global environment.

Pacific and Atlantic Ocean Oscillations

Given the spatial extent of oceans on planet Earth, there is little doubt that the surface temperatures of these water bodies dominate the global temperature. Considering that coastlines and, to a certain extent, some inland areas exhibit maritime climates, it becomes clear that the ocean influence reaches beyond the limits of the ocean basins.

When we think of heat content, oceans have as much heat in their lower 2.5 m (meters) than does the entire atmosphere. However, ocean heat content is rather large all the way down to the thermocline, with an average depth of about 100 m. Considering that the oceans take in about 85% of the heat absorbed by the entire Earth, ocean temperature has to be a good measure of global temperature.

The two largest oceans in the global climate system are the Pacific and Atlantic, although the Indian Ocean is a significant player. The largest ocean, the Pacific, would therefore be expected to dominate global ocean and air temperatures. There are cases where an anomalously warm Atlantic would offset some of the effect of a cool Pacific, although the Pacific signal would still be there. Take the period of time from about 2003 through 2009, when the Pacific entered its cool cycle (oscillation between cold and warm cycles is referred to as the Pacific Decadal Oscillation [PDO]). While land-based instrumental air temperature measurements have continued to detect a warming trend, these data could be warm biased, given that the "global" network is dominated by climate stations in Europe and the United States, particularly the Eastern United States. A recent survey of these climate stations found that 85% of them fail to meet the U.S. National Atmospheric and Atmospheric Administration standards for reliable climate measurements, with some of the violations being extreme, such as temperature sensors located immediately next to concrete buildings, air conditioning and heating vents, mechanical room exhaust fans, and so on. Not only that, but there has been a significant instrumentation change over the past 20 yrs., with studies showing the new Maximum Minimum Temperature Systems (MMTS) to be recording overnight low temperatures several degrees higher than the previous Cotton Region Shelter alcohol in glass thermometers. Given that daily mean temperatures are computed as simple averages between 24-hour maximum and minimum temperatures, this could have quite an effect. Also, many of the traditionally rural stations have been encroached on by urban environments, which has warmed the local climate considerably. In a recent study, it was found that while urban climate stations continue to observe rising temperatures, rural stations in the same regions continue to measure nearly constant, if not falling, temperatures. Both NASA satellites and NOAA ocean buoys that record temperatures over the oceans have detected a consistent cooling trend since 2003, so while some land areas appear

to be warming, rural land areas and oceans seem to be cooling. When all is said and done, global temperature appears to have begun cooling in 2003 at the same time the Pacific Ocean entered its cool phase. This cooling trend has continued to this day.

The Pacific Decadal Oscillation (PDO) exhibits a low-frequency periodicity of about 30 yrs. Prior to about 1977, the PDO was in its cold phase just as it has been at the time of this entry being written, which means that the Pacific Ocean as a whole was colder than average, with a predominance of strong La Niña events in the tropical Pacific. During a cold PDO, El Niño events do occur, as they must in reaction to the conclusion of a La Niña event, but they are not as frequent and not as strong. During a warm PDO, strong El Niño events dominate, which increases global temperatures, while weak and infrequent La Niña events are not uncommon. The El Niño/La Niña cycle, called the El Niño Southern Oscillation (ENSO), is well known, and a proper understanding requires recognition that the increase in sea levels due to a long period of a La Niña event necessitates the Pacific warm pool to "slosh" back toward the east, creating an El Niño environment.

By 1977, the PDO shifted dramatically to its warm phase, an event still called the "Great Pacific Climatic Shift." From this time until the early 21st century, temperatures were much warmer than normal, since this was the PDO warm phase. Two of the most spectacular El Niño events occurred within this period: the 1982–1983 event and the record 1998–1999 event. Strong La Niña events were few and far between during this 25-yr. period.

Now that the PDO has shifted to its cold cycle, global temperatures, except those over land, have come down consistently at the same time as CO_2 has increased dramatically. This recent 6-yr. cooling cycle was not picked up by any of the climate models that suggest a runaway greenhouse effect and drastic global warming. So while CO_2 continues to increase at a dramatic rate, global temperatures controlled by the Pacific Ocean have cooled. Clearly, there are problems with these climate models, as alluded to earlier.

The Pacific Ocean is not the only oceanic consideration. The Atlantic Ocean also has cool and warm phases, with a periodicity of 20 to 40 yrs. This modulation is called the Atlantic Multi-decadal Oscillation (AMO). During the mid 1920s through the late 1960s, the AMO was in its warm phase, and with all that heat available, hurricane frequency was rather large. By about 1970, the AMO entered its cold phase, and hurricane frequency decreased dramatically. However, in 1995, the AMO reentered its warm phase, and hurricane frequency picked up again dramatically. In 2009, the AMO appeared to be peaking in its warm mode. How long this will continue is difficult to say, but some predict it will begin to decline in about the year 2015. Until then, we can expect a higher-than-average frequency of hurricanes, and hurricanes that are more powerful than average. The only factors that will limit them during some years are El Niño events, when the upper-level wind shear is too strong to allow full development.

As it turns out, however, a warm AMO favors a blocking high near Greenland during the cold season, which amplifies Eastern U.S. troughs. With the polar jet ushering colder than normal air from the northwest, due in part to a cold PDO, then frequent snowstorms should be the rule. During cold AMO events, the upper-level troughs do not become as amplified, and a Greenland block does not exist, so the pattern is very progressive. The East Coast sees its share of snowstorms during this phase, but they move through quickly, leaving only light snow. And even though both oceans will be cold during about 2020 to 2030, which should result in much colder-than-average global temperatures, the milder Pacific air in a progressive flow across the continental United States will likely keep temperatures near to slightly above normal in the East Coast region.

David L. Arnold

See also Adiabatic Temperature Changes; Air Masses; Anthropogenic Climate Change; Atmospheric Circulation; Atmospheric Composition and Structure; Atmospheric Energy Transfer; Atmospheric Moisture; Atmospheric Pressure; Chinooks/Foehns; Climate Types; Climatology; Coriolis Force; Fronts; Hurricanes, Physical Geography of; Monsoons; Thunderstorms; Tornadoes; Wind

Further Readings

Bothmer, V. (2007). *Space weather.* New York: Springer-Verlag.

D'Arrigo, R. D., Jacoby, G. C., Free, M., & Robock, A. (1999). Northern Hemisphere annual to decadal temperature variability for the past three centuries: Tree-ring and model estimates. *Climatic Change, 42,* 663–675.

Houghton, H. G. (1985). *Physical meteorology.* Cambridge: MIT Press.

Marshall, J., & Plumb, R. A. (2008). *Atmosphere, ocean, and climate dynamics: An introductory text.* Burlington, MA: Elsevier/Academic Press.

Robock, A. (1979). The "Little Ice Age": Northern Hemisphere average observations and model calculations. *Science, 206,* 1402–1404.

ATOLL

The term *atoll* is derived from the Maldivian word *atholu* and refers to a coral reef that forms a ring around a central lagoon. Above this coral rim are typically numerous small islets that are composed of coral sand, gravel, and aggregate. These islets, often known by the Polynesian word *motu*, are low lying and on average may be no more than 3 m (meters) above sea level. There are more than 250 atolls in the world; most occupy the warm tropical waters of the Pacific and Indian oceans, but some are found as far south as the Elizabeth and Middleton reefs in the Tasman Sea. Some atolls are very small, such as Rose Atoll in the Central Pacific, which has a combined land and lagoon area of only 7 km^2 (square kilometers), whereas one of the largest atolls in the world, Kwajalein in the Marshall Islands, has a lagoon area of 2,304 km^2 (see photo).

Atolls have long been of interest to geographers in terms of how they form, as well as the challenges faced by those who live in such isolated and unique environments. This entry will briefly review debates surrounding atoll formation, and it will discuss issues faced by atoll-dwelling communities.

Some atolls of the Marshall Islands

Source: William James Smith Jr. Reprinted with permission.

Atoll Formation

Charles Darwin first hypothesized the subsidence theory for the origin of atolls in 1842. Convinced of the existence of seabed subsidence to counterbalance tectonic uplift, his theory of subsidence suggested that atolls are a product of an evolutionary sequence of coral reef formation around the perimeter of subsiding oceanic seamounts and volcanoes. These fringing reefs maintain growth close to the sea's surface, the optimum depth for coral growth being less than 25 m. Over time and keeping pace with the subsiding landmass, these fringing reefs gradually evolve into barrier reefs and with the final submergence of the landmass, result in an atoll (Figure 1). Darwin's theory was supported by the results of drilling in Bikini Atoll in 1947 and Eniwetok Atoll in 1951, both in the Marshall Islands, which found volcanic rock beneath reef limestone at depths of 1,600 and 1,405 m, respectively.

Although Darwin's explanation is widely promulgated, there have been two theories put forth as alternatives to his model of atoll formation by subsidence. Reginald Daly's "glacial control theory" took into account the effects of sea level rise as a consequence of historical glaciations and suggested that atolls were the product of weathering and erosion of marine terraces at times of low sea level. In contrast to Daly's proposition, the "karst control" or "antecedent karst" theory developed by Edward Purdy suggested that at times of low

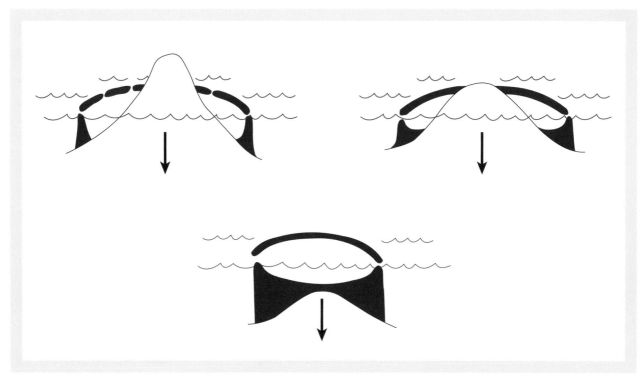

Figure 1 Subsidence theory of atoll formation as a consequence of subsiding landmass

Source: Adapted from Scoffin, T. P., & Dixon, J. E. (1983). The distribution and structure of coral reefs: One hundred years since Darwin. *Biological Journal of the Linnean Society, 20,* 13.

sea level, the exposure of the marine terraces to rainwater would actually result in karst topography. The rim of the marine terrace would result in rainwater runoff from the edge of the marine terrace and concentration of rainwater at the center. These differential rates of solution experienced at the rim and center would result in a depression at the center of the platform that, with subsequent sea level rise, would flood. Coral growth on the peripheral rim would thus form the barrier reef, and flooding of the interior depression in the central lagoon (Figure 2). Debates over atoll formation are ongoing.

Atoll Environments and Social Issues

Atoll environments, with their limited land area, poor soils, and restricted freshwater supplies, tend to support limited terrestrial flora and fauna and have very low levels of species endemism. Marshall Sahlins, most notably, suggested that the social structures of atoll-dwelling communities reflected the distinctive ecological conditions of atoll environments. Although now widely criticized, his argument linked degree of social stratification to supply and distribution of resources and modes of production; atoll-dwelling communities with limited resources would generally be expected to have less hierarchical and more egalitarian social structures than their high-island counterparts.

Atolls retain a special place in popular imagery and are often described as "natural," "pristine" environments and isolated and exotic "tropical paradises." In contrast, however, their geographical remoteness from Western centers has resulted in atolls continuing to play important roles as nuclear testing grounds for Western military forces. The U.S. military, for example, conducted more than 23 nuclear tests on Bikini Atoll between 1946 and 1958, claiming that such tests and the resultant displacement of the entire Bikinian population were necessary for the good of humankind. French nuclear testing on Mururoa and Fangataufa atolls in French Polynesia continued up until 1995, although claims that such tests

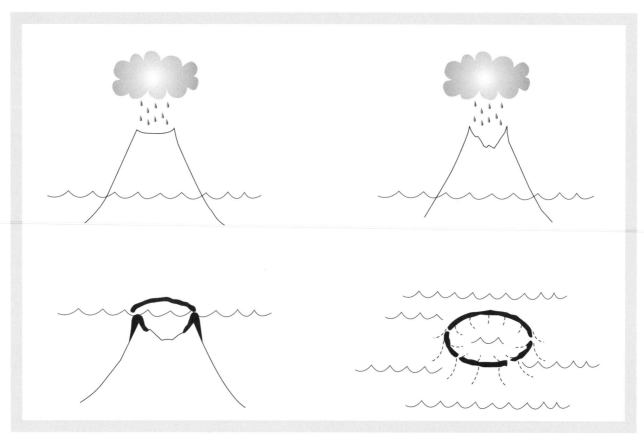

Figure 2 Karst control theory of atoll formation as a consequence of sea-level rise

Source: Adapted from Woodroffe, C. D. (2008). Reef-island topography and the vulnerability of atolls to sea-level rise. *Global and Planetary Change, 62*, 80.

were damaging to the environment continue to be dismissed by French authorities. U.S. military testing of its missile defense systems continues to this day on Kwajalein Atoll.

Atolls and Climate Change

Atolls are deemed to be particularly at risk from events associated with climate change such as sea-level rise, increasing frequency of extreme weather events, and rising sea-surface temperatures. The perceived vulnerability of atolls to such climate change–related phenomena is in part due to their physical characteristics, such as the high ratio of coastline to land area, low relief, and susceptibility to coastal erosion. In addition, many atolls support high population densities, which, combined with low levels of infrastructure, present significant challenges to the residents of such island nations as Tuvalu and the Maldives, in terms of whether their environments will be able to sustainably support human settlements.

Charlotte N. L. Chambers

See also Climate Change; Coral Reef; Coral Reef Geomorphology; Global Sea-Level Rise; Island Biogeography; Islands, Small; Karst Topography; Oceanic Circulation

Further Readings

Barnett, J., & Adger, W. N. (2003). Climate dangers and atoll countries. *Climatic Change, 61*, 321–337.

Daly, R. A. (1910). Pleistocene glaciation and the coral reef problem. *American Journal of Science, 30*, 297–308.

Darwin, C. (1962). *The structure and distribution of coral reefs.* Berkeley: University of California Press. (Original work published 1842)

Davis, J. S. (2007). Scales of Eden: Conservation and pristine devastation on Bikini Atoll. *Environment and Planning D: Society and Space, 25,* 213–235

Purdy, E. G. (1974). Reef configurations: Cause and effect. In L. F. Laporte (Ed.), *Reefs in time and space* (Special Publication 18, pp. 9–76). Tulsa, OK: Society of Economic Paleontologists and Mineralogists.

Sahlins, M. D. (1958). *Social stratification in Polynesia.* Seattle, WA: American Ethnological Society.

 # AUTOMOBILE INDUSTRY

Motor vehicle production constitutes an important industrial sector for many countries. More than 70 million *light vehicles* (defined as passenger cars and light trucks but excluding heavy-duty trucks) were produced worldwide in 2006. Production exceeded 1 million vehicles in 2006 in 16 countries on every continent: Brazil, Canada, China, France, Germany, India, Iran, Italy, Japan, South Korea, Mexico, Russia, Spain, Thailand, the United Kingdom, and the United States. During the first decade of the 21st century, global production of light vehicles was increasing at an annual rate of 4%.

With the growth of motor vehicle production worldwide has come an increasing concentration of producers. The two leading motor vehicle producers, Toyota and General Motors (GM), together manufactured one fourth of the world total. Eight other companies shared another one half of the market. Three dozen companies, mostly based in Asia, split the remaining one fourth, in part through joint ventures with one or more of the top 10 companies.

Small-scale commercial production of motor vehicles began in the United States and several European countries in the 1890s. Early carmakers located in places with expertise in producing key components, especially gasoline engines and bodies. Also critical was proximity to sources of capital willing to invest in a risky new industry with uncertain prospects.

Under the influence of the Ford Motor Company, the geography of motor vehicle production was set during the 1910s. Ford's low-cost Model T held more than one half of the market worldwide, not just in the United States; minimizing transport-related production costs was one way that Ford kept the Model T price low. Ford employees calculated that the optimal spatial arrangement for motor vehicle production was to assemble vehicles near the consumers.

As a bulk-gaining product, the cost of shipping finished vehicles to consumers was much higher than the cost of packing parts into freight cars and trucks. In addition, finished motor vehicles were fragile, high-value items subject to costly damage if shipped long distances. Consequently, Ford produced most of its parts in Michigan but actually assembled its cars at three dozen final assembly plants around the United States near major metropolitan areas where most of the customers were clustered. When GM replaced Ford as the leading carmaker in the 1920s, it too produced most of its parts in Michigan and shipped them to assembly plants near major population centers.

Ford and GM, which dominated worldwide motor vehicle production during the first half of the 20th century, extended to other countries the strategy of assembling vehicles where they were sold. Rather than relying primarily on exports from the United States, the two companies opened assembly plants throughout Europe, Latin America, and Asia to meet—as well as to stimulate—growing local demand.

The changing distribution of motor vehicle sales and production worldwide has reinforced the long-standing geographic paradigm. As more consumers around the world have the income to purchase motor vehicles, manufacturers open assembly plants in these places. Thus, production has grown rapidly in Asia during the early 21st century primarily because demand from consumers in the region has grown rapidly. For example, as sales of light vehicles increased in China from 2.4 million in 2001 to 7.5 million in 2006, demand was accommodated by increasing production in China from 2.4 million in 2001 to 7.3 million in 2006. On the other hand, stagnant production levels in North America during the period reflected stagnant demand in a mature, long-saturated market.

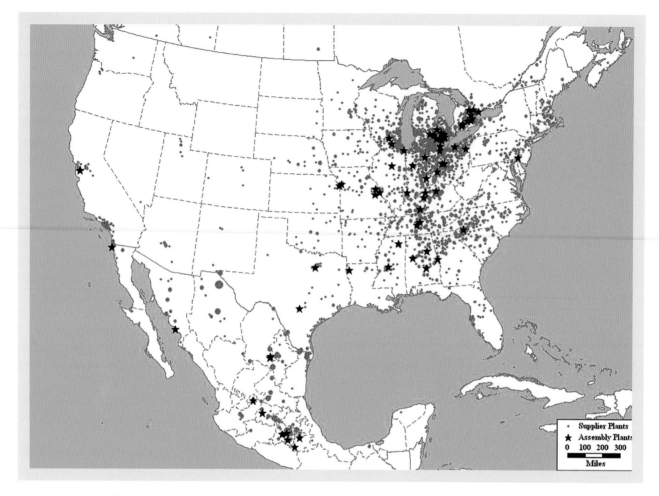

Figure 1 Auto alley in the United States. The restructuring of the automobile industry has produced new geographies stretched between the traditional core and new centers in the South.

Source: Updated by author from Klier, T., & Rubenstein, J. (2008). *Who really made your car? Restructuring and geographic change in the auto industry* (p. 13, Figure 1.3). Kalamazoo, MI: W. E. Upjohn Institute for Employment Research.

The principal exception to the assemble-where-sold model in the early 21st century was the net movement of several million vehicles from Asia to North America. Japanese carmakers sold 7.5 million vehicles in North America in 2006 but produced only 5 million of them there and so met the excessive demand through imports.

The build-where-sold paradigm has been more flexible for parts than for final assembly plants. Parts account for 70% of the value added in the production of motor vehicles. Carmakers were once highly vertically integrated; that is, they built most of their parts in-house. But the leading carmakers, notably, Ford, GM, and Toyota, converted most of their in-house parts operations into independent companies, primarily in the late 1990s.

Under mounting pressure from carmakers to reduce prices for their parts, independent suppliers have adopted global sourcing strategies. More than one fourth of parts used at U.S. final assembly plants were imported in 2006. Mexico, Canada, and Japan together accounted for three fourths of imported new vehicle parts. China ranked a distant fourth and was capturing an increasing share. China was poised to become one of the world's largest producers of new vehicle parts, but most of the parts were destined for domestic producers.

The location of motor vehicle plants has been changing within North America and Europe. In North America, most motor vehicle production has clustered in a region known as auto alley, a

700-mile-long corridor between the Great Lakes and the Gulf of Mexico, for the most part following interstate highways I-65 and I-75 (Figure 1). Nearly all U.S. final assembly plants built since 1980 have been in auto alley, and nearly all plants outside auto alley have been closed. A location in auto alley minimizes shipping costs to the North America market as a whole, as well as ensuring proximity to parts suppliers.

Within auto alley, most new facilities have been located at the southern end, while the northern end has had to endure most of the closures. The southern drift within auto alley is primarily a function of the shift in market share from the "Detroit Three" U.S.-owned carmakers (Chrysler, Ford, and GM) to foreign-owned carmakers. Foreign-owned companies have preferred to locate facilities at the southern end of auto alley, primarily to avoid communities traditionally associated with the Detroit Three. Detroit's three plants and many of their suppliers have a unionized workforce, whereas unions are rare in foreign-owned plants.

Within Europe, motor vehicle production stagnated during the early 21st century in the traditional manufacturing centers, such as France, Germany, Italy, and the United Kingdom, whereas the Czech Republic, Poland, and other former communist Eastern European countries had sharp increases. Plants in Eastern Europe were close enough to ship to Western European markets yet had much lower wage rates.

The principal uncertainty in anticipating the future geographic distribution of motor vehicle production is the impact of changing fuel sources. Hybrid engines in North America and diesel engines in Europe—early-21st-century bridge technologies—did not fundamentally alter the geography of production. However, vehicles powered by truly alternative sources, such as batteries or fuel cells, may have a more profound spatial impact.

The prospect of commercializing alternative fuel vehicles has set off a scramble among the major motor vehicle producers, large parts suppliers, software providers, and electronics specialists to control the architecture of motor vehicle technology. Just as motor vehicle production clustered in Michigan in the early 20th century to take advantage of local expertise in gasoline engines, so may motor vehicle production in the 21st century locate near the successful creators of postpetroleum technology.

James M. Rubenstein

See also Automobility; Deindustrialization; Economic Geography; Fordism; Industrialization; Industrial Revolution; Manufacturing Belt

Further Readings

Flink, J. (1988). *The automobile age.* Cambridge: MIT Press.
Klier, T., & Rubenstein, J. (2008). *Who really made your car? Restructuring and geographic change in the auto industry.* Kalamazoo, MI: W. E. Upjohn Institute for Employment Research.
Liker, J. (2004). *The Toyota way.* New York: McGraw-Hill.
Maynard, M. (2003). *The end of Detroit.* New York: Currency Books.
Rubenstein, J. (2001). *Making and selling cars: Innovation and change in the U.S. automotive industry.* Baltimore: Johns Hopkins University Press.
Womack, J., Jones, D., & Roos, D. (1990). *The machine that changed the world.* New York: Rawson.

 # AUTOMOBILITY

Automobility refers to the economic, social, and spatial impacts of the automobile. From its modest roots as a "horseless carriage," a plaything for affluent Europeans, the automobile emerged to become arguably the defining technology of the 20th century. Automobiles encapsulated the era's infatuation with speed and power, epitomized mass mobility, reworked cities and rural areas alike, transformed the lives of millions, if not billions, and dramatically reconfigured the structure and meanings of individual and social time and space.

The origins of the automobile lay in Europe, particularly France (hence the numerous French terms, e.g., *chassis, chauffeur*). In 1884, Karl Benz built the first self-propelled vehicle in Germany. Early automobiles were fitted with solid rubber tires, which made transport slow and uncomfortable, until 1888, when John Dunlop invented the air-filled pneumatic tire, which absorbed shocks. The mass production and mass consumption of cars, however, was a distinctively American phenomenon. Henry Ford's introduction of the moving assembly line at the Highland Park factory near Detroit in 1913 initiated a vast revolution in the world of work and production, taking mass production to new heights of productivity and profitability. By pioneering a form of production centered on vertically integrated firms and enormous economies of scale, Ford developed an entire ensemble that became the model for most forms of manufacturing throughout much of the 20th century. (A similar system was adopted by Alfred Sloan of General Motors, who also introduced planned obsolescence.) The process worked spectacularly, generating millions of jobs and relatively high wages (and unions) and raising standards of living.

It was in the United States that the automobile, like the bicycle, first came into widespread use, democratizing access to transportation to an unprecedented degree. In 1895, only four automobiles were registered in the entire country, although this number grew to 8,000 by 1900 and to 458,000 by 1910. The automobile soon displaced rail as the dominant intercity carrier of goods and people. What Ford called the "family horse" evolved into the Model T, the first automobile to gain mass acceptance, whose price dropped in half between 1913 and 1916, initiating a privatization of transportation unparalleled in history.

Sharp gender differences characterized early automobile ownership: Driving was an exclusively male phenomenon, and women took to the road in large numbers only after World War II. The invention of the electric starter in 1912 helped bring a few women into the automobile market, but they remained the exception.

Although most urbanites still used light rail transit in the 1920s, it was not long before the private car was no longer a luxury but a necessity for the middle class. Spurred by Henry Ford's Model T, mass ownership of automobiles in the United States grew stupendously. Whereas the Model T cruised at 30 kilometers per hour, by the

1930s, improvements in engines and chassis and the introduction of leaded gasoline doubled this speed. Farmers turned to cars because of their ability to negotiate unpaved roads and urbanites because of the convenience they offered. Within two generations, almost universal availability of motor transportation became the norm. In the United States, the number of registered automobile owners rose from 1 million in 1912 to 10 million in 1921, 30 million in 1937, and 60 million in 1955. The American experience was quickly emulated by Canada and several Western European countries. However, no European country attained the high levels of U.S. auto ownership, in part due to lower incomes, better mass transit systems, and higher gasoline taxes. European cities, in consequence, assumed a very different form from the sprawling, low-density, multinucleated model prevalent in North America.

With the internal combustion engine, it was the infrastructure rather than the vehicle that presented the greatest constraint to continued reductions in travel time. The internal combustion engine allowed investments to be spread out among users rather than be concentrated on producers, as with railroads. The result was far more flexible geographies that emerged along the interstices of fixed rail lines. Trucks, for example, offered unprecedented flexibility in the spatial decisions of firms, as they allowed perishable and low-valued goods to be distributed quickly and efficiently. Yet the building of roads and highways is fraught with political as well as economic predicaments. In the United States, political pressures from the burgeoning middle class for improved roads accompanied the widespread adoption of cars, leading to rounds of parkways in the 1890s on the outskirts of numerous metropolitan areas. The 1920s witnessed the construction of numerous parkways, which were typically curved and allowed for only limited velocity. Within cities, streets changed from places of pedestrian travel and public life to those whose primary function was to serve as open space to arterials for vehicles.

Mass automobile ownership triggered an enormous wave of time-space compression in the form of suburbanization, the quintessential spatial fix of late modern capitalism and a reversal of the long-standing drift of people from rural areas to cities. (Some observers attribute the decline of dense, tightly knit urban neighborhoods in the 1920s and the parallel increase in alienation to the rise of the automobile and associated suburbanization.) On the urban periphery, automobiles allowed the radial trunk lines extending from urban cores to be supplemented by networks of circumferential trips. Suburbanization for the middle class became synonymous with upward social mobility, home ownership, and mass consumption. Indeed, the rise of the single-family home and all that it represents in many ways can be attributed to the shift to an automobile-oriented culture. The low residential densities of suburban areas in American cities were typically one quarter of those found in European ones. However, the freedom of choice offered by the automobile came at the price of depriving others: The decentralization of people and economic activities rendered public transportation systems, and those who depend on them, significantly disadvantaged, a cleavage marked by race/ethnicity as much as class. The result was that even the poor became heavily automobile dependent.

Concomitantly, the automobile led to numerous cultural, perceptual, and psychological changes. The generation that first learned to drive on a mass basis in the 1920s may be the only one to appreciate these changes fully, for their predecessors never mastered the art of driving, and their successors took it for granted. Learned doctors warned that the human body was not designed to withstand speeds of 60 miles per hour, and young men boasted if they had traveled at such velocities. Driving combined Western notions of freedom, mobility, privacy, and effortless mastery of enormous power, immersing drivers and passengers into a series of ever-changing spaces that rushed by rapidly. With its drivers and passengers ensconced within their private bubbles, the automobile reaffirmed and deepened the deeply individualistic character of bourgeois life. Indeed, automobile driving generated its own, unique set of feelings, perceptions, and culture (e.g., drive-through services), all of which are enmeshed in racialized, gendered, and national feelings of movement, mobility, identity, and embodiment. Many of the skills and practices of driving became so deeply internalized as to be unconscious.

Much later, the automobile continues to be a decisive force in the structure and restructuring of urban space. In many economically advanced countries, ownership among the middle class approaches near universal levels. The popularity—and necessity—of cars remains highest in the United States, which, with its underfunded mass transit system, has the world's highest rate of automobile ownership. Even in Europe and Japan, however, with their extensive rail and bus networks, the bulk of the population relies on automobiles to navigate space. Convenient, affordable, rapid, and safe automobile transit in the 20th century was, in turn, tied to the availability of cheap gasoline.

As the automobile became the defining symbol of 20th-century mobility, roads displaced rail as the primary object of state intervention in transportation. If the distinctive symbol of 19th-century urbanism was the boulevard, its 20th-century equivalent was the highway. The highway system, like the railroads, represented a major round of investment in the infrastructure, a spatial fix that both facilitated and imprisoned subsequent rounds of capital accumulation. In 1930s Europe, only dictators like Hitler or Mussolini possessed the necessary powers to force highway systems through the landscapes of private land ownership and impose a centralized transportation network. Italy under Mussolini initiated the *autostrade* network in 1924, the world's first roads exclusively designed for high-speed motor traffic. In 1930, the Germans undertook the *autobahnen* system connecting the country's major cities, an effort militarized by the Nazis. But it was the U.S. Interstate Highway System, the largest public works project in history, that epitomized this modality of movement. In requiring national defense to obtain legislative authorization, the interstates also epitomized Cold War politics. Automobile companies, construction companies, earth-moving equipment manufacturers, real estate firms, labor unions, and assorted other business interests supported the project enthusiastically. The system that eventually materialized over four decades (mid 1950s to mid 1990s) included even-numbered routes that ran east-west and odd-numbered ones that ran north-south (Figure 1).

The impacts of the Interstate Highway System are difficult to exaggerate. Its very construction

Figure 1   The U.S. Interstate Highway System, built between the late 1950s and early 1970s, exemplified the country's enormous dependence on the automobile for transportation of people and goods.

Source: Adapted from Moon, H. (1994). *The Interstate Highway System.* Washington, DC: Association of American Geographers.

displaced countless numbers of people in largely minority, impoverished inner-city communities. Together with the millions of cars that made use of it daily, it gave birth to a new landscape of industrial parks, office complexes, shopping malls, motels, fast food, and gasoline stations. Average interurban travel times declined by 10% within the first decade. The new road networks accelerated the flight of manufacturing from the Manufacturing Belt to the emerging Sunbelt, the suburbanization of people and firms, and the dispersal of industry to nonmetropolitan areas. With the option of both enhanced access to downtowns and life in low-density peripheries, commuting distances, but not times, for many suburbanites increased. Highway interchanges became intensive sites of commercial development in their own right. The evacuation of the middle class rose to new heights, with attendant crises for the governments and neighborhoods of the urban core. The

highway system thus completed the long-running transition of the American city into a completely automobile-dependent spatial formation, including massive suburbanization, a process that has continued well into the 21st century.

Barney Warf

See also Automobile Industry; Fordism; Mobility; Suburbs and Suburbanization; Time-Space Compression; Transportation Geography; Urban Spatial Structure; Urban Sprawl

Further Readings

Featherstone, M. (2004). Automobilities: An introduction. *Theory, Culture and Society, 21,* 1–24.

Flink, J. (1988). *The automobile age*. Cambridge: MIT Press.

Hugill, P. (1982). Good roads and the automobile in the United States, 1880–1929. *Geographical Review, 72*, 327–349.

Moon, H. (1994). *The Interstate Highway System*. Washington, DC: Association of American Geographers.

Newman, P., & Kenworthy, J. (1999). *Sustainability and cities: Overcoming automobile dependence*. Washington, DC: Island Press.

Scharff, V. (1991). *Taking the wheel: Women and the coming of the motor age*. New York: Free Press.

 # AVALANCHES

Avalanches occur as rapid, gravity-driven accelerations of different materials downslope at high rates of speed. All avalanches are caused by an overburden of material that is too massive and unstable for the slope that supports them. At least five different types occur, each type with numerous subtypes or relationships to other types: (1) snow avalanches, (2) ice avalanches, (3) slush avalanches, (4) debris avalanches, and (5) rock avalanches. Because of their common high velocities, all avalanches are greatly threatening to life, limb, and property, and the lives of tens of thousands of people have been lost in the many notorious occurrences. Because of the steep slopes and high potential energies of mountains that can convert their potential energy into high kinetic energy wherever snow, ice, slush, debris, and rock are detached from cliffs through climatic or seismic events, many mountain areas are the locations of all five of these types of mass movement. People who live in such areas are generally well aware of these natural hazards and face such risks with a certain fortitude, based on an understandable desire to beat the odds of what they hope are rare events. But, of course, the gambling odds are all too often not in favor of the people or their infrastructures in mountain regions where any or all types of avalanches can be common.

Snow Avalanches

Snow falls and accumulates on mountain slopes as snowpack wherein the commonly light and fluffy flakes of powder snow pile up, become unstable in various ways, and eventually flow downhill as a snow avalanche. Because of the ubiquity of snow, and therefore of snow avalanches, in almost all high mountain regions, the science of analysis and prediction of such events has become quite sophisticated. Mountain weather, snow formation and snow pack, avalanche formation, and avalanche terrain, motion, and effects are all important elements in the prediction of snow avalanches wherein the elements of stability evaluation and snowpack observations enable better avalanche forecasting.

Avalanche-relevant mountain weather first concerns the deeper snows and relatively milder temperatures of the maritime snow climates of coastal mountains, as opposed to the lighter snows and colder climes of interior ranges, both of which types can set up different conditions of snow instability. Temperature conditions are an essential element in the consideration of avalanches—both while the snow is falling, being redeposited, or both, as well as much later in the heat exchange at and within the snowpack. The snow itself falls in many different forms of crystal type, from the light and fluffy powder flakes to the heavy, coarse graupel, or hail, with an enormous variation in between. Heat exchange at the snow surface occurs as heat enters or leaves the snowpack by conduction or convection of radiation as well as by condensation resulting from diffusion of water vapor. Metamorphism of the snow crystal results. Formation of feathery crystal of unstable surface hoar frost (e.g., frozen dew on a surface) can be a problem, especially where buried by later snowfalls. The melting and refreezing of moisture between snow crystals in contact with each other causes the formation of grain bonds, or sintering, which are also crucial elements in snow strength, so that the temperature, temperature gradients, grain geometries (i.e., orientation of grains in a deposit), and pore-space (i.e., gaps between grains in a deposit) configuration can all figure in avalanche formation. The highest crystal growth rates occur where there are large temperature gradients, higher temperatures, and large spaces between crystals. This produces unstable angular or faceted (i.e., grains with one or more smoothed surfaces) grains or depth hoar.

In addition to temperature and pressure in snowpack metamorphism as a factor in snow-avalanche formation, horizontal and vertical winds also control the amounts and distributions of snow as primary precipitation events or as secondary redistributions through blowing and drifting. Snow is eroded in acceleration regions and deposited in deceleration regions, which produces lee-zone (i.e., in the leeward side of an obstacle) accumulations, cross-loading, deposition in gullies and notches, and unstable cornices (cantilevered snow structures formed by drifting snow) built out as overhangs off ridge tops, all of which can become potential avalanche sites as a result.

Snow-avalanche formation types are generally classified as loose snow avalanches that generally start at a point or the more dangerous slab avalanches that can propagate laterally across a slope. Loose snow avalanche formation occurs because such snow has little if any cohesion between flakes. A local loss of cohesion can result in a small movement that will propagate in a downhill direction into a bigger mass that can incorporate other snow layers of different moisture contents. Slab avalanches generally develop from a weaker layer such as faceted snow or depth hoar that fails and allows fractures of the snow to propagate both upslope and across slope to release the avalanche. The slope angles can control the types of snow avalanches so that dendritic (i.e., like the branching pattern of tree roots) or stellar snow crystal shapes have the highest stable angle of repose (up to 80°), while this may decrease to 35° for more rounded forms. As the water content increases, the angle of repose decreases, so that slush avalanches can occur on slopes of less than 15°. The normal range of slopes for release of slab avalanches is ~25° to 55°.

Avalanche terrain is the area in mountains where certain features exist that warn of an avalanche hazard. An avalanche area is one in which various times, types, and geographies of snow avalanches occur throughout one or more avalanche paths. Any given avalanche path will have its own starting zone, track, and runout zone. Starting zones or zones of origin of avalanches are commonly higher on mountain ridges, slopes, gulleys, and other collection points for snow. Slope angles there range from 90° to 60° for small snow sloughs, because enough snow can never accumulate to form large avalanches. Dry snow types range from 30° to 60°, small slabs from 45° to 55°, multisized slabs from 45° to 35°, and large slabs or wet, loose snow avalanches from 35° to 25°. Infrequent wet snow avalanche types occur from 25° to 10°. Other controls include orientation to the wind, orientation to the sun, forest cover, underlying ground surface, altitude, slope dimensions, and crown and flank positions of fracture lines according to local terrain features. The track or zone of transition through which the moving snow mass flows is the area of highest velocity. These areas are either open slopes or channels, and the typical slopes are 30° to 15°, with significant deceleration on slopes below 10°. The runout zone or zone of deposition is where the avalanche comes to a stop and drops its load of snow and any entrained debris or trees and so forth. Determination of runout distances in avalanche hazard analysis includes long-term observations of avalanche deposits, observations of damage to vegetation, structures, and the ground surface, and historical records. Avalanche frequencies can also be determined by analysis of the disturbed vegetation.

Ice Avalanches

Rapid avalanches of ice in mountains are generally caused when a glacier terminus (snout or end) is perched high on cliffs above, and as its front collapses through forward advance to the cliff edge, ice masses can be precipitated into the air or down steep slopes as huge and rapid masses of ice blocks. In some cases, where rock materials are incorporated as well and the internal ice melts, the moving mass can become a kind of rapid and highly mobile debris or rock avalanche. The two classes of ice avalanches resulting are thus based on size.

The smaller type is the result of *calving* or falling of ice blocks caused by internal flow (creep) and possibly also sliding of the glacier over its bed to a cliff edge where blocks break off and roll or bounce down the cliff as an analog to a rockfall. In many of the higher mountains in the world, these ice avalanches accumulate again at the cliff bottom to reconstitute a kind of glacier detached from its accumulation zone higher on the mountain.

An avalanche in the Caucasus Mountains

Source: iStockphoto.

The larger, rare types of ice avalanche occur where a huge piece of the glacier breaks off to form a massive ice avalanche, analogous to a large rapid landslide or rock avalanche, which can be greater than 10^6 m^3 (cubic meters) in size. These are enormously destructive, having buried whole towns and thousands of people in the Andes, Alps, Caucasus, and elsewhere. Some of them have rock entrained in the failure as well, so that a continuum between large ice avalanches and rock avalanches probably exits.

Slush Avalanches

These are a class of wet slab avalanches that are most common at high latitudes, generally due to rapid onset of spring snowmelt when the sun returns after long, dark winters. Starting zones can be 5° to 40°, but only rarely >25° to 30°. The snowpack is commonly partially or completely water saturated, and the bed surface of the failure is fairly impervious to that water. Depth hoar frost is common at the base of the snow cover, and failure release into a slush flow or "slusher" is commonly associated with intense snowmelt or heavy rain. Slush avalanches have exceptionally high densities, in some cases >1,000 kg (kilograms)/m^3, so that impact forces are among the highest and most destructive of any snow avalanches known.

Debris Avalanches

As their name implies, debris avalanches are usually mixed up masses of soil, gravel, rocks, trees, houses, cars, and whatever random materials

On May 31, 1970, a massive debris avalanche in Yungay, Peru, caused by a 7.9-magnitude earthquake, scoured this valley from wall to wall, burying the town of Yungay and killing most of its population. Some people were saved by climbing Cemetery Hill (shown here) to escape the avalanche.

Source: © Lloyd Cluff/Corbis.

exist on mountain slopes that can be mobilized and accelerated downslope in torrential rainstorms, rapid snowmelt, or collapse of water dams. Rainfall intensity or duration, along with the prior rainfall and general soil moisture conditions, is also a strong control in the triggering of debris avalanches.

Most well-vegetated mountain slopes have a thin veneer of sediment and soil on them that has been draped over them by glaciers, wind, and other processes and that is produced by long-term weathering of the bedrock beneath. Debris avalanches are common in topographic concavities or hollows in first-order watersheds. This geometry is conducive to the accumulation of colluvium and other sediment as well as the convergence of underground water necessary to cause the failure. Failure of this surficial cover is triggered by an unusual amount of water being delivered abruptly to the slope so that the water first infiltrates into the soil and sediment, saturates it to fill all pore spaces, and then, through return flow, begins to come back to the surface again lower down the slope. This greatly increased weight of water on the slope in the soil, coupled with the hydrostatic water pressure upward inside the slope cover and the seepage pressure of the return flow out onto the surface, lifts the sediment and soil mass and mobilizes it into the debris avalanche. Such rapid movements have a dominance of inertial forces with a high sediment concentration in a granular flow.

Rock Avalanches

These rapid inertial granular flows are caused when a large volume of mostly dry rock fragments derives from the collapse of a slope or cliff and moves for a long distance, even on gentle slopes. These *strurzstrom* failures (i.e., failures that exhibit much greater horizontal movement than initial vertical drop) can be initiated by a sudden seismic acceleration, a melting of internal permafrost in the rock joints, a debuttressing of the slope when nearby glaciers melt downward and remove support, or perhaps a simple freeze-thaw that wedges the rock mass outward. Once started, the fall height or sliding distance can impart a high velocity, which produces dilation and reduction of internal friction. These highly mobile masses of rock fragments can flow for long distances, apparently owing to a low effective coefficient of friction produced by some combination of fluidization produced by entrained air, vacuum-induced upward flow of air from the airfoil shape of the upper flow surface, trapped basal air-layer lubrication, buoyant dust suspensions, acoustic fluidization, or high-frequency vibration, steam generation from frictional heating, and various other hypothesized mobilization mechanisms. The result is that these long-runout-zone landslides are incredibly destructive, move with great rapidity, and obliterate everything in their path. In fact, as the size of the falling rock mass increases, the ratio of the fall height to the travel distance decreases, so that lower distances of fall can lead to further distances of travel. Lateral and frontal rides can develop when the rapidly traveling mass comes to an abrupt stop. These enigmatic features have also been observed on the Moon and Mars, where the gravity and atmospheres are very different, yet the physics of their motion must somehow be similar to produce near-identical surficial features.

John F. Shroder

See also Earthquakes; Geomorphology; Glaciers: Mountain; Ice; Landforms; Landslide

Further Readings

Easterbrook, D. J. (1999). *Surface processes and landforms.* Upper Saddle River, NJ: Prentice Hall.
McClung, D., & Shaerer, P. (1993). *The avalanche handbook.* Seattle, WA: Mountaineers Books.
Pudasaini, S. P., & Hutter, K. (2007). *Avalanche dynamics: Dynamics of rapid flows of dense granular avalanches.* New York: Springer.

AVIATION AND GEOGRAPHY

In the more than 100 years since the Wright brothers' first flight at Kitty Hawk, North Carolina, aviation has become a significant influence on human geography. The world's air carriers now account for most intercontinental and transcontinental passenger trips and a rapidly growing share of shorter-haul movements, and they carry about 40% of world trade by value. Air transportation has become more important as it has become more affordable due both to technological change and, especially in the past few decades, to the deregulation and privatization of the airline industry. The resulting increased share of the world's travelers and cargo moved by air has affected patterns of urban and regional economic development and has helped enlarge the spatial scale of everyday life—as is evident in the global traffic in everything from tourists to fresh-cut flowers. At the same time, however, inexpensive and widespread air mobility has emerged as a potent force affecting the environment.

An early theme of geographic research on air transportation concerned the development of airline networks. In the 1950s, geographers examined the structure of airline networks in the United States, for instance, as a way of learning about the broader structure of the American urban hierarchy. Today, the scale of interest is global. The world's airlines operate networks within which the primacy of alpha world cities such as London, New York, and Tokyo is evident. Conversely, cities in the periphery tend to be more difficult and expensive to reach, and that disadvantage, among many, perpetuates their poverty.

The architecture of an airline network is not, however, solely attributable to the relative wealth and power of the cities linked. Aircraft technology is another important consideration; advances in the capacity, speed, and range of commercial airliners have encouraged time-space compression, the stretching of nonstop linkages, and the consolidation of traffic at a relative handful of major hubs. In the past decade, airlines have begun flying several aircraft models whose range is long enough to connect almost any two cities on the

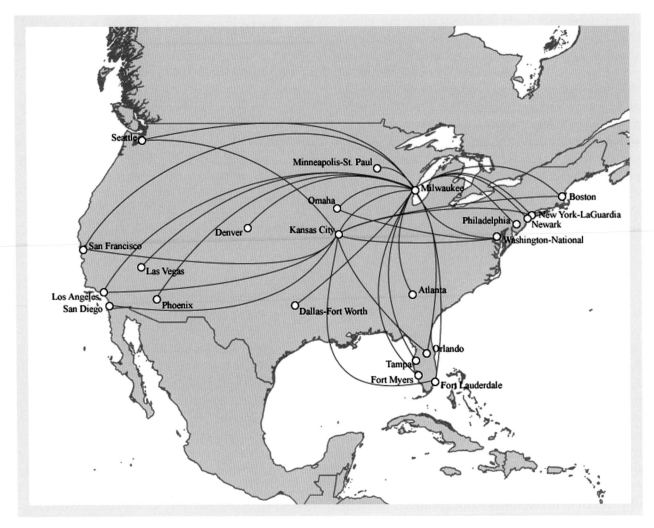

Figure 1 Midwest Airlines network: an example of a hub-and-spoke network

Source: Author.

planet nonstop. Among these new planes is the Airbus A380, which not only has a very long range but is also the largest commercial jet in service. The A380, which has lower costs per passenger-kilometer due to economies of scale, is expected to fly on routes linking the world's most important hubs, especially in Asia. Interestingly, the maiden commercial flight taken by the "Superjumbo" linked Singapore and Sydney, making it the first major jetliner to debut on a route involving neither the United States nor Europe.

Singapore and Sydney are examples of cities that have benefited greatly from the privatization and deregulation of the airline industry. Since the 1970s, many formerly state-owned carriers, such as Singapore Airlines and Australia's

Qantas, have been at least partially privatized, and governments have loosened regulations that once limited the number of airlines on particular routes, the capacity they offered, and the airfares and air cargo rates they levied. Geographers have analyzed the consequences of these changes across the world. In the United States, deregulation in 1978 led to the elaboration of hub-and-spoke systems with many "spoke" cities feeding traffic into and being fed in turn by traffic from one of several major hubs. Midwest Airlines (Figure 1), for instance, is able to offer higher frequency service to more destinations by consolidating traffic at its two hubs.

Research has shown that the largest hubs have gained from their superior accessibility. Delta Air

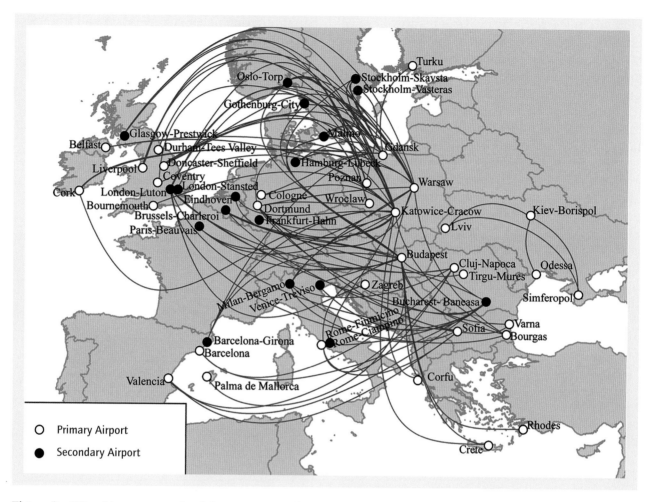

Figure 2 Wizz Air: an example of the greater emphasis on point-to-point services in a low-cost carrier's network
Source: Author.

Lines' hub in Atlanta has helped make the city's airport the busiest in the world in terms of passenger volume, which in turn has had a catalytic effect on Atlanta's economic development, especially in the vicinity of the airport (and the same is true in cities that have important air cargo hubs, such as Memphis with its FedEx hub). On the other hand, Delta's dominance of flights to and from the city has made the city one of several "megahubs," where high levels of market concentration have muted the effect of deregulation on fares. A number of spoke cities, too, with few competitors, have been high-cost "pockets of pain."

The more general pattern, however, has been for competition to increase in deregulated markets. Many of the new players in the industry are low-cost carriers (LCCs), which have costs that are, on average, less than half those of full-service network carriers such as Delta. An interesting spatial element common to many LCCs is the use of inexpensive secondary airports. For instance, the Hungary-based airline Wizz Air (Figure 2) uses airports such as London-Luton instead of the more costly London-Heathrow. Several other important aspects of LCCs are illustrated by Wizz Air. First, LCCs often have a stronger emphasis on point-to-point services instead of hub-and-spoke operations—though many LCCs do have hubs. Budapest, Katowice, Warsaw, and Gdansk are more important than other cities in the Wizz Air network, but there are many nonstop services that bypass these cities. Second, Wizz Air has taken advantage of deregulation in Europe to operate routes that would have been impossible a few decades ago. Although it is a

Hungarian carrier, Wizz Air operates several hubs in Poland and flies domestic routes in the Ukraine. The very existence of Wizz Air is testament to liberalization, since Hungary had only one state-owned airline for decades. Finally, the irreverent name of the carrier reflects the democratization of air travel that LCCs and their low fares have helped engender.

The very popularity of air travel may be a problem, however. The environmental sustainability of LCCs and the air transportation system generally have been called into question by geographers. Aviation has significant environmental costs at a variety of scales, including the global. In fact, the world's airlines account for an estimated 2% of anthropogenic climate change, but that share is projected to grow, and with it, so will the concerns about the "hypermobility" afforded by inexpensive air transportation. Low fares and air cargo rates have made it easier to vacation farther from home, to eat fresh produce year-round, to operate globally dispersed production networks, and even to commute between countries; but in aviation's second century, these practices are likely to come under increasing pressure, from higher energy prices and from policies intended to protect the environment.

John Bowen

See also Accessibility; Mobility; Time-Space Compression; Transportation Geography; World Cities

Further Readings

Derudder, B., Devriendt, L., & Witlox, F. (2007). Flying where you don't want to go: An empirical analysis of hubs in the global airline network. *Tijdschrift voor Economische Geografie, 98,* 307–325.

Graham, B. (1995). *Geography and air transport.* New York: Wiley.

Solberg, C. (1979). *Conquest of the skies: A history of commercial aviation in America.* Boston: Little, Brown.

Zook, M. A., & Brunn, S. D. (2006). From podes to antipodes: Positionalities and global airline geographies. *Annals of the Association of American Geographers, 96*(3), 471–490.

B

🌐 BARRIER ISLANDS

Barrier islands are an important part of the natural, economic, and recreational landscape of the North American coast. These elongated and detached bodies of unconsolidated sand or gravel lie above the high-tide level and are separated from the mainland by a lagoon, a marsh, or a sound. The islands are most prominent on wide and gently sloped trailing edge coasts, although they can form on collision coasts where there is an adequate supply of sediment. The islands typically occur in chains separated from neighboring islands by inlets, which are spaced depending on the relative strength of wave and tide forces. Long (wave-dominated) barriers tend to form where tidal range is small, and waves are able to close breaches that form during storms (e.g., from Southern Texas to the Northwest Florida coast). In contrast, mixed-energy barriers form through a combination of wave and tide forcing, leading to short barriers that are wider on one end (a drumstick) and separated by stable inlets with large sand shoals (e.g., the Georgia coast).

Theories of barrier island formation include (a) the subaerial growth of near-shore bars, (b) the long-shore progradation and eventual breaching of spits, and (c) the submergence of coastal dune ridges during the last glacial lowstand. Most barrier islands are less than 7,000 years in age, suggesting that they formed during a period when sea level was gradually rising, most probably as a combination of all the above mechanisms. Once formed, the morphology, composition, and stability of a barrier island depends on the relative importance of waves and tide, the availability of sediment, and relative sea-level rise. If there is a sufficient supply of sediment from offshore or alongshore, the island progrades seaward, leading to a regressive barrier, able to maintain itself against rising sea level. However, most modern barrier islands are retreating landward in response to both global (eustatic) sea-level rise and local subsidence at a rate that depends on the slope of the continental shelf. Despite an almost constant rise in sea level, barrier island transgression (rollover) tends to only occur during extreme storms capable of breaching the foredune and distributing sediment to the back of the island in the form of washover fans (see second photo).

The ability of coastal dunes to control the rate and timing of transgression depends on the ability of dunes to redevelop in height and extent from sediment delivered from the beach or washover fans by the wind following storms. Barrier islands affected by frequent storm events tend to be low in profile and are susceptible to island overstepping (island inundation). A lower frequency of storms or a large supply of sediment leads to larger dunes that can recover quickly following a storm, thereby limiting washover and promoting island rollover. A change in the rate of sea-level rise or the frequency of hurricanes and

Santa Rosa Island, a wave-dominated barrier island in northwest Florida, photographed in 2004

Source: Author.

Santa Rosa Island in northwest Florida following Hurricane Ivan (September 2004). Shown are washover fans extending into the sound behind the island. This transfer of sediment is part of the barrier's transgression.

Source: Author.

tropical storms and a reduction in sediment supply through the construction of groins and jetties may lead to more rapid transgression and shoreline erosion.

Chris Houser

See also Coastal Erosion and Deposition; Coastal Hazards; Coastal Zone and Marine Pollution; Delta; Dunes; Geomorphology

Further Readings

Dean, C. (1999). *Against the tide: The battle for America's beaches*. New York: Columbia University Press.

Sallenger, A. (2000). Storm impact scale for barrier islands. *Journal of Coastal Research, 16*, 890–895.

 ## BARROWS, HARLAN (1877–1960)

Harlan H. Barrows was an American geographer best known for his interpretation of geography as human ecology. He expressed this view in his 1922 presidential address to the Association of American Geographers. In Barrows's opinion, simply describing what exists in the landscape was not enough. He believed geography should focus on the actions people take and the adjustments they make to live in their environments. In other words, Barrows thought it was important to focus on what people do to survive and thrive in their environments rather than looking for how the environment influences them.

In Barrows's day, many believed that cultures were influenced almost entirely by the environment. This belief, called *environmental determinism*, does not consider the fact that people make decisions and adapt to their environments. Instead, determinists look for environmental factors that influence or "determine" culture. Barrows held that geographers should study the interactions people have with the environment rather than looking for influences. He thought environmental determinists searched for simple answers, seeking influences in the environment that do not really exist. His colleague, Ellen Churchill Semple, was a well-known environmental determinist and one of the first female geographers.

In Barrows's view, geography was largely about the relationships people had with their environments. He believed that this approach would help geographers discover the complexity of human-environmental relationships. Achieving this goal of discovering human and environmental relationships required geographers to balance knowledge of the natural world with knowledge of the social world, especially history and economics. Barrows differentiated history and historical geography. In his view, history starts at the past for the purpose of learning about the past, but historical geography starts with the present and should study how relationships with the environment progressed. Barrows was more concerned with the present than the past.

Born in Armada, Michigan, he earned his BS degree from the University of Chicago in 1903. That same year, the University of Chicago established the first university geography department in the United States. In 1907, Barrows was hired as a geography instructor and became a professor in 1914. He was promoted to chair of the department of geography in 1919 and maintained that position until stepping down and becoming an emeritus professor in 1942. Barrows also worked on numerous natural resource projects for the federal government. A common theme for much of this work concerned the conservation of water resources, which he included in reports to the U.S. Bureau of Reclamation and the U.S. National Resources Committee.

Jessey Gilley

See also Anthropogeography; Environmental Determinism; Human Ecology; Semple, Ellen Churchill

Further Readings

Barrows, H. (1923). Geography as human ecology. *Annals of the Association of American Geographers, 13,* 1–14.

Chapell, J. (1971). Harlan Barrows and environmentalism. *Annals of the Association of American Geographers, 61,* 198–201.

Koelsch, W. (1969). The historical geography of Harlan H. Barrows. *Annals of the Association of American Geographers, 59,* 632–651.

BASIN AND RANGE TOPOGRAPHY

Basin and range topography is a type of landscape where ridges and mountains alternate with depressions in a somewhat regular pattern. Such topography occurs wherever diastrophic processes have produced faults that have uplifted blocks and dropped down intervening basins. These features are not tied genetically in any way to arid areas but many of the best-known examples do occur in such regions, which allows for greater recognition of them partly because of good exposures of the causative structural processes. Many of the basins have entirely interior or endorheic drainage so that no water drains out to the sea. Such landscapes occur in the well-known Great Basin geomorphologic province of the southwestern United States and northwestern Mexico, as well as the dry littoral regions of Chile and Peru, parts of Iran, Western Afghanistan, and Southern Pakistan, in Central Asia, and in parts of the Sahara Desert in Northern Africa. Additional examples are known from northeastern Japan and the Gulf of Corinth, Greece.

The primary genetic attributes characteristic of basin and range topography are down-dropped *graben* rock structures and uplifted *horst* block rock structures that are bounded with tensional normal faults. In addition are numerous cases of tilted fault blocks that are lifted up on one side and dropped down on the other into what are termed *half-grabens*. In either case the graben and half-graben basins are generally filled in with sediment derived from the erosion of the nearby uplifted horst blocks or mountains. In arid regions, most of these graben or basin-filling sediments are comprised of alluvial fan sands and gravels, with the basin centers being largely alluvial plains down which ephemeral stream channels flow, or lacustrine (lake) infillings left as former water bodies dried up and became dry or *playa* lakes, the most well-known of these playa lakes is Lake Bonneville that existed during the Pleistocene Ice Age. Lake Bonneville was a large lake centered on what is now the State of Utah, which dried up as the climate warmed and dried after about 10,000 years ago to leave the shrunken remnant of what is now the Great Salt Lake, as well as the Bonneville Salt Flats that are used for attempts at land speed records.

The cause of faulting of basin and range structure and topography in the American Southwest is crustal extension or stretching of the rocks of the crust, perhaps several tens of millions of years ago. In the course of the disruption of the crust, huge outpourings of lava occurred, which left extensive lava-capped tilted fault blocks at the edges of the half-grabens. The cause for this crustal extension is not known with certainty but apparently when the great slab of the North American tectonic plate was pushed up on top of the San Andreas transform fault zone, a compressional gap developed. This seems to have reduced the compressional forces that the Great Basin area was under and allowed extension of the crustal rocks toward the slab gap and out more rapidly to the west. The result was the extensive north-south normal fault system that allowed the crust to fragment so extensively.

John F. Shroder

See also Climate: Dry; Diastrophism; Geomorphology; Landforms

Further Readings

McPhee, J. (1982). *Basin and range.* New York: Farrar, Straus & Giroux.

BATTY, MICHAEL (1945–)

Michael Batty is one of the most renowned and forward-thinking, analytical-planning researchers of his generation, and it is from this perspective that he has made enduring contributions to GIScience. He graduated in Town and Country Planning from the University of Manchester (United Kingdom) in 1966 and earned a PhD from the University of Wales in 1984. He has had a successful and varied career on both sides of the Atlantic Ocean: After holding junior faculty positions at the universities of Manchester, Reading (United Kingdom), and Waterloo (Canada), he became professor of city and regional planning at Cardiff University (United Kingdom) in 1979, where he also served (between 1983 and 1986) as dean of the School of Environmental Design. In 1990, he moved to the State University of New York at Buffalo in order to serve as site director of the National Center for Geographic Information and Analysis. He returned to the United Kingdom in 1995 to establish and direct the interdisciplinary Centre for Advanced Spatial Analysis (CASA) at University College London, where he also serves as Bartlett Professor of Planning. Since 1981, he has also worked as editor-in-chief of the journal *Environment and Planning B: Planning and Design*. He was elected a fellow of the British Academy in 2001 and was awarded a CBE (Commander of the British Empire) for services to geography in 2004.

Many of his contributions to the development of GIScience are summarized in the 2003 CASA-edited collection, *Advanced Spatial Analysis: The CASA Book of GIS*. The most recent summary of his path-breaking contributions to the understanding of the size, shape, scale, dimension, and functioning of city systems can be found in his 2005 monograph, *Cities and Complexity*. His 1976 book, *Urban Modelling*, defined the agenda for quantitative analysis of city systems over the following 25 years, and his 1994 book *Fractal Cities* provides an early yet rich statement of ideas concerning the way in which cities grow and change. Other key research has focused on the development of agent-based models in GIScience and the laws that characterize population growth.

Paul A. Longley

See also Agent-Based Models; Cellular Automata; Complexity Theory; GIScience; Models and Modeling; Spatial Analysis

Further Readings

Batty, M. (1976). *Urban modelling*. Cambridge, UK: Cambridge University Press.

Batty, M. (2005). *Cities and complexity: Understanding cities with cellular automata, agent-based models, and fractals*. Cambridge: MIT Press.

Batty, M. (2006). Rank clocks. *Nature, 444*, 592–596.

Batty, M. (2008). The size, scale, and shape of cities. *Science, 319*, 769–771.

Batty, M., & Longley, P. (1994). *Fractal cities: A geometry of form and function*. London: Academic Press.

Longley, P., & Batty, M. (2003). *Advanced spatial analysis: The CASA book of GIS*. Redlands, CA: ESRI Press.

BAYESIAN STATISTICS IN SPATIAL ANALYSIS

Bayesian maximum entropy (BME) is a group of spatial analysis techniques that are used to study natural systems (physical, biological, health, social, financial, and cultural) and their attributes. Methodologically, spatial BME analysis is a synthesis of stochastics theory with epistematics principles. Stochastics theory involves random fields (spatial and spatiotemporal; ordinary and generalized). Epistematics introduces a fusion of evolutionary concepts and methods from brain and behavioral sciences (e.g., intentionality, adaptation, teleology of reason, complementarity, and prescriptivity) in real-world problem solving. Consequently, instead of a dry presentation of spatial analysis within a hermetically sealed mathematics discourse, the BME approach focuses on the basic inquiry process that investigates the problem's conceptual background and knowledge status, and presents it within a general methodological context that accounts for objective reality-human agent interactions and introduces quantitative tools of mathematics in an environment of realistic uncertainty.

BME analysis and techniques have certain important advantages, both theoretical and computational, compared with mainstream spatial statistics methods (including statistical regression, standard Bayesian inference, Markovian analysis, Kalman filters, Gaussian process, and geostatistical kriging techniques). The BME techniques (ordinary and generalized) incorporate various kinds of knowledge bases (core and site specific) in a rigorous and unified framework rather than in an ad hoc and artificial manner (which is the case, for example, with many statistical regression methods). The BME core knowledge base may include physical laws, scientific models, theoretical equations, social structures, logical rules, and reasoning principles; they assume spatial coordinate systems that accommodate Euclidean and non-Euclidean metrics; and the site-specific knowledge base includes hard and soft data, secondary information, empirical relationships, fuzzy sets, and engineering charts. BME techniques can study systems and attributes with heterogeneous space-time dependence patterns; they make no restrictive assumptions concerning estimator linearity and probabilistic normality (nonlinear estimators and non-Gaussian laws are automatically incorporated rather than been restricted by the linear/linearized estimators of mainstream spatial statistics); and they can be used in the spatiotemporal domain too. Also, BME analysis can consider uncertain yet valuable information at the estimation (prediction) points themselves, when available; it provides a complete predictive probability density function at every spatial point rather than just an attribute estimate (in this way, more than one possibility is available at each spatial point, as far as estimation is concerned); and it incorporates higher-order spatial moments (e.g., effect of skewness on spatial analysis).

BME techniques have been used in a wide range of scientific disciplines, including medical geography, epidemiology, earth and atmospheric sciences, biology, human exposure, global health, ecology, and environmental engineering. In these applications, BME analysis has been shown to rigorously and effectively process multisourced uncertainty sources (conceptual, technical, ontologic, and epistemic); to account for measurable and categorical variables; and to work in the case of multiple attributes (vector or co-BME) and different space-time supports (functional BME). Noticeably, the BME techniques are particularly suitable for multidisciplinary real-world projects.

A number of computer software packages exist for the implementation of spatial and spatiotemporal BME statistics concepts and techniques in practice. One such package is the SEKS-GUI software library (Spatiotemporal Epistematics Knowledge Synthesis & Graphical User Interface; see http://homepage.ntu.edu.tw/~hlyu/software/SEKSGUI/SEKSHome.html). A comprehensive user's manual is available at the same Web site, which addresses a broad audience ranging from the novice spatiotemporal modeling user to the field expert.

George Christakos

See also Geostatistics; Heuristic Methods in Spatial Analysis; Quantitative Methods; Spatial Statistics

Further Readings

Bogaert, P. (2002). Spatial prediction of categorical variables: The BME approach. *Stochastic Environmental Research and Risk Assessment, 16,* 425–448.

Christakos, G. (2000). *Modern spatiotemporal geostatistics.* New York: Oxford University Press.

Christakos, G. (2009). *Treatise on epistematics: An evolutionary framework of real-world problem solving.* New York: Springer.

D'Or, D., & Bogaert, P. (2003). Continuous-valued map reconstruction with the Bayesian Maximum Entropy. *Geoderma, 112,* 169–178.

Douaik, A., Van Meirvenne, M., & Toth, T. (2005). Soil salinity mapping using spatio-temporal kriging and Bayesian maximum entropy with interval soft data. *Geoderma, 128,* 234–248.

Lee, S.-J., Balling, R., & Gober, P. (2008). Bayesian maximum entropy mapping and the soft data problem in urban climate research. *Annals of the Association of American Geographers, 98*(2), 309–322.

Orton, T., & Lark, R. (2007). Estimating the local mean for Bayesian maximum entropy by generalized least squares and maximum likelihood, and an application to the spatial analysis of a censored soil variable. *Journal of Soil Science, 58,* 60–73.

Savelieva, E., Demyanov, V., Kanevski, M., Serre, M. L., & Christakos, G. (2005). BME-based uncertainty assessment of the Chernobyl fallout. *Geoderma, 128,* 312–324.

 # BEHAVIORAL GEOGRAPHY

Human spatial behavior is not always obviously rational or efficient. Geographers interested in understanding spatial choice realized that spatial behavior could not be described using solely aggregated data with broad assumptions of rationalism and maximum efficiency. Confronting these limitations, behavioral geography evolved as a theoretical framework and methodological approach for investigating spatial behavior.

Behavioral geography is the study of human perceptions, cognition (e.g., internal mental process such as recognition and recall), and contextual interpretations of space, and ultimately choice as it relates to spatial decisions and spatial problem solving. Emphasis is on individuals as the unit of study paired with human-environment interaction assumptions inherent in constructivist philosophy. The primary objective remains true to its analytical ancestor of traditional location analysis, that is, a search for explanation or process of individual spatial choice with the intent of generalizing to social groups.

Theoretical Basis

The foundations for modern behavioral geography are derived from three main sources: (1) economic theories of location and spatial analysis, (2) theories of cognitive acquisition of spatial concepts, and (3) theories of learning. The psychologists Piaget and Inhelder proposed a hierarchy of spatial concepts based on human development: topological, projective, then Euclidean. Geographers modified this idea in various ways. For example, Reginald Golledge proposes an anchor point theory of spatial cognition that postulates that spatial knowledge is gained through a hierarchical process of acquiring first landmark knowledge, followed by route knowledge, and then configurational (or survey) knowledge. Using hierarchical theories for the cognitive development of spatial concepts, behavioral geographers have explored route choice and strategies for learning new routes and places.

Cognitive mapping, introduced by the psychologist Edward Tolman, greatly influenced research in behavioral geography. Cognitive mapping is useful as means to explain internal processes for storage and retrieval of spatial information and as an external method of analysis to understand individuals' perceptions of spatial information. Geographers such as David Stea, Peter Gould, and David Montello used cognitive maps to explore spatial bias, cultural influence, orientation skills, perceived importance of locations, and how these factors influence behavior such as distance estimation, wayfinding, and perception of natural hazards.

Constructivism, proposed by Piaget as a general learning theory, has influenced behavioral geography broadly. The primary assumption of constructivism is that new knowledge (learning) is built on existing knowledge. Individuals' interpretation of new knowledge is always influenced by culture, prior experiences, language, social interaction, attitudes, and motivation. Constructivism emphasizes the importance of context and individual construction of knowledge.

Behavioral geography is characterized by four main assumptions as described by John Gold. First, "space" is composed of both a real, physical, "objective" environment and a perceived, metaphysical, subjective environment. Real and perceived environments interact constantly setting a stage that determines human perception of the next "fact." As a result, a second assumption is that individuals not only respond to their environments but also change their environments. Third, the unit of study starts at the individual level and transitions to social groups. Behavioral geography is a bottom-up process of inquiry. Finally, behavioral geography is multidisciplinary. Cognitive psychology, education psychology, urban planning, anthropology, sociology, and recently cognitive neuroscience have contributed ideas and methods to behavioral geography.

Methods

Location theories based on aggregated quantitative data did not adequately predict or explain real-world spatial choices. Purely quantitative and stochastic methods intentionally exclude complex social and cultural variables. Because social and cultural variables are critical to understanding spatial behavior, mixed methods are employed in an effort to get at these complex and dynamic variables. Recognizing that spatial choices and preferences are likely to change over time, research is often centered on understanding process rather than on predicting specific outcomes.

Gathering information on variables such as perceptions, attitudes, and values requires direct interaction through methods such as surveys, interviews, and focus groups. Resulting data are frequently in noninterval or nonnumerical formats such as binary responses (yes/no), Likert scales, sketches, pictures, video and audio recordings, field notes, and interview notes. Sample size is often small with nonrandom or selective sampling that does not fit parametric assumptions. Nonparametric statistics, qualitative analysis, quantitative analysis, and spatial analysis are selected and weighted according to the specific research question or problem. Research emphasis is on applications in the real world rather than in a lab under controlled conditions. In general, methods in behavioral geography are applied in nature; they focus on large-scale and real-world space, use mixed methods for data collection and data analysis, and use probabilistic models of spatial analysis.

Criticisms

Perhaps, there is too much emphasis in behavioral geography on disaggregated individual information. If individuals are truly unique in their spatial decision making, as extreme constructivism and postmodernism suggest, generalization is impossible. Behavioral geographers assume that humans share the same "real" world, even though individual perceptions vary. They also assume that perceptions and knowledge are shared and taught to others in the same setting or social group so that within that group generalizations are plausible.

A second criticism is aimed at the contradictions between the importance of context and the methods employed to understand behavior. Surveys, interviews, focus groups, and other forms of qualitative data collection can be intrusive, and they likely alter the context and perceptions that behavioral geographers seek to understand.

In spite of these criticisms, behavioral geography has made positive contributions. Examples include advances in visualization methods, cognitive cartography, spatial cognition, hazards research, geography education research, marketing research, geographic information science (GIScience), and transportation models. With the increasing need to understand spatial behavior, behavioral geography will continue to gain momentum as a theoretical framework and methodological approach.

Sandra Metoyer

See also Environmental Perception; GIScience; Golledge, Reginald; Mental Maps; Phenomenology; Spatial Cognition

Further Readings

Gold, J. (1983). *An introduction to behavioral geography.* New York: Oxford University Press.

Golledge, R. (1981). Misconceptions, misinterpretations, and misrepresentations of behavioral approaches in human geography. *Environment and Planning A, 13*(11), 1325–1344.

Golledge, R. (2008). Behavioral geography and the theoretical/quantitative revolution. *Geographical Analysis, 40,* 239–257.

Golledge, R., & Stimson, R. (1997). *Spatial behavior: A geographic perspective.* New York: Guilford Press.

National Research Council. (2006). *Learning to think spatially: GIS as a support system in the K-12 curriculum.* Washington, DC: National Academies Press.

BERKELEY SCHOOL

The Berkeley School of geography, also sometimes called the *California School*, was a movement of geographic analysis that originated with Carl Sauer and the University of California, Berkeley (UC Berkeley) in the early to mid 20th century. It was a major branch of cultural geography concerned with how humans interact with and change their landscapes. The Berkeley School helped diminish the popularity of environmental determinism and placed an emphasis on historical human culture and behavior. As such, it was heavily aligned with UC Berkeley's history and anthropology departments. Scholars at the Berkeley School believed that by studying historical cultures and their impacts on the landscape, they could better understand modern human impacts. Unlike its counterpart in the Midwest, the Berkeley School was also more concerned with theoretical studies as opposed to practical applications of scholarship such as urban planning or resource management. Thus, both its topics of study and its methods differed significantly from those of Midwestern American geographers. The Berkeley School was significant because it represented a new way to study geography and also gave the subject prominence on the U.S. West Coast.

Origins

The Berkeley School began in 1923 when Carl Sauer moved from the University of Michigan to UC Berkeley. Prior to Sauer's arrival, geography at Berkeley was an independent department but was housed with geology; most of its courses were technical in nature and focused on physiography and geomorphology. William Morris Davis and his ideas of the cycle of erosion dominated the department at the time. When Sauer arrived, he sought to restructure the curriculum with the help of John Leighly, a student from Michigan, and separated the various types of study within geography. "Physical Geography" became the first course students took, but it was followed by a new course in cultural geography,

which later became the main component of the Berkeley School of thought. Cultural geography as it was studied at Berkeley was regional in focus and was concerned with the density and development of populations, the character of the world's peoples, economic systems, specific regional cultures, and how people shape their physical environment. As cultural geography developed at Berkeley, it later became more concerned with the past and was aligned with the university's history and anthropology departments.

Characteristics of Study

Those studying in the Berkeley School used Sauer's dimension of time to study places, people, and their interactions with the landscape. In studying these things, however, Berkeley scholars did not rely on the opinions of city planners and politicians but instead conducted their own extensive fieldwork, which frequently involved long-term observations of people in their day-to-day lives. In addition to fieldwork, Berkeley scholars also examined cultural history and humanized landscapes with archival research. The regions and cultures studied in the Berkeley School were often outside the United States and other English-speaking areas and much of the research to come out of the school was focused on, though not limited to, Latin American studies.

Another important characteristic of the Berkeley School was its location. Because it was on the West Coast and located far from the concentration of geography departments in the Midwest, it was relatively isolated from other universities studying geography. This isolation allowed the school to develop and reinforce its own set of ideas and characteristics of study that significantly differed from those of other universities—most important the "practical" aspects of education prevalent in the Midwest. The theoretical ideas important in the Berkeley School were then reinforced through generations of students as the school's isolation allowed Sauer to hire many of his former students to work in the department on receiving their degrees.

Key People and Dissertations

Apart from Sauer, the Berkeley School has a number of important people who either studied or worked there, or both. Peveril Meigs and Lauren Post were the first students of the Berkeley School, and Meigs graduated in 1933 after completing his dissertation titled *The Dominican Mission Frontier of Lower California: A Chapter in Historical Geography*. Post received his PhD in 1937 with his dissertation titled *The Cultural Geography of the Prairies of Southwest Louisiana*. One of the most famous geographers to come out of the Berkeley School though was Charles Warren Thornthwaite, who completed his PhD in 1930 with his dissertation *Louisville, Kentucky: A Study in Urban Geography*. The titles of these students' works are significant because they emphasize cultural and historical geography common in the Berkeley School. Following these works were projects such as the 1944 *Historical Geography of Michoacán* by Dan Stanislawski and *The Peoples and Economy of Begemden and Semyen, Ethiopia* of 1956 by Frederick John Simoons. In addition to the historic and cultural aspects, these projects clearly represent the emphasis on places outside the United States.

Critiques and Decline of the Berkeley School

Like many academic movements, the Berkeley School experienced several analytical critiques during its height. Critics said that its research methods were static and believed that the school was concerned only with material human landscape artifacts such as farms. As geography became increasingly theoretically oriented in the late 20th century, the Berkeley School's popularity declined as the "New Cultural Geography" took hold.   •

Amanda DeeAnn Briney

See also Cultural Geography; Environmental Determinism; Sauer, Carl

Further Readings

Cosgrove, D., & Jackson, P. (1987). New directions in cultural geography. *Area, 19*(2), 95–101.
Entrikin, J. (1984). Carl O. Sauer, philosopher in spite of himself. *Geographical Review, 74*(4), 387–408.
Kenzer, M. (Ed.). (1987). *Carl O. Sauer: A tribute.* Corvalis: Oregon State University Press.
Price, M., & Lewis, M. (1993). The reinvention of cultural geography. *Annals of the Association of American Geographers, 83*(1), 1–17.
Sauer, C. (1974). The fourth dimension of geography. *Annals of the Association of American Geographers, 64*(2), 189–192.

BERRY, BRIAN (1934–)

One of geography's most well-known and productive scholars, Brian Berry has played an enormously influential role in urban and economic geography, primarily as the steadfast defender of traditional quantitative modeling.

Born in 1934 to working-class parents in England, Berry defied the confines of the British class system to rise to the topmost tiers of academia. He completed a BS in economics at the London School of Economics in 1955, where he was exposed to historical geography and introduced to the quantitative modeling of spatial phenomena. Immediately thereafter, he traveled to the University of Washington in Seattle just as the geography program there initiated the quantitative revolution in American geography. Berry thus formed one of William Garrison's cadre of "space cadets," along with Duane Marble, William Bunge, Michael Dacey, Arthur Getis, Richard Morrill, John Nystuen, and Walter Tobler, arguably the discipline's most successful and famous single cohort of students.

Three years later, armed with a PhD—at age 22—he began the first of a long list of academic positions at prestigious institutions, including the University of Chicago (1958–1976) and Harvard

University (1976–1981), where he served as the Frank Backus Williams Professor of City and Regional Planning, chair of the PhD program in urban planning, director of the Laboratory for Computer Graphics and Spatial Analysis, and as a faculty fellow of the Harvard Institute for International Development. From 1981 to 1986 he served as dean of the School of Urban Public Affairs at Carnegie-Mellon University in Pittsburgh. Beginning in 1986, he taught at the University of Texas, Dallas, where, in 1991, he became the Lloyd Viel Berkner Regental Professor of Political Economy and, in 2006, dean of the School of Social Sciences. He is the recipient of numerous awards and medals. In 1975, he became the first geographer and youngest social scientist ever elected to the National Academy of Sciences. In 1977–1978, he served as president of the Association of American Geographers. He also served as a founding coeditor and editor-in-chief of the journal *Urban Geography* from 1980 to 2006.

Berry's lengthy publication record—including more than 500 books, articles, and other publications—has earned him enormous recognition as one of geography's most fecund scholars. He advocated for a discipline that was self-consciously nomothetic in outlook and positivist in epistemology, thus emphasizing the need for general laws of explanation, quantitative methods, and rigorous empirical testing of hypotheses. Throughout his long career, he subscribed to a paradigm that privileged the abstract over the concrete, deduction over induction, and the universal over the specific. Berry emphasized the use of models as a means to simplify and shed light on the bewildering complexity of the world. He was instrumental in the adoption of multivariate statistics in the discipline. His early papers stressed the applicability of central place models of urban systems and detail studies of retail shopping patterns. Subsequent work on market centers and retailing was very influential in geography and business and economics. He also delved into the rank-size distributions of cities, hierarchal diffusion processes, and the impacts of transportation systems. In addition, Berry had a long-standing interest in urban morphology and urban problems such as inner-city poverty. Over time, Berry's

works came to be characterized by an increasing concern for the role of public policy.

Berry's later career focused on the dynamics of regional development in different national contexts. He conducted extensive work in India, Australia, Indonesia, and other countries. Comparative analyses of urbanization bridged these national contexts. Urban trends in the United States received considerable attention as well, including the question of counterurbanization. A persistent theme was the relation of demographic shifts and migration to regional change. This phase of his career was characterized by a mounting interest in issues of globalization, particularly the ways in which multinational corporations intersected with state policies to shape urban growth around the world. In the 1990s, Berry turned his focus to the role of long cycles or Kondratief waves and their relation to regional development and political relations. Subsequent work viewed utopian communities and turmoil associated with the periodic restructuring brought about by long waves.

As the embodiment of positivism and the quantitative revolution, Berry's intellectual position came under mounting criticism from the 1980s onward. A newer generation of geographers attuned to political economy and social theory began increasingly to question the relevance of abstract, ahistorical models and the unrealistic neoclassical logic of utility maximization. This schism was exemplified in a famous debate between Brian Berry and the Marxist geographer David Harvey. As a result of academic geography's shift to political economy, Berry appeared to many observers as increasingly conservative and disconnected from the field. However, never one to give up, Berry is known to this day for the enthusiasm and commitment with which he advocates his views, and whatever philosophical differences some geographers may have with him, he remains widely respected.

Barney Warf

See also Central Place Theory; Location Theory; Logical Positivism; Models and Modeling; Quantitative Revolution; Spatial Analysis; Urban Geography; Urban Hierarchy; Urban Spatial Structure

Further Readings

Berry, B. (1967). *Geography of market centres and retail distribution*. Englewood Cliffs, NJ: Prentice Hall.

Berry, B. (1973). *The human consequences of urbanisation: Divergent paths in the urban experience of the twentieth century*. New York: St. Martin's Press.

Berry, B. (1976). *Urbanisation and counter-urbanisation*. London: Sage.

Berry, B. (1980). Creating future geographies. *Annals of the Association of American Geographers, 70*, 449–458.

Berry, B. (1991). *Long-wave rhythms in economic development and political behavior*. Baltimore: Johns Hopkins University Press.

Berry, B. (2002). Clara voce cognito. In P. Gould & F. Pitts (Eds.), *Geographical voices: Fourteen autobiographical essays* (pp. 1–26). Syracuse, NY: Syracuse University Press.

Clark, G. (2004). Brian Berry. In P. Hubbard, R. Kitchin, & G. Valentine (Eds.), *Key thinkers on space and place* (pp. 47–52). London: Sage.

Warf, B. (2004). Troubled leviathan: The contemporary U.S. versus Brian Berry's U.S. *Professional Geographer, 56*, 85–90.

 # BHOPAL, INDIA, CHEMICAL DISASTER

On the night of December 3, 1984, the toxic chemical methyl isocyanate leaked from the premises of the U.S.-based Union Carbide's pesticide plant in Bhopal, the capital city of the central Indian state of Madhya Pradesh. Union Carbide had set up its plant amid a dense "squatter settlement" and had stored methyl isocyanate in violation of the 1975 Bhopal Development Plan. The leaked chemical engulfed this community and left a trail of fatalities. The number of people affected is disputed, with estimates suggesting that 2 to 10 thousand died in the immediate aftermath of the accident and that an additional half million were permanently maimed. Environmental scholars argue that the disaster had all the characteristics of what the sociologist Charles Perrow has termed

a *normal accident* and note that the probability of a normal accident occurring could have been minimized had Union Carbide meticulously designed its technological system and scrupulously maintained the plant's safety checks.

Scholars also note the perpetuation of a "second disaster" in Bhopal, a term that highlights two further facets of the disaster: (1) the travails of a protracted legal battle that secured paltry monetary compensation but failed to secure adequate health care facilities for the original victims of the disaster and (2) the production of a second generation of victims through the failure to adequately clean up the premises of the old factory, which houses 425 tons of toxic waste. The failure of Union Carbide and the Indian government to undertake environmental remediation in nearly 24 years has contaminated soil and water and has led to high incidence of birth defects among children born into the surrounding community. Yet a comprehensive study that would determine the nature and extent of danger posed by the exposure to toxic waste has not been commissioned. Such bureaucratic apathy coupled with willful corporate inaction has amplified the magnitude of suffering caused by the Bhopal disaster and turned it into an enduring disaster.

A committed set of activists and survivors continue to fight for justice for Bhopal victims: Coalitions such as the "International Campaign for Justice in Bhopal" are aggressively campaigning to force Dow Chemical (which acquired Union Carbide in 2001) to assume responsibility for toxic waste and undertake environmental remediation. This campaign relies on activists and experts located in various parts of the globe to exert pressure on media to report the everyday disaster of Bhopal and force Dow to act. The ongoing tragedy of the Bhopal disaster has spurred new forms of environmental justice activism. Furthermore, the enactment of various environmental legislations, such as the publication of the Toxics Release Inventory in the United States, and calls for greater corporate accountability, such as Greenpeace International's elaboration of the "Bhopal Principles on Corporate Accountability," are the legacy of this disaster. Bhopal, the site of the world's worst industrial accident, remains important for human geographers interested in the study of globalization and corporate

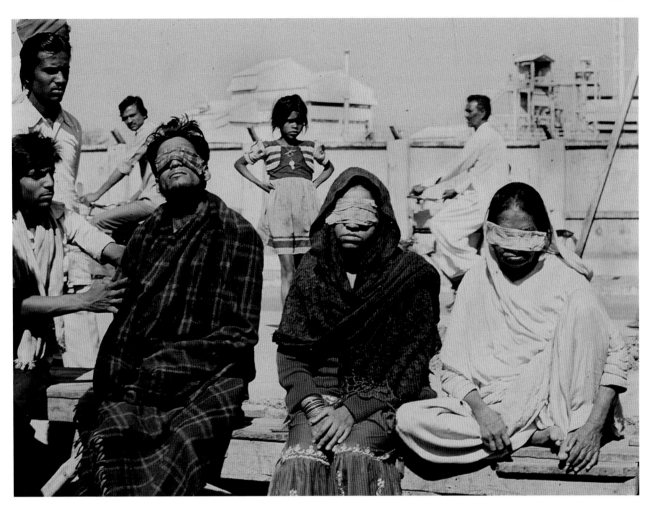

A December 1984 file photo shows victims who lost sight after poison gas leak from a pesticides plant in the central Indian city of Bhopal squatting in front of the U.S. Union Carbide factory (shown in background).

Source: AFP/Getty Images.

accountability, the evolution of global environmental activism, and new social movements.

Rajyashree Reddy

See also Chernobyl Nuclear Accident; Disaster Preparedness; Environmental Justice; Natural Hazards and Risk Analysis

Further Readings

Fortun, K. (2001). *Advocacy after Bhopal: Environmentalism, disaster, new global orders.* Chicago: University of Chicago Press.

Jasanoff, S. (2007). Bhopal's trials of knowledge and ignorance. *Isis, 98,* 344–350.
Rajan, R. (2001). What disasters tell us about environmental violence: The case of Bhopal. In M. Watts & N. Peluso (Eds.), *Violent environments* (pp. 380–398). Ithaca, NY: Cornell University Press.

BIBLICAL MAPPING

Biblical mapping relates to the various uses that have been made of maps in seeking to understand or explain the Bible. It embraces maps found in

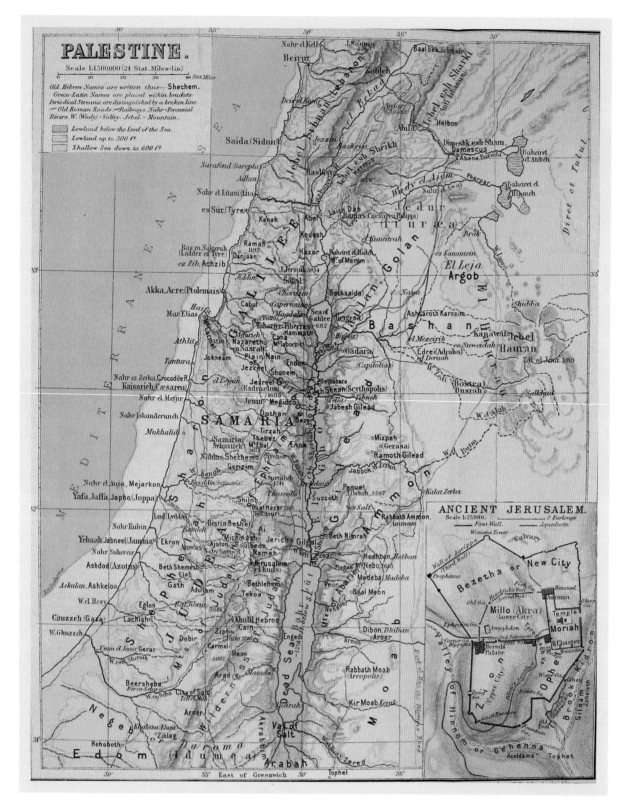

Figure 1 This map by C. Conder, typical of those in many early-20th-century Bibles, is a summary of 19th-century Holy Land research. It combines the topographical precision of surveyors with the onomastic conjectures of biblical scholars.

Source: Original is in the Founders' Library, Lampeter.

bibles, in special biblical atlases, stand-alone maps, and maps in tracts explaining biblical passages. Given that (a) the narratives that compose, and the events related within, the Judeo-Christian scriptures took place in, or are located within, the areas of Palestine, the Levant, Egypt, and Mesopotamia and that (b) most of those religious texts present their story within a historical form, it is not surprising that they have attracted attempts at mapping since a very early period after their canonization among both Jews and Christians. When events, places, pieces of land, and movements of individuals (e.g., Jesus going from Galilee to Jerusalem) or groups (e.g., the story of the "exodus" from Egypt) have deep significance to a group, those places will be mapped; and, thereby, the place of the narratives' audience is related to the events (see Figure 1).

Whether or not the authors of the Bible used maps in creating their texts is uncertain—several use language that closely coheres with ancient extant maps—but since Eusebius of Caesarea (ca. 260–ca. 340 BC), maps have been used to represent, clarify, or explain the biblical scenes: the 6th-century mosaic map in Madaba, Jordan, being an example inspired by Eusebius. With the passing of time, among both Jewish and Christian scholars, maps became an integral part of the exegetical process whereby geography was called on to solve textual problems (e.g., the 9th-century map in BNF [Backus-Naur Form, a context-free syntax] lat 11561 explicating the Book of Joshua); and later maps became part of biblical apologetics. With the arrival of print, maps became part of the apparatus of Bibles, while biblical/historical maps became a fixed feature of atlases from the time of Ortelius.

Still today maps are almost invariably included in printed Bibles, while biblical maps in atlases have evolved to become the specialist subcategory of atlases: the "Bible atlas." These atlases range in style from works for children to elaborate coffee-table books and in perspective from sectarian propaganda to works of critical scholarship. One common feature of most biblical mapping is that the maps combine the best available physical information (nowadays derived from geodetic survey and satellite imaging) with scholarly conjecture as to locations and distributions, working on the assumption that the world as we map it is the world as their ancient authors imagined it.

Thomas O'Loughlin

See also Ortelius; Religion, Geography and; T-in-O Maps

Further Readings

Bartlett, J. (2008). *Mapping Jordan through two millennia*. London: Maney.

Delano-Smith, C. (1991). Geography or Christianity? Maps of the Holy Land before A.D. 1000. *Journal of Theological Studies, 42,* 143–152.

Delano-Smith, C., & Morley Ingram, E. (1991). *Maps in Bibles 1500–1600: An illustrated catalogue.* Geneva: Librairie Droz.

Laor, E. (1986). *Maps of the Holy Land: Cartobibliography of printed maps, 1475–1900.* New York: Alan R. Liss.

Nebenzahl, J. (1986). *Maps of the Holy Land.* New York: Abbeyville Press.

O'Loughlin, T. (2005). Map and text: A mid ninth-century map for the Book of Joshua. *Imago Mundi, 57*(1), 7–22.

Rainey, A., & Notley, R. (2006). *The sacred bridge.* Jerusalem: Carta.

BIODIVERSITY

The study of biological diversity has a considerable legacy in ecology and geography. The concept refers generally to the sum total of all genes, species, and ecosystems in a particular region. How many bird species occur in a particular area, for example, and how does this feature change in response to evapotranspiration or latitude? The term *biodiversity*, a contraction of "biological diversity," did not appear until the 1980s and was a response to increasing scientific concern with ecosystem degradation and the species extinction crisis. Biodiversity as a field of interest thus has come to encompass not only understanding species numbers and patterns but also ecological interactions, conservation status, and relevant management strategies.

The question of how many species exist on Earth has long intrigued scientists. Although the number of plant and animal species is fairly well-known in the mid and higher latitudes, the number of taxa (distinct rank of organisms) in the tropics, especially of insects and microorganisms, will never be fully enumerated. Estimates place the total number of biota (plant and animal life) at between 4 million and 20 million species, although only 1.7 million species have been documented to date. The beetles (*Coleoptera*), with about 24% of all known species, are far and away the most species-rich taxon. Compare this group to plants, which make up 14%, and birds and mammals, which together make up only 2.7%, of the world's species.

Measures and Maintenance of Biodiversity

There are various means of measuring and conceptualizing biodiversity, and each possesses its own merit depending on the objective of the research. *Alpha* diversity refers to the number of species within a particular area or plot. In a 10-meter × 10-meter plot in a chaparral plant community, the alpha diversity of shrubs is simply a compilation of the number of shrub species encountered in the area. *Beta* diversity, on the other hand, refers to the change in species from one plot or area to another in an ecosystem. Also known as species turnover, beta diversity reveals much more about underlying diversity than simply a numbers count. For example, in a 1-ha (hectare) coniferous forest plot, a total of seven species of trees are identified. In a second 1-ha plot a few kilometers away, seven coniferous tree species are also recorded. However, six of the seven species encountered in the first plot are different from those encountered in the second plot. Thus, whereas the alpha diversity of the two plots is equal, the high beta diversity (difference in species composition between the two plots) suggests that the overall diversity of the ecosystem is quite high. A third type of diversity is referred to as *gamma* or landscape-level diversity. This is similar to beta diversity but refers to the total biodiversity of the various communities within a landscape. In this case, for example, we could be measuring the biotic diversity of the Columbia River watershed. Finally, *epsilon* diversity describes species richness at the broadest possible scale, that is, a region that encompasses many landscape types.

An important dimension of biodiversity at the local scale is community evenness. In this case, the issue is not only the number of species but also the number of individuals represented by each of the respective species. Two of the traditional means of measuring this feature are the Simpson and the Shannon-Weiner Indices. For example, two tropical forest plots are found to have equal alpha diversities, 25 tree species each, and an equal total number of trees, 100. However, in one plot, the 100 trees are equally divided among the 25 species, with 4 trees in each species. In the second plot, one species is represented by 76 individual trees, whereas the other 24 species are represented by only 1 tree each. The biodiversity of the first plot, as measured by community evenness, is thus much higher than the second plot.

Maximum species diversity in a particular ecosystem was long held to be a function of environmental stability. This approach followed the Clemensian ideal of community succession and species recovery following habitat disturbance from fire, flood, landslide, and other natural and anthropogenic action. This notion fell out of favor, however, and was replaced by the "intermediate disturbance hypothesis." In this case, the highest biodiversity is recorded neither in recently and heavily disturbed habitats, which are dominated by a few hardy pioneer species, nor in relatively undisturbed communities, in which a few highly competitive species have come to dominate. Rather, the sites with the highest biodiversity are patchy and moderately disturbed, and they include the myriad taxa associated with each of the various successional stages, from pioneer to old growth.

Latitudinal Biodiversity Gradients

During the European voyages of exploration, colonial naturalists reported that taxonomic diversity appeared to increase along a meridian from the poles toward the tropics. This geographical gradient in species richness, elaborated early by Alexander von Humboldt, holds for most types of terrestrial organisms, including birds, mammals, amphibians and reptiles, insects, and vascular plants. It also was found to apply to marine and freshwater fishes and arthropods, corals, mollusks, and other aquatic life-forms. For example, the number of bird species in the Canadian Arctic ($\approx 55°$ N) ranges from 60 to 100 species, whereas in Panama and Costa Rica ($\approx 10°$ N latitude) the number rises to more than 600 species. Likewise, a 1-ha plot of Northern California coniferous forest will yield around 15 species of trees, whereas the same sized area in Brazil's Atlantic Coastal forest will register well over 400 tree species. This latitudinal cline exists at other taxonomic ranks as well, including numbers of families and genera. There are of course some biota for which this pattern does not hold; the diversity of penguins, for example, does not increase from Antarctica toward the Congo River basin, nor does ursine biodiversity grow along a line from Alaska to Costa Rica. Species with parasitic life histories tend not to conform to a latitudinal pattern, nor do aquatic plants. Nevertheless, the latitudinal biodiversity gradient has proved remarkably consistent in terrestrial, freshwater, and marine ecosystems, and in both hemispheres.

Numerous hypotheses have been proposed to explain this phenomenon, but none has proven completely satisfactory. It remains to this day the most widely debated biogeographical pattern. Possible mechanisms range from competition, predation, and patchiness to environmental stability, productivity, and stochastic processes. One of the first to speculate on this biodiversity gradient was Alfred Russell Wallace, the cofounder of the theory of evolution via natural selection. He suggested that because the tropical and subtropical regions of the world had escaped the devastating effects of Pleistocene glacial action, they represented in effect "biodiversity museums." Lacking high extinction rates, these stable latitudes accumulated huge numbers of species of plants and animals over the course of evolutionary time. Contrarily, the arctic and temperate regions, he posited, existed in a state of perpetual recovery from the massive species extinctions that would have accompanied the various glacial advances. Although paleoecology and other lines of research strongly dispute the notion of tropical environmental stability during the glacial era, recent work continues to support the idea that the tropics represent a global font of biological novelty.

The biodiversity of any particular location represents a fluctuating equilibrium between species diversification or arrival and species extinction or departure. These variables, in turn, depend in part on available space. At the scale of continental-sized landmasses, there is more real estate in the tropical latitudes than in the temperate and arctic zones. Greater landmass, in turn, translates to more geographical barriers, greater likelihood of population isolation, and enhanced rates of speciation. Furthermore, larger areas equate in principle to greater numbers of individuals per species, thus providing a buffer against extinction. This follows in part from MacArthur and Wilson's theory of island biogeography, which associates area and number of individuals per species with the stochastic extinction rate and consequent species richness (alpha diversity). In the case of the latitudinal gradient in species diversity, this species-area relationship favors higher taxonomic diversification, lower extinction rates, and overall elevated levels of biodiversity near the equator. Although the "geographical area hypothesis" is tenable in the tropical realm, it falls short as an explanatory mechanism in the temperate and arctic zones, where similarly sized biomes exhibit significantly different levels of species richness.

The "productivity hypothesis" suggests that the latitudinal biodiversity gradient results largely from differences in net primary productivity (photosynthesis). This follows because solar intensity and duration and (often) mean annual precipitation are tightly linked to latitude. Net primary productivity, in turn, which is mostly a function of temperature and precipitation, frequently correlates with species diversity patterns. Thus, for example, hot deserts and tundra exhibit both low primary productivity due to drought and low temperatures, respectively, and relatively few numbers of species. Warm and wet equatorial latitudes, on the contrary, which provide ideal conditions for primary productivity, are also home to the greatest numbers of species of plants and animals. It may be, therefore, that higher productivity provides for greater populations per species, and consequently lower extinction rates. Moreover, the opportunities for niche partitioning between taxa with similar ecological requirements would seem to be greater in ecosystems with higher available

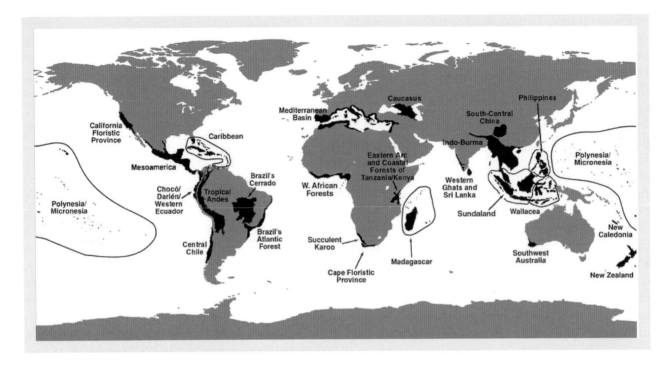

Figure 1 Twenty-five biodiversity hotspots

Source: Mittermeier, R., Mittermeier, C., da Fonseca, G., Kent, J., & Myers, N. (2000). Biodiversity hotspots for conservation priorities. *Nature, 403,* 853–858. Copyright © 2000, Nature Publishing Group.

biomass. While accounting for biodiversity patterns in many biomes, the productivity hypothesis also has weaknesses. For example, it is not always the case that number of individuals per species is correlated with productivity. Rarity is quite common in highly productive ecosystems. In addition, contrary to expectation, some of the highest net primary productivity ecosystems on Earth, such as temperate zone estuaries, maintain quite low levels of species richness.

Unlike the other explanations, the relatively recent "geometrical constraints hypothesis" does not require that environmental gradients be associated with latitudinal gradients. In this case, species ranges are assumed to be randomly distributed between physical or physiological barriers, that is, bounded domains. In a world characterized by barriers, species richness becomes a stochastic function of the number and types of barriers. According to the hypothesis, if species' latitudinal ranges are randomly shuffled, they will tend to overlap more in the mid-domain area. These mid-domain areas in turn correspond with lower latitudes. Although this model has encouraged considerable recent scholarship, it fails to explain many latitudinal species gradients. Clearly, the last word on this vexing question has yet to be heard.

Biodiversity Hotspots

Biodiversity hotspots are biomes that exhibit both a high degree of species endemism (restricted to a particular region) and diversity (i.e., possesses at least 0.5% of the world's known vascular plant species), and they are critically endangered. Twenty-five terrestrial hotspots have been proposed, covering a total of only 1.4% of the Earth's surface (see Figure 1). Of these, 15 occur in tropical rain forests, and 5 are in Mediterranean-type climates. These 25 regions are believed to contain 44% of all known plant species and 35% of all known mammals, birds, reptiles, and amphibians. Later, the terrestrial hotspots were supplemented with 11 marine hotspots, mostly hyperdiverse coral reef ecosystems.

The motivation behind the biodiversity hotspot concept is both biogeographical and political. In the first instance, it derives from the notion of congruence of regions of high species diversity and endemism among different types of biota. Thus, a region that has an unusually high number of endemic species in the pine family, for example, is also likely to exhibit high species richness for other organisms, such as arachnids, salamanders, or butterflies. However, given that the geographical ranges of most biota in species-rich lower latitudes are barely known, these ideas are still fairly speculative. Moreover, the knowledge that ongoing climate change is forcing plant and animal migration in ways that are barely understood calls into obvious question whether today's hotspots will be around in the future. Nevertheless, hotspots represent high priorities for protection, all things considered. Because funds for conservation are limited, protecting biomes that are significantly rich in biodiversity on a global scale, and that retain only a fraction of their original vegetation, clearly constitutes a sensible strategy.

Robert Voeks

See also Biogeography; Biota and Climate; Biota Migration and Dispersal; Crop Genetic Diversity; Ecosystems; Extinctions; Great American Exchange; Island Biogeography; Species-Area Relationship

Further Readings

Gaston, K. (1996). Biodiversity: Latitudinal gradients. *Progress in Physical Geography, 20,* 466–476.
MacDonald, G. (2003). *Biogeography: Introduction to space, time and life.* New York: Wiley.
Myers, N., Mittermeier, R., Mittermeier, C., da Fonseca, G., & Kent, J. (2000). Biodiversity hotspots for conservation priorities. *Nature, 403,* 853–858.
Willig, M., Kaufman, D., & Stevens, R. (2003). Latitudinal gradients of biodiversity: Pattern, process, scale and synthesis. *Annual Review of Ecology and Systematics, 34,* 273–309.

BIOFUELS

A biofuel is a solid, liquid, or gaseous fuel made from crops or trees and their residues, human and animal wastes, and algae. In many parts of the world, biofuels are the primary or only source of fuel. Concerns about global climate change, high oil and gas prices, and energy security have increased interest in biofuels over the past decade. Consequently, key questions have been raised regarding biofuels utilization:

- To what extent are biofuels economical? Because the market for biofuels is still emerging, and long-term oil prices are unknown, it is difficult to calculate the true cost of biofuels. Further complicating the situation is the wide mix of subsidies and taxes for both biofuels and petroleum products.
- What are the impacts of biofuels on food prices? While it is clear that not all biofuels compete with foodstuffs, there is concern that agricultural crop-based feedstocks such as corn do.
- What are the environmental impacts of increased biofuel production? Current biofuel production practices typically require extensive inputs, such as fertilizers and water, which have damaging environmental and ecological side effects. Two other key issues that must be considered are the impacts stemming from land use changes that often occur with increased biofuel production and the ratio of energy return on investment in biofuel production.
- How carbon neutral are biofuels? While biofuels allow for a closed-carbon cycle, initial land use change (e.g., turning forests into agricultural land) can emit significant amounts of carbon dioxide into the atmosphere. Another concern is the quantity of petroleum products required to harvest and process some biomass feedstocks.

Current technological and economical research is primarily focused on the creation of alcohol (especially ethanol) and biodiesel as fuel for vehicles. Ethanol and other alcohols can be used as substitutes in gasoline engines, although some engines require modification. Biodiesel is used in diesel engines. Pure ethanol (biodiesel) is commonly referred to as E100 (B100), and a blend of 20% ethanol (biodiesel) and 80% gasoline (diesel) is E20 (B20). In 2006, global ethanol production was 10.6 billion gallons, 90% of which was produced in Brazil and the United States. One billion gallons of biodiesel were produced in 2006—75% by the European Union (EU) and 13% by the United States. However, Malaysia and Indonesia have begun investing in biodiesel production and will likely become much larger producers in the next decade. Case studies of Brazil, the United States, Germany, and Malaysia are presented below.

Brazil

Brazil, South America's largest energy user, is the world's second-largest ethanol producer (42%) and the largest ethanol exporter. The ethanol is produced from sugarcane, and Brazil has been criticized for allowing rain forests to be converted into cropland. Still, record-high oil prices and increased acceptance of ethanol on the world market indicate that Brazil will continue to be a major ethanol producer.

Brazil has produced ethanol for many years. As early as the 1930s, laws required that gasoline be blended with 5% ethanol, and during World War II this percentage was as high as 42%. However, the decline in oil prices during the 1950s and 1960s resulted in the blend being lowered to 3% to 7%. The 1973 oil embargo by the Organization of Petroleum Exporting Countries resulted in the creation of the National Alcohol Program (Proalcool), a governmental program designed to increase ethanol production. Proalcool provided both investment and subsidies for sugar production, while at the same time gasoline was taxed heavily to decrease demand. Proalcool was very successful in the implementation of ethanol-powered engines.

Market forces (in particular a steep decline in the oil prices coupled with a rise in sugar prices) during the 1980s along with new government policies combating inflation led to the elimination of subsidies and a sharp decline in ethanol production. Meanwhile, worldwide automotive standardization meant that most new cars that were produced had gasoline-based engines. By 2001, only about 1% of car engines produced in Brazil were ethanol fueled. Interest in ethanol increased again during the 1990s, with concerns over global warming and the signing of the Kyoto Protocol leading Brazil to once again implement fuel-mixing standards. Since 1998, all gasoline fuel sold in the country has been E24. Over the past two decades, Brazil has been a leading designer of flex-fuel engines, which can use any blend of ethanol and gasoline without requiring modifications.

Ethanol production in Brazil nearly doubled between 2000 and 2007, increasing from 200,000 barrels per day to 390,000 barrels per day. Projections for 2009 were at 530,000 barrels per day. Currently, about half of the nation's sugarcane crop is used for ethanol production. The primary export markets are the United States and Europe, although both heavily tariff imported ethanol.

Brazil is expected to become a net oil importer by the end of the decade, and there is now strong interest in biodiesel as well as ethanol. In 2008, Brazil's major oil company, Petrobras, announced plans to build three biodiesel plants. In addition, government regulation went into effect requiring all diesel to be B3, which may be raised in the future.

United States

Ethanol has been used as an automotive fuel in the United States since the late 1800s. When Ford's Model T was introduced in 1908, it was able to use both ethanol and gasoline as fuel. Use of ethanol increased during World War I, but the introduction of lead into gasoline quickly replaced ethanol as an oxygenate. The United States is today the world's largest ethanol producer and was responsible for 46% of the world's ethanol output in 2007. Ethanol production has grown steadily since 2000, when 1.6 billion gallons were produced. In 2007, 6.5 billion gallons were produced.

Interest in ethanol has revived in recent years because of concerns over climate change, a desire to reduce foreign oil imports, and strong governmental support—there are currently several federal programs that support ethanol production. Perhaps the most important incentive is the Renewable Fuel Standard, created by Congress, which mandated that 36 billion gallons of ethanol be produced in 2022 (of which 16 billion gallons must be from cellulosic ethanol). Another important incentive is a 51-cent-per-gallon excise tax exemption on fuel sales. Finally, several states have begun to require an ethanol blend (typically E10) in lieu of the oxygenate MTBE (methyl tertiary butyl ether), owing to concerns over groundwater pollution from the latter.

The primary feedstock for ethanol production in the United States is corn, which is contentious for two reasons. First, some people believe that the rapid increase in ethanol production, now accounting for 30% of all corn production, is competing with food supplies. This is questionable, since most corn is used for animal feed, and the by-product of ethanol production, distillers grains, is a valuable feed product. Still, with a worldwide increase in all staple food prices since the mid 2000s, this debate is likely to continue. Concerns also have been raised about the environmental impacts that increased ethanol production has caused due to the large area of land needed and the amount of fossil fuels and fertilizers required to grow, harvest, and process the corn. Nevertheless, nearly 80 new ethanol plants are currently being built in the Midwest.

Interest in cellulosic ethanol has increased rapidly over the past decade, and several commercial plants are being constructed. Likely feedstocks include corn stover, wheat straw, switchgrass, short-rotation woody crops, and forest residues. Cellulosic ethanol does not require high-intensity inputs like fertilizer, does not require land use conversion, and does not compete with foodstuffs. However, the economic viability of cellulosic ethanol is still in question, and current processing technology requires large volumes of water. The cellulosic ethanol market has the potential to surpass the volumetric supply of corn ethanol.

Germany

Germany is the largest biodiesel producer in the world, producing about half of the EU's biodiesel, which itself accounts for 75% of the world's market. Recent years have seen a strong increase in the amount of biodiesel produced, primarily from rapeseed oil, as the government has given strong support to the industry through investment and an excise duty reduction. When leaded gasoline was outlawed in 1996, more than 1,000 pumps were converted to distribute B100. Germany produced about 3 million tons in 2006, compared with 290,000 tons in 2000, and annual growth is expected to continue to be as high as 20% to 30% over the next few years. Germany has about 25 biodiesel production plants, the largest of which produces 165,000 tons per year. The largest constraint the industry faces is land availability.

In 2007, an EU commission published the Renewable Energy Road Map, which sets targets for biomass utilization. At the same time, a new German government began eliminating tax breaks on biodiesel (and vegetable oil), which will completely end in 2012. As a result, production fell in 2007–2008. Still, biodiesel is seen as part of the strategy toward energy independence, and pure biodiesel is now being marketed for trucks instead of cars. The automobile industry's support for this program is a significant change.

Malaysia

Malaysia produced around 550,000 tons of biodiesel from palm oil in 2006. During the past two decades, Malaysia became the largest palm oil producer and exporter in the world. As of 2006, it is responsible for 47% of the world's palm oil production. There has been an explosive growth of palm plantations in the past two decades— from less than 120,000 acres of palm plantations in 1970 to around 10 million acres by 2005—and they now account for more than 10% of the nation's land area. In 2006, Malaysia produced a total of 16.5 million tons of palm oil, 90% of which was exported. Globally, palm oil accounted for 33% of global vegetable oil supply whereas soy accounted for 31%. Soy is also commonly used to produce biodiesel (e.g., in the United States), but it requires 7 to 10 times more land to produce an equivalent amount.

The government of Malaysia has shown great interest in continued biodiesel production output growth and works closely with industry. Consequently, around 100 commercial plants have been planned with the first ones having begun production in 2007 and 2008. All diesel sold in Malaysia is B5, although the government has not shown signs of increasing this to a richer blend. Exports of palm oil as a diesel substitute are expected to grow, both to southeast Asia and to Europe.

Despite the economic success of palm oil, several environmental issues have been raised. The effects of palm oil plantation growth had caused deforestation of more than 2.4 million acres, with the rest of the growth being absorbed by a decrease in other crops, such as cocoa, rubber, and coconut. With 64% of Malaysia remaining forested, future expansion will likely be into forested areas. Because of land use changes, decreasing biodiversity is an issue. Awareness of the rapid land use change has brought about some positive results. However, although part of the biofuel industry had tried to counter the effects, without strong government intervention, it is expected that the tropical rain forest deforestation will continue. Indonesia has a palm oil industry of comparable size with Malaysia's and is experiencing many of the same environmental challenges.

Conclusion

Biofuel demand worldwide is expected to grow due to fears of climate change, sustained high oil and gas prices, and the desire for localized and diverse energy supplies. Liquid biofuels, especially ethanol and biodiesel, hold great promise as partial alternatives to gasoline and diesel, but questions remain as to their economic viability and environmental and social costs. Because large-scale biofuel production is still an emerging market, countries must offer strong government support if they wish to invest in biofuels.

Nicholas H. Johnson

See also Agricultural Land Use; Climate Policy; Energy and Human Ecology; Energy Policy; Energy Resources; Renewable Resources; Sustainable Production

Further Readings

Bomb, C., McCormick, K., Deurwaarder, E., & Kåberger, T. (2007). Biofuels for transport in Europe: Lessons from Germany and the UK. *Energy Policy, 35,* 2256–2267.

Gardner, B., & Tyner, W. (2007). Explorations in biofuels economics, policy, and history. *Journal of Agricultural & Food Industrial Organization, 5*(2), Article 1.

Solomon, B., Barnes, J., & Halvorsen, K. (2007). Grain and cellulosic ethanol: History, economics, and energy policy. *Biomass and Bioenergy, 31,* 416–425.

Tan, K.-T., Lee, K.-T., Mohamed, A., & Bhatia, S. (2009). Palm oil. *Renewable & Sustainable Energy Reviews, 37,* 554–559.

Worldwatch Institute. (2007). *Biofuels for transport.* London: Earthscan.

BIOGEOCHEMICAL CYCLES

Biogeochemistry is the chemistry of Earth's surface. Or, in other words, it is physical geography through the lens of chemistry. Akin to the study of water moving through the hydrologic cycle, a biogeochemical cycle examines how a particular chemical element moves through the physical geographic spheres (i.e., the atmosphere, biosphere, lithosphere, and hydrosphere). Viewed from the integrative perspective of the biosphere, this entry explores several of the major biogeochemical cycles, with a special emphasis on how these cycles affect—and are affected by—human society.

Biosphere and Biogeochemical Cycles

During the past 3.5 billion years of Earth's history, the biosphere has come to play an increasingly dominant role in shaping the chemistry of Earth's surface, making the study of Earth surface chemistry inherently the study of *bio*geochemistry. Indeed, as noted by James Lovelock and Lynn Margulis in their Gaia hypothesis, it was the biosphere that shaped the biogeochemistry of the earth surface into a chemical composition most favorable for the continued presence of life on earth. Most important, the biosphere changed the earth's atmosphere from an inhospitable, carbon-dioxide-rich, hot atmosphere into the hospitable, oxygen-rich, temperate atmosphere of the present.

This dominant role of the biosphere has been further solidified with the rise of one biotic species—humans. In particular, the advent of the agricultural and industrial revolutions has caused human actions to vastly increase many elemental fluxes and pools within the most important biogeochemical cycles.

Earth's Intersecting Biogeochemical Cycles

Analysis of a biogeochemical cycle may focus on any given chemical element found on Earth, but it is most commonly applied to the abundant elements that play an integral role in the functioning of the Earth system. The elements with a lighter atomic weight are most abundant in the solar system, and those light elements, which were soluble in the early seas, formed the basis of modern biogeochemistry. The key effect of the biosphere on the biogeochemistry of Earth's surface can be traced to the evolution of the photosynthetic metabolic pathway. Photosynthesis represents the critical link between the biogeochemical cycles most important for humans and the functioning of the Earth system. The simplified form of this crucial reaction of photosynthesis is

$$6H_2O + 6CO_2 + \text{Light energy} \rightarrow C_6H_{12}O_6 + 6O_2.$$

It is vital to note that photosynthesis must be sustained by a consistent supply of mineral nutrients. Thus, in addition to linking biogeochemical cycles of carbon (C), oxygen (O), and hydrogen (H), photosynthesis causes the biogeochemical cycling of mineral nutrients to also play key roles in the functioning of the Earth system. Of these mineral nutrients, nitrogen (N) is the most important. After nitrogen, photosynthesis requires an abundant supply of several other macronutrients. All these remaining macronutrients are weathered from rocks, thereby forming an important biogeochemical link between the biosphere and lithosphere.

In short, photosynthesis is the nexus of the major biogeochemical cycles. Given this role of photosynthesis, a plant's stoichiometry, or ratio of the concentration of its chemical elements, provides a useful means for roughly assessing the relative importance of the various biogeochemical cycles within the earth system. By dry mass, the average plant is 45% carbon, 45% oxygen, 6% hydrogen, 1.5% nitrogen, 1.0% potassium, 0.5% calcium, 0.2% magnesium, 0.2% phosphorous, 0.1% sulfur, and 0.1% silicon. Biogeochemical cycles of other "micronutrients" such as iron (Fe) or manganese (Mn) can also be important to examine in instances where a shortage of one of these rarer elements may adversely affect the growth or health of a plant, animal, or human. This entry proceeds by concentrating on the major biogeochemical cycles organized here as (a) the hydrogen and oxygen cycles, (b) the carbon cycle, (c) the nitrogen cycle, and (d) cycling of rock-derived nutrients.

Hydrogen and Oxygen Cycles

Not surprisingly, among the most abundant elements on Earth are hydrogen and oxygen, the elements that constitute water (H_2O). Water is the medium by which life on earth evolved and is sustained. The biogeochemical cycling of H_2O, more often referred to as the hydrologic cycle, is highly coupled to the biogeochemical cycling of other elements. For instance, because water limits photosynthesis and soil microbial transformations, the carbon and nitrogen cycles proceed more slowly in drier regions of the Earth. Water is also an important agent of weathering and erosion, causing it to also strongly affect the cycling of rock-derived elements such as phosphorus and calcium.

The cycling of hydrogen is also important in its own right. For example, an increase in hydrogen ions increases the acidity of a soil, and this increase has subsequent effects on the mobility and plant availability of nutrients such as calcium or nitrate. Through the burning of fossil fuels and resulting deposition of acids (e.g., "acid rain"), humans have increased acidity over large regions, causing deleterious effects for soil and water quality.

Finally, in addition to its participation in the hydrological cycle, the cycling of oxygen is also inextricably linked to the global carbon cycle. As mentioned above, if photosynthetic organisms did not synthesize molecular oxygen (O_2) from carbon dioxide and water, the 21% abundance of oxygen in the atmosphere would not occur. The global cycle of oxygen is quite similar to carbon. Like carbon, most of the annual fluxes of oxygen are between the atmosphere and biosphere, but each year a small fraction of Earth's oxygen is weathered from or buried in rocks.

Carbon Cycle

Carbon forms "the building blocks" of life on Earth. Its abundance and its tetravalent chemical properties allow it to form the ubiquitous complex polymers (e.g., DNA, carbohydrates, proteins) that are the hallmark and foundation of life. Despite this importance, it is only recently that the carbon cycle has garnered societal interest. Since the late 18th century, carbon-rich fossil fuels have enriched human society by providing energy, food, and mobility to an exponentially growing human population. Yet through our understanding of the global carbon cycle, we now realize that these treasures of fossil fuels come at the peril of unprecedented climatic instability.

Fossil fuels form during rare moments in space and time when dead biotic remains become buried in sediments before they can fully decompose. Historically, the amount of carbon entering the ocean through dissolution and entering the biosphere through photosynthesis tends to be balanced with the carbon returned to the atmosphere by natural processes of combustion, decomposition, and respiration. Combustion releases heat energy while rapidly oxidizing organic carbon into atmospheric CO_2. Microbial decomposition returns carbon to the atmosphere as carbon dioxide (CO_2) and methane (CH_4). Respiration is the reverse chemical process of photosynthesis, thereby producing CO_2.

The carbon cycle also includes a substantial abiotic component. Atmospheric carbon dioxide naturally dissolves in cool surface ocean water. Indeed, the vast oceans contain almost 56 times more carbon than the atmosphere. The slow precipitation of limestone ($CaCO_3$) on the ocean floor provides the only long-term storage for carbon and forms the largest pool of carbon on earth. Limestone slowly releases this carbon back into the active carbon cycle when it is weathered or converted to concrete by humans (a process that also adds CO_2 to the atmosphere).

As humans increasingly produce cement, cut down forests, and combust fossil fuels, we tip the natural balance of the carbon cycle. In 2007, the natural biotic and oceanic sinks of carbon could only take in about half of the carbon humans emitted. As of 2007, human emissions are still rapidly increasing and have caused the atmospheric CO_2 content to reach its highest levels in the past 650,000 years and probably in the past 20 million years. This high level of atmospheric CO_2 is extremely important. Since CO_2 is a greenhouse gas that helps our atmosphere retain heat, our emissions of CO_2 are radically altering Earth's energy budget and climate. Thus, through our extensive participation in the carbon cycle, we must adapt to a global climate that will be unprecedented in human history.

Nitrogen Cycle

As the key limiting nutrient for photosynthesis and biotic growth, nitrogen is central to the functioning of all plants and animals on Earth. Within photosynthesis, nitrogen is a key element within the chlorophyll used to harvest light energy from the sun, as well as the Rubisco enzyme that catalyzes the carboxylation of atmospheric CO_2 into the photosynthate (i.e., $C_6H_{12}O_6$) that ultimately sustains life on Earth. Finally, due to its importance in animal protein, nitrogen serves to limit biotic growth at all higher trophic levels.

At first glance, this limiting nature of nitrogen for life on Earth is curious. In its N_2 molecular form, nitrogen makes up 78% of Earth's atmosphere. Yet while life on Earth is indeed surrounded by nitrogen, most organisms cannot directly access this inert atmospheric N_2 gas. All biologically available, or "reactive nitrogen" (N_r) must first be "fixed." This "fixing," or breaking apart the strong triple bond between atoms in the N_2 molecule, can only be done by two natural sources: (1) lightning and (2) nitrogen-fixing bacteria, such as those associated with the rhizomes of legumes such as soybeans. Through the fairly recent advent of the Haber-Bosch process for synthesizing ammonium nitrate fertilizer (NH_4NO_3) from atmospheric N_2, humans can also fix nitrogen. In fact, when this fertilizer production is added to the N_r created by planting nitrogen-fixing crops, as well as the N_r released from the combustion of fossil fuels, humans now contribute almost half of the biologically available nitrogen (N_r) produced worldwide each year. Indeed, this enormous yearly anthropogenic supplement of N_r has helped continue to feed the world's growing population. However, it has also resulted in widespread instances of overfertilization, leading to a host of environmental problems.

The adverse effects of anthropogenic fixation of nitrogen occur because once N_2 is fixed into N_r, it stays in this biologically available form for a long time. In fact, each molecule of N_r cycles through the world's terrestrial or marine ecosystems an average of almost 38 times. This nitrogen "cascade" can only be stopped through a microbial process of denitrification or through burial in marine sediments. Presently, human additions of N_r greatly outstrip denitrification or burial, and Earth's ecosystems are gaining greater than 20 metric megatons of N_r per year.

A poignant example of the adverse effects of all this excess N_r is the vast "dead zone" in the Gulf of Mexico at the delta of the heavily fertilized Mississippi River basin. In these locations dramatically affected by human actions, whole ecosystems are shifting toward those species, which thrive under nutrient-rich conditions. Often, these shifts to a nitrogen-rich ecosystem come at the detriment of the clean water or productive fisheries needed to sustain human life. Thus, our environmental policies must now also aspire to better manage our substantial interaction with the nitrogen cycle.

Cycling of Rock-Derived Nutrients

Rock-derived nutrients may be taken up by plants only after they are slowly weathered from the underlying parent material. Most important among these rock-derived nutrients is phosphorus, a nutrient essential to the nucleotides (i.e., DNA and RNA) that carry and reproduce genetic information. While phosphorus is the primary limiting nutrient in many terrestrial, freshwater, and marine ecosystems, the cycling of other rock-derived nutrients such as calcium, magnesium, and sulfur are also essential to life.

Geography of Biogeochemical Cycles

Much of the study of biogeochemistry has been aspatial in its focus. Great efforts have been undertaken to quantify biogeochemical budgets at scales ranging from the globe to a small watershed. With this understanding to build from, it is only recently that researchers have begun to examine the great spatial heterogeneity within Earth's biogeochemical cycles. Exploring spatial patterns of biogeochemical cycling requires a holistic approach. The biogeochemical cycles of different elements are inherently coupled and are affected by many interacting "statelike" environmental factors (e.g., climate, relief). Moreover, maps derived from remote sensing instruments such as MODIS or AVIRIS now quantify meaningful spatial patterns in biogeochemical cycles, such as marine carbon sequestration or forest canopy nitrogen content. As humans further dominate the functioning of their sustaining ecosystems, the understanding gained by these views into the geography of biogeochemical cycling will become paramount for renewing sustainable societies.

Brenden E. McNeil

See also Acid Rain; Atmospheric Composition and Structure; Atmospheric Pollution; Biota and Soils; Carbon Cycle; Climate Change; Ecosystems; Gaia Theory; Global Environmental Change; Hydrology; Imaging Spectroscopy; Land Use and Cover Change (LUCC); Nitrogen Cycle; Nutrient Cycles; Phosphorus Cycle; Remote Sensing; Rock Weathering; Soils

Further Readings

Canadell, J., Le Quere, C., Raupach, M., Field, C., Buitenhuis, E., Ciais, P., et al. (2007). Contributions to accelerating atmospheric CO_2 growth from economic activity, carbon intensity, and efficiency of natural sinks. *Proceedings of the National Academy of Sciences, 104*(47), 18866–18870.

Galloway, J., Townsend, A., Erisman, J., Bekunda, M., Cai, Z., Freney, J., et al. (2008). Transformation of the nitrogen cycle: Recent trends, questions, and potential solutions. *Science, 320*(5878), 889–892.

Likens, G., & Bormann, F. (1995). *Biogeochemistry of a forested ecosystem.* New York: Springer-Verlag.

Lovelock, J. (1979). *Gaia: A new look at life on earth*. Oxford, UK: Oxford University Press.

Schlesinger, W. (1997). *Biogeochemistry: An analysis of global change*. San Diego, CA: Academic Press.

Sterner, R., & Elser, J. (2002). *Ecological stoichiometry: The biology of elements from molecules to the biosphere*. Princeton, NJ: Princeton University Press.

Vitousek, P., Mooney, H., Lubchenco, J., & Melillo, J. (1997). Human domination of earth's ecosystems. *Science, 277*, 494–499.

BIOGEOGRAPHY

Biogeography is the study of the geographical distribution of living and fossil plants and animals as a result of ecological and evolutionary processes. Biogeography analyzes organism-environment relations through change over space and time, and it often includes human-biota interactions. The main questions explored by biogeographers deal with organism patterns to understand the underlying processes. Biogeographers ponder questions such as why is a species present in a given area? Conversely, if a species is not present, then why is it missing from the area? What are the historical and ecological factors that help determine where a species occurs? What are the effects of evolution and plate tectonics? How have humans altered geographic distribution of organisms? The science of biogeography has been revitalized in the past 60 years due to our understanding of plate tectonics, mechanisms limiting distributions, island biogeography theories, and mathematical and technological tools.

Current work in biogeography uses spatial patterns of organisms, past and present, to determine ecological processes. Biogeographers use experimental testing and quantification of biotic interactions. Vegetation dynamics is the primary focus for approximately half of the biogeographic research conducted by U.S. geographers. Other major focuses include ecosystem structure and function, zoogeography, paleoecology, and development of new biogeographic methodology. In particular, mapping and modeling spatial patterns of abundance and distribution of species of plants and animals have greatly advanced with geographic information systems and remote sensing technology.

Development of Biogeography

To better understand the current field of biogeography, it is important to explore the foundations and history of the science. Biogeography is a synthetic study, which is based in part on the subjects of community ecology, geology, systematics, evolutionary biology, and paleontology. The development of the subject of biogeography may be broken into four historical periods.

1600–1850: The Age of Reason

Early studies of organisms' geographic distributions were focused on descriptive studies with historical explorations. These scientists focused on documenting spatial patterns of organisms, emphasizing on the effects of climate, latitude, and altitude. Comte de Buffon (1707–1788), also known as Georges-Louis Leclerc, determined that distant regions with similar climate and similar-appearing vegetation have different animal species. This is now referred to as Buffon's Law. He is also the author of *Histoire Naturelle*, a 44-volume natural history encyclopedia. Carl Linnaeus (1707–1778) studied the plants and animals spread from Mount Ararat in Turkey to explore the idea of the biblical flood. As a result of documenting elevational zones of Ararat, he came up with the idea of biomes defined as major ecological communities. In addition, Carl Linnaeus is considered the father of the science of taxonomy, which is the science of classification.

This time period is also known as a great age for exploration. Johann Reinhold Forster (1729–1798) was the naturalist on James Cook's second Pacific voyage in 1778. He advanced biogeography by creating global biotic regions for plants. Forster noted the higher-species diversity in the tropics, as well as species diversity being correlated with island size. Alexander von Humboldt (1769–1859) created a botanical geography that was foundational to the field of biogeography. He determined that plant vegetation types are strongly correlated with local climate to create latitudinal belts of vegetation. Moreover, he developed elevational vegetation zones for the Andes in South America.

1850–1900: Evolution by Natural Selection

The idea of evolution based on natural selection greatly altered the way species distributions were explained. Charles Darwin (1809–1882) is most famous for publishing *The Origin of Species*, outlining his idea of evolution through natural selection. Natural selection occurs when individuals in a population either do not survive equally well or do not breed equally well, or both due to inherited differences. Evolution in turn can be thought of in two ways: (1) microevolution and (2) macroevolution. In microevolution, evolution is considered as changes in the genetic composition of a population with the passage of each generation. For macroevolution, evolution is the gradual change of organisms from one form into another, with the origins of species and lineages from ancestral forms. For an example, Darwin studied the variations in mockingbirds on different Galapagos Islands. This divergent evolution is a diversification over evolutionary time of a species into several different species, commonly referred to as adaptive radiation.

Alfred Russel Wallace (1823–1913) is also famous for independently developing the idea of evolution by natural selection, based on his work in Indonesia. He found that the species on Sumatra and Java were very different from nearby New Guinea, even though the climates were similar. Wallace's study of biota in southeast Asia showed geographic distance is not equal to taxonomic similarity, and the boundary area between these islands is now referred to as *Wallace's Line*. Wallace is also considered to be the originator of zoogeography, which is the biogeography focused on animals. Wallace integrated geological, fossil, and evolutionary information to consider paleoclimate influences distributions, developing six great biotic regions.

Other notable contributions to biogeography during this period include mapping biotic regions and understanding limiting factors. Philip Lutley Sclater (1829–1913) advanced the subject of biogeography with his defining terrestrial biotic regions for birds and marine regions for marine mammals. Justus Liebig (1803–1876) changed the way scientists viewed restrictions on organisms away from a focus on total resources available with his law of the minimum. The law of the minimum states that the scarcest resource (or limiting factor) in the environment makes it difficult for a species to live, grow, and reproduce.

1900–1950: Continental Drift and Ecology

Themes in biogeography in the first half of the 20th century focused on links to paleontology, centers of species origins, and the biological species concept. The emphasis in the science of biogeography was on evolution, history, dispersal, and mechanisms of survival. The greatest impact on biogeography in this period was the theory of continental drift in 1912 and 1915 by the German geologist Alfred Wegener (1880–1930). Before the theory of plate tectonics, it was difficult for biogeographers to explain certain patterns of species distributions with the assumption that land masses were fixed in their geographic positions. Wegener's theory was actually not widely accepted until the 1960s when proof of continental drift came from a series of linear magnetic anomalies on either side of the Mid-Atlantic Ridge. With the acceptance of the continental drift theory, biogeographers could now explain the disjunct biogeographic distribution of present-day organisms found on different continents but having similar ancestors. Species can interact as continents collide. Subsequently, when the continents separate, they take their new species with them. Biogeographers now ponder how plate tectonics may have affected the evolution of life. In turn, biogeographers offer evidence for plate tectonics such as dispersal of species via such corridors as the Bering land bridge or widely separated ("disjunct") species distributions that can't be explained by dispersal; for instance, *Nothofagus* (southern beech) trees, which only occur in Southern South America and in New Zealand.

In addition to historical explanations of organism distributions, biogeographers also examined ecological reasons for spatial patterns. Theories on ecological succession were formally developed in the late 1800s and early 1900s to show predictable and orderly changes in the composition or structure of ecological communities. In 1899, Henry Cowles published his study of stages of vegetation development on dunes along Lake Michigan. In 1916, Frederic Clements published his famous theory of vegetation development focusing on gradual changes over time to best fit the local conditions. His climax theory of vegetation-dominated plant ecology was later largely replaced by other models, notably by Henry Gleason's 1926 concept of distribution of plants depending on the individual species rather than Clements's idea of plant associations. In 1934, Christen Raunkiaer (1860–1938) helped change the way biogeographers classified species with life forms based on ecological rather than taxonomic classification. In 1935, Sir Arthur Tansley (1871–1955) refined the term *ecosystem* to mean the whole complex natural unit in a system consisting of all plants, animals, and microorganisms (biotic factors) in an area functioning together with all the nonliving physical (abiotic) factors of the environment.

1950–Present: Ecological and Historical Theories

Since 1950, the field of biogeography has been revitalized with advances in ecological and historical theories focused on phylogenetic classification to related different species, mechanisms limiting geographic distribution, and distances and size influencing number of species in an area. During this period, the concept of new species arising due to geographic isolation was developed by Ernst Mayr (1904–2004). Mayr is also well-known for defining the "biological species concept" as potentially interbreeding to produce fertile offspring. In addition, Mayr helped define the term *cladistics* to refer to classifications, which only take into account geneology, based on evolutionary ancestry. Cladistics, or phylogenetic classification, views a species as a group of lineage-connected individuals, compared with the traditional Linnaean taxonomy, which focused on the similarities between species. Cladograms are created based on the order in which different groups branched off from their common ancestors, arranged with the most closely related species on adjacent branches of the phylogenetic tree.

Theories also expanded during this time period on how a species can occur in widely geographically separated areas and the mechanisms that limit these distributions. In 1958, Leon Croizat (1932–1982) published his concept of "vicariance biogeography" to explain disjunction of multiple species due to the growth of barriers instead of via dispersal. Croizat's works include *Manual of Phytogeography* (1952), *Panbiogeography* (1958), and *Space, Time, Form* (1964). Robert Harding Whittaker (1920–1980) proposed a new method to analyze limits to plant distributions by comparing species abundance with environmental gradients. His gradient analyses approach focuses on abiotic factors such as light, water, temperature, and soil nutrients in plant communities.

Biogeography during this period moved from observational to predictive studies with the theory of island biogeography. In 1963, R. H. MacArthur and E. O. Wilson hypothesized that species richness of an area could be predicted to explain distributions. The theory states that if one knows the rates of colonization and extinction of an island, then it is possible to predict the number of equilibrium species that area could support. They based the species richness prediction on two factors: (1) distance of the island from a mainland source of species for a colonization pool and (2) the size of the island for available habitat and its variety of ecological niches. With these two factors, MacArthur and Wilson predicted the number of species the area could maintain, as well as the turnover rate for the area. According to island biogeography theory, small and distant islands have a lower number of species that can be maintained compared with large and near islands. The theory also states that there would be a turnover of the species as new species colonize and old species go extinct, but the number of species overall should achieve an equilibrium number. This theory has been applied to other nonisland areas that act like islands due to habitat fragmentation, such as nature preserves and national parks.

Spatial Distributions of Organisms

Modern biogeography explores spatial patterns in the geographic variation of individuals and populations, including genetic, physiological, and morphological variations. Organisms are often studied in biomes, consisting of distinct flora and fauna related to climate, soil, and geological factors. Biogeographers analyze species ecological niches (both fundamental and realized), defined as the total requirements for resources and physical conditions. The foundation of biogeographic distribution patterns follows the first rule of geography, namely that closer equates to more similar (referred to as "spatial autocorrelation"). When relating species niche to geographic distributions, key interactions and adaptations to consider include (a) stress, with regards to climate, predation, and availability of symbiosis (close association between species); (b) competition, both between and within species, which exclude some species from their fundamental niches; and (c) disturbance, which occurs less predictably and causes a greater change in the environment than stress. Biogeography seeks to answer why and how species distribution patterns are a response to historical and ecological limiting factors, dispersal mechanisms, and human influences.

Vicariance Biogeography

Geographic distributions may be separated into disjunct populations by historic events that create barriers, called *vicariant events*. Disjunct distribution occurs when two or more closely related taxa live today in widely separated areas. Barriers that split a continuous distribution into disjunctions are created by many processes, including changing in the distribution of land via continental or tectonic shifts, volcanism and mountain building, shifts in river patterns, glacial cycles, climate change, and human alterations to landscapes. Barriers may be physical (such as a mountain), physiological (such as freshwater vs. saltwater blocking for aquatic organisms), and/or ecological-behavioral (e.g., predators). For example, one vicariance-splitting taxons' ranges occurred from historical climate change during glacial periods, resulting in the drying of tropical rain forest in South America into smaller fragments of rain forest refugia surrounded by grassland. Disjunct distributions from vicariant events are supported by fossil evidence and by the geography of present species that cannot be explained by natural dispersal. For example, *Nothofagus* (southern beech) trees occur today in such widely separated regions as Southern South America and New Zealand, which can be explained by historic plate movement but not dispersal. In another example, disjunct populations of southern migrations were created during glacial ice advancements in North America, with isolated populations cut off as ice retreats (partly because soil was removed by glaciers). Organisms from temporary or fluctuating environments (such as seasonal variations in temperature) typically are less separated by barriers. For instance, crossing mountain ranges are less of a barrier to species from temperate climates (adapted to cold winters) compared with species from tropical climates.

Geographic isolation from vicariant events leads to reproductive isolation. Even if the disjunct population reunites in time, the geographic isolation may have already resulted in the groups no longer being able to interbreed (especially in animal species). In other words, when barriers geographically isolate populations over time, a different evolutionary lineage might occur with new species being created (referred to as *allopatric* speciation). Another reason for its occurrence may be because different geographic regions have different selective pressures, such as temperature, rainfall, predators, and/or competitors. In addition, even if the environments separated by a barrier are not very different, the populations may differentiate because different genetic combinations and mutations occur by chance (see Figure 1).

Dispersal Biogeography

Disjunct biogeographic distributions may also be created by dispersal events. Dispersal can be defined as movement away from one's point of origin. The possible results of dispersal include extending *within* current range by colonizing new habitat in range or colonizing *distant* location across a major physical barrier of unfavorable habitat. Individuals may move great distances through unsuitable habitats by traveling through corridors, flying over hostile environments, being blown or floating through sweepstake events (such as a hurricane). Dispersal agents typically are wind, water, rafting, or animals. Dispersal biogeography studies distribution patterns of organisms, emphasizing dispersal capabilities as well as ecological properties of species to evaluate origins of taxa in a biota. Dispersal mechanisms affect rate of species movement across landscape, as well as the efficiency of species in colonizing new areas.

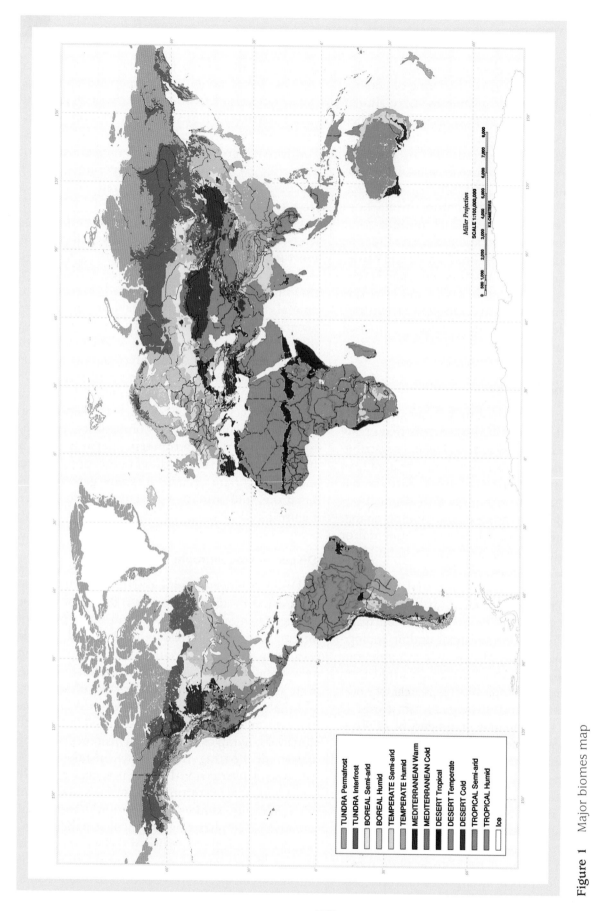

Figure 1 Major biomes map

Source: U.S. Department of Agriculture, Natural Resources Conservation Service, Soil Survey Division, World Soil Resources, Washington, D.C.

TUNDRA Permafrost
TUNDRA Interfrost
BOREAL Semi-arid
BOREAL Humid
TEMPERATE Semi-arid
TEMPERATE Humid
MEDITERRANEAN Warm
MEDITERRANEAN Cold
DESERT Tropical
DESERT Temperate
DESERT Cold
TROPICAL Semi-arid
TROPICAL Humid
Ice

Miller Projection
SCALE 1:100,000,000

0 500 1,000   2,000   3,000   4,000   5,000   6,000   7,000   8,000
KILOMETERS

Modes of dispersal vary from jump dispersal (e.g., the movement of gypsy moth *to* North America) to diffusion (e.g., gypsy moth spread *within* North America). Jump dispersal includes traveling over long distances across inhospitable habitat—in other words, long-distance dispersal mostly by organisms that can fly or swim. Jump dispersal events are rare, cover large distance, and are considered "surprising" events. These long-distance dispersals can explain large discontinuous distributions of some organisms, as well as taxonomic similarity of distant biotas and populations. Most animal and all plant jump dispersals are passive, although occasionally some animals have active long-distance dispersals. Those individuals who succeed at jump dispersal have the ability to (a) travel long distances, (b) withstand unfavorable conditions during passage, and (c) establish viable population on arrival.

Pathways of dispersals can be broken into the categories of corridors, filters, and sweepstakes. Corridors are routes that permit spread of taxa through continuous favorable habitat with relatively little risk. These routes may in turn link larger areas of habitat. Famous corridors include past land connections created by sea level changes, such as the Bering land bridge connecting Asia and North America. In contrast, a "net" or "filter" is a route that contains patches of suitable habitat interspersed between larger areas of unsuitable habitat. Filters can act as a barrier for some taxa, blocking or slowing passage of some organisms. A classic example is the Isthmus of Panama filtering the dispersal between North and South America starting around 3.5 million years ago. The "sweepstakes" pathway of dispersal refers to chance dispersal across a major barrier—in other words, a long shot for dispersal, which usually involves accidents, low probability, and unusual means of travel. For instance, sweepstakes dispersal includes birds caught in hurricanes, seeds traveling in upper atmosphere winds, and animals floating on driftwood (called "Noah's Ark" if there is an assemblage of organisms deposited en masse).

Human Impacts of Distributions

In addition to barriers and dispersal events, biogeography also explores species distribution patterns affected by humans, whether in limiting or expanding ranges. Humans decrease other species ranges through habitat destruction, hunting and commercial exploitation, polluting environments, deliberately or accidentally introducing competing nonnative species, changing historic disturbance regimes, and causing the loss of ecological partners. Species most likely to have their geographic ranges limited by humans include species that have relatively few offsprings that are nurtured for a long time ("K" species), specialists who are selective about their niche conditions, economically valuable species, naturally rare species, species with naturally restricted ranges, species sensitive to pollutants, and species in competition with humans for habitat or resources. People also increase species geographic ranges, either by introducing species to a new geographic region or by expanding ranges by favoring some species that can adapt to human landscapes (weeds, agricultural species, species adapted to urban areas). Introduction of species by people into a new area may be deliberate or accidental. Although most introduced species fail to establish viable population, those that succeed typically have harmful long-term effects, competing with native species or transmitting diseases.

Conclusion

Biogeographers seek to develop theories to explain past, present, and changing future composition of plants and animals distribution patterns in order to better understand the processes of how organisms interact with our planet. Biogeography is both interdisciplinary within other subfields of geography as well as other disciplines, such as ecology, geology, and biology. An integrative approach helps biogeographers provide a holistic understanding of the diversity of life and make recommendations for the conservation of biological diversity.

Joy Nystrom Mast

See also Adaptive Radiation; Biome: Boreal Forest; Biome: Desert; Biome: Midlatitude Deciduous Forest; Biome: Midlatitude Grassland; Biome: Tropical Deciduous Forest; Biome: Tropical Rain Forest; Biome:

Tropical Savanna; Biome: Tropical Scrub; Biome: Tundra; Cultural Ecology; Darwinism and Geography; Ecological Regimes; Ecoregions; Ecosystems; Exotic Species; Extinctions; Forest Fragmentation; Great American Exchange; Invasion and Succession; Island Biogeography; Islands, Small; Landscape Ecology; Nature; Political Ecology; Species-Area Relationship; Tree Farming

Further Readings

Cox, C., & Moore, P. (2005). *Biogeography: An ecological and evolutionary approach*. Malden, MA: Blackwell.

Lomolino, M., & Heaney, L. (Eds.). (2004). *Frontiers of biogeography*. Sunderland, MA: J. H. Sinauer.

Lomolino, M., Riddle, B., & Brown, J. (Eds.). (2006). *Biogeography* (3rd ed.). Sunderland, MA: J. H. Sinauer.

Lomolino, M., Sax, D., & Brown, J. (Eds.). (2004). *Foundations of biogeography: Classic papers with commentaries*. Chicago: University of Chicago Press.

MacArthur, R., & Wilson, E. (2001). *The theory of island biogeography*. Princeton, NJ: Princeton University Press. (Original work published 1967)

MacDonald, G. (2003). *Biogeography: Introduction to space, time, and life*. New York: Wiley.

Quammen, D. (1997). *The song of the Dodo: Island biogeography in an age of extinction*. New York: Simon & Schuster.

 # BIOME: BOREAL FOREST

Boreal forests, also known as *taiga* in Russia, stretch across a large zone in the Northern Hemisphere, spanning North America and Northern Eurasia, and represent nearly one third of the global forested areas. Formed under harsh climatic conditions, these forested ecosystems provide crucial ecological and societal benefits and strongly influence many global processes, including carbon storage and release and, subsequently, concentration of greenhouse gases in the atmosphere. Boreal forests are dominated by coniferous tree species that are frequently replaced by broad-leaved deciduous trees after an occurrence of disturbances such as fire, insect infestation, or logging. Although poor in dominant-tree biological diversity, these forests are the largest intact ecosystem on Earth and sustain considerable populations of large land mammals such as moose, elk, and bear. Under the observed warming trends, boreal forests show signs of a shift in vegetation composition and an increase in extent and severity of natural disturbances.

Definition and Geographic Extent

Boreal forest received its name after the Greek god of northern winds Boreas, indicating the geographic position of this biome in high northern latitudes. No equivalent belt exists in the Southern Hemisphere since large land masses at similar latitudes are not found there. The geographic boundaries of the boreal biome are frequently defined by climatological variables representative of the temperature-limited growing conditions. Several temperature-related metrics are commonly used to define the exact northern and southern boundaries of boreal forests. One of the approaches is to define the southern boundary of the biome at the average yearly +5 °C isotherm (i.e., a line connecting points of equal temperature) and the northern boundary at the average yearly −5 °C isotherm. Another approach uses mean monthly July +18 °C and +13 °C isotherms for southern and northern boundaries, respectively. Yet another approach defines the boreal zone using the length of the period with mean daily temperatures of at least 10 °C and draws the southern boundary at less than 120 days and the northern boundary at approximately 30 days. The area within these boundaries typically has 50 to 100 frost-free days per year. The amount of precipitation in the boreal forests depends on their proximity to the zone of oceanic influence and varies from 200 to 2,000 millimeters per year. However, relatively low temperatures and a short warm period lead to low evaporation rates and thus to generally humid climate even within areas that receive little precipitation.

A different approach to defining boreal forests is to use the dominance of a particular vegetative form. Under this approach, boreal forests are defined as tree-dominated landscapes in high northern latitudes with prevalence of needle-leaved species in primary (climax) forests. Although this approach is very different from the climatological definitions described above, the overall global extents of boreal biome produced by various schemes are similar and reflect a strong relationship between climate and vegetation distribution. In North America, the majority of boreal forests are found between 45° N and 55° N, whereas in Northern Eurasia boreal forests are rarely found below 55° N. The vegetation-based definition extends the northern boundary of boreal forest in Eurasia above 70° N; however, most ecologists agree that there it represents a forest-tundra transitional zone rather than traditional boreal forests. Boreal forests are also found south of their broad geographic boundaries presenting Alpine extensions of the biome range or mountain taiga. These extensions can penetrate deep into general ranges of other biomes stretching along mountains as in the example of the Appalachians.

Vegetation, Wildlife, and Major Ecosystem Processes

The limitations imposed on vegetation growth in the boreal zone by relatively low temperatures are offset by the length of the day during the warm season. The maximum day length varies from 16 hrs. (hours) of sunlight at the southern boundary of the boreal zone to 24 hrs. at the northern, promoting vigorous vegetation growth during summer. Trees are the dominant life form in the boreal forest. Primary or climax forests are dominated by coniferous species such as spruce (*Picea*), fir (*Abis*), pine (*Pinus*), and larch (*Larix*). The majority of coniferous species are evergreen with the exception of larch that drops its needles during the cold period of the year, adapting to extremely harsh winters of Central and Eastern Siberia. Following a disturbance, the primary forests enter a successional stage frequently dominated by broad-leaved deciduous trees such as alder (*Alnus*), birch (*Betula*), and aspen (*Populus*). Early colonizers, these species subsequently create favorable conditions for reestablishment of shade-tolerant coniferous trees. The "broad-leaved deciduous coniferous" succession is typical for many but not all boreal forests. For example, light coniferous

larch stands frequently regrow as larch bypassing the broad-leaved tree dominance stage. Even with postdisturbance successional landscape mosaic of coniferous and broad-leaved deciduous stands, boreal forests show little diversity in main tree species. In addition to the dominant tree tier, boreal forests may contain shrubs, dwarf shrubs, herbaceous cover, lichens, and mosses, although the composition and abundance of those vegetation types differ in various stands and zones.

Boreal forests occupy areas of the most recent ice-age glacial cover and commonly overlay permafrost-affected soils. The distribution of permafrost is not uniform and ranges from completely permafrost-free soils and scattered permafrost in interior Alaska and Canada to widespread solid permafrost of Central and Eastern Siberia. Poor soil drainage in combination with vigorous growth of vegetation, low temperatures, and short frost-free period facilitates accumulation of partially decomposed organic material and formation of peatlands. Although peatlands develop in many biomes, nearly 80% of all peatlands worldwide are found within the boreal zone. Boreal peatlands and trees are estimated to account for nearly 40% of global terrestrial carbon pool and therefore play a crucial role in the global carbon budget. In addition to their prominent role in carbon storage, peatlands also receive and store atmospherically deposited pollutants, including nitrogen, sulfur, and heavy metals. Despite great accumulation of organic layer, nutrient availability in soils within the boreal biome is generally low.

Although the biological diversity of dominant tree species in the boreal biome is low, boreal forests now represent the largest intact ecosystem on Earth and therefore represent a highly valuable biodiversity conservation resource. The spatial extent of these forests sustains very large populations of large herbivores, including woodland caribou (*Rangifer tarandus*), moose (*Alces alces*), and elk (*Cervus elaphus*), as well as smaller herbivorous boreal species such as snowshoe hare (*Lepus americanus*) and beaver (*Castor Canadensis*). In turn, the availability of large numbers of prey species sustains considerable populations of fur-bearing predators, including fairly common species such as brown bear (*Ursus arctos*), lynx (*Lynx* spp.), grey wolf (*Canis lupus*), as well as

Forest area with mountains in the background in Northern Ontario, Canada

Source: Ken Chiu/iStockphoto.

the endemic and critically endangered Amur (Siberian) tiger (*Panthera tigris altaica*), and various members of the weasel family such as wolverine (*Gulo gulo*), marten and sable (*Martes* spp.), and mink (*Neovison* spp. and *Mustela* spp.). Boreal forests contain hundreds of threatened plant and animal species, included in the World Conservation Congress (IUCN) red list, and over a dozen of the United Nations Educational, Scientific and Cultural Organization (UNESCO) World Heritage sites.

Natural and Human-Caused Disturbances

Boreal forests are affected by many disturbances that play an important role in regulating forest succession and maintaining diversity and mosaic structure of boreal forests. The most prevalent natural disturbance is wildland fire that is an integral part of the boreal ecosystem functioning. Mean fire-return interval in the boreal zone, defined as an average time required to burn an area equivalent to the entire ecosystem, is approximately 139 yrs. (years) with a range between 50 and 270 yrs. The severity of fire impacts on the forest vary from low-impact surface fires, which rarely kill mature trees, to full mortality stand replacement fires. Peat fires are more frequent in continental zones of West-Central Canada and Western Siberia than in maritime boreal zones. Peat fires are characterized by different burning patterns compared with forest fires and are more frequently smoldering than flaming. As a result of smoldering burning of deep organic layers, peat fires release massive amounts of greenhouse gases and other pollutants into the atmosphere. Subsurface smoldering of peatland fires also makes it extremely difficult to control them as fire may spread beneath the surface layer without being detected. Unlike forests that turn into a net carbon sink during early successional stages 10 to 30 yrs. following the fire, peat accumulation is a very slow process, therefore burned nonforested peatlands remain a source of carbon emissions for a long time after fire occurrence.

Insect infestation is the second most widespread natural disturbance agent in the boreal forest. The major threat is posed by defoliating insects, such as spruce budworm and forest tent caterpillar that feed on foliage during various life stages, and borers, such as bark beetles, which bore into the tree and generally complete their life cycle under the bark. While the impact from defoliators can range from partial and temporary loss of tree canopy to tree mortality following a prolonged infestation period, bark beetle infestation almost always leads to complete tree mortality during the life cycle of one generation. Unlike fire impacts, the full extent of insect infestation may not be immediately apparent, and it may take several seasons to define the extent and quantify the amount of forest loss. Insect infestation-weakened or killed forests are frequently subsequently affected by fire, removing the dead standing biomass and resetting successional forest growth. Although not quite as prevalent as fire and insect infestation, disease and wind throw also affect boreal forest and may lead to stand replacement.

Commercial timber harvest is the leading agent of human-caused (or anthropogenic) disturbances in the boreal forest. Clear-cut logging is the most frequent form of timber harvesting, which results in removal of forest cover similar to stand replacing natural disturbances. In addition to commercial timber harvesting, boreal forests overlay major oil, gas, and coalfields and are affected by energy development projects and other accompanying development. Peat mining presents the third major anthropogenic disturbance leading to significant modification of boreal ecosystem functioning. Postharvesting dehydration and mineralization of peat changes soil and water qualities and water tables and subsequently results in a turnover in species composition and vegetation structure. Large-scale peat harvesting also has a pronounced impact on wildlife and certain bird species, leading to long-term replacement of typical peatland fauna with generalist species.

Cultural Significance of the Boreal Biome

In addition to their prominent role as a major economic resource in industrial forestry, including timber harvesting and pulp and paper production, mining, and oil and gas extraction, boreal forests present an important resource for tourism, trapping, hunting, and recreation. The cultural significance of boreal forests is especially high because many of the world's remaining northern indigenous cultures live in the zone of boreal forests. Many native people of the boreal biome maintain their traditional practices and lifestyle that has survived since the most recent Ice Age. The genetic biodiversity, remoteness, and vast ecological resources of boreal forests support the central way of life for many indigenous groups and allows for preservation of their cultural identity. Many of these cultures have become vulnerable under the impact of ongoing climatic, economic, and cultural changes in the surrounding areas making it particularly hard for them to maintain an active role in the contemporary society while retaining close association with the traditional way of life.

Tatiana Loboda

See also Agroforestry; Biodiversity; Biota and Climate; Carbon Cycle; Deforestation; Forest Fragmentation; Forest Land Use; Forest Restoration; Peat; Sustainable Forestry; Tree Farming

Further Readings

Goldammer, J., & Furyaev, V. (Eds.). (1996). *Fire in ecosystems of boreal Eurasia.* Dordrecht, Netherlands: Kluwer Academic.

Kasischke, E., & Stocks, B. (Eds.). (2000). *Fire, climate change, and carbon cycling in the boreal forest: Vol. 138. Ecological studies.* Heidelberg, Germany: Springer.

Shugart, H., Leemans, R., & Bonan, G. (Eds.). (2005). *A systems analysis of the global boreal forest.* New York: Cambridge University Press.

Wieder, R., & Vitt, D. (Eds.). (2006). *Boreal peatland ecosystems: Vol. 188. Ecological studies.* Heidelberg, Germany: Springer.

BIOME: DESERT

The desert biome is broadly described as any land that receives less than 500 mm (millimeter) of rainfall per year, including tropical/subtropical deserts falling within ± 30° latitude of the equator, as well as temperate deserts falling between 30° and 50° latitudes (North and South) with an annual temperature less than 10 °C. This biome covers more of the earth's land surface than any other biome—roughly one third of all land is desert. Most of this biome falls within Africa, Asia, and the Middle East.

The word *desert* is derived from the Latin word *desertus*, meaning "deserted," reflecting many people's first impression of deserts as being barren and devoid of life. Contradictory to such first impressions, the desert biome has a remarkably high diversity of species, second only to the rain forest biome, but it has the lowest productivity of any biome. It is perhaps this disparity that has intrigued geographers and naturalists for centuries and led scientists to base many environmental theories on desert systems. Nowhere else on earth are ecological and geomorphic processes more visible.

Categorization of Deserts

In addition to being broadly categorized into tropical/subtropical and temperate, deserts are categorized according to their level of aridity, traditionally by assessing annual precipitation. However, moisture input should not be considered without factoring in moisture loss, particularly in the limited moisture systems of deserts. Furthermore, precipitation in deserts is too variable to be the sole basis of categorization. Most geographers measure the aridity of a desert according to the ratio between annual precipitation and evapotranspiration (the combined loss of moisture through evaporation and transpiration by plants). These criteria were developed by UNESCO in the 1970s, but several similar formulas have been developed over the past century.

The extent to which mean annual precipitation (P) falls below mean annual evapotranspiration values (ET), reflects a level of aridity that falls into one of three categories: (1) semiarid, (2) arid,

and (3) hyperarid. This measurement is sometimes altered to include other factors. For example, calculations of this ratio are often made for set periods of time to account for climate variability, and some formulas factor in the water storage capacity of plants and soil. Nevertheless, categorizations are most commonly based on the following:

- *Semiarid deserts:* P/ET value <0.5 and mean annual precipitation 200 to 500 mm.
- *Arid deserts:* P/ET value <0.20 and mean annual precipitation between 80 to 200 mm.
- *Hyperarid deserts:* P/ET value <0.03 and mean annual precipitation <80 mm.

The delineation of different deserts is challenging at best, and satellite images now show us how the boundaries of deserts can shift in just 1 year— the sands of the Sahara Desert in North Africa continually shift and China's desert land expands every year due to desertification processes. In addition, a desert that is considered arid one year could be semiarid in another due to a rare spike in rainfall. To overcome these variations, there is an important temporal element to the categorization of deserts, usually requiring decadal measurements at minimum. Other measurements have also been used to help identify where a desert area begins and ends, such as vegetation distribution, geomorphology, or landscape type.

Semiarid Deserts

Semiarid deserts are a common type of desert within the desert biome, covering roughly 14% of Earth's land surface—this percentage doubles if one includes dry subhumid areas (see Mediterranean Biome). The features of semiarid deserts vary considerably from dry savanna ("steppe" in temperate zones) to shrubland to dry woodland. Because of their widespread distribution and varying land cover, many semiarid deserts are not delineated as "deserts" but are, nonetheless, included in the desert biome.

All semiarid deserts are vulnerable to desertification processes, mainly because more human settlements occur in these less arid deserts due to the agricultural potential. Unfortunately, land use practices in these areas are often unsustainable,

Wildebeest migrating across semiarid plains of East Africa

Source: J. S. Lalley, 2008.

and some of the worst droughts and famines ever experienced have been in semiarid deserts. This observation also means that pristine semiarid deserts are harder to come by than pristine arid to hyperarid deserts.

Flora and Fauna of Semiarid Deserts

Semiarid deserts are often referred to as "fringe deserts" and will harbor species that live on the edge of both higher- and lower-rainfall areas. Florae that are characteristic of semiarid deserts include drought- and fire-resistant plants (particularly grasses), plants with thorns, and plants with deep penetrating roots that tap into ground water. Fauna is characterized by migrating large mammals and migrating birds.

Movement is key to survival in semiarid areas. In tropical semiarid deserts, animals will quite literally follow the rain. In temperate semiarid deserts movement is further dictated by seasonal temperatures. Well-known examples of such movement include the mass migration of wildebeest in East Africa and pronghorn sheep in North America (see wildebeest photo).

Location of Semiarid Deserts

Temperate, Semiarid Deserts (or "Steppes")

Asia: peripheral areas of the Gobi Desert, Tibetan Plateau, Pamir Plateau, Turanean lowland, Ust Urt Plateau

North America: Colorado plateau, dry grasslands from Texas to Montana and North Dakota

South America: parts of the Patagonian Desert and bordering areas

Tropical/Subtropical Semiarid Deserts

Africa: Northern and Eastern Kalahari Desert, Karoo Desert, Sahel Desert, Western Somali-Chalbi Desert and bordering areas of East Africa

Asia: Deccan Plateau

Australia: all areas bordering the great arid deserts of the interior

Livestock overgrazing in a desert

Source: J. S. Lalley, 2008.

Middle East: most of the Middle East outside arid and hyperarid deserts

North America: the southern and eastern boundaries of Chihuahuan Desert

South America: Chaco, Monte Desert, most areas east of the Atacama Desert, Caatinga Scrubland (Brazil)

Arid Deserts

Arid deserts are considered "true deserts" because they have extremely water-limited systems with desert-adapted flora and fauna. They cover 15% of the earth's land surface, and this percentage is increasing due to the desertification of semiarid deserts and exceeding climate variability worldwide (see livestock photo). Human population growth has meant that human settlements have spilled into the nonarable lands of arid deserts, and while desert dwellers have traditionally stuck to desert-sensitive livelihoods, most modern inhabitants of arid deserts have not. Hence, arid deserts are becoming increasingly vulnerable to human impacts. Such impacts can be profound where overgrazing leads to extreme dust storms, water tables readily collapse leading to contamination of already limited freshwater resources or where the loss of one species can lead to a dramatic and cascading loss of others.

Flora and Fauna of Arid Deserts

Arid deserts are home to a unique variety of desert-adapted species. Vegetation is characterized by plants with high water-retention capabilities, such as succulents and cacti, and with small, unpalatable and/or spiny leaves that resist water loss and browsing by animals. Plants are often tolerant of saline soils, and they often have shallow roots for the quick absorption of rare rain events or extremely deep roots for tapping into ground water. Drought-resistant grasses are also typical of arid deserts, but growth is usually patchy (see springbok photo). In fact, most vegetation growth in arid deserts is patchy, which is largely dictated by soil conditions (nutrient content, salinity, porosity) and by water availability. In turn, water availability is dictated by water-collecting and water-channeling features, such as soil depressions,

Springbok grazing on patchy grasses of the arid Kalahari Desert

Source: J. S. Lalley, 2008.

rock crevices, or certain soil types with greater water-holding capacity. Patch dynamic theory was developed to explain this phenomenon.

Given the extreme temperatures commonly found in arid deserts, much of the animal life seeks refuge in burrows and/or is nocturnal. Reptiles and insects are usually the most diverse group of animals in arid deserts, while mammal life is less diverse and typically characterized by rodents and small carnivores (see chameleon photo). Exceptions are found in the large antelope inhabiting the Gobi and Arabian deserts, and many charismatic large mammals in Africa's Kalahari Desert (e.g., zebras, giraffes, lions). The birdlife of arid deserts is often found in the form of ground-dwelling birds and small, insect-eating birds. Bright coloration is rare.

Location of Arid Deserts

Temperate Arid Deserts

Asia: Gobi Desert, Ordos Desert, Tengger Desert

Middle East: Karakum Desert, Taklimakan Desert, Caspian Lowland Desert, Central Iranian Plateau, Syrian Desert

North America: Great Basin, Colorado Plateau

South America: Patagonian Desert and Monte Desert

Tropical Arid Deserts

Africa: Kalahari Desert, Karoo Desert, peripheral zones of the Sahara Desert

Asia: Thar Desert

Middle East: Arabian Desert, Judean Desert

North America: Chihuahuan Desert, Sonoran Desert, Mojave Desert

Australia: Great Sandy Desert, Tanami Desert, Simpson Desert, Great Victorian Desert, and Gibson Desert

Hyperarid Deserts

Hyperarid deserts of the world have long been misinterpreted as desert wastelands, devoid of conspicuous vegetation, typically covered to some degree by sand dunes, and usually lacking any surface water or easily accessed ground water.

A desert-adapted chameleon

Source: J. S. Lalley, 2005.

They are the deserts most frequently pictured as typical deserts, with wind-swept sand dunes and barren gravel plains (see sand dunes photo). They cover only 8% of Earth's surface and are sparsely populated by humans, yet most people are familiar with images of these intriguing "wastelands." Because of their low human population, hyperarid deserts of the world are the most pristine of all deserts. Where human activities have ensued, these deserts are highly vulnerable to change: A single vehicle track across a hyperarid desert can last several hundred years.

Flora and Fauna of Hyperarid Deserts

Hyperarid deserts are known for having high rates of endemism among their flora and fauna, meaning the species occur in that desert and nowhere else on earth. For example, Southern Africa's hyperarid Namib Desert has an astonishing diversity of wildlife with a third of its vertebrates being endemic. The old age of all the deserts in this region have led to a high diversity of species, but it is the long isolation of the Namib Desert that has led to its highly endemic fauna.

The vegetation of hyperarid deserts includes scarce outcrops of succulents and cacti—as in arid deserts, these outcrops are dictated by soil characteristics and water availability. Microflora, such as algae and lichens, can be found in great abundance on soil surfaces, filling in a niche usually dominated by grass in less arid environments. These species can survive by absorbing atmospheric sources of moisture such as dew (or fog in coastal deserts).

Dew, fog, and water from vegetation and prey are the key sources of moisture exploited by the animal life of hyperarid deserts. Reptiles and insects are known to bask in fog, allowing moisture to condense on their bodies for later ingestion. They will also strategically sit on or under plants and allow leaves to collect the moisture for them. As in arid deserts, mammals inhabiting hyperarid deserts are typically nocturnal and small in size. However, there are many notable exceptions such as the large antelope (oryx) found in the hyperarid Namib Desert and Arabian Desert or the elegant gazelles of the Persian Desert.

Location of Hyperarid Deserts

Temperate Hyperarid Deserts

Asia: parts of the Gobi Desert, Ala-Shan Desert, Tsaidam Desert

Middle East: Syrian Desert (southeast)

Tropical/Subtropical Hyperarid Deserts

Africa: Central Sahara Desert, Namib Desert

Middle East: Arabian Desert, Southern Iranian Plateau, Negev Desert

North America: parts of the Mojave and Sonoran Deserts

South America: Atacama Desert, Sechura Desert, Guajira Desert

Jennifer S. Lalley

See also Biome: Tropical Savanna; Biome: Tropical Scrub; Climate: Dry; Desertification; Desert Varnish; Environmental Management: Drylands; Xeriscaping

Sand dunes in a hyperarid desert

Source: J. S. Lalley, 2009.

Further Readings

Laity, J. (2008). *Deserts and desert environments.* Oxford, UK: Wiley-Blackwell.

Mares, M. (1999). *Encyclopedia of deserts.* Norman: University of Oklahoma Press.

Meigs, P. (1953). World distributions of arid and semi-arid homoclimates (Arid Zone Research No. 1). In *Arid zone hydrology* (pp. 203–209). Paris: United Nations Educational, Scientific and Cultural Organization.

Middleton, N., & Thomas, D. (1997). *World atlas of desertification.* London: Arnold.

Whitford, W. (2002). *Ecology of desert systems.* London: Academic Press.

World Resources Institute. (2002). *Drylands, people, and ecosystem goods and services: A web-based geospatial analysis.* Washington, DC: Author. Retrieved November 2, 2009, from www.wri.org/publication/drylands-people-and-ecosystem-goods-and-services-web-based-geospatial-analysis

 # BIOME: MIDLATITUDE DECIDUOUS FOREST

Deciduous forests can be classified into two distinct biomes: (1) the deciduous midlatitude biome and (2) the tropical deciduous biome. Although in both biomes the trees lose their leaves during a dormant season, tropical deciduous trees experience dry conditions and subsequent leaf loss while deciduous midlatitude trees lose their leaves because of the cold season. The deciduous midlatitude forest is anything but homogeneous. Flora and fauna, soil types, and climate differ along a latitudinal and/or an east-west gradient. Tree genera are very similar between the European, North American, and Asian deciduous, midlatitude forests but diversity varies. Disturbances are a vital part of this biome.

Structure and Composition

Extending throughout the Eastern United States, much of Europe, parts of Eastern Asia, Southern

South America, Eastern Australia, and New Zealand, the deciduous midlatitude biome is characterized by broad-leaf trees that undergo a dormant season due to low temperatures. North American, European, and Asian deciduous midlatitude forests are similar in structure and composition, although diversity in European deciduous forests is much lower compared with its American counterpart. The difference in diversity is mainly due to the position of mountain ranges in Europe and North America and glacial advances on both continents. Mountain ranges run north to south in North America and east to west in Europe. Mile-high glaciers advanced south during the last glacial maximum, displacing the deciduous tree species in North America and Europe. Trees were able to migrate south in North America but the deciduous tree species in Europe encountered a barrier. Many deciduous species went extinct in Europe, resulting in lower diversity.

Considering the forest composition, the genera of deciduous trees in the different forests of North America, Europe, and Asia are, in general, the same with some combination of maple (*Acer* spp.), chestnut (*Castanea* spp.), beech (*Fagus* spp.), ash (*Fraxinus* spp.), oak (*Quercus* spp.), basswood (*Tilia* spp.), and elm (*Ulmus* spp.), among others present in different forests. Some evergreen trees, such as pine (*Pinus* spp.) and hemlock (*Tsuga* spp.), and evergreen shrubs, such as holly (*Ilex* spp.), are also present in the deciduous midlatitude forest.

In addition to the similarity in the forest composition, the structure in deciduous midlatitude forests of North America, Europe, and Asia is also similar. In general, four layers exist in these forests. Mature canopy trees (20 to 30 m [meters] tall) dominate the overstory. Subcanopy trees are not as tall as mature canopy trees but still receive direct sunlight at the top of the tree's canopy. Subcanopy trees usually fill a gap produced by a dead canopy tree and will eventually grow to become part of the mature canopy trees. Understory trees only receive minimal sunlight through the canopy and consist mainly of shade-tolerant saplings. Finally, the ground layer consists of tree seedlings and shrubs, ferns, and herbs.

Several animals are associated with the eastern deciduous forest in North America, among them mammals such as the white-tailed deer (*Odocoileus virginianus*), gray squirrel (*Sciurus carolinensis*), American beaver (*Castor canadensis*), and black bear (*Ursus americanus*). Birds in this biome include owls, crows (*Corvus* spp.), hawks, pigeons, and jays among many others. A once very common bird in North America, which became extinct in the late 1800s and early 1900s, is the passenger pigeon (*Ectopistes migratorius*), and before its extinction, it took days for a flock to pass by. Billions of passenger pigeons existed when European settlers came to North America. However, due to overexploitation by hunting and loss of habitat, their numbers declined. Because the passenger pigeon is a social creature that requires large flocks to live, it did not survive. Today, the most prominent animal in the eastern deciduous forest in North America is probably the white-tailed deer. Brought to very low numbers in the early 1900s due to hunting pressures and habitat destruction, the population rebounded to pre-European levels, exceeding those numbers by several million today.

Climate, Soil, and Light Availability

Mean annual temperature and average annual precipitation have a wide range in the deciduous, midlatitude forest, with a mean annual temperature between 5 °C and 20 °C and the average annual precipitation ranging between 50 and 250 cm (centimeters). Mean January temperatures can vary from less than −10 °C in, for example, in northeastern North America to above freezing temperatures in the southeastern United States. Most of these differences relate to the latitudinal gradient in the Eastern United States. Therefore, the plant and animal distribution in the Eastern United States deciduous midlatitude forest is heterogeneous, and, in general, nine different associations can be distinguished. An association is a forest region that is often named after its most dominant tree species. For example, the maple-basswood association in southeastern Minnesota and southwestern Wisconsin is dominated by maple and basswood.

Soils also vary with latitudinal gradient. However, major soil types in the deciduous, midlatitude biome are alfisols and ultisols. Alfisols are sometimes described as the soil type of the deciduous forest because the soil depends on the leaf litter of the broad-leaf trees. This soil is relatively fertile and important for agricultural purposes. Alfisols are usually found in cooler regions, such as the northern part of the deciduous forest in Eastern North America. Ultisols, on the other hand, are found in warmer regions, such as the southeastern United States and southeastern Asia. Ultisols are well leached and therefore have relatively low fertility. However, they can support relatively productive forests. Because the plants in areas where ultisols predominate receive their nutrients from organic matter and not the soil itself, the fertility of these soils depends on the biological recycling system, that is, the nutrient cycle.

Compared with the tropical rain forest, where about 2% of the light that is intercepted at the top of the canopy reaches the forest floor, the deciduous midlatitude forest floor receives approximately 10% during the summer and 70% during the winter. During summer, the understory can be dense with seedlings, herbs, and ferns in these forests. Nevertheless, the understory receives less light than the overstory, and trees have adapted to this discrepancy. For example, leaves on the same tree can be quite different based on their position on the tree. Leaves in the canopy (sun leaves) are thicker, have less chlorophyll, and are smaller in area compared with so-called shade leaves that can be found lower on the trunk of the tree. One reason for this is the fact that a smaller surface area in sun leaves helps reduce water loss. At low light levels, shade leaves are often more efficient at photosynthesizing, whereas at high light levels, the sun leaves are more efficient.

Another way that trees have adapted to the discrepancy between light levels in the overstory and understory is with shade tolerance. Tree species differ based on their tolerance or intolerance of shade. Maples, for example, are very shade tolerant, that is, they can germinate and survive in the shade. Oaks are more intermediate in shade tolerance, that is, they need medium light levels to germinate and survive in the understory. Birches are on the other end of the spectrum and need high light levels to thrive. This characteristic is also indicative of the successional stage of the forest. For example, a forest that has mainly shade-intolerant species probably just experienced a major disturbance such as a crown fire. Crown fires kill all trees, open up space, and free nutrients for new plant growth to become established. For a variety of reasons, usually shade-intolerant species become established. However, those species cannot grow under their own canopy and shade-tolerant species will germinate. If there is no other major disturbance, the shade-intolerant species will senesce and eventually fall over, creating a canopy gap into which an existing shade-tolerant tree species will grow, and gradually replace the shade-intolerant canopy.

If it was not for the dormant season, the productivity of deciduous midlatitude biome could be the highest compared with other biomes. During summer, productivity is higher in the deciduous forest than in any other biome, including the rain forest because days are longer. Photosynthesis, however, does not occur when temperatures are too cold. Trees and shrubs become dormant by losing their leaves. Before losing their leaves, the hardwood trees in northeastern North America experience the so-called Indian summer where the leaves show spectacular fall colors ranging from red to yellow. The reason for this color change in the leaves is the loss of the green chlorophyll, which masks the red and yellow pigments that are already in the leaves. With the loss of chlorophyll, those pigments become visible. If they did not lose their leaves, the soft tissue that is exposed would be damaged by the freezing conditions. Timing of leaf loss and leafing out varies between species and by year. However, the process of leafing out in one stand or forest happens in about the same time frame of 2 to 4 weeks every year. The same is true for the timing of losing leaves. After losing their leaves, most trees in the deciduous midlatitude forest require a chilling period, that is, a certain amount of time of freezing temperatures, before trees leaf out again. This ensures that trees only leaf out after winter is over. For example, sugar maples (*Acer saccharum*) in northern parts of North America require approximately 100 consecutive days of chilling temperature for buds to form.

Disturbances

Disturbances are a major part of the deciduous midlatitude forest. Commonly, 5% to 10% of a forest is influenced by natural disturbances and mortality. Those disturbances usually result in canopy gaps. Natural disturbances, such as fire, windthrow, insects, and pathogens, and human-induced disturbances, such as invasive species, logging, and forest clearing, shape forest ecosystems with regard to structure, composition, and functional processes. Of the primary (deciduous) forest, that is, forests that have not been altered by humans, which existed in the Eastern United States before European settlement, 99% has been cut or otherwise been altered during the past 150 yrs. (years). Often also named old-growth forest when no major disturbance has killed the canopy trees in recent centuries, primary forest is very rare in deciduous forests around the world.

Although most of the deciduous forest in the Eastern United States has been cleared for agriculture or logged, some of the resultant agricultural land has been found to have only marginal economic value. The eastern deciduous forest in the United States was decimated by the coming front of people clearing land for agriculture. By 1850, approximately 50 million hectares had been cleared and converted to agriculture. That trend reversed in the 20th century when many farms converted back to forest because they were abandoned, resulting in an increase in forested land in the contiguous United States in the past 100 yrs. In addition, primary forests cleared for their wood have been replaced by second-growth forest. This second-growth forest is structurally very different from the primary forest as species and structural diversity are much lower than in the primary forest.

Natural and human-induced disturbances have different effects on forest structure, composition, and functional processes. An example of a human-induced disturbance was the introduction of the fungus chestnut blight (*Cryphonectria parasitica*) to the United States with the arrival of chestnut lumber or trees in the early 1900s. The American chestnut was one of the most dominant trees in Eastern North America, but the blight eliminated all mature American chestnut (*Castanea dentate*) trees in the Eastern United States within 30 to 40 yrs. Today, only small American chestnut seedlings and saplings exist in the understory but are killed before they mature. This blight was introduced from Eurasia, where trees have resistance to this fungus because they evolved with it.

A native pathogen or insect, on the other hand, does not usually kill its host species completely. Host trees might die in an insect outbreak, for example, the spruce budworm, which defoliates balsam fir (*Abies balsamea*) and spruce (*Picea* spp.) in the boreal forest. However, this defoliation is a way for the forest to rejuvenate. While older, mature trees are killed, younger and smaller trees survive and repopulate the forest after the outbreak. The forest tent caterpillar (*Malacosoma disstria*) is a native insect that defoliates deciduous trees in Eastern North America. However, here the insect does not necessarily kill the trees in an outbreak but reduces growth significantly. The trees usually recover after an outbreak. This is probably the case because the trees have had a chance to adapt to the insect by process of evolution.

Julia Rauchfuss

See also Agroforestry; Biogeography; Biome: Boreal Forest; Biome: Tropical Deciduous Forest; Deforestation; Forest Fragmentation; Forest Land Use; Forest Restoration; Invasion and Succession; Nutrient Cycles; Social Forestry; Soils; Sustainable Forestry

Further Readings

Archibold, O. (1995). *Ecology of world vegetation.* New York: Chapman & Hall.

Braun, E. (1950). *Deciduous forests of Eastern North America.* Philadelphia: Blakiston.

Kricher, J., & Morrison, G. (1998). *A field guide to eastern forests: North America.* Boston: Houghton Mifflin.

MacDonald, G. (2003). *Biogeography: Space, time and life.* New York: Wiley.

Yahner, R. (2000). *Eastern deciduous forest: Ecology and wildlife conservation.* Minneapolis: University of Minnesota Press.

BIOME: MIDLATITUDE GRASSLAND

The midlatitude grasslands are not easily defined. Most grasslands are areas of low relief and low rainfall, the combination of which protects them by reducing erosion disturbances that allows rapid invasion by trees and shrubs. Much of the former grasslands in higher-rainfall regions were removed by cultivation. The midlatitude grasslands lie mostly between 25° and 50° of latitude in both hemispheres. The grass species of the midlatitude grasslands come from rich sources both in the tropics and high latitudes. The species did not arrive together and many grasslands have been in place for thousands to millions of years. New grass species invade along disturbance corridors, such as valley walls or bottoms. The dominant grass species use a variety of seed dispersal mechanisms. Some are adapted to transport by wind, while others move in the digestive tracts of animals or adhere to the hides of animals. This entry will discuss the nature of the midlatitude grasslands, the grasses of these regions, human interactions with and maintenance of grasses, and animals that use the grasslands.

At the global scale, scant and uneven data limit our descriptions to higher-taxonomic levels, such as tribes. Available data do not permit examination of taxa by abundance. We can compensate for this deficiency to some degree by looking at the diversity or dominance of some taxa in the total grass flora of different regions. The most widely available data are species lists for various areas of the world. Hartley and colleagues have done this for the seven taxonomic groups that contain more than 75% of the global taxonomic diversity of grasses. These most important tribes are divided into two distinct photosynthetic types that evolved as adaptations to distinctly different environments. *Agrosteae*, *Aveneae*, *Festucae*, and the largest genera, *Poa*, comprise C_3 species. This photosynthetic pathway, common in other plants, is adapted to high-atmospheric carbon dioxide and cool summer temperatures. For these grasses, production declines above 20 °C (68 °F). They have migrated to the midlatitudes from higher latitudes in both hemispheres. They also dominate in areas of wet winters and dry summers. This seasonal adaptation allows these grasses to complete their growth cycle before the summer heat limits production. Domesticated species from this group include rice (*Oryza* spp.), wheat (*Triticum* spp.), oats (*Avena sativa*), barley (*Hordeum* spp.), and rye (*Secale cereale*). These and other domesticated grasses have larger seeds that are presented high on the plant and, therefore, easily harvested and handled than plants with seeds that are small designed for transport by the wind.

A later evolution, the C_4 (warm season) photosynthesis grasses, adapted to higher temperatures and lower carbon dioxide, such as occurred during glacial maxima, by enriching the CO_2 taken in the stomata and feeding it to a deeper C_3 tissue layer. Production declines for these grasses above 35 °C (95 °F). *Eragrosteae* (now regrouped with *Chlorideae*), *Paniceae*, and *Andropogoneae* are dominated by species that use C_4 photosynthesis, although a few of the *Paniceae* are C_3 species. The highest species diversity of these grasses is found within 30° latitude of the equator. Figure 1 illustrates the dependence of the midlatitudes on the adjacent tropical and high-latitude source areas for the species that migrated into midlatitude grasslands. Domesticated species from this group include corn (maize; *Zea mays*) and sorghum (*Sorghum bicolor*).

The Grasses

While extant grasslands are a good place to find some grass species, they are often not the places of greatest species diversity. Of the more than 9,000 grass species known worldwide, only about 1,400 occur in the United States. Fewer than 400 of those are known to grow somewhere in the Midcontinent Plains grasslands. Ten percent of these 400 species account for more than 99% of the grass biomass in the American Plains. Biomass of the 41st-ranked species is less than half of 1% that of the most abundant *Andropogon gerardii*.

Many of the higher-taxonomic groups of grasses are not well represented in the American Plains flora. In the Poeae tribe, only 18 of the 500 species known worldwide (3.6%) are documented in the Plains. Only two of those (*Poa pratensis* and *Puccinellia nutaliana*) rank among the top 41 Plains grasses. Seventeen species of the vast *Chloridoideae* subfamily are

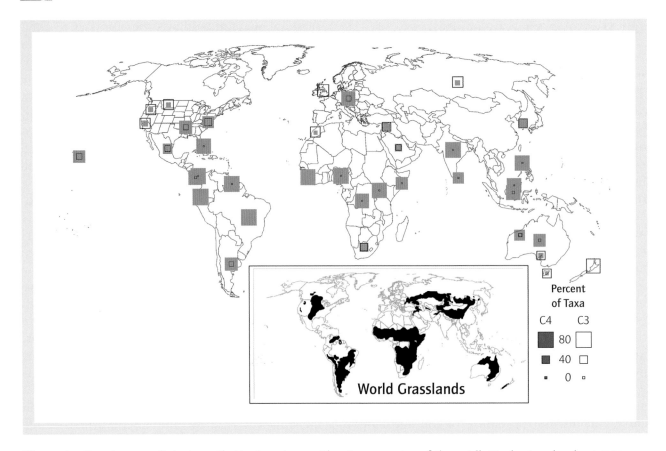

Figure 1   Dominance of photosynthetic grass types. The grass species of the midlatitude grasslands comes from areas of species richness in the high latitudes and the tropics.

Sources: Map created by author from data in Hartley, W. (1950). Studies on the origin, evolution and distribution of the Gramineae. *Australian Journal of Agricultural Research*, 1, 355–373; Hartley, W., & Slater, C. (1960). Studies on the origin, evolution and distribution of the Gramineae III: The tribes of the subfamily Eragrostoideae. *Australian Journal of Botany*, 8(3), 256–276.

represented in the Plains. The *Andropogoneae* and *Paniceae* tribes are represented by 6 and 5 species, respectively, in the top 41 Plains grasses. Of the 63 worldwide genera in the *Aveneae* tribe, only one species is known in the Plains, and it ranks among the top 41. Of the top 41 Plains grasses, five each are from the *Titiceae* and *Stipeae* tribes. Data are not available to show the worldwide affinities of 10 of the top 41 Plains grass species.

The dominance patterns of a grass genus, tribe, or subfamily are reflected in the species abundance patterns in the American Plains. For example, the species of the *Chloridoideae* tribe are concentrated in the Southwestern Plains. In contrast, *Paniceae* tribe species are more abundant in the Southeastern Plains.

The broad geographic pattern of *Andropogoneae* dominance suggests a tropical Asian origin with a connection to North America through tropical Africa and South America. Six of the top 41 Midcontinent Plains grasses (big, little, and silver bluestem, cane beardgrass, eastern gamagrass, and Indiangrass) are members of the *Andropogoneae* tribe, and the abundance of these species are all higher in the southeastern Plains.

Other genera dominate coastal grasslands, but species vary with salinity. The *Spinifex* of Australia, a wide variety of coastal species in South Africa, and *Spartina* in North America are all C_4 grasses. Bamboos (*Bambusa*) and related species of *Phragmites* and *Arundinaria* are C_3 grasses and are important components of some coastal grasslands. Coastal grasslands are seriously threatened by development and potential rise in sea level.

Many of the world's grasslands are of considerable antiquity. We do know that members of the *Stipeae* tribe have a long presence in the American Plains, dating back to 10 million years ago. Most of these are related to living forms that are more common in Central and South America, such as *Piptochaetium*, *Stipa*, and *Nassella*. Of these fossil genera, there are 40 species known in the United States. Twenty-five are found in the Plains, and 4 are among the top 41 grasses.

Human Interactions and Maintenance

Most of our food comes from a few grasses that were domesticated in the Eastern Hemisphere: rice, wheat, oats, barley, and rye. Sorghum was domesticated in either India of Africa. Corn (maize) stands virtually alone as an American grass domesticated for human nutrition. Wild rice (*Zizania aquatica*) that has long been harvested from natural stands now joins corn as a New World domesticated grain crop. These grasses directly and indirectly feed human populations. Most animals grown for meat, fiber, and dairy products are a product of domestic and nondomestic grasses that support huge grazing economies throughout the world.

Grasses are also used directly as fiber for baskets or furniture, as structural elements or thatching for dwellings, and as covering for animal sunshades. Most notable of the construction grasses are the bamboos, *Arundinaria* sp., *Phyllostachys* sp., *Bambusa* sp., and *Dendrocalamus* sp. Only giant cane (*Arundinaria gigantea*) is native to the North American midlatitudes and has been used for fishing poles, pipe stems, and baskets. This very tall grass can achieve a diameter of 5 centimeters and probably was important in constructing Native American lodges. Its range, however, did not extend west of the forest regions of the country. Common reed (*Phragmites australis*) is more widespread and found throughout the Plains in wet areas. It grows as tall as 4 m and has been used as lattice in adobe construction, as arrow shafts, and as fiber for mats, screens, thatching, cordage, and carrying nets.

In recent decades, interest in grasslands has focused on animal habitat, recreation, and environmental esthetics, including a momentary cure for claustrophobia. However, attention is often focused toward the sky, flowers, people, and their architecture. Infrequently, attention to grasslands focuses on the dominant or formerly dominant plant family. In short, grasses do a lot for humans, but you must work hard to appreciate them.

Soils developed under long periods of grassland cover are nutrient rich. While the low-relief terrain protects grasslands from erosion disturbances that facilitate woody species invasion, it facilitates the greatest threat to grassland persistence, the agricultural implements that take advantage of the rich soil to grow crops with requirements similar to those of native grasses. Overgrazing damage to grasslands is accentuated during droughts that limit forage for grazers. Early season grazing also harms some species. Dominant species are very tolerant of fire, which limits the invasion of many woody plant species, but overgrazing can compound the effects of fire. Many areas of grasslands no longer have weather patterns that support reestablishment of the current dominant species; therefore conservation is the only way to ensure their continued existence.

Animals That Use the Grasslands

Ruminant animals (those with four stomachs) are the most visible and characteristic animals that evolved in the grasslands. They include rhinoceros, camels, bison, gazelles, sheep, goats, antelope, and giraffes. These are joined by nonruminant horses, asses, elephants, and zebras. These animals have the ability to digest large amounts of roughage and travel long distances to water. Most will also browse shrubs and trees.

Dwight A. Brown

See also Biogeography; Climate: Midlatitude, Mild; Climate: Midlatitude, Severe; Prairie Restoration; Prairies

Further Readings

Barkley, T. (Ed.). (1977). *Atlas of the flora of the Great Plains.* Ames: Iowa State University Press.

Brown, D., & Brown, K. (1996). Disturbance plays key role in distribution of plant species. *Restoration & Management Notes, 14*(2), 140–147.

Hartley, W. (1950). Studies on the origin, evolution and distribution of the Gramineae. *Australian Journal of Botany, 8*(3), 256–276.

Hartley, W., & Slater, C. (1960). Studies on the origin, evolution and distribution of the Gramineae: III: The tribes of the subfamily Eragrostoideae. *Australian Journal of Agricultural Research, 1,* 355–373.

Hitchcock, A., & Chase, A. (1950). *Manual of the grasses of the United States* (2nd ed., U.S. Department of Agriculture Miscellaneous Publication No. 200). Washington, DC: USDA.

BIOME: TROPICAL DECIDUOUS FOREST

Tropical deciduous forest is a major vegetation type that occurs in association with seasonally wet and dry or monsoon climates. These forests vary significantly in their composition and structure in relation to their geographic location in the tropical realm, rainfall and soil conditions, and history of disturbances. The transition from evergreen to deciduous forests is gradual in response to the occurrence of a water deficit, the length of the dry season, and the amount of rain received during the rainy season. Tropical deciduous forest, in contrast to tropical rain forest, includes trees that lose their leaves synchronously and remain without leaves for varying amounts of time. In contrast to tropical savanna, tropical deciduous forests maintain a greater dominance by trees (>50%) across the landscape.

Generalized attributes of tropical deciduous forests include a closed to open tree canopy that may be tall (>20 m [meters]) with a single trunk or low in stature (<10 m) with a coppiced trunk that produces multiple stems close to the ground surface. Leaves are broad but often smaller and thicker than those typical of tropical rain forest, and the forest type varies greatly in the adaptations these trees have to the seasonal climatic conditions. Semievergreen forests have a high forest canopy with deciduous trees that remain dormant for a short period of time and an understory that remains evergreen (see the first photo). These forests can support a high diversity of lianas and understory vegetation. Dry deciduous forests occur as low (<8 m) closed-to-open woodland with trees that lose their leaves and remain dormant during long dry seasons of 3 to 6 mos. (months; see the second photo). Evergreen tropical rain forest can maintain itself only if the drought period is very short, 1 to 2.5 mos., and rainfall is very high, 2,500 to 3,000 mm (millimeters). Tropical deciduous forest transitions to tropical thorn scrub under conditions of very low rainfall (<500 mm) and a dry season that exceeds 6 to 8 mos.

Tropical deciduous forests occur between approximately 5° and 20° north and south of the equator, and extend into the subtropics where rainfall is distinctly seasonal and there is only a slight chance of frost. Seasonal rainfall patterns coincide with the migration of the sun's direct rays between the tropics of Cancer and Capricorn, bringing precipitation under high sun (summer) conditions. In South and Central America, they are most diverse in Mexico and Bolivia and along the Pacific Coast of Central America, but "dry" forests also occur between the Cerrado savannas and Caatinga dry scrublands of Northeast Brazil, and on Caribbean islands where mountains (e.g., the Blue Mountains in Jamaica) or other islands block the receipt of rainfall from the Northeast trades. Seasonal monsoon forests occur in southwest India, Sri Lanka, and Indochina. A small region in Northern Australia is distinctly seasonal with summer rainfall conditions, but evergreen *Eucalyptus* species dominate on that continent and maintain an evergreen forest canopy. Tropical deciduous forests are particularly important as a vegetation type in Africa, where they occur under two wet and dry seasons in equatorial East Africa, under one wet and dry season south of the equator from Tanzania to Southern Africa, and in Western Madagascar. These forests differ in their tree species composition, including the *Acacia-Commiphora* bushland of East Africa, the *Brachystegia*-dominated

Semievergreen forest along the Tana River, Kenya

Source: K. E. Medley.

miombo woodlands of Tanzania and south-central Africa, and the mopane (*Colophospher-mum mopane*) woodlands of South Africa. Whereas one species of baobob (*Adansonia digitata*) occurs in tropical deciduous forest on the continent of Africa, at least seven species are recorded in the dry forests of Western Madagascar (the third photo). Tree genera with very broad geographic ranges show high diversity in different woodland types.

Seasonal rainfall supports tree growth when it can infiltrate and be stored deep in the soil profile. Trees lose their leaves to conserve water during the long dry seasons, but the storage of water at depth supports the onset of flowering and leaf production during high temperatures that precede the onset of rains. Soil conditions, therefore, are also important to the development and structure of tropical deciduous forest. Where an impermeable soil layer prevents infiltration, tropical savanna grasslands predominate, such as the llanos of Venezuela. Moreover, the trees that occur

in tropical deciduous forest vary distinctly between highly weathered deep lateritic soils and those locations where clays (i.e., black cotton soils) occur. In East Africa, tropical deciduous forest (bushland) occurs as a diverse mosaic of different plant communities in relation to the distribution of these soils. In Southern Africa, "sour veld" and "sweet veld" distinguish the woodland types on more acidic lateritic soils and more alkaline soils, respectively. Moreover, the structure of tropical deciduous forest may be greatly altered by disturbances. For example, the destruction of trees by elephants in the Tsavo ecosystem of East Africa at least partially explains changes over time in the distribution of woodlands and grasslands. Fire has been shown to maintain miombo woodlands in Tanzania and certainly influences its composition and structure.

Tropical deciduous forest, especially when the biome includes the broad vegetation transition between tropical rain forests and tropical savanna, supports a diverse fauna that show

Tropical deciduous forest in East Africa

Source: K. E. Medley.

unique adaptations to the seasonally wet conditions. Among the large mammals, browsers are most significant, where giraffe rely on the leaves of the canopy trees, and the kudu and gerenuk browse the understory vegetation. Mammal population densities remain low, at least in part because of the lower availability of forage during the dry season and in Africa the continued persistence of tsetse fly. Certain bird species are particularly adapted to fruit production in dry forests, such as the diverse population of hornbills that rely on the fruits of *Commiphora* in East Africa. As with the tropical savanna, termites are critical to the redistribution of nutrients and plant establishment.

Tropical deciduous forest occurs in regions that experience a long dry season and typically have poor infertile soils that limit settlement and farming. When human populations expand into these regions, such as on the continent of India and in northeast Brazil, forest cover is quickly degraded to a grass or scrubland and production is typically unsustainable. Pastoral activities, particularly for cattle or goats, can be sustained under low population densities, but the distribution of the tsetse fly in Africa also occurs in this region thereby limiting human activities. Tropical deciduous forests, particularly in Africa, provide a major source of fuel wood and charcoal. Tree coppicing can sustain much extraction, but when trees are cut at the base, regeneration is not possible and the composition and structure of the forest will change. The construction of roads facilitates access to woodland regions, supporting an export market for charcoal that significantly affects woodland conservation. Some of the most important tropical trees for timber are extracted from tropical deciduous or semievergreen forest, such as mahogany (*Swietenia Mahogani*) in South America and teak (*Tectona grandis*) in southeast Asia. Valuable timber and nontimber resources support the importance of establishing sustainable management practices for this regionally diverse tropical biome.

Tropical deciduous (dry) forests currently occupy about 1% of the land surface, but the

Adansonia (baobob) species in the tropical deciduous forests of Western Madagascar

Source: K. E. Medley.

relationship between the actual and potential distribution of this vegetation type remains unclear. Dense forest areas occur in South America and on the Yucatan peninsula of Mexico, but globally its distribution is very fragmented, often occurring as a mosaic with tropical savanna and degraded lands cleared by human activities. Fires can be employed to expand grazing lands with a direct impact on the distribution of forest, just as heavy extraction of fuel wood or charcoal production can degrade the structure of existing woodlands. The degree to which projected climate changes in seasonal rainfall regimes might influence the persistence of closed forest cover forms a special scientific concern especially in Central and South America. Tropical deciduous forest is among the most threatened biomes because of its small and fragmented distribution and the complex influences that both "natural" and "human" disturbances have on its potential distribution.

Kimberly E. Medley and
Christine Mango Mutiti

See also Agroforestry; Biogeography; Biome: Tropical Rain Forest; Biome: Tropical Savanna; Biome: Tropical Scrub; Deforestation; Forest Degradation; Forest Fragmentation; Forest Land Use; Forest Restoration; Nomadic Herding; Social Forestry; Sustainable Forestry

Further Readings

Miles, L., Newton, A., DeFries, R., Ravilious, C., May, I., Blyth, S., et al. (2006). A global overview of the conservation status of tropical dry forests. *Journal of Biogeography, 33,* 491–505.

Rodgers, W. (1996). The miombo woodlands. In T. R. McClanahan & T. P. Young (Eds.), *East African ecosystems and their conservation* (pp. 299–325). New York: Oxford University Press.

Walter, H. (1971). *Ecology of tropical and subtropical vegetation.* Edinburgh, UK: Oliver & Boyd.

BIOME: TROPICAL RAIN FOREST

The plant geographer A. F. W. Schimper first used the term *tropical rain forest* in 1898 to characterize the broad-leaved evergreen trees that dominate the wet equatorial regions of the world. Under tropical wet conditions, a structurally diverse forest occurs that is typically high stature, closed canopy, species rich, structurally diverse, and unique in the adaptations that different plants and animals have to the environmental conditions. Rain forest can be described by these general attributes and also by the unique conditions that occur across continents and within continents along environmental gradients.

The Physical Environment

The potential geographic distribution of tropical rain forest is pantropical within the equatorial latitudes of 10° N and 10° S. These forests occur on the continents of South America, Africa, and Asia, and extend into the subtropical realm in Eastern Central America and the Caribbean, southwestern India and Sri Lanka, and Northern Australia. Day lengths hardly change from 12 hours during the spring and fall equinox, noon sun angles are always high, and the sun rises and falls quickly in the horizon. Temperature conditions are always warm, above 18 °C, and warm moist equatorial air rises to support daily convective showers and constantly moist conditions. Tropical rain forests experience no frost; daily temperature differences show greater ranges than the monthly means. Monthly precipitation always exceeds water needs (potential evapotranspiration) and can average more than 100 mm (millimeters) per month or an excess of moisture. Rainfall averages 2,000 to 3,000 mm in South America, 1,500 to 2,500 mm in Africa, and >3,000 mm in southeast Asia.

Tropical rain forests extend beyond the equatorial zone where the periods of moisture deficit are very short and the rainy periods are exceedingly wet, such as those brought by the monsoons to southwest India, southeast Asia, and West Africa, and those brought by the northeast and southeast trade winds to Central America,

the Eastern Caribbean, the South Pacific, and Australia. Moisture deficits, however, will eventually limit broad-leaved evergreen trees and transition tropical forests toward dominance by trees that are semievergreen or deciduous during the dry season. Tropical rain forests and tropical deciduous forests capture the broad diversity of forest types that occur across the tropical realm where temperatures are constantly warm and suitable for plant growth.

Biogeographical Patterns of Diversity

Plant and animal occurrences in the tropical rain forests show a relationship with their biogeographical distribution among tropical regions. By far the greatest land area (>50%) and species diversity occurs in the New World tropics, which include South and Central America. Tropical rain forest occurs in the Amazon River basin over a 3,000-km (kilometer) distance from the Andes to the Atlantic, along the Orinoco River in Venezuela, and in north coastal Guyana, Suriname, and French Guiana. The rain forests of the western Andes in Peru and Ecuador are among the wettest in the world, which then extend northeastward along the east coast of Central America and include the northeastern exposed Caribbean islands (e.g., Jamaica, Hispaniola, Puerto Rico, Dominica, and Trinidad). In Africa, tropical rain forests occur in the Congo River basin and along a narrow coastal zone in West Africa. More than 60% of the Democratic Republic of Congo is in rain forest, and some of the most diverse rain forests occur in Eastern Madagascar. In southeast Asia, tropical rain forests straddle the equatorial zone between Indochina, Malaysia, Indonesia, and the Philippines. One of the most famous biogeographical boundaries occurs across Indonesia, Wallace's line, distinguishing the rain forests of southeast Asia from Australia and New Guinea along a plate boundary that merged about 50 million years ago (mya). Moreover, the isolated distribution of the Atlantic forest from Recife to São Paulo in Brazil and the coastal and Eastern Arc forests of East Africa suggest a historically broader distribution of rain forests across these continents.

Tropical rain forests are among the most biologically diverse vegetation types on Earth, and there are both historical and ecological explanations for their patterns of diversity. Broad-leaved evergreen trees arose in the early Tertiary period more than 70 mya and experienced geographic shifts in their distribution and periods of diversification and extinction over time. For example, tropical forests expanded northward during the Eocene (about 50 mya) into North America, and expanded and contracted significantly on the South American and African continents during interglacial and glacial periods, respectively. Woody plants and some of the most primitive woody plant groups (*Magnioles* and *Laurales*) are especially diverse in the tropics. Historical climatic changes at least partially explain the high species diversity of the New World tropics, lower diversity on the continent of Africa, and the occurrences of unique biological attributes among the regions. Smaller high-canopy primates with tails in the New World, larger primates without tails and with an opposable thumb in Africa, the very primitive lemurs of Madagascar, and the persistence of some very old plant species in the more equable climate of southeast Asia are just some examples.

Ecological Structure, Function, and Dynamics

Biological diversity also corresponds with the high vertical complexity of the forests. Tropical rain forest typically maintains five structural layers: (1) an open emergent layer at >50 m (meters) of very tall tropical trees with broad and flattened crowns exposed to open sunlight, (2) an open to closed canopy layer of large trees at about 30 m, (3) an open subcanopy tree layer, (4) an open sapling and shrub layer, (5) and a very infrequent herb layer. Dense undergrowth or more "jungle" conditions only occur in open disturbances such as forest edges and within tree-fall gaps. High sun angles contribute to the vertical complexity of the structural layers, providing sunlight at different levels for lianas (woody vines), epiphytes that germinate and grow on the surfaces of trees, and hemiepiphytes that first germinate in tree crevasses but then extend stems to root in the ground and often strangle their host plants for structural support. Canopy and emergent trees

Tall emergent trees in tropical rain forests often have buttressed trunks to maintain stability.

Source: Author.

show different shade tolerances, influencing their ability to establish and recruit to the canopy in tree-fall gaps that vary in their size and the amount of light available through the opening.

Competition for light, nutrients, and reproduction dictate the special adaptations and growth forms of the plants. For example, large tropical trees often have buttresses at the base of their trunks, which keep trees stable while maintaining shallow root systems that can acquire nutrients when they are released by decomposing leaves. The upper canopy of tropical rain forests can be exceedingly diverse where trees, lianas, and epiphytes compete for available light. More than 90% of all lianas occur in tropical wet environments where they can be as diverse as the trees. Large flowers, cauliflory where flowers germinate

Smooth leaves with drip tips and cauliflory, the production of flowers on the main stems, are just some of the special adaptations of tropical rain forest plants.

Source: Author.

on the main stems of understory trees, and large fleshy fruits best ensure specialized pollinators and dispersal mechanisms. Large entire leaves and very shade-tolerant herbs maintain productivity at low light levels in the forest understory. Light intensities can be as low as 0.1% to 0.5% at the forest floor. Drip tips on the leaves shed excess moisture. None of these plants need to become dormant in the continually warm and wet conditions for growth, and often demonstrate red nodding leaf shoots when the formation of supporting tissue and chlorophyll lag behind during rapid shoot elongation. These specialized adaptations restrict habitat preferences and allow for high species packing in small areas. For example, tree diversity can be as high as 300 species measured in a 1-ha (hectare) plot in the Western Andes and almost always is above 40 tree species in 1-ha plots. While tropical rain forests are well known to have the highest diversity of growth forms for different kinds of plants, trees typically account for more than 60% of the plant diversity.

High primary production and plant diversity in the tropical rain forest provides resources to support a high faunal diversity. Invertebrates that feed on litter and contribute to the very high rates of decomposition and rapid release of nutrients make up about 60% of the animal biomass. Less than 20% of the faunal biomass relies on leaves and much less on the fruits and flowers. Tropical plants, especially in lowland environments where faunal diversity is highest, produce secondary compounds that are either toxic or indigestible to animals. The vertical stratification and specialized adaptations of insects, birds, and mammals are critical to the reproduction of plants and their sustained diversity. Many plants rely on a certain insect, bat, or bird to ensure pollination, and the large fruits produced by tropical trees often depend on a certain mammal for dispersal. For example, the production of Brazil nuts can only be successful in closed tropical forest, where the agouti can ensure dispersal and establishment. The tropical rain forest also maintains interdependencies among animals as a strategy to gain access to limited food resources. For example, ant swarm followers are a group of birds that follow large colonies of army ants and rely on the insects that are stirred up as the ants move through the forest litter.

Equatorial rain forests for the most part coincide with some of the oldest, most weathered landscapes in the world. For example, both the Amazon and Congo rivers flow across the continental shields of South America and Africa, respectively, where ancient rocks form the substrate for vegetation growth. Warm moist conditions promote exceedingly high rates of decomposition and chemical weathering. Upland tropical forests in these regions typically occur in oxisols, which are soils characterized by the mobilization and accumulation of iron and aluminum oxides to form a dense laterite that is exceedingly low in organic matter and nutrients. Soils are covered by a shallow litter layer that rapidly decomposes and releases its nutrients. The depth of weathering in the soil profile can be as high as 20 m. Shallow roots and epiphytic-growth forms acquire nutrients in this environment from stem flow off the trees before the rainwater reaches the soil surface. In contrast, where younger volcanic soils occur, such as in southeast Asia, nutrient availability can be much higher and support greater productivity.

A Diversity of Plant Community Types

Tropical rain forests, defined by their distribution in tropical wet environments, also show a diversity of forest communities adapted to local physical-environmental conditions. Mangroves dominate estuarine environments across the equatorial tropics where temperatures are always warm. Along the river basins, seasonally flooded forests can occur that are adapted to high water conditions, just as low-lying areas support swamp forests that differ in their composition and stature. Particularly unique are the "heath" forests that develop in large regions of tropical spodisols, such as the *igapo* forests of South America and the *kerangas* of southeast Asia. Tropical spodisols form in coarse-textured sands that are exceedingly low in nutrients and high in organic acids (pH <5.5). These forests are low in stature, have thickened small leaves that are high in tannins, and support many carnivorous plants or symbiotic plant and ant relationships that are specialized to acquire nutrients. For example, the Rio Negro (black water river) flows through these soils and gains a brown color from the tannins in the decomposing forest vegetation.

Tropical rain forest occurs typically at lowland elevations below 900 to 1,200 m, but environmental conditions suitable for forest growth can extend above sea level to around 4,000 m. Along an altitudinal gradient, mean temperature decreases and available moisture increases to create a lower-stature, exceedingly moist montane forest. These "cloud forests" differ in their tree species composition and are covered from the canopy to the ground with epiphytic plants, ferns, mosses, and club mosses that acquire their moisture directly from the atmosphere under continually wet conditions. Other forests of unique composition may be restricted to locations where a different rock type is exposed. For example, *Leptospermum*-dominated forest on Mt. Kinabalu, Sabah, Malaysia, only occurs on ultrabasic soils.

Heath forests, such as the Kerangas in Sabah, Malaysia, occur on nutrient-poor tropical spodisols.

Source: Author.

Human Relations With Tropical Rain Forests

Human relationships with the tropical rain forest are particularly complex in how they vary through time and differ by region in response to local and extra-local forces. Under low-population densities, hunter and gathering societies rely on their indigenous knowledge of biodiversity and how to extract resources to meet their livelihoods. In contrast, agricultural societies typically employ shifting cultivation systems that remove and burn the vegetation to establish crops temporarily in the remaining nutrients. With increasing population densities, the expansion of fields into mature forests and the intensity of production on open lands

Montane cloud forests on the Eastern Arc mountains in Kenya (A) and on Mt. Kinabalu in Sabah, Malaysia (B)
Source: Author.

greatly affects forest area and the potential for forest regeneration. These local forces on forest resources are further intensified by extra-local demands for timber and land resources. Tropical deforestation rates exceed rates of regrowth, with an estimated loss of >50% of the original land area. Moreover, certain regional forces, such as the resettlement of populations and pasture development along the Amazonian highway, fuelwood and charcoal production in Africa, and the international export of high-quality timber from the rain forests of southeast Asia and West Africa intensify local impacts and magnify global deforestation statistics. While the rates of forest clearing declined in recent years, countries such as Brazil, Peru, Democratic Republic of Congo, and Indonesia rank among the top 10 countries with highest rates of deforestation. Unlike temperate forests, specialized requirements for tree reproduction, small species ranges, and soil infertility predict much higher rates of extinction and degradation with tropical deforestation. Conservation action for tropical rain forest focuses on better understanding the value of species diversity, validating indigenous rights and local participation in conservation plans, and providing global support for reserves that are compatible with national interests for development.

Kimberly E. Medley

See also Biodiversity; Biogeography; Biome: Tropical Deciduous Forest; Biome: Tropical Savanna; Biome: Tropical Scrub; Biota and Climate; Biota and Soils; Climate: Tropical Humid; Deforestation; Forest Degradation

Further Readings

Primack, R., & Corlett, R. (2005). *Tropical rain forests: An ecological and biogeographical comparison.* Malden, MA: Blackwell.

Richards, P. (1996). *The tropical rain forest: An ecological study* (2nd ed.). Cambridge, UK: Cambridge University Press.

Terborgh, J. (1992). *Diversity and the tropical rain forest.* New York: Scientific American Library.

Walter, H. (1971). *Ecology of tropical and subtropical vegetation.* Edinburgh, UK: Oliver & Boyd.

BIOME: TROPICAL SAVANNA

Tropical savanna or savanna is a global vegetation type where grasses are continuous and important but occasionally interrupted by trees and shrubs. The term *savanna* originated as a description of treeless plains in South America but now describes the global distribution of this mixed vegetation type. The type of woody plants and how they are arranged can vary significantly in relation to the grasses, creating a diverse mosaic landscape across regions where tropical savanna occurs and at the local scale within each savanna. These differences sometimes dictate how the vegetation is classified and named, identifying gradients of change from savanna grassland with no trees or shrubs to parklands or woodlands with a greater abundance of trees, and to thickets and scrub with an absence of trees. Tropical savanna for the most part overlaps geographically, particularly in Africa, with the distribution of tropical deciduous forest and tropical scrub (see tropical savanna photo).

Biogeographical Patterns of Diversity

Tropical savanna is one of the largest global vegetation types (biomes), potentially occurring over about 20% of the land surface in the tropics and subtropics between 30° north and south of the equator in South America, Africa, South Asia, and Australia. Major savannas are mapped in South America, where densely wooded grasslands dominate in the "cerrados" of northeastern Brazil, the grass-dominated "campo" in southeast Brazil, and

the "llanos del Orinoco" in Venezuela. In Africa, tropical savanna is most extensive, predominating across a broadband between the west-central equatorial tropical rain forest to the Sahara Desert in the north and Kalahari Desert in the south, continuing across plains in equatorial East Africa, and dominating much of the landscape in Western Madagascar. When mapped with the distribution of tropical deciduous forest and tropical thornscrub, tropical savanna covers approximately 65% of the African continent. Much of the Indian subcontinent is also mapped as savanna and small patches occur in southeast Asia and on the islands of the East Indies and the Pacific. In Australia, tropical savanna occurs in the north and extends southeast between the more arid central plains and the humid tropical and subtropical forests along the east coast.

Tropical savanna, therefore, shows a disjunct distribution across the continents in the tropical realm. While savanna vegetation may be similar in its overall structure as a mixed system with grasses and woody plants, the composition of plants species that occur in the region can also influence its appearance. For example, evergreen *Acacia* trees dominate the savannas of the new world tropics in South America, just as evergreen eucalyptus trees occur across a broad moisture gradient on the continent of Australia. In Africa, evergreen *Acacia* trees can predominate, but much of the region shows a mixed composition with some evergreen trees (e.g., *Acacia* species, *Balanites aegyptica*), deciduous trees (*Commiphora* species, *Delonix elata*, *Sclerocarya birrea*), and evergreen (e.g., *Croton*) or deciduous (e.g., *Combretum*) shrubs. In Africa, baobob (*Adansonia digitata*) is a prominent deciduous tree in savanna that varies its deciduous habit in relation to changes in the occurrence of rains across the region. The plants that make up the tropical savanna are much more diverse than temperate grasslands. About half of the plant species on the African continent occur in this vegetation type.

Savanna Climatic Conditions

Much of the composition and structural diversity in tropical savanna can be attributed to moisture changes across an equally broad region characterized by the tropical to subtropical wet-dry climate.

Tropical savanna in the Masai Mara, Kenya

Source: K. E. Medley.

Along a climatic gradient between tropical wet and tropical-subtropical dry, moisture deficits first emerge as the occurrence of a dry season and then change in available moisture by extending the length of the dry season and reducing the amount of rainfall received during the wet season or seasons. Moist savanna along this gradient occurs in regions receiving more than 700 mm (millimeters) of rainfall with a dry season of less than 3 mos. (months) and dry savanna can occur in regions that receive as little as 300 mm with a dry season of more than 6 mos. The primary production of savanna grasses shows a direct relationship with the amount of rainfall. Seasonal rainfall and dry seasons in the tropics are dictated by the migration of the sun's direct rays or subsolar point from the equator to the Tropics of Cancer and Capricorn and concurrent shifts in the positioning of the equatorial low and associated convective showers. Moreover, tropical wet-dry conditions can be intensified in some regions by strong monsoonal winds that bring moisture from the ocean toward the continent, particularly in West Africa and the Indian subcontinent, during high sun (summer) conditions and dry continental winds during low sun (winter) conditions. East Africa occurs as an anomaly, characterized by a northeast monsoonal flow from October to December and a southeast monsoonal flow from March to May that bring "short" and "long" rains to the continent, respectively, in contrast to a single wet and dry season. Rainfall seasonality is a key attribute of tropical savanna. The dry season or seasons extend from 3 to 9 mos. with a mode of 5 to 7 mos., and rainfall always occurs during the growing season. The most profound vegetation changes that occur along this climatic gradient are observed in the woody plants, with a corresponding shift from evergreen trees, to deciduous trees, to evergreen shrubs, and to deciduous shrubs along a humid to arid gradient. These changes in the types of plants, however, vary significantly among the continents where tropical savanna occurs and within savanna regions in relation to soil conditions and disturbance regimes.

Ecological Adaptations of Tropical Savanna

Vegetation structure in tropical savanna can be mostly explained by the contrasting adaptations that woody plants and grasses have under a seasonal moisture regime and their competitive

Acacia-dominated savanna on black-cotton soils in Kenya

Source: K. E. Medley.

relationships. Most of the grasses are perennial, growing from underground stems or rhizomes and living 4 to 8 years. Much of the plant biomass occurs below the ground, where grasses maintain a dense fibrous root system at the ground surface and to a depth of about 1 meter. Grasses are well adapted to quickly absorb the rains as they fall and infiltrate through the soil near the surface. In contrast, woody plants produce deep taproots that rely on the deep infiltration of precipitation and its storage at depth. Because of these contrasting adaptations, soil conditions that influence the availability of water under seasonal precipitation regimes are critical to the competitive ability of grasses and woody plants and their relative dominance across savannas. Grasses predominant in regions where soils are fine textured or underlain by an impermeable layer that keeps moisture closer to the surface and available to the fibrous roots systems that grasses maintain. For example, the "arecife" layer of impermeable iron hydroxides (a lateritic crust) that occurs beneath the llanos of Venezuela and an impermeable soil layer of volcanic deposits in the southern Serengeti of Tanzania, explain the broad expanse of pure grasslands across those regions. In contrast, coarse-texture soils typical of riparian corridors often support savanna woodlands with a dense canopy of trees. In Africa, more coarsely textured lateritic soils support a different composition and structure of trees and grasses than the fine-textured clays or black-cotton soils in low-lying areas (see acacia savanna photo). At a local scale within a savanna type, termite mounds can also alter the structure and nutrient content of soils in a way that allows for the local establishment of trees and shrubs. The isolated occurrences of trees across a savanna often correspond with the distribution of abandoned termite mounds.

Disturbance Ecology and Derived Savannas

Tropical savanna, particularly in East Africa, often occurs in regions that possess climatic conditions also suitable for the development of dense woodland or scrub. The geographic distribution of tropical savanna under these conditions is not climatically determined but occurs in response to soil conditions, such as those described above or "derived" in response to disturbances that reduce the competitive ability of woody plants. A prominent disturbance that reduces tree cover and promotes the production of grasses is grazing (the consumption of grasses) and browsing (the consumption of woody foliage) by large mammals (see arid bush savanna photo). Anteaters in South America and the marsupials of Australia influence

Gerenuk browse in more arid bush savanna

Source: K. E. Medley.

the ecology of tropical savanna on those continents, but the diversity and impact by large herbivores is a prominent influence in Africa that at least partially explains the broad expanse of this vegetation type across the region.

The highest diversity and abundance of large mammals in the world occur in the tropical savanna of East Africa and are a profound ecological disturbance. Large mammals and grassland coevolved and show some interesting examples of coadaptations. Savanna mammals migrate in response to seasonal changes in grassland production, and species diversity is maintained by their efficient use of different plant resources. Elephants consume large quantities of coarse grasses, shrubs, and trees and the seeds that pass through their digestive tract are dispersed over broad distances. Zebra, through their consumption of coarse grasses that are high in cellulose and lignin, provide a grass composition more suitable for smaller ruminant species such as gazelles. Woody plants develop both physical (thorns) and chemical (indigestible compounds) defenses to browsing. Grasses, with their source of growth at the base of the stem, and woody plants,

with an ability to produce multiple stems on browsing, can adapt and may even respond favorably (i.e., overcompensate) to the grazing/browsing activities of large mammals. An interesting example of the interplay between large mammals and tropical savanna occurred when the Rinderpest epidemic reduced wildlife herds by more than 90% after 1890, and woodlands expanded over much of the region until the mid 1900s. The heavy poaching of elephants in the 1980s had a similar effect, allowing for the expansion of woodland in some regions. The structure of tropical savanna and its species composition is often determined by the coadaptations of plants and animals, and the ecological separation is maintained by the animals as they migrate through or reside in certain regions.

Fire can also reduce woody plants in tropical savanna, but like grazing disturbances, its natural occurrence is unclear in relation to the very long history of human activities in the region. Humans evolved in the tropical savanna region of East Africa more than 2 million years ago and remain a significant influence on the composition and structure of this vegetation type. Pastoral populations

Heavily grazed savanna in the Masai Mara National Game Reserve, Kenya

Source: K. E. Medley.

rely on livestock that heavily graze the tropical savanna, and their livelihood strategies include migration and fires to ensure the availability of forage resources. Human conflicts arise when the conservation of wildlife or agricultural activities compete with local strategies for sustainable pastoral livelihoods. For example, much of the tropical savanna in East Africa is now protected in large reserves that limit land resources for grazing livestock and intensify the competition between agricultural and pastoral populations (see photo above). At the same time, there are also examples of cooperative ventures that support both the production of "natural" savanna resources and human livelihoods under community-based natural resource conservation initiatives for the region. Sustainability of tropical savanna relies on understanding the physical-environmental conditions and disturbances that influence and thereby maintain savanna productivity and the opportunities for sustained human livelihoods across the region.

Kimberly E. Medley and
Christine Mango Mutiti

See also Biome: Desert; Biome: Midlatitude Grassland; Biome: Tropical Deciduous Forest; Biome: Tropical Rain Forest; Biome: Tropical Scrub; Biota and Climate; Biota and Soils; Climate: Dry; Community-Based Natural Resource Management; Wildfires: Risk and Hazard

Further Readings

Archibold, O. (1995). *Ecology of world vegetation.* New York: Chapman & Hall.

Boulière, F. (Ed.). (1983). *Ecosystems of the world: Vol. 13. Tropical savannas.* Oxford, UK: Elsevier Scientific.

Gichohi, H., Mwangi, E., & Gakahu, C. (1996). Savanna ecosystems. In T. McClanahan & T. P. Young (Eds.), *East African ecosystems and their conservation* (pp. 273–298). New York: Oxford University Press.

Shorrocks, B. (2007). *The biology of African savannahs.* Oxford, UK: Oxford University Press.

Walter, H. (1971). *Ecology of tropical and subtropical vegetation.* Edinburgh, UK: Oliver & Boyd.

BIOME: TROPICAL SCRUB

On every continent except Antarctica, there are areas with dry or desert climates that receive enough rainfall to support shrubs as a dominant plant growth form. The land cover of these places consists of various types of shrublands or woodlands, which can be collectively known as "scrub." Tropical scrub refers both to this climate zone as found in tropical and subtropical latitudes and also to the typical scrub vegetation type of those latitudes, characterized by dominance of multistemmed woody plants that are 1 to 5 m (meter) tall. It may also be known as *brush*, *bush*, or *thicket*, depending on local usage. Slightly higher woody plants with thorns, form the closely associated "thorn woodland."

Globally, tropical scrub and its variants can be found in large areas of Africa (including Madagascar), Southern Asia (India, Thailand), Australia, Mexico and Central America, the Andes Mountains, Eastern Brazil, and enclaves in Venezuela and elsewhere. It may be localized in pockets of suitable climate in mountainous areas and is also positioned on the dry sides of tropical and subtropical islands. It is found where mean annual precipitation is on the order of 20 to 50 (70) centimeters/year, with mean annual temperatures from as low as 12 °C and as high as 30 °C.

One way to conceptualize the biogeography and distribution of tropical scrub is in relationship to its relative position on a theoretical environmental gradient associated with annual precipitation. Such a gradient would extend from vegetation formations such as tropical rain forests and tropical deciduous forests to tropical and subtropical deserts, following a gradient of wet to dry climates. Scrub would be situated in climate zones somewhat moister than the deserts, while drier than the conditions found in tropical deciduous forests. However, its position along this idealized moisture gradient is also shared with tropical savanna, a vegetation formation wherein dominance is typically shared between both grasses and scattered trees. Climate on its own is not enough to predetermine the dominant vegetation in a particular area. Instead, the influence of disturbances such as fire and the complicating effect of edaphic or soil-related factors also must be taken into account.

The shrubs and short-statured trees that dominate tropical scrub are often deciduous, losing their leaves as an adaption to extended dry seasons when little or no rain may fall for 4 to 8 months a year. The evergreen species often have small leaves to minimize water loss. In addition, many species have thorns, providing protection from herbivores. These plant species also have varying degrees of adaptation to fires. They may have some combination of a fire-resistant bark, the ability to resprout following limb loss, or seeds cued to grow on recently burned substrate. By definition, tropical scrub does not have a continuous graminoid (grass or sedge) layer, so it is less prone to annual or semiannual fires than in many tropical savannas because fuel loads are less. Some tropical scrub may include cacti or other succulent plants that are well adapted to survival in seasonally dry environments.

Tropical scrub offers an excellent opportunity to examine evolutionary convergence globally. This phenomenon occurs when unrelated species develop similar appearances and adaptations to the prevailing climatic conditions found in a certain biome type in different parts of the world. Because tropical scrub is found worldwide, there are lineages of plants that have independently evolved suitable adaptations and so are generally similar in size, morphology, and physiology, despite not being close relatives. The same would be true for animals specialized in using scrub habitats, including a variety of bird, lizard, and insect species.

The interaction of soil nutrients, chemistry, and structure with vegetation dynamics is universal. For example, the *cerrado* savannas and *caatinga* woodland/scrub of Brazil are atop poor-nutrient soils derived from the ancient and weathered Brazilian shield. The role of soil characteristics may override climatic controls, allowing for tropical scrub vegetation to persist in wetter environments than is the norm. New Caledonia offers examples of such vegetation, locally known as *maquis*, wherein harsh serpentine soil conditions favor shrubs and other tolerant species in early stages of succession following disturbance or with recurrent fires. Recovery to rain forest is prolonged and depends on colonization and facilitation by *Araucaria* conifers.

Tropical scrub can also result from persistent disturbance caused by strong winds or large animals. For example, scrub vegetation in Southern Africa may persist in landscape mosaics that include riparian woodland forest and extensive grass-dominated savannas. This persistence is due to feedbacks apparently resulting from increasing elephant and impala populations. The elephants, which are megaherbivores, preferentially feed at 1 to 3 m heights and may knock down and kill even quite large trees. The impala and other antelope feed at lower levels and may browse on tree seedlings, preventing tree establishment. The actions of these herbivores are concentrated near rivers and watering holes, especially during the long dry seasons, thus tending to increase total area in shrubland. Scrub is often found on ridges in otherwise tall forests located in the rough topography of tropical mountains. In this case, it is wind action that results in smaller stature woody plants being the dominants, perhaps with additional feedbacks from soil characteristics and herbivores.

Of course, in many places in the world, humans also have become a further ecological control on vegetation structure and dynamics. People may act in such a way as to increase, decrease, or eliminate disturbances that affect vegetation. They may impose land uses that create unique conditions, for example, through novel fire regimes that have an altered seasonality, intensity, and frequency than would be found under conditions wherein only climate regulated fire behavior and influences. They may introduce cattle, sheep, or goats, which alter plant species dominance and may promote woody plant establishment in grasslands. They may selectively remove or favor certain species considered valuable or nuisances. The floristic and faunistic elements of tropical scrub that are of use to people may become exhausted due to overharvest. Woody plants valuable as timber or as polewood, or animals hunted for their meat, may be locally eliminated.

Due to human influences, over time, some areas of tropical deciduous forest may begin to resemble tropical scrub, without emergent or tall trees. On the other hand, some forms of land use in tropical savannas, such as certain kinds of grazing systems, or as associated with the suppression of seasonal fires, may in fact promote establishment and invasion by shrub species, thus shifting the savanna into a form of scrub. Encroachment of shrubs into both grasslands and desert vegetation has been observed worldwide in recent decades. This woody plant invasion may be causally related to increased amounts of carbon dioxide in the atmosphere, which could give a competitive advantage to the woody plants that can overtop and shade grasses.

As a result, collectively, human impacts would tend to increase tropical scrub. In landscapes occupied by people for many millennia, such as in India (Eastern and Western Ghats) or in the Andes Mountains (intermontane valleys from Colombia to Bolivia), large areas that would appear to be in a tropical deciduous forest climate zone, in fact have scrub vegetation. The loss of large forest trees, however, may have happened 2,000 to 4,000 years ago in some of these landscapes given the antiquity of human settlements, although both colonial and recent time periods also were moments in history with extensive extraction of natural resources. Only a subset of the original animals from the dry forests is able to continue to thrive in the shorter and more open habitat of scrub.

A conceptual model of the dynamism of tropical scrub would thus need to include all these features. A climatic regime with a pronounced dry season is essential. But then the composition and size of area with scrub vegetation will be a function of the degree of alteration due to land use. Seasonally dry forests may be "desertified" and converted to shorter stature vegetation through deforestation and through burning. Tropical savannas may disappear as woody shrubs come to dominate through altered fire regimes and as promoted by the influence of increased background carbon dioxide levels. Land use and large animal impacts could further alter conditions through herbivory and by changing soils through compaction and erosion. This model makes predictions for vegetation dynamics and also might permit the development of landscape management strategies when ecological restoration is a goal.

Vulnerable species found in tropical scrub may need special conservation efforts to survive in a changing world. Although scrub as a habitat type may be on the increase globally, due to shrub encroachment and forest, grassland, or desert degradation, there will still be species that need protection or management due to their value or limited dispersal or reproductive capacities. In other cases, land managers are worried about exotic plant and animal species that alter ecosystem properties and lessen the value of the habitat for native species. This kind of biotic invasion is of particular concern in Australia, and also in the thousands of smaller tropical islands, all of which are susceptible. Some grassland areas in tropical Australia are now occupied by a shrub species from India, creating an anthropogenic scrub.

In summary, tropical scrub denotes an important semiarid climate type that predisposes vegetation to be dominated by shrubs and small trees with adaptations to drought. Because it is present in low latitudes around the world, and has existed for millions of years in some form, there are native species of plants and animals that have converged in their appearance and in their adaptations for survival. Some human land uses and some natural combinations of poor soils and fire may further increase the surface area covered by some type of shrubland. Tropical scrub is of interest for conservationists and biogeographers, but it is also an object of concern for those land managers who need to control the degree of dominance by shrubs in order to promote values for pasturelands, for fire control, or for ecological restoration.

Kenneth R. Young

See also Biome: Tropical Deciduous Forest; Biome: Tropical Rain Forest; Biome: Tropical Savanna; Human-Induced Invasion of Species

Further Readings

Asner, G., Elmore, A., Olander, L., Martin, R., & Harris, A. (2004). Grazing systems, ecosystem responses, and global change. *Annual Review of Environment and Resources, 29,* 261–299.
Bond, W., & Midgley, G. (2000). A proposed CO_2-controlled mechanism of woody plant invasion in grasslands and savannas. *Global Change Biology, 6,* 865–869.
Brown, J., & Carter, J. (1998). Spatial and temporal patterns of exotic shrub invasion in an Australian tropical grassland. *Landscape Ecology, 13,* 93–102.

Enright, N., Rigg, L., & Jaffré, T. (2001). Environmental controls on species composition along a (maquis) shrubland to forest gradient on ultramafics at Mont Do, New Caledonia. *South African Journal of Science, 97,* 573–580.
Furley, P. (1999). The nature and diversity of neotropical savanna vegetation with particular reference to the Brazilian cerrados. *Global Ecology and Biogeography, 8,* 223–241.
Jayakumar, S., Ramachandran, A., Bhaskaran, G., & Heo, J. (2009). Forest dynamics in the Eastern Ghats of Tamil Nadu, India. *Environmental Management, 43,* 326–345.
Joubert, D., Rothauge, A., & Smit, G. (2008). A conceptual model of vegetation dynamics in the semiarid Highland savanna of Namibia, with particular reference to bush thickening by *Acacia mellifera. Journal of Arid Environments, 72,* 2201–2210.
Kellman, M., & Tackaberry, R. (1997). *Tropical environments: The functioning and management of tropical ecosystems.* London: Routledge.
Mehta, V., Sullivan, P., Walter, M., Krishnaswamy, J., & DeGloria, S. (2008). Ecosystem impacts of disturbance in a dry tropical forest in Southern India. *Ecohydrology, 1,* 149–160.
Moe, S., Rutina, L., Hytteborn, H., & du Toit, J. (2009). What controls woodland regeneration after elephants have killed the big trees? *Journal of Applied Ecology, 46,* 223–230.
Shahabuddin, G., & Kumar, R. (2007). Effects of extractive disturbance on bird assemblages, vegetation structure and floristics in tropical scrub forest, Sariska Tiger Reserve, India. *Forest Ecology and Management, 246,* 175–185.
Van Auken, O. (2004). Shrub invasion of North American semiarid grasslands. *Annual Review of Ecology and Systematics, 31,* 197–215.

BIOME: TUNDRA

The word *tundra* originates from the Finnish *tunturi,* which means "completely treeless heights." This is how the word is applied broadly all over the world to areas of higher altitude than the tree line. However, the geographically largest and most significant tundra areas are those north of the latitudinal tree line in Eurasia and North America. These tundra areas are mainly situated in lowlands.

The tundra as a biome is relatively young, having developed in the early Pleistocene. However, the characteristic floras of tundras developed earlier, probably during the Miocene-Pliocene, in the highlands of Central Asia and in the Rocky Mountains of North America. In the same period (late Tertiary), present tundra areas were covered by various types of mixed and coniferous forests, now associated with the boreal forest region. Circumpolar regions of tundra-steppe environments developed during the Pleistocene. Rapid expansion of these areas during this period may have been associated with the success of such now-extinct macrograzers as mammoths.

A relative mild climate with spruce forests north of their present limit occurred in the last interglacial (Eem). During the last glaciation (Wisconsin), major parts of the Northern Eurasian and American continents were covered by the Eurasian and Laurentide ice sheets, respectively. There were, however, nonglaciated pockets extending far north, as in parts of Beringia and Northern Yukon Territory. In such nonglaciated areas, the diversity of animal and plant species and the general development of the soil and plant communities have been shown to be no more complex than comparable glaciated land masses nearby. This has lent support for the theory that at present arctic tundra ecosystems, including soils, are in equilibrium with the prevailing climate. Thus, they have been considered quite stable, although many areas have been deglaciated for only 3,000 to 8,000 years. Whether this apparent stability now is threatened with a warming climate is the subject of considerable current research. The following two sections review briefly the basic biogeographical and soil formation characteristics of the tundra region.

Biogeographical Subcategories

From a biogeographical viewpoint, the Arctic is often defined as the lands beyond the climatic limit of tree growth in upland habitats between river drainages. These areas have often, in particular in North America, been considered to consist of only two floristic units, tundra and

polar deserts. Furthermore, in the Western Hemisphere, the application of basic, diagnostic characteristics for the division of the Arctic into simple subdivisions based on the degree of closedness of the vegetation has been widespread. Traditionally, scientists of the Soviet Union have identified a larger number of biogeographical subzones than North Americans.

Andreev and Aleksandrova, for example, identified 13 arctic vegetation types and five subzones within the tundra zone based on the species composition and characteristics related to life form and migration history. In addition, they identified a number of longitudinal provinces across the Eurasian continent. Part of the reason for the varying use of subzones by North American and Eurasian scientists is probably real differences in the physical geography of the two continents. In Eurasia, most of the land north of the tree line is continental, with groups of islands in the Arctic Ocean. The climate shows a gradual northward shift over contiguous land masses, which provides the basis for major subzones or "belts" of specific vegetation types. In contrast, the American continent has a different geomorphology: No unbroken land mass extends to 78° N as in Eurasia and translongitudinal mountain ranges, sea barriers, and ice caps combined with their respective climatic influences cause a coarse-grained mosaic vegetation pattern rather than arrangements in belts or zones. Bliss and Matveyeva provide a useful overview of tundra subcategorization still used widely and also in connection with modern impact studies of climate change. Table 1 shows the areal extent of the subcategories described in the following sections.

Shrub Tundras

In North America, shrub communities are dominated by *Betula nana* and various species of *Salix*. The ground cover includes *Carex* and *Eriophorum* spp., numerous dwarf shrubs, grasses, and forbs; and an abundance of lichens and mosses. The most common dwarf shrubs belong to the genus *Ledum*, *Vaccinium*, *Empetrum*, *Rubus*, *Arctostaphylos*, and *Cassiope*. The most common mosses include species of *Hylocomium*, *Aulacomnium*, *Polytrichum* and, figuring prominently, *Sphagnum*. Important lichen families are *Cladina*, *Cetraria*, and *Cladonia*. The tall shrub canopy is

Vegetation Type	Alaska	Canada	Greenland, Iceland	Eurasia	Total Area
Low Arctic					
Tall shrub	0.018	0.026	0.018	0.112	0.174
Low shrub	0.09	0.264	0.032	0.896	1.282
Tussock, sedge-shrub	0.126	0.088	0.036	0.672	0.922
Wet sedge	0.104	0.176	0.04	0.56	0.88
Semidesert	0.018	0.325	0.014	—	0.358
Ice caps	—	—	0.776	—	0.776
High Arctic					
Wet sedge	0.004	0.096	—	0.032	0.132
Semidesert	—	0.72	0.093	0.192	1.005
Polar desert	—	0.64	0.127	0.080	0.847
Ice caps	—	0.144	1.031	0.016	1.191
Total land	0.36	2.336	0.368	2.544	5.6
Total land plus ice caps	0.36	2.48	2.167	2.56	7.567

Table 1   Areal extent ($\times 10^{12}$ square meters) of various tundra types in different regions of the Arctic

Source: Bliss, L. C., & Matveyeva, N. V. (1992). Circumpolar Arctic vegetation. In F. Chapin, R. Jefferies, J. Reynolds, G. Shaver, & J. Svoboda (Eds.), *Arctic ecosystems in a changing climate* (pp. 281–300). San Diego, CA: Academic Press.

40 to 60 cm (centimeters) high, and the heath shrubs and forbs are 10 to 20 cm in height; the cryptogams provide more or less complete ground cover. This type of vegetation is widespread in Arctic Alaska and in many areas of Arctic Canada except for the central and eastern parts. Shrub tundra in Eurasia extends on rolling uplands across much of Siberia. Again the taller shrubs are dominated by varying subspecies of *Betula nana* and species of *Salix*. The ground cover includes *Carex* spp. and heath (low) shrubs, *Ledum*, *Vaccinium*, *Empetrum*, *Arctostaphylos*, *Dryas*, and *Cassiope* spp. There is a continuous moss cover comprising *Hylocomium*, *Tomenthypenum*, *Aulacomnium*, and *Dicranum* spp. Lichens *Cladina*, *Cladonia*, and *Cetraria* spp. occur within the mosses. The tallest shrub (up to 2 m [meters]) forming thickets are *Alnus fruticosa*, which are important in many places from Ural to Chukotka. Also, thickets of *Salix* spp., 1 to 2 m tall, occur mainly in drainages and along river banks.

Tussock and Sedge-Dwarf Shrub Tundras

In the vast areas of tussock and sedge-dwarf shrub tundra, many of the shrubs mentioned above are still present, but they do not form a canopy and they occur almost exclusively in depressions, on raised polygons in mires or along river banks. The tussock and sedge-dwarf shrub tundra corresponds largely to what scientists of the Soviet Union refer to as "typical tundra."

In the Siberian Arctic, the sedge-dwarf shrub tundra is most widespread. Here, the main species are mosses *Hylocomium*, *Tomenthypnum*, and *Aulacomnium* spp. and species of lichens *Cladina* and *Cladonia*. Dwarf shrub species of the genus *Vaccinium*, *Salix*, and *Cassiope* are common as are *Carex*, *Ptilidium*, and *Dryas* spp. in the ground layer. In northwestern Siberia, *Dryas octopetala* dominate many dry communities.

Tundras dominated by *Eriophorum vaginatum* tussocks with *Carex* spp., along with common dwarf shrub species and an abundance of mosses and lichens occupy large areas in western parts of North America, particularly Alaska. Again dwarf shrubs of *Betula* and *Salix* are common along with numerous forbs, grasses, and an abundance of mosses, including species of *Sphagnum*, *Hylocomium*, *Dicranum*, *Aulacomnium*, and *Tomenthypnum*. Common lichens include *Cetraria*, *Cladonia*, *Cladina*, and *Thamnolia* spp. Tussock tundra is more limited in the eastern parts of Arctic America and West Siberia. However, large tracks of land dominated by *Eriophorum vaginatum*, along with the common heath shrubs and an abundance of lichens and mosses, are found in East Siberia and Chukotka.

Wet Tundras

Wetland plant communities in North America dominate on the coastal plain of Alaska and in the flat coastal areas in the Yukon. They extend on islands in the Mackenzie River Delta and eastward on the Tuktoyaktuk Peninsula. Peat in these communities reaches considerable depths typically 1 to 5 m. Only a shallow (20–50 cm) active layer develops in these cold, wet, soils.

The dominant sedges are in particular species of *Carex* and *Eriophorum*. Grasses include *Arctagrostis*, *Dupontia*, and *Arctophila* spp. Mosses are abundant, including species of *Aulacomnium*, *Calliergon*, *Ditrichum*, *Drepanocladus*, *Hylocomium*, *Meesia*, *Tomenthypnum*, and *Sphagnum*. Various dwarf shrubs are common along the rims of the mires and on raised hillocks. Sedge-dominated mires occur in lowlands across the Canadian Shield but in limited extent.

Wet tundras form a prominent part of the Siberian tundra. They are especially well developed in the central part of the Yamal Peninsula and in the lowlands of the Yana, Indigirka, and Kolyma river basins. Various types of mires have been identified in the Russian Arctic, but all are dominated by species of *Carex*, *Eriophorum*, *Caltha*, and *Comarum* as well as moss species *Drepanocladus*, *Meesia*, *Calliergon*, *Polytrichum*, *Sphagnum*, and *Cinclidium*. Dense thickets of *Betula nana* are typical for large peat hillocks up to 30 m in diameter. *Dryas*, *Vaccinium*, *Salix*, and *Betula* spp. typically grow on the rims of polygonal mires. Mires similar to the above but without the abundance of shrub species *Betula*, *Salix*, and *Vaccinium* are also common in the low and high arctic communities, particularly in the Northwest Territories of Canada and in Greenland.

Polar Semideserts

The tundras described so far, except for mires, are exclusive to what Bliss and Matveyeva identify as the Low Arctic. The polar semideserts are in the High Arctic, which is exemplified by a number of structural and floristic changes in vegetation. These changes include a shift from the predominance of low shrub (*Betula*, *Salix*), dwarf shrub (heath species), and cottongrass-tussock-dwarf shrub tundras to an open vegetation dominated by cushion plants (*Dryas* and *Saxifraga* spp.), prostrate shrubs of *Salix arctica* and rosette species of *Saxifraga*, *Draba*, and *Minuartia*.

Along a south-north transect in the continental Siberian true tundra, the boreal elements of the low shrub species are the first to disappear. Subsequently, other dwarf shrub species decrease significantly in the transition to high arctic tundra. Arriving on the "true tundra," as named by Russian scientists, the dominating species are *Salix polaris*, *S. arctica*, the graminoids *Alopecurus alpinus*, *Deschampsia borealis*, and *Luzula confuse*. Comparable tundras in North America have been called *polar semideserts*.

As mentioned earlier, the above major vegetation types tend to appear in zones in the Eurasian Arctic and in a mosaic in North America (including Greenland). However, all over the tundra regions on a local scale these vegetation types are mixed, depending on local climatic, hydrologic, and topographic features. The photograph (next page) shows a typical tundra valley in northeastern Greenland where in the lower parts wet tundra dominates and is surrounded by typical sedge shrub tundra that is replaced with polar semidesert-type vegetation as moving upslope with almost barren land beyond. This transition, occurring within a few hundred meters or less, is a common feature of tundra environments in landscapes with rolling hills.

Tundra Soil Formation and Characteristics

The tundra soils are young, typically dating back less than 12,000 years except in some areas that were not ice-covered during the latest glaciation. In some local sectors, the landscape has been icefree only during relatively recent time. In addition to disappearance of the ice cover itself, major areas of the Arctic have undergone isostatic adjustment resulting in emerging landforms. A large area surrounding Hudson Bay and extending

The Zackenberg valley in northeastern Greenland in mid August. The variation in vegetation types over short distances depending on hydrology and local topography is clearly visible. Muskoxen can be seen in the background.

Source: Charlotte Sigsgaard.

northward to the arctic islands was depressed well below sea level during the Pleistocene, as was Northern Scandinavia and the lower courses of major rivers of Siberia. Some emerged landforms have accumulated organic-rich sediments, which, coupled with flat terrain and poorly developed drainage patterns, have resulted in sluggish surface drainage and formation of extensive wetland conditions. The presence of permafrost further restricts soil and plant development. Consequently, decomposition, release of nutrients, and synthesis of secondary minerals from weathering of clay all progress very slowly.

The soil formation process of podzolization is limited to well-drained tundra soils with a deep active layer. Where dwarf shrub species predominate, weakly developed podzols (Spodosols) are found. Less-well-developed soils of uplands and dry ridges are the arctic brown soils (Inceptisols). The most common group of soils in the Low Arctic region includes the tundra soils (Inceptisols) underlying cottongrass-dwarf shrub and some sedge communities of imperfectly drained habitats. These soils form under the process of gleization. Poorly drained lowlands where soils remain saturated all summer accumulate peat. These wet and highly organic tundra soils belong to the Histosol group.

The drier arctic soils, podzols, and arctic brown soils show some translocation of humus and iron, with iron-enriched B2 horizons and weakly eluviated A2 horizons in the podsols. Surface layers tend to be acidic (pH 6–4) and low in available nutrients but quite well drained above the permafrost. Inceptisols (arctic tundra soils) are less well drained but show generally similar pH characteristics and low nutrient availability. They contain B horizons that have sub-angular to angular structures, are grayish in color, and include iron oxide mottles. Histosols of poorly drained lands are acidic and are similar to arctic tundra soils in having limited translocation of minerals into the B horizon.

Torben R. Christensen

See also Biome: Boreal Forest; Biota and Climate; Biota and Soils; Climate: Midlatitude, Severe; Poles

Further Readings

Aleksandrova, V. (1980). *The Arctic and Antarctic: Their division into geobotanical areas.* Cambridge, UK: Cambridge University Press.

Andreev, V., & Aleksandrova, V. (1981). Geobotanical division of the Soviet Arctic. In L. Bliss, O. Heal, & J. Moore (Eds.), *Tundra ecosystems: A comparative analysis* (pp. 25–34). Cambridge, UK: Cambridge University Press.

Arctic Climate Impact Assessment. (2005). *Arctic Climate Impact Assessment: Scientific report.* Cambridge, UK: Cambridge University Press.

Bliss, L. (1981). The evolution and characteristics of tundra. In L. Bliss, O. Heal, & J. Moore (Eds.), *Tundra ecosystems: A comparative analysis* (pp. 5–46). Cambridge, UK: Cambridge University Press.

Bliss, L., & Matveyeva, N. (1992). Circumpolar Arctic vegetation. In F. Chapin, R. Jefferies, J. Reynolds, G. Shaver, & J. Svoboda (Eds.), *Arctic ecosystems in a changing climate* (pp. 281–300). San Diego, CA: Academic Press.

Chernov, Y. (1985). *The living tundra.* Cambridge, UK: Cambridge University Press.

Linell, K., & Tedrow, J. (1981). *Soil and permafrost surveys in the Arctic.* New York: Oxford University Press.

Tedrow, J. (1977). *Soils of the polar landscapes.* New Brunswick, NJ: Rutgers University Press.

BIOPHYSICAL REMOTE SENSING

Remote sensing is a collection of spatiotemporal views (images) of Earth's surface from vantage points, high above Earth. Such a collection has become instrumental to monitor and predict the future biophysical changes on Earth's surface that human beings will need to overcome. By using remote sensing techniques that involve the interpretation of reflective and absorptive properties of electromagnetic radiations (EMR), image interpreters are able to predict sea surface temperatures, wind patterns, the health of vegetation, possible occurrences of fire based on the structures of various plant canopy, and many other environmental factors. The EMR is absorbed and reflected by both the biological and physical objects of the Earth.

The solar radiation distribution among the vegetation is a function of canopy structure that strongly affects the productivity of the canopy and influences the surrounding biophysical environment. The productivity of the plant canopy is related to the well-being of humankind. The recent advances in biophysical remote sensing has made it possible to include all the biological and physical processes influencing the microclimate and atmospheric exchange characteristics of terrestrial biosphere over a range of spatial and temporal scales. The vegetation, the principal biosphere component, tightly couples with the radioactive, meteorological, hydrological, and biological processes. There is interconnectedness among atmosphere/cloud, wind, water, topography, soil moisture, and land cover types in the biosphere, and the biophysical remote sensing helps monitor such interconnectedness.

Biophysical remote sensing helps display patchy warm-and-cool patterns of the atmospheric/cloud conditions on the images. A dark image (dark signature) indicates relatively cool conditions while the bright signatures indicate relatively warm conditions.

Wind produces characteristic patterns of smears and streaks on images resulting in lighter and darker color signatures. Often, these signatures extend over vast areas. For example, wind velocity is typically lower downwind due to

This MODIS Terra image, acquired in August 2, 2006, shows most of Greece, with its jagged coastline, and many islands and peninsulas.

Source: Jeff Schmaltz/NASA.

obstructions and reduces cooling effects resulting in relatively bright images; sheltered areas are generally warmer than terrain exposed to windy conditions during the winter. Theoretically, warmer areas generate relatively brighter images than cooler areas.

Another biophysical component, water, behaves differently than a land surface. During the daytime, water bodies have a cooler surface temperature than soils and rocks, and water bodies appear darker on the daytime image. Since water and ground surface temperatures reverse themselves during the day and nighttime, the thermal signature of water bodies becomes a reliable index of the timing of image. Warmer signatures on water bodies as compared with the adjacent terrain indicate nighttime image acquisition, while relatively cooler signatures indicate daytime imagery.

Scattered rain showers produce parallel lines on the image, just like the lines developed from a malfunctioned scanner. A heavy overcast layer (dense cloud) reduces thermal contrasts between terrain objects because of reradiation of energy between the terrain and cloud layer but the resulting thermal contrast becomes relatively low.

Topography plays a dominant role in remote sensing as solar heat and shadow vary with the timing of images. If images are taken in the nighttime, the thermal properties of images are displayed due to the reradiation of surface temperatures. If the images are taken during the daytime, the ridges and slopes facing the south and east reveal brighter signatures due to solar illumination, while areas facing the north are shaded from solar illumination and display dark (cool) signatures. On the nighttime image, topographic features are mostly eliminated as only the geologic features (thermal characteristics of objects) influence the image appearances. The total and spectral global solar irradiance absorbed by the vegetation, canopy layer, spectral reflectance, and transmittance for each discrete wavelength at various exposures can be calculated by using Equation 1 and merging satellite images with digital elevation data.

$$R_f = \cos(A_f - A_s)\sin(H_f)\cos(H_s) + \cos(H_f)\sin(H_s), \quad (1)$$

where R_f = relative radiance value of a facet in the image merged with an elevation layer, A_f = the facet of the aspect, A_s = sun's azimuth, H_s = sun's altitude, and H_f = facet's slope.

Another important component of biophysical remote sensing is the soil moisture that influences electromagnetic radiation. Wet or damp soil is cooler than dry soil, both during day and nighttime. The presence of water in the soil increases thermal inertia making the soil comparable with the inertia of rocks. The infrared image of geologic faults and fractures displays differently during the evaporative cooling process. Interpreters at times may confuse man-made built-in infrastructure with geological outcrops. Some geological outcrops and built-in urban areas often show warmer (brighter) surface "heat islands" than the relatively cooler surrounding nonurban countryside or geological outcrops of moist surfaces (rocks or soils). Often, the nature of electromagnetic radiation from soil, urban areas, and vegetation become difficult to distinguish because dry vegetation, such as crop stubble in agricultural areas, appears warm on nighttime imagery similar to the daytime imagery of the dry soil or built-in areas. Likewise, the dry vegetation insulates the ground to retain heat and results into the warm nighttime signature.

The most important component of biophysical remote sensing is the property of individual green vegetation, which exhibits very distinct thermal responses. Green deciduous vegetation has a cool signature on daytime images and a warm signature on nighttime images. During the day, leaves

stomata open for transpiration lowering leaf temperatures. As a result, vegetation generates a cool signature relative to the surrounding environment. At nighttime, the insulating effects of leafy foliage and the high water content of the leaves retain heat resulting in warm nighttime temperatures; however, this situation varies with the species. For example, leaves of coniferous species have thick cuticular walls and smaller stomatal openings; as a result, the high nighttime and low daytime radiant temperature and their low leaf water contents do not affect the appearance of the images. Reflective properties of other vegetation also differ due to leaf structures and the water content of the leaves. Collenchymatous and parenchymatous cells (broad and young leaves) will hold more water than sclerenchomatous (older) cells. The proportion of total and spectral global irradiance absorbed by individual canopy layers may affect the global climate but such effects vary greatly as a function of topography and solar zenith angle.

Vegetation growing in various topographic conditions displays specific spectro-physiographic characteristics, often resulting in spectrally different classification results. Many researchers use the normalized differential vegetation index (NDVI) to determine species-specific vegetation vigor based on red- and near-infrared bands of satellite images (Equation 2).

$$NDVI = \frac{X_{nir} - X_{red}}{X_{nir} + X_{red}}, \quad (2)$$

where X_{red} represents sensor's visible red and X_{nir} represents near-infrared bands of a satellite image. For example, the Douglas fir growing at >2,000 meters elevation displays an NDVI >0.5 on spring image and <0.9 on a landform that is moderately moist with a slope of 10° to 35°. To further analyze the effects of topography on plant growth, remotely sensed images are combined with biophysical factors, such as digital elevation data (DEM) and Tasseled-Cap bands (brightness, greenness, and wetness images). DEM derivatives, the topographic relative moisture index and landform layers are generated in geographic information systems, and used a scale index to determine relative slope position, slope angle, slope shape, and slope aspect for accurate classifications.

These biophysical properties are analyzed at topographic-bathymetric (x = longitude, y = latitude, and z = elevation) locations based on the mixing of the color and spectral signature of features. Since remotely sensed images alone cannot represent all the features at planimetric (x and y) locations, additional topographic-bathymetric features, such as slope, aspect, and elevation from a digital elevation model are needed to identify the appropriate feature locations.

Advances in biophysical remote sensing has made it possible to estimate leaf area index (LAI), percentage of green cover, biomass as well as the percentages of photosynthetically active radiation absorbed by the leaf. Many researchers use LAI to measure the physical quality of life index assuming the green environment as a proxy to healthy and prosperous living. Further advances in biophysical remote sensing suggest that the use of the enhanced vegetation index to determine areas with high vegetation density by taking into account the canopy structure of vegetation rather than just the chlorophyll concentration. Scientists have developed the Fire Weather Index in addition to the NDVI to identify areas that are at fire risk. Biophysical remote sensing is imperative to determine the canopy reflectance indices that link crops to individual wavelengths and for calculating the spectral vegetation indices. Additionally, advances in the biophysical remote sensing have developed the green simple ratio index and green normalized difference vegetation index by estimating vertical temperature profiles of Earth's atmosphere based on the narrow bands in the near- and mid-infrared measures. Such information is not only useful in meteorology but also useful to estimate water surface temperature, chlorophyll concentration on various types of leaves, ground surface temperatures, and the level of moisture present in the atmosphere.

In summary, biophysical remote sensing has been instrumental for vegetation analyses, including climate changes, biogeochemical cyclical and hydrological modeling, land resource monitoring, land use planning, agricultural crop monitoring, yield forecasting, deforestation, and pollution and public health issues. It involves the study of soil moisture, spectral signature features, chlorophyll contents of leaves, vegetation biomass, and surface temperature. Biophysical remote sensing can be used to analyze various types of forest covers and to analyze the effects of vegetation on climate. Biophysical remote sensing is also useful to measure the land surface temperature and vegetation distribution from space.

Keshav Bhattarai

See also Remote Sensing

Further Readings

Boyd, D. S., Phipps, P. C., Duane, W. J., & Foody, G. M. (2002). Remote monitoring of the impact of ENSO-related drought on Sabah Rainforest using NOAA AVHRR Middle Infrared reflectance: Exploring emissivity uncertainty. In G. M. Foody & P. M. Atkinson (Eds.), *Uncertainty in remote sensing and GIS* (pp. 119–142). New York: Wiley.

Chang, K.-T. (2008). *Introduction to geographic information systems* (4th ed.). Toronto, Ontario, Canada: McGraw-Hill.

Gabban, A., San-Miguel-Ayanz, J., & Viegas, D. X. (2008). A comparative analysis of the use of NOAA-AVHRR NDVI and FWI data for forest fire risk assessment. *International Journal of Remote Sensing, 29*(19), 5677–5687.

Harvey, L. (2000). *Climate and global environmental change.* Harlow, UK: Prentice Hall.

Jensen, J. R. (1983). Biophysical remote sensing. *Annals of the Association of American Geographers, 73*(1), 111–132.

Lillesand, T. M. (1996). A protocol for satellite-based land cover classification in the upper Midwest. In J. M. Scott, T. H. Tear, & F. Davis (Eds.), *Gap analysis: A landscape approach to biodiversity planning* (pp. 103–118). Bethesda, MD: American Society for Photogrammetry and Remote Sensing.

Liu, W., Song, C., Schroeder, T. A., & Cohen, W. B. (2008). Predicting forest successional stages using multitemporal Landsat imagery with forest inventory and analysis data. *International Journal of Remote Sensing, 29*(13), 3855–3872.

Piwowar, J. M. (2005). Digital image analysis. In S. Aronoff (Ed.), *Remote sensing for GIS managers* (pp. 287–336). Redlands, CA: ESRI Press.

Sabins, F. F. (2000). *Remote sensing: Principles and interpretation.* New York: W. H. Freeman.

BIOREGIONALISM

Bioregion denotes the conflux of cultural and ecogeographical features in human-defined territories. Bioregionalism has been promoted by geographers as a pragmatic research framework for understanding, and discovering solutions to problems within human-ecological relations. In many regards, bioregionalism is a comfortable fit for a discipline long concerned with the human place in larger ecological systems. Bioregional thought has a strong normative character, rooted in the North American environmental social movements of the 1970s. Essentially interdisciplinary, bioregional thought views communities of land and life as overlapping and internested. As a political philosophy, it proposes that political units need to be defined according to this combination of features and that decisions would be more socioeconomically and ecologically sustainable if they were taken at smaller, local levels than is the case in the present state-centric model. At the ethical level, it stresses attention to the human place within, and responsibility toward, the natural world.

The interdisciplinary nature of bioregional theory renders it suitable for addressing the complex challenges of developing sustainability. Indeed, bioregional approaches have spanned ecological science, conservation biology, social theory, geography, ethics, and political philosophy.

Criticism of bioregionalism has focused on its alleged lack of theoretical rigor, romanticism, or an unduly restrictive conception of place. Taking a holistic stance drawn implicitly from the epistemological insights of poststructuralism, bioregional scholars respond that all regions are continuous with one another; that no portion of territory on Earth is entirely discrete from the next. According to this view, global ecology, like landforms, is characterized by continuities and transitions. Similarly, there is constant cultural and economic continuity and exchange between both adjacent and (ostensibly) geographically separated human communities, all of which are simultaneously embedded in greater natural systems. The task of bioregionalists is to highlight the intensities within this continuity on which human politics may be based and ecology depends.

The following are some bioregional axioms:

1. The cumulative impacts of industrial civilization on the biosphere are rolling back many of Earth's evolutionary achievements of the past several million years.

2. The territorial assumptions inherent to the global industrial political system must be radically reformed.

3. Decentralization to smaller units defined by a combination of ecogeographical and cultural factors is essential.

4. Human politics must emphasize a commitment to place in order to promote the healthy human-environment interactions that are the foundation of healthy communities.

5. Humans bear ethical responsibility toward the nonhuman world.

William Hipwell

See also Biosphere Reserves; Carrying Capacity; Deep Ecology Movements; Ecological Economics; Ecoregions; Environmental Ethics; Nature; Nature-Society Theory; Political Ecology

Further Readings

Aberley, D. (Ed.). (1993). *Boundaries of home.* Gabriola Island, British Columbia, Canada: New Society.

Alexander, D. (1990). Bioregionalism. *Environmental Ethics, 12,* 160–173.

Ankersen, T., Regan, K., & Mack, S. (2006). Towards a bioregional approach to tropical forest conservation: Costa Rica's greater Osa bioregion. *Futures, 38,* 406–431.

McGinnis, M. (Ed.). (1999). *Bioregionalism.* London: Routledge.

McTaggart, W. (1993). Bioregionalism and regional geography. *The Canadian Geographer, 37,* 307–319.

Rajeswar, J. (2002). Development beyond markets, and bioregionalism. *Sustainable Development, 10,* 206–214.

Sale, K. (1991). *Dwellers in the land: The bioregional vision* (Rev. ed.). Gabriola Island, British Columbia, Canada: New Society.

BIOSPHERE RESERVES

Biosphere reserves are geographic areas recognized by the United Nations Education, Scientific and Cultural Organization (UNESCO) and intended to demonstrate how people can live and work in harmony with the natural environment. Prior to designation, these areas are subject to strict scrutiny by UNESCO. Advocates for biosphere reserve status must demonstrate the special ecological, cultural, and social characteristics of the local areas. Although certain locations within a biosphere reserve area may be subject to national or subnational legislation, there is no new regulation imposed by the recognition of the region by UNESCO.

Initially conceived in the 1960s, the first biosphere reserves were created in 1976 under UNESCO's Man and the Biosphere program established in 1971. Indeed, many reserves were created in the 1970s. After a quiet period during the 1980s and 1990s, there has been an increase in the rate of designation internationally since 2000. As of May 2008, there were 531 biosphere reserves in 105 countries.

Biosphere reserves are intended to demonstrate three functions: (1) environmental protection, (2) logistical provisioning for scientific research and education, and (3) sustainable resource use. Over the years, the emphasis placed on each of these functions has changed. In the late 1960s, the greatest emphasis was placed on environmental protection through scientific research and the application of its results. Promoters of biosphere reserves, typically natural scientists, believed that modern science would help local people establish rational methods of resource use that would help them conserve the world's biodiversity. By the early 1980s, UNESCO indicated that biosphere reserves were to serve as field laboratories, which would help find solutions to the problems facing local populations, and indeed, the global community.

During the 1980s, however, there was a growing realization that ecosystem conservation was directly connected to development. This realization was given greater impact with the report of the World Commission on Environment and Development, *Our Common Future*, in 1987. Particularly in developing countries, there was a growing idea that biosphere reserves should be used to promote "ecodevelopment" strategies that would also meet the basic needs of local communities. There was also concern that natural science research should be accompanied by social science research that would find ways to improve cooperation and communication among researchers, resource managers, and local residents living in and near biosphere reserves. Thus, the "sustainable resource use" function became a higher priority.

This priority was given greater force with the Seville Strategy of 1995, an outcome of the Second UNESCO International Conference on Biosphere Reserves. The strategy noted that the purpose of biosphere reserves was to include social and cultural dimensions into environmental management. The strategy also suggested that biosphere reserves should help people who live and work within them by demonstrating how to attain a sustainable future. These considerations illustrate an increased emphasis on addressing cultural and development considerations in both the establishment and the management of biosphere reserves.

To address the three functions, biosphere reserves contain three zones: (1) a core that must be protected, typically by national legislation; (2) a buffer where research and recreation use compatible with ecological protection are allowed; and (3) a transition zone where sustainable resource use is practiced. In some countries, the outer zone is also referred to as an *area of cooperation*. Thus, biosphere reserves retain some form of protected area at their core, but they must also incorporate adjacent areas and the inhabited surrounding "working landscapes" to demonstrate how they integrate conservation with sustainable development. Biosphere reserves are typically established on the basis of watersheds or other landscape-level features that extend beyond the boundaries of local human communities.

Although designation of biosphere reserves does not confer changes in property rights or regulation, biosphere reserves are subject to periodic review whereby each biosphere reserve is reviewed for UNESCO once each decade. This review requires an assessment of whether the biosphere reserve is fully functional, meaning whether all three functions are being addressed. The intention of the review is to encourage each reserve to be sensitive to all three objectives, namely, (1) conservation, (2) sustainable development, and (3) scientific provisioning, that formed the basis for designation.

Contemporary research in biosphere reserves places greater emphasis on socio-ecological systems than in the past. With reference to socio-ecological systems, biosphere reserves currently serve as experiments in environmental governance, community-based conservation, sustainable development, and adaptation and resilience. All such research points to the interconnection of humans and natural landscapes. For example, UNESCO now believes that biosphere reserves should demonstrate innovations in environmental governance and can help meet international obligations, such as Agenda 21, the Convention on Biological Diversity, and the Millennium Development Goals. Thus, biosphere reserves offer researchers an opportunity to study complex drivers of change and stressors on ecological and social systems and to identify and test strategies for adaptation and resilience. This kind of research links biosphere reserves to the goals and objectives of the Millennium Ecosystem Assessment and the related discourse of sustainable development.

Despite these lofty opportunities, many biosphere reserves struggle because they are not well supported financially or logistically by their local or national governments. For example, ongoing financial support is somewhat uneven. In some countries, individuals are hired to pursue conservation and sustainable development objectives, but they are typically minimally staffed with small budgets. In other countries, biosphere reserves rely on a volunteer labor to advance their interests, making their progress slow and sometimes halting. In general, the success of biosphere reserves in meeting their mandate is predicated on the initiative and skills of a small number of people who often spend years dedicated to demonstrating the links between environmental protection and sustainable development.

Maureen G. Reed

See also Biodiversity; Bioregionalism; Environmental Ethics; Environmental Management; Indigenous and Community Conserved Areas; Nature; Nature-Society Theory; Species-Area Relationship; Sustainable Development

Further Readings

Francis, G. (2004). Biosphere reserves in Canada: Ideals and some experience. *Environments, 32*(3), 3–26.

Pollock, R., Reed, M. G., & Whitelaw, G. (2008). Steering governance through regime formation at the landscape scale: Evaluating experiences in Canadian biosphere reserves. In K. Hanna, D. Clark, & S. Slocombe (Eds.), *Transforming parks* (pp. 110–133). London: Routledge.

UNESCO. (1995). *The Seville strategy for biosphere reserves.* Paris: Author.

UNESCO. (2000). *Solving the puzzle: The ecosystem approach and biosphere reserves.* Paris: Author.

UNESCO. (2005). *Biosphere reserves: Benefits and opportunities.* Paris: Author.

World Commission on Environment and Development. (1987). *Our common future.* New York: Oxford University Press.

BIOTA AND CLIMATE

The word *biota* is a collective term, simply meaning "living organisms." Strictly speaking this includes human beings, but here the focus is on nonhuman living things. *Climate* can be defined as the average weather conditions of a place, determined over a long enough period of time to ensure that the full range of conditions normally experienced is taken into account, while unusual conditions happening within the measurement period do not unduly influence the average. Climate is therefore commonly calculated from weather conditions over a period of 20 to 50 years and includes temperature, precipitation, daylight, and wind, among other variables. Biogeography focuses directly on the study of patterns of life in both space and time. The biotic environment is an important part of "the environment" more generally, which is the central element of physical geography. Both present and past climate affect which organisms can live in a place, but the strength of this influence varies depending on the type of organism and the medium in which it lives (land or water).

The study of the present climate's influence on organisms is a strong focus of what is often called "environmental biogeography," while the influences of past climate—in tandem with the movement of tectonic plates—are a major focus of "historical biogeography." Increasingly, the distinction between these two subdisciplines of biogeography is breaking down, as it seems clear that a good explanation of patterns of life in the world today requires elements of both. Furthermore, it is increasingly appreciated that, as well as being strongly affected by climate, living organisms can affect climate and therefore climate change at all spatial scales. Clearly, relationships between biota and climate are a very important part of geography. This entry first discusses how living organisms affect climate and examines the feedbacks between the two, using the relationship between climate and species richness as an example. The entry concludes with an overview of current research directions and priorities in the field more generally.

The Influence of Climate on Biota

For living organisms, climate can be considered part opportunity and part threat. Some aspects of climate are usable resources (opportunities), especially water and daylight—most important, those wavelengths of light that constitute photosynthetically active radiation, which plants can convert into food. Water is vital to life for three main reasons: (1) it can carry other substances in solution or suspension, (2) it is one of the few substances that is liquid at most temperatures experienced around the world, and (3) it is one of the raw materials for photosynthesis. Other aspects of climate, most important temperature, represent ambient conditions rather than resources. Temperature affects the rates at which chemical processes happen within the bodies of living organisms. Ambient conditions also interact with resource inputs, the most important interaction being between temperature and water. If it is too cold or too hot, water is not available to living organisms because it is in the form of ice or water vapor, neither of which is of any use for chemical processes within organisms' bodies. Beyond affecting water availability, extremes of temperature are directly stressful to organisms. Water expands on freezing, meaning that organisms without adaptations to cope will suffer lethal rupture of cell walls and other important parts of their internal structures, such as freezing pipes bursting. Increasingly high temperatures tend to be associated first with the denaturing of enzymes, which are vital to living organisms, and second, at more extreme levels, with the direct destruction of the living tissues.

Evolution largely concerns organisms adapting to their environment, with climate a key part of this. For most of the history of life on Earth, the climate has been warmer than it is today. Probably for this reason, combined with the physical properties of water, frost is dangerous to most organisms. Among plants, various strategies exist for coping with frost, including shutting down over the winter (e.g., deciduous trees) and surviving winter as either seeds (e.g., annual plants) or underground structures, with the above-ground parts dying. For animals, this means that frost is not only directly dangerous but also often associated with food shortages. There may also be water shortages, when most water in the environment is locked up in ice. Animals' strategies for coping with frost include hibernation, the use of burrows, and storing food for the winter. Even in warmer climates, water is often in short supply, either seasonally or generally (e.g., deserts). Again, both plants and animals have developed numerous ways of coping with water shortage. For example, the "tree" *Welwitschia mirabilis*, found in the Namib Desert and in a family of its own, has a large underground trunk. Above-ground are only two thick leathery leaves that grow continually for up to about 2,000 years. This plant is adapted to grow in places with exceptionally low annual rainfall: Its leaves, being at ground level, are less desiccated by winds. Its leaves also contain special structures that obtain moisture from dew, providing the bulk of its water supply. It photosynthesizes using the CAM (crassulacean acid metabolism) pathway, which allows it to open its stomata only at night, thereby minimizing water loss. This example illustrates that there are many different ways of adapting to the environment. Because most adaptations have an energetic cost, they trade-off against each other, and the balance of fitness can shift between species with only minimal variation in climate. Thus, each species has a unique geographical range, which has been shaped by the responses of its individuals and populations

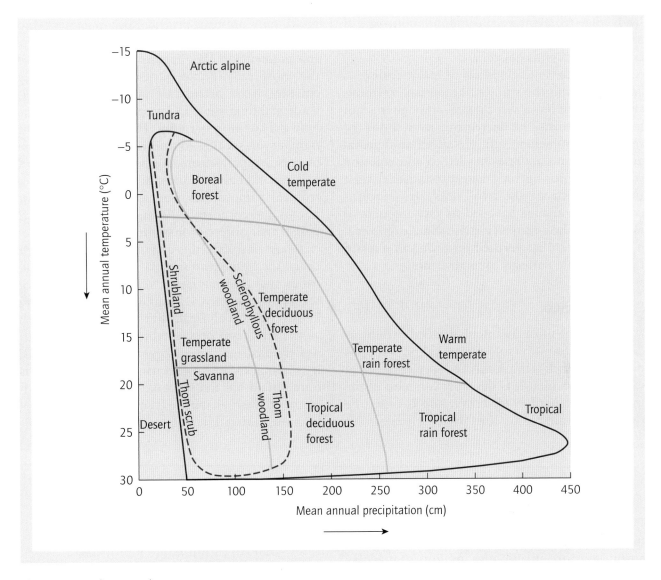

Figure 1 A climograph

Source: Brown, J. H., & Lomolino, M. V. (1998). *Biogeography* (2nd ed., Figure 5.12, p. 111). Reprinted with permission from Sinauer Associates.

Note: The higher temperatures are represented lower on the vertical axis.

to their environment during its evolutionary history. The role of climate in determining species' ranges is so fundamental that the dominant method of predicting biotic responses to climate change is currently "climatic envelope modeling," in which the climatic limits of each species' current range is determined; then, the ranges of all the species are redrawn according to the predicted future climate.

Climate-plant relationships are the most important link between biota and climate for two main reasons. First, photosynthesis is responsible for all the food used by almost all living organisms (propagated via "food webs"). Second, the inefficient transfer of energy between links in the food webs ("trophic levels") means that plants have far more biomass than animals. A key part of biogeography, therefore, involves relationships between spatial patterns of climate and of plants, globally. Climate varies around the world, the most important variation being in temperature and precipitation, for the reasons explained above. A very useful starting point for biogeography is a climograph (Figure 1). This maps "biomes" (broad vegetation types) in climate space—that is, on axes' defined by average temperature and precipitation—and is an excellent demonstration of the close links between biota and climate.

Much of the emerging field of macroecology is concerned with the study of broadscale relationships between biota and climate—though it is not limited to this. The increasing availability of regional, continental, and global environmental and biological databases has led to an explosion of studies focusing on emergent statistical relationships between the two. Similarly, concerns over climate change have led to vast amounts of research on the effects of climate change on biota. Much research has focused on trying to explain the broadscale patterns of biodiversity.

Species Richness

Perhaps the most commonly studied variable is "species richness" (sometimes called "species density"), which is the number of species recorded in a specified area. It is the most common of the many measures of biodiversity, and is often the only one available at the broadest scales, for which only presence-absence species data tend to be available. The well-known "latitudinal diversity gradient" (LDG, Figure 2) describes one of the most remarkable patterns that repeatedly emerges from broadscale studies of this sort. Understanding the LDG is often considered the holy grail of biogeography. The pattern is that the number of species (or other taxonomic units) increases rapidly from the poles to the equator. Correspondingly, the most diverse ecosystems are tropical rain forests (see montane rain forest photo) and coral reefs. Because we do not know, within two orders of magnitude, how many species exist on Earth (this gap in knowledge is often called the "Linnaean shortfall") and because distributional information is very poor for most of the approximately 1.7 million species that are known (the "Wallacean shortfall"), patterns such as the LDG are only established for particular, well-studied, groups of organisms. Nonetheless, it is remarkable how consistently the LDG is found in all but the narrowest and most idiosyncratic of these groups (e.g., penguins and seals), both on land and in water.

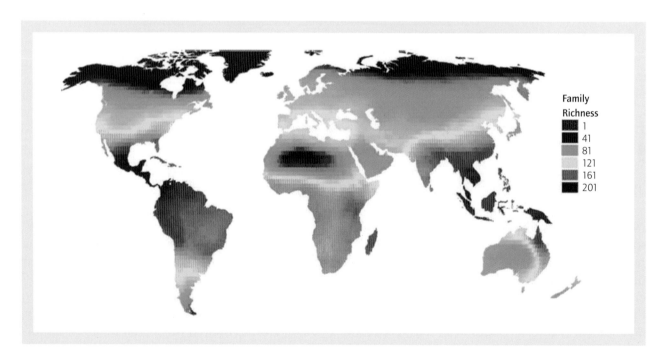

Figure 2 The latitudinal gradient of angiosperm family richness: number of families in 220-km × 220-km grid cells

Source: Francis, A. P., & Currie, D. J. (2003). A globally consistent richness-climate relationship for angiosperms. *American Naturalist, 161*, 523–536. Copyright © 2003 The University of Chicago Press. Used with permission.

A tropical montane rain forest in Cusuco National Park, Honduras. Tropical montane rain forests often harbor many endemic species, which are species found nowhere else.

Source: Author.

Broadscale spatial species-richness patterns are rarely purely latitudinal. For instance, in Australia, the richness of most organisms peaks around the edges of the continent, especially the eastern edge (Figure 3). This is almost certainly because the climate near the east coast of Australia is wetter than in the interior of the continent. Studies repeatedly find that the strongest correlate of species-richness patterns at broad scales is some measure of the climate—usually a measure of energy, such as temperature or potential evapotranspiration, or a measure of water availability, such as rainfall, or a mixture of both, such as actual evapotranspiration. As a result, it is generally accepted that climate is one of the main causes of variation in species richness at this scale. However, exactly how climate causes the patterns

One of the many species endemic to Cusuco National Park is the large frog *Plectrohyla exquisita*.

Source: Author.

remains a very controversial question. Some of the leading theories focus on direct effects of contemporary climate on within-cell processes or on the availability of food. Others stress the role of past climatic changes, with species' niches being conserved to a considerable extent over geological time ("niche conservatism"), or evolutionary rates being faster in warmer conditions. Others invoke increasing importance of climate toward the poles, with biotic-abiotic interactions accounting for much of the high tropical diversity. Another claims the importance of a larger contiguous area of comparable climate in the tropics resulting in the species richness in that area. Yet another stresses the roles of refugia from harsh climatic conditions, such as during glacial periods, with these refugia often resulting in barriers to reproduction that lead to more species via allopatric speciation, especially in the tropics. One important logical distinction in this debate is between climate as a direct cause of increased biodiversity and climate as a constraint on diversity. In all probability, climate acts in both ways.

Spatial patterns of species richness at narrower scales tend to be less latitudinal and less strongly correlated with climate, but this may well be largely an artifact of the amount of climatic variation that is present in a smaller study area. One situation in which considerable climatic variation occurs over relatively short distances is along a steep altitudinal gradient. This is for two main reasons. First, temperature rapidly decreases with increasing altitude. Second, precipitation often increases greatly with altitude (up to a point) because cooler air can hold less water vapor. Along altitudinal gradients, rapid changes in species richness tend to occur. There are two very commonly reported shapes of species richness–altitude relationship. The first is called "monotonic," in which species richness consistently decreases at higher altitudes, whether linearly or via some other decreasing function. The second is usually referred to as "hump-shaped" and occurs when species richness is relatively low at the bottom of the gradient, initially rising to peak at midaltitude, and then falling at the highest altitudes. As with the LDG, there are numerous competing hypotheses to explain the species-altitude relationships, including the decreased area of high-altitude zones, but again, climatic changes with altitude tend to correlate well with the observed patterns and feature strongly in most explanations. Many argue that topography is best seen as a landscape-scale climatic variable when trying to account for spatial patterns of biodiversity. It certainly tends to be among the variables most strongly correlated with species richness at this scale.

The marine equivalent of the altitudinal gradient is the depth gradient of the oceans. Less is known about this relationship, as deep sampling of the oceans remains sparse. Initial indications have suggested that both monotonic decreases in species richness with depth and hump-shaped relationships are found here. In the

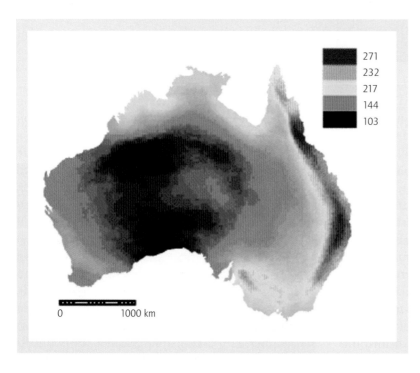

Figure 3 Bird species richness in Australia, in 27.5-km × 27.5-km grid squares

Source: Hawkins, B. A., Diniz-Filho, J. A. F., & Soeller, S. A. (2005). Water links the historical and contemporary components of the Australian bird diversity gradient. *Journal of Biogeography, 32,* 1035–1042.

sea, water availability changes little with latitude, longitude, or depth (some gradients in salinity can cause slight variation in the availability of water to some organisms). Water temperature does vary, not least with depth, and probably plays a part in causing depth gradients of species richness. Perhaps the most important aspect of climate here, though, is light, which rapidly attenuates to minimal levels with increasing depth in the oceans.

At the narrowest spatial scales, such as field plots, climate is also often implicated in species distribution patterns and therefore in species-richness patterns. The main reason is the angle of incidence of the sun: Where topography is steep, the aspect of a place (the direction it faces) can be quite important. For example, some tropical plant species have been shown to favor east-facing slopes because they receive direct sunlight in the morning, before the clouds have built up. In higher latitudes, the more important distinction tends to be between south-facing slopes, which may face the sun nearly at the perpendicular, and north-facing slopes, which may receive sunlight only at the most acute angles, if at all. However, fine-scale climatic variables such as aspect are often only weakly correlated with species richness. In such circumstances, aspect may be more strongly related to species identities and community composition than to species richness.

A complication arises at fine scales, in that microclimate can be an important variable affecting species richness, identity, and other ecological variables. Yet microclimatic variation is often largely determined by trees, such that broadscale climate affects biota, which affects fine-scale climate, which affects biota.

Other Key Biotic Variables

While many, many studies have been devoted to relationships between climate and species richness (and other measures of biodiversity), much research has also linked climate to species' traits, such as body size, abundance, range size, growth form, dispersal, and phenology. Among the ways in which climate affects dispersal is the manner in which it forms barriers to dispersal, which lead to isolation of populations and ultimately to speciation. In particular, high mountains are very effective barriers to the dispersal of most species because

of the extreme climatic conditions at high altitudes. Conversely, tropical valleys can be formidable barriers for many cool-adapted species of tropical mountains, leading to high rates of endemism in parts of Central and South America (see the montane rain forest photo). "Phenology" refers to the timing of the different stages or events in the life cycle of organisms, such as the onset of bud burst or flowering in plants. The Royal Botanical Gardens in Kew (London) has been keeping records of the date of the spring bloom of more than 100 plant species for more than 50 years. Recent trends have been for spring bulbs to come up and flower earlier—and consistently so over the past two decades or so (except 2009). One major concern is that the phenology of plants and the insects that pollinate them, or feed on them, are becoming decoupled as the climate changes. This can happen because the phenology of some species is triggered mainly by daylight, while that of others is triggered mainly by temperature or rainfall. Although climate has always changed in the past, sometimes rapidly, and therefore species have clearly managed to cope with such decoupling, the concern is that multiple stresses may reduce their ability to do so. Multiple stressors result from the fact that human actions are not only changing climate but are also reducing and fragmenting habitats, polluting them, introducing alien species, and so on.

The Influence of Biota on Climate

Living organisms have long been known to be powerful influences on many aspects of the environment, such as soil, hydrology, geology (many rocks are the remains of living organisms), and atmospheric composition. It has long been known that the composition of the Earth's atmosphere would be very different were it not for living organisms, especially those that photosynthesize. After all, oxygen is a highly reactive element and would not remain long in the atmosphere without a continual supply. The stability of the composition of the atmosphere is also due in large part to the actions of plants. More controversially, some have argued that ocean salinity and the Earth's surface temperature have also remained so remarkably constant for so long because of the influence of living organisms. Quite recently, it has been appreciated that broadscale rainfall is also affected by vegetation. For example, about half of the annual rainfall of Amazonia is recycled by the vegetation: Rainfall is intercepted by leaves and branches, or is taken up by the roots, and is returned to the atmosphere in great quantities via evapotranspiration, mostly to fall again as rain.

Amazonia is now known to be a major sink for anthropogenic carbon dioxide (CO_2) emissions: The forests have been increasing in biomass. Another vital biota-related carbon sink is peatlands, where plants take up CO_2, but the normal decomposition processes do not work fully so carbon is locked up in the resulting peat. This process is reversed by humans when we burn peat and fossil fuels and when humans drain peatlands for agriculture.

Current and Future Research Directions

The practical and theoretical implications of the relationships between biota and climate are very far-reaching. The feedbacks—climate affecting biota and biota affecting climate—make the task of understanding the Earth system all the more challenging. Global circulation models need to include the feedbacks between biota and climate more effectively than they have done so far. One result of increased atmospheric CO_2 (an aspect of climate) is the acidification of the oceans, as the CO_2 dissolves in seawater, becoming carbonic acid. This is increasingly one of the greatest areas of current concern, in relation to oceanic ecosystems and the continuing supply of ecosystem services from the ocean. Land use planning is increasingly concerned with the conservation of nature; at all scales this needs to include good understanding of relationships between biota and climate. Thus, a fast-growing subdiscipline is "conservation biogeography," in which biogeographic theory is applied to conservation and land use planning practice.

Finally, it is worth noting that, with the increasing incorporation of spatial and temporal scale and spatial analysis in studies of relationships between biota and climate and with the increasing use of geographic information systems in such studies, what was largely the preserve of biologists and ecologists is increasingly becoming geographical.

Richard Field

See also Biodiversity; Biogeochemical Cycles; Biogeography; Biome: Tropical Rain Forest; Biota and Soils; Biota and Topography; Climate Change; Climate Types; Climatology; Ecological Regimes; Ecological Zones; Ecoregions; Ecosystems; Ecotone; Species-Area Relationship

Further Readings

Cleland, E. E., Chuine, I., Menzel, A., Mooney, H. A., & Schwartz, M. D. (2007). Shifting plant phenology in response to global change. *Trends in Ecology and Evolution, 22,* 357–365.

Field, R., Hawkins, B. A., Cornell, H. V., Currie, D. J., Diniz-Filho, J. A. F., Guégan, J. F., et al. (2009). Spatial species-richness gradients across scales: A meta-analysis. *Journal of Biogeography, 36,* 132–147.

Huston, M. A. (1994). *Biological diversity.* Cambridge, UK: Cambridge University Press.

Lomolino, M. V., Riddle, B. R., & Brown, J. H. (2006). *Biogeography* (3rd ed.). Sunderland, MA: Sinauer Associates.

Lovejoy, T. E., & Hannah, L. (Eds.). (2005). *Climate change and biodiversity.* New Haven, CT: Yale University Press.

Malhi, Y., & Phillips, O. L. (Eds.). (2005). *Tropical forests and global atmospheric change.* New York: Oxford University Press.

McCain, C. M. (2009). Vertebrate range sizes indicate that mountains may be "higher" in the tropics. *Ecology Letters, 12,* 550–560.

Whittaker, R. J., Willis, K. J., & Field, R. (2001). Scale and species richness: Towards a general, hierarchical theory of species diversity. *Journal of Biogeography, 28,* 453–470.

BIOTA AND SOILS

At the Elkhorn Slough Visitor's Center at the head of Monterey Bay, California, is a diorama that shows a shorebird standing in a bucket of mud taken from the adjacent slough. After asking visitors to estimate how many and what kinds of organisms might live in this much mud, the diorama offers an estimate of 500 billion bacteria,

500 million diatoms, 50,000 protozoa, 50,027 worms, 5,000 crustaceans, 39 clams—plus the bird. While we cannot vouch for the ballpark accuracy of the estimate, what we know about soil organisms and numbers suggests it may be low. But in pondering the diorama, several questions are raised. Does slough mud qualify as soil? If so, what is soil? How many organisms, and what kinds, live in soil? And finally, what roles do biota play in forming soil and its biomantle? This entry addresses these questions.

Does Slough Mud Qualify as Soil?

Slough mud with its myriad organisms does qualify as soil, but it qualifies more so when it exists in an undisturbed state, before being slurried as a bucket of mud. While the conventional idea that soil forms subaerially on land has a long tradition, the idea that it might also form underwater has long been espoused, since the early 19th century. In recent years, terms and expressions such as *submerged soils*, *submarine soils*, *ocean soils*, *freshwater soils*, and *subaqueous soils* follow these views and are now garnering validity for such soils. However, while the early and later observers used the word *soil*, they more likely were thinking about its uppermost part where most organisms dwell—the part more recently conceptualized and named "soil biomantle."

What Is Soil?

Views and paradigms of soil—including what soil is—have been in flux, not only for agronomists, soil scientists, and pedologists but also for archaeologists, biologists, ecologists, foresters, geographers, geologists, oceanographers, and many others. Such a community of scientists must deal with a concept of soil that covers a wide agenda—an ecologically based one that embraces both applied and pure science research, transcends multiple disciplines, and encompasses all environmental contexts. Those who study soil or who work with it in any way must come to grips with understanding the central pathways of soil formation. A definition of soil that is compatible with such an agenda is a necessary step.

In advance of these tasks—to better accomplish them and to gain some understanding of the entrenched mind-sets that might resist such a broadened approach to soil—we must briefly examine the gardener-farmer-agronomic practical traditions that underlie the conventional views of soil and that link traditional soil science dubiously to pedology, the two disciplines that claim soil as an object of study. Soil science has a broad societal practical agenda that includes crop yields, soil amendments (e.g., fertilizers, pesticides, herbicides, fungicides), plant diseases, soil survey, soil classification, soil genesis, soil quality, soil and natural resources conservation and management, extension/outreach services, and so on. The domain of pedology is narrower and less practical, with a focus on a scientific understanding of how soil forms, its properties, its depth (thickness), its ecological functionality, and other aspects—subtle though very basic distinctions.

Two Gatekeeper Soil Documents

For most soil scientists, agronomists, horticulturalists, gardeners, and farmers, the answer to the question "What is soil?" may be simply that soil is what plants grow in. This was a common definition in the 19th- and 20th-century literature on soil in the agricultural sciences, and still is. Several similar, though more specific and lengthier, definitions are given in recent editions of two gatekeeper soil compendia, the *1993 Soil Survey Manual* and the *1999 Soil Taxonomy*, both hereafter referred to as *Manual* and *Taxonomy*. These two treatises are basic and standard reference works for soil scientists and pedologists, and both have become conceptual and operational "reference bibles" for practitioners. The concepts and language within them play a mentoring role for both soil and nonsoil specialists, who consult them when dealing with any aspects of soil. Both volumes foster, in the first instance, a practical agronomic agenda, where soil survey, classification, and mapping—not necessarily ecologic or biospherical understanding—are the stated goals. This practical focus defines the traditional core domain of soil science and is what separates it from pedology, whose main focus is a more purely scientific study of soil.

The soil definitions in both documents are intimately linked to a late-19th-century model of soil formation commonly known as the "five factors," or "clorpt" model of soil formation. In the model, Soils form as a *function* of *cl*imate, *o*rganisms, *r*elief, *p*arent material, and *t*ime: or S = f(clorpt). Built-in provisos limit the depth of soil on land to 2 m (meters) for purposes of survey and classification—a strictly utilitarian (and arbitrary) contrivance—whereas depth of rooting on land for many plants is actually much deeper. For example, the roots of some phreatophytes in midlatitudes extend down to 18 m and to 85 m for plants in certain drought-prone tropical areas, such as northeastern Brazil. Also, according to the clorpt definition, for substrate to be considered soil, it must show evidence of horizonation or layering due to pedogenesis, and it must be capable of supporting rooted plants in natural environments. Substrates that lack both are considered "not-soil" (or nonsoil).

The definition includes shallow water substrates as soil, but only if plants are rooted in them, and then only to 2.5-m water depth. However, rooted plants in marine environments may grow to water depths of 10 to 40 m, occasionally even to 90 m. In sum, soil depth in *Manual* and *Taxonomy* is arbitrarily set at 2 m on land and 2.5 m underwater—all for the practical need to survey, map, and classify soils. Such arbitrariness, while understandable for its practical purposes, has created conceptual confusion and even misinformation for those in the Earth and environmental sciences, especially in explaining many-meters-thick biomantles in tropical soils and their far deeper subsoils and weathering zones.

Rooted plants can grow in many different environments in totally nonsoil contexts—as long as moisture is available. For example, plants grow in air (e.g., on building roofs, parapets, gutters, concrete walls, abandoned autos, as epiphytes on power lines, tree trunks, and branches), as free-floating plants in water (aquatic hydrophytes, some with dangling water roots), on animals (e.g., three-toed sloths and others), in pure quartz and gypsum sand, on rocks, under certain translucent quartzose-type rocks—and even *within* some porous rocks, on freshly exposed bedrock cores of bulldozed mountains, on fresh unweathered quarry spoil, on vertical walls of bedrock quarries, and so on.

While no question exists regarding the important role of plants in soil formation, a problem exists with the traditionally lopsided "clorpt" emphasis on plants to the near exclusion or marginalization of other biota. Many members of *all* life-form groups are involved in forming Earth's soil, with bioturbating animals particularly effective in texturally sorting soil into layers, as will be shown. But, in light of active space explorations and the search for life on other planets, should a definition of soil be exclusively Earth-bound?

Universal Definition of Soil

So what is soil and how should we define it insofar as soil technically is the "skin" of landforms and an integument of planets? Now, the best scientific definition is one that is most useful and comprehensible to a broad spectrum of scientists and the lay public. It especially should be simply expressed, concise, scientifically sound, easily explainable, and broad enough to cover all cases, not just a few or most cases. In sum, it should be clear, simple, logically sound, and have universality.

One definition that fits these criteria, being applicable to soil on Earth and all other generally lithic-composed planets and their satellites, is as follows: Soil is substrate at or near the surface of the Earth and similar such bodies that are altered by biological, chemical, and/or physical agents and processes.

The definition is brief and to the point. The inclusion of "or near" acknowledges that soil on at least one planet—Earth—has variable depth, sometimes great depth, as in some humid tropical areas. "Or near" also covers subaqueous soils, for such soils, even in deep water, are still relatively near Earth's planetary surface. The two words *or near* also embrace the notions of "suspended soils," and "soil body extensions." Both notions cover temporary aboveground pockets of organic-rich soil that teem with organisms in rain forest canopies and soil temporarily incorporated into nests by birds, wasps, termites, and other arboreal lifeforms in their aboveground living-reproductive activities. The words *or near* also cover bodies of wind-deposited soil on rooftops and other human structures in which plants, insects, and other soil lifeforms live, as well as temporary islands and rafts of vegetation-bound soil that floats on rivers (e.g., Paraná), estuaries (e.g., of the Rio de la Plata), lakes (many), and deltas (e.g., of the Danube and Okavango). They also cover soil washed from surface horizons downward into bedrock cavities, caves, joints, fractures, and vugs in limestone and other rocks.

The definition also focuses attention on the realization that only one celestial body—Earth—has all three processes operating more or less surficially and that because ours is such a different planet from all others—with the combination of a just-right distance to our star, water, and life, *especially life*—simple and complex surface and near-surface life must make the difference. Specifying simple and complex life—where life ranges from single-celled simple prokaryotes (whose cells lack nuclei), such as archaea and bacteria, through more complex single- and multicelled eukaryotes (whose cells have nuclei), such as protoctists, fungi, plants, and animals—leaves open the possible existence of deeper prokaryotic biospheres on Earth and other planets, as has been proposed by several distinguished scientists. What we do know is that Earth's subaerial and subaqueous near-surface soil-biomantle continuum and *its* planetary integument does literally teem with simple to complex forms of life representing all known groups of organisms, with their numbers probably more than indicated in the slough diorama.

The definition also accommodates the fact that the term *soil* is being increasingly applied to extreme substrate environments on Earth and to substrates of other planets (the current technical and popular literature has statements about "cave soil," "lunar soil," "Martian soil," etc.). The definition also reminds us that soil is not simple and implies that on Earth it is an ecological entity—thus, free from the conceptual limitations imposed by the agronomic-inspired "clorpt"-linked definition.

How Many Organisms Are There and What Kinds Live in Soil?

The question of how many organisms, and what kinds, live in Earth's soil intrigues scientists of all stripes, and probably we will never know total numbers, only ballpark estimates. However, evidence suggests that the estimate of biota living in the bucket of slough mud mentioned in the first paragraph of this entry is probably low, perhaps way low. In a paper presented at a geological symposium titled "The Solar Stew: The Search for Ingredients of Life and Biomarkers in Our Solar System," Johnson and Johnson gleaned data from various sources and offered an estimate of the kinds and numbers of invertebrate organisms that likely live in 1 g (gram) (not a bucket) of typical humid midlatitude soil and its biomantle. The data are as follows: bacteria (10^{8-9}), actinomycetes (10^{5-8}), fungi (10^{5-6}), micro-algae (10^{3-6}), protozoa (10^{3-5}), nematodes (10^{1-2}), and other invertebrates (10^{3-5}). Because the estimates were based on such a tiny amount of soil, 1 g (less than half a thimble full), many larger organisms, including more familiar ones such as ants, termites, worms, bugs, and so on, were of necessity unspecified in the "other invertebrates" category. The gram-sized data, of course, also did not include vertebrates, which as a group have a huge collective impact on soil.

Other research suggest that insects alone, apart from all other soil organisms, represent more than half of all known living organisms and potentially represent more than 90% of the differing life-forms on Earth. This estimate is particularly notable if, as has been averred by others, 95% of all insect species reside in soils at some period in their lives. Clearly a staggering number of organisms have co-evolved with Earth's soil and are essential and integrated parts of it.

What Roles Do Biota Play in Forming Soil and Its Biomantle?

Ours is indeed a very different planet, with a very different soil, most notably because it has a biomantle brimming with biota that other planets lack. The biomantle is the "epidermis" of Earth's soil and can be defined at several levels. In simplest terms, it is the surficial zone of biodynamic activity, with bioturbation dominant. Bioturbation under some dense forests is exemplified by floralturbation (e.g., root growth expansions, tree uprootings), whereas in grasslands-shrublands and most other landscapes, even including many forests, faunalturbation—animal bioturbation—is the more dominant process. More broadly, the biomantle is the bioturbated and sorted, organic-rich, often darker upper part of soil, including its surface, where most biota are conceived, born, grow, live, burrow, reproduce, die, and become

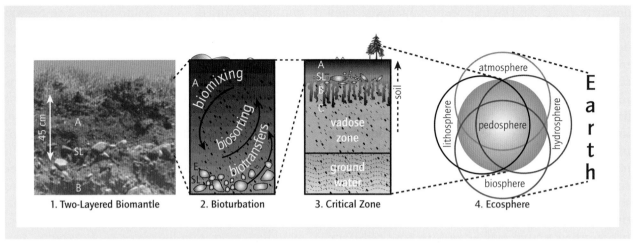

Figure 1 Photo of two-layered biomantle (45 centimeters thick) near Carmel, California, and its formation and place in soil, the critical zone, and the ecosphere

Source: Authors.

reassimilated. To this end, the soil biomantle functions variously as the global composting, mixing, sorting, particle comminuting, recycling, and nutrient-supplying epidermal layer of soil.

Animal bioturbations involve three main subprocesses: biomixing, biotransfers, and biosorting, as indicated in Figure 1. Now, depending on which organisms are doing the bioturbating and what soil particle sizes are present, either a one- or two-layered biomantle will be formed. Figure 1 illustrates how this process works.

In Figure 1, Part 1, A = topsoil, SL = stonelayer, and B = subsoil: (A + SL = biomantle) + B = soil. Part 2 shows major bioturbational subprocesses: A = topsoil, E = leached horizon, and SL = stonelayer (A + E + SL = biomantle). Part 3 shows soil in the upper vadose zone of Earth's critical zone: (A + B + SL = biomantle) + B + C = soil. Part 4 shows Earth's ecosphere, with its five great interacting subspheres.

Let us assume the dominant bioturbator in a given soil is a pocket gopher (*Geomyidae*), a small but supremely burrowing subterranean rodent endemic to North America (similar convergent burrowers in other families occupy five other continents). Let us also assume that the soil in which the animal lives contains coarse gravels. Now, because gopher burrows are about 4 to 6 centimeters in diameter, all larger particles (cobbles, human artifacts, boulders, etc.) will, as soil is bioturbationally removed from under them, settle downward under gravity. These coarse gravels concentrate as a stonelayer at the base of the biomantle—the main zone of bioturbation. A biosorted, porous, two-layered gravelly biomantle thus results, which consists of an upper layer of fluffy, permeable, biomixed soil with small gravels that overlies an even more porous and permeable basal stonelayer, similar to the one displayed in Figure 1—which actually portrays just such a soil produced mainly by gophers. Conversely, if the soil in Figure 1 *lacked* gravels, or if the particles (including gravels and artifacts) were all *smaller* than the burrow diameters of pocket gophers, then the result would have been a fluffy, porous, biomixed one-layered biomantle (i.e., one *without* a basal stonelayer). If the dominant bioturbators were invertebrates such as ants, termites, and worms, as is typical in many humid temperate and tropical soils, and if the particle sizes were all small, such as sands and silts (no gravels), then a fluffy, biomixed, one-layered biomantle (i.e., without a basal stonelayer) would also result. But, if we litter the surface of soil with gravels (i.e., pebbles, cobbles, and artifacts), then wait for some years, decades, or millennia—a two-layered biomantle would result. Notably, the more porous and permeable stonelayer, which results from such animal biosorting, commonly functions as a French drain for soil-water throughflow.

Bioturbation in fact is what produces the abundant pore spaces, porosity, and permeability that is typical of biomantles. Interconnected pore spaces, together with the basal stonelayer, function as an effective porous and permeable subsurface reservoir and aquifer for soil-water storage, transmission, and capillary movements. The biomantle thus plays a key process role in the dynamics of soil-water circulation in the Critical Zone as implied in Figure 1, and as embedded functions in the new subfield of hydropedology.

Thicknesses and Whole Soil Biomantles

In the midlatitudes and subtropics, the soil biomantle is relatively thin (±1–2 m) and includes the conventional A and E soil horizons, whereas in the tropics, the biomantle is often much thicker (±2–6 m) and may include other horizons. And as indicated, if soils are gravelly, a stonelayer invariably forms at, and defines, the biomantle base. Thus, if present, stonelayers of most midlatitude and some subtropical soils tend to lie at shallow depths (≤1–2 m), whereas in most tropical and some subtropical soils, they generally occur more deeply (≥2–6 m).

In some deserts and drylands of the world (e.g., Sahara, Atacama, Patagonia, Namib, Simpson, Taklamakan), where moisture is sparse or rare and weathering minimal or nonexistent and where both subsoil and underlying zone of weathering are absent, the biomantle constitutes the entire soil. Whole soil biomantles may be also observed, indeed are common, on moderately steep and very steep slopes of many mountainous areas where rates of erosion exceed those of weathering, but where a soil biomantle nevertheless is present.

Computer Animation

Using digital animation programs, computer models can now be constructed that demonstrate how biomixing, biotransfers, and biosortings interact to produce one- and two-layered biomantles. In one example that displays several predictive aspects of the soil biomantle model, a gravelly soil evolves under intense bioturbation into a two-layered biomantle, whose basal stonelayer accumulates cobbles, pebbles, plus human artifacts and other objects that were dropped on, or fell on, the soil surface in earlier times. Then, if the environment changes such that erosion exceeds the rates of bioturbation, the biomantle

gradually thins, until the stonelayer becomes exposed as a surface pavement.

Donald L. Johnson and Diana N. Johnson

See also Biodiversity; Biogeochemical Cycles; Biogeography; Biota and Climate; Biota and Topography; Carbon Cycle; Ecosystems; Ecotone; Geomorphology; Peat; Soils; Species-Area Relationship

Further Readings

Darwin, C. (1881). *The formation of vegetable mould through the action of worms, with observations on their habits.* London/New York: J. Murray/ Appleton. (Facsimiles produced in 1982 and 1985 by University of Chicago Press; available at http:// darwin-online.org.uk/content/frameset?itemID=F1 357&viewtype=side&pageseq=1)

Demas, G. P., & Rabenhorst, M. C. (2001). Factors of subaqueous soil formation: A system of quantitative pedology for submerged environments. *Geoderma, 102,* 189–204.

Dittmer, H. J. (1948). A comparative study of the number and length of roots produced in nineteen angiosperm species. *Botanical Gazette, 109,* 354–358.

Domier, J., & Johnson, D. L. (2007). *Biomantle and soil thickness processes: Computer animation.* Retrieved October 1, 2009, from https://netfiles. uiuc.edu/jdomier/www/temp/biomantle.html

Fey, M. V. (2009). *Soils of South Africa.* Singapore: Craft Printing International.

Gold, T. (2001). *The deep hot biosphere.* New York: Copernicus Books.

Hartog, C. den. (1970). *The sea-grasses of the world.* Amsterdam: North-Holland.

Hole, F. D. (1981). Effects of animals on soils. *Geoderma, 25,* 75–112.

Johnson, D. L. (1990). Biomantle evolution and the redistribution of earth materials and artifacts. *Soil Science, 149,* 84–102.

Johnson, D. L., Ambrose, S. H., Bassett, T. J., Bowen, M. L., Crummey, D. E., Isaacson, J. S., et al. (1997). Meanings of environmental terms. *Journal of Environmental Quality, 26,* 581–589.

Johnson, D. L., Domier, J. E. J., & Johnson, D. N. (2005). Animating the biodynamics of soil thickness using process vector analysis: A dynamic denudation approach to soil formation. *Geomorphology, 67*, 23–46.

Johnson, D. L., Maxwell, T. A., Haynes, C. V., Jr., & Johnson, D. N. (2007). Saharan biofabric: A biomarker for planetary soils? *Geological Society of America, Abstracts With Programs, 39*(6), 126.

Kevan, D. K. M. (1962). *Soil animals*. London: H. F. & G. Witherby.

Lacey, E. A., Patton, J. L., & Cameron, G. N. (Eds.). (2000). *Life underground: The biology of subterranean rodents*. Chicago: University of Chicago Press.

Lin, H. S., Bouma, J., Wilding, L., Richardson, J., Kutilek, M., & Nielsen, D. (2005). *Advances in hydropedology: Advances in agronomy* (Vol. 85). San Diego, CA: Elsevier.

Margulis, L., & Schwartz, K. V. (1999). *Five kingdoms* (3rd ed.). New York: W. H. Freeman.

Morras, H., Moretti, L., Picolo, G., & Zech, W. (2009). Genesis of subtropical soils with stony horizons in NE Argentina: Autochthony and polygenesis. *Quaternary International, 196*, 137–159.

Nardi, J. B. (2003). *The world beneath our feet: A guide to life in the soil*. New York: Oxford University Press.

Nepstad, D. C., de Carvalho, C. R., Davidson, E. A., Jipp, P. H., Lefebvre, P. A., Negreiros, G. H., et al. (1994). The role of deep roots in the hydrological and carbon cycles of Amazonian forests and pastures. *Nature, 372*, 666–669.

Robinson, W. O. (1930). Changes occurring in submerged soils. *Soil Science, 30*, 197–217.

Russell, R. S. (1977). *Plant root systems*. New York: McGraw-Hill.

Schenk, H. J., & Jackson, R. B. (2002). Rooting depths, lateral root spreads and below-ground/above-ground allometries of plants in water-limited ecosystems. *Journal of Ecology, 90*, 480–494.

Soil Survey Division Staff. (1993). *Soil survey manual* (USDA Handbook No. 18). Washington, DC: U.S. Department of Agriculture.

Soil Survey Staff. (1999). *Soil taxonomy* (NRCS Handbook No. 436). Washington, DC: U.S. Department of Agriculture.

Van Breemen, N., & Buurman, P. (2002). *Soil formation* (2nd ed.). Dordrecht, Netherlands: Kluwer.

 # BIOTA AND TOPOGRAPHY

Biota and topography have always played an integral role in geographic approaches and studies. Biota (modern Latin, from Greek *biotē*, "life")—that is, living organisms of a particular region, habitat or geological period and in extension their influence on the physical environment—is explicitly taken into account in biogeography. Biogeography is a subdiscipline of geography that studies the spatial distributions of living organisms and the underlying causes for these differentiations. More recently, biota have been considered major agents contributing to the definition of the Earth's surface topography by the geographic subdiscipline of biogeomorphology. Biogeomorphology is an emergent subdiscipline at the interface between ecology and geomorphology; it promotes the development of new interdisciplinary concepts and models better adapted to the study of topography as an outcome of complex interactions and feedbacks between biota and physical processes.

Topography (via late Latin from Greek *topographia*, from *topos*—"place"—and *graphia*—"writing") characterizes the geometry of landscapes at diverse scales (e.g., plain, mountain, hill, coast, alluvial fan, dune, river channel, river bar). The geometry of landforms and their situation in space (e.g., spatial extension, roughness, altitude, inclination, orientation) at a given place and at a given moment translates and determines their functioning, relative to their physical, ecological, and anthropogenic components.

Topography: Definition by Substrate Cohesion and Physicochemical Agents

Earth surface topography is characterized by a substantial spatial heterogeneity at diverse spatiotemporal scales. Spatial and temporal distributions of topographic variations are not random. Landforms are dynamic features that result from self-organization processes of matter and energy, commonly revealing geometric and dynamic constants—or attractor domains—representing optimum physical solutions that minimize energy dissipation according to endogenous or exogenetic morphogenetic forces (e.g., tectonics, gravity, water flow, wind). For this reason, specific landforms are generally recognizable and can be distinguished from their surrounding environment—that is, from other landforms. It is possible to classify landforms based on their topography and complementary geomorphic characteristics such as sediment physicochemical properties. In many cases, the identification of the family of processes that are at work or, in the case of inherited landforms, were at the origin of the landforms and their topography, is also possible. Erosional and depositional landforms on the Earth surface can be perceived as a characteristic signature of a certain combination of interactions between the material (e.g., sediments, rocks, ice) and the physical constraints applied on it (e.g., tectonics, gravity, water flow, wind, frost, chemical alteration). However, topographic signatures vary greatly because of the many possibilities of combinations between the fundamental and basic morphostructures and processes. The high variability of topography on the Earth's surface is also related to the frequency, the magnitude, and the timing of the physical constraints (i.e., the disturbance regime and its historical pathways) that is applied on the material.

The three main components that together define the variety of topographic features need to be distinguished:

1. The material

2. The endogenous geothermal force that is at the origin of tectonics and the uplift of landscapes

3. The exogenetic weathering agents that induce sediment erosion, transportation, and deposition on Earth's surface

The first component—the material—defines landform cohesiveness, that is, the ability of the substrate to resist deformation or desegregation when a mechanical or a chemical constraint is applied. The second component—tectonics—produces the most important variations in topography on the Earth through uplifting and deforming rocks and landforms, generating heterogeneity in topography at the global scale. The third component—sediment erosion, transportation, and deposition—redistributes matter in space and produces characteristic topographic signatures through the dissipation of kinetic energy associated with matter flows. For example, (a) wind produces dunes with a structural organization depending on a delicate balance between wind regime and sediment grain size; (b) water flow produces landforms such as hydrographic networks, rivers, or intertidal channels and alluvial fans (see alluvial bar photo); (c) gravity is at the origin of characteristic topographic signatures such as hill slopes, screes (accumulations of primarily angular clasts that lie at an angle of around 36° beneath exposed free rock faces or cliffs), solifluction lobes (tongue-shaped lobes that are the result of slow gravitational downslope movement of water saturated materials); and (d) frost and the freezing of water produce geometrical forms at the surface such as polygons delimited by stones.

Topography: Significant Control Through Biota

Geomorphologists recently identified biota as an additional major agent shaping topography at various spatiotemporal scales on the Earth surface. The influence of life is quite apparent over short timescales, for example, pedogenesis. Substrate erosion and stabilization and river and dune dynamics are controlled by living organisms that directly mediate chemical reactions, disintegrate rocks, modify soil structure, texture and cohesion, and indirectly control sediment dynamics by modifying flows of matter and kinetic energy. The effects of living organisms over geologic time affect the climate, which control exogenetic (pertaining to processes occurring at or near the surface of the Earth and to the landforms produced by such processes) Earth surface processes and topographic changes.

Since the beginning of the 1980s, biogeomorphology has provided an increasing number of studies demonstrating the importance of living organisms in driving or modulating Earth surface processes and landform dynamics. This relatively new field of investigation, which adds the biological dimension (i.e., the biological dynamics of single organisms, populations, or communities) to the exclusive consideration of chemical and physical geomorphic factors and agents, clearly led to conceptual improvements in geography through the explicit consideration of the structural and

Effect of vegetation on topography on an alluvial bar of the River Allier, France

Source: Authors.

functional linkages between biology, ecology, and geomorphology. Very recent conceptual developments in biogeomorphology pointed out that the main difference between biota—as a morphogenetic agent—and weathering and tectonics is related to life as an evolving (i.e., in the biological sense) force which may in many cases produce very characteristic adjusting topographic features.

The regions on the Earth where interactions between biota and topography may occur are restricted to the bioclimatic contexts that favor the development of life. Deserts and polar regions remain dominantly shaped by physical and chemical processes. In tropical, subtropical, and temperate climatic zones, biota may commonly play a major role in modulating micro- and macrotopography while the climatic conditions (temperature, humidity, solar radiation) for the development of biodiversity and biomass are optimized. However, a large range of different configurations exist,

encompassing particular bioclimatic contexts (e.g., within semiarid, Mediterranean, or alpine climatic regions) and very particular systems such as terrestrial-aquatic interfaces (e.g., fluvial and coastal systems), which concentrate flows of matter and energy and stimulate primary production through favorable environmental conditions (e.g., the availability of water and nutrients for plants). Influences of biota on topography can be significant from very local to regional scales and may vary significantly. Direct influences are observed when organisms dig or burrow galleries in soils or build mounds or other structures at topographic surfaces. Indirect influences occur through the modulation by living organisms of sediment erosion, transportation, and deposition produced by water flow or by wind. Biotic influences are autogenic when landforms and an associated topography are modified or created in situ through the organisms themselves as, for example, in the case of coral reefs.

Control of Topography Through Biota Since the Precambrian Era

Microorganisms such as cyanobacteria and microscopic algae were the first living organisms to significantly modulate the topography of Earth as early as 3,500 mya (million years ago), during the Precambrian era. Prokaryotic and eukaryotic organisms present in the oceans started to build large-scale reefs on continental margins, and they drastically modified the atmosphere's composition by producing in the long term huge quantities of oxygen. The mineral material exposed to geomorphic agents has adjusted its properties with the emergence of life. Two thirds of the known types of minerals on Earth are directly or indirectly linked to biological activity. This is a consequence of the oxygen-rich atmosphere of the Earth, which is the product of photosynthesis by microscopic algae in the oceans. The evolution of organisms with shells and mineralized skeletons also generated thick-layered deposits of minerals such as calcite, which would be very rare on a lifeless planet. The atmospheric change contributed to modify climate, encompassing carbon dioxide (CO_2) concentrations, temperature, and precipitation, which represent fundamental morphogenetic agents shaping the Earth surface topography. The first organisms that started to dig and burrow the substrate in the ocean about 542 mya, at the Proterozoic-Phanerozoic transition, also modified very significantly the microtopography via pellet production, track formation, and a variety of types of construction, for example, mounds and pits. These changes in ocean surface and subsurface microtopography had very important consequences on hydrodynamics—sedimentary and biochemical cycles. It oriented the evolution of life in the oceans and led to an important increase of biodiversity in the Cambrian era while habitat morphology and physicochemical properties started to diversify.

Plant Control of Topography on Continents

A fundamental step of the history of the influence of biota on topography on Earth is associated with the colonization of land by life—and especially plants—from the beginning of the Silurian era (470–430 mya). The first organisms, such as bacteria, lichens, and then plants, which colonized terrestrial-aquatic interfaces, started to modulate topography by altering rocks and by modifying fine sediment dynamics. In particular, plants acted as a fundamental morphogenetic agent through the modification of substrate cohesion by roots and surface roughness with their canopies at terrestrial-aquatic interfaces between oceans and continents and within fluvial systems. Vegetation evolved from the Devonian to the Carboniferous (380–290 mya) and from small plants to complex ecosystems, dominated by trees with heights of 30 m. At the global scale, trees in particular contributed to directly modify the surface topography by their own structures—that is, trunk and canopies—which represent physical habitats for many other kinds of organisms such as insects, birds, and mammalians. From the Devonian on, plants spread efficiently to the inside of continents; they colonized various bioclimatic and geomorphic systems and also contributed to modify topography by increasing rock alteration, soil formation, and sediment retention and trapping, at a global scale. Plants are considered as the main agent of direct creation and indirect modulation of topography on the Earth surface in various systems, encompassing terrestrial-aquatic interfaces, hill slopes, and sand dune systems. While vegetation developed strategies to colonize continents, it offered, through the biologic function of primary production and through facilitation processes associated with physical habitat modification, the possibility for microorganisms, insects, and animals to colonize lands and to spread into continents. Some of these organisms also contributed in varying proportions to modify and sometimes to create topography.

Control of Topography by Microorganisms, Insects, and Animals

Microorganisms act as an important morphogenetic agent on Earth's surface by altering or protecting rocks from erosion. Lichens may provide good protection against weathering agents; or, on the contrary, they may contribute to the acceleration of rock erosion. Insects also contribute to the modification of topography in different ecosystems and through diverse means. For example, termites build mounds, which

A typical North American beaver dam in Alaska. Beaver dams form ponds, wetlands, and meadows, which increase biodiversity and improve overall environmental quality.

Source: John J. Mosesso/NBII.gov.

Caddisfly larvae also contribute to significantly modify topography in streams by increasing the stability of fine channel bed sediment in many rivers, worldwide. Ants modify soil microtopography by digging galleries; these dynamics modify substrate properties and contribute to control water infiltration and erosion processes on large surfaces and in the long term.

Small and large animals also represent important morphogenetic agents on the Earth surface. For example, as already observed by Charles Darwin, earthworms deeply modify topography over very large scales through their activity of bioturbation contributing to soil formation. Earthworm activity contributes to the elevation of horizontal topographic surfaces on historical and geological timescales (10^1 to 10^4 years). Bioturbation also stimulates downslope creep and increases the susceptibility of soil to erosion. In the long term, this biological activity significantly contributes to smooth topography—to level hills and to fill up entire valleys. Crayfishes may also significantly modify topography in river channels by affecting gravel and/or sand structures and sediment transportation in stream channels. The net effects on below-ground topography of burrowing animals such as shrimps and crabs, at terrestrial-aquatic interfaces, and prairie dogs, tuco-tucos, and pocket gophers, in terrestrial contexts, are spatially extensive. In systems where bioturbation is frequent, animal transportation is a key process that determines the vertical structure and texture in the soil profile, the surface microroughness, and the resulting hydrodynamics associated to water flow and to wind. North American beavers induce very significant topographical changes in rivers. Beaver dams trap large quantities of sediment and maintain a complex mosaic of landforms. Entire North American valley floors have been raised many meters by beaver pond sediments. Even large animals such as alligators and the African elephant may induce very important topographic modifications in river channels by removing huge quantities of sediment. Large animals such as the antelope can also affect significantly slope dynamics in arid systems by trampling along preferential pathways across scarp slopes; this results in scarp recess and initiates cutback development.

directly modify topography and act as roughness elements in plains and fluvial corridors. Such roughness elements alter local infiltration rates and represent points for clastic and chemical sedimentation and coalescence. Termite mounds can initiate, in certain environmental contexts, for example, in the Okavango delta, the construction of islands by cumulative sediment and nutrient accretive processes. Pioneer shrubs and trees, which colonize these mounds, then contribute to the increase of transpiration. This causes calcite and silica to precipitate, eventually inducing vertical and lateral growth and island expansion.

Manhattan from the Empire State Building in New York City in 2008

Source: Steven K. Martin.

Control of Topography Through Human Activities at a Global Scale

Humans have become a keystone engineering species modifying both directly and indirectly topography from the microscale to the global scale. Human societies have directly modified Earth surface geomorphic patterns through land use practices since the Neolithic. The important modification of vegetation cover, in particular in temperate, subtropical, and tropical regions, led to drastic topographic changes because of the increase or the attenuation of hill slope erosion processes. This process is illustrated, for example, in the Mediterranean basin in which several ancient ports (e.g., Ephesus) are now disconnected from the sea. This phenomenon is explained by intensive erosion crises linked to the deforestation of entire Mediterranean catchments, having caused massive sediment deposition within and several kilometers downstream of the ancient ports. Recent anthropogenic modifications of stream flow regimes and alterations of sediment transfer from production zones to oceans also induced important topographic changes within fluvial corridors and on coasts, generally associated with erosional landforms. Human infrastructures, encompassing towns and megalopolises, also directly and very significantly modified Earth's surface topography at regional scales. Concentration of buildings represents the most recent significant direct modification of surface topography on Earth's surface, which has enormous regional impacts on aerial hydrodynamics and microclimate (see New York City photo).

One of the most significant topographic changes at the global scale, indirectly associated with biota, may occur in the near future through human activity that contributes to global warming and climate change. This change could be caused by modification of the frequency, intensity, and distribution of morphogenetic agents such as water flow and wind, induced by regional changes of precipitation and temperature regimes. Topographic changes will also be associated with the modifications of the structure and functioning of ecosystems caused by global change, for example, the loss of vegetation in some regions may have dramatic consequences on erosion and subsequent depositional processes.

Dov Corenblit and Johannes Steiger

See also Biodiversity; Biogeochemical Cycles; Biogeography; Biota and Climate; Biota and Soils; Biota Migration and Dispersal; Carbon Cycle; Geomorphology; Human Dimensions of Global Environmental Change; Landforms; Species-Area Relationship

Further Readings

Butler, D. R. (1995). *Zoogeomorphology: Animals as geomorphic agents*. Cambridge, UK: Cambridge University Press.

Carter, N. E. A., & Viles, H. A. (2005). Bioprotection explored: The story of a little known earth surface process. *Geomorphology, 67,* 273–281.

Corenblit, D., Gurnell, A. M., Steiger, J., & Tabacchi, E. (2008). Reciprocal adjustments between landforms and living organisms: Extended geomorphic evolutionary insights. *Catena, 73,* 261–273.

Darwin, C. (1881). *The formation of vegetated mould through the action of worms with observation of their habitats*. London: Murray.

Dietrich, W. E., & Perron, J. T. (2006). The search for a topographic signature of life. *Nature, 439,* 411–418.

Goudie, A. S. (1993). Human influence in geomorphology. *Geomorphology, 7,* 37–59.

Goudie, A. S. (Ed.). (1994). *The encyclopedic dictionary of physical geography*. Oxford, UK: Blackwell.

Hazen, R. M., Papineau, D., Bleeker, W., Downs, R. T., Ferry, J. M., McCoy, T. J., et al. (2008). Mineral evolution. *American Mineralogist, 93,* 1693–1720.

Jones, C. G., Lawton, J. H., & Shachak, M. (1994). Organisms as ecosystem engineers. *Oikos, 69,* 373–386.

Meysman, F. J. R., Middelburg, J. J., & Heip, C. H. R. (2006). Bioturbation: A fresh look at Darwin's last idea. *Trends in Ecology and Evolution, 21,* 688–695.

Naylor, L. A., Viles, H. A., & Carter, N. E. A. (2002). Biogeomorphology revisited: Looking towards the future. *Geomorphology, 47,* 3–14.

Phillips, J. D., & Lorz, C. (2008). Origins and implications of soil layering. *Earth-Science Reviews, 89,* 144–155.

Taylor, T. N., & Taylor, E. L. (1993). *The biology and evolution of fossil plants*. Englewood Cliffs, NJ: Prentice Hall.

Viles, H. A. (Ed.). (1988). *Biogeomorphology*. Oxford, UK: Blackwell.

Viles, H. A., Naylor, L. A., Carter, N. E. A., & Chaput, D. (2008). Biogeomorphological disturbance regimes: Progress in linking ecological and geomorphological systems. *Earth Surface Processes and Landforms, 33,* 1419–1435.

 BIOTA MIGRATION
AND DISPERSAL

In general, migration and dispersal both refer to the movement of plants or animals over distances beyond their regular home range. Migration is typically seasonal, long distance, and repeated annually, such as the movement of birds between breeding and wintering grounds. Dispersal is more permanent, varies in distance, and is only done once or a limited number of times in the life of an organism. The timing, distance, and effects of movement are intrinsically linked to the habitat features being moved from, through, and to.

Animal migration takes an individual from one physical and ecological environment to another, typically, in search of resources or mates or in response to an environmental shift. The most common examples include birds that breed in northern latitudes during the Northern Hemisphere summer and then travel south toward or into the tropics before the weather turns cold (e.g., Purple Martin, Mississippi Kite). Ungulates and other large mammals in the African savannas travel in enormous herds tracking water during the dry season, creating amazing spectacles of mass migration. Rather than seasonal, migration could take a lifetime to complete its cycle. Pacific salmon, for example, are born in freshwater streams, travel to sea for most of their lives, and then to their natal stream to spawn and then die. Migration can take two lifetimes. In some insect species, a parental generation migrates, reproduces, and dies, and the offspring return.

The triggers to initiate this movement behavior are often linked to either seasonal cues such as day length or resource availability such as access to water or food. Several species of migratory birds (e.g., the European quail and the white-crowned sparrow) have been shown to respond to circadian rhythms with changes of flight activity, suggesting day length as a cue for their seasonal migration. Insectivorous birds respond instead to a reduction in the abundance of insects as a cue for travel. If the target of the migration is finding a mate or mating, as in the case of the salmon, then the cue may be the reaching of sexual or hormonal maturity. In the case of the following resources, migration may be less structured and more nomadic, without return to the original area specifically. This type of movement is seen in the large mammals of the African savannas (e.g., the wildebeests and the zebras) as they follow the availability of water and forage.

As opposed to migration, dispersal is typically permanent and unidirectional. It is a phenomenon of both plants and animals and is the main mechanism for gene flow among populations. The dispersal process involves three somewhat distinct stages: (1) emigration, (2) transfer, and (3) settlement. Emigration, or departure from the predispersal area or group, often but not always, occurs at or near sexual maturity. Movement or transfer between sites or social groups, typically, involves travel through unsuitable or unavailable habitat that can include increased mortality risk. Finally, settlement or immigration in a secondary area or group may be as simple as haphazardly landing somewhere and surviving if it is suitable or sequentially investigating several areas and then choosing the most suitable. The dispersing individual can be entirely passive in this process, as

seeds are expelled from a parent plant and then carried on the wind. Or the individual could be entirely active, as are female birds that leave the parental nest and investigate the territories of several males before choosing one with whom to nest and mate.

Dispersal is further defined by the breeding experience of the dispersing individual. Natal dispersal is the movement of a propagule from the place of birth to the site of its first breeding, often engaged at or near the age of sexual maturity. Breeding dispersal is the movement of an individual between subsequent breeding sites and is somewhat less common. Some researchers further distinguish successful dispersal as resulting in successful breeding at the secondary site. In terms of monitoring gene flow and the effects of it on the fitness of populations, movement without breeding is inconsequential.

Primary hypotheses for the evolution of dispersal include avoidance of inbreeding depression, avoidance of kin competition, and response to habitat variability. In most vertebrate species, only one sex disperses while the other is philopatric (i.e., it remains in the natal area). This sexual dimorphism would effectively reduce the likelihood of inbreeding. The mating system of the species would then dictate which sex disperses. For example, in mammals, which are typically polygynous with males competing for females and females investing more parental care than males, males are expected to disperse and females are expected to be philopatric. However, whether this reduction in inbreeding was the selective force for the evolution of dispersal or a fortunate side benefit is debated. An alternate selective advantage of dispersal is the reduction in competition with kin for resources, including mates.

Habitat variability could also lead to selection for dispersal. In temporally or spatially variable habitats, the situation in which an individual is born is likely to either change (causing temporal variation) or be better elsewhere (causing spatial variation). Thus, the probability of finding an improved situation postdispersal would add selective advantage.

Both migratory and dispersal behavior appear to be under strong genetic control with plasticity that responds to environmental variation. The selective advantage of the behavior is suggested by its risk. In almost all species in which it can be quantified, an increased risk of mortality is experienced during transition between sites. For this behavior to have persisted, the benefit must outweigh this cost.

Elizabeth R. Congdon

See also Adaptive Radiation; Biodiversity; Biogeography; Biota and Climate; Biota and Soils; Biota and Topography; Climate Change; Extinctions; Human-Induced Invasion of Species; Species-Area Relationship

Further Readings

Clobert, J., Danchin, E., Dhondt, A. A., & Nichols, J. D. (Eds.). (2001). *Dispersal.* Oxford, UK: Oxford University Press.
Cullock, J. M., & Kenward, R. E. (2002). *Dispersal ecology.* Malden, MA: Blackwell.
Dingle, H. (1996). *Migration: The biology of life on the move.* Oxford, UK: Oxford University Press.

 # BIOTECHNOLOGY AND ECOLOGICAL RISK

Since the late 1970s, advances in biotechnology have allowed for the unprecedented crossing of genes between species, and over the past decade many of these genetically modified organisms (GMOs) have been released into ecological systems. The environmental release of GMOs has taken place at landscape and test plot scales, predominately involving crop and tree varieties, although livestock and fish species have also been modified and are on the verge of commercialization in some countries. While GMOs mainly offer production benefits, it is largely accepted that biotechnology is still in its infancy and may cause unanticipated ecological risks, which requires further study that may be aided by the holistic discipline of geography.

The terminology regarding biotechnology and risk is controversial and confusing. Industry often claims that biotechnology has existed for millennia and suggests that selective plant breeding is one example of this. More colloquially, the media and public usually equate biotechnology with recombinant DNA and transgenesis, which has created a breakdown in language among stakeholders regarding this technology. Academics have tried to introduce new terminology to differentiate between "traditional" (e.g., selective breeding) and "modern" (e.g., GMO crops) biotechnology, although its uptake has been met with mixed success. Similarly, the term *risk* is also contested and means different things to different stakeholders. Natural scientists define risk quantitatively as the "probability of harm" and use "risk assessment" to predict and avert potential problems that may be associated with GMOs. Social scientists define risk more broadly and use "risk analysis" to also include people's risk perceptions in the evaluation of biotechnology. Indeed, that the language over biotechnology and risk is in dispute speaks to the larger controversy regarding the promise and peril of the technology itself.

Applications of Modern Biotechnology

Since the mid 1990s, "first generation" GMO crops such as corn, rice, soybean, and canola have been commercially available, and many farmers across the globe have adopted these varieties, which offer operational benefits predominately due to herbicide, insect, and pathogen resistance. "Second generation" crops that are supposed to offer end-user benefits are now being created and deployed, such as plants that contain nutritionally enhanced and pharmaceutical traits. The forestry industry has also developed herbicide, insect, and pathogen resistant trees, as well as varieties with low lignin content that makes processing easier. Various GMO animals have been developed, tested, and are nearing commercialization, such as "super fish" that have been modified for unregulated growth (which speeds rearing of market-size fish). While applications of modern biotechnology can offer benefits, many people argue that GMOs do not have biological antecedents and therefore may pose unique ecological risks.

Ecological Risk

The primary ecological concerns regarding the release and use of GMOs is that they might spread, become invasive, and harm nontarget organisms and biodiversity overall. Importantly, the degree to which a GMO may be considered an ecological risk depends on several factors, including the species, the introduced trait, the scale of release, and whether the organism is present in a natural or managed ecosystem. Although prerelease studies have identified possible ecological risks, society's understanding of this issue has been informed most by North America's commercialization of GMOs, particularly crops, which many observers consider a "living experiment."

Since commercialization, GMO crops in North America have spread in the environment through cross-pollination and direct seed movement, which has "contaminated" the non-GMO cultivars. Canadian farmers can no longer grow non-GMO canola because herbicide-tolerant (HT) varieties have outcrossed with conventional and organic seed supply. Similarly, GMO corn has escaped into non-GMO varieties and maize landraces in the United States and Mexico, respectively. That industry and regulators have been unable to effectively manage the spread of this genetic material indicates that gene escape happens, and therefore the "coexistence" between GMO and non-GMO crops will be difficult, if not impossible.

Another concern is that GMO crops might cross-pollinate with genetically related weed species and this may increase the fitness advantage and "invasiveness" of these plants. These so-called superweeds occur in very low frequencies and have yet to cause serious problems, although proper stewardship will be required to mitigate risk, especially in areas with genetic compatibility between crop and weed species. However, in agricultural fields, GMO crops that are HT have become problematic weeds, by shattering their seeds and growing in "volunteer" fashion in subsequent years. Studies show that HT volunteers are more costly and difficult to control, can become resistant to multiple herbicides (i.e., "trait-stacked") through cross-pollination with different HT crops, and may undermine soil-conserving farming practices that use herbicide instead of tillage to control weeds.

Despite the commercial use of GMO crops, few studies have investigated the long-term ecological risks associated with this technology. A major concern is that GMO crops will harm biodiversity, particularly nontarget organisms, and recent studies seem to confirm this. In 2003, U.K. scientists released the findings of the "Farm-Scale Evaluations," which were the first and only large-scale studies investigating HT crops and their overall impact on agricultural biodiversity. The studies found that HT canola and beet were too effective and by significantly reducing weeds also reduced nontarget weed-dependent arthropods. Scholars interpreting the findings believe that arthropod-dependent birds will also likely be reduced by the efficacy of these HT crops. Indeed, studies like this suggest that more research is required.

Geography and Holism

It must be emphasized that the ecological risks associated with GMOs cannot be generalized. This is why holistic research on this topic is important, and geography is an ideal discipline to achieve this. Both physical and human geography research on GMOs—from landscape mapping to stakeholder surveys—will continue to inform society regarding biotechnology and the ecological risks involved.

Ian J. Mauro

See also Agricultural Biotechnology; Agrofoods; Biotechnology Industry; Environmental Impacts of Agriculture; European Green Movements; Genetically Modified Organisms (GMOs); Intergovernmental Environmental Organizations and Initiatives

Further Readings

Conner, A. J., Glare, T. R., & Nap, J. P. (2003). The release of genetically modified crops into the environment: Overview of ecological risk assessment. *Plant Journal, 33,* 19–46.

Mauro, I. J., & McLachlan, S. M. (2008). Farmer knowledge and risk analysis: Postrelease evaluation of herbicide-tolerant canola in Western Canada. *Risk Analysis, 28,* 463–476.

BIOTECHNOLOGY INDUSTRY

Biotechnology may be defined as the application of molecular and cellular processes to solve social, scientific, and environmental problems, develop new products and services, or modify living organisms to carry desired traits. Arising after the discovery in 1973 of recombinant DNA, biotechnology has been a rapidly growing industry worldwide, with extensive linkages to agriculture, health care, energy, and environmental sciences. In 2005, the U.S. biotech industry (excluding medical equipment firms) consisted of roughly 1,500 firms that employed 400,000 people and generated $64 billion in output. There is a wide range in the size of firms in this industry, including single proprietorships and firms of more than 500 employees; the mean national annual salary in the industry is $62,500, which is well above the national average.

Pharmaceutical firms, which tend to be much larger than biotechnology companies, form the major market for biotechnology products. Large pharmaceutical firms are reliant on biotech clusters for innovative drug solutions, and human therapeutics thus account for the vast bulk of the biotechnology industry's revenues. Other applications are found in agriculture, manufacturing, and veterinary medicine. Many biotech firms enter into alliances with drug manufacturers, who may provide venture capital in return for marketing rights after the product is commercialized.

Venture capital is critical to making basic research in biotechnology commercially viable. Most small biotech firms lose money, given the high costs and the enormous amounts of research necessary to generate their output and the long lag between research and development (R&D) and commercial deployment (generally in the order of 12–15 years of preclinical development). Only 1 in 1,000 patented biotechnology innovations leads to a successful commercial product, and it may take 15 years. Venture capitalists may invest in many biotech firms, and one biotech firm may receive funding from several venture capitalists. Above all, venture capitalists look for an experienced management team when deciding in which companies they are willing to invest. Venture capitalists often provide advice and professional contacts and serve on the boards of directors of young biotech firms. As a biotech firm survives and prospers, its relations with investors often become spatially attenuated—that is, venture capitalists gradually withdraw from day-to-day direct management.

There has been extensive state involvement in establishing biotechnology complexes since the industry began. Because of its rapid growth as well as its demonstrated and potential technological advances, many national science policies target it as a national growth sector. The survival and success of biotechnology firms is heavily affected by federal research funds, primarily through institutions such as the National Science Foundation and the National Institutes of Health. Other federal offices such as the Small Business Technology Transfer, Small Business Innovation Research, Environmental Protection Agency, and the Food and Drug Administration are also significant. Federal policies regarding patents and intellectual property rights, subsidies for medical research, and the national health care programs are all important. Roughly 83% of U.S. state and local economic development agencies have targeted the industry for industrial development. State-level determinants are also critical, including regulatory policies, educational systems, taxation, and subsidies.

Biotechnology firms tend to cluster in distinct districts, and place-based characteristics are essential for the industry's success in innovation. Europe, for example, hosts the BioValley Network situated among France, Germany, and Switzerland. In Britain, Cambridge has assumed this role. Similarly, Denmark and Sweden formed Medicon Valley. Geographically, the U.S. biotechnology industry is currently dominated by a small handful of cities, particularly Boston, San Diego, Los Angeles, San Francisco, New York, Philadelphia, Seattle, Raleigh-Durham, and Washington, D.C., which together account for three fourths of the nation's biotech firms and employment. All these cities have excellent universities with medical schools and state-of-the-art infrastructures (particularly fiber optics and airports) and offer an array of social and recreational environments (see Figure 1).

Biotechnology firms tend to agglomerate for several reasons. In an industry so heavily research intensive, the knowledge base is complex and rapidly expanding, expertise is dispersed, and innovation is to be found in networks of learning rather than individual firms. This observation is at

Figure 1 Geography of U.S. biotechnology centers. Like many knowledge-intensive sectors, biotechnology firms tend to concentrate in large metropolitan areas.

Source: Author.

odds with the popular misconception that such firms are the products of heroic individual entrepreneurs. In a highly competitive environment in which the key to success is the rate of new product formation, and in which patent protections lead to a "winner take all" scenario, the success of biotechnology firms is closely related to their strategic alliances with universities and pharmaceutical firms. Although many biotechnology firms engage in long-distance partnering, these tend to be complementary, not substitutes, for colocation in clusters where tacit knowledge is produced and circulated face-to-face, both on and off the job.

Because pools of specialized skills and a scientifically talented workforce are essential to the long process of research and development in biotechnology, an essential element defining the locational needs of biotech firms is the location of research universities and institutions and the associated supply of research scientists. Most founders of biotech companies are research scientists with

university positions. Regional human capital may be measured by examining the prevalence of bachelor's degrees, which indicate a region's educational attainment, and the location and size of regional universities that grant PhD degrees in biology and related fields. However, to a large extent, local labor shortages can be mitigated through in-migration. Because knowledge is generated and shared most efficiently within close loops of contact, the creation of localized pools of technical knowledge is highly dependent on the detailed divisions of labor and constant interactions of colleagues in different and related firms. Successful biotechnology firms often revolve around the presence of highly accomplished academic or scientific "stars," with the requisite technical and scientific skills but also the vision and personality to market it. Often such individuals begin in academia and move into the private sector.

Barney Warf

See also Agglomeration Economies; Agricultural Biotechnology; Biotechnology and Ecological Risk; Clusters; Economic Geography; Innovation, Geography of; Knowledge Spillovers; Research and Development, Geographies of

Further Readings

Cooke, P. (2004). The accelerating evolution of biotechnology clusters. *European Planning Studies, 12,* 915–920.

Delaney, E. (1993). Technology search and firm bounds in biotechnology: New firms as agents of change. *Growth and Change, 24,* 206–228.

Zucker, L., Darby, M., & Brewer, M. (1998). Intellectual human capital and the birth of U.S. biotechnology enterprises. *American Economic Review, 88,* 290–306.

Image of Biruni from a 1973 stamp issued by Afghanistan

Source: Public domain.

 # BIRUNI (973–1048)

Abu 'l-Rayhan Muhammad ibn Ahmad al-Biruni, one of the greatest Muslim astronomers, geographers, and scientists, was an advocate of empirical science and a practitioner of the experimental and comparative methods.

One of the most original and prolific medieval scientific writers in the natural sciences, al-Biruni was born in the city of Kat (now Biruni, in Uzbekistan), the ancient capital of Khorezm, south of the Aral Sea. Political history played an important role in his life and career. Biruni was a native speaker of the Khorezmian language (related to Persian), but the body of his work is in Arabic, including poetry; he also knew some Greek, Sogdian, Hebrew, and Syriac and learned Hindi and Sanskrit while residing in India. His early education was supervised by the learned emir of Khorezm, Abu Nasr Mansur ibn 'Iraq al-Ja'di, an astronomer and mathematician. The young Biruni constructed astronomical instruments and globes, conducted astronomical observations to determine the ecliptic angles and coordinates of various towns in Khorezm, and described a solar eclipse. In the early 990s, war forced Biruni to flee to Rayy (near Tehran); in about 998, he moved to Gurgan (Jurjan), on the Caspian Sea, where his mentor was the Christian astronomer and physician Abu Sahl 'Isa al-Masihi. In 997, he recorded in Kat an observation of a lunar eclipse for purposes of determination of the longitude. From Gurgan, Biruni engaged in correspondence with Abu 'Ali Ibn Sina (Avicenna). During his stay in Gurgan, Biruni authored *Kitab al-athar al-baqiya* (*The Chronology of Ancient Nations*), the astronomical treatise *Spherics*, two books about the religious movement of the Qarmatians, and several works (now lost) on natural history, astronomy, and astrology. Altogether, Biruni is known to have authored, or have ascribed to him, about 150 works.

In 1004, Biruni returned to Khorezm, dividing his time between politics and science cultivated within the circle of scholars of "Al-Ma'mun's Academy" at Urgench; this group included Avicenna and was named after the Khorezm-shah Abu 'Ali Ma'mun (992–999). Astronomical observations conducted during this period formed the basis of Biruni's theory of the solar motion; he also wrote a history of Khorezm. In the 1017 attack on Khorezm by the army of Sultan Mahmud of Ghazna, Biruni and his teacher Ibn 'Iraq were taken captive and taken to Ghazna, where he remained until his death in 1048. Biruni's appointment in Ghazna was as court astrologer; in Ghazna, Biruni composed some of his most significant works. In 1025, he completed *The Geodesy*, an astronomical-geodetic treatise *Kitab tahdid nihayat al-amakin li-tashih masafat al-masakin*, or *The Determination of the Coordinates of Positions for Corrections of Distances Between Cities*. In 1029, he finished *Kitab al-tafhim li-awa'il sina'at al-tanjim*, or *The Book of Instruction in the Elements of the Art of Astrology*, which contains 530 questions and answers related to astronomy, geography, chronology, and the theory of the astrolabe and goes considerably beyond the information required for astrological predictions.

From Ghazna, Biruni traveled to India, and in about 1030 he completed a monumental treatise, *Kitab Tahrir ma li'l-Hind min maqula maqbula fi'l-'aql aw mardhula*, sometimes called *Ta'rikh al-Hind* (*Alberuni's India*) or simply *India*. It contains vast and detailed historical, ethnographical, scientific, and cultural data and is distinguished by a remarkable lack of religious bias. In India, Biruni was able to measure degrees of latitude as well as determine the coordinates of many of the cities he visited. He learned about Indian astronomy and mathematics, translated several Sanskrit works into Arabic, and, in turn, translated into Sanskrit Euclid's *Elements*, Ptolemy's *Almagest*, and his own treatise on the astrolabe. Between 1035 and 1037, Biruni completed his major work, *al-Qanun al-Mas'udi fi hay'at wa-l-nujum*, or *Encyclopedia of Astronomical Sciences*, dedicated to his patron Mas'ud ibn Mahmud (1030–1040). This encyclopedic work contains the traditional (but expanded) astronomical tables compiled by his predecessors and a thorough discussion of theoretical questions of chronology, trigonometry, astronomy, and geography, with experimental arguments and mathematical proofs. Late in life, Biruni completed his *Kitab al-jamahir fi ma'rifat al-jawahir* (*The Book of Most Comprehensive Knowledge on Precious Stones*) and *Kitab al-saydana fi-l-tibb* (*Book of Pharmacology in Medical Science*).

Biruni is often called the most original and critical thinker among the medieval Muslim scholars, but sometimes his inclination to deal with empirical evidence as opposed to theoretical speculation is perceived as a limitation. In the field of geography, Biruni demonstrates both his empirical knowledge and his expertise in astronomy and mathematics. Mathematical geography is extensively treated in *Geodesy* and in Books IV to VI of the *Encyclopedia of Astronomical Sciences*. Among the questions discussed are the rotation of the Earth around its axis, determination of the latitude and longitude, the azimuth (especially of the *qibla*), the radius of the Earth, and the method and mistakes in measuring the degree of the meridian. These books also include material on chorography and physical geography, which are also dealt with in some of the questions and answers of the *Astrology* and in parts of *India*. In particular, he discussed the different methods behind different systems of dividing the inhabited world: nine parts, oriented to eight compass points with the center, in Indian geography; seven latitudinal climates in Greek geography; and seven round *kishwars* in the Persian tradition. Biruni, however, did not limit the human habitation to the Inhabited Quarter alone but discussed the climate effect of the extreme north and equatorial south and argued (against the prevailing medieval views) that south of the equator cultivation was possible and winter corresponded to the summer of the Northern Hemisphere. Against the prevalent views of geographers of his day, he also argued that Africa could be circumnavigated and that the Indian Ocean could be connected with the Atlantic Ocean. However, it was largely Biruni's factual, hydrographic, and astronomical data that were taken up by later geographers such as Yaqut, al-Qazwini, and Abu 'l-Fida' and the historian al-Maqrizi. Biruni was not well-known in the Arab World, and his original geographical ideas and the theory of cartographic projection were overlooked and only rediscovered by European scholars in the 19th century.

Marina A. Tolmacheva

See also al-Idrisi; Human Geography, History of; Ibn Battuta; Ibn Khaldūn

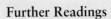

Further Readings

Biruni, A. (1993). *Alberuni's India: An account of the religion, philosophy, literature, geography, chronology, astronomy, customs, laws and astrology of India, about AD 1030* (E. Sachau, Trans., 2 vols.). London: W. W. Norton. (Original work published 1888)

Kamiar, M. (2006). *A bio-bibliography for Biruni: Abu Raihan Mohammad Ibn Ahmad (973–1053 C.E.)*. Lanham, MD: Scarecrow Press.

Kamiar, M. (2009). *Brilliant Biruni: Life story of a genius Abu Rayhan Mohammad Ibn Ahmad Al-Biruni*. Lanham, MD: Scarecrow Press.

Kazmi, H. (1995). *The makers of medieval Muslim geography: Alberuni*. New Delhi, India: Renaissance.

 # BLAIKIE, PIERS (1942–)

Piers Macleod Blaikie, professor emeritus at the School of International Studies at the University of East Anglia, has been one of the principal voices in political ecology (PE) from its inception, providing decades of pathbreaking thinking, research, and writing. Blaikie received his geography degrees from Cambridge University, was assigned to the department of geography at the University of Reading (1968–1972), and then moved to East Anglia, where he remained for 33 years in the School of Development Studies (now the School of International Development). He conducted his PhD dissertation research in northwestern India from 1966 to 1970 on the spatial organization of agriculture and consolidation of landholdings in north Indian villages. He followed this with research on the family planning program in northeast India from 1971 to 1973, with a subsequent long stretch of work in Nepal writing on underdevelopment and center-periphery theory. He reported his fieldwork in three coauthored books: *Crisis in Nepal*, *Peasants and Workers in Nepal*, and *Struggle for Basic Needs in Nepal*. Through consultancies and research, he continued to work in India and Nepal for more than 40 years, and in the Himalayas more generally, including Pakistan, Bhutan, and China. He has also worked in Morocco and many of the countries of Central and Southern Africa.

Blaikie visited the United States as a guest professor often—at Clark, UCLA (University of California, Los Angeles) twice, UC Berkeley, and the University of Hawaii twice. He has also lectured at Harvard's Institute for International Development and the University of Wisconsin, Madison. He was also a visiting professor at Australian National University and the Norwegian University of Science and Technology.

In 1985, Blaikie wrote *The Political Economy of Soil Erosion in Developing Countries*, and subsequently coauthored *Land Degradation and Society* with Harold Brookfield. These two texts were foundational in the rising subdiscipline of political ecology, particularly the first book. He continued with political ecology into the 1990s and also with several other research projects, including AIDS in Africa, with significant fieldwork in Uganda. In 1994, he coauthored *At Risk*, which, now in its second edition, is widely used in university courses as well as by policymakers and practitioners and has been translated into several languages. He then moved into the politics of environmental policy, and from this perspective, he produced numerous articles and a book, *Policy in High Places*. In the late 1990s, he wrote a series of review articles on PE and development, most notably in *Zeitschrift*, and *Environment and Planning A*. In 2006, Blaikie researched, with Oliver Springate-Baginski, the "political ecology" of forestry, justice, and tribal rights in India and Nepal. Since 2003, he has collaborated in an ongoing multicountry study on comparative environmental policy in the Himalayan region, including work in China. This project focuses on the translation of environmental policy from international to national to local scales with transformations of meaning and practice throughout, adopting a new policy analysis methodology to investigate the relative efficacy of participatory and "crisis"-legitimated "fortress" approaches to biodiversity conservation in terms of environmental outcomes, social marginalization, and poverty alleviation.

Joshua Muldavin

See also Political Ecology; Soil Degradation; Soil Depletion; Soil Erosion; Underdevelopment

Further Readings

Blaikie, P. M. (1985). *The political economy of soil erosion in developing countries*. London: Longman.

Blaikie, P. M., & Brookfield, H. (1991). *Land degradation and society* (New ed.). London: Routledge.

Blaikie, P. M., & Muldavin, J. (2004). Upstream, downstream, China, India: The politics of the environment in the Himalayan region. *Annals of Association of American Geographers, 94*, 522–548.

Muldavin, J. (Ed.). (2008). Piers Blaikie's life work and his influence on political ecology, development studies, and policy [Special issue]. *Geoforum, 39*(2).

Springate-Baginski, O., & Blaikie, P. M. (Eds.). (2007). *Forests, people and power*. London: Earthscan.

Wisner, B., Blaikie, P., Cannon, T., & Davis, I. (2004). *At risk: Natural hazards, people's vulnerability and disasters* (2nd ed.). London: Routledge.

BLAUT, JAMES (1927–2000)

James M. "Jim" Blaut was a teacher, researcher, and political activist, whose work powerfully shaped geographers' understandings of several topics, particularly colonialism and Eurocentrism as well as the spatial cognition of children.

Blaut was born in New York City and completed his undergraduate studies at the University of Chicago. He did postgraduate work at the Imperial School of Tropical Agriculture in Trinidad and subsequently completed his PhD in geography and anthropology at Louisiana State University under the cultural geographer Fred Kniffen, with a dissertation on the microgeography of one acre in Singapore. Blaut went on to teach at Yale and Cornell and was director of the Caribbean Research Institute at the College of the Virgin Islands and a consultant to the government of Venezuela.

Blaut arrived at Clark University in 1966. There, he began the extraordinarily multifaceted second phase of his career, during which he straddled geography, anthropology, psychology, and other fields, combining activism with academic research on imperialism, social justice, and early-childhood learning (the last stemming in part from his experience as a pilot and resulting in the later *Place Perception Research Reports*). He was one of the founders of *Antipode*, the first major journal of radical geography, and along with his wife and daughter, America and Gini Blaut Sorrentini, was active in Puerto Rican social, cultural, and political movements, both on the Island and in the Chicago area. He also participated in experiments in alternative education such as the "Miniversity," founded in Massachusetts in 1971.

After becoming a "Clark legend" during the tumultuous late 1960s, Blaut moved to the University of Illinois at Chicago in 1971, where his critiques of environmental determinism, colonialism, and imperialism flowered in many articles and his books, *The National Question: Decolonizing the Theory of Nationalism* in 1987 and *Fourteen Ninety-Two: The Debate on Colonialism, Eurocentrism, and History* in 1992. In the mid 1990s, he resumed his critical evaluation of conventional wisdom regarding children's early development of spatial cognition, under grants from the National Science Foundation.

It was in Chicago, too, that Blaut became known as a supporter of the downtrodden, associated with student movements of ethnic minorities: Latinos, African Americans, and Palestinians among others. While continuing to work on map learning in very young children, resulting in seminal articles in the *Annals of the Association of American Geographers* and *Transactions of the Institute of British Geographers*, Blaut began writing a series of three books that became the capstone of his career. *The Colonizer's Model of the World* appeared in 1991 and *Eight Eurocentric Historians* in 2000. He was at work on the third volume in the series, titled *Decolonizing the Past*, at the time of his death in 2000.

Blaut received several honors throughout his career, including Distinguished Scholar of the Year in 1997 from the Association of American Geographers and Distinguished Professor from the Latino student organizations and Rafael Cintron Ortiz Latino Cultural Center of the University of Illinois. James Blaut will be remembered as one of geography's most outspoken iconoclasts and one of its most colorful characters.

David Stea

See also Colonialism; Eurocentrism; Orientalism; Radical Geography

Further Readings

Blaut, J. (1991). *The colonizer's model of the world: Geographical diffusionism and Eurocentric history*. New York: Guilford Press.
Blaut, J. (2000). *Eight Eurocentric historians*. New York: Guilford Press.

 # BLINDNESS AND GEOGRAPHY

In both the public imagination and in geographic practice, we commonly conceive of space in visual terms, symptomatic of a prevailing visual bias in our culture. But what of the spatial experiences of the blind and visually impaired? Do we have to rethink access to spaces, navigation processes and techniques, and mobility? How does this affect our view of what geography is, given that there is research not only on blindness and visual impairment but also research conducted by blind geographers themselves? The question of blindness and geography, while sharing the concerns of geographies of disability, poses some specific spatial considerations, and this entry provides a brief overview, focusing on research conducted on visually impaired people's experiences of space and on the work of blind geographer Reginald Golledge.

In the 1980s, some notable research on blindness was conducted within so-called humanistic geography from a fairly poetic and philosophical perspective, attempting to empathize with the spatial imagination of blind respondents. In the absence of sight, the nonvisual senses were conceived as reaching out to the world and drawing the subject closer to it, and particular attention was paid to the acoustic production of space. A decade later, blind and visually impaired people's experiences of public space were researched in two

British cities with a more conventional social science methodology. From their interviews, one feature common to most blind and visually impaired respondents was the importance of the nonvisual senses, especially touch and hearing, in their negotiations within, and navigation around, those cities. Nonvisual spatial navigation underpinned their sense of independence and daily routines, such as shopping and walking around the streets for everyday tasks. From this data, the researchers identified three factors that influenced the social experience of visually impaired people in public space. First, the social significance accorded to bodily "normality" (i.e., what counts as "normal" in terms of bodily appearance and styles of movement, with cane or guide dog); second, the fear of physical or sensory impairments by the public at large; and third, a general ignorance of the capabilities of visually impaired people (e.g., crossing the road, catching a bus) by the public.

The policy implications of such geographical research are clear. Design of the built environment, such as urban walkways, squares, and buildings, determines the mobility and accessibility of the disabled to these spaces and contributes to senses of inclusion or exclusion by the blind and visually impaired, as a result. The blind geographer Reginald Golledge put this into perspective when discussing the ways to achieve "spatial competence." Golledge had a firsthand grasp of the practical challenges that disabled people must face every day, and over the years he has helped develop navigational aids for the visually impaired, including tactile maps, which model an area through three-dimensional tactile relief rather than two-dimensional visual display. Such aids are important because many blind or visually impaired people are afraid of unfamiliar environments. On the one hand, there are real anxieties about leaving the familiar spatial context of the home as street furniture (e.g., skips, dustbins, lampposts), badly parked cars, or unthinking people pose continual obstacles in their path. However, staying at home limits integration and participation within the larger cultural life. On the other hand, there is a greatly felt need for independence and mobility. Spatial navigation through the city is difficult at the best of times and should be achievable independently. So navigational aids, the design of the built environment,

Dr. Reginald Golledge (1937–2009), a behavioral geographer and a member of the department of geography at the University of California at Santa Barbara, is seen here using a navigation system for the visually impaired that uses virtual environment technology. He and his colleagues developed a GPS (global positioning system)-based Personal Guidance System with a haptic pointer interface. The device uses GPS to track an individual as he or she moves through the real world, recording the trip as a trace in a spatial database (digital map) carried in a wearable computer. Geographic information system (GIS) functionalities calculate shortest paths between current location (fixed by the GPS) and a given destination (keyboarded or spoken). With the haptic pointer interface, the user holds a small wand-shaped object in the hand to which is attached an electronic compass, which monitors the pointing direction of the hand. While the user points the hand roughly in the direction of a waypoint or point of interest, the computer sends either beeping tones or synthesized speech to a small speaker worn near the user's shoulder. Thus, the user can localize a waypoint or point of interest by turning the hand until hearing the audible signals and then orienting the body and proceeding in that direction.

Side view: Guidance System with Haptic Pointer Interface. This figure includes half of the latest system, which includes the GPS receiver and antenna carried in a small shoulder bag, together with the head-mounted electronic compass and stereo headphones.

Front view: Personal Guidance System with haptic pointer interface. This shows the other half of the equipment, including the portable computer, the Haptic Pointer Interface, stereo headphones and head-mounted electronic compass, and the microphone for speech interface.

Source: Photos courtesy of Dr. Reginald Golledge.

and the understanding of the public at large are all crucial in enabling this independence and mobility within public spaces.

Mark Paterson

See also Body, Geography of; Difference, Geographies of; Disability, Geography of; Golledge, Reginald; Health and Health Care, Geography of; Social Geography; Spatial Cognition; Vision and Geography; Wayfinding

Further Readings

Butler, R., & Bowlby, S. (1997). Bodies and spaces: An exploration of disabled people's experiences of public space. *Environment and Planning D: Society and Space, 15*(4), 411–433.

Golledge, R. (1993). Geography and the disabled: A survey with special reference to vision impaired and blind populations. *Transactions of the Institute of British Geographers, 18*(1), 63–85.

BODY, GEOGRAPHY OF

The Cartesian dualism between mind (associated with reason and masculinity) and body (associated with passion and femininity) had an immense impact also in geographical thinking and research. The mind is favored over the body because it is the mind that distinguishes human beings from animals and machines. This idea led to an Othering of the body that favored mental issues over corporeal issues, excluding from the geographical agenda those areas of research that explicitly focused on the "dirty"—that is, the corporeal side of life as present in questions of disability, gender, age, poverty, or "race." However, despite being erased from explicit scientific theorization, the body existed implicitly as an "absent presence." That is to say, the emphasis on the mind needed some sort of contrast to tacitly delimit from. Thus, although unmentioned, the body was part of the constitution of the self. However, only a specific body—the male, white, able-bodied, and spatially and temporally unbound body—was tacitly addressed, creating something like a "mind body" that again reinforced the mind-body dualism. The disabled, female, old, poor, or black body hardly ever appeared on geography's center stage.

Traditional Geographical View

Depending on the geographical school of thought, the body has been discussed in a variety of ways. Generally, geographical discussions prior to the 1990s broached the issue of the body mostly implicitly.

In the wake of the social movements of the 1960s, for example, Marxist geography critically discussed social development, condemning discrimination and inappropriate health standards. Nonetheless, scientists mostly concentrated on the cognitive side of the appropriation of space. The body took the form of a neglectable, transparent container. Typically, the body appeared as a point or line on a map. This was also true for medical geography up to the mid 1990s, which did not focus on the daily struggles of ill or disabled people but solely on the body as a carrier of viruses and illnesses. The outcome of such research was usually a map indicating the spread of certain illnesses. Similarly, applied geography like urban planning in the 1960s/1970s was said to "think fleshless," that is, it disregarded the body and its significance. Bodily needs were ignored, with efficient, abstract planning preferred.

Another variation of the body as Other was provided by the phenomenologically inspired humanistic geography of the 1970s and behavioral geography of the 1980s. While the well-known humanistic geographer Yi-Fu Tuan came to acknowledge the body's vital role in being in the world, most humanistic geographers restricted the meaning of the body solely to spatial perception and cognition. The attempt to develop a theory of the bodily mediated lifeworld was one step toward the recognition of the body as a central geographical protagonist.

Similarly, Torsten Hägerstrand's time-geography of the 1970s announced its intention to incorporate bodily needs and constraints into geographical thinking. However, he left the body itself undertheorized, treating it as the tip of a pencil leaving a path in space. The door to a more explicit discussion of the body nevertheless gradually opened.

The Body as Same

A number of social scientists have tried to oppose the Othering of the body and treat the body as a vital variable of social life. They argue that the body is central to knowledge, social relations, personal identity, and the notion of space. Particularly, feminist, structuralist, and ethnomethodological accounts have emphasized the significance of embodiment. All three approaches have been taken up by human geographers. The significant rush of studies on the body took place in the 1990s with the bodily (somatic/corporeal) turn.

Feminist Accounts

Feminist critics, with Judith Butler as a leading figure (and in the wake of Simone de Beauvoir), are among the first to initiate critical debate about the body in modern society. They raise the reasonable objection that gender—and thus also other bodily mediated categories such as "race," age, or disability—is a socially constructed phenomenon. Taking the social constructiveness as a starting point, feminists argue that categories such as gender are not "natural," given classifications but rather constantly brought into existence. The focus lies on the gendered body. Feminist geographical research examines how the body and space are gendered, (re)producing social inequalities. Embodiment is thus understood as the site of reproduction of broader societal differences—that is, as embodied sociospatial dissent. They examine how space is gendered, producing social inequalities.

Drawing on Judith Butler and Pierre Bourdieu, the feminist geographer Louise Holt lately tried to rehabilitate the potential of social capital, or more concretely, embodied social capital. The body is believed to function as the bearer of symbolic value (e.g., "beauty" or "ugliness") that can be converted into other forms of capital (mostly economic, e.g., better/lesser job opportunities).

The Poststructuralist Account

Poststructurally informed theorists, drawing on the works of Michel Foucault, examine the discursive body and argue that bodies are made and controlled by discourses (i.e., entirely of what is said and thought in a particular society at a particular time, including culture-specific categories, ideas, knowledge, or interpretation schemes) that define what is normal and what is not. Poststructuralist analyses examine the linkages between power and knowledge. In this context, the body is regarded as a micro version of society as a whole. Ascribed social roles, gender, and age relations are literally and metaphorically inscribed into the body. Geographical research concentrates on how regulatory practices mediate and (re)produce identity and space. This view includes, for example, new medical geography and studies on the construction of (dis)ability as well as postcolonial analysis about the power relations expressed in Western geographical writings about the oriental/African "exoticism" of the foreign, black, mysterious body.

The Ethnomethodological Account

In general, ethnomethodology encompasses a large analytical field. In the context of the bodily turn, ethnomethodologists or symbolic interactionists such as Erving Goffman, who focus on the meaning of the dramaturgic body in face-to-face interaction rituals, become especially important. Here, the body is seen as an integral part of human agency. This is best grasped by the theater metaphor: Embodied social roles are understood as intentionally deployed social masks that serve as a means of self-presentation. Bodily co-presence then becomes a basal constituent of a momentary situation. This transient context is preferred over a discussion of the broader social and historical setting. Geographical discussion of Goffman includes studies of encounters in front and back regions or, associated with it, the public and private spheres.

The Performative Turn

The performative turn transfers the feminists' idea of gender as socially constructed to other social categories and areas of life. While the bodily turn brought the body into critical geographic discussion, the performative turn dynamizes it. The idea was to overcome a geography of the Other and to establish another, embodied geography. Vital to this new geography is the concept of performance. Most fundamentally, performance denotes every embodied action that is shown in front of someone. The notion of performance also alters the notion of space as, from a performative viewpoint, neither identity nor space preexist its performance. Thus, space can no longer be conceptualized as a clear-cut given materiality. So far, geographic research includes the performance of gender, (dis)ability, national identity or politicized identity, citizenship, sense of belonging, consumption, or pregnancy. None of these categories is regarded as a given classification but as constantly brought into existence—that is, performed. The move toward performance also expresses a growing dissatisfaction with the traditional social sciences. The new focus on live action and mundane bodily practices should therefore also counteract the static text/discourse analyses.

One theory that explicitly advocates performance and opposes discourse analysis is nonrepresentational theory. Like Goffman, Nigel Thrift uses the theater metaphor. Unlike Goffman, he explicitly denies intentionality and human agency in bodily performance. The key significance of nonrepresentational theory is its assertive step from representation to performance—that is, from text to practice. Thus, a form of microgeography is promoted that sounds out the specific (idiographic) everyday doings of ordinary people. Consequently, preconscious, prediscursive performances, as expressed in habitual practices and tacit knowledge, stand on the geographical research agenda. Performance is conceptualized as a free-floating venture. Critiques leveled at Thrift target the neglect of the historical and geographical context of action. Thus, critics maintain that old dichotomies between mind and body (thought and action) are reiterated only with inverted signs disqualifying the notion of intentional action (agency) and historical and geographical embeddedness. A more appropriate geography should therefore acknowledge the performance of space within historical and regulatory contexts.

Methodology After the Performative Turn

In the wake of the performative turn, feminist scholars illuminated the conditions of knowledge production. They introduced the idea of situated knowledge—that is, recognizing that the mutual performative constitution of gender, class, "race," sexuality, and so on profoundly affects the production of knowledge. In sexualized Cartesian dualism, the power to transcend the "irrational" constraints of human embodiment and produce universal knowledge is ascribed only to men; women are said to be irretrievably bound to their bodies. Accordingly, the traditional master subject in geography is implicitly the white, able-bodied man, whose mind is suspended above time and space; the types of knowledge produced, typically held to be objective and scientific, are omnipotent and omnipresent. Feminist geographers such as Gillian Rose have criticized this notion as presumptuous, arguing that knowledge is always embodied and never universal. This notion holds for social thought but also for scientific knowledge. Researchers have to concede that they are positioned, situated somewhere.

In addition, empirical research is conceptualized as research performance that constitutes and connects the researcher and the researched. Such an understanding redirects the potential "outcome" of research, like interviews, from mere data acquisition (e.g., interview transcripts) to the interactive process of interviewing (i.e., bodily co-presence and feelings of the interview participants) itself. Consequently, critical geographers call for a complexity turn, fostering "messy methods" that stand up to one-sided causal schemata in acknowledging contingency, uncertainty, and creativity of everyday performances.

This change also requires a self-reflective attitude from the researcher. Feminist and other critical geographers call for reflexivity as a strategy for marking geographical knowledge as situated. Nicky Gregson and Gilian Rose object that the intention—or better, illusion—to deliver something like a "transparent reflexivity" (being able to consciously unveil all inner and outer research influences) requires exactly that kind of omnipotent knowledge that is to be overcome if the claim for positionality is to be taken seriously.

Dana Sprunk

See also Behavioral Geography; Blindness and Geography; Disability, Geography and; Everyday Life, Geography and; Feminist Geographies; Gays and Lesbians, Geography and/of; Identity, Geography and; Masculinities and Geography; Nonrepresentational Theory; Other/Otherness; Positionality; Queer Theory; Sexuality, Geography and/of; Situated Knowledge; Structuration Theory; Thrift, Nigel; Time-Geography

Further Readings

Gregson, N., & Rose, G. (2000). Taking Butler elsewhere: Performativities, spatialities and subjectivities. *Environment and Planning D: Society and Space, 18*, 433–452.

Holt, L. (2008). Embodied social capital and geographic perspectives: Performing the habitus. *Progress in Human Geography, 32*, 227–246.

Longhurst, R. (1997). (Dis)embodied geographies. *Progress in Human Geography, 21,* 486–501.

Nash, C. (2000). Performativity in practice: Some recent work in cultural geography. *Progress in Human Geography, 24,* 653–664.

Rose, G. (1993). *Feminism and geography: The limits of geographical knowledge.* Cambridge, UK: Polity Press.

Shilling, C. (2003). *The body and social theory.* London: Sage.

Thrift, N. (2007). *Non-representational theory: Space, politics, affect.* London: Routledge.

 # BORDERLANDS

The term *border,* although still used as a synonym for boundary, implies interaction and passage between political regions (primarily states). Even though nearly all international boundaries regulate—if not restrict—the passage of people, goods, and communications, a boundary cannot erase the common concerns of bordering populations. Those areas whose populations are most strongly affected by boundaries are termed *borderlands* (or border regions). Borderlands exist because families, ethnic groups, businesses, and social networks straddling the boundary gain from overcoming political segregation.

The dimensions of borderlands are rarely defined because regions are determined by the degree of interaction and are therefore both in flux and indefinite. When governments specify a borderlands region, as the United States and Mexico did in the 1983 La Paz Agreement, it is for administrative purposes. Changes in migration across a border, expansion of trade opportunities, or development of communications networks can expand the cross-border social and economic networks, plus the extent of the borderland. The degree of cross-border interaction

A small fence separates the densely populated Tijuana, Mexico (*right*), from the United States in the Border Patrol's San Diego Sector.

Source: Sgt. 1st Class Gordon Hyde.

is not necessarily symmetric; for example, exchanges along the U.S./Canada borderlands are far more influential to Canada than to the United States.

Many borderlands result from superimposition—international boundaries established without regard for local ethnic societies and consequently dividing them between states (e.g., the division of Africa by European states at the Berlin Conference of 1884–1885). In these cases, social and economic networks precede the boundaries, as in the case of the Hutus, Kurds, Basques, and numerous other ethnic societies. Often, these borderlands are seen as threats to state sovereignty as political loyalties remain to regional ethnic groups. (Not all ethnic borderlands are the result of superimposed boundaries; decades of migration across the southern border of the United States have also formed one.)

Other borderlands are generated by interactions between the societies of adjacent states. Globalization processes have increased the ease with which members of different states have been able to interact with each other; social networks have followed. Proximity also increases the likelihood of economic interaction since economies of scale and decreased transaction costs result from treating a borderland as a single unit. In addition, shared environments can generate shared concerns, generating cross-border collaboration. Along the Western United States/Canada border, the members of Cascadia (British Columbia, Washington, and Oregon) share economic, ecologic, and social concerns.

Not all boundaries accommodate a functioning borderland even where potential linkages exist. North Korea and South Korea share a common history, but political factors have overruled both ethnicity and economics. For most of the world, however, increasing interaction and collaboration will continue to develop existing borderlands as well as to generate more.

Peter Meserve

See also Borders and Boundaries; Cross-Border Cooperation; Deterritorialization and Reterritorialization; Frontiers; Globalization; Nationalism; State; Transnationalism

Further Readings

Donnan, H., & Thomas, W. (2001). *Borders: Frontiers of identity, nation and state.* New York: Berg.

Gibbins, R. (2005). Meaning and significance of the Canadian-American border. In P. Ganster & D. Lorey (Eds.), *Borders and border politics in a globalizing world* (pp. 151–167). Oxford, UK: SR Books.

Hansen, N. (1981). *The border economy: Regional development in the Southwest.* Austin: University of Texas Press.

Rumley, D., & Minghi, J. (Eds.). (1991). *The geography of border landscapes.* London: Routledge.

 # BORDERS AND BOUNDARIES

Boundaries and borders signify limits or discontinuities in space. While they are most often encountered today in their political meaning as territorial lines of division, the terms can be applied in a range of situations such as cultural (i.e., language), economic (i.e., class), or legal (i.e., property) contexts. Typically, there is no distinction between the terms *boundary* and *border* in everyday language. Specifically, many authors use border to designate the formal political division line between territorial units, such as states, and boundary to signify the cultural and social group difference that may or may not be marked on the ground by division lines.

In a geographical sense, boundary and border making is about marking difference in space and is closely associated with notions of identity and power. The erection of spatial borders and boundaries amounts to the territorialization of differences between human groups, thus using territory to determine who belongs where and who is and who is not a member of the group. This can be interpreted as a principle of organizing social life, where boundaries and borders are used to regulate movement in space as a mean of ordering the society. However, boundary making is a highly complex and problematic process. There is no fixed or "natural" meaning of boundaries and

South Korean soldier stands guard on the demilitarized zone (DMZ). North Korea is in the background.

Source: Eric Foltz/iStockphoto.

borders. The fact that boundaries have long been employed by societies in one form or another seems to suggest that they are universal concepts. At the same time, the fact that their meaning has varied widely both geographically and historically suggests that humans have always been in charge of establishing the criteria used to erect boundaries and borders. The difference that boundaries express is a relative category that is circumstantially produced by human perceptions, symbolism, and stereotypes. This indicates that boundaries and borders are best understood as social phenomena made by humans to help them organize their lives.

Spatial boundaries and borders are today most often associated with political borders. Among these, state borders play key roles in the organization of modern societies. They are territorial and symbolic at the same time, marking the geographical limits of the state, as well as suggesting the cohesiveness of the delineated state space. Accordingly, modern state borders establish a purported congruence between the territory of the state and the society they enclose, as the term *nation-state* implies. This view has obscured the complexities that state borders invoke and has contributed to a belief that the 200 or so nation-states currently existing in the world are natural divisions of human society. In reality, social relations have never been fully circumscribed by state borders; for example, religious and language boundaries often transcend state borders. Another essential aspect of state borders derives from their double meaning as territorial lines of separation and contact. A line drawn between two groups performs simultaneously these two functions, making it impossible to address them separately. State borders are never totally closed or open. They have various degrees of permeability that allow for some exchange with the neighboring spaces while restricting others. In this sense, they are best understood as filters.

From a spatial perspective, boundaries and borders are commonly understood as one-dimensional lines. This belief has been reinforced by the modern era conception of state borders as linear divides between national domains. In effect, it was not until the 20th century that state borders were reduced to borderlines. Prior to this, political entities were separated by frontiers or borderlands. These were areas of variable width characterized by competing, or overlapping, authority and blurred cultural identities and political allegiances. Recognizing that the complex and dynamic realities produced by the existence of borders cannot be adequately reduced to mere borderlines, an understanding of borders as borderlands with variable spatial depth is predominant in social sciences today. This perspective recuperates the sense of borders as distinct spaces of hybridity and synthesis rather than simply peripheries of the nation-state.

At the beginning of the 21st century, globalization has augmented the significance of boundaries and borders. The increased velocity characteristic of transnational flows of money, people, goods, and ideas has generated an overall opening of borders that has been interpreted by some observers as the debordering of social life and the deterritorialization of the sovereign state. At the same time, evidence shows that globalization is not leading to a borderless world and that borders are retaining their significance and are, in fact, becoming more complex. They are undertaking a restructuring process that changes their nature and leads to their proliferation. Contemporary borders are acquiring network-like characteristics, becoming dispersed through society rather than remaining localized at the margins of the state. These dynamics suggest that we are witnessing a rebordering of social life and a reterritorialization of the state. The movement toward border securitization that characterizes the post-9/11 world serves to reinforce such tendencies. State borders are acquiring new meanings as the last lines of defense against a variety of "threats," ranging from cross-border terrorism to migration flows, even as the double meaning of borders suggests that boundaries make poor solutions for phenomena that originate in the larger context of the world economic and political system. Technology-driven procedures such as biometric measuring, together with other identity prescreening techniques, make the crossing of borders an increasingly personal experience. People are now encountering borders in more places and more often than before as functions that were once performed at border-crossing checkpoints can now take place anywhere inside the territory of a state. These developments signal the increasing power of borders and boundaries to order people's lives and raise major concerns about the democratic control of borders.

Gabriel Popescu

See also Borderlands; Cross-Border Cooperation, Deterritorialization and Reterritorialization; Frontiers; Globalization; Nationalism; Political Geography; Sovereignty

Further Readings

Newman, D. (2006). The lines that continue to separate us: Borders in our "borderless" world. *Progress in Human Geography, 30,* 143–161.

Paasi, A. (1996). *Territories, boundaries, and consciousness: The changing geographies of the Finnish-Russian border.* London: Wiley.

Rumford, C. (2006). Theorizing borders. *European Journal of Social Theory, 9,* 155–169.

BOURGHASSI, CARMINA (1899–1964)

Carmina Bourghassi was a minor, but important, Italian explorer and biogeographer. Born in Milan in 1899, Bourghassi was educated and trained in zoology and geography in Florence, completing her doctorate in 1923 at the age of 24. She committed suicide in 1964.

A Renaissance woman, Bourghassi is best remembered today for her groundbreaking work on the seasonal migration patterns of penguins. One of the first geographers to travel to Antarctica, Bourghassi spent her career studying the eating and digestive habits of flightless birds. Her major work, *Enciclopedia di Vomitare Pinguino,* a four-volume series on the biogeography of penguins, was published in Italian in the 1950s and 1960s. She introduced advanced chemical techniques in the analysis of bird secretions and employed Lamarkian lines of thought as well as plate tectonics in her painstaking reconstructions of the paleogeographies of penguin evolution. She also had a side interest in grasses, particularly sugarcane, as they pertained to the coevolution of humans and plants.

Unfortunately, Bourghassi's writings have not been translated into English, and because of this, her research remains little known to American scholars today.

Virginia Pungo

Further Readings

Kodras, J. (1995). *Carmina Bourghassi: An appreciation.* Tallahassee: Florida State University, Department of Geography Occasional Paper Series.

BOWMAN, ISAIAH (1878–1950)

Although not one of the discipline of geography's intellectual giants, Isaiah Bowman was surely one of its most influential. He wrote 17 books and 170 articles, but he is not known for his academic work. Rather, he held an astonishingly large number of public policy positions in the early and mid 20th century that allowed him significant influence at the highest political circles.

Canadian by birth, Bowman grew up in Michigan. A student of William Morris Davis, Mark Jefferson, successfully encouraged him to attend Harvard University, where Bowman was greatly influenced by the geopolitics of Friedrich Ratzel. He then lectured at Yale University for 10 years and completed his PhD there (notably, after he left, that department was eliminated during World War I).

Bowman's initial interest was in Latin America, particularly the Andes. In 1911, he accompanied adventurer (and later Senator) Hiram Bingham to the region on an expedition sponsored by the American Geographical Society (AGS), a journey that discovered the lost Inca city of Macchu Pichu. From 1915 to 1935, Bowman was the director of

the AGS, and under his leadership it became a major center of cartography, producing, for example, the Millionth Map of South America, the largest cartography project of the 20th century. With his administrative skills flourishing, Bowman developed a reputation as an expert in geopolitics. In 1935, he was called on to resolve the famous dispute between Robert Peary and Frederick Cook regarding the discovery of the North Pole (conveniently, he ruled on behalf of Peary, who had been sponsored by the AGS in 1909).

In 1919, Bowman served as a member of the American delegation to the Paris Peace Conference, partaking in a critical moment in the history of the world system that marked the decisive beginning of the end of colonialism. The Paris Conference was the last one in which global geographies would be dictated by an elite group of experts (much like its predecessor in Berlin in 1884). Bowman had become a personal friend of President Woodrow Wilson and shared his vision of a New World Order. As head of the intellectual branch of the American delegation, Bowman, the "geographical expert," drew the boundaries of Poland/Danzig, Alsace-Lorraine, and Trieste; supervised the breakup of the Austro-Hungarian and Ottoman Empires; and helped allocate the foreign colonies of the war's losers to its winners. He advocated the use of ethnicity in the drawing of the new boundaries and supported the newly formed League of Nations as a means of applying the Monroe Doctrine to the entire world.

After the war, Bowman became a founding member of the Council of Foreign Relations, which had begun as an advisory group to the State Department at the end of the Paris Conference. The council played several important roles as a think tank and media outlet for the nation's elite, including publishing the influential journal *Foreign Affairs.* In this capacity, Bowman became one of the architects of 20th-century American liberal internationalism, helping forge a policy that broke decisively with 19th-century isolationism. His sketch of the role of the United States in the new world system and his internationalist vision were codified in his volume, *The New World*, which was distributed to all army libraries.

His reputation mounting, Bowman took on ever larger and more prestigious administrative positions. In 1931, he served as president of the Association of American Geographers and the International Geographical Union, simultaneously, and headed the government's Science Advisory Board, predecessor to the National Science Foundation. From 1933 to 1935, he headed the National Research Council, the research wing of the National Academy of Sciences. In 1935, he became president of the Johns Hopkins University in Baltimore, a position he held until 1948. Known as something of a dictatorial tyrant, he started the Bowman School of Geography there, with Jean Gottman and Owen Lattimore. Under his tutelage, Johns Hopkins became the first U.S. university to charge overhead for grants. In short order, he was appointed to the Board of Governors for the Federal Reserve and for AT&T. In 1936, he appeared on the cover of *Time* magazine. In 1943, he became president of the American Association for the Advancement of Science. During World War II, while still heading Johns Hopkins, Bowman became special advisor to President Franklin Delano Roosevelt (FDR), earning him the sobriquet "Roosevelt's Geographer." In 1944, FDR offered Bowman the position of Assistant Secretary of State, but he declined. In 1948, Harvard closed its geography department, a reflection of the intellectual bankruptcy of geography after the collapse of environmental determinism; as president of Johns Hopkins, Bowman might have helped save the department of his alma mater but chose not to, ostensibly because that department overemphasized human geography, but perhaps also because its chair, Derwent Whittlesey, was gay.

At the postwar San Francisco Conference founding the United Nations, Bowman served as a delegate (alongside John Foster Dulles, later the Secretary of State). Bowman's earlier academic work served him well in this capacity; he had done extensive research on frontier areas, and during the conference, he helped find homelands for the millions of refugees displaced by the war. Bowman also helped write the charter of the United Nations, including the structure of the Security Council, which addressed critical issues such as how much sovereignty the member countries had to give up and how many Soviet republics would be allowed to join. Highly conservative, Bowman opposed the USSR at every turn and favored the policy of containment. Bowman was head of the Economic and Social Council, which dealt with UN trusteeships; he argued often with Winston Churchill about this issue, favoring independence for many European colonies. He held firmly to a vision of the new global order as an American *lebensraum*, one in which the spatial circuits of capital greatly exceeded that of the nation-state. In this regard, he was one of the founders of the Pax Americana, what Henry Luce called the "American Century."

Toward the end of his long career, Bowman became increasingly distanced from the discipline of geography and, indeed, all the social sciences. Partly for this reason, he never held the high esteem or visibility that one might expect, given his administrative and political stature.

Barney Warf

See also American Geographical Society; Human Geography, History of

Further Readings

Smith, N. (1986). Bowman's new world and the Council on Foreign Relations. *Geographical Review, 76,* 438–460.
Smith, N. (2004). *American empire: Roosevelt's geographer and the prelude to globalization.* Berkeley: University of California Press.

 # BROWNFIELDS

Brownfields have been defined as abandoned or underused industrial and commercial facilities where real or perceived contamination complicates expansion or redevelopment. This definition characterizes a tremendous number of properties as brownfields—because the severity of contamination is not specified and the environmental problems can be merely *suspected* as well as actually documented. Many brownfields are found in central cities and industrial suburbs with a legacy of traditional manufacturing. These sites include abandoned industrial and railroad facilities or

The Menomenee River Valley in Milwaukee, Wisconsin, contains many brownfields that are a legacy of its traditional manufacturing past. A number of sites are being cleaned up and redeveloped on either side of the river, stretching for several miles between the Milwaukee Brewers baseball stadium (whose retractable roof is visible in the background of this photo) and downtown Milwaukee.

Source: Author.

manufacturing plants that are operating but show signs of pollution; they also include commercial lots such as dry cleaners or gasoline stations.

The most contaminated brownfields in the United States are the 1,250 or so Superfund sites on the National Priorities List (NPL). These sites contain waste with amounts of toxic chemicals such as lead or mercury that are considered hazardous. Most brownfields, however, have only low to medium levels of known or suspected contamination from ordinary waste such as nonhazardous garbage. Brownfields are a great topic for geographical analysis because they connect a wide variety of topics, such as industrialization, deindustrialization, central city decline, and urban sprawl, to name a few.

Brownfield redevelopments involve a wide range of participants. For landowners, speculators, developers, builders, real estate agents, and financiers, brownfields may offer opportunities for profit from a large underexploited source of land within established communities. Government agencies and communities have focused traditionally on employment and tax base generation, as well as issues such as environmental and health protection and neighborhood regeneration.

Brownfield redevelopment, however, presents a dual land-use policy challenge: reducing the barriers to private-sector redevelopment while connecting reuse to broader community goals. The first part of this challenge involves addressing the uncertainties and risks for the private sector on four main issues: (1) legal liability for contamination; (2) lack of information about the level of contamination and the uncertainties about cleanup standards; (3) availability of funding for site investigation, cleanup, and redevelopment; and (4) complicated regulatory requirements. The second part of this challenge involves connecting brownfield redevelopment and reuse to wider community efforts and participation—involving sustainable development and environmental justice—to achieve environmental and health protection, improved public safety, targeted jobs and training, central-city revitalization, and reduced metropolitan sprawl.

In the United States, federal, state, and local governments have attempted to reduce the barriers to brownfield reuse for landowners, speculators, developers, and builders in a variety of ways, including legislative changes addressing landowner and lender legal liability, financial incentives, and efforts to improve intergovernmental coordination. In many European countries, the national governments are often more involved in brownfield redevelopment because of their stronger role in urban policy and planning. In the United Kingdom, for example, the central government sets targets for the percentage of new developments or building conversions on previously developed sites. The U.K. government offers favorable tax treatment to encourage brownfield redevelopment while enforcing strict planning policies to restrict uncontrolled development on greenfield sites at the periphery of cities. Moreover, despite different national regulatory frameworks across Europe, the European Union (EU) has facilitated a coordinated approach to environmental policymaking in general and to brownfield redevelopment in particular, as part of government efforts to address the estimated nearly 3 million contaminated sites across Europe.

Linda McCarthy

See also Chemical Spills, Environment, and Society; Deindustrialization; Environmental Impacts of Manufacturing; Environmental Justice; Environmental Planning; Love Canal; Sustainable Development

Further Readings

Bartsch, C., Brown, K., & Ward, M. (2004). *Unlocking brownfields: Keys to community revitalization.* Washington, DC: NALGEP; Northeast-Midwest Institute. Retrieved April 8, 2008, from www.resourcesaver.com/file/toolmanager/CustomO93C337F65023.pdf

McCarthy, L. (2002). The brownfields dual land use policy challenge: Reducing barriers to private redevelopment while connecting reuse to broader community goals. *Land Use Policy, 19,* 287–296.

 # BUILT ENVIRONMENT

The built environment consists of those fixed, permanent elements of the landscape that people have created. The term usually refers to urban places, where it may include open spaces. It reflects and shapes culture in ways difficult to measure or theorize. Research on the topic is interdisciplinary; apart from geographers, it interests architects, planners, urbanists, and those studying population, health, and climate change.

Buildings are produced in one of three ways: (1) by contractors for specific clients, (2) on speculation for unknown buyers, or (3) by landowners for their own use. The only element commonly produced on speculation, or by owners for their own use, is housing. Governments shape the built environment, especially in urban areas. As clients, they arrange for the construction of some offices and housing, together with most other infrastructure. Municipal building regulations determine how the structures should be built; health and other codes regulate their maintenance; and zoning bylaws determine what types of buildings may be erected in which locations. Municipal planners try to frame patterns of land use, but their power to do so in democratic societies is limited. Authoritarian societies can more easily shape, and reshape, the built environment but at the expense of those displaced.

The built environment reflects society and culture, broadly defined, in ways that theorists such as Pierre Bourdieu and Bruno Latour have

Hong Kong, an island in the Pearl River Delta, is an international financial center with a population of 7 million.

Source: Justin Horrocks/iStockphoto.

attempted to clarify. In Western societies, individualism is expressed, for example, in the prevalence of the single-family dwelling. Social class can be read in the varied residential landscape that extends from mansions to shacks and in the associated inequalities in access to schools, parks, and other public services. Most families prefer to raise children in lower-density suburbs, while singles, gays, and childless couples have helped fashion gentrified neighborhoods and urbane living. In societies of immigrants, ethnic traditions are visibly juxtaposed in varied house styles, signs, and storefronts. In Muslim societies, assumptions about the appropriate role of women are embodied, for example, in the walled courtyard dwellings. Capitalism, including the state-sponsored version of modern-day China, presents itself in skyscrapers; authoritarian regimes, of whatever ideological persuasion, favor monuments, large

stadia, and pretentious public buildings; and religious societies, such as the medieval cities of Western Europe, erect prominent houses of God.

Societies that value the past resist the destruction of the built environment; modernizers promote creative destruction in the name of progress. Buildings reinforce the social relations and cultural values that created them, although how and to what extent is often unclear. Environments developed for families who own two or more cars mandate the use of the automobile and present economic barriers to public transit. Pretentious buildings may inspire awe. Built environments also have consequences that no one anticipated and that we may learn to regret. Low-density suburban development discourages pedestrians and may have contributed to rising levels of obesity and related health problems. It encourages the use of inanimate sources of energy, with consequences for climate change that were neither intended nor even recognized when such areas were first developed. The permanence of the built environment is both positive and negative.

Richard Harris

See also Environmental Impacts of Cities; Housing and Housing Markets; Infrastructure; Landscape Interpretation; Land Use; Social Justice; Spatial Fix; Suburbs and Suburbanization; Urban and Regional Development; Urban Geography; Urban Land Use; Zoning

Further Readings

Gieryn, T. F. (2002). What buildings do. *Theory and Society, 31,* 35–74.

Harvey, D. (1973). *Social justice and the city.* London: Arnold.

Koskela, L. (Ed.). (2008). Developing theories of the built environment [Special issue]. *Building Research and Information, 36*(3).

Moore, J. B., & Glandon, R. P. (Eds.). (2008). Built environment and public health [Special issue]. *Journal of Public Health Management and Practice, 14*(3).

 # BUNGE, WILLIAM (1928–)

William Bunge, American geographer, is accurately self-described as a quantitative analyst, spatial theorist, radical humanist, and Marxist geographer, and even at times as "Wild Bill." Bunge did not operate well within the constrained modes of conventional academia, but his stellar significance in world geography is firmly based on a number of extraordinary theoretical and empirical contributions and on a fierce determination to participate in efforts to fight injustice and change society.

Bunge was born in LaCrosse, Wisconsin, of German American heritage and with possibly a 1/16th Ojibwa background. Despite his radicalism, he served in the U.S. Fifth Army during the Korean War (1950–1952), teaching atomic warfare at the Chemical, Biological and Radiological Warfare School. He received his MA in 1955 at the University of Wisconsin under Richard Hartshorne but shifted to the rival University of Washington for his PhD in 1960, as a prominent member of the pioneering "quantitative and scientific" geographers. His dissertation, *Theoretical Geography*, was published in Sweden in 1962 and is regarded as a forthright exposition that a theory for understanding the Earth's surface needed to be based on conceptions of geometric and topologic mathematics. That this great work could not find a publisher in the United States was a testament to a pattern of suppression of Bunge's philosophical and political radicalism.

He taught at the University of Iowa from 1960 to 1961 but was not retained, owing to his unconventional behavior. He taught later at Wayne State University in Detroit from 1962 to 1969 but was again fired for his incompatibility with normal academic rules. Subsequently, Bunge had a few visiting positions but mainly became an independent scholar. In 1970, he left the United States and has since lived in Canada, mainly in Quebec.

He founded the Detroit Geographical Expedition in 1968, an intense and astounding exercise in urban fieldwork. Out of this 2-year effort came perhaps Bunge's greatest personal and intellectual contribution, *Fitzgerald: The Geography of a Revolution* in 1971, detailing the operations of racism in a Detroit community; this was a new and provocative regional geography. The work is based on interviews, visual representations (photos, maps, and charts) combined with Bunge's intellectual and radical conceptions. The very idea of mapping variables never shown before—such as rat bites or hit-and-run deaths—was revolutionary. The fieldwork reinforced Bunge's intense concern for the health and survival of children. In *The Geography of Human Survival* of 1973, Bunge turns more broadly geographical and political, stressing on the inhumanity of dependence on machines, including those of war. *The Nuclear War Atlas*, published in 1988, also received significant acclaim and again ingeniously combines maps, graphs, and text.

Richard Morrill

See also Quantitative Revolution; Radical Geography

Further Readings

Bunge, W. (1962). *Theoretical geography* (Lund Studies in Geography). Lund, Sweden: Lund University.

Bunge, W. (1971). *Fitzgerald: The geography of a revolution*. Cambridge, MA: Schenkman Books.

Bunge, W. (1973). The geography of human survival. *Annals of the Association of American Geographers, 63*, 275–295.

Bunge, W. (1979). Fred K. Schaefer and the science of geography. *Annals of the Association of American Geographers, 69*, 128–132.

Bunge, W. (1988). *Nuclear war atlas*. New York: Blackwell.

BUSH FALLOW FARMING

Bush fallow farming currently occurs mainly in the humid tropics of Africa, South and Central America, southeast Asia, and parts of Oceania. In these regions, average temperatures for the coolest month are above 18 °C and annual precipitation exceeds potential evapotranspiration. Temperatures and moisture conditions in turn affect soil and vegetation characteristics. Soils are generally heavily weathered and lacking in basic nutrients, but on these soils grow a great biodiversity of plants in forest and savanna biomes. Bush fallow farming is a response and adaptation of farmers to this environment. It can be defined as an agricultural land use system and set of practices that is based on the rotation of land between different uses rather than a single permanent use, to achieve several agronomic goals. It involves the rotation of land between cultivation and fallow to create favorable agro-ecological conditions, such as regenerating soils through a vegetation-soil nutrient cycling. Following widely varying periods of cultivation of different crops, a farm plot is allowed to rest or remain fallow, and for soils, in particular, to regenerate as vegetation grows and returns large quantities of biomass to the soil—all this while another plot is brought into cultivation. Some authors have used the term *shifting cultivation* as a synonym for this type of farming, while others have suggested that shifting cultivation be limited to situations of land rotation that involve the shifting of homes of farmers when cultivation shifts to a new land patch.

A key to understanding all farming systems is to be conscious of variety within them and the changes they undergo over time. With the exception of the land rotation criterion, there are many varieties of bush fallow farming, as the ecological, social, economic, cultural, demographic, and technological conditions in which they occur vary. Tropical crops cultivated in bush fallow farming vary greatly in number and biodiversity. They include rice, maize, millet, sorghum, groundnuts, yams, taro/cocoyam, cassava/manioc, plantain, and a wide variety of vegetables and fruits, often intercropped in various combinations. Bush fallow farming occurs in areas of land abundance and those with less abundant land, in areas of low to medium population densities to those with high population densities, and in areas of varying cropping-fallow regimes, ranging from a few months of fallow to many years of fallow. Fallow periods may be shorter, the same length, or longer than the period of cultivation. Human-assisted fallow regeneration occurs in some practices, while greater reliance is placed on natural successional fallow growth for restoring soil fertility in others. The fallow vegetation is varied. Some bush fallow farming involves heavy tillage of soils, while others have minimum tillage. As commercialization and markets have expanded, the sizes of fields and farms, the land titles and rights, the types of labor arrangements, the technologies and inputs employed, and the proportion of output sold on markets have evolved. Given such complexity and variety, considerable caution is warranted when generalizations are made about such a varied system of farming.

Yet in spite of such complex variations in bush fallow farming, there is a dominant discourse that derides it as the preeminent threat to tropical forests and their massive biodiversity. Bush fallow farming is viewed as synonymous with swidden cultivation/slash-and-burn cultivation, where following a fallow period, temporary agricultural fields are cleared and prepared for cultivation by burning the vegetation. This burning, it is argued, destroys tree seeds, seedlings, saplings, and leads to the loss of many soil nutrients, such as carbon and nitrogen. Soil erosion and leaching exacerbates nutrient loss. Soil fertility thus declines rapidly with cultivation, leading farmers to abandon plots and clear forests for more fertile plots. Unfortunately, forest and soils are not allowed to regenerate long enough before land is put to cultivation, leading to detrimental ecological change. The loss of forest and their biodiversity is said to accelerate with the high rates of population growth and rising population densities in the tropical developing world. The loss of forest carbon sinks and their biodiversity, increasing carbon dioxide, and global warming lead to the verdict that the ecological impact of bush fallow farming is negative and that modernization is needed. Agronomic shortcomings in terms of the low productivity of bush fallow farming are also given as further justification for change.

Another discourse is more sensitive to the variations within bush fallow farming and acknowledges both its advantages and problems. On the positive side, some bush fallow farming is viewed as an ecologically sound, affordable system of farming—a traditional form of agroforestry—for small-scale farmers in the tropical developing world who can hardly afford modern agricultural inputs. Burning in this system leads to the release of large stores of nutrient ions in the ash, such as potassium and phosphorous, from the biomass into the soils. As land is temporarily abandoned for fallows long before soils are exhausted during cultivation, fallow vegetation rapidly accumulates nutrients in its biomass. The mineralization of litterfall, in particular, replenishes soil nutrients, leading to a vegetation succession process and the natural rehabilitation of forests. While population pressure, problems of access to land, and pressures of meeting rising costs of living have led to the intensification of production, it has led to the decline in the period of fallow and biodiversity degradation in some bush fallow farming areas; others have adapted through practices that nurture a vigorous growth of forests to meet agronomic objectives, such as soil regeneration and other social objectives. Practices include fire-avoidance strategies such as mulching rather than burning the cleared vegetation. One influential argument is that in parts of West Africa, forests are actually increasing as a result of such tree biodiversity nurturing practices. The generalization about the productivity shortcomings of such biodiversity-friendly practices compared with modern agricultural practices is also questioned.

Louis Awanyo

See also Agricultural Land Use; Agriculture, Preindustrial; Agrobiodiversity; Agroforestry; Deforestation; Shifting Cultivation

Further Readings

Nye, P., & Greenland, D. (1960). *The soil under shifting cultivation*. Bucks, UK: Commonwealth Agricultural Bureau.
Richards, P. (1987). *Indigenous agricultural revolution: Ecology and food production in West Africa*. Boulder, CO: Westview Press.

BUSINESS CYCLES AND GEOGRAPHY

Capitalism is a society and economy notorious for its instability over time. The history of capitalist economies is replete with boom and bust periods, epochs of rapid growth, high profits, and low unemployment followed by periods of crisis, economic depression, bankruptcies, and high unemployment. The Great Depression of the 1930s was the most spectacular example of capitalism's anarchical and self-destructive tendency. There are many theories of capitalism's cyclical behavior, including, for example, those from Karl Marx, Joseph Schumpeter, and Simon Kuznets. Taming business cycles has long been a central objective of Keynesian economic policy. This entry focuses on what is perhaps the most famous form of business cycles, Kondratiev waves, as well as how geographers have linked multiple cycles of investment and disinvestment to the evolution of local landscapes.

Kondratiev Waves

The most famous depiction of business cycles is that of Kondratiev cycles, named after the Soviet economist Nikolai Kondratiev (1892–1938), who first identified them in the 1920s. Examining historical data on changes in output, wages, prices, and profits, Kondratiev hypothesized that industrial countries of the world experienced successive waves of growth and decline since the beginning of the Industrial Revolution. Based on the emergence of key technologies and industries, these long cycles have a periodicity of roughly 50 to 75 years' duration. The reasons that underlie this duration of these waves reflect the long-term trends in the rate of capital formation and depreciation; as fixed capital investments reach the end of their useful economic life, the drag on productivity they create generates incentives to search for new technologies.

The first Kondratiev wave, at the dawn of the Industrial Revolution, arose on the heels of the textile industry and lasted from roughly 1770 to the 1820s, when the West was swept by a series of recessions, bankruptcies, and bank failures. The second wave, focused on railroads and the iron industry originated in the 1820s, peaked in the 1850s, and ended in the great round of consolidation in the 1880s and 1890s, particularly the depression of 1893, the second worst in world history. The third wave, associated with Fordist industries and mass production of goods such as automobiles, but also including electricity and chemicals, arose at the end of the 19th century, peaked around World War I, and collapsed suddenly during the Great Depression of the 1930s. The fourth Kondratiev, which was propelled by World War II, peaked in the 1960s, corresponded to the postwar wave of growth, and included major propulsive industries such as petrochemicals and aerospace; it ended with the petroshocks of the 1970s. Many believe that we live in the midst of a fifth wave, starting in the 1980s and centered on services and information technology.

Joseph Schumpeter, a famous German economist, explained Kondratiev's observation in terms of technical and organizational innovation. Schumpeter suggested that long waves of economic development are based on the diffusion of major technologies, such as railways and electric power. Throughout capitalist history, innovations have significantly bunched at certain points in time, often coinciding with periods of depression that accompany world economic crises. His very famous phrase "creative destruction" spoke of the birth of new forms of production and organization out of the ashes of the old, and has come to encapsulate capitalism's simultaneous enormous vitality and propensity to annihilate forms of social life.

Simon Kuznets described Kondratiev cycles in terms of successive periods of recovery, prosperity, recession, and depression. The upswing of the first cycle was inspired by the technologies of water transportation and the use of wind and captive water power; the second, by the use of coal for steam power in water and railroad transportation and in factory industry; the third, by the development of the internal combustion engine, the application of electricity, and advances in organic chemistry; and the fourth, by the rise of chemical, plastic, and electronics industries after World War II. In the present world economic crisis, with higher energy costs, lower profit margins, and decline of the Fordist forms of production, many ask whether a fifth wave is under way.

Some scholars argue that a fifth Kondratiev cycle began in the 1980s and is associated with information technology based on microelectronic technologies, including microprocessors, computers, robotics, satellites, fiber-optic cables, and information handling and production equipment, including office machinery and fax machines. The importance of information technology results from the convergence of communications technology and computer technology. Communications technology involves the transmission of data and information, whereas computer technology is concerned primarily with the processing, analysis, and reporting of information. Such technologies allow for extraordinarily low costs of storing, processing, and communicating information. In this perspective, the late 20th century was a prolonged period of social adaptation to the growth of this new technological system, which is affecting virtually every part of the economy, not only in terms of its present and future employment and skill requirements but also in terms of its future market prospects.

At the front of the contemporary wave are business services such as advertising, purchasing, auditing, inventory control, and financing. The defining characteristic of these new services is that they create and manipulate knowledge products in almost the same way as the manufacturing industries that peaked in earlier rounds transformed raw materials into physical products. These services have become the salient forces of the new postindustrial society and the information economy. These are the leading forces that are restructuring the geography of manufacturing, because they are the basis of productivity increases—technological innovation, better resource allocation through expert systems, increased training, and education. As a result, the new geography of world cities emerged: a command and control economy centered on world cities such as New York, Los Angeles, London, Paris, Tokyo, Hong Kong, and Singapore.

Business Cycles and the Spatial Division of Labor

The spatial dimensions of business cycles are complex and important, revealing that uneven development in time and space are two sides of the same coin. Uneven development in space occurs through the specialization of production in different areas, including the comparative and competitive advantages that regions enjoy at different moments in time. Given the fluidity of capitalism, however, there is no reason for a region or country to enjoy its advantages in production indefinitely. Capital, labor, and information move across space, continually changing the conditions of profitability in different places. As Doreen Massey argued in a famous volume, *Spatial Divisions of Labor*, uneven development in time is manifested when a region's or nation's comparative advantage is created and lost as capital creates and destroys regions over successive business cycles. The loss of comparative advantage makes a region attractive to firms: It offers pools of unemployed, and hence cheap, labor and infrastructure and often other advantages as well. In short, regions abandoned by capital may be ready to be recycled for a new use.

Over different business cycles, several industries may locate in one region, each leaving its own imprint on the local landscape. Each industry constructs a labor force, invests in buildings, and shapes the infrastructure in ways that suit its needs (and profits). From the perspective of each region, therefore, business cycles resemble waves of investment and disinvestment. Each wave, or Kondratiev cycle, deeply shapes the local economy, landscape, and social structure and leaves a lasting imprint on a region that is not easily erased.

For example, the textile industry in New England created its industrial landscape in the 19th century, ranging from small mills located on streams to the large factories in Lowell, Massachusetts, or Manchester, New Hampshire. These landscapes, including the people who inhabited them, persisted long after the industry abandoned New England for the South in the early 20th century. For many years, New England was a relatively poor part of the country, with high unemployment rates. By the 1980s, however, a new wave of production had centered on the region—the electronics industry. Firms producing computer hardware and software found the human resources of the Boston metropolitan region attractive, including the famous Route 128 corridor, and the local geography of this industry was shaped to no small extent by the residues of earlier ones.

In short, each set of investment/location decisions in a region is prestructured by earlier sets of decisions. Thus, as their comparative advantage changes, regions accumulate a series of different imprints: Each wave is shaped by and transforms the vestiges of past waves. Thus, regions are unique combinations of layers of investment and disinvestment over time. Such an approach explains the unique characteristics of places through their economic histories. Because capitalism constantly reproduces spatial inequality by diverting capital from low-profit to higher-profit regions, individual places are perpetually open to the lure of new forms of investment and vulnerable to the risk of being abandoned by capital. This view allows us to integrate the specifics of regions with broader understandings of capitalist processes.

Barney Warf

See also Crisis; Economic Geography; Massey, Doreen; Palimpsest; Uneven Development

Further Readings

Berry, B. (1991). *Long-wave rhythms in economic development and political behavior.* Baltimore: Johns Hopkins University Press.

Massey, D. (1984). *Spatial divisions of labor: Social structures and the geography of production.* New York: Methuen.

 # BUSINESS GEOGRAPHY

Business geography integrates geographic analysis, reasoning, and technology to improve business decisions. This ability to enhance business decisions distinguishes business geography from the traditional explanatory frameworks of economic and urban geography. Business geography, moreover, goes far beyond merely the application of geospatial technologies to business requirements. Business geography combines a keen understanding of geospatial technologies and business systems and operations, which together can significantly improve real-time, real-world business decisions.

Business geography has its roots in the traditional location analysis models and applications, mainly siting retail enterprises. These locational techniques emanated from analogue gravity modeling and transformed over the decades into more complex probabilistic expressions of distance-decay interactions of retail consumer demand. Of these advanced models, David Huff's is the most well-known. Although location analysts today incorporate geospatial technology, data sets, and analytic techniques, many analysts also continue to rely on their qualitative judgments honed from practitioner experiences of applying the science and art of business geography.

Business geography has evolved to provide solutions to complex spatial management objectives, including, for example, real-time spatial tracking of products from manufacture to consumer purchase, to containers, to vehicles, and to employees. These solutions move beyond barcodes to radio frequency identification (RFID) tags and satellite global positioning system (GPS) triangulation. Personal privacy concerns about RFID tags and a constant need for updated GPS data raise questions regarding these two tracking systems.

Private sector applications of spatial forecast modeling, geographic information systems (GISs), remote sensing, and geovisualization work is growing remarkably and is expected to overtake the government as the largest geospatial employment market by 2020. Early business adopters of these integrated "geographic management systems" in enterprises such as real estate, marketing, transportation, banking, insurance, and communications have realized substantial cost efficiencies.

Business Geography in Academia

The Association of American Geographers' Web site (aag.org) explicitly enumerates professional positions that pertain to spatial analysis of shopping habits, regional sales characteristics, route delivery management, and real estate appraising. Yet most geography departments in the United States have not developed undergraduate or graduate curricula needed to position themselves and their students for the vigorous employment growth in the business sector. An analysis of the curricula of U.S. geography departments identified only five offering business geography programs. Several Canadian geography departments, following the strong European (particularly British) tradition, have robust business geography programs. Moreover, only five U.S. colleges of business have GIS courses in their curricula. Although the expansion of geotechnologies, particularly GIS, will produce up to $100 billion in the private sector in the near term, geography departments and colleges of business have not recognized the huge market in the U.S. economy for college graduates with a combination of geospatial skills and business knowledge. However, professional initiatives such as the globally recognized Certified Commercial Investment Manager (CCIM) program, with its rigorous geospatial real estate market analysis, will undoubtedly change the thinking in colleges of business regarding the analytical power of geotechnologies and the growing employment opportunities in business geography.

Business Geography Academic Curriculum

Most U.S. geography departments are located in academic colleges focusing on the liberal arts or science and have little affinity for business. Nevertheless, the following model suggested for a business geography undergraduate degree program, operating within each university's general education requirements and melding into a department's fundamental knowledge and skills framework, will guide geography students into the burgeoning business geography market:

1. Geography Major
   a. *Core:* business geography (including project management topics), economic geography, locational analysis
   b. *Geography techniques and methods:* introductory cartography, introductory GIS, advanced GIS, quantitative analysis/spatial statistics
   c. *Internship/practicum:* emphasizing the use of spatial analysis in business operations
2. Business Minor
   a. Marketing, retail, real estate, management (particularly operations or systems management), economics, finance, or entrepreneurship; or individually tailored programs that include courses from more than one of these fields

The Future of Business Geography

The future employment demand is optimistic for geography and business graduates who have geospatial skills if the forecast by the U.S. Department of Labor is accurate in identifying geotechnology as one of the three most important emerging fields, along with nanotechnology and biotechnology. College graduates who have knowledge of spatial analysis and skills in geospatial techniques, as well as an understanding of business concepts and operational frameworks, will be exceedingly marketable. Yet few geography departments in the United States currently have curricula and faculty that relate well to business contexts, much less offer degrees in business geography. If geography departments do not act to fill the substantial future market for business geographers, then colleges of business will.

Lawrence E. Estaville and Grant Thrall

See also Business Models for Geographic Information Systems; Cost-Benefit Analysis; Dangermond, Jack; E-Commerce and Geography; Economic Base Analysis; Economic Geography; Environmental Impact Assessment; GIS Implementation; Location Theory; Real Estate, Geography and; Regional Economic Development; Retail Trade, Geography of

Further Readings

Damowicz, E. (2005). Retail trade area analysis using the Huff model. *Directions Magazine*. Retrieved April 20, 2008, from www.directionsmag.com/article.php?article_id=896&trv=1

Estaville, L. E. (2007). Colleges of business and GIS. *Journal of Real Estate Literature, 15,* 441–448.

Estaville, L. E., Keys-Mathews, L., Brown, B. J., & Strong, W. R. (2005). Business geography: Development of a curriculum model. *Papers of the Applied Geography Conferences, 28,* 292–300.

Estaville, L. E., Keys-Mathews, L., Brown, B. J., & Strong, W. R. (2006). Educating business geographers. *Geospatial Solutions.* Retrieved April 25, 2008, from www.geospatial-solutions.com/geospatialsolutions/article/articleDetail.jsp?id=325930

Gewin, V. (2004). Mapping opportunities. *Nature, 427,* 376–377.

Richardson, D., & Solis, P. (2004). Confronted by insurmountable opportunities: Geography in society at the AAG's centennial. *Professional Geographer, 56*(1), 4–11.

Thrall, G. I. (2002). *Business geography and new real estate market analysis.* New York: Oxford University Press.

 # BUSINESS MODELS FOR GEOGRAPHIC INFORMATION SYSTEMS

Business use of geographical information systems (GIS) is growing as spatial costs are reduced and more spatial information becomes available. GIS can be used by business customers, suppliers, internal analysts, decision makers, managers, and executives. A number of models are available to explore, analyze, predict, and strategize with spatial information. Examples are geological modeling of the Earth's fossil deposits for the petroleum industry, identification of "hot sales zones" for the real estate industry, siting models of outlets in fast foods, and decision support in insurance for pricing of policies. This entry covers the business topics of spatial decisions support, enterprise applications of GIS, Web and mobile spatial applications, spatial data, and business GIS strategies.

GIS to Support Business Decisions

Spatial information can assist in decision making. Decision support systems (DSSs) are in widespread use to guide managers, analysts, and specialists by recommending the optimal decision. A spatial decision support system (SDSS) does all this and includes spatial boundary layers, spatial modeling, and analysis features.

An example of an SDSS is a large insurance firm that insures for property, casualty, and workers' compensation. It can decide how much insurance exposure is allowable at a single location, conduct risk analysis to analyze properties for property insurance, and make spatial decisions in the event of catastrophes. When Hurricane Katrina struck in 2005, the insurance firm used government and commercial topographic and flood maps to model the depth of water in the affected metropolitan areas and better predict the subsequent extent of damages.

Enterprise Applications of GIS

Large and medium-sized firms often use an Enterprise Resource Planning (ERP) system that integrates into one system the firmwide processing for marketing and sales, customer relationship management (CRM), finance and accounting, human resources, manufacturing, inventory, and supply chain. GIS software can be connected to the ERP system to provide spatial analysis capabilities. Although ERPs embed GIS as a feature, most of them allow connection with GIS through remote function calls (i.e., the ERP invoking functions from the GIS and vice versa),

third-party connectors, and/or middleware. Chico's, a medium-sized women's apparel chain, has integrated enterprise CRM with GIS. CRM features enable personalized customer care, which tracks customer shopping patterns spatially and forecast by geographic area, based on historical georeferenced sales information—that is, how customers would respond to a promotion. The advantage of connecting ERP and GIS is that there is a much greater amount of attribute information from the functional areas of the whole corporation available for spatial analysis.

Web and Mobile Spatial Applications for Business

An expanding aspect of GIS in business is Web-based and mobile-based spatial applications. For a Web-based architecture, one or more spatial Web servers combine with application and data servers to provide mapping capabilities to browser-based end users. This has the advantages of a service area worldwide, greater ease of use, and improved access for users. However, this architecture requires that the business either acquire and support the servers and spatial networking architecture internally or make often costly arrangements with an outside provider to provide it. For instance, Lamar Advertising, one of the largest outdoor advertising sign firms in the world, combines its own server information with a commercial spatial Web service to provide 2D (two-dimensional) and 3D maps of prospective sites to its U.S. sales managers, who in turn use these maps in sales presentations to customers and their advertising consultants. It allows Lamar managers to gain easy access to maps and more quickly respond to sales opportunities.

Mobile map services are also expanding as 3-G (third-generation) and 4-G mobile devices with GPS become prevalent worldwide. The GIS capability is mobile browser based. The advantages of mobility are that field personnel can have map and spatial analysis available as they move around and when they need it. For example, spatial mapping is prevalent for field maintenance employees of middle- and large-sized utilities. They update their maintenance information in real time at field sites, as well as visualize utility networks and other assets in fieldwork.

Managing Spatial Data

Underlying business spatial applications are databases that store both attribute and boundary layer information. Most business GIS desktop software is connected to external databases for attribute information while storing the boundary information as specialized files or geodatabases. The principles of database design and management are essential in organizing and maintaining these databases. For servers providing Web-based or mobile-based GIS, the databases likewise reside on specialized database servers. Some firms such as Oracle have provided spatial database products that integrate attribute and boundary data and include spatial functionality. Another option for businesses is to use spatial data warehouses, which provide long-term archiving of time slices of all the firm's attribute and spatial information. They are useful for spatial trend analysis, data mining, and long-term record keeping.

Models of GIS and Business Strategy

Since GIS is a competitive factor in industries such as fast food, real estate, and transportation, firms need to consider management models of how to best leverage GIS to competitive advantage. One model that is useful is the GIS-IT business alignment model. As discussed by Pick in 2008, businesses can leverage GIS competitively better by aligning business processes with GIS processes and IT processes. This is best seen in companies that have corporate goals to provide spatial services or products. For many of these firms, the GIS is aligned to achieve the business goals for these services and products, and IT is also aligned with the GIS and business goals. Case studies have shown that the alignment is sometimes precluded if the GIS and IT departments are merged or are working closely together or if business management does not communicate well with the GIS unit.

James B. Pick

See also Business Geography; Database Management Systems; Web Service Architectures for GIS

Further Readings

Drummond, J., Billen, R., João, E., & Forrest, D. (Eds.). (2007). *Dynamic and mobile GIS*. Boca Raton, FL: CRC Press.

Pick, J. (2005). *Geographic information systems in business*. Hershey, PA: Idea Group.

Pick, J. (2008). *Geo-business: GIS in the digital organization*. New York: Wiley.

 # BUTTIMER, ANNE (1938–)

Anne Buttimer, a world famous geographer, gained exceptionally high regard through her work in social and humanist geography, her understanding of the human life experience, and her work on planning and sustainable development, all within a conception of geography as a social science.

Buttimer was born in rural Ireland, educated (BA 1957, MA 1959) at the National University of Ireland, and destined for service in the Dominican Sisters of Providence. She moved to Seattle in 1960 to obtain a Washington State teaching certificate at Seattle University. She joined the PhD program at the University of Washington, completing her dissertation in 1965. She taught initially at Seattle University but moved to the University of Glasgow as a lecturer in urban studies, followed by a long tenure at Clark University—from postdoctorate to professor, from 1970 to 1981—during which she was at times a visitor to Lund University in Sweden. She moved to Lund as a research fellow in humanities and social sciences from 1982 to 1988, where she met and married engineering professor B. Broberg. After 2 years at the University of Ottawa, she returned to Ireland in 1991, where she remained till her retirement in 2003, as professor and head of the department of geography at University College Dublin.

She has held numerous professional positions—on editorial boards; as a member, secretary, or chair of commissions; and as officer of professional organizations, such as the Association of American Geographers, the Irish National

Committee for Geography, the Institute of British Geographers, and the Acadameia Europaea. Especially significant has been her many roles in the International Geographical Union (IGU), beginning in 1973 and culminating in her service as president of the IGU from 2000 to 2004. She has received many awards and honors, including two honorary doctorates.

The major themes of Buttimer's research and writing have been (1) social geography and social space, (2) *genres de vie* or understanding the life world of people in places, (3) dialogue—orally derived biographies of practicing geographers, (4) the philosophy of a humanistic social science, and (5) humanist planning and sustainability. Her initial work was to help establish social geography within American and British geography and develop the concept of social space within the emerging social scientific geography, challenging simplistic and mechanical modeling and arguing that understanding social space required incorporation of cultural diversity and competing individual values.

A major aspect of Buttimer's social geography was her reinvigoration of concepts of *genres de vie*, or life worlds, as social space experienced and created, not just perceived by individuals and groups. At Lund, she went beyond the use of the developing field of time-geography at the daily level to that of people's lifetimes. This led to her important and intellectually valuable dialogue project, which has resulted in hundreds of fascinating interviews concerning the actual practice of geography, which occupied much of her time at Lund.

A parallel theme throughout her career was her concern with the philosophy of geography, ccharacterized by a spirit of reconciliation and collaboration and incorporating a humanist concern for people's different experiences and values into a social scientific geography, not rejecting the quantitative tradition but broadening its perspective and capability. Yet another related theme has been a concern with effective planning—a practice in need of a humanistic, social science understanding. This began with work in the slums of Glasgow and continued with work in IGU commissions and current work in sustainable development.

Anne Buttimer remains a person inspired by her early affiliation with the Dominican order, leading a life of service and dedication to an ideal of humanism as knowledge integration.

Richard Morrill

See also Existentialism and Geography; Humanistic Geography; Phenomenology; Vidal de la Blache, Paul

Further Readings

Buttimer, A. (1976). Exploring the dynamics of lifeworld. *Annals of the Association of American Geographers, 66,* 277–292.

Buttimer, A. (1983). *Practice of geography*. London: Longman.

Buttimer, A. (1990). Geography, humanism and global concerns. *Annals of the Association of American Geographers, 80,* 5–34.

Buttimer, A. (1993). *Geography and the human spirit*. Baltimore: Johns Hopkins University Press.

CADASTRAL SYSTEMS

Cadastral systems consist of official written records and the corresponding mapped registries of land. They include information such as boundaries of parcels, precise location, value, ownership, and a record of interests in the land (e.g., rights, restrictions, and responsibilities). Cadastral systems are thus composed of two main components: (1) the land registries and (2) the cadastral parcel map based on surveying (Figure 1). Together, they form the legal basis of land ownership in many parts of the world and are used when partitioning the land, when establishing and documenting legal rights to land, for maintaining legal title records, and in property tax assessment by government.

Cadastral systems have a profound influence on geography: The patterns created by cadastral partitioning schemes are among the longest-lasting physical characteristics of the landscape and are difficult to reshape once they have been implemented. We live among and still see evidence of cadastral systems developed hundreds, sometimes thousands, of years ago.

The *Cadastre*, or Parcel, in Land Administration

The basic building block of land administration systems is the cadastral lot or parcel, sometimes called the *cadastre*. Each parcel is given a unique code or identifier, such as an address, map coordinates, or a lot number shown on a survey plan or map and cross-referenced to the land registries (Figures 2A and B). Cadastral maps are typically large scale, ranging between 1:500 and 1:10,000 (1 inch equals between approximately 42 and 830 feet; see Figure 3). In the United States, most real property can be described in one or more of three ways: (1) by metes and bounds (using a combination of references to local landmarks such as roads, trees, or geological features; cardinal directions; measurements in feet and inches; and latitude and longitude coordinates); (2) by means of the PLSS (Public Land Survey System, in which property is identified by its location within a grid subdivided by successive grids and referenced to standard meridians); and (3) by platting (lots and blocks).

Geodetic surveying, the profession most closely associated with cadastral systems, includes boundary surveying, land information systems (LIS) and geographic information systems (GIS), hydrography, photogrammetry and remote sensing, minerals and mining surveying, cartography, and geodetic networks and reference systems. The Fédération Internationale des Géomètres (FIG, the International Federation of Surveyors) was established in 1878 as an international body representing all surveying disciplines.

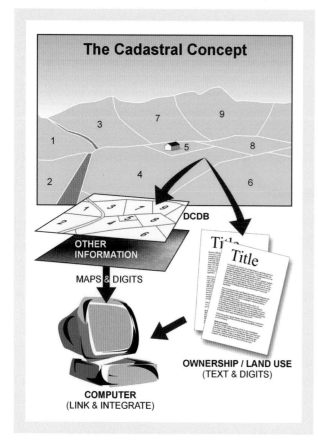

Figure 1 The two main components of the cadastral system: the written land registry and the surveyed parcel map

Source: International Federation of Surveyors, FIG Commission 7. (1995). *Statement on the cadastre* (Publication No. 11). Copenhagen, Denmark: Author. Reprinted with permission.

History of Cadastral Systems

The word *cadastral* is attributed both to the Greek *katastikhon*, a list or register, and to the late Latin *capitastrum*, a register of the "poll tax," which was a per capita "head tax." It is thought that both the Egyptians and the Mesopotamians used cadastral systems starting about 2300 BC, and although few concrete examples of these cadastral maps exist today, we know that these civilizations had the technical and administrative expertise to conduct cadastral surveys. The Romans were masters at land partitioning and incorporated cadastral systems into their overall governance from at least 1600 BC, using them to grant lands, administer their empire in Europe, Asia, and North Africa, and assess and collect tax revenues (Figure 4).

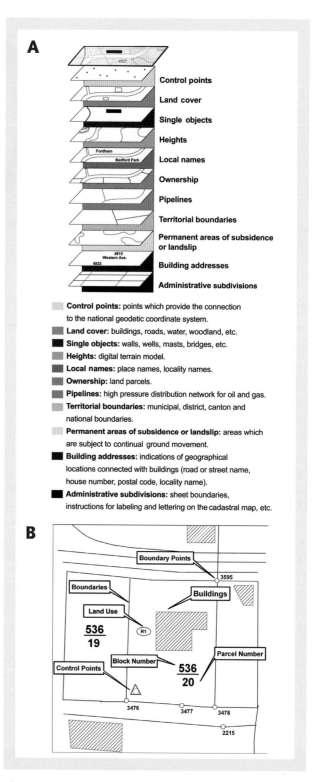

Figure 2 The construction of a cadastral mapping system

Source: Figures redrawn by Brian Morgan, Urban GISc Lab, City University of New York, Lehman College, from *Cadastral template—Country data: Switzerland.* (2003). Retrieved December 19, 2009, from Cadastral Template Web site: www.cadastraltemplate.org/countrydata/ch.htm.

Figure 3 Example of a traditional cadastral map

Source: From Cadastral Template: A Worldwide Comparison of Cadastral Systems. (2003). *Cadastral template—Country data: Switzerland*. Retrieved December 19, 2009, from www.cadastraltemplate.org/countrydata/ch.htm.

After the fall of the Roman Empire and the beginning of the feudal era, comprehensive mapping of real property was largely discontinued, although isolated historical events highlight the continued importance of recording real property information. For instance, the Domesday Book, commissioned by William the Conqueror of England in 1086, was a population census as well as a record of real property and livestock ownership and valuation for a large part of the country; it was instrumental in consolidating royal power and raising tax revenues for the crown.

In the nascent capitalist nations of 16th- and 17th-century Europe, land reemerged as a private commodity, and it once again became vital to administer this property by mapping and recording information about ownership and rights held. Private property took the place of feudal and communal property, coinciding roughly with the European "age of exploration" and the wholesale acquisition of new continents full of land, which

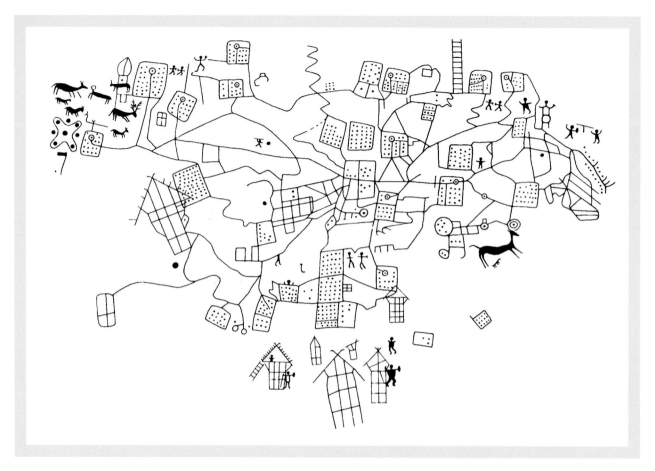

Figure 4 Early cadastral map from Northern Italy. Streams, irrigation channels, and roadways are delineated as lines. A circle with a dot represents a well, and a rectangular piece of land with regular dots is a cultivated field.

Source: From Larsson, G. (1991). *Land registration and cadastral systems: Tools of land information and management*. New York: Longman Scientific and Technical. Copyright © 1991 by Longman. Reprinted by permission of Pearson Education Limited.

also had to be surveyed, mapped, recorded, and distributed. The birth of modern cadastral mapping took place in this period, and in most of North and South America, Australia, India, and other colonized regions, this process was in conflict with the indigenous populations' beliefs about land ownership. In the United States, both the Land Ordinance of 1785 and the Homestead Act of 1862 laid the foundations of American land policy by surveying, mapping, and giving away millions of acres of land, most of which was already inhabited or otherwise used by the original populations. In 18th- and 19th-century Western Europe, cadastral surveys served primarily as a means of collecting taxes, although by the 19th century, they also assumed importance in providing security of private property rights.

Importance and Major Societal Benefits of Cadastral Systems

The manner in which a society perceives its relationship to land is emblematic of its general worldview. In many cultures today, land forms the basis of wealth, and therefore, it is thought necessary to thoroughly inventory it and document its value and ownership in order to make effective decisions about its use, access, and control.

Many governmental agencies and private firms directly depend on reliable information about land. Functions such as urban and rural land use planning and zoning; environmental analysis and protection; natural resources management; parks and recreational management and planning;

Figure 5   A 19th-century European cadastral map: Village of Pielnia, Poland, 1852

Source: Public domain.

housing and preservation; economic development; emergency preparedness and disaster relief; real estate markets; municipal finance and property tax assessment and collection; infrastructure maintenance and construction; and many social services depend on the existence of accurate and up-to-date information on land at the parcel level.

Having comprehensive cadastral information has administrative and legal benefits. For instance, it allows the establishment of a land market, permitting land rights to be bought, sold, mortgaged, and leased; and it enables easier and better-protected real property transactions, resulting in less litigation, such as property boundary disputes. Thus, cadastral systems can be seen as a stabilizing influence on a nation's financial and legal systems, promoting a sense of security, protection, and fairness.

Ongoing Issues: Reform, Equity, and New Technologies

The definition and distribution of land rights are undertaken by most societies and are based on the society's social, political, and religious structures. These rights, in turn, influence how labor and capital are invested in the land. There are a number of common land tenure forms, such as the Anglo-American, the Roman-French-German, the Communist/Socialist (state and cooperatives), Islamic, and community. In developing countries, communal land tenure is often converted to individual tenure in an effort to increase economic productivity, to modernize, and to promote nation building. This change from communal to individual land tenure also occurs where planned economies are supplanted by market economies, and it usually requires a completely new cadastral land registration system to be instituted.

The establishment or reform of cadastral systems and their associated legal structures is a complicated matter, presenting special problems. Standardizing cadastral systems and creating an international cadastral template may not be feasible, due to the many models that currently exist and countries' different land management needs. In many places, it is critical for cadastral systems to incorporate customary and "informal" rights (e.g., the traditional right to harvest woodland resources, rights to common pasture grounds, or seasonal rights to certain land uses). In other places, "3D" (three-dimensional) rights and responsibilities are of paramount importance, which include underground mineral rights, air rights, and riparian and groundwater rights, in addition to the typical "2D" surface land rights (Figure 6).

Over the past decade, advances in geographic technologies have made it likely that many more countries can have cadastral systems based on common concepts, allowing data exchange and interoperability. Cadastral systems can be improved through innovations in GIS, new geospatial databases, integration with computer-aided data (CAD) and aerial photos, automated scanning, updating with mobile GIS, online search and mapping capabilities, new measurement methods, new ways to access large imagery collections, browser-based editing, and citizen access to this information and functionality. However, there are ethical issues involved: The disproportionate availability and access to geotechnologies and the expertise necessary to use them can result in inequity, and the associated costs of technology may be burdensome for developing nations.

On the other hand, there are a number of ethical reasons to establish new cadastral systems and undertake cadastral reform, including poverty mitigation and gender equity. There is a strong correlation between poverty and access to land. Through an expansion of rights to own land, poverty and inequity can be alleviated. Land mapping is one way to formally document land rights that may have been held by custom or traditionally but not legally. Cartographic and geodetic information helps modernize the cadastral structure and establishes a database of land information, which can augment a legal case for ownership. When there are constraints on access to land, residents have difficulties obtaining financing. For example, in many permanent slum areas with marginal housing and tenuous ownership, credit is usually only available for people who hold formal title to the land. There are often no cadastral records of these areas, or the record does not reflect the actual users/residents. Something as mundane as structure numbers or a viable street numbering system can enable proper cadastral registration and help secure financing for individuals or for property owners in common.

Development of cadastral systems can also play an important role in gender equity, since in many countries women traditionally have limited rights of ownership and inheritance. Yet this is

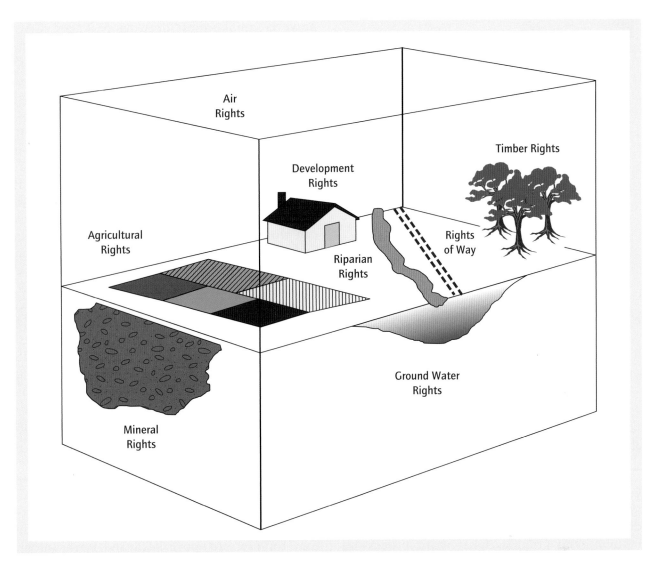

Figure 6 The three dimensions of land rights. Cadastral systems may also record and map three-dimensional rights attached to the land, such as underground mineral rights, air rights, and riparian and groundwater rights.

Source: From Dale, P., & McLaughlin, J. (1988). *Land information management: An introduction with special reference to cadastral problems in third world countries* (p. 3). Oxford, UK: Clarendon Press. Copyright © by Clarendon Press, 1988. Based on Platt, R. (1975). *Land use control* (Resource Paper No. 75-1). Washington, DC: Association of American Geographers.

sometimes by custom and not by law, so instituting a cadastral system and official recording of women's ownership on titles and deeds can have the effect of clarifying joint or individual ownership, ensuring rights in decision making about property management, and providing protection of property in case of divorce or death of a spouse.

In developing countries, land administration can be a critical element in the wider economic development agenda, and in both developing and developed countries, cadastral systems are crucial to support decision making for a sustainable environment. As Will Rogers, the American humorist, famously said, "Land—they ain't making any more of the stuff."

Juliana A. Maantay

See also Cartography, History of; GIS in Land Use Management; Surveying; Township and Range System; Zoning

Further Readings

Dale, P., & McLaughlin, J. (1988). *Land information management: An introduction with special reference to cadastral problems in Third World countries.* Oxford, UK: Clarendon Press.

Dale, P., & McLaughlin, J. (2000). *Land administration (spatial information systems).* Oxford, UK: Oxford University Press.

Kain, R., & Baigent, E. (1992). *The cadastral map in the service of the state: A history of property mapping.* Chicago: University of Chicago Press.

Larsson, G. (1991). *Land registration and cadastral systems: Tools of land information and management.* New York: Longman Scientific and Technical.

Price, E. (1995). *Dividing the land: Early American beginnings of our private property mosaic.* Chicago: University of Chicago Press.

CAD SYSTEMS

Computer-assisted (or aided) drafting (CAD) systems are used to produce maps, diagrams, drawings, plans, and charts. CAD systems are to maps and graphics what word processing packages are to the written word: an easier, digital method for creating and editing documents, which were once drafted by hand. CAD systems are used extensively to map infrastructure (e.g., roads, utilities, property, telecommunication networks) as well as for architecture plans and renderings, mechanical designs, and drawings used in civil and industrial design. Although geographers use these systems largely for mapping geospatial information, CAD systems are remarkably versatile and are used for applications as diverse as designing microchips; shipbuilding; aircraft, aerospace, and satellite design; automobile and engine design and rendering; textile and fashion design; robotics; and computer-assisted manufacturing.

CAD systems, such as Illustrator or Corel-Draw software for drawing, offer great flexibility in representing graphical and spatial entities; however, unlike most drawing packages, CAD systems are designed for drafting especially complex and highly precise maps and plans. CAD systems are also very closely related to vector-based geographic information systems (GIS) and often provide inputs or outputs for GIS analysis. Early CAD systems focused on ease of graphic design and editing and, in so doing, relied on Cartesian coordinate systems and lacked database interfaces for linking graphic and spatial features to nonspatial attributes—both features essential to GIS. Contemporary CAD systems can work with geographical coordinate systems and readily permit database linkages between spatial and nonspatial data, meaning that there is no longer a sharp divide between CAD and GIS. GIS systems usually offer a wider range of analytic, geospatial, and modeling options than do CAD systems, particularly in the domain of topological modeling, but even these types of GIS functionality can be built into CAD systems using add-on software, scripts, and macros. In the areas of terrain modeling, surface modeling and rendering, and three-dimensional modeling, mapping, and animation, GIS and CAD are moving ever closer together because their modeling, visualization, and computational models are closely interrelated.

CAD, vector-based GIS, and drawing software share many concepts that make it relatively easy for users to move back and forth among systems. Phenomena are represented graphically as points, lines, and areas/shapes. Features of particular types (e.g., roads, streams, or census boundaries) are organized into common layers or levels. All the systems offer tremendous flexibility in how the data are symbolized, for example, by hue, value, texture, size, orientation, and shape. Principles of effective cartographic and visual symbolization and display apply just as strongly to CAD systems as they do to GIS and drawing software.

CAD files are used widely to archive and exchange geospatial information. Commonly used file types that can be found on the Web are .dwg and .dxf (AutoCAD formats) and .dng (Microstation/Intergraph format), which cannot be readily imported into most GIS.

Other related terms are *computer-assisted mapping* (CAM), *computer-aided design* (CAD), *automated mapping/facilities management* (AM/FM), *computer-assisted drafting and design* (CADD), and *computer-assisted cartography.*

Kenneth E. Foote

See also GIS Implementation

Further Readings

Jones, C. (1997). *Geographical information systems and computer cartography.* New York: Prentice Hall.

 # CÂMARA, GILBERTO (1956–)

Gilberto Câmara is a Brazilian researcher in the areas of geographical information science, spatial databases, spatial analysis, and environmental modeling. He was director for Earth observation (2001–2005) and general director (2006–2010) of Brazil's National Institute for Space Research (INPE) and is internationally known for defending open access to geospatial data and software. Câmara is responsible for setting up the free and open policy for remote sensing images from the China-Brazil Earth Resources Satellites (CBERS) worldwide and for creating the Remote Sensing Data Center of INPE, which has made decades of imagery freely available on the Internet. This made Brazil the world's largest distributor of remote sensing imagery (2005–2008). He was also responsible for setting up a methodology for real-time detection and monitoring of deforestation in the Brazilian Amazon rain forest.

Câmara is a professor in INPE's graduate programs in remote sensing and computer science and has published several books and more than 100 papers in conference proceedings and journals. He led the development of SPRING, a free GIS that was innovative in the use of the concepts of object-oriented modeling within a GIS environment, and of TerraLib, an open-source GIS library. These contributions have made him

one of the leaders in the establishment of the infrastructure of geospatial data and the GIS community in Brazil.

Pedro Ribeiro de Andrade

See also GIScience; GIS Software; Remote Sensing

Further Readings

Câmara, G., Fonseca, F., Monteiro, A., & Onsrud, H. (2006). Networks of innovation and the establishment of a spatial data infrastructure in Brazil. *Information Technology for Development, 12*(4), 255–272.
Câmara, G., Souza, R. C., Freitas, U. M., & Garrido, J. C. P. (1996). SPRING: Integrating remote sensing and GIS with object-oriented data modelling. *Computers and Graphics, 15*(6), 13–22.
Câmara, G., Vinhas, L., Queiroz, G. R., Ferreira, K., Monteiro, A. M. V., Carvalho, M., et al. (2008). TerraLib: An open-source GIS library for large-scale environmental and socio-economic applications. In B. Hall (Ed.), *Open source approaches to spatial data handling.* Berlin, Germany: Springer.

 # CANADIAN ASSOCIATION OF GEOGRAPHERS

The Canadian Association of Geographers (CAG)/L'Association canadienne des géographes (ACG) is the national organization of geographers in Canada and for geographers outside Canada who have scholarly interests in Canada. The professional association includes members from universities as well as both public and private sectors. Its mission is to encourage geographic research, promote geographic education, and recognize geographic excellence through various awards. In addition, the organization is dedicated to cooperation and participation with other national and international geographic organizations and interdisciplinary organizations.

In 1950, a group of geographers representing the Canadian government's Geographical Branch and several Canadian universities met to discuss the creation of a Canadian organization of geographers. They contacted other geographers in Canada and invited them to an inaugural meeting in May 1951, which was held at McGill University in Montreal. Although the precise number of geographers who participated in the meeting is not known, various sources suggest that number to have been about 60. By the end of the first day, the CAG was formally organized and officers elected; on the second day, eight papers were presented, seven of which were subsequently published in the first volume of *The Canadian Geographer.*

The CAG publishes *The Canadian Geographer*, a quarterly journal and the principal means through which it disseminates geographic research. In addition to this journal, the CAG publishes its *Newsletter* six times a year. The *Newsletter* is the venue in which information about professional jobs in geography, profiles of nominees for various elected positions, and information on study groups, regional divisions, and geographers in the news are made available to CAG members. The organization also publishes an annual directory that provides information about each geography department within Canada, its faculty, and the current research projects and recent publications of faculty members.

Although most of its members are associated with universities, the CAG is committed to geographic education at all levels. Since its establishment, the CAG has promoted geography in Canada through publication of its journal, newsletters, and annual meetings. The CAG hosts a national meeting each year. The event attracts geographers from across the country and many others from the United States and Europe. Both students and faculty members alike present at these meetings. The CAG has five regional divisions: (1) Western Division (British Columbia, Alberta, Yukon, and the Northwest Territories); (2) Prairie Division (Manitoba, Saskatchewan, and Nunavut); (3) Ontario Division; (4) Quebec Division; and (5) Atlantic Division (New Brunswick, Nova Scotia, Prince Edward Island, and Newfoundland). These divisions also host annual meetings in their respective regions and often share those meetings with corresponding regional divisions of the Association of American Geographers.

In addition to its regional divisions, the CAG also has 15 study groups that provide a forum for dis-

cussion and promotion of specific subdisciplines of geography. The groups are (1) Women and Geography; (2) Environment and Resources; (3) Tourism and Recreation; (4) Economic and Social Change; (5) Marine Studies and Coastal Zone Management; (6) Health and Health Care; (7) Indigenous Peoples; (8) Rural Geography; (9) Geomorphology; (10) Public Policy; (11) GIS; (12) Geographies of Asia; (13) Geographic Education; (14) Diversity, Migration, Ethnicity, and Race; and (15) Historical Geography. These groups organize special sessions at national and regional meetings and many have a presence on the Web and publish newsletters.

The CAG is governed by an executive committee, which includes a president, a vice president, a secretary-treasurer, and nine councillors, and by a short constitution with 13 articles. It is represented by university representatives at 49 geography departments across Canada. The offices of the CAG are housed at McGill University.

Dawn S. Bowen

See also Association of American Geographers

Further Readings

Canadian Association of Geographers: www.cag-acg .ca/en/index.html
Kobayashi, A. (2001). Truly our own: Canadian geography 50 years after. *The Canadian Geographer, 45*, 3–7.

 # CANCER, GEOGRAPHY OF

Within the fields of medical geography and spatial epidemiology, a specific topic is the geography of cancer, which includes the study of spatial and temporal patterns of different types of cancer for purposes of surveillance, prevention, and etiologic inference.

Cancer Maps

Maps are important tools in the geographic study of cancer; however, when producing and interpreting cancer maps, it is essential to consider how cancer is measured and displayed. Incidence, mortality, survival, prevalence, and stage of diagnosis are common measures of cancer disease outcomes. Typically, these measures are presented as rates, standardized by population size, age, sex, and race, as a means of accounting for the underlying demographic differences between regions. Since cancers are relatively rare diseases, data are often grouped across many years to provide sufficient stability to the rates. Some maps are descriptive (e.g., for purposes of surveillance), while others are used in spatial analyses undertaken to detect statistically significant patterns or possible associations with potential risk factors. An important issue is the spatial resolution of the cancer data. Depending on the purpose of the map, data may either be aggregated to some geographic level, for example, national, census units, postal districts, and so on, or the data may be mapped using the locations of individual persons (e.g., location of residence). In most applications the resolution is often determined by the availability of data, which is driven by confidentiality concerns, resulting in very few spatial studies of individual-level cancer. As a consequence, the modifiable areal unit problem (MAUP) must be considered, necessitating tests of whether results are consistent across different scales and levels of aggregation; without considering the MAUP, results should be interpreted cautiously.

Spatial Analysis of Cancer

Beyond descriptive maps of cancer, researchers use spatial analyses to quantify and explain spatial and temporal patterns of cancer. First, it is essential to determine if observed patterns are simply random or if there are true geographic differences. Common spatial analytic algorithms include Kulldorff's Spatial Scan and Cuzick and Edwards's Nearest Neighbour Test, although others also exist. If the patterns are statistically significant, researchers may try to identify some underlying factors that could explain the observed patterns. Such factors include smoking, diet, lifestyle, genes, environment, health care resources, and differences in cancer registration. If the spatial analyses point to certain possible risk factors,

a new hypothesis may be developed and tested, for example, by using spatial regression analysis such as Geographically Weighted Regression (GWR) or in a more structured epidemiologic study. One example would be the relationship between large concentrations of arsenic in drinking water and the high cancer mortality rates in Taiwan. This association was first suggested based on aggregated data on cancer mortality, which coincided with aggregated data on arsenic concentrations. Later, several case-control and cohort studies supported this hypothesis at the individual level. It should be noted that it is essential that any hypothesis originating from aggregated data be tested at the individual level due to the ecological fallacy.

Geographic studies of cancer, by and large, have generated little etiologic insight. This may be attributable to the neglect of individual mobility information. In the analysis of chronic diseases such as cancer, causative exposures may occur over a long time, and the disease may be manifested only after a lengthy latency period. During this latency period, individuals may move from one place of residence to another. This can make it difficult to detect patterns of cases in relation to the spatial distribution of their causative exposures without having information on residential mobility. Building on Torsten Hägerstrand's concepts, space-time mobility paths can now be visualized using temporal GIS technology. Statistical algorithms have recently been developed for evaluating space-time clustering in residential histories. This approach is theoretically appealing but largely untested, and it is not yet known whether it will reveal insights into disease etiology.

Examples

Cancer rates and types vary across continents, countries, regions, and even local neighborhoods; furthermore, differences also occur over time. The publication *Cancer Incidence in Five Continents* by the International Agency for Research on Cancer (IARC), for example, presents data on cancer incidence in more than 200 populations, showing significant international differences. Furthermore, noticeable differences in cancer incidence and mortality are even observed across very similar

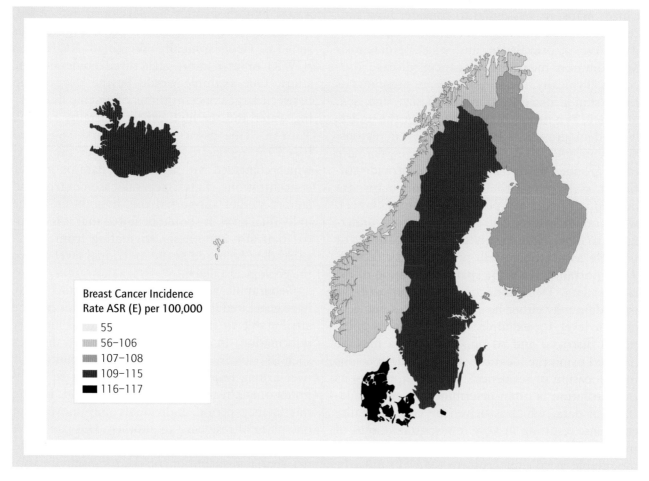

Figure 1   Female breast cancer incidence rates in the Nordic countries, 2003

Source: Map created by R. Baastrup from data in Engholm, G., Ferlay, F., Christensen, N., Bray, F., Gjerstorff, M. L., Klint, Å., et al. (2009). *NORDCAN: Cancer incidence, mortality and prevalence in the Nordic countries* (Version 2.4). Association of Nordic Cancer Registries, Danish Cancer Society. Retrieved December 18, 2009, from www.ancr.nu.

Note: ASR (E) = age-standardized rates (Europe).

countries, such as Sweden and Denmark (see www-dep.iarc.fr/NORDCAN.htm). Figure 1 shows the 2003 female breast cancer incidence rate in the Nordic countries of Europe (Iceland, Faeroe Islands, Norway, Sweden, Denmark, and Finland). The data are calculated using age-standardized rates (ASR), which enable calculation of the rates that the populations would have if they had a common age structure, thereby making it possible to compare rates across populations of different age structures. In this example, the European standard is used. The map shows noticeable differences in incidence across similar countries. Figure 2 shows the temporal trend in breast cancer incidence in the Nordic countries.

In these examples, the denominator for the incidence rate was drawn from the population as a whole, which can lead to problems if estimates of the underlying population are inaccurately estimated. For example, since census surveys are often only conducted once a decade, changes in the population distribution in the intercensus years can lead to inaccurate population estimates. One way around this problem is to map the proportions surviving or diagnosed at an early stage, using the incidence population as the denominator. In this way, both the numerator and the denominator are drawn from the same data set, typically a cancer registry. Maps of survival or stage of diagnosis provide information different

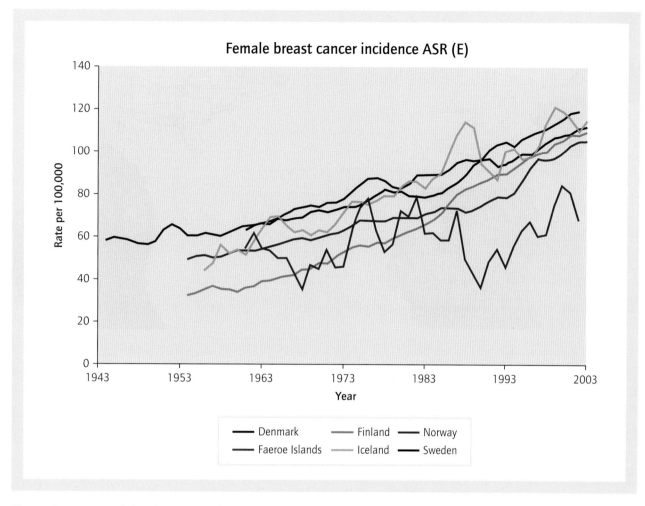

Figure 2　Temporal development in female breast cancer incidence rates in the Nordic countries

Source: Figure created by R. Baastrup from data in Engholm, G., Ferlay, F., Christensen, N., Bray, F., Gjerstorff, M. L., Klint, Å., et al. (2009). *NORDCAN: Cancer incidence, mortality and prevalence in the Nordic countries* (Version 2.4). Association of Nordic Cancer Registries and Danish Cancer Society. Retrieved December 18, 2009, from www.ancr.nu.

from that provided by incidence rate maps, but such maps are also valuable. As an example, Figure 3 shows the proportion of breast cancer survivors in Michigan State House legislative districts from 1985 to 1993, using data from the Michigan State Cancer Registry.

In a similar manner, this same registry data can be used to map proportions of early-stage breast cancer in Michigan. Figure 4 displays a cluster of early-stage breast cancer diagnosed from 1994 to 2002, detected using SatScan.

Residence at the time of diagnosis is an appropriate geographic unit for gaining insights about why survival or stage of diagnosis differs from place to place. The etiologic factors responsible

for cancer incidence, however, are best detected incorporating individual mobility histories because of cancer's latency period. Detecting spatial patterns in the mobility of cancer cases is a pivotal area of current research in the geography of cancer.

*Rikke Baastrup and
Jaymie R. Meliker*

See also Carcinogens; Disease, Geography of; Ecological Fallacy; Geographically Weighted Regression; GIS in Health Care; Health and Health Care, Geography of; Medical Geography; Modifiable Areal Unit Problem; Spatial Analysis; Time-Geography

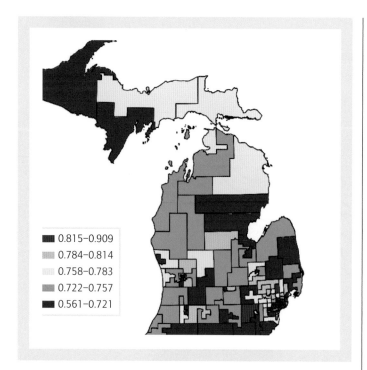

Figure 3 Proportion of female breast cancer survivors out of all breast cancer cases in Michigan State House legislative districts, 1985–1993

Source: Map created by J. R. Meliker from data in Meliker, J. R., Jacquez, G. M., Goovaerts, P., Copeland, G., & Yassine, M. (2009). Spatial cluster analysis of early stage breast cancer: A method for public health practice using cancer registry data. *Cancer Causes and Control, 20,* 1061–1069. Copyright © 2009 by Springer Science + Business Media.

Figure 4 Cluster of high proportion of early-stage female breast cancer cases out of all breast cancer cases in Michigan, 1994–2002

Source: Map created by J. R. Meliker from data in Meliker, J. R., Jacquez, G. M., Goovaerts, P., AvRuskin, G. A., & Copeland, G. (2009). Breast and prostate cancer survival in Michigan: Can geographic analyses assist in understanding racial disparities? *Cancer, 115,* 2212–2221. Copyright © 2009 Wiley-Blackwell.

Further Readings

Jacquez, G., Meliker, J., AvRuskin, G., Goovaerts, P., Kaufmann, A., Wilson, M., et al. (2006). Case-control geographic clustering for residential histories accounting for risk factors and covariates [Electronic version]. *International Journal of Health Geographics, 5*(32). Retrieved January 12, 2010, from www.ehjournal.net

Kulldorff, M., & Nagarwalla, N. (1995). Spatial disease clusters: Detection and inference. *Statistics in Medicine, 14,* 799–810.

Meliker, J., Slotnick, M., AvRuskin, G., Kaufmann, A., Jacquez, G., & Nriagu, J. (2005). Improving exposure assessment in environmental epidemiology: Application of spatio-temporal visualization tools. *Journal of Geographical Systems, 7,* 49–66.

Parkin, D., Bray, F., Ferlay, J., & Pisani, P. (2005). Global cancer statistics, 2002. *CA-A Cancer Journal for Clinicians, 55,* 74–108.

Pickle, L., Waller, L., & Lawson, A. (2005). Current practices in cancer spatial data analysis: A call for guidance [Electronic version]. *International Journal of Health Geographics, 4*(3). Retrieved January 12, 2010, from www.ehjournal.net

Vieira, V., Webster, T., Weinberg, J., Aschengrau, A., & Ozonoff, D. (2005). Spatial analysis of lung, colorectal, and breast cancer on Cape Cod: An application of generalized additive models to case-control data [Electronic version]. *Environmental Health, 4*(11). Retrieved January 12, 2010, from www.ehjournal.net

Waller, L., & Turnbull, B. (1993). The effects of scale on tests for disease clustering. *Statistics in Medicine, 12,* 1869–1884.

CARBONATION

Carbonation is the chemical reaction that occurs when the mineral calcite, or calcium carbonate ($CaCO_3$), comes in contact with an acid. In the natural environment, where rotting organic matter occurs in soils, the soil water reacts with the carbon dioxide from the organic matter to produce carbonic acid:

$$H_2O + CO_2 \leftrightarrow H_2CO_3 \qquad (1)$$

In turn, the weak carbonic acid that is produced then disassociates or comes apart into smaller atomic particles and compounds:

$$H_2CO_3 \leftrightarrow H^+ + HCO_3^- \qquad (2)$$

These materials can react in turn with the calcium carbonate. In this case, the hydrogen ion, H^+, is a proton, which is an extremely small particle from the hydrogen nucleus that can work its way into many mineral particles and disrupt them. In the case of the disruption of the solid calcium carbonate mineral or the limestone bedrock, the disassociated ions of the carbonic acid disrupt the atomic matrix and recombine to form the new compound of calcium bicarbonate, which is then dissolved in water.

$$CaCO_3 + H^+ + HCO_3^- \leftrightarrow Ca^+ + (HCO_3^-)^2 \qquad (3)$$

In this fashion, the carbonate rocks of the world (sedimentary limestone and dolostone, or the crystalline rocks of metamorphic marble and igneous carbonatite) are subject to the carbonation reaction, and caves or caverns result. As the rocks dissolve, the rock is carried away in the water in solution as ions of Ca^+ and HCO_3^-. Thus, the solid calcium carbonate rock goes into solution and is carried away in underground waters as $Ca(HCO_3)^2$, to eventually end up in the ocean as one of the many kinds of salts in the sea. Organisms in the sea such as coral, bivalves, and other sea life use this dissolved calcium carbonate to make their reefs and shells.

The double arrows in the above reaction equations (Equations 1, 2, and 3) indicate that these reactions are reversible, that is, they can go in both directions, with the result that the carbon dioxide (CO_2) gas that is dissolved in liquids can come out of the solution as the effervescence or "fizz" of carbonated beverages. Thus, in situations in which surface waters containing a high proportion of dissolved calcium carbonate are agitated in rapids as they flow over an uneven streambed, they can lose some of their dissolved CO_2 to the atmosphere and force the reactions to move $CaCO_3$ back out of solution and into precipitation as solid carbonate rock. This is the same process inside caves wherein the carbonate-bearing water moves through the fissures and pore spaces until it arrives on the surface of the underground opening, loses some CO_2, and thus produces *stalactites* that hang from the ceiling, *stalagmites* growing up from the floor, and various forms of flowstone, dripstone, and other cave *speleothems* that are produced in these carbonation reactions. Thus, first carbonation reactions produce caves, mostly below the water table in the limestone bedrock, and then, once the water table drops and exposes the new cave to the atmosphere, the process reverses, and the cave is then in-filled back again with new *travertine* calcium carbonate.

John F. Shroder

See also Carbon Cycle; Caverns; Geomorphology; Groundwater; Karst Topography; Rock Weathering

Further Readings

Ford, D., & Williams, P. (2007). *Karst hydrogeology and geomorphology*. New York: Wiley.

CARBON CYCLE

The natural cycle of carbon is one of the biogeochemical cycles that link Earth's *atmosphere*, *hydrosphere*, and *lithosphere* with continental and oceanic *biota*. Together, these spheres and biota form the largest natural system: the *ecosphere* or *global ecosystem*. Detailed information on the size of carbon pools in the global ecosystem and on the direction, and magnitude of exchange fluxes are important parameters used in the construction of global carbon cycle models and in the forecasting

of atmospheric carbon dioxide concentration and changes in the atmospheric *greenhouse effect* and climate.

Biota

During their formation from protoplanetary dust cloud, Earth, Venus, and Mars captured similar percentages of carbon. Due to differences in their history, however, the concentrations of carbon in the modern atmospheres of these terrestrial planets vary widely. The atmospheres of Venus and Mars consist mostly of CO_2, with concentrations of 96.5% and 93.5% by volume, respectively. The atmosphere of Earth, in turn, is a 99% mixture of nitrogen and oxygen; the concentration of carbon dioxide is very low (~0.0000289%, or 289 ppm [parts per million] by volume in the middle of the 19th century and 383.86 ppm in 2007). Unlike Mars and Venus, most of the carbon on Earth is found not in the atmosphere but in the lithosphere in the form of carbonate deposits and organic carbon (Table 1). This striking difference in carbon distribution between Earth and other terrestrial planets can be explained only by the activity of living organisms.

Reservoir	Carbon Content
Atmosphere (at 384 ppm [parts per million]; 2007 data)	804
Phytomass (living biomass of plants)	610
Soil organic matter	1,580
Oceanic biota	2
Dissolved and suspended organic matter in the ocean	1,830
Dissolved inorganic carbon in the ocean	36,000
Organic carbon in the lithosphere	12,000,000
Carbonates in the lithosphere	94,000,000

Table 1 Carbon pools in modern Earth

Sources: Adapted from Budyko, M., Ronov, A., & Yanshin, A. (1987). *History of the Earth's atmosphere*. New York: Springer-Verlag; Houghton, R. (2005). The contemporary carbon cycle. In W. Schlesinger (Ed.), *Biogeochemistry* (pp. 473–513). Amsterdam: Elsevier Science.

Note: Some estimates vary in large by as much as 25%.
Units are in gigatons (Gt) of carbon (1Gt = 10^{12} kilograms).

In photosynthesis, green plants (including marine phytoplankton) produce carbohydrates (organic matter). The net primary productivity (the rate of annual biologic fixation of carbon by continental and oceanic biota) is about 85 to 90 Gt/yr. (gigatons per year) (Table 2). More than 99% of this organic matter is decomposed and returned to the atmosphere (Table 2). Depending on the molecular weight of specific organic compounds, however, the rate of their decomposition varies from less than a year to hundreds and even thousands of years.

The biomass of marine organisms is two orders of magnitude smaller than the biomass of terrestrial organisms, but the productivity of marine biota is only two times smaller than the productivity of all terrestrial organisms (see Tables 1 and 2).

Lithosphere

A common form of organic carbon in geologic sedimentary rocks is kerogen, which is insoluble in water, a carbohydrate of irregular structure and very high molecular weight. In the past few hundred million years, a rather small portion of kerogen was transformed inside the lithosphere into various types of fossil fuel: coal, oil, and natural gas. Overall, however, sedimentary geologic rocks contain most of their carbon not as organic matter but as carbonates (Table 1). Carbonates are mineral sedimentary rocks formed in the process of weathering of original igneous (volcanic) rocks. The process of chemical weathering is regulated by the reaction of carbon dioxide with silicate minerals in soil. This reaction can be accelerated by root respiration of vascular plants as well as by activity of soil bacteria. As a result of chemical weathering, calcium and/or magnesium ions are released from soil into water, while carbon dioxide is transformed into bicarbonates:

$$CaSiO_3 + 2CO_2 + H_2O \rightarrow Ca^{2+} + 2HCO_3^- + SiO_2$$

Dissolved products of this reaction are carried to the ocean by river runoff. In oceanic waters, calcium and magnesium can be recombined with bicarbonates to form carbonate deposits:

$$Ca^{2+} + 2HCO_3^- \rightarrow CaCo_3 + CO_2 + H_2O$$

Fluxes	Gigatons per Year
Net primary production of continental biota	60
Root respiration	18.5
Flux to top of soil with organic litter	41.5
Carbon mineralization on top of soil	38.3
Formation of new humus (fast and slow cycling)	2.5
Oxidation of humus	2.2
Flux of terrestrial carbon to the ocean as atmospheric aerosol and dust; discharge of organic carbon to the ocean with continental runoff	1.0
Net primary production of oceanic plankton	25–30
Oxidation of organic carbon in water column	20–25
Flux to suspended and dissolved oceanic carbon reservoir	5
Sedimentation of organic carbon at the oceanic floor	1–3
Oxidation of organic carbon at the oceanic floor	0.9–2.9
Preservation of organic carbon in the lithosphere	0.02
Preservation of carbonate carbon in the lithosphere	0.15
Exchange of carbon dioxide between the ocean and the atmosphere	90–100
Net intake of anthropogenic CO_2 by the ocean	1.6–3.2

Table 2 Major fluxes in the global carbon cycle

Sources: Adapted from Kobak, K. I. (1988). *Biotic components of carbon cycle* (p. 248). Leningrad, Russia: Hydrometeoizdat; Sundquist, E. T. (1985). Geological perspective on carbon dioxide and the carbon cycle. In E. Sundquist & W. S. Broecker (Eds.), *The carbon cycle and atmospheric CO₂: Natural variations Archean to present* (Geophysical Monographs No. 32; pp. 5–60). Washington, DC: American Geophysical Union.

Note: Some estimates vary in large by as much as 50%. This uncertainty is larger for some fluxes, such as the rate of organic carbon sedimentation and the rate of organic carbon oxidation at the ocean floor.

The resulting reaction (below) demonstrates the net flux of carbon dioxide from the atmosphere to geologic deposits (*Urey reaction*):

$$CaSiO_3 + CO_2 \rightarrow CaCo_3 + SiO_2$$

This reaction demonstrates that an increase in the atmospheric CO_2 should lead to an increase in the accumulation of carbonate sediments. In other words, the Urey reaction represents an important negative feedback mechanism (also called a geochemical *thermostat*) that may have kept the atmospheric greenhouse effect within limits suitable for life. Indeed, the high level of volcanic activity in early geologic epochs caused a significant emission of CO_2 into the atmosphere. The high CO_2 concentration, in turn, caused an increase in chemical weathering and transportation of carbon dioxide to geologic sediments. Thus, the Urey reaction helped reduce the ancient greenhouse effect and kept the planet relatively cool. The opposite is also true. As volcanic activity slowed down, the rate of chemical weathering and removal of CO_2 from the atmosphere had to decline.

More than 90% of carbonate deposits on Earth were formed by living organisms. The rate of biogenic carbonate preservation (formation of $CaCO_3$ in the right side of the Urey reaction) is a function of many external physical and chemical parameters, such as temperature, saturation of oceanic waters with carbonates, and some others. Every year, oceanic and continental deposits accumulate 0.1 to 0.2 Gt of mostly biogenic carbon (Table 2), which is practically equal to the rate of carbon dioxide emission due to volcanic activity and, at the same time, 25 to 50 times smaller than recent anthropogenic emissions of carbon into the atmosphere.

Ocean

Oceanic waters constitute the second largest (after the lithosphere) pool of carbon (Table 1). Unlike the lithosphere, however, oceanic carbon is present in dissolved form and is readily available for exchange with the atmosphere.

In seawater, inorganic carbon exists in four major forms: (1) dissolved carbon dioxide (CO_2), (2) carbonic acid (H_2CO_3), (3) carbonate (CO_3^{2-}), and (4) bicarbonate (HCO_3^-) ions. All these forms

of carbon participate in a series of interrelated chemical reactions. The resulting chemical equilibrium in seawater can be presented in the following equation:

$$CO_2 + H_2O \leftrightarrow H_2CO_3 \leftrightarrow$$
$$H^+ + HCO_3^- \leftrightarrow H^+ + CO_3^{2-}$$

The intensity of carbon dioxide exchange between ocean and atmosphere depends on the chemical and physical properties of surface oceanic waters. The concentration of dissolved inorganic carbon in surface waters, in turn, is regulated by the exchange with deeper layers and by the activity of oceanic biota. Marine life significantly decreases the concentration of dissolved inorganic carbon and essential nutrients such as nitrogen and phosphorus in the mixed layer of ocean (upper 50–100 meters). Often, this effect is called *biologic pump*. Biota absorb carbon from the mixed layer in the process of photosynthesis and when, under the force of gravity, organic matter descends into deeper layers, where it degrades. The speed of organic matter descent is directly proportional to the size and density of marine organisms. At the current rate of net primary production, marine biota cause a reduction of equilibrium with the atmospheric concentration of CO_2 in the mixed layer of ocean of about 300 ppm.

Atmosphere

The share of atmospheric carbon in the global carbon cycle is small compared with that of other carbon reservoirs (Table 1). Therefore, the atmospheric concentration of CO_2 can be changed drastically in response to relatively small disturbances in other reservoirs such as ocean, soil, and geologic sediments or continental and oceanic biota.

Over millions of years, changes in the concentration of atmospheric carbon were caused by slight deviations from the balance between two major fluxes: (1) volcanic activity and (2) chemical weathering of volcanic rocks.

During the past 100 million years, the atmospheric concentration of carbon dioxide declined from more than 3,000 ppm in the Late Cretaceous period to 400 to 500 ppm in the Pliocene.

The main cause of such change was gradually diminishing volcanic activity. The appearance of angiosperm plants at the end of the Cretaceous period, however, might have enhanced chemical weathering in the soil and contributed to a decrease in the atmospheric carbon dioxide concentration.

At the timescale of thousands of years, the interaction of atmospheric carbon with carbon pools in the oceans, biota, and soil seems to be more important than the relatively weak fluxes of chemical weathering and volcanic activity. During the past few hundred thousand years, atmospheric carbon dioxide concentration has exhibited almost periodic changes, from 200 to 300 ppm, coinciding with the major glacial-interglacial epochs.

During the 20th century, atmospheric carbon dioxide concentration demonstrated a consistent growth proportional to the rate of fossil fuel burning and to the rate of deforestation. At present, the atmosphere receives ~7 Gt/yr. of anthropogenic, which is 25 to 50 times greater than natural volcanic activity. The total amount of known fossil fuel reserves is ~5,000 Gt of carbon, or six times the amount of CO_2 in the modern atmosphere. Since the beginning of the Industrial Revolution, the combustion of fossil fuel has been by far the most important factor influencing changes in the atmospheric carbon dioxide. During the past few decades, however, the biosphere has demonstrated some visible increase in photosynthesis (especially in the circumpolar forests of the Northern Hemisphere). This increase, however, is still at least an order of magnitude smaller than the rate of CO_2 production through combustion of fossil fuel. Therefore, the growth of atmospheric CO_2 most likely will continue as long as humans continue to burn fossil fuel.

Andrei G. Lapenas

See also Anthropogenic Climate Change; Atmospheric Pollution; Biogeochemical Cycles; Biota and Climate; Carbonation; Carbon Trading and Carbon Offsets; Climate Change; Climate Policy; Climatology; Ecosystems; Energy Models; Energy Policy; Global Sea-Level Rise; Greenhouse Gases; Sedimentation; Symptoms and Effects of Climate Change; Volcanoes

Further Readings

Berner, R., Lasaga, A., & Garrels, R. (1983). The carbonate-silicate geochemical cycle and its effect on atmospheric carbon dioxide over the past 100 million years. *American Journal of Science, 283,* 641–683.

Bolin, B., & Eriksson, E. (1959). Changes in the carbon dioxide content of the atmosphere and the sea due to fossil fuel combustion. In B. Bolin (Ed.), *The atmosphere and the sea in motion* (pp. 130–142). New York: Rockefeller Institute Press.

Budyko, M., Ronov, A., & Yanshin, A. (1987). *History of the Earth's atmosphere.* New York: Springer-Verlag.

Houghton, R. (2005). The contemporary carbon cycle. In W. Schlesinger (Ed.), *Biogeochemistry* (pp. 473–513). Amsterdam: Elsevier Science.

Huggett, R. (1999). Ecosphere, biosphere, or Gaia? What to call the global ecosystem? *Global Ecology and Biogeography, 8*(6), 425–431.

Petit, J., Jouzel, J., Raynaud, D., Barkov, N., Barnola, J., Basile, I., et al. (1999). Climate and atmospheric history of the past 420,000 years from the Vostok ice core, Antarctica. *Nature, 399,* 429–436.

Volk, T. (1987). Feedbacks between weathering and atmospheric CO_2 over the last 100 million years. *American Journal of Science, 287,* 763–779.

Wigley, T. & Schimel, D. (2000). *The carbon cycle.* Cambridge, UK: Cambridge University Press.

CARBON TRADING AND CARBON OFFSETS

The carbon trade involves the exchange of either *emission allowances* or *carbon offsets*. An *allowance* is a right to emit a unit of greenhouse gases (GHGs)—typically defined as 1 metric ton of carbon dioxide (tCO_2) or equivalent quantity of other GHGs. An *offset* is created when a project or policy achieves a reduction of GHG emissions relative to a *baseline* scenario (defined as emissions that would have occurred if no project or policy was undertaken). Both allowances and offsets are accepted, through various institutional arrangements, such as fungible units of trade. These forms of exchange have been introduced to facilitate GHG emission reductions through the use of market mechanisms.

In creating these markets, society has placed a price on CO_2 and related pollutants, which are linked to all aspects of human activity. Since people all over the globe emit GHGs (albeit in highly differentiated ways), carbon markets span spatial scales and link disparate world regions. These markets have received attention from a range of geographic sub-disciplines and will likely continue to be the objects of study into the future.

Theoretical Foundations of Emissions Trading

Carbon trading can be traced to the mid 20th century, when economists began to examine the issue of social costs of economic activity. These *externalities* represent a form of market failure that arises when an economic exchange among one set of actors leads to impacts among other actors who are not participants in the exchange. Impacts may be positive, negative, or a combination, and environmental pollution is a classic example of a negative externality.

Prior to the introduction of market-based environmental regulation, the primary means of reducing pollution was through government-imposed limits. This approach was criticized because it restricts economic activity and does little to promote innovation. Alternative approaches were proposed, predicated on the theory that although pollution has negative social costs, so does the restriction of economic activity that causes pollution. Based largely on the work of economist Ronald Coase, the theory holds that the problem is reciprocal rather than one-sided. For example, if pollution emitted by A causes damages to B, the initial reaction may be to limit A's pollution. However, reducing the damages on B inflicts harm on A by adding abatement costs or forcing cuts in A's production.

Theoretically, there is an "optimal" level of pollution at which the net benefits to society can be maximized. In economists' jargon, this is the level of pollution at which the marginal social benefit of an additional unit of pollution equals the marginal social cost. However, finding an optimal level of pollution is more complex than this simple description implies. First, in Coase's original formulation, both A and B were single actors incurring individual costs and benefits. In reality, there may be multiple As releasing pollutants with similar impacts on B. There may also be many Bs, each incurring costs that can be difficult to enumerate or attribute to a single polluter. Furthermore, A may deploy technical experts and litigators who challenge B's claims. Additional complications arise if B is displaced spatially or temporally from A. Moreover, the impacts of pollution are typically transmitted through air or water, which lack well-defined property rights. GHGs possess all these complicating factors, which is one reason why consensus on mitigation has been so elusive.

Nevertheless, the notion that pollution can have an optimal level has gained traction and facilitated the establishment of emissions trading in several ways. First, finding an optimal level of emissions puts a price on pollution, which can be used to guide policy formation. It also defines a "cap" on total emissions. This cap is subdivided among polluters so that each receives the right to emit a fixed quantity of pollution. This right can assume the form of private property, which creates the opportunity to sell, trade, or even "bank" it for future use.

In the context of climate change, the opportunity for trade is bolstered because GHGs are long-lived pollutants. Longevity and broad spatial dispersion create the sense that all emissions are identical. However, the costs of reducing emissions vary across different economic sectors and geographical regions. Therefore, net costs to society are minimized if actors subject to emission caps are permitted to trade their emission rights. Those with lower abatement costs prefer to reduce emissions below their cap to sell excess rights, while those with high costs prefer to purchase rights from others rather than reducing their own emissions.

These theories were linked to policy in the 1970s with the U.S. Clean Air Act (CAA) and operationalized with cap-and-trade of sulfur dioxide (SO_2) in the early 1990s. The success with SO_2 was invoked during negotiations for the Kyoto Protocol several years later.

Current Carbon Markets

In 2007, 2.7 billion metric tons of carbon emission reductions worth US$60 billion were traded in several distinct carbon markets. Most trading occurs in *allowance-based markets*, where governments grant allowances to polluting firms to create a cap on GHG emissions. *Project-based markets* exist in parallel with allowance markets and supply them with alternative sources of emission reductions in the form of *carbon offsets*. Offsets are created through activities that reduce GHG emissions relative to a baseline scenario. *Voluntary markets* are a particular type of project-based market which caters to buyers interested in meeting self-imposed emission reduction goals rather than state-imposed mandates.

The 2007 trade was split roughly 60-40 between allowances and offsets. The allowances are largely traded within the European Union's Emissions Trading Scheme (EU-ETS). Offsets are primarily derived from the Clean Development Mechanism (CDM) of the Kyoto Protocol, which is a scheme to promote mitigation projects in developing countries. Voluntary markets also rely on offsets and constituted about 4% of the trade in 2007.

The ETS is the largest active market. There are also several national and subnational schemes active or coming online, including Australian, Canadian, U.S., and possibly Japanese markets. Each has distinct sets of actors, governing institutions, and rules creating demand for offsets among participating players. There are also differences in the types of offsets that are acceptable for trade in each market. Markets are currently distinct, but there are discussions about the ways in which they could be linked in the future to create a single global carbon market. For this to happen, existing differences would need to be smoothed over. Table 1 describes some of the characteristics of existing markets, and Figure 1 illustrates the recent growth in volume and revenue in these markets.

Carbon Trading Under the International Climate Regime

The Kyoto Protocol is the impetus for the bulk of existing carbon markets, with Kyoto instruments collectively representing more than 95% of market activity. The Protocol introduced three "flexible mechanisms" linked to the carbon trade. Article 17 explicitly introduces emissions trading into the climate regime. Articles 6 and 12 define Joint Implementation (JI) and the CDM, respectively. Both are project-based mechanisms. JI supports projects in developed countries that

Table 1 Key characteristics of existing carbon markets

Market	Participating Countries or Subnational Units	Unit of Trade	Allocation Mechanisms	Enforcement Mechanisms	Trade in 2007 Million Metric Tons CO_2 (tCO_2)	Trade in 2007 Million US$
Allowance-based markets						
European Union's Emissions Trading Scheme (EU-ETS)	27 European Union (EU) member states and several non-EU members (e.g., Norway, Iceland, and Liechtenstein)	EU Allowance Unit (EUA) equivalent to 1 tCO_2	National Allocation Plans (NAPs)	The European Commission can reject NAPs, impose fines on noncompliant entities, and publicly name noncompliers; accepts some types of certified emission reductions (CERs)	1,600	44,000
Regional Greenhouse Gas Initiative (RGGI)	*Participants:* Northeastern U.S. states: Maine, New Hampshire, Vermont, New York, New Jersey, Delaware, Massachusetts, Maryland, Rhode Island *Observers:* Pennsylvania, D.C., and several Canadian provinces	CO_2 allowance (1 tCO_2)	Auction	Existing government laws: EPA acid rain rules require CO_2 reporting. EPA Clean Air Interstate Rule requires equipment capable of CO_2 reporting	NA	NA
Chicago Climate Exchange (CCX)	Range of actors covering industry, municipalities, universities, offset providers, offset aggregators	CCX Carbon Financial Instrument (CFI) contract (100 tCO_2 equivalent). There are CFI allowances and offsets	Legally binding, voluntary annual greenhouse gas (GHG) reductions	Legally binding compliance regime, providing independent, third-party verification	24	104
New South Wales (NSW) GHG Reduction Scheme	New South Wales, Australian Capital Territory	Two types of allowances (1 tCO_2); transferable NSW GHG Abatement Certificates; nontransferable Large User Abatement Certificates	Companies must meet benchmarks relative to share of powermarket	Independent Pricing and Regulatory Tribunal (IPART) can impose fines	19.9	NA
Project-based schemes						
Clean Development Mechanism (CDM)	Signatories to the Kyoto Protocol defined as non–Annex 1 countries	CERs (1 tCO_2)	Any non–Annex 1 signatory can host projects	Use of methodologies approved by the CDM executive board; third-party verification	947	18,900
Joint implementation (JI)	Signatories to the Kyoto Protocol defined as Annex 1 countries	Emission Reduction Unit (ERU): 1 tCO_2	Any Annex 1 signatory can host projects	Joint Implementation Supervisory Committee (JISC) supervises verification process	38	512

Source: Based on data from Capoor, K., & Ambrosi, P. (2008). *State and trends of the carbon market 2008.* Washington, DC: World Bank.

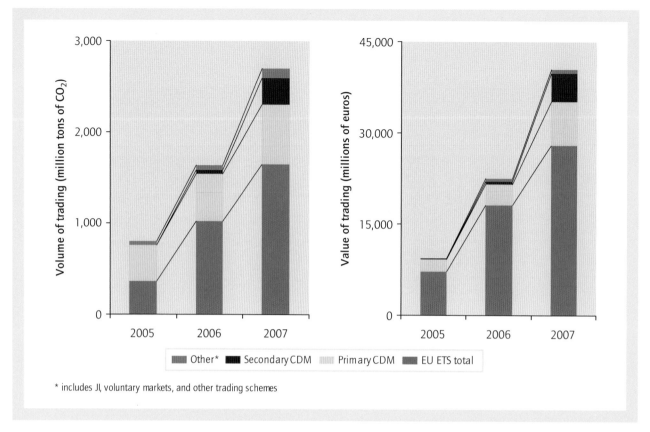

Figure 1   Activity in carbon markets showing trading volume (*left*) and value (*right*) from 2005 to 2007

Source: Figures created by author based on data from Røine, K., Tvinnereim, E., & Hasselknippe, H. (Eds.). (2008). *Carbon 2008: Post-2012 is now*. Washington, DC: PointCarbon.

were in transition to market economies at the time the Protocol was being drafted. CDM supports projects in developing countries that do not have commitments under the Protocol. Figure 2 depicts the countries and sectors that had been most active in CDM and JI from inception to mid 2008.

Critiques of Carbon Markets and Emissions Trading

Markets for carbon and other pollutants have been subject to numerous critiques, several of which have stemmed from geographic disciplines. Some critics challenge the fundamental validity of market mechanisms as a form of environmental governance, while others accept the potential of carbon markets to reduce emissions in theory but are critical of the ways in which these markets are implemented and operationalized.

Critiques of the Theoretical Validity of Carbon Markets

Market-based approaches to pollution control involve the creation of property rights to use a fraction of the atmosphere's finite ability to absorb heat-trapping GHGs. This "atmospheric sink" had been a common pool resource prior to the establishment of a market, and some challenges to the theoretical validity of carbon markets question the idea that such a common pool resource can be privatized. The exhaustible nature of the atmospheric GHG sink raises questions concerning the allocation of rights within and between generations.

Other challenges maintain that the privatization of the atmospheric sink and the commodification of landscapes' abilities to store or accumulate carbon exemplify neoliberal encroachment into environmental governance. This stance, originating from political ecology, critical geography, and

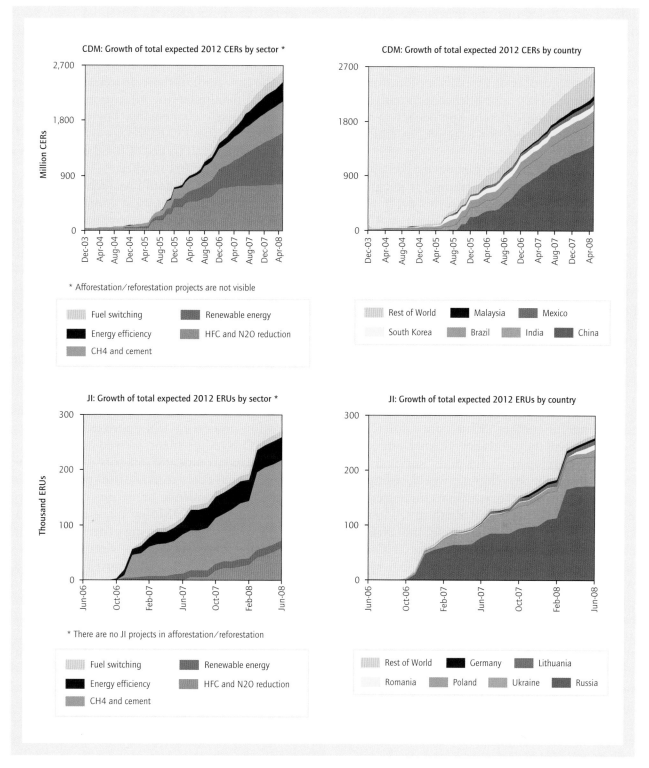

Figure 2   Growth in carbon offsets expected in 2012 from the CDM (*top*) and JI (*bottom*) by sector (*left*) and country (*right*)

Source: Figures created by author based on data from Fenhann, J. (2008, June). *CDM and JI pipeline overviews*. Retrieved July 25, 2008, from the CD4CDM—UNEP Risø Centre Web site: http://cd4cdm.org.

Notes: CERs = certified emission reductions; ERUs = emission reduction units. These are the names of the fungible offsets created by projects in the CDM and JI, respectively. Each is equal to 1 ton of CO_2.

allied fields, holds that the hegemonic nature of the "free-market environmentalist" approach to pollution control simultaneously appropriates and dilutes attempts to prevent the severe environmental consequences of capitalist production and consumption. They also silence alternative discourses proposing different approaches to GHG abatement, grounded, for example, in ethical or ecocentric arguments.

Further challenges to carbon markets attack the notion that emissions trading is geographically neutral. As discussed previously, 1 ton of CO_2 from a coal-fired power plant in China, from a cement factory in Mexico, and from a sport utility vehicle in the United States are identical from the narrowly defined perspective of atmospheric chemistry. However, the systems of production causing emissions are not at all equivalent. Critics claim that equating emissions from different sectors and world regions to render them fungible commodities hides the social and political context from which emissions originate.

Trading pollution allowances also reduce emissions from one sphere of activities and allow them to continue in another. Some critics differentiate between what they consider "luxury emissions" from the developed world and "survival emissions" from the developing world, using this dichotomy to argue that emissions trading encourages countries in the global South to sell low-cost carbon offsets to high-emitting industrialized countries, which perpetuates existing GHG-intensive activities in an inequitable global system. Another aspect of this critique is aimed specifically at markets that permit offsets generated by carbon sequestration, for example, afforestation and reforestation. Here, critics take issue with the notion that CO_2 fixed in living biomass is exchangeable with emission reductions from industrial processes. They claim that biophysical differences render the two incompatible and also point to the serious implications of land use change and, in particular, plantation forestry for local communities and indigenous populations.

Critiques of the Implementation of Carbon Markets

Critiques of carbon markets also arise from analysts who do not challenge the theoretical potential of these markets but take issue with the ways in which these markets are put into practice. Critics have taken issue with the setting of emission caps and allocation of allowances under the cap. Both of these processes are subject to political manipulation. Allocation has proven to be particularly challenging. For example, the first phase of the EU-ETS was overallocated, which led to a crash in prices for the allowances that were traded in that market.

Ideally, allowances should be auctioned so that they go to the actors that most value them. Auctioning also creates a flow of revenue to the state, which assists in defraying the costs of establishing the market and creating necessary institutional oversight. In practice, many markets have been established via "grandfathering" of allowances, in which firms are *given* rights to emit that are proportional to their past emissions, which creates a large windfall for polluting firms.

Additional critiques have been leveled at the monitoring and verification in the production and trade of carbon offsets. To be traded in compliance markets, offsets are subjected to a rigorous verification process by UN-approved third-party verifiers. This requirement, viewed as essential to uphold the environmental integrity of the system, adds additional costs to any offset project. Even with these requirements, numerous charges have been leveled at the CDM that question the validity of the offsets that are being generated. Voluntary markets are subject to similar critiques currently and have fewer safeguards. In response, numerous standards are under development to introduce some degree of verification to voluntary markets.

Some critics hold that barriers to implementation of carbon trading are so severe that cap-and-trade systems should be scrapped in favor of a carbon tax. Theoretically, a tax on emissions should have the same effect as a functioning cap-and-trade system. However, proponents of the tax claim that taxes are preferable because they are more transparent, less susceptible to manipulation, and immune to market volatility. Taxes are also preferable when there is uncertainty in abatement costs. Such taxes have been implemented in several European countries but with many loopholes. Carbon taxes are also problematic because they do not guarantee a specific level of emission reductions. Moreover, once implemented, taxes are difficult to adjust if further mitigation becomes necessary. Taxes may also be political nonstarters under certain circumstances. This point is particularly true in the United States, where market-based policies are preferred.

Robert Bailis

See also Carbon Cycle; Climate Change; Climate Policy; Ecological Economics; Environmental Management; Market-Based Environmental Regulation; Neoliberal Environmental Policy; Political Ecology; Social and Economic Impacts of Climate Change; United Nations Environmental Summits

Further Readings

Bailey, I. (2007). Neoliberalism, climate governance, and the scalar politics of EU emissions trading. *Area, 39,* 431–442.

Bumpus, A. G., & Liverman, D. M. (2008). Accumulation by decarbonization and the governance of carbon offsets. *Economic Geography, 84,* 127–155.

Capoor, K., & Ambrosi, P. (2008). *State and trends of the carbon market 2008.* Washington, DC: World Bank.

Coase, R. H. (1960). The problem of social cost. *Journal of Law and Economics, 3,* 1–44.

Fenhann, J. (2008, July). *JI and CDM pipeline overviews.* Retrieved July 25, 2008, from CD4CDM—UNEP Risø Centre Web site at http://cd4cdm.org

Lohmann, L. (2006). *Carbon trading: A critical conversation on climate change, privatization, and power.* The Hague, Netherlands: Dag Hammarskjöld Foundation.

McCarthy, J., & Prudham, S. (2004). Neoliberal nature and the nature of neoliberalism. *Geoforum, 35*(3), 275–283.

McKibbin, W. J., & Wilcoxen, P. J. (2002). The role of economics in climate change policy. *Journal of Economic Perspectives, 16*(2), 107.

Voß, J. P. (2007). Innovation processes in governance: The development of "emissions trading" as a new policy instrument. *Science and Public Policy, 34*(5), 329–343.

CARCINOGENS

Cancer is a series of diseases characterized by abnormal cell growth and spread of cancerous cells through the body. A carcinogen is a substance that can cause cancer in humans or animals. Sometimes substances promote cancer or aggravate cancers but do not cause them. These substances also may be called carcinogens. Carcinogens include both naturally occurring substances (radon gas) and substances created as a result of production (the components of tobacco). Lists of known carcinogens are maintained by national and world agencies, such as the National Cancer Institute, the International Agency for Research on Cancer, the National Toxicology Program, and the World Health Organization.

Geographers have been involved in the study of cancer because there are obvious geographical differences between nations, within states and provinces in the same nation, among adjacent local governments, and even within neighborhoods. This entry highlights several of the most important contributions made by epidemiologists and geographers.

At the international and intranational scales, researchers have gained important insights about carcinogens by comparing the cancer incidence and mortality rates of immigrants. For example, a Polish epidemiologist compared cancer rates of Polish citizens who lived in Poland, Polish migrants to the United States, and the progeny of Polish migrants who lived in the United States. Residents of Poland had notably higher stomach cancer rates than the progeny of Polish migrants to the United States. Migrant studies are not confined to international comparisons. For instance, within the United States, the cancer death rates of Southern-born African Americans were compared with those of their counterparts born elsewhere in the United States. Southern-born African Americans had higher digestive, urinary, and respiratory cancer rates, irrespective of whether they lived in the South or had migrated to other locations in the United States.

Migrant studies such as these provide clues to field- and laboratory-based cancer researchers, allowing them to investigate the reasons for the different rates. For example, investigators studied the food preferences and eating habits of African Americans residing in Harlem in New York City who were born in the Northeast, in the South, and in the Caribbean and found marked differences in their food preparation and consumption patterns. The United States has experienced a substantial increase in immigration in places such as California, New York, and New Jersey, and these places are rich environments for cancer-oriented migrant studies. These so-called migrant studies are inherently geographical and appeal to those who are interested in the study of risk factors for cancer, including food preferences, cooking and food preservation practices, tobacco and alcohol use, workplace exposures, and many others.

A second important area of cancer research that geographers have contributed to is cancer cluster investigations. Every year, residents of some neighborhoods report that there are an excessive number of cancers in their communities. To verify that a cluster exists, scientists must review every case, map the data, and compare the cancer cases with the number expected by chance. If there is indeed an excess, then public health experts will try to understand why the rates are higher than expected. Typically, no cause can be found, leading often to considerable public angst.

There are reasons for our failure to determine the carcinogens responsible for clusters. One is that the exposures are no longer present; that is, cancer normally takes two to four decades to appear. The source may be gone, or the individual who has contracted the cancer may have been exposed in an entirely different part of the country. Another reason is that cancer is a stochastic process. We should expect thousands of cancer clusters to occur in any one year across the United States at the neighborhood scale. This also means that there will be thousands of neighborhoods that have relatively few cancers. The important question is whether an excess of cancer cases is observed for many years, in which case it is highly improbable and warrants a detailed investigation. Geographers have played a role in cluster investigations.

Both migrant studies and cluster investigations depend on reliable data. The best way of accumulating reliable data for cancer studies is through proactive surveillance. Because of their understanding of spatial attributes, geographers can make important contributions to the geography of cancer surveillance. Envision a database that includes the locations of people, known carcinogenic agents and their concentrations in the environment, and the distribution of incident cancers. If routinely updated, the database would allow state officials to identify locations where the number of cancers seems to be in excess. Armed with powerful GIS tools, they could map the distribution of the cases, review the geography of risk factors, and decide what steps, if any, are warranted. If they did this over an entire state, then the limited resources that state health and environmental protection departments have for cancer cluster investigations could be allocated on the basis of the best information. The alternative, which we heretofore have relied on, is to wait for local residents, local health department officials, physicians, and other people to identify what they perceive as a cluster and report those perceptions to state health and environmental officials. By the time the observations are made, more people may be exposed, budgets for investigation may be limited, and the locations that most warrant investigation may not be addressed. Hence, some scientists strongly believe that proactive surveillance will become much more common and the most important basis on which locations with high cancer rates or clusters of a specific kind of cancer will be determined and followed up. The surveillance approach to analyzing the geography of cancer implies that geographers who are well trained in GIS, medical geography, and epidemiology can become increasingly important to understanding carcinogenic processes as they play out across the landscape.

Michael R. Greenberg

See also Cancer, Geography of; Disease, Geography of; Ecological Fallacy; GIS in Health Care; Health and Health Care, Geography of; Medical Geography; Risk Analysis and Assessment

Further Readings

Doll, R., & Peto, R. (1982). *The causes of cancer.* New York: Oxford University Press.

Greenberg, M. (1983). *Urbanization and cancer mortality*. New York: Oxford University Press.

Greenberg, M., & Schneider, D. (1995). The cancer burden of Southern-born African-Americans: Analysis of a social geographical legacy. *Milbank Quarterly, 73*(4), 599–620.

Nasca, P., & Pastides, H. (Eds.). (2001). *Fundamentals of cancer epidemiology*. Gaithersburg, MD: Aspen.

 # CARRYING CAPACITY

Carrying capacity is best known today as a means of expressing the finitude of Earth's resources relative to the human population. However, the concept has been used in a remarkably wide range of fields beyond demography and geography, including anthropology, ecology, population biology, range and wildlife management, engineering, medicine, and law. Considered across all these uses, *carrying capacity* may be defined as the maximum or optimal amount of some X that can be conveyed or supported by some encompassing entity or place Y. The cogency and utility of this definition depend on the X and Y in question, however, and an understanding of its complex history is necessary to evaluate carrying capacity as a concept.

In the 1840s, when the term originated, *carrying capacity* referred to the mechanical or engineered attributes of manufactured objects or systems, specifically ships. It arose in disputes over the tariffs and duties applied to steamships, as compared with the sailing vessels for which earlier laws and practices had been designed. Later in the 19th century, the concept was extended to things such as commuter rail systems, electrical lines, lightning rods, irrigation ditches, and pipelines.

The term *carrying capacity* was applied to living organisms and natural systems beginning in the 1870s. Examples include how much weight pack animals could haul, how much pollen certain bees could carry, and how much floodwater a bayou could transport. In the late 1880s, the term began to be used to describe the number of livestock a given quantity and quality of land could support. This marked an important reversal, as an earlier "Y" became an "X" being "carried," in a more figurative sense, by the environment in which it lived. In the 1920s and 1930s, early game managers applied this concept to quail and deer, and in the 1940s, British colonial administrators applied it to human populations in Africa as part of resettlement and agricultural development schemes. Also in this period, biologists and chemists conducted laboratory experiments in which invertebrates and microorganisms were found to display sigmoid population growth curves when provided a stream of nutrients under controlled environmental conditions. This observation led to the rediscovery of Pierre-François Verhulst's logistic equation (showing population growth and decline as an S-shaped curve), which would provide subsequent ecologists with a means of quantifying and modeling carrying capacity.

After World War II, two additional types of carrying capacity concepts emerged concurrently. One retained fauna as its object but transformed the epistemological basis of carrying capacity from inductive and applied to deductive and theoretical. This ecological concept has been poorly supported by empirical research, however, and is now seen to have only limited, heuristic value. The other shifted the object of the concept to humans and expanded its scale to regions, continents, and the entire globe, giving rise to the neo-Malthusian sense of *carrying capacity* that pervades the general use of the term today. When applied to living populations, however—and especially humans—static conceptions of carrying capacity are empirically unsupportable, while variable ones are theoretically incoherent.

Nathan F. Sayre

See also Neo-Malthusianism; Population, Environment, and Development; Population and Land Degradation; Population and Land Use

Further Readings

Sayre, N. (2008). The genesis, history, and limits of carrying capacity. *Annals of the Association of American Geographers, 98,* 120–134.

CARTOGRAMS

A cartogram is a map on which the underlying geography has been distorted to convey the distribution of some thematic variable. If the distorted features are linear, such as the lines of a subway system modified to show travel times, then the cartogram is termed a *linear*, or *distance*, cartogram. Without a qualifier, *cartogram* typically refers to the value-by-area cartogram, in which polygonal features, such as state or country boundaries, are reduced or enlarged based on a variable such as population (see Figure 1). Such area cartograms take two major forms—contiguous and noncontiguous—and have been used primarily within the domains of political cartography and epidemiology.

The first use of the area cartogram technique is difficult to pinpoint, but early examples were produced in France and Germany in the late 19th and early 20th centuries. Early cartograms were produced using manual and mechanical methods. In the 1960s, Waldo Tobler pioneered the use of computer programs in cartogram production, and several fully specified algorithms have been produced since then.

Area cartograms are typically classified as either contiguous or noncontiguous, though a full range of designs exists between these extremes. Fully contiguous, "rubber sheet" cartograms distort the shapes of units but maintain all border relationships from the original geography. Noncontiguous cartograms preserve the shapes of features, simply creating gaps between units to accommodate scaling. Both major types have

Figure 1 Counties are sized by population and colored by the 2008 presidential election results. This contiguous cartogram was produced using a diffusion-based algorithm borrowed from particle physics.

Source: Map created by Michael Gastner, Cosma Shalizi, and Mark Newman of the University of Michigan.

distinct advantages and disadvantages, with the former design known for its compactness and the latter for its easily recognizable features.

For a cartogram to be effective, readers must be able to accurately recognize and estimate the areas of its features. The former task is hindered by the distorted, ballooned shapes found on many cartograms. Most cartographers attempt to maintain certain critical shape points along feature edges as visual cues to the original geographic space. Some, though, have taken the radical step of abstracting enumeration units to circles or squares, leaving only arrangement as a clue to the identity of cartogram features.

Area estimation in information graphics is typically poor, with most readers underestimating the size of symbols. Cartograms, which use the visual variable of area to convey information, are at a particular disadvantage. Rather than the circles and squares found on most proportional symbol maps, readers must estimate the areas of complex, sometimes multipart polygons.

Another factor in the effectiveness of cartogram communication is the distribution of the thematic variable chosen to map. A variable with a high correlation to land area will likely lead to an uninteresting cartogram. Cartograms are most effective at showing variables whose spatial distributions diverge markedly from land area.

Zachary Forest Johnson

See also Cartography; Choropleth Maps; Isopleth Maps; Map Design; Map Visualization

Further Readings

Dent, B. (1999). *Cartography: Thematic map design.* Boston: McGraw-Hill.

Olson, J. (1976). Noncontiguous area cartograms. *The Professional Geographer, 28*(4), 371–380.

Slocum, T., McMaster, R., Kessler, F., & Howard, H. (2005). *Thematic cartography and geographic visualization.* Upper Saddle River, NJ: Prentice Hall.

Tobler, W. (2004). Thirty-five years of computer cartograms. *Annals of the Association of American Geographers, 94*(20), 58–73.

CARTOGRAPHY

The term *cartography* is derived from the Greek words *chartis* (map) and *graphein* (to write). Cartography is commonly defined as the art and science of mapmaking, although these dimensions are not mutually exclusive. The "art" component of cartography refers to the aesthetic or design aspects of maps and mapmaking. Similar to artwork, each map is unique, and maps may be critiqued similarly to pieces of art. An artistic step in the mapmaking process includes the creation of an overall design layout for a map. The "science" component of cartography refers to the science and mathematics necessary for accurate and effective map production. Examples of the science aspect of cartography include the use of map projections to transform the spherical surface of the Earth to a flat map and the development of an experiment with human participants to determine which color combinations on a map are perceived most effectively. A person who compiles, designs, and produces maps of any type is referred to as a *cartographer.* This entry describes the types of maps and the variety of ways in which they are used and provides a brief history of Western cartography. It then reviews the elements of maps and explains the steps involved in creating them. In exploring how map users perceive and interpret maps, cartographers have conducted research in the areas of perception and cognitive studies. These studies, along with technological innovation, have led to a variety of new applications of cartography in today's world. The entry examines some of these advances and concludes with comments on the cultural and ethical issues associated with cartography.

Categories of Maps and Map Uses

Throughout human history, maps have been created for a variety of uses and purposes. One of the earliest uses of maps was for navigation or wayfinding purposes, and this remains a common use of maps today. For example, a nautical chart used by the crew of a ship for navigation by sea and a subway map used by a tourist to travel around a city are both examples of maps used for navigation purposes. In addition, maps are commonly employed for displaying or visualizing

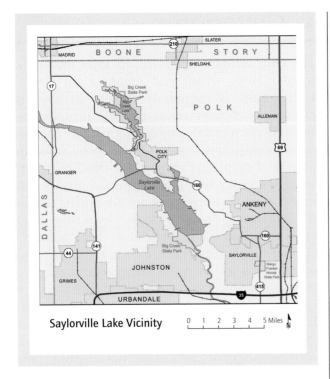

Figure 1   Example of a reference map

Source: Map created by author based on data from Environmental Systems Research Institute. (2008). *ESRI data and maps 9.3*. Redlands, CA: ESRI Press.

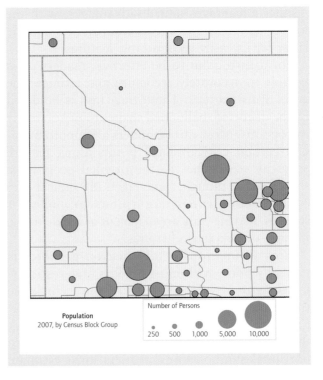

Figure 2   Example of a thematic map (total population by census block group) for the same area as Figure 1

Sources: Map created by author based on data from U.S. Census Bureau and Environmental Systems Research Institute. (2008). *ESRI data and maps 9.3*. Redlands, CA: ESRI Press.

geographic trends or patterns in data. For example, a map of population density for provinces in a country would display geographic patterns of high- and low-density population in the country. Some maps depict geographic change over time, such as weather maps on television that forecast trends in daily temperature change over a week for a location. Other maps are used primarily for the management of resources, for example, maps used by employees of a city to manage infrastructure such as utility or water lines. Maps are also commonly used to assist in the decision-making process, such as a map that displays possible building locations for a new business based on proximity to prospective customers.

Maps produced by cartographers may be categorized in many different ways depending on their purpose or use. One common classification is to divide maps into two categories, reference maps and thematic maps. Reference maps (Figure 1) display the location of major geographic features such as cities, political boundaries, transportation networks, or physical features

(e.g., rivers, lakes, and mountain ranges). Examples of reference maps include the topographic maps published by the U.S. Geological Survey and the British Ordnance Survey. Thematic maps (Figure 2), commonly referred to as *statistical maps*, display a particular theme, topic, or attribute. Examples include a map of per capita income for states in the United States and a map of languages spoken in countries within a region of the world. Thematic maps are constructed from a base map (e.g., state boundaries) and thematic or statistical data (e.g., per capita income for each state). Maps may also be classified in any number of other ways, such as by specific theme or topic, map scale, time period, or geographic area.

A Brief History of Western Cartography

The earliest known maps that survive are in the form of paintings, drawings, or engravings on

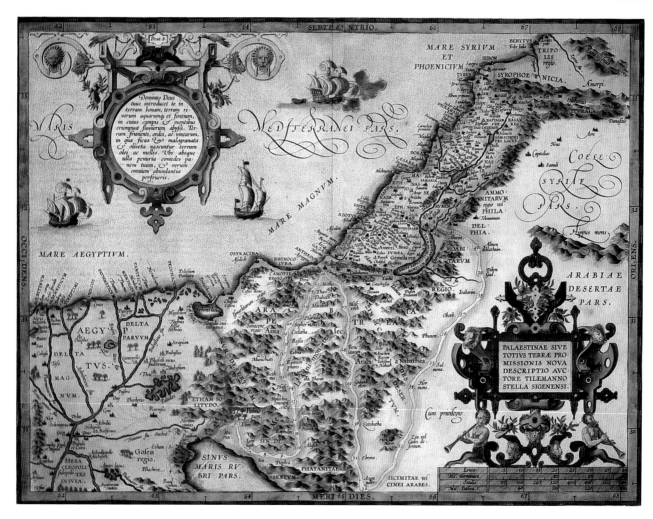

Figure 3   Ortelius's Map of Palestine and the Holy Lands, 1570. This map is from Abraham Ortelius's *Theatrum Orbis Terrarum*, which was the first atlas that produced a uniform series of maps of the world.

Source: Hemispheres Antique Maps and Prints (http://betzmaps.com/AS-168.html).

cave walls, clay tablets, animal hides, or parchments in ancient and indigenous civilizations. A painting of a village on the wall of a cave discovered in Catal Huyuk (modern Turkey) from the 7th century BC is widely regarded as the earliest known map. Babylonian maps dating to the 4th and 5th centuries BC depict the layout of ancient cities and the location of geographic features such as rivers and hills. The ancient Greeks (4th century BC) contributed several important advancements to cartography. For example, Eratosthenes (276–195 BC) devised a method of measuring the angle of the sun that he used to estimate the circumference of the Earth to within 100 miles of its actual circumference. The Greek geographer

Claudius Ptolemy (AD 90–168) published several maps in his eight-volume *Geographia*, including a world map with lines of latitude (parallels) and longitude (meridians) clearly indicated. Ptolemy and the Greeks also devised a number of map projections, some of which are still in use today.

Mapmaking during the Roman Empire focused on cadastral or property ownership maps that were used for tax assessment purposes by the Roman emperors. A prominent example of a Roman map is the Peutinger Table, a map of the road network that connected territories administered by the Roman Empire.

The Middle Ages were characterized by *mappae mundi*, including T-O maps (named for their depiction of an O-shaped world divided into three continents by rivers and seas that form a "T"), which incorporated Christian doctrine and world-views into maps of the world. As coastal exploration of the Mediterranean increased during the 13th century, Portolan charts emerged to display the location of coastal ports and cities.

Several important advancements were made in cartography during the Renaissance. Prince Henry the Navigator (1394–1460) established a center for cartography in Portugal; seafaring European nations sponsored explorations around the globe during the Age of Exploration, beginning in the 15th century. In 1569, the Dutch cartographer Gerardus Mercator (1512–1594) introduced a map of the world based on a new map projection that would bear his name and would prove invaluable for sea navigation. The Dutch cartographers Abraham Ortelius (1527–1598; see Figure 3) and Jodocus Hondius (1563–1612) published the first modern atlases of the world based on the geographic discoveries of the Age of Exploration.

Thematic mapping arose in the 18th century and accompanied the rise in national censuses and statistics collected by government organizations. In the United States, the academic study of cartography in universities developed in the early part of the 20th century and accelerated following World War II.

Map Scale, Generalization, Coordinate Systems, and Map Projections

Regardless of the type of map created by the cartographer, all maps are reductions and generalizations of the Earth. The map scale, which may be expressed on the map as a representative fraction (e.g., 1:24,000), as a verbal statement (e.g., "one inch represents two miles"), or as a graphic known as a bar scale, indicates how much the Earth has been reduced to fit on the map. The map scale influences the size of the geographic area displayed on the map as well as the amount of detail that may be displayed. Large-scale maps are those that display a relatively small geographic area but may display much detail. For example, a map of a city is a large-scale map that displays a relatively small geographic area but in much detail, with the inclusion of features such as streets, bus stops, and points of interest. In contrast, a small-scale map displays a larger geographic area but with less detail. For example, a map of a country or continent is an example of a small-scale map.

Generalization is the process whereby the cartographer selects the appropriate amount of detail to display on the map according to the map scale and the purpose of the map. An example of generalization would be the selection of cities with population figures above a certain number or the simplification of the appearance of a river by eliminating the smallest bends. Coordinate systems are necessary in mapmaking for plotting accurately the location of geographic features on the map. A coordinate system defines an x (horizontal) and y (vertical) coordinate that is referenced to the Earth's surface and then may be displayed on a map. *Latitude and longitude* is an example of a coordinate system that is used commonly in mapmaking.

A common challenge in cartography is the transformation of the mostly spherical Earth on a flat map surface. This challenge is overcome through the use of a map projection, a geometric or mathematical transformation of the Earth from a three-dimensional (3D) to a 2D surface. Hundreds of map projections have been developed to date, yet only a small number are used commonly in mapmaking today. Due to the necessity of depicting the mostly spherical Earth on a flat surface, every map projection distorts at least one of the following: area, angles, distances, or directions. Equivalent or equal-area map projections preserve the relative sizes of landmasses displayed on the map, whereas conformal map projections preserve the angular measurements. Equidistant map projections preserve the distance measurements between two points on the map, whereas azimuthal projections preserve the measurements of direction between two points. The selection of an appropriate map projection is an important step in mapmaking, and the purpose of the map often influences the type of map projection that is chosen. For example, a cartographer would choose an equivalent map projection for a map intended for comparing the size of countries or continents.

The Cartographic Process

A cartographer follows a series of steps for producing a map—these steps are commonly referred to as the *cartographic* or *mapmaking process.*

Typically, the first step is envisioning the goal or purpose of the map and identifying the map's intended audience. The type of map (e.g., paper for printing or digital for display on a computer) and format (or size of the map) are also determined by the cartographer at an early stage. A suitable map projection is selected, and all data are added to the map. Any additional data that are presented on the map must also be gathered. Sources of data for maps may include other maps, statistical data (e.g., population data), aerial photographs or satellite images, archival data, or field surveys.

Once all the data have been gathered, the cartographer must decide how to represent or symbolize the data on the map. Geographic features on maps are represented as either points (e.g., a city), lines (e.g., a river), or areas (e.g., a country). Each geographic feature is represented or symbolized with graphic variables, known as the *visual variables*, such as color, value, size, texture, pattern, or shape. If the map is a thematic map that displays quantitative data, several types of symbolization are possible that the cartographer may choose from, depending on the characteristics of the data. For example, *choropleth* maps depict data by symbolizing each enumeration or area unit with a shade of a color that represents a defined range or class of data. A *graduated* or *proportional symbol* map depicts quantitative data with symbols such as circles or squares that are sized according to the data value at each location on the map, as in Figure 2. Once a method of symbolization is chosen, the overall design of the map is clarified. The objective is to create a map that is balanced, harmonious, and aesthetically appealing to map users. Map elements such as a scale, title, legend/key, and north arrow are added to the map, each positioned to give an appealing appearance to the map.

After a completed draft of the map is produced, it is common for the cartographer to request feedback from typical users of the map. An important goal of the cartographic process is to determine if a given map communicates effectively or not. Feedback is used by the cartographer to improve the map, if necessary, prior to final production or distribution. Such feedback may be collected in a variety of ways, using both formal and informal methods. Formal methods of feedback may include structured interviews, focus groups, or questionnaires conducted with typical users.

Cognition and Perception in Cartography

Cartographers are keenly interested in how map users interpret, perceive, and remember information presented on maps, and for this reason, cognitive science is commonly integrated into cartographic research. In particular, cartographers study the *perception* (the processes associated with how humans process and interpret vision and other sensory information) and *cognition* (the processes associated with human thought and memory) of maps through structured experiments. In his book *The Look of Maps* (1952), the American cartographer Arthur Robinson established the foundation for a scientific approach to the study of map design and called for perceptual and cognitive studies in cartography to ensure effective map design. Such perceptual and cognitive studies have focused on numerous topics related to maps, such as the estimation of the sizes of map symbols (e.g., proportional circles, text, etc.) or the memory or recollection of geographic patterns displayed on thematic maps. Typically, such studies have been conducted in carefully controlled laboratory experiments. Such experiments may include the use of eye-tracking instruments that record data about what people look at on maps and how much time they spend viewing various aspects of the map. Collectively, the results from such perceptual and cognitive research help cartographers understand how people interpret maps and may be used to create better guidelines for making maps.

Technology and Cartography

Technological change and innovation have had an important influence on the development of many aspects of cartography, from the earliest days of mapmaking to the present. Technological change has affected both the type of instrumentation used in the compilation of maps by cartographers and the means of reproducing or disseminating maps to others. For example, the invention of celestial instruments such as the sextant and astrolabe allowed cartographers to plot latitude coordinates on maps and nautical charts more accurately. The magnetic compass, theodolite, and other surveying instruments improved the positional accuracy and geometry of features on maps. The development of the printing press in the 15th century was an important innovation that allowed for the mass production of maps. The technology used for printing maps has evolved from methods such as engraved plate printing to lithography to digital printing today.

The computer revolution has had a significant impact on mapmaking by transforming maps into a digital medium and automating many steps in the cartographic process. Digital maps may be displayed by map users on personal computers or portable electronic devices such as cellular phones or car navigation systems. Global positioning system (GPS) technology has revolutionized mapmaking through increased accuracy and precision of the coordinates that are displayed on maps. The development of the World Wide Web (WWW) has created a new method for distributing and viewing maps; Web-based maps may be either static maps that may be viewed on a Web page or interactive/dynamic map displays. In addition, the WWW has provided a framework for easy distribution of data that may be used for the creation of maps. Advances in aerial photography and satellite imagery have had implications for cartography as well, as these both serve as important data sources that are often used in the compilation of maps.

Modern Cartography

Although maps have been compiled by hand-drawn or analog techniques for many centuries, modern cartography is almost entirely digital in nature and is completed with the assistance of computers. Geographic information systems (GIS), graphic illustration, and computer-aided drafting (CAD) software are commonly employed by cartographers for map production. An impact of the computer revolution in cartography is what Morrison (1997) described as the "democratization of cartography," which is the increased production of maps by those with little or no cartographic training due to the easy availability of computer software that may be used for map production purposes.

The term *geovisualization* refers to interactive or dynamic maps that are developed today, such as those on Web sites that allow a user to change the appearance of features that are displayed on the map. For example, an interactive map may allow a map user the option of zooming in/out, modifying the colors displayed on the map, or changing the method by which data are symbolized on the map. Animated maps are a common method for displaying geographic change over time on maps. Data exploration mapping software is used to extract geographic patterns from large data sets that may involve the comparison of multiple variables on the map. As computer graphics technology has improved, so has the capability of cartographers to create 3D and virtual reality map displays that depict a more realistic representation of the Earth. As more focus is placed on the use of interactive and dynamic maps due to the technological changes in cartography, much attention has been focused on the design of the user interfaces and understanding how people interact with such displays.

Social, Cultural, and Ethical Aspects of Cartography

Since maps are used by many people for different purposes, there are a number of social, cultural, and ethical issues related to map production and use. Historically, maps have been produced primarily by those with political power or influence and, therefore, may intentionally or unintentionally express the views or opinions of the mapmaker. Cartographic ethics calls for the accurate presentation of information on maps; however, maps may misrepresent the real world as a result of error or carelessness by the cartographer. The increased availability of highly detailed, Web-based maps and satellite imagery, used for applications such as navigation or real estate purposes, has raised privacy issues in recent years. A cultural issue related to mapmaking includes human interpretation of colors and symbols on maps, which may have various meanings or connotations in different cultures.

John Kostelnick

See also Argumentation Maps; Cartograms; Cartography, History of; Choropleth Maps; Color in Map Design; Coordinate Geometry; Coordinate Systems; Coordinate Transformations; Countermapping; Dasymetric Maps; Data Classification Schemes; Datums;

Dot Density Maps; Dynamic and Interactive Displays; Earth's Coordinate Grid; Ecological Mapping; Electronic Atlases; Environmental Mapping; Exploratory Spatial Data Analysis; Flow Maps; Gazetteers; Indigenous Cartographies; Isopleth Maps; Land Use and Land Cover Mapping; Latitude; Longitude; Map Algebra; Map Animation; Map Design; Map Evaluation and Testing; Map Generalization; Map Projections; Map Visualization; Mental Maps; Multimedia Mapping; Portolan Charts; Resource Mapping; Self-Organizing Maps; Trap Streets; Typography in Map Design; Virtual and Immersive Environments; Virtual Globes

Further Readings

Brewer, C. A. (2005). *Designing better maps: A Guide for GIS users*. Redlands, CA: ESRI Press.

Dent, B. D., Torguson, J. S., & Hodler, T. W. (2009). *Cartography: Thematic map design* (6th ed.). Boston: McGraw-Hill.

Krygier, J., & Wood, D. (2005). *Making maps: A visual guide to map design for GIS*. New York: Guilford Press.

MacEachren, A. M. (1995). *How maps work: Representation, visualization, and design*. New York: Guilford Press.

Monmonier, M. (1996). *How to lie with maps*. Chicago: University of Chicago Press.

Morrison, J. L. (1997). Topographic mapping for the twenty-first century. In D. Rhind (Ed.), *Framework for the world* (pp. 14–27). Cambridge, UK: Geoinformation International.

Robinson, A. H. (1982). *Early thematic mapping in the history of cartography*. Chicago: University of Chicago Press.

Robinson, A. H. (1952). *The look of maps*. Madison: University of Wisconsin Press.

Robinson, A. H., Morrison, J. L., Muehrcke, P. C., Kimerling, A. J., & Guptill, S. C. (1995). *Elements of cartography* (6th ed.). New York: Wiley.

Slocum, T. A., McMaster, R. B., Kessler, F. C., & Howard, H. H. (2009). *Thematic cartography and geovisualization* (3rd ed.). Upper Saddle River, NJ: Prentice Hall.

Snyder, J. P. (1993). *Flattening the Earth: Two thousand years of map projections*. Chicago: University of Chicago Press.

Wilford, J. N. (2000). *The mapmakers* (Rev. ed.). New York: Random House.

CARTOGRAPHY, HISTORY OF

Existing histories of geography and cartography in the Western world have generally focused on one or the other of these disciplines, thereby missing the significance of their interdependent development. Maps are an integral and integrating element for geography. While much of the practice of cartography operates independently of geography, work in geography has often been explicated best by maps. The exploration and mapping of the world have been widely documented, as, for example, in coffee-table books illustrated with early maps, but the developing map of the world (from Waldseemüller's map naming America in 1507 to Heezen and Tharp's depiction of the ocean floor in 1977) is only a small part of the broad picture. The focus in this entry is on the ways in which maps have been used in many cultural environments over the millennia. This approach unites geography and cartography by tracing the uses of maps as tools in human interaction with the spatial environment.

The history of cartography is not just a timeline of significant maps and associated major events. It is more important to understand how maps have been used and what they have contributed to the cultures that produced them and how they reflect the characteristics and activities of the times and places in which they were created and employed.

Dual Functions of Maps

Maps have two general functions. One is to help people navigate from one place to another. At the most basic level, maps for wayfinding and navigation consist of the individual's personal mental (cognitive) maps. Simple wayfinding maps may be easy to make (such as a sketch map directing a dinner guest to your house) or may be sophisticated simplifications of a complex environment (such as the Vignelli New York subway map of 1977), but both are understandable at a glance. More complex navigation tasks and transportation systems require more complex mapping systems—both to gather the data for navigation maps and to then incorporate them in the complex systems of tools used for navigation (such as global positioning system [GPS]-based air, sea, or land transport navigation systems).

The second use of maps is for environmental management. As a group, maps in this category are highly varied and very large in number. The mental (cognitive) maps in this group range from the simple "image" that a person has of his or her neighborhood to the opposite extreme of the scientist who examines complex arrays of mapped data, gaining insights into relationships or searching for a solution to a problem. In both of these examples, the activity is one of visualization; the maps are manipulated in the mind. However, map use for environmental management may also involve precise measurement, complex data analyses and problem-solving procedures, and elaborate technical processes (such as in civil engineering planning and construction).

Prehistoric and Ancient Origins of Cartography

Defining a map by its function and manner of use encompasses maps in a broader spectrum of familiar formats than the conventional map drawn or printed on paper. The first practice of cartography began as mental (cognitive) maps, long before the creation of the oldest surviving map artifacts, and such cognitive maps still inform the spatial behavior of humans and animals. The patterns of stars in the sky—the celestial sphere—were among the first external maps that humans saw and sought to understand (Figure 1), as evidenced by the fact that Stonehenge and other prehistoric monuments were aligned to mark seasonal astronomical events. Prehistoric hunters followed animal tracks and paths—maps on the ground—and they depicted hunting scenes in rock carvings and cave paintings, some of which can be interpreted as maps. Ancient peoples—migratory and nomadic, foragers and hunters—learned the signs and locations of environmental resources and features and used them as landmarks to guide their travels.

When groups of humans established long-term settlements and became sedentary, new aspects of mapping became critical. The organization of the community—the relationships among living spaces, public spaces, utilities, basic resources, and, above all, gardens—became complex, requiring environmental management. Organizing the infrastructure of these early permanent communities required maps. In Egypt, the boundaries of

Figure 1 Nebra sky disk. A metal disk from ca. 1600 BC found in Northern Germany is obviously a celestial map. Interpretations range from cosmological to proto-astronomical: the sun god crossing the sky or the angles of the setting sun at the summer and winter solstices.

Source: Kaulins, A. (2009). *The sky disk of Nebra: Evidence and interpretation*. Retrieved June 27, 2009, from www.scribd .com/doc/12363758/The-Sky-Disc-of-Nebra-Evidence-and-Interpretation-English-Version-DOC.

fields cleared by the annual flooding of the Nile River were reestablished by surveying, but any maps involved would have been drawn on fragile papyrus and have not survived. The earliest environmental management maps survive among the archeological remains of the civilizations of the ancient Middle East, where cuneiform writing on clay tablets was first invented (Figure 2). Maps in the same medium were created there, the oldest about 2500 BC, to record land transactions, plot the course of a canal, or lay out the ground plan of a temple. These appear to be the earliest surviving maps drawn to scale, constructed by measurement and meant to be commensurable.

From Mesopotamia has also come the earliest surviving cosmological map (600 BC). Typical of cosmological maps produced by various cultures in being ethnocentric, it is centered on a homeland shown at a larger scale and in greater detail, while the surrounding areas shrink, losing detail and becoming distorted toward the edges of the

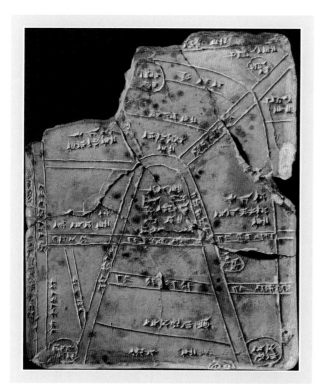

Figure 2 Clay tablet showing Nippur, the capital of Sumeria, about 1500 BC, with the surrounding fields along a bend in a river, presumably the Euphrates, labeled in cuneiform script.

Source: Courtesy of the Penn Museum, object B13885, image #14210.

known world. Egypt, where tomb paintings were an important preparation for the afterlife, also produced maps with cosmological significance. A "Book of the Dead" often included in tombs was intended to serve as a route map enabling the deceased's soul to reach heaven within 24 hours following death.

Ancient navigation in the Mediterranean Sea has left no wayfinding maps, although the archeological evidence of shipwrecks reveals the area as a crucible for the early development of seafaring technology: ships, sails, and navigation techniques. Sailing directions passed down by word of mouth; the movements of the sun, moon, and stars; the directions of the seasonal winds; the color of the water; the composition of the sea bottom; the types of birds and fish; and other environmental signs contributed to each mariner's mental map. Navigation techniques combined piloting from point to point along the coast with the increasing use of dead reckoning to sail

courses across open water. A panoramic wall painting found during the excavation of a house (dated about 1550 BC) in the seaport of Akrotiri, on the island of Thera, depicts a voyage from the Egyptian Nile delta to the Greek islands.

Similarly, mental maps and orally transmitted information guided overland wayfinding. In some cases, there were well-traveled and marked paths. In other instances, in ventures into the unknown, there was only the intuition of the explorer about the landscape. The so-called imaginary features that were plotted on early maps were the result of imagineering, as the mind of the mapmaker extended the available environmental information, relying on experience and geographical knowledge, to "map the unknown."

With the development and spread of Western civilization, the practice of cartography changed. While the basic cognitive system is still fundamental to nearly all map activities, developments in societal organization, science, and technology during the past two millennia have radically altered mapmaking and use.

Over a period of more than 600 years, Greek cartographers created a legacy of cartographic concepts and processes, culminating in the work of Claudius Ptolemy in the 1st century AD (Figure 3). His ideas for the creation of maps based on projections and the use of the graticule to locate places on maps were revolutionary when reintroduced to Europe early in the Renaissance.

Medieval and Renaissance Cartography

Lost to Europe during the feudal era, Ptolemy's writings were preserved in the Middle East, thanks to the Islamic practice of copying classical texts. For more than a millennium during the Middle Ages, the principal map of the world in Western civilization was the schematic tripartite (also called "T-in-O") map (of the *orbis terrarium* [the world]), closely identified with the Roman Catholic Church. Following the collapse of the Roman Empire and the division of Europe into small kingdoms, it was the doctrinaire Christian world schema that prevailed. The resulting maps, deriving much of their geographical information and perspective from the Bible, usually showed Earth as a disk, oriented with east at the top and centered ethnocentrically on Jerusalem (Figure 4).

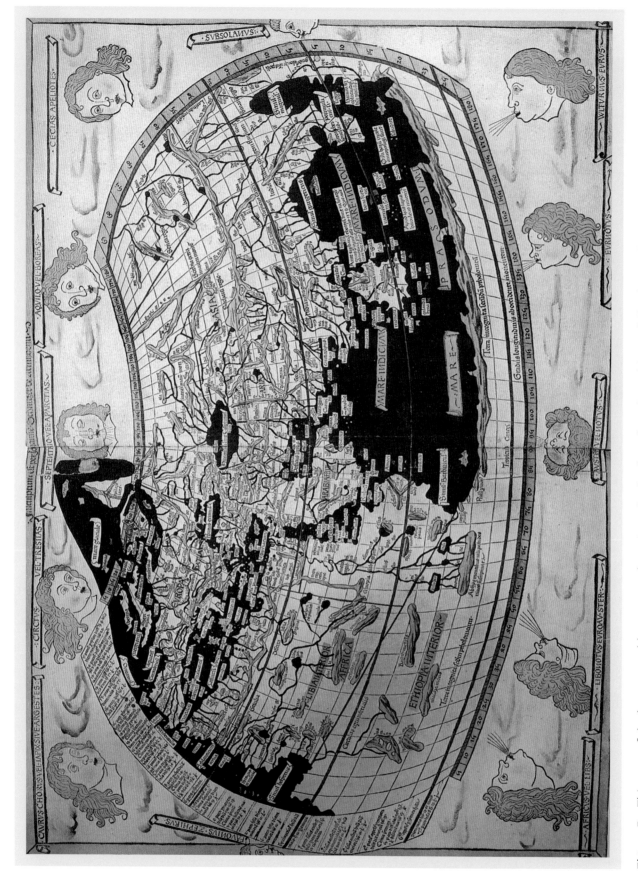

Figure 3 This map of the known world, on a pseudoconic projection based on the writings of Claudius Ptolemy (ca. 150), the map uses parallels and meridians not only to locate places on the Earth's surface but also to suggest the sphericity of the Earth.

Source: Shirley, R. W. (1984). *The mapping of the world: Early printed world maps, 1472–1700* (Cartographica 9, Plate 20). London: Holland Press.

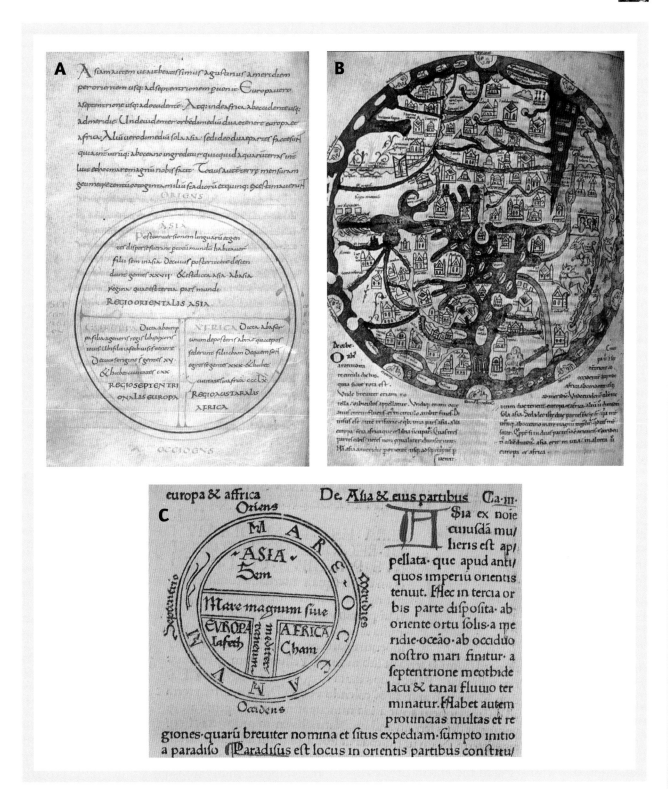

Figure 4   Three versions of the T-in-O map from the Etymologia of Isidore of Seville (560–636). Figures 4A and 4B are 7th- and 11th-century manuscripts, respectively, and illustrate the changing Christian cosmography, while Figure 4C is the first map ever printed, in 1472.

Sources: Figures 4A and 4C: Nebenzahl, K. (1986). *Maps of the Holy Land* (Figures 3 and 4). New York: Abbeville Press; Figure 4B: Harvey, P. D. A. (1991). *Medieval maps* (Figure 16). Toronto, Ontario, Canada: University of Toronto Press.

The mapmaking skills of Islamic geographers were brought to Europe in 1154, when al Idrisi completed for King Roger, the Norman conqueror of Sicily, a systematic description of the known world (extending west to east from Northern Africa and Spain to Asia and north to south from Europe to Africa) (Figure 5). Known as the *Book of Roger*, it was illustrated by a world map and sectional maps. Several centuries later, a manuscript of Ptolemy's *Geography* was brought to Italy from Constantinople and translated into Latin in about 1405. After the advent of printing in the mid 15th century, it was published repeatedly in Europe, at first uncritically and later, as scholars acknowledged that the content of Ptolemy's maps was out-of-date, amended by adding modern maps.

The introduction of printing and the commercial publication of maps were key to the democratization of map use, a process ongoing since the 15th century. Books and graphic images, including maps, had been the province of the wealthy and the literate elite in the manuscript era. Reproduction by woodcut and copper engraving also made maps accessible to the growing middle classes from the 15th century onward.

During the 16th century, the center of map printing shifted from Italy to the Netherlands. The *Theatrum Orbis Terrarum* of Abraham Ortelius in 1570 became the first of a succession of atlases expanding the limited world view of Ptolemy with the results of Europe-based global explorations. In these Renaissance atlases, environmental management expanded its scope to cover the entire Earth, systematically presented in a hierarchy of scales, ranging from city maps to world maps. These atlases were designed to "provide insight through visual methods." One could see and, without direct experience of a place or an environment, gain familiarity with some remote part of the world. This is, to use the modern terminology, *map visualization*. Decorative suites of maps, painted on the walls of palaces, such as the Vatican, impressed visitors with the extent of the ruler's vast domain. Visualization is a vicarious experience, unlike individual mental maps based on direct experience of the environment.

Navigation in the Mediterranean Sea, the age-old heart of the European trading network, generated a new type of navigation map in the late Middle Ages—the portolan chart (Figure 6). Its origin may have been associated with the introduction of the magnetic marine compass to Europe by way of trade routes from China, but it also grew out of centuries of Mediterranean nautical activity. The rhumb lines (lines of constant compass direction) radiating from compass roses on portolan charts worked effectively for navigation, despite distortions of direction and distance, in the Mediterranean Sea and along the coastal areas of Europe and North Africa. However, expansion of navigation into more remote areas of the Earth during the 16th century required maps with accuracy of direction over great distances. The solution was provided by Gerardus Mercator in 1569 and Edward Wright in 1599 (Figure 7). The Mercator projection, with its correct representation of angles across the entire map (conformality or orthomorphism), provided the navigator (as well as the surveyor and others who needed a map providing the correct representation of angles) with an efficient way to plot courses, especially when used in combination with the ancient Greek gnomonic projection, which portrayed great circles as straight lines.

National Topographic Mapping

While Renaissance atlases and sea charts provided a global cartographic perspective, an even more challenging cartographic mission, born in the 16th century, was the detailed topographic mapping of the countries of Europe. The impetus for this shift came from the emergence of national governments, aided by significant encouragement from other elements of society. For the ruling royalty, there was the desire to know the exact extent of the kingdom for purposes of administration and taxation. Similar activities were carried out at other times in other areas of the world, notably in China and Japan.

For scientists, the shape and size of the Earth was a subject for speculation and study. Seeking and mapping mineral resources fueled the growth of industry and commerce. The military general staff was concerned with national defense, and terrain was a critical factor in the movement of troops and artillery. Detailed measurement of land areas became part of the mapping process, as the mathematics underlying land surveying,

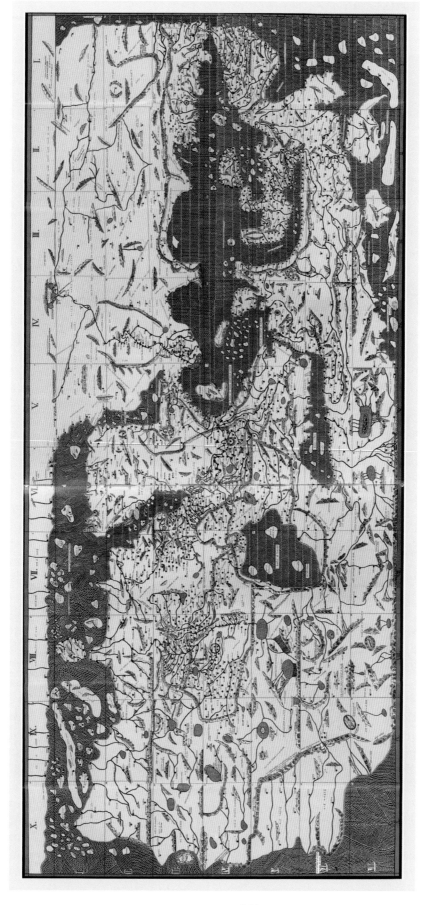

Figure 5 The world map in the *Book of Roger* (1154) perpetuates ancient Greek knowledge (such as the seven climatic bands, clima). Sectional maps (indicated on the world map) prefigure the systematic treatment of geography in modern atlases.

Source: Bibliotheque nationale de France (MSO Arabe 2221)/Wikimedia Commons, http://en.wikipedia.org/wiki/File:TabulaRogeriana.jpg.

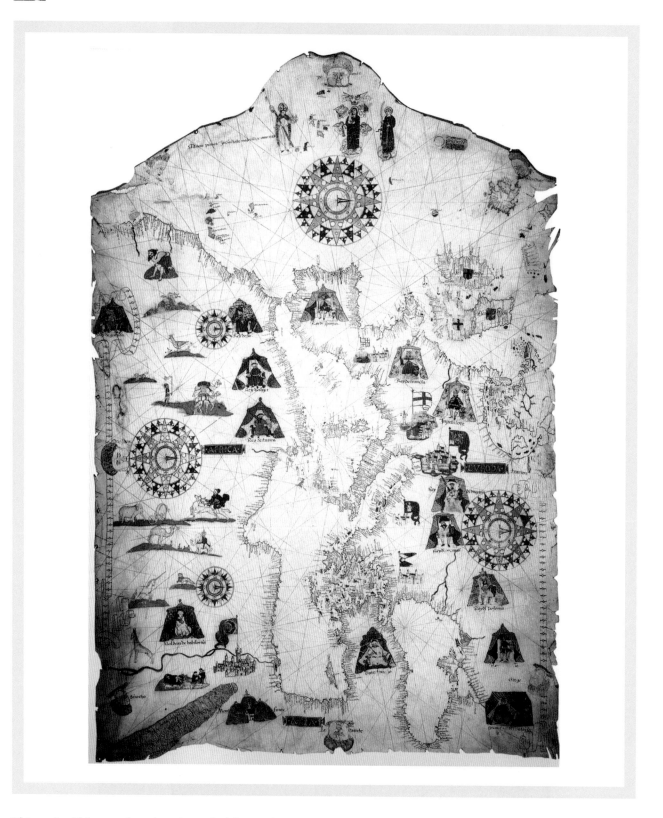

Figure 6 This portolan chart intended for Mediterranean and North Atlantic navigation was drawn in 1559 by a Catalan chart maker, who oriented this sea chart with west at the top to fit the proportions of his vellum (treated-sheepskin) drawing surface.

Source: Virga, V., & Library of Congress. (2007). *Cartographia: Mapping civilization* (Plate 32, p. 39). New York: Little, Brown.

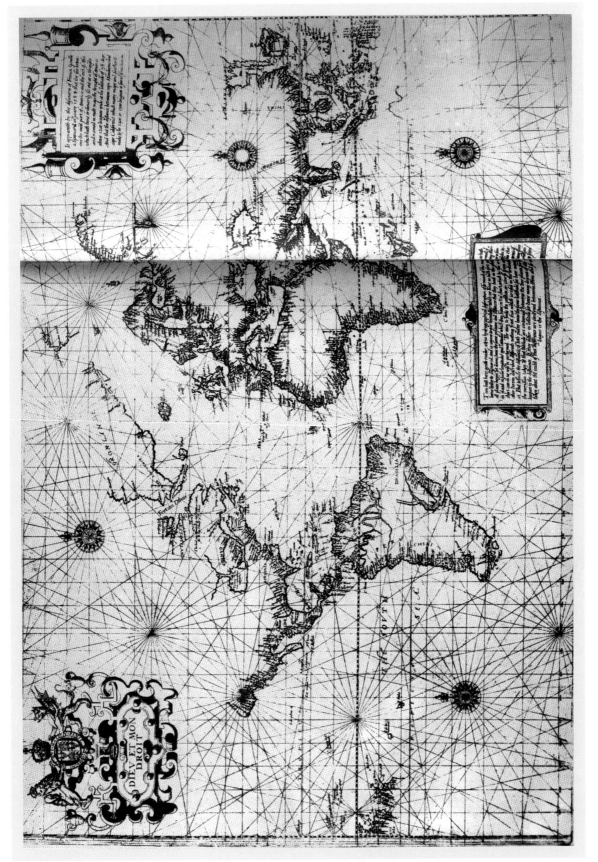

Figure 7 This world chart on the Mercator projection was used to illustrate Richard Hakluyt's book *The Principal Navigations* in 1599 and has been attributed to Edward Wright, who published tables for plotting courses on the projection, published in 1569.

Source: Ehrenberg, R. (2006). *Mapping the world: An illustrated history of cartography* (pp. 110–111). Washington, DC: National Geographic.

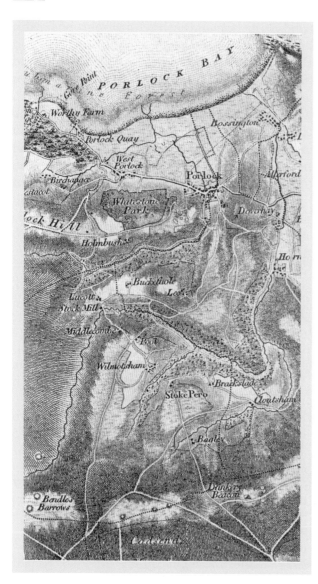

Figure 8   This detail is from an early Ordnance Survey 1-inch sheet printed from an engraved copper plate. A part of Exmoor in the hachuring style typical of the 1820s.

Source: Winterbotham, H. S. L. (1939). *A key to maps* (2nd ed.; Plate II, facing p. 70). London: Blackie.

particularly trigonometry, had been developed. However, the measurement and representation of land surface form remained a problem.

Earlier maps portraying land surface form had used pictographic symbols to represent the major features of the landscape. That approach gradually yielded during the late 14th century to a more integrated portrayal, with profiles used to represent the surface. Using the slope data that could be gathered most efficiently at the time, hachures (a short line used for shading and denoting surfaces in relief [as in map drawing] and drawn in the direction of slope) became the standard method to represent slope on 18th- and 19th-century military maps and made their way into early editions of national topographic map series throughout Europe, such as the "one-inch" Ordnance Survey of England (see Figure 8). These arrangements of upslope-downslope lines used different approaches (graphic contrasted to metric) to indicate steepness. At the same time, new methods of measuring terrain elevation provided data for mapping contour lines (isohypses or lines of equal elevation derived from the earlier concept of isobaths or lines of constant depth first used on a Dutch river chart in 1584) (Figure 9). Recognition of the contour line as the ideal metrical representation of the landscape for engineering and other commensurable uses led eventually to the adoption of contours as standard on topographic maps by every national mapping organization.

In the digital era, the database for the digital elevation model (DEM) has become the foundation for developing maps of the land surface. This numerical representation of the surface, based on elevation, provides commensurable representations of the land surface in the form of contour lines. There is still the problem that the contour line is not an effective *visual* representation of land surface form. The map user cannot visualize easily the characteristics of the surface from contour lines without significant training, so the search for the best graphic representation continues. The invention of lithography and the photographic screen in the 19th century, followed by computer graphics processes in the mid 20th century, have made it possible to produce terrain shading more easily on maps. Today, topographic maps of all types are often used as base maps, forming a geographical background of terrain and infrastructure on which other themes or types of information are displayed or draped (Figure 10). This role for the topographic map was several centuries in the making.

Thematic Maps

Until the 18th century, in Europe the main task of geography was seen as collecting enough data for

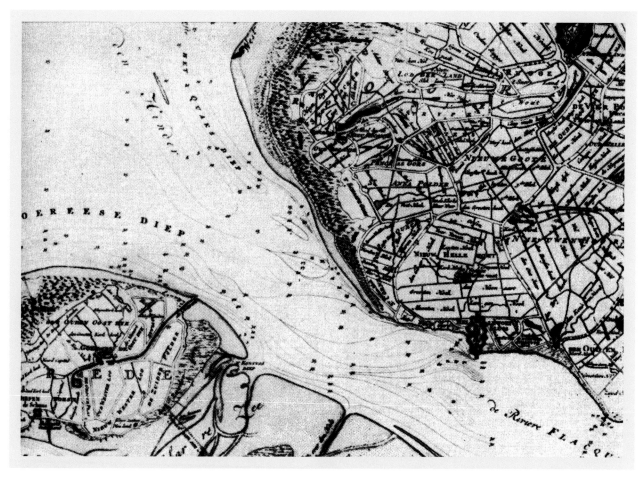

Figure 9   When Pieter Bruinsz surveyed the River Spaarne in 1584, he plotted the earliest known lines of equal depth (isobaths). This map is an early printed version of this technique. A map by Nicolaas Cruquius (1729–1731), with plotted isobaths.

Source: de Dainville, F. (1970). From the depths to the heights. *Surveying and Mapping, 30,* 393.

description of the terrestrial infrastructure. The aim of cartography, as a tool of geography, was to map the world in increasing detail, focusing on infrastructure. Where were the corridors for travel on land, rivers, and seas? Where were the nodes of population—the cities and towns? Where were the seacoasts, environmental barriers where the technological shift of trade goods between sea and land transport occurred? What was the shape of the terrain? And so on.

While the data-gathering and descriptive function of geography remained central, it gradually expanded to include the climatic, biotic, and sociocultural environments. In addition, theoretical consideration of geographical patterns and spatial relationships among different categories of data began to be explored and mapped. Early examples include Dud Dudley's 1665 map predicting the pattern of occurrence of iron and coal deposits encircling the town of Dudley in England; Edmond Halley's maps of magnetic variation (1700, Figure 11), trade winds (1688), and a predicted eclipse (1715); Alexander von Humboldt's early-19th-century map of temperatures (1817) and a diagram showing the variation of vegetation with altitude on the Ecuadoran volcano, Chimborazo (1807); moral maps by Balbi and Guerry relating the levels of education across France to the incidence of crime (1829); and John Snow's map relating the distribution of cholera cases to the location of the Broad Street pump in London (1865).

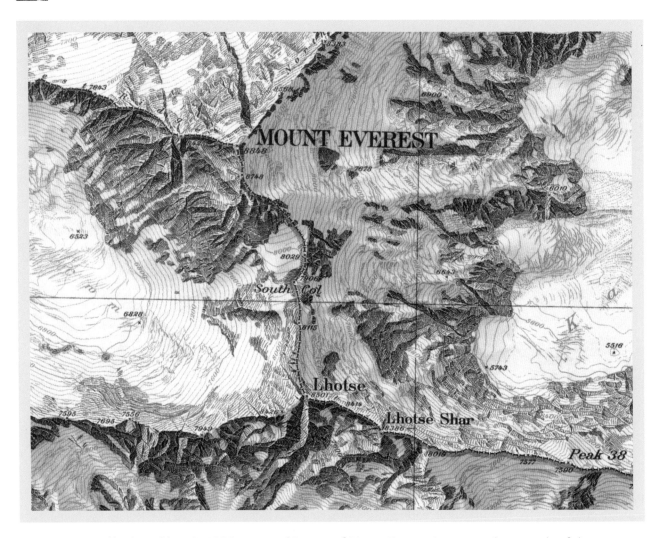

Figure 10 Bradford Washburn's 1988 topographic map of Mount Everest is a masterly example of the combined use of terrain shading and contour lines (isohypses) to depict complex mountainous terrain.

Source: Ehrenberg, R. (2006). *Mapping the world: An illustrated history of cartography* (p. 225). Washington, DC: National Geographic.

During the period from the 1820s through the 1840s, there was much innovation in the methods for symbolizing such thematic information on maps. The term *thematic map* was first used in the 20th century, replacing earlier names for this type of map (such as "special" map and "statistical" map). Into this category, cartographers placed maps representing data other than general infrastructure ("geographical" maps) and land surface form ("topographic" maps). Many types of thematic data can be mapped, some visible and tangible (such as land use or land cover) and others invisible and intangible (such as average annual temperature or median household income). During the 19th century, many thematic maps and even some thematic atlases were created. New scientific disciplines emerged, and the geographers, geologists, sociologists, and demographers (the latter called political economists) became interested in presenting their data on maps. Both individual scientists and national governments were collecting data and making thematic maps. For example, the U.S. Bureau of the Census conducted decennial (and other) censuses and also produced many series of maps (Figure 12).

German geographers, in particular, produced atlases and maps in books and journals that

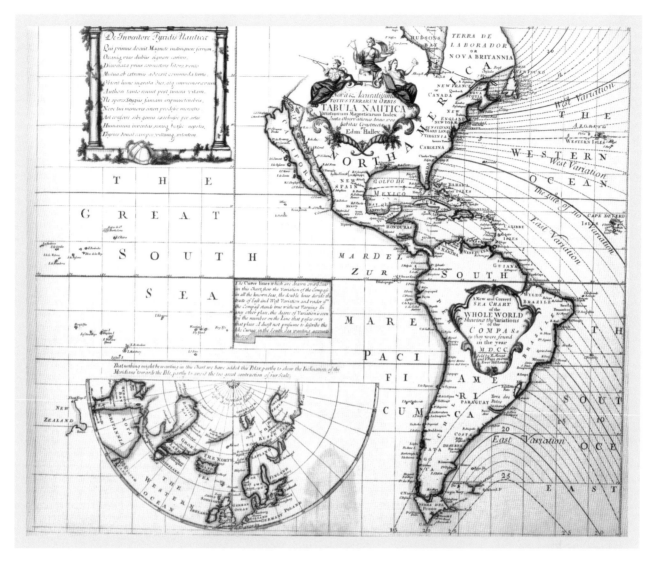

Figure 11 Edmond Halley's 1700 chart of the Western Atlantic Ocean was the first published map to include lines of equal magnetic declination.

Source: Clark, J. O. E. (Ed.). (2005). *100 maps: The science, art and politics of cartography throughout history* (p. 53). New York: Sterling.

provided the opportunity for the user, through visualization processes, to develop an understanding of the relationships between human activities and the characteristics of the physical environment. Thematic maps became a major component of national atlases.

Cartography in the 19th and 20th Centuries

The latter half of the 19th century saw extensive growth in the development of map projections. Prior to 1800, only 32 map projections had been created (the earliest being those of the Ancient Greeks). During the 19th century, 53 new projections were used. More than 180 were added in the 20th century. While most were created by geographers and cartographers, others were developed by cartophiles from other professions. Many sought a projection that would provide a better map of the world than the Mercator projection, and many pseudocylindrical projections were created. This activity continues as needs for new perspectives of different parts of the world arise.

The 19th and 20th centuries saw not only remarkable innovations in the science and technology of cartography but also the creation of new

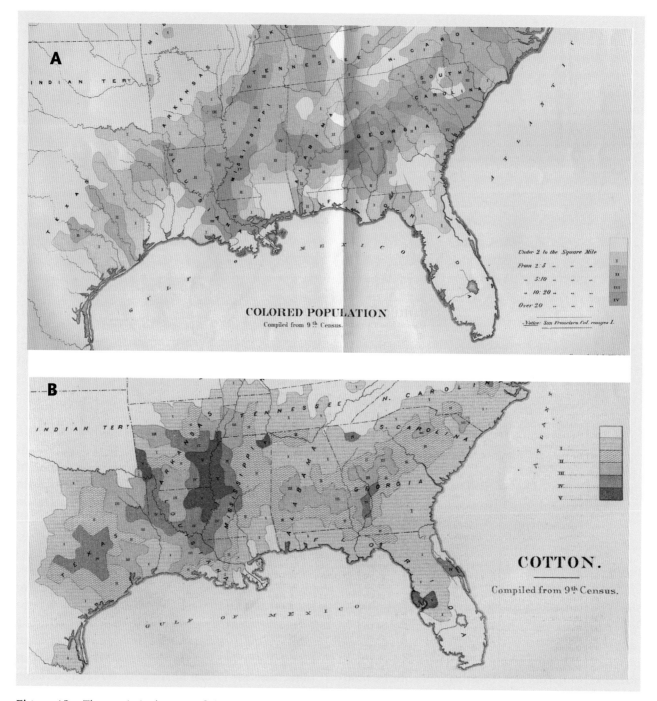

Figure 12 The statistical maps of the 1870 U.S. Census were innovations in both data gathering and presentation. Figures 12A and B use "degrees of density" for the colored population and an index of "relative power" for the levels in the production of cotton.

Source: Walker, F. A. (1872). *Colored population and cotton (Ninth Census): Vol. 3. The statistics of wealth and industry of the United States*. Washington, DC: Government Printing Office (maps preceding pp. 77 [12a] and 160 [12b]).

Notes: Recognizing the ineffectiveness of many procedures for presenting cropland data on maps, the director of the 1870 Census, Francis Walker, employed an index of "productive power." Since no single element satisfactorily measures productive power of a crop, the index relates the quantity of a crop to both the number of inhabitants and the number of acres of improved land. The derived values for each county, when mapped, indicate "the important and the insignificant sections of the country distinctly from each other" (here in six ordinal classes, I, II, etc.). To compute these values, the number of bushels, tons, or pounds produced in each county was divided by the number of inhabitants and by the number of acres of improved land in the county; these two quotients were multiplied, and the square root of the product was taken as the measure of productive power of the county with respect to that crop.

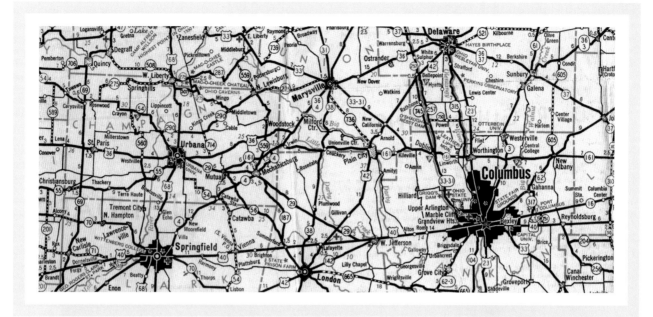

Figure 13 Portion of the State of Ohio Official Road Map, 1938, showing numbered "hard surfaced" and "principally stone or gravel" highways and "good connecting roads," along with unincorporated villages and incorporated municipalities.

Source: State of Ohio, Department of Highways. (1938). *Ohio 1938: Official highway map*. Columbus, OH: Author. (Original scale 1:633,600.)

types of maps for Environment Waikato map uses. In the realm of navigation, aeronautical charts and automobile road maps (Figure 13) were added; these and railroad maps are complemented by the variety of tourist maps that became available. In the realm of environmental management, weather maps and the news and information maps promulgated by the media not only provide spatial details for military campaigns and natural disasters but also emphasize the importance of location. Aerial photographs and satellite images have assumed the role of the base map (Figure 14). All these have been accompanied by the introduction into the cartographic process of new printing and publication processes. The 20th century saw the introduction of photomechanical process color printing, microfilming, blueline printing, photocopying, and color proofing, each offering the cartographer a new approach to map design and production (Figure 15). The cartographer gained control over the design of the map, which had been the province of the engraver and the printer for more than five centuries. When, near the end of the 20th century, the analog procedures were replaced with comput-

er-based mapping systems, the overall design of the map became the result of the default values embodied in the computer software. With the rise of the Internet, Web-based mapping services provided even more personalized maps, and the emboldened map user now deals directly with the available software.

While technical manuals and reference volumes dealing with the practice of cartography were already common at the beginning of the 20th century, textbooks emerged first in German in 1912 and then in English (Erwin Raisz, *General Cartography*) in 1938. Textbook development expanded after World War II, promoting study and research dealing with cartographic concepts and processes. Today, textbooks are available in many languages, taking many different approaches to the development and production of all types of maps.

Shortly after the mid 20th century, the rise of academic cartography in geography departments generated psychology-based studies aimed at making symbolization of data and map design more effective (Figure 16). Based on the principles of visual perception, this work expanded to include

Figure 14   Using data from the SPOT and Landsat systems, cartographic annotations in vector format from a resource management database were added. Each step in the process was carried out using the components of Intergraph's MGE software system.

Source: Intergraph Corporation. (1991). *Kuwait City*. Huntsville, AL: Author. (Original scale 1:100,000.)

Figure 15 Progressive stages of engraving and printing presented in the order they occur in this reproduction of a painting. Combinations of four inks can produce millions of colors. Produced by the Peerless Engraving & Colortype Company (Chicago)

Source: Flader, L. (Ed.). (1927). *Achievements in photo-engraving and letter-press printing* (p. 329). Chicago: American Photo-Engravers Association.

It is important to recognize that significant scientific and technological advances occurred in 20th-century cartography outside the sphere of "geographical cartography." Mapping activities during World War II involved both historians and geographers, working with a diversity of cartographic products used for navigation and environmental management. By mid century, the basic features on Earth had been well mapped. Improvements continue, and additions to the coverage of large-scale national topographic map series and the extension of the geodetic network continue (both now supported by the global positioning system [GPS]). A new global mapping reference system was completed (World Geodetic System [WGS] 84). Digital databases, from the Digital Chart of the World (DCW; 1:1,000,000) to the hundreds of "local" databases, provide data for environmental management at many different government levels. Attention has turned to more precise study of problems such as environmental pollution and sustainability, preparation for natural disasters, global disease distribution, and more. All these can be integrated, in one way or another, into a database and can contribute to geographic information systems (GIS), and they are related to other elements of the environment. As individual maps, these are all used for environmental management. In the case of the examples used here, the management involved is the handling of information, for understanding and knowledge.

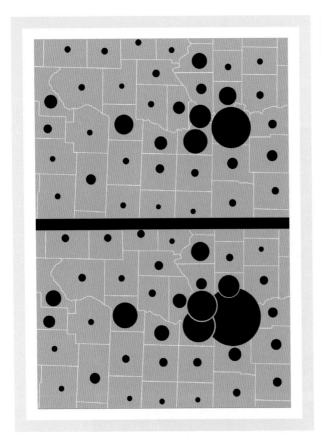

Figure 16 The concepts and processes of psychophysics were explored by cartographers throughout the 20th century. Here, the visual underestimation of circle-size differences is countered by rescaling the sizes, a more efficient representation of the data.

Source: McCleary, G. F., Jr. (1981). How to design an effective graphics presentation. In P. A. Moore (Ed.), *How to design an effective graphics presentation* (Vol. 17, pp. 15–64). Cambridge, MA: Harvard College.

The 21st Century

It is clear that the maps produced by a culture mirror that culture. In various indigenous cultures around the world, the traditional role of environmental management has been to preserve and perpetuate the relationship of humans and environment. In Western civilization, maps have been created and used not simply for navigation and environmental management but rather for navigation and environmental exploitation. It is time to make maps for environmental sustainability. Computer-based GIS are likely to assist in finding solutions to the looming global environmental problems, offering not only the means of carrying out effective environmental management using maps but also a renewed role for the

work in cognition as well as graphic design and communication (Figure 17). The attention to this user-oriented research was soon diverted toward experimentation with the new digital technology. Continuing ignorance of the basics of graphic design; of the fundamental elements of statistical data representation, particularly the visual variables; and of the considerable resources from the literature in graphic design, psychology, and human factors has perpetuated the problems associated with efforts to create truly effective user-friendly maps. Unfortunately, the use of poor design procedures has been sustained by tradition and convention, as well as the aforementioned default options of the computer software.

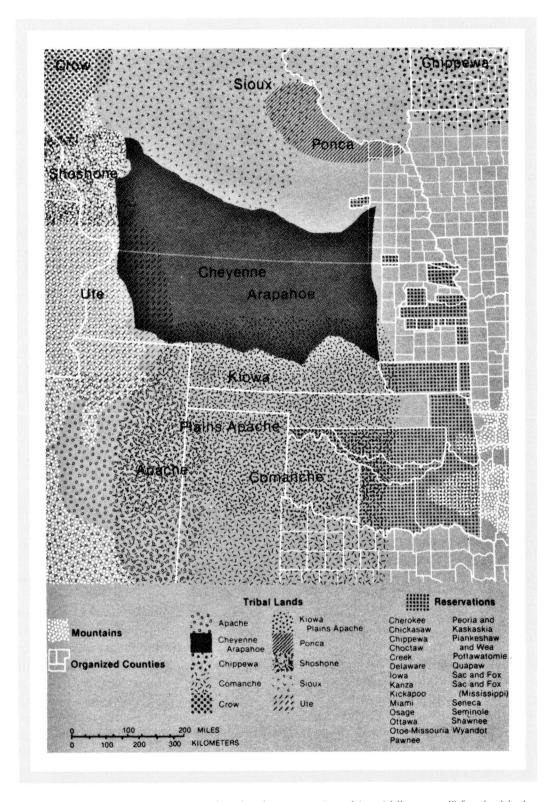

Figure 17   In this map, tint screens are used to develop a gray "graphic middle ground" for the black and white symbols conveying the environmental confrontations (*in white*) experienced by the Cheyenne (and other indigenous groups) during the 19th century.

Source: McCleary, G. F., Jr. (1990). Pursuing the Cheyenne: Mapping tribes, trails, the 1857 expedition and the battle of Solomon's Fork. *Meridian, 4*, 3–28.

continuing partnership of cartography and other environmental sciences. To achieve this, there are several problems that must be resolved.

The first is the extension of the GIS software beyond rudimentary analytical capabilities that are only the beginning of a comprehensive set of spatial analytical procedures. The technology has to mature beyond paying lip service to the study of relationships and reach a new level of sophistication. Second, to manage the environment, "you gotta know the territory." This is the only way to sort out and understand the numerical data and the resulting maps. It is not as simple as Bacon (and others) looking at the world map to gain a perspective on plate tectonics, which occur in geological time. Time is so short in many environmental management situations that long before there is any clear understanding of the problem, a decision has been made and implemented by public officials, based on limited mental (cognitive) maps (and other understandings). Too often, they have cleared the forest and built the highway intersection and shopping mall, and the environment has lost again. The joint expertise of geographers and cartographers is needed more than ever to create the GIS, generate the maps, and interpret the environmental factors and processes they depict.

George F. McCleary Jr. and Karen S. Cook

See also al-Idrisi; Biblical Mapping; Cadastral Systems; Cartography; Digital Terrain Model; GIS, History of; Human Geography, History of; Indigenous Cartographies; Map Projections; Map Visualization; Mental Maps; Mercator, Gerardus; Ortelius; Portolan Charts; Ptolemy; T-in-O Maps; Waldseemüller, Martin

Further Readings

Bertin, J. (1983). *The semiology of graphics: Diagrams, networks, maps.* Madison: University of Wisconsin Press. (Original work published 1967)
Harley, J. B., & Woodward, D. (Eds.). (1987). *The history of cartography.* Chicago: University of Chicago Press.

Harvey, P. D. A. (1980). *The history of topographical maps: Symbols, pictures and surveys.* London: Thames & Hudson.
Lynch, K. (1960). *The image of the city.* Cambridge: MIT Press.
Raisz, E. (1938). *General cartography.* New York: McGraw-Hill.
Snyder, J. P. (1993). *Flattening the Earth: Two thousand years of map projections.* Chicago: University of Chicago Press.
Wallis, H. M., & Robinson, A. H. (1987). *Cartographical innovations: An international handbook of mapping terms to 1900.* Hertfordshire, UK: Map Collector.

CASTELLS, MANUEL (1942–)

Manuel Castells is a prominent urban sociologist who has made major interdisciplinary impacts through his theorizing of various contemporary global socioeconomic transformations.

Born in a Catalan family in Spain in 1942 and educated in France, Castells exemplifies a global theorist of contemporary societies. His expertise ranges from urban sociology and social movements to theories of technology and network society, and he has been influential in a wide range of social science disciplines, including anthropology, urban planning, economic geography, and telecommunications. Castells is known for a combination of talents rarely gifted to a single individual: an ability to interpret, analyze, and incorporate real-world examples around the globe; an ability to craft theories by combining rigorous inductive and deductive methods; and a commitment to the masses as the instigators and primary drivers of social change. His hallmark extensive international coverage of topics and rigorous comparative analysis are based on his firsthand observations and experiences gained through appointments and lectures he conducted in 40 countries, primarily in Europe, the Americas, and Asia. Castells's lifetime interests have been to uncover the factors and agents that bring about revolutionary changes in society, and he has done so through engagement in research on urban politics, social movements, and information technologies.

Exiled by the Franco regime in the early 1960s, Castells began his intellectual journey in Paris under the guidance of Alain Touraine, an esteemed urban sociologist. Castells's dissertation and his first book, *The Urban Question: A Marxist Approach* (published in English in 1977), focused on urban politics, planning, and policy. The book is remarkably suggestive of analytical traits that are evident in his later work. Castells developed a global interpretation of urban structures by incorporating historical analysis and examining data from American, French, and Latin American cities, as well as those in the socialist world. His goal was to demystify and expose ideologies of the dominant classes through analysis of their social practices and develop a Marxist urban theory, which he claimed to have generated questions yet lacking sufficiently specific analytical framework. The outcome is an extremely dense writing that combines rich theoretical explorations with extensive empirical research that illustrates the underlining forces behind the ideology of the urban question.

At the University of California at Berkeley, which became Castells's longest intellectual home (1979–2003), he expanded his inquiry on the urban question into social movements by developing a cross-cultural theory of urban social change. In *The City and the Grassroots* (1983), Castells examined grassroots mobilizations in San Francisco's Mission district; in the gay community around Castro Street; in squatter settlements in Lima, Mexico City, and Santiago de Chile; and in citizens' movements in Madrid. Although somewhat overshadowed by the fame of his later work, this book is still regarded as a groundbreaking work in social movements research.

Subsequently, Castells shifted his focus to technology as a driving force of social change, with a particular emphasis on information technologies and their impacts on cities (as depicted in his books *The Informational City* in 1989, in which the concept of "space of flows" is first elaborated, and *Technopoles of the World* coauthored with Hall in 1994). The trilogy on *The Information Age: Economy, Society and Culture* (1996 to 2004) is considered Castells's seminal work, and it has been translated into 22 languages. He analyzed the socioeconomic transformations of globalization, seen as outcomes of the information technology revolution. In the first volume of the trilogy, titled *The Rise of the Network Society*, Castells focused on the socioeconomic aspects of informationalism. In the second volume, *The Power of Identity*, Castells reconceptualized identity, social movements, and the state in the new global order and explored religious fundamentalism, patriarchalism, and information politics. The third volume, *End of Millennium*, covers broad contemporary sociopolitical transformations, ranging from the collapse of the Soviet Union to the rise of the fourth world, the global criminal economy, the Asian economic crisis, and European unification.

Castells also published works that focused on social changes in a digitalized world (see, e.g., *The Internet Galaxy*, 2001, and *The Information Society and the Welfare State: The Finnish Model*, with P. Himanen, 2002). But it is the trilogy that showcases the best of Castells's skillful grand theorizing of the highest sophistication; he does so by compiling an enormous amount of quantitative and qualitative evidence to support his claims, by presenting the most comprehensive global coverage humanly possible by an individual scholar, and by mobilizing an in-depth and multidisciplinary knowledge of theory in capturing and illuminating fundamental social changes.

At the University of California, Berkeley, Castells was known to his students as a methodological whiz, being able to instantaneously process and translate any potential research topic presented by students—often still broad and undefined—to one with an elegant research design, complete with a compelling question driven by a series of theoretically informed, rigorous yet answerable hypotheses. In fact, in spite of his voraciousness and openness to theoretical experiments, Castells was simultaneously feared and admired by students for demanding rigorous research design and competent social analysis. What made Castells truly remarkable was his dedication to data collection. Castells once even remarked that becoming a competent empiricist—in the sense that research should be based on empirical data rather than critical theoretical reflection alone—is often more difficult than engaging in grand theorizing. His sensitivity to cultural nuances, combined with his ability to generate grand theories by incorporating empirical details in a truly bottom-up manner, was clearly enabled by his extensive interdisciplinary knowledge of sociological theories, but it also came out of his innate ability to generate and articulate compelling questions and craft them into sophisticated, well-informed hypotheses.

For someone so deeply involved in social activism of the political left, Castells's departure from Marxism was never absolved by the Marxist community. Particularly frustrated were those who expected a manifesto, or at least a prescriptive message, from the trilogy. Yet Castells's relationship with Marxism is clear from his earliest work, which was underscored by a desire to refine Marxist analysis and resolve its analytical weaknesses (*The Urban Question*) and contradictions (*The City and the Grassroots*). Castells in *The City and the Grassroots* criticized both Marxism and Leninism for their inability to incorporate urban social movements and declared, "So although Marixist theory might not have room for social movements other than the historically predicted class struggle, social movements persist. So experience was right and Marxist theory was wrong on this point" (p. 299). In the *End of Millennium*, he wrote, "Theory and research, in general as well as in this book, should be considered as a means of understanding our world, and should be judged exclusively on their accuracy, rigor, and relevance" (p. 390). Furthermore, his firsthand knowledge of those who suffered under the Soviet regime led to the conviction that Marxism as an intellectual project resulted in considerable human sufferings. In *End of Millennium*, he explicitly cautions against ideology:

> What is to be done? Each time an intellectual has tried to answer this question, and seriously implement the answer, catastrophe has ensued. . . . I have seen so much misled sacrifice, so many dead ends induced by ideology, and such horrors provoked by artificial paradises of dogmatic politics that I want to convey a salutary reaction against trying to frame political practice in accordance with social theory, or, for that matter, with ideology. (pp. 389–390)

Ultimately, Castells's unwavering concern for the social well-being of the people has won over his commitment to a particular intellectual project. His commitment to real-world experiences is what makes Castells a world-renowned intellectual.

Yuko Aoyama

See also Marxism, Geography and; Space of Flows

Further Readings

Castells, M. (1977). *The urban question: A Marxist approach*. Cambridge: MIT Press.

Castells, M. (1983). *The city and the grassroots: A cross-cultural theory of urban social movements*. London: Edward Arnold.

Castells, M. (1989). *The informational city: Information technology, economic restructuring and the urban-regional process*. London: Blackwell.

Castells, M. (1996). *The rise of the network society: Vol. 1. The information age: Economy, society and culture*. London: Blackwell.

Castells, M. (1997). *The power of identity: Vol. 2. The information age: Economy, society and culture*. London: Blackwell.

Castells, M. (1998). *End of millennium: Vol. 3. The information age: Economy, society and culture*. London: Blackwell.

Castells, M. (2001). *The Internet galaxy: Reflections on the Internet, business, and society*. New York: Oxford University Press.

Castells, M., & Himanen, P. (2002). *The information society and the welfare state: The Finnish model*. New York: Oxford University Press.

CAVERNS

Caverns occur under the surface of the Earth wherever soluble rocks allow them to develop large openings. Limestone or calcite-rich ($CaCO_3$) bedrock is the most common host rock, as well as dolostone rocks composed of the mineral dolomite ($CaMg(CO_3)_2$). The evaporite rocks of gypsum ($CaSO_4$) and rock salt ($NaCl$) are also subject to solution and can develop underground openings large enough to be considered caverns. In a few places, caverns have developed in other rock lithologies, such as where a calcite-cemented sandstone has also been subjected to solution. Several natural processes other than the dissolving of the bedrock can also produce caves, such as ocean waves and lava flows, but these are almost never large enough to constitute a true cavern, which has an implication of being of great size.

Caverns are almost always also of the solutional type, wherein the rock is dissolved, either partly in the upper vadose (undersaturated with free air surfaces) zone above the water table or more commonly beneath an unconfined water table in the bedrock as a water-filled phreatic cavern. Only when the water table is lowered, usually when the climate changes to a more arid one and the water table drops or when a nearby stream erodes downward enough to allow the water to drain out of the caverns, does air enter the opening and the cavern become part of the vadose realm. Vadose caverns can also pass from phreatic to vadose and back again many times in response to water table fluctuations associated with varying climates or tectonics. In addition to those formed in association with unconfined water tables, some caverns are formed entirely under conditions of confined or artesian water.

Speleogenesis, or the origin of caves and caverns, is a process wherein surface water (H_2O) infiltrating through the soil reacts with rotting organic matter to gain a bit of carbon dioxide (CO_2) and becomes carbonic acid (H_2CO_3). This weak acid in the groundwater trickles slowly along joints and fissures in the bedrock to come in contact with the calcium carbonate

Mammoth Cave in Kentucky is the longest known limestone cavern with about 367 miles of surveyed passageways. Mammoth Cave National Park was named a World Heritage Site in 1981 and is a core area of the International Biosphere Reserve.

Source: National Park Service.

($CaCO_3$) or the calcite mineral of the limestone, thereby dissolving it in the carbonation reaction. Mammoth Cave in Kentucky is this kind of cavern. Gypsum is 10 to 20 times more soluble in water than is carbonate rock, so caverns in that rock can form more quickly. Similarly, rock salt is even more soluble than gypsum, so large caves can form in it. Neither gypsum nor rock salt are very common, except in arid areas, with the result that caverns developed in these rocks are far rarer than those in limestone. Furthermore, because rock salt is so mobile underground and moves upward in salt domes or diapirs, where it is loaded by overlying rock, any openings that have formed in it deep underground tend to anneal or be resealed and infilled by the flowing salt. Only where a surface salt deposit is subjected to a falling water table or where a salt diapir penetrates the surface into the vadose zone and forms an uplifted region do caverns form and remain open. In these cases, the rainwater simply dissolves the salt, so that the brine flows back out onto the surface at a lower level.

Alternatively to solution in water, a rarer process occurs underground to produce caverns wherein some source of hydrogen sulfide (H_2S) gas, such as oil or natural gas, or perhaps also volcanic gases, reacts with the groundwater to produce a gas of sulfuric acid (H_2SO_4), which rises from below to produce large openings. Carlsbad Caverns and Lechuguilla Cave in New Mexico are believed to have been produced in this unusual fashion.

Caverns commonly have a huge variety of speleothems (a secondary mineral deposit formed in a cave) in them that are the product of deposition after the initial dissolution produces the cavity in the first place. Once limestone caverns leave their underwater, phreatic-formation stage, where the rock is removed in solution and enters the vadose realm with a new gaseous atmosphere, they are then subjected to renewed infilling by speleothems. The varieties of speleothem infillings of caverns, such as stalactites, stalagmites, dripstome, flowstone, helectites, cave pearls, and myriad other forms, are truly awe inspiring and commonly very beautiful. Generally, in limestone bedrock, the speleothems will be of a similar composition because the carbonation reaction can go in either direction: toward dissolution of the rock in the cavern-forming phreatic stage or toward deposition of calcite speleothems in the cavern-refilling vadose stage. In some cases, however, where the underground water in a limestone region has much dissolved calcium sulfate ($CaSO_4$) in it, perhaps as the result of nearby igneous intrusions, the speleothems can be crystals of gypsum. The Cave of Crystals in Chihuahua, Mexico, for example, has crystals of selenite gypsum up to 11.3 m (meters) long, which are among the longest in the world. In arid and semi-arid climates, speleothems in caverns are mainly composed of gypsum, whereas in temperate, humid, or tropical regions, carbonate formations are more predominant.

The size of caverns has been measured in most parts of the world. The published sizes change frequently with every new expedition that revises the survey. The longest known limestone cavern is Mammoth Cave, Kentucky, with about 591 km (kilometers) of chambers, whereas the longest in gypsum is the Optimisticheskaya Cave in West Ukraine, at about 230 km. The longest cavern in salt is in a salt diapir in Iran on Qeshm Island in the Persian Gulf at 6.58 km. The deepest cavern is Kubera (Voronja) at 2,191 m in Abkhazia, Georgia, in the West Caucasus Mountains. The largest underground chamber by surface area is the Sarawak Chamber, which measures 600 m by 415 m on a side and is about 80 to 100 m high. It is in the Lubang Nasib Bagus Cavern, in the Gunung Mulu National Park in Malaysia on the island of Borneo. The deepest pit in the world of caverns is that of Vrtoglavica in Slovenia at 603 m deep.

John F. Shroder

See also Carbonation; Groundwater; Karst Topography

Further Readings

Ford, D., & Williams, P. (2007). *Karst hydrogeology and geomorphology.* New York: Wiley.

 # CELLULAR AUTOMATA

Cellular automata (CA; singular: cellular automaton) are dynamic discrete systems that operate in space and time on a uniform and regular lattice of cells. These cells are driven by rules that describe interactions at the local level to produce patterns at the global level. Depending on the problem context, the cells can also be rectangular, triangular, or hexagonal.

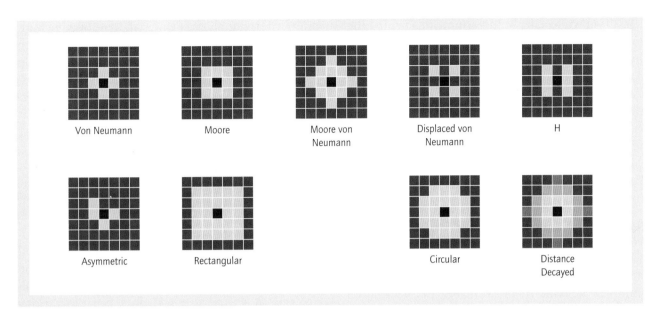

Figure 1 Different cellular automata neighborhoods

Source: Author.

History

The mathematical description of CA was partially postulated by D'Arcy Thompson (1917), then further developed by Von Neumann (1966) in his work on self-reproducing systems. The work of Von Neumann was motivated by Alan Turing's groundbreaking developments in algorithms and digital computing as well as Stanislaw Ulam's work on the growth of crystals in the 1940s and 1950s. CA concepts were further explored by Conway's *Game of Life* and popularized by Gardner in a 1970 article published in the journal *Scientific American*. Later, CA studies were linked to the theory of complex systems and were further elaborated by Wolfram (1984). Since then, CA theory and applications have gained rapid popularity in various scientific fields, such as physics, medicine, chemistry, biology, ecology, forestry, geography, and atmospheric and Earth sciences, among others.

Mathematical Formalism

The main components of CA are the cell spaces represented by a regular grid; cell states, S; neighborhood of the cell, N; functions of cell transition rules, R; and discrete time increments, ΔT. The cell state $S(x, y)_T$ at time T can be represented as function F depending on CA elements at previous time $T - 1$ and can be formalized as

$$S(x, y)_T = F\{S(x, y)_{T-1}, N_{T-1}, R_{T-1}, \Delta T\},$$

where $S(x, y)_{T-1}$ is the cell state at location (x, y).

Each discrete time step represents the CA model iteration. During the iteration, an update of the state of each cell in the grid is provided using transition rules and the states of other cells in its local neighborhood. Figure 1 shows the various types of neighborhoods used in the CA modeling procedures. The Von Neumann, Moore, extended Moore, displaced von Neumann, Moore von Neumann, and H neighborhoods are the more popular ones that are used in modeling studies. The shape and the size of the neighborhoods may vary, from those with symmetric, asymmetric, rectangular, and circular shapes and distinct values to those that consider cells from different locations in the neighborhoods with gradual values of different importance based on distance-distance decay of a given variable.

The transition rules form an important part of the CA model. They define the way in which the cell will develop over time and hence mimic the overall change process of the cells over space and time. The rules can be deterministic, probabilistic, stochastic, or fuzzy and, therefore, imply a bottom-up modeling process from the local scale to global scales. The model calibration and validation procedures are key phases of the CA model development. Calibration and validation allow fine tuning of the model parameters, and model testing determines how closely the final model can simulate or represent the spatial patterns found in the real world.

Cellular Automata and Geography

Geographic phenomena are inherently dynamic and complex and can therefore be studied using complex systems theory. During the geographic change process, elements of the system may evolve, interact, and bifurcate over space and time. These interactions are usually nonlinear and self-organizational, so CA theory is suitable to represent them.

Using cells and neighborhoods to represent the dynamics of geographic systems as spatial models can be traced back to the development of diffusion models by Hagerstrand (1953), the probabilistic model of residential growth by Chapin and Weiss (1968), and Schelling's (1969) segregation model. Later, Tobler (1970) introduced strict cellular automat rules for cellular geography, which was further theorized by Couclelis (1985). Since then, CA was increasingly used in geography.

Geographic Applications of Cellular Automata

The main geographical applications of CA were in land use change and urban growth studies at local or regional scales. Structure formation related to micro- and macrodynamics were explored in early research efforts in 1990s and further improved to include the modeling and representing of urban form and urban growth, population dynamics, rural residential settlements, sustainable growth, and, even, urban gentrification. CA models of land use change were developed to account not only for urban land use dynamics but for forest and agricultural dynamics as well. Many mathematical approaches were used to make CA models closer to reality, such as

multicriteria analysis, principal components analysis, genetic algorithms, and neural and Bayesian networks. Attempts have been made to develop CA models that operate on irregular spatial tessellations with Voronoi diagrams, use graph theory, and develop urban cadastral lots as cells (Figure 2).

Cellular Automata and Geographic Information Systems

Advances in satellite data collections, raster geographic information systems (GIS), and increased geospatial data availability also amplified the attractiveness of CA as a modeling approach. GIS deals with various spatial data representations and can handle, analyze, and manipulate different geospatial data sets, but it is, however, known as a static tool. The CA-GIS fusion provided the extended capability of GIS to use both spatial and temporal dimensions and allowed the analysis and modeling of inherently dynamic and complex geographic processes. The Idrisi GIS software contains modules that allow the development of simple CA models. However, most studies rely on a loose coupling mechanism with the GIS software to implement various CA modeling approaches. The most common stand-alone CA modeling approaches, tools, and environments are, for example, IDUEM (Integrated Dynamic Urban Evolutionary Model), GEONAMICA, SLEUTH (slope, land use map, excluded area, urban area, transportation map, and hillside area model), UrbanSim, and iCITY (irregular city). One drawback is that they are customized based on the model design or data, making them difficult to adapt for new case studies. Some researchers argue that CA models are for analysis but not for prediction of possible simulated urban and land use patterns. Others have implemented CA models for planning contexts and policy-making purposes or for public and stakeholder involvements in land use decision making.

Suzana Dragicevic

See also Complex Systems Models; Complexity Theory; Distributed Computing; Dynamic and Interactive Displays; GIScience; Neogeography

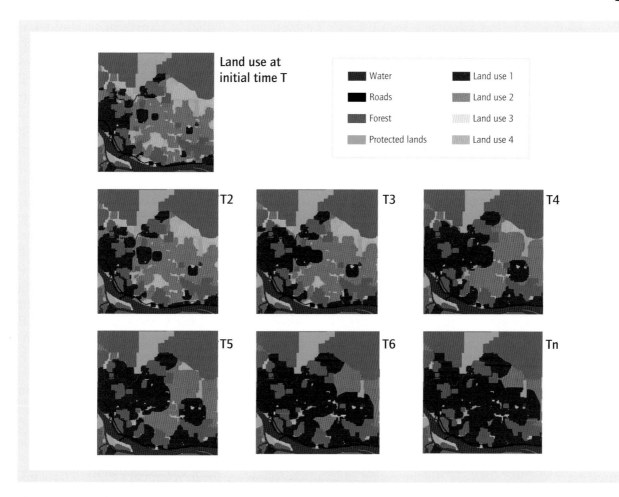

Figure 2 Simple cellular automata model of land use change over time
Source: Author.

Further Readings

Batty, M. (2005). *Cities and complexity*. Cambridge: MIT Press.

Maguire, D., Batty, M., & Goodchild, M. (2005). *GIS, spatial analysis and modelling*. Redlands, CA: ESRI Press.

Portugali, J. (2006). *Complex artificial environments: Simulation, cognition and VR in the study and planning of cities*. Berlin, Germany: Springer-Verlag.

White, R., & Engelen, G. (2000). High-resolution integrated modelling of the spatial dynamics of urban and regional systems. *Computers, Environment & Urban Systems, 24*, 383–400.

Wolfram, S. (1984). Computation theory of cellular automata. *Communications in Mathematical Physics, 96*, 15–57.

CENSUS

A census is a data collection activity that is usually periodic and is typically conducted by a governmental entity for the purpose of obtaining statistical information and producing data for the entire population. A census contrasts with a survey, which uses sampling methods and obtains information from a subset of rather than the entire population. In the United States, the term *census* is commonly associated with the Decennial Census of Population and Housing, which takes place every 10 years and is conducted by the U.S. Census Bureau for the purpose of counting the population to apportion the seats in the House of Representatives among the states and to provide small-area data necessary for legislative redistricting.

Many countries conduct periodic censuses for their populations; their design, the data collection methods, and the data produced vary. Not all censuses focus on counting populations and housing units. For example, in addition to a census of population and housing, the U.S. Census Bureau conducts an economic census, which collects information on the characteristics of businesses, and a census of governments, which collects data on the finances and administrative characteristics of state and local governments.

Censuses are critical to and depend on the discipline of geography. Census activities rely heavily on geographic databases and geographic information system software as well as cartographic operations. Census data, based on the areas defined and maintained by the U.S. Census Bureau and statistical agencies in other countries, are widely used throughout the geospatial industry, in geography programs in academia, and throughout the public and private sector, wherever geography is practiced. This entry focuses on the U.S. Census of Population and Housing.

Article 1, Section 2, of the U.S. Constitution requires enumeration of the nation's population every 10 years. The United States has conducted censuses every decade since 1790; the scope of the census, the methods used, and the data tabulations produced have evolved over time.

Census 2000 collected basic information (name, household relationship, sex, age, Hispanic or Latino origin, race, and tenure—whether a home is owned or rented) for all people and housing units and detailed information on demographic, social, housing, and economic characteristics for people and housing units through a sample survey conducted as part of the decennial census. The 2010 Census collected basic information for people and housing units. However, detailed information was collected through the American Community Survey (ACS). The ACS is an ongoing survey that has replaced the sample survey used for Census 2000 and previous censuses and produces timely demographic, social, housing, and economic data (Table 1).

Like previous censuses, the 2010 Census required a full range of preparations that included the testing of concepts that are the basis for census questions, improvements to address lists, various promotional activities, a variety of technological improvements, the dissemination of data products, and other activities. Twelve regional offices located in major cities throughout the United States provided critical support to local operations, including hiring and data collection. The Census Bureau made significant efforts to develop effective partnerships with local communities and organizations to encourage full participation and to communicate a commitment to respecting privacy and protecting the confidentiality of the data collected. Title 13, United States Code, is the basis for the Census Bureau's commitment to safeguarding data. Most addresses received an English language questionnaire, although a bilingual Spanish-English questionnaire was mailed to selected areas. Then, to assist people who do not speak English well, the Census Bureau provided translated forms in five languages (Spanish, simplified Chinese, Vietnamese, Korean, and Russian), telephone questionnaire assistance in those same languages, and language assistance guides in 59 languages. Promotional and outreach materials were provided in many languages.

The geographic tools and techniques used for census taking have expanded over the years, especially following the development of the Topologically Integrated Geographic Encoding and Referencing (TIGER)-automated mapping database for the 1990 Census. Modernizing TIGER and the Census Bureau's Master Address File, the source of millions of addresses used to deliver census forms, were critical objectives for Census 2010. Mobile computing devices were used to collect data and verify addresses. The 2010 Census operations also included a Local Update of Census Addresses program, which allowed state, local, and tribal governments to complete and correct address files for their communities, and a nationwide address-canvassing operation. Continuing geographic challenges for the Census Bureau in preparing for the 2010 Census included improving address information, particularly in rural areas, collecting new street features, acquiring boundary updates from all governments, and improving quality measurement of address and spatial data and geographic processes.

Census data for geographic areas such as census blocks, census tracts, incorporated places, and American Indian Reservations are of interest to a

Demographic characteristics	Economic characteristics
Age	Income
Gender	Food stamps benefit
Hispanic origin	Labor force status
Race	Industry, occupation, and class of worker
Relationship to householder (e.g., spouse)	Place of work and journey to work
	Work status last year
Social characteristics	Vehicles available
	Health insurance coverage[a]
Marital status and marital history[a]	
Fertility	**Physical characteristics**
Grandparents as caregivers	
Ancestry	Year structure built
Place of birth, citizenship, and year of entry	Units in structure
Language spoken at home	Year moved into unit
Educational attainment and school enrollment	Rooms
Residence 1 year ago	Bedrooms
Veteran status, period of military service, and VA	Kitchen facilities
Service-Connected Disability Rating[a]	Plumbing facilities
Disability	House heating fuel
	Telephone service available
	Farm residence
	Financial characteristics
	Housing tenure (owner/renter)
	Housing value
	Rental costs
	Selected monthly owner costs

Table 1   Subjects included in the American Community Survey

Source: U.S. Census Bureau. (2008). *A compass for understanding and using American community survey data: What general data users need to know.* Washington, DC: Government Printing Office.

Note: a. Marital History, Department of Veterans Affairs (VA) Service-Connected Disability Rating, and Health Insurance Coverage are new for 2008.

variety of geospatial data users. Defined anew for each Decennial Census of Population and Housing or updated by a periodic Boundary and Annexation Survey, these areas form the basis for geographic tabulations that help identify needs for housing, economic development, new schools, job training, or other goals critical to national, regional, state, local, or tribal needs. Each year, more than $400 billion in U.S. federal funds is distributed on the basis of census data, and every new release of census data is accompanied by media coverage of the latest demographic or housing trends affecting the nation's communities. Census data products are available through American FactFinder and other data access tools that can be accessed through the Census Bureau's Web site, www.census.gov.

Nancy K. Torrieri

See also Census Tracts; GIS, History of; Population Geography; Redistricting; United States Census Bureau

Further Readings

Peters, A., & MacDonald, H. (2004). *Unlocking the census with GIS.* Redlands, CA: ESRI Press.
Steffey, D. L., & Bradburn, N. M. (Eds.). (1994). *Counting people in the information age.* Washington, DC: National Academy Press.

CENSUS TRACTS

Tracts are regions of U.S. census geography that originated as analogs of urban neighborhoods. They retain some of that purpose today, and for that reason, tract limits are partially defined by population characteristics in addition to political boundaries and population. Tracts are preferable to other regions because the tabulations are more complete than those of smaller areas and are more easily compared than wards, zip codes, or other administrative regions not defined by population.

First proposed in 1906 by the demographer Walter Laidlaw, census tracts were intended to be a permanent framework to simplify tracking demographic changes. Although the census bureau agreed in principle to provide tract-level tabulations for eight cities in 1910, there was no immediate interest outside New York City. Demand had grown enough by 1940 so that the census bureau adopted the tract as an official tabulation area, including them in decennial tabulations, and took control over the definition of their boundaries. Tract coverage was extended into the suburbs following World War II, and most metropolitan areas were covered by 1980 and the entire nation, in 1990. Although tracts have always been intended to provide stable census geography, it has proven impossible to retain stability against the background of a highly fluid population. For this reason, efforts have been made to keep them comparable from one enumeration to the next by preferring changes that simplify reaggregation, especially by splitting a high-population tract or merging multiple low-population tracts.

Most changes in tract geography result from the population redistribution; however, some changes result from changes in the criteria defining tracts, and there have been two general overhauls of the system since 1940. Some basic tract criteria have been constant since 1910; a particularly important criterion is that every attempt should be made so that they are compact and follow visible and identifiable features but do not cross state or county boundaries.

The Census Bureau redefined tracts nationwide in 1960 to standardize the populations of existing tracts and to impose a uniform decimal numbering system. For the second reorganization in 2010, the bureau established new standards meant to address the problems of low-population tracts resulting from extending the system across the country. The newer criteria define a tract as having either a minimum population of 1,200 persons (down from 1,500) or encompassing a minimum of 480 housing units. Persistent low-population areas, including parks, airports, or other unoccupied land, may now be designated as belonging to a "special land use area" so long as it exceeds a minimum size of 1 mi.^{2} (square mile) of urban land or 10 mi.^{2} outside an urban region.

Jason Bryan Jindrich

See also Census; United States Census Bureau

Further Readings

U.S. Bureau of the Census. (1994). Census tracts and block numbering areas. In *Geographic areas reference manual* (pp. 10-11–10-17). Washington, DC: Author.

CENTERS OF DOMESTICATION

Centers of domestication refer to areas of agricultural origins and innovations. Cultural geographers emphasize that particular local groupings of domesticated plants constituted the basic cultivated species for each regional agricultural development zone, or "center" of domestication. These hearths of domestication were located mostly in tropical or subtropical regions in areas of marked diversity of useful plants or animals and diverse topography, where the first sedentary societies developed. Regions of domestication could be quite large. One of the primary regions of crop domestication, the southwestern Asia area (which includes Mesopotamia) stretches from northwest India to the Caucasus Mountains and into Turkey, reaching over 40° N.

Plant domestication refers to the process of artificial selection by humans, whereby plants with desirable traits are selected and cultivated for food, fiber, medicines, and other needs. The domestication of plants began more than 10,000 yrs. (years) ago as an experimental process that was partly accidental as well as deliberate. While there is debate over the precise definition of domestication, plants that have been domesticated over time by humans become more, if not entirely, dependent on human cultivation for their survival. Domestication was a lengthy process in which plants and animals were raised by humans, and the development of village-based agricultural economies probably did not occur until around 5,000 yrs. ago. Smallholders all over the world continue to domesticate wild species today.

Chili peppers in South America. The distinctly different shapes and colors of these chili peppers provide evidence of their artificial selection over time, a key part of the domestication process.

Source: Author.

At first, plants with desirable traits were most likely simply protected from destruction in the wild. As humans learned more about them, these protected plants were cultivated (cared for), and then, as knowledge progressed, they were actually planted. Thus, the farming of wild plants emerged as a practice before the actual domestication of these species took place. Einkorn wheat has been considered the world's oldest domesticated crop (about 10,500 BP), although recent discoveries indicate that figs were domesticated 11,400 BP and place rye at 13,000 BP. All three plants were domesticated in the Fertile Crescent, considered to be the oldest of the world's independent centers of plant domestication.

Plants may be selected for their desirable traits (a cultigen), such as larger, tastier, or easily harvested edible parts (e.g., maize, manioc, avocados), though some domesticated plants were essentially no different in morphology from their wild counterparts (e.g., teff, fonio, pecans). Plants must be studied to understand if or how humans were involved in their development, but a good indicator of a domesticated species that has undergone artificial selection is one in which the edible parts exhibit different shapes or colors, such as the different varieties of chili peppers in the photograph. Today, most domesticated plants used for commercial purposes look and even taste significantly different from their wild counterparts.

How did all this come about? New genetic data help confirm that a multitude of factors contribute to the domestication of plants across space and time, but scientists have also known for some time that the earliest humans needed to eat plants to survive. Plant materials were an extremely important source of carbohydrates for hunter gatherers, as well as fats where or when game animals were scarce. Although these foraging societies existed long before the sedentary, farming societies, most were probably well aware of ways to maintain patches of wild crops. For example, with wild roots, not all the root or roots would be harvested but a part would be left in the ground to regenerate, allowing foragers to return and harvest the roots again. This may be considered a type of predomestication cultivation, although we can assume that over time the better or hardier root specimens were selected or survived. However, natural mutations outside human control also make certain species more compatible to human cultivation and help the domestication process. For example, humans noticed mutations in wild crops such as wheat or legumes, where the seeds remained on the stem when ripe rather than falling to the ground. These seeds were selected for harvest, and over time the mutation became the primary or sole source of seed for the crop. Natural fires caused by lightning helped make certain plants edible. Along with the use of fire, the establishment of clearings, climate change, and human consumption of plant species all aided the process of domestication.

We must remember that humans dispersed seeds by simply discarding them, by defecation, and through the early exchange of seeds, all of which promote artificial selection and the domestication process. It was only natural for humans to select the larger and sweeter fruits from trees and to take them elsewhere to eat. These conditions still prevail in traditional agricultural and hunter-gatherer societies today.

The terms *cultivation*, *domestication*, and *agriculture* should be used with care when discussing the development of crops and agriculture. Cultivation refers to the effort humans put into planting, tending, or caring for a plant, while domestication means that plants have been changed genetically, morphologically, or in their growth behavior by humans. Domesticated plants tend to depend on human cultivation for their survival over time. Agriculture is defined as the cultivation of domesticated plants as crops (cultivars) in systems that require some working of the soil.

The processes and applications of plant and animal domestication are fundamental examples of how humans have adapted to different natural environments and the cultural landscapes they create over time. The dog is considered to be the oldest domesticated animal, approximately 14,000 yrs. (years) ago in East Asia, and was an important hunting animal for humans. Animal domestication influenced plant domestication, such as the sheep and goats in the Fertile Crescent, which helped clear vegetation for crops and

could eat parts of the domesticated plants that were not eaten by humans. There were six major centers of agricultural origin in the world, along with several other regions where agriculture developed independently. The early diffusion of domesticated crops and animals across regions has made it difficult to define the original spatial patterns of domestication.

Plant and animal domestication was a primary research topic and interest of the father of American geography, Carl Sauer (1889–1975), and greatly influenced his thinking about cultural landscapes, a concept that he promoted and developed beginning in the early 20th century. Sauer believed that there were perhaps 11 independent centers of plant domestication, including the Fertile Crescent (wheat, rye, peas, lentils, figs), Mesoamerica (where maize was domesticated from teosinte), East Asia (rice), and the highlands of South America (potato, quinoa, peanuts, manioc). It was the tropical forests of southeast Asia and South Asia that he believed were the oldest centers of domestication, for root crops such as yams more than 14,000 yrs. ago. Sauer's beliefs differed from most other scientists' during the 1950s, who found little evidence to support his ideas because root crops quickly rotted away in humid environments, while archeologists had found thousands of seeds from grains in the arid Middle East that were at least 10,000 yrs. old. New evidence now confirms that forest dwellers had cultivated squash in Ecuador at least 10,000 yrs. ago and that forest fruits such as avocados may have been cultivated in Mexico far earlier than was previously known. The tall, common peach palm (*Bactris gasipaes*), native to the humid lowlands of Central and South America, has never been found in its wild, undomesticated state, also suggesting that cultivation took place quite early and often. Sauer noted that animals domesticated in southeast Asia, such as dogs, pigs, chickens, and ducks, were household animals that were tended to mostly by women, as compared with the herd animals originating in Southwest Asia; emphasizing the major role played by women in the development of agriculture.

Where traditional agriculture is practiced, smallholders still cultivate wild species in their gardens, and the process of domestication continues much as it had evolved centuries ago. These farmers will protect useful wild species that regenerate in their gardens or will collect wild-planting stock to grow there. They will also exchange wild-planting stock with other farmers. The Amazon basin is a region where a large number of wild forest species are promoted by agronomists and even governments as a way to promote agricultural development in an environmentally friendly way. Recent programs promoting camu camu (*Myrciaria dubia*) planting in Peru are an interesting example of initial or incipient domestication. Seeds are collected from wild camu camu trees and simply distributed to farmers, usually without any selection for traits. Areas next to the wild stands were selected first for cultivation because the fruit trees have the best chance of growing well in the same habitat. The small, water-loving trees produce a tasty fruit that is very high in vitamin C. As confidence grew in the ability to cultivate the species, new fields were established farther away from the wild trees (Figure 1). This is a classic example of in situ and adjacent domestication, a process that cranberry growers followed in North America. While new evidence shows both human settlements and plant domestication arising earlier than researchers had thought, we also have much to learn about the efforts of today's farmers to domesticate little-known but promising wild plant species.

Jim Penn

See also Agriculture, Preindustrial; Domestication of Animals; Domestication of Plants; Hunting and Gathering

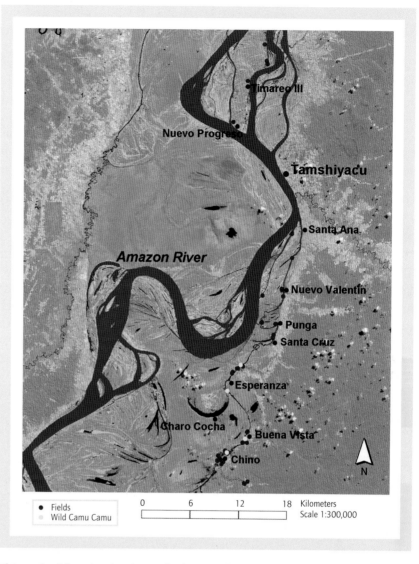

Figure 1 Mapping in situ and adjacent domestication. The close proximity of cultivated fields of camu camu trees to natural patches of wild camu camu is a recent example of in situ and adjacent domestication of this promising species.

Source: From Penn, J. W., Jr. (2008). Non-timber forest products in Peruvian Amazonia: Changing patterns of economic exploitation. *Focus on Geography, 51*(2), 22. Copyright © 2008 American Geographical Society. Reprinted by permission of the author.

Further Readings

Balter, M. (2007). Seeking agriculture's ancient roots. *Science, 316,* 1831–1835.

Penn, J. W., Jr. (2006). The cultivation of camu camu (*Myrciaria dubia*): A tree planting programme in the Peruvian Amazon. *Forests, Trees, and Livelihoods, 16*(1), 85–101.

Prance, G., & Nesbitt, M. (Eds.). (2005). *The cultural history of plants*. New York: Routledge.

Sauer, C. O. (1952). *Agricultural origins and dispersals: The domestication of animals and foodstuffs*. New York: American Geographical Society.

Sauer, J. D. (1993). *Historical geography of crop plants: A select roster*. Boca Raton, FL: CRC Press.

 ## CENTRAL BUSINESS DISTRICT

The central business district (CBD) is the core area of a city, where specialized business services are concentrated. The CBD is essentially an American phenomenon, initially appearing in the early 1800s and evolving over time into the high-rise skyline characteristic of many large cities today. Other terms widely used synonymously for the CBD are *downtown*, *city center*, and *central city*. The American CBD differs from business centers in other countries in several ways. European cities have long histories during which growth spreads outward from the center. Residential land uses are common in European city centers, where the wealthy, upper-class citizens reside. High-rise office buildings are generally built in the peripheral areas. Many American cities, on the other hand, were initially constructed with a business district in the center of the city. Residential land uses in downtown areas of most American cities are rare.

The reason for the development of the CBD lies in the high accessibility afforded by the convergence of transportation networks into the city

Salt Lake City, Utah, with the Wasatch Range in the background

Source: © Joseph Sohm/Visions of America/Corbis.

center, which is central to the agglomeration economies there. This accessibility resulted in high land values as businesses were attracted to office buildings for accounting and management functions. These activities necessitated that ancillary services such as banking, finance, insurance, and law firms also be located in the immediate vicinity. Nearby hotels, restaurants, and high-end retailers attended to the needs of the professionals, their clients, and the office workers employed in the CBD.

Several technological innovations allowed a further intensification of CBD land uses. The skyscraper and electric elevator permitted the construction of ever taller buildings, while subways facilitated the rapid movement of people into and out of the CBD. High-rise office buildings provided enormous amounts of office space on a relatively small footprint on the ground. As land became more and more valuable, the CBD actually contracted into a smaller area because of the abundant office floor space. Zoning took on a vertical orientation with the peak land values at the street level, where retailing could provide the greatest revenues per square foot.

Suburbanization has caused the CBD in many cities to stagnate as office functions decentralized and retailing migrated to distant shopping malls. The edge and the "centerless" city phenomena have also challenged CBD dominance of business services. Anchored by government offices and trophy buildings housing corporate headquarters, the CBD still remains a major provider of specialized business services. But other than a few cities such as New York, Chicago, Boston, and San Francisco, which have active night lives, the CBD has become a daytime employment destination with vacant buildings and streets at night.

Norman Carter

The UBS Tower (above, center) is a 651-foot-tall skyscraper located at One North Wacker Drive in Chicago, Illinois. The 50-story structure, started in 1999 and completed in 2001, was the first major office skyscraper to be built in Chicago since 1992. Chicago's central business district is bounded on the west and north by the Chicago River, on the east by Lake Michigan, and on the south by Roosevelt Road.

Source: Eric Mathiasen.

See also Agglomeration Economies; Chicago School; Suburbs and Suburbanization; Urban Land Use; Urban Spatial Structure; Urban Sprawl; Vance, James

Further Readings

Fogelson, R. (2001). *Downtown: Its rise and fall, 1880–1950*. New Haven, CT: Yale University Press.

Ford, L. (1994). *Cities and buildings: Skyscrapers, skid rows, and suburbs*. Baltimore: Johns Hopkins University Press.

Vance, J., Jr. (1971). Focus on downtown. In L. S. Bourne (Ed.), *Internal structure of the city* (pp. 112–120). New York: Oxford University Press.

CENTRAL PLACE THEORY

Central place theory, developed by the German geographer Walter Christaller as his doctoral dissertation, *Central Places in Southern Germany* in 1933, was a highly influential set of models widely used to describe and analyze urban hierarchies in the mid 20th century.

Essentially, central place theory is a model of city systems that posits them as retail centers (central places) that distribute goods and services to their surrounding hinterlands. Like many models of spatial analysis, it assumes a featureless isotropic plain in which population density and transport costs are equal in all directions. Each good has a threshold, or minimum market size, as well as a range, or maximum distance consumers will travel to purchase it (Figure 1). The size of a given good's range and threshold vary with transportation costs, or the friction of distance, as well as the willingness of consumers to travel to obtain a given good or service. As the effective price increases with distance from the central place, demand declines accordingly. The model is thus compatible with the neoclassical views of economics that privilege consumer demand in the analysis of social and spatial structures.

Different goods and services have different thresholds: inexpensive, frequently purchased, and everyday necessities have low thresholds

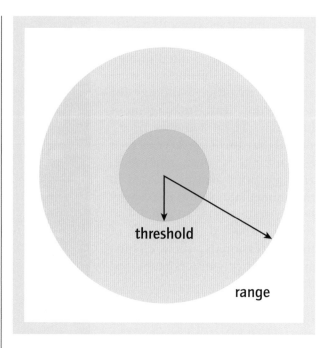

Figure 1 Range and threshold. In central place theory, each good has a minimum market size (threshold) and a maximum distance (range) that consumers will travel to purchase it.

Source: Ingolf K. Vogeler, University of Wisconsin–Eau Claire, http://www.uwec.edu/Geography/Ivogeler/w111/urban1.htm.

(e.g., convenience stores, groceries, eating and drinking establishments, dry cleaners, and other personal services), while costly, infrequently used goods and services, including luxuries, have much larger ones (e.g., jewelry, specialized medical services). The degree of specialization of a good or service is thus directly proportionate to the willingness of consumers to travel to obtain it: Increasingly specialized products attract consumers from progressively longer distances. A hierarchy of goods and services leads to a hierarchy of central places to distribute them. Central place theory thus ranks urban places on the basis of their ability to distribute goods that range from "low-order" to "high-order" ones. The order of a central place is determined by the highest-order goods that it distributes to consumers around it.

The resulting hierarchy of central places— the urban hierarchy—reflects the imperatives of market forces over space, which tend to maximize the number of centers, maximize the number of consumers served, and minimize the aggregate distances that consumers must travel to

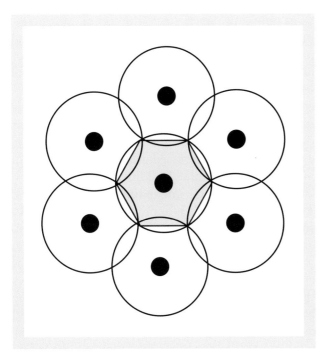

Figure 2 Formation of hexagonal market areas. Overlapping circular hinterlands are divided in half as consumers travel to the closest one, leading to hexagonal, nonoverlapping ones.

Source: Ingolf K. Vogeler, University of Wisconsin–Eau Claire, http://www.uwec.edu/Geography/Ivogeler/w111/urban1.htm.

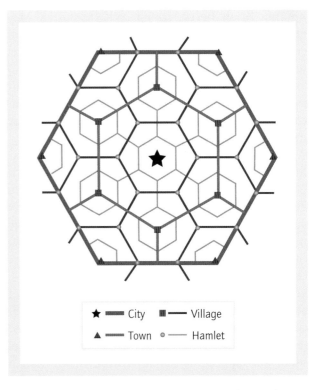

★ ▬ City   ▦ ▬ Village
▲ ▬ Town   ○ ▬ Hamlet

Figure 3 Geometry of the central place hierarchy. The nested hexagonal market areas reflect different levels of the urban hierarchy.

Source: Ingolf K. Vogeler, University of Wisconsin–Eau Claire, http://www.uwec.edu/Geography/Ivogeler/w111/urban1.htm.

purchase goods of varying levels of specialization. Large cities distribute both low- and high-order goods, the latter over increasingly large hinterlands. Higher-order centers are more widely spaced, with larger populations, serve large market areas, and are encountered less frequently on the landscape. Thus, metropolitan areas, the highest order, offer the greatest diversity of goods and services and hinterlands that extend over vast distances. Conversely, progressively lower-order centers are nested within the market areas of higher-order centers. Small, rural hamlets, for example, historically formed central places serving low-density agricultural regions. Progressing up the urban hierarchy, one encountered increasingly larger but infrequent places with a mounting diversity of goods and services.

In the classical central place model, which has several variations, consumers minimize their transport costs to the closest city that offers the goods they desire, and producers maximize profits. On an isotropic plain, market areas form circles around each center. However, as consumers located between competing centers opt for the closest one, overlapping market areas are divided by lines (Figure 2). Assuming no spatial variations in transport costs or population, a hexagonal series of market areas should emerge, with lower-order ones nested within the hinterlands of higher-order ones (Figure 3). Hexagonal market areas became the defining feature of central place theory and were justified on grounds that they "packed" space—that is, they served all clients without overlapping market areas, more efficiently than any other shape. The model was thus decisively static in nature and offered a fixed, frozen geometrical portrait of cities.

Central place theory became extremely popular in geography in the 1950s and 1960s as part of the emerging positivist school of thought, which emphasized modeling. It formed the dominant paradigm within urban geography for almost a generation. For example, urban geographers, notably Brian Berry, sought out hexagonal systems of market areas in the Midwest, which most closely resembled an isotropic plain. The model was also used in urban planning ventures in Israel and the Netherlands. Positivist theoreticians built increasingly more complex—and unrealistic—central place models that attempted to incorporate uncertainty, entropy, and evolution over time.

Ultimately, central place theory was abandoned as human geography shifted decisively into political economy and social theory in the late 20th century. It was widely criticized for its overly simplistic assumptions; the absurdly unrealistic reduction of cities to shopping centers; the neglect of social relations in favor of individual consumer choice and the resulting silence concerning class, gender, consciousness, and power; the absence of any account of production and the spatial division of labor; and its ahistorical portrayal of urban networks. Today, the once dominant model of urban geography remains little more than a historical curiosity.

Barney Warf

See also Berry, Brian; Christaller, Walter; Location Theory; Positivism; Regional Science; Spatial Analysis; Urban Hierarchy

Further Readings

Berry, B. (1967). *Geography of market centers and retail distribution.* Englewood Cliffs, NJ: Prentice Hall.

King, L. (1984). *Central place theory.* Beverly Hills, CA: Sage.

CHEMICAL SPILLS, ENVIRONMENT, AND SOCIETY

The use of chemicals and chemical products in the 21st century is comprehensive and widespread, and modern society depends on them. The chemical industry plays an important role in the industrial sectors of the economy, and many different chemicals are important in manufacturing, such as in pulp and paper, steel, aluminum, auto making, oil refining, and so on. In the agricultural production sector, chemicals are used in pesticides and as important ingredients in fertilizers. In the food sector, chemicals play a major role for processing foods. The transportation sector depends on fossil fuel products (gasoline and diesel). In our daily lives, we find chemicals everywhere. Some examples are flame retardants used in textiles in airplanes and public buildings and cosmetics, cleaning detergents, and paints used in homes. The use of chemicals is somewhat paradoxical. In smaller amounts and/or for certain purposes, chemicals are needed and effective. However, chemicals can also cause serious environmental and health-related problems. This entry reviews the scope of chemical use today, describes the harms that can result from chemical spills, and examines the potential harm associated with their regular use. It then discusses some of the paradoxical ways in which modern chemical use involves both benefits and costs.

There are 90 natural chemical elements, and in total, 111 chemical elements are known. The most common chemical elements (percentage by weight) in the Earth's crust are oxygen (45%), silicon (27.2%), aluminum (8.3%), ferrous (6.2%), and calcium (4.7%). The number of chemical compounds has grown dramatically since the end of World War II. Some of the most widespread and commonly used chemical compounds are petroleum products. In the North American and European markets, about 75,000 different chemical products can be found. These consist of about 13,000 different chemical compounds. Global production of chemicals has increased from 1 million metric tons/yr. (per year) in 1930 to 400 million metric tons/yr. of chemicals today, according to the Swedish Chemical Agency. Some of these chemicals are less harmful to health and environment, but we lack knowledge of the effects of many of these compounds, and there is a tendency to replace harmful compounds with new compounds that have not been sufficiently tested.

Many harmful compounds and products are being produced and transported despite large uncertainties about their effects on human health and the environment, and thus the massive use of chemical compounds could be better controlled. Chemicals are also related to environmental

A worker uses a water hose to clean Highway A5 near Frankfurt, Southwestern Germany, September 22, 2004. About 20 metric tons of methylamine, an inflammable and poisonous chemical, was spilled onto a road near the city of Frankfurt after an accident involving a hazardous goods transporter. Emergency services were attempting to contain the chemical, which was seeping into the ground and the nearby river Nidda.

Source: Kai Pfaffenbach/Reuters/Corbis.

problems such as depletion of the ozone layer, acidification of soil and water, pollution of the seas, and concentration of long-lived organic chemical compounds such as PCBs (polychlorinated biphenyls), DDT (dichloro-diphenyl-trichloroethane), and dioxins.

Chemical Spills

Chemical spills are accidents that occur in the extraction process, in industrial processes, or in transportation, as well as through overuse or misuse in industry, agriculture, and society and uncontrolled leakage to the atmosphere, ground, and water from industrial processes, agriculture, the energy sector, and consumption of goods and their transport. Indeed, a chemical spill can be one of the most serious environmental problems.

Chemical spills can be categorized in many ways. In chemistry, the main distinction is between organic and inorganic compounds. Organic compounds are carbon based, and in this group are the petroleum products. The inorganic compounds are of mineral origin, often metals of different kinds. There is also a group of semi-organic compounds such as methyl mercury. When it comes to transportation, storage, and chemical spills and risk assessments, it is important to distinguish between acids and bases, between compounds with health risks and/or environmental risks, between flammable and nonflammable compounds, between stable and nonstable compounds, and between gaseous and liquid compounds.

Waste Management

Waste management often suffers from poor institutional capacity, improper treatment, large long-term environmental risks, and human health risks, especially in the developing world. In the modern industrialized world, a "filter" preventing misuse of chemicals—environmental assessments and health declarations, improved waste management, and so on—has been developed by better control, research, legislation, and certification. This has meant that the chemical product life cycle is better controlled. However, one major problem in industrialized countries is old waste disposal or storage sites. Many people consider these sites as "ticking environmental time bombs," and some of the sites contain materials and chemicals that are unstable and hazardous. Waste treatment is both better and safer today and is generally built on the idea of recycling and safe storage.

Transboundary dumping of hazardous waste is still a problem. Taking care of hazardous waste is costly; one alternative is to export the waste. The export of waste from the industrialized world to developing countries in Asia and Africa is regulated in the Basel Convention on the Control of Transboundary Movements of Hazardous Wastes and Their Disposal. The Basel Convention has been adopted by 170 countries and came into force in 1992. This treaty aims to regulate transboundary movements of hazardous waste and to ensure that hazardous waste and other waste are managed in an environmentally safe manner. Approximately 338 million metric tons of hazardous and other wastes were generated in 2001.

Even if the control of chemicals is satisfactory, there is always a risk of illegal transport, treatment, storage, and dumping of hazardous wastes, especially when improper disposal can be very profitable. Often, it is a question of how the hazardous waste is defined and declared. For example, the same hazardous waste could be declared as materials for recycling or as raw materials for industrial production.

Chemical Use and the Development of Modern Society

There are many challenges and paradoxes to face in relation to the use of chemicals. The use of chemicals and the risk of handling chemicals are very closely related to the development of modern society. For instance, by using pesticides in agriculture (in combination with other measures), many nations have been able to increase production in the agricultural sector, which has resulted in better nutrition conditions for a growing global population. On the other hand, there is a risk of overuse of pesticides, which can be harmful to the environment and to agricultural workers over the long term. This paradox could be called the "development paradox." A second paradox is related to research and replacement (the "replacement paradox"). When people find that a compound is unhealthy and environmentally hazardous, there is a tendency to replace it with another chemical compound that seems to be better but in fact sometimes is not. A third paradox relates to the fact that people use large amounts of energy to extract, enrich, produce, and transport chemical compounds around the world, followed by great effort to collect, store, and handle the residues as hazardous wastes. All these steps take place over time and space. One element is extracted in a specific place, transported to the manufacturer, which in turn sends it to the consumer market. Waste management could be equally described in terms of time and space. These material flows in time and space are still rather unexplored but are important to describe and analyze.

Olof Stjernström

See also Agrochemical Pollution; Brownfields; Heavy Metals as Pollutants; Industrial Ecology; Mining and Geography; Petroleum; Polychlorinated Biphenyls (PCBs); Risk Analysis and Assessment

Further Readings

Basel Convention. (2008). *The Basel Convention on the control of transboundary movements of hazardous wastes and their disposal.* Retrieved December 1, 2008, from www.basel.int/convention/about.html

Ibitayo, O. (2008). Transboundary dumping of hazardous waste. In *The encyclopedia of Earth.* Retrieved December 1, 2008, from www.eoearth.org/article/Transboundary_dumping_of_hazardous_waste

Scott-Andersson, Å., Stjernström, O., & Fängsmark, I. (2005). Use of questionnaires and an expert panel to judge the environmental consequences of chemical spills for the development of an environment-accident Index. *Journal of Environmental Management, 75,* 247–261.

CHERNOBYL NUCLEAR ACCIDENT

A serious accident occurred on April 26, 1986, at 1:23 a.m. at Unit 4 of the Chernobyl Nuclear Power Plant near Pripyat in the Ukrainian Soviet Socialist Republic. The power plant was located near the border of Ukraine and Belarus. A series of explosions sent a plume of highly radioactive fallout of iodine and cesium into the atmosphere. Cesium-137 deposition spread throughout Europe, and the radioactive plume reached the U.S. West Coast on May 6, 1986. Official word of the accident first reached the Western world when Sweden announced on April 28 that workers at its Forsmark nuclear plant had radioactive particles on their clothing and the source was traced back to Chernobyl. Initially, former Soviet ruler Mikhail Gorbachev downplayed the incident's severity, which is considered the worst nuclear power plant accident in history.

The Chernobyl accident was caused by a combination of human error and design flaws. It was initiated when power plant technicians conducted an experiment on the emergency core cooling system during its shutdown procedure. First, the emergency core cooling system was turned off, and 205 of 211 fuel control rods were removed from the reactor core. Following this, automatic safety devices that shut down the reactor when

Inspectors just outside the Chernobyl Nuclear Power Plant's crumbling west wall take water and soil contamination samples and conduct land surveys to repair and build new access roads. Twenty years after the Chernobyl nuclear accident in 1986, radiation monitoring occurs on a regular basis on the grounds.

Source: Gerd Ludwig/Corbis.

water and steam levels fall below normal and turbine stops were turned off because engineers did not want to spoil the experiment. Next, an additional water pump to cool the reactor was turned on. However, with only a 7% power output and an extra drain on the power system, water did not reach the reactor. Finally, since the reactor power output was lowered too much, it was very difficult to control. The result was a massive and catastrophic power excursion and steam and hydrogen explosions. These tore off the steel and concrete top of the reactor, exposed the core, and dispersed large amounts of radioactive particulate and gaseous debris, allowing oxygen to contact the superhot core, containing almost 2,000 tons of combustible graphite moderator. The graphite moderator fire increased the radioactive particle emissions. Much of the nuclear fuel supply in the reactor core eventually melted.

There has been great controversy over the number of people who died from the Chernobyl accident, either directly or later through acute radiation syndrome or thyroid and other cancers. The official death toll stands at 56, while the estimated number of eventual fatal cancers ranges from 4,000 (World Health Organization) to 93,000 people (Greenpeace). Millions of other people became sick from Chernobyl, and many thousands have lost their homes and livelihoods. From 1986 to 1992, about 336,000 people were evacuated and resettled because of the accident. In addition, millions of acres of farmlands and forests have become contaminated, especially in Ukraine, Belarus, and Russia. Overall, 4.5 million people were affected. In December 2000, the Chernobyl Nuclear Plant was fully decommissioned.

Barry D. Solomon

See also Nuclear Energy

Further Readings

Marples, D. R. (2004). Chernobyl: A reassessment. *Eurasian Geography and Economics, 45,* 588–607.
Medvedev, Z. A. (1992). *The legacy of Chernobyl.* New York: W. W. Norton.

CHICAGO SCHOOL

In the 1920s and 1930s, Chicago emerged at the forefront of American urban analysis. As the model of the rapidly growing, industrialized city populated by streams of immigrants, Chicago became the prototypical example of the industrial American city. The University of Chicago played a major role in disciplines such as economics, sociology, and geography. Within this context, the Chicago School of urban studies arose, which was enormously influential in sociology and geography for the next several decades. The Chicago School is credited with the first systematic attempt to understand the dynamics of urban areas, including social change, urban planning, and territoriality.

The origins of the Chicago school lay largely with Robert E. Park, a former journalist turned teacher. In 1925, Park, Ernest Burgess, and Roderick McKenzie published *The City*, a famous collection of interpretive essays about urban life, a volume that both summarized and inspired a long tradition of urban ethnography. Chicago School practitioners inaugurated the tradition of detailed case studies, ranged far and wide over the city, studying, among other things, the wealthy, immigrants, the destitute and homeless, dance halls, gangs, criminals, and prostitutes in an attempt to draw rich and detailed portraits of urban life. In the process, they irrevocably fused the study of space and the study of society.

The first paradigm of urban structure offered by Chicago School theorists, particularly McKenzie, centered on the metaphor of the city as urban jungle, a view derived from the social Darwinism prevalent in the early 20th century. For example, the displacement of one ethnic group by another in a given neighborhood was framed as a process of invasion and succession, a view that drew directly from ecological studies of how one plant species displaced another through successive stages in the evolution of ecosystems. Later, this biological metaphor was dropped in the face of criticism that it lacked a coherent account of social relations and naturalized the inequality of urban areas. Throughout the Chicago School's worldview, competition appears repeatedly as a driving force behind ethnic and class segregation.

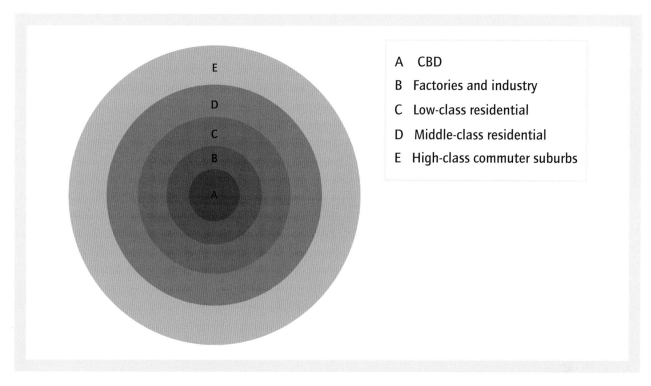

A   CBD
B   Factories and industry
C   Low-class residential
D   Middle-class residential
E   High-class commuter suburbs

Figure 1   Concentric ring model. Burgess's concentric ring model focused on socioeconomic differences that arose with distance from the CBD.

Source: Author.

Chicago School theorists also drew on the urban sociology of Frederick Tönnies and notions such as *Gemeinschaft* and *Gesellschaft* to examine daily life in light of the massive rural-to-urban migration then characteristic of most U.S. cities. In this view, urbanization represented the annihilation of mythologized rural communities, in which everyone knew everyone else. In contrast to the idealized image of small towns in which everyone ostensibly was intimately connected to everyone else and presented the same sense of self under all contexts, urbanization was held to decompose these traditional bonds and erode the foundations of mutual trust. Cities, it was held, were not conducive to the formation of a sense of community. Louis Wirth, in particular, advocated a desolate but compelling view of city life as structured around population size, density, and heterogeneity. Size or total population, he held, created a climate that was inherently predatory, utilitarian, uncaring, and commodified: Strangers were rare in small towns but the norm in large cities. Density, he argued, led people to

be physically but not emotionally close; indeed, alienation was the norm. Finally, social and cultural heterogeneity, manifested in the diverse lifestyles found in large cities, undermined the common values necessary to the success of healthy communities. Wirth concluded that these factors were responsible for the high rates of urban crime and other social pathologies ranging from suicide to psychoses. (Subsequent work has rectified this stereotype by pointing to the high crime rates often found in many small cities and that large cities often have healthy, vibrant urban neighborhoods.)

Three Models of Urban Social Structure

The most famous products of the Chicago School are three models of urban social structure often described in introductory sociology and geography textbooks. The first, proposed by Earnest Burgess in 1927, was the concentric ring model (Figure 1), which, extrapolating from the specific instance of Chicago, viewed the city as a series of

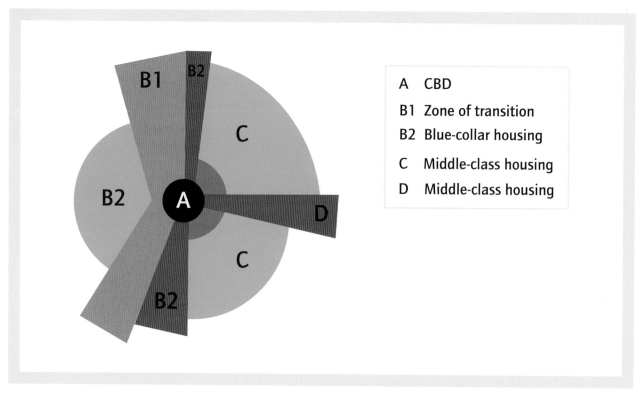

Figure 2 Hoyt's sector model. The sector model emphasized variations in land values along different transport routes.

Source: Author.

rings of varying size centered on the *central business district* (CBD), a term coined by the Chicago School. Adjacent to the CBD was a zone of factories and warehouses, sometimes called a *zone of transition*. Moving outward, this was followed by the "zone of workingmen's homes," that is, working-class, blue-collar communities. Yet farther out were the medium- and then the high-income belts of suburbia.

Burgess observed that cities tend to expand horizontally as the wealthy had new homes constructed on the urban periphery; as they came to occupy these, the relocation of families outward set off a chain of vacancies that reverberated across the urban landscape, as less well-off families in turn occupied the cast-off mansions of the rich, a process known as *filtering*, or the *trickle-down* theory of housing supply. Moreover, Burgess observed a paradox: Low-income residents in cities lived on expensive, accessible land near the urban core, whereas more well-to-do inhabitants of wealthier rings occupied the less expensive

land. This puzzle, he noted, was explained by the population density curves characteristic of cities, which decline exponentially with distance from the CBD. The poor, crowded into dense communities in the urban core, collectively generate high aggregate rents that create relatively high rates of profit in inner-city areas, whereas the low-density environments of the wealthier classes reduce the profitability of the less accessible periphery.

In contrast to the rigid geometry of the Burgess model, Homer Hoyt, an economist and another influential Chicago School theorist, proposed the sector model of urban growth in 1939 based on an empirical analysis of 142 cities (Figure 2). In this view, rather than concentric rings, urban growth occurred along transportation lines centered on the CBD. Once parts of the central city acquired distinctive uses, they radiated outward. High-income land uses played a determining role in shaping the rest of the city, growing along waterfronts, in high-altitude areas, or toward other high-income neighborhoods, and

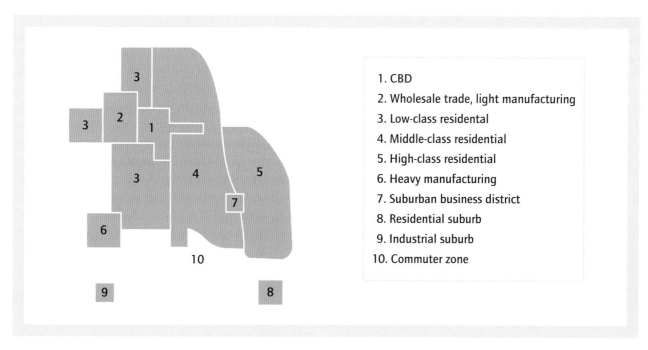

1. CBD
2. Wholesale trade, light manufacturing
3. Low-class residental
4. Middle-class residential
5. High-class residential
6. Heavy manufacturing
7. Suburban business district
8. Residential suburb
9. Industrial suburb
10. Commuter zone

Figure 3   Harris and Ulman's multiple-nuclei model. The multiple-nuclei model attempted to reflect the complexity of urban space as it was structured around multiple nodes of economic activity.

Source: Author.

other uses filled in the spaces between them. Rather than belts, socioeconomic groups occupy sectors or wedges. The wealthy, with the greatest ability to pay, outcompeted less advantaged groups for the most desirable locales, generally far from the disamenities of factories or railroad lines. Low-income groups found themselves confined in zones relatively far from the wealthy.

The third of the Chicago School trilogy, proposed by Chauncey Harris and Edward Ullman in 1945, was called the *multiple-nuclei model* (Figure 3). Essentially, this view attempted to rectify the perceived simplistic shortcomings of the previous two. Rather than a single city center, it maintained that American cities had become polycentric, with many nuclei, around which land uses were organized in a complex quilt. Rather than a single overarching logic, this perspective maintained that certain land uses would repel one another, while others might be mutually attractive.

Impacts of the Chicago School

Although the Chicago School began to diminish in importance shortly after World War II, its ideas were carried into urban sociology and geography for many years afterward. For example, social ecologists in the 1960s, armed with multivariate statistical methods and census data, argued that each of the three classic models effectively captured a different aspect of urban social space. Thus, family status was distributed in rings per Burgess's concentric ring model, reflecting the dynamics of the family life cycle. Economic class was held to occur in sectors, conforming to Hoyt's theory of land use. Finally, ethnicity was theorized to reflect the dense nucleations of different immigrant groups, as proposed by the multiple-nuclei model.

The Chicago School essentially defined American urban analysis throughout the 20th century. More recently, however, attempts by urban political economists to reveal the complex dynamics of globalization and immigration in a postindustrial context have yielded the so-called Los Angeles School, which takes the low-density, polynucleated, postindustrial Southern California metropolis as its point of departure.

Barney Warf

See also Central Business District; Ethnic Segregation; Hoyt, Homer; Los Angeles School; Neighborhood; Urban Geography; Urban Land Use; Urban Spatial Structure

Further Readings

Bulmer, M. (1984). *The Chicago School of sociology: Institutionalization, diversity, and the rise of sociological research.* Chicago: University of Chicago Press.

Dear, M. (Ed.). (2002). *From Chicago to L.A.: Making sense of urban theory.* Thousand Oaks, CA: Sage.

Park, R., Burgess, E., & McKenzie, R. (Eds.). (1925). *The city.* Chicago: University of Chicago Press.

Wirth, L. (1938). Urbanism as a way of life. *American Journal of Sociology, 44,* 1–24.

CHILDHOOD SPATIAL AND ENVIRONMENTAL LEARNING

Understanding the spatial and environmental learning of children is a critical task for geography as well as several subdisciplines of geography. This entry presents some of the important developmental issues related to the spatial cognition and environmental learning of children and examines some of the implications for pedagogy.

Challenges in Teaching Young Children

Young children who are just starting school can have a difficult time with multidimensional information. Children of this age have shown some difficulty with cartographic-scale spaces (extensive spaces such as large regions, nations, continents, and the world) as well as with making transitional judgments between large- and small-scale spaces. Young children have shown limits in their ability to process and store information, an issue that must be considered when developing appropriate activities in any subject area.

Initiating geographic learning from the immediate and the visible allows the child to conserve cognitive resources that might otherwise be required for visualization of the problem or environment. Teachers and developmental researchers can often be surprised by the insights that children exhibit when given time to consider more deeply a specific spatial or geographic problem. Starting from a basis that is familiar to the child will help that child pursue more complex concepts than would be possible in abstract or foreign situations. Using a small number of geographic skills and concepts to start with, the teacher can help children develop an understanding of geography from perspectives that young children can easily grasp. The cumulative accretion of knowledge is essential; once the building blocks of geographic knowledge and problems solving are laid down, more complex concepts and skills can be introduced.

The majority of cognitive development occurs during early and middle childhood. Much of the material for later development has its foundation in cumulative knowledge based on earlier development and learning. Once the fundamental geographic concepts are introduced to teachers, opportunities to ask geographic questions and discover the answers flourish. Many of these techniques rely on a geographic basis of knowledge and can first be introduced through lessons and activities associated with maps and aerial photographs. Once a basis is established, students are well prepared for more advanced geographic topics that will produce geographically literate people.

Developmental and Learning Theories

The basis of environmental and geographic learning is established during childhood. The way children acquire geographic knowledge is quite distinct from the way humans learn as adults, and it shapes the nature of their geographic understanding and how they experience the world. Both developmental and learning theories play a meaningful role in the understanding of children's geographies and the nature of childhood spatial learning. From a developmental point of view, it is important to consider the range of abilities a child of a certain age might possess while selecting appropriate content and methods for communicating that content and facilitating learning.

Consideration of developmental issues is only part of the picture; learning theory must also be considered. Although the consideration of learning theory is important from the onset of education, it begins to usurp developmental issues around the age of 12 or 13. By this time, the majority of cognitive skills have emerged, and students begin to rely on problem solving and knowledge acquisition as the main tools for developing intelligence and understanding. Furthermore, working memory capacity, cognitive processing, and attention have all reached levels equal to or near those of adults. Although there are many other developmental issues beginning to emerge at this time (puberty, social developmental, etc.), the majority of the cognitive systems are close to completely developed (allowing for individual differences in rates of development and the emergence of cognitive abilities).

There are many important fundamental geographic and spatial concepts that children must understand to learn the full breadth of geography or to appreciate in depth a single geographic topic. Before considering these specifically, it is important to recognize the existence of cognitive resources that support the development of all skills, spatial and nonspatial (language, mathematics, etc.). These resources include working-memory capacities for both storage and information processing and current knowledge concerning a situation. The latter necessarily forces one to consider a variety of learning theories, including, but not limited to, activity-based learning, situated cognition, and constructivism. With development, the capacity and efficiency of cognitive resources improve while a child grows, develops, and experiences the world. Experience also helps children develop in specific ways, particularly with respect to a knowledge base in experiential domains, such as large parts of geography.

A small number of geographers over the past 30 years have attempted to examine the building blocks of geography, sometimes called *spatial primitives*. Among the identified primitives are location, identity, distance, direction, and orientation. Some of these can be considered second-order, or derived, primitives, because they rely on other primitives for their structure and definition. For instance, to measure the distance between two places, a person must know the location and identity of each. Spatial primitives represent some of the fundamental concepts that must be understood to grasp more complex geographic topics and processes and as such are a good place to start.

Piaget and Bruner

Almost all discussions of cognitive development begin with the Swiss developmental theorist Jean Piaget, particularly in the spatial domain. Piaget provided a general framework on which a substantial amount of later work is based. Although Piaget and his colleagues outlined the stages through which children develop on their way to adulthood, the greatest contribution of their work is most likely the description of the qualitative changes in thought that occur during development rather than the timing of stages and the various specific structural and operational changes that were described. Other research that is important to geographic learning and thought is that of Jerome Bruner, whose cognitive growth theory was based on how children represent the world. Changes in the mode of representation could be understood with respect to how the use of language changes over time during childhood. Bruner's theory included three representational modes: enactive, iconic, and symbolic. The latter is critical to one's ability to understand models and symbols, a central geographic concept.

Contemporary Developmental Theory

More contemporary developmental theory has focused on information-processing models of cognitive development, such as those developed by Robbie Case (executive processor and automatization), Robert Siegler (attention and encoding), and David Klahr (generalization). In each of these theories, the child is viewed as a strategist who is always trying to construct a solution for a problem. Other, more specific research has produced concrete understanding in areas specific to geographic education.

Scott Bell

See also Children, Geography of; Geography Education; Spatial Cognition; Wayfinding

Further Readings

Allen, G. L., Kirasic, K. C., Siegel, A. W., & Herman, J. F. (1979). Developmental issues in cognitive mapping: The selection and utilization of environmental landmarks. *Child Development, 50,* 1062–1070.

Bruner, J. S. (1966). *Studies in cognitive growth: A collaboration at the Center for cognitive studies.* New York: Wiley.

Case, R. (1985). *Intellectual development: Birth to adulthood.* Orlando, FL: Academic Press.

Cornell, E. H., Heth, D., & Rowat, W. L. (1992). Wayfinding by children and adults: Response of instructions to use look-back and retrace strategies. *Developmental Psychology, 5,* 755–764.

DeLoache, J. S., & Burns, N. M. (1993). Symbolic development in young children: Understanding models and pictures. In C. Pratt & A. F. Garton (Eds.), *Systems of representation in children: Development and use* (pp. 91–112). Chichester, UK: Wiley.

DeLoache, J. S., Kolstad, V., & Anderson, K. N. (1991). Physical similarity and young children's understanding of scale models. *Child Development, 62,* 111–126.

DeLoache, J. S., Miller, K. F., & Rosengren, K. S. (1997). The credible shrinking room: Very young children's performance with symbolic and nonsymbolic relations. *Psychological Science, 8,* 308–313.

Klahr, D. (1992). Information processing approaches. In R. Vasta (Ed.), *Six theories of child development: Revised formulations and current issues* (pp. 133–185). London: Jessica Kingsley.

Klahr, D., & Wallace, J. G. (1976). *Cognitive development: An information processing view.* New York: Wiley.

Mayer, R. E. (1992). *Thinking, problem solving, cognition* (2nd ed.). New York: W. H. Freeman.

Piaget, J., & Inhelder, B. (1967). *The child's conception of space.* New York: W. W. Norton.

Siegler, R. S., & Alibali, M. W. (2004). *Children's thinking* (4th ed.). Upper Saddle River, NJ: Prentice Hall.

CHILDREN, GEOGRAPHY OF

The geography of children is a subdiscipline that examines the relevance of space and place to the study of childhood. Research in this area examines the unique experiential, political, and ethical experiences of this social group. The geography of childhood was born out of the fields of environmental psychology, urban planning, Marxist geography, behavioral geography, and geographic education. While Roger Hart discussed an explicit geography of childhood in his 1979 book, *Children's Experience of Place,* its modern practitioners did not publish widely until the 1990s.

Early research examined how children used, perceived, and made sense of space and place for the purposes of developing theories on children's spatial learning and planning environments that take into consideration children's unique experiences. Recent studies apply a wide range of disciplines to the field, such as sociology, psychoanalysis, and feminist geography, to examine diverse topics such as children's participation in community development and environmental management and children's identity formation in relation to the cultural practices of a particular place.

Childhood as studied by geographers varies widely in its definition and concept. Some research is based on biological classifications, such as age and other developmental considerations, while other work addresses cultural and socially constructed definitions of childhood. Researchers acknowledge the unique experiences of children, as opposed to adults, and some argue that children have historically been marginalized by society, rendering their actions and agency as a threat to the moral order of the adult world. Since the cultural turn in geography, researchers have come to view children as active agents in the construction of their lives and environments. This has led to a fundamental shift in ideology toward childhood, one that complicates adult-child relations and challenges what some refer to as *adult spatial hegemony.*

The status of research on children's geographies in the early 21st century is quite comprehensive (Table 1), including topics such as youth

Topics	Disciplines and Subfields	Key Concepts and Theories
Youth subcultures, hybrid identities, cross-cultural studies of youth, places of resistance, representation, contestation	Cultural geography, regional geography, anthropology, sociology	Resistance theory, identity construction, representation, subcultural analysis, cultural reproduction, peer cultures, cultural capital
Childhood as marginalized "other," childhood as subordinate, the role of space and place in identity formation and construction, the control of space by dominant groups	Feminist geography, women's studies, social psychology, psychoanalysis	Identity construction, adult spatial hegemony, ideology of childhood, social marginality, "othering" of childhood
Children's spatial or place behavior, children's use of public and private space, children's spatial ecologies	Behavioral geography, time geography, environmental psychology, social psychology	Time-activity patterns, spatial range, ecology of human development, children's territories, adult spatial hegemony, independent mobility
Children's understanding of space, place, and human-environment interactions; children's image of space and place; children's place attachment	Geography education, developmental psychology, environmental psychology, environmental education	Place attachment, cognitive learning theory, place value, spontaneous geography, social reproduction, spatial learning, situated learning
Impact of race, class, and gender on children's access and use of space and place; impact of globalization on children and youth; role of youth in the labor market; children's identities in a changing economy; impact of gentrification on youth identities; and use of public space	Social geography, human geography, sociology, economics	Marxism, capitalism, structuralism, agency and structure, childhood as a social structure, social biography, lifecourse, social reproduction, local-global, political economy, sociospatial identity, street literacy, gentrification
The role of physical hazards (traffic, pollution) and social hazards (molestation, violence) in children's use of and access to space, children's relationship to the "natural" world	Hazards geography, environmental geography, environmental studies, planning	Risk management, terror talk
Historical and place-based perspectives on children and youth, childhood in perspective of time and space	Historical geography, history, sociology, psychoanalysis	Social construction, ideology of childhood, psychogenic theory of history (regression, projection), structural notion of childhood, production and reproduction of childhood, moral order and/or panics
Designing space and place for and with children	Applied geography, planning, architecture, sociology, urban studies	Sustainable development, new urbanism
Growing up in cities; how the design of cities relates to children's use, image, and representations of place	Urban geography, urban studies, sociology	Social ecology, modernization theory, political economy, gentrification
Children's rights, children's participation in general and specifically in planning neighborhoods and communities	Political geography, political science, planning	Children's rights, children's participation

Table 1 Summary of research on children's geographies. The spatiality of children has been approached from numerous conceptual angles.
Source: Author.

subcultures, spatial representations and sites of resistance, the impact of gentrification on children's play spaces, the impact of globalization on children's participation in the labor force, the role of physical and social hazards in children's spatial freedom and practices, designing environments for and with children, and children's rights in international law. The importance of children's geographies as a subdiscipline is further witnessed in newly created academic journals devoted to the topic, such as *Children's Geographies; Children, Youth and Environments*; and the *Journal of Youth Studies*.

Pamela Wridt

See also Childhood Spatial and Environmental Learning

Further Readings

Aitken, S. (1994). *Putting children in their place.* Washington, DC: Association of American Geographers.

Aitken, S. (2001). *Geographies of young people: The morally contested spaces of identity.* New York: Routledge.

Bartlett, S., Hart, R., Satterthwaite, D., de la Barra, X., & Missair, A. (1999). *Cities for children: Children's rights, poverty and urban management.* London: Earthscan, UNICEF.

Hart, R. (1997). *Children's participation: The theory and practice of involving young citizens in community development and environmental care.* London: Earthscan.

Katz, C. (2004). *Growing up global: Economic restructuring and children's everyday lives.* Minneapolis: University of Minnesota Press.

Matthews, M. (1992). *Making sense of place: Children's understanding of large-scale environments.* Savage, MD: Harvester Wheatsheaf.

Moore, R. (1990). *Childhood's domain: Play and place in child development.* Berkeley, CA: MIG Communications.

Skelton, T., & Valentine, G. (Eds.). (1998). *Cool places: Geographies of youth cultures.* New York: Routledge.

CHINOOKS/FOEHNS

The chinook is a warm, dry, and sometimes gusty downslope wind that blows leeward of the Rocky Mountains. The term *chinook* is an aboriginal word meaning "snow eater," because of the rapid melt of snow that almost invariably accompanies its occurrence in winter. It is a type of wind that occurs in many parts of the world where large orographic barriers are oriented more or less perpendicular to the prevailing direction of atmospheric circulation. Alternative names used to identify this phenomenon include the *foehn* in Europe, the *ibe* in Central Asia, the *Zonda* in Argentina, the *Berg* in South Africa, and the *Nor'wester* in New Zealand. The environmental consequences of chinooks include increased wind erosion, reduced air quality, vegetation mortality, and adverse health-related effects.

The onset of chinook conditions is typified by a rapid increase of temperature and wind speed and a corresponding decrease of humidity (Figure 1). Additional distinguishing characteristics include wind flow oriented perpendicular to the orographic barrier, a dry adiabatic lapse rate, and leewave clouds.

Synoptic conditions commonly associated with chinooks involve a surface ridge of high pressure on the windward side of an orographic barrier and a trough of low pressure on the lee side, which produces a steep pressure gradient and strong cross-barrier flow. Changes in air temperature and humidity during a chinook are generally ascribed to the forced ascent of warm, moist air over the barrier. The rising air cools and expands at the dry adiabatic lapse rate (≈ 9.8 °C/km [per kilometer]) until the dewpoint is reached, after which the air continues to cool at the saturated adiabatic lapse rate (≈ 6 °C/km). Air descends and warms at the dry adiabatic lapse rate on lee slopes, arriving warmer than it was at equivalent elevations on the windward side. In some cases, this change in air temperature can be dramatic. At Pincher Creek, located in southwestern Alberta, Canada, R. W. Turner reported an increase of 25.5 °C in 1 hour.

The explanation for increased wind speed down the lee side is not straightforward and is a topic of ongoing investigation. One of the leading hypotheses involves hydrodynamic theories of fluid

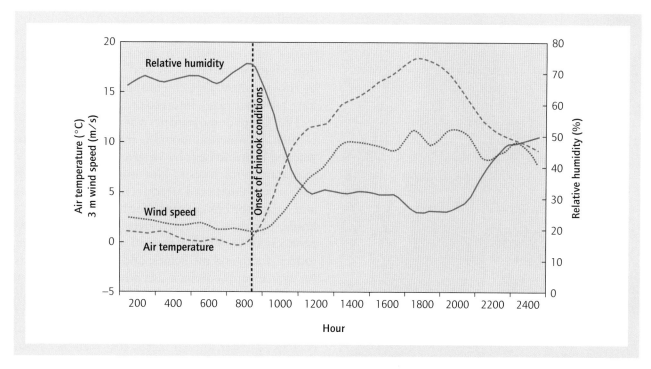

Figure 1 Chinook conditions in Saskatchewan, Canada (50°11′ N, 109°12′ W), on April 1, 2005
Source: Author.

mechanics, analogous to the behavior of water flowing over obstacles in river channels. As air ascends an orographic barrier, it accelerates toward the crest, where it transitions to supercritical flow and continues to accelerate down the leeslope. At some distance leeward of the barrier, the flow decelerates and produces a hydraulic jump. The latter typically corresponds to the position of the chinook arch, which is a narrow leewave cloud band oriented parallel to the orographic barrier.

Christopher Hugenholtz

See also Adiabatic Temperature Changes; Wind; Wind Energy; Wind Erosion

Further Readings

Brinkmann, W. A. R. (1971). What is a foehn? *Weather, 26,* 230–239.

Durran, D. R. (2003). Downslope winds. In J. Holton, J. Pyle, & J. Curry (Eds.), *Encyclopedia of atmospheric sciences* (pp. 644–650). Oxford, UK: Elsevier.

Nkemdirim, L. C. (1996). Canada's chinook belt. *International Journal of Climatology, 16,* 441–462.

CHIPKO MOVEMENT

The Chipko movement in India is regarded as the icon of ecofeminist movements across the globe, whose primary goal has been the preservation of forests. The word *chipko* in Hindi literally means "to stick to." The movement came into the limelight in March 1974, when Gaura Devi of Reni village, in the Garhwal region of the Indian Himalayas (then in the state of Uttar Pradesh and currently a part of the state of Uttarakhand), accompanied by 27 other women, decided to embrace the trees of their village, challenging axe-wielding lumber workers to axe them down before they cut down the trees. This act of courage and nonviolence was emulated across the Kumaon and the Garhwal regions of the Indian Himalayas, as well as in other areas. This resistance was directed at the unjust practice of tree felling, a manifestation of the close relations between the state and commercial contractors. The movement gained national and international media coverage, forcing the Indian state to change its policy and impose a 15-year ban on tree felling. To date, environmental activists are often referred to as *tree huggers,* the

This February 19, 1982, photo shows Indian women of the Chipko movement embracing trees to prevent them from being cut down.

Source: India Today Group/Getty Images. Used by permission.

term being a literal translation of the Chipko movement of India.

There are many reasons for the immense popularity and significance of the Chipko movement. First, it was spearheaded by women who were largely illiterate subsistence peasants, representing grassroots democracy; second, it was an environmental movement protesting the deforestation of the Himalayas; third, it resorted to Gandhian methods of peaceful resistance and nonviolence against a postcolonial state that collaborated with capital interests against its own citizens. It also helped demolish the myth that poor or oppressed people are incapable of looking beyond their immediate material needs and practical interests by showing how they can mobilize to protect their long term, strategic interests.

The Chipko movement was led primarily by women, because traditionally, the women of Garhwal and Kumaon have had a special bond with their mountains and forests. Not only do the women have a deep emotional attachment to the mountains and regard the Himalayas as their *maiti*, or mother's home, but they also rely extensively on the forest products of wood for firewood and fuel; leaves for fodder, food, and medicines; and rivulets and rivers for water supply (which are adversely affected by deforestation). Although the Chipko movement is regarded as an ecofeminist movement, highlighting the symbiotic relationship between women and ecology, a lesser known fact is that men were also actively involved. Indeed, the basic idea was initiated by two men, Gandhian activists Chandi Prasad Bhatt and Sunderlal Bahuguna.

Anupam Pandey

See also Ecofeminism; Feminist Political Ecology

Further Readings

Dankelman, I., & Davidson, J. (1988). *Women and the environment in the third world* (pp. 57–60). London: Earthscan.

Guha, R. (1994). *The unquiet woods: Ecological change and peasant resistance in the Garhwal Himalayas.* Delhi, India: Oxford University Press.

Rangan, R. (2001). *Of myths and movements: Rewriting Chipko into Himalayan history.* London: Verso Press.

Shiva, V. (1989). Women in the forest. In *Staying alive: Women, ecology and development* (chap. 4, pp. 55–95). London: Zed Books.

Shiva, V. (1991). *Ecology and the politics of survival.* Newbury Park, CA: Sage.

Sinha, S., Gururani, S., & Greenberg, B. (1997). The "new traditionalist" discourse of Indian environmentalism. *Journal of Peasant Studies, 24,* 65–99.

CHLORINATED HYDROCARBONS

Chlorinated hydrocarbons (CHCs) are a group of organic molecules that consist of chlorine, carbon, and hydrogen atoms. While some CHCs occur naturally, many of these compounds are synthesized for industrial purposes or are formed as disinfection by-products (DBPs). As the result of accidental spills and improper disposal methods, CHCs are among the most frequently detected hazardous chemicals at abandoned waste sites, in refuse disposal areas, and in industrial and municipal wastewaters. These compounds persist in the environment, and many bioaccumulate in the food chain. Some CHCs form separate, dense nonaqueous-phase liquids (DNAPL) with very low solubilities in water. However, the solubilities of DNAPL CHCs are orders of magnitude greater than their typical drinking water limits, making DNAPL CHCs long-term sources of groundwater contamination. While the human and environmental health risks of chlorinated hydrocarbons are compound specific, exposure to most CHCs has been linked to cancer. Studies continue to develop environmental remediation techniques for individual CHCs.

There are three major types of CHCs: aliphatic, aromatic, and heterocyclic. Aliphatic CHCs are those in which chlorine atom(s) are bonded to a straight or branched carbon chain. Some of the more well-known aliphatic CHCs are trihalomethanes (THMs) such as chloroform (CHCl$_3$; Figure 1), chlorinated methanes such as dichloromethane (DCM; CH$_2$Cl$_2$), and chlorinated ethenes such as trichloroethene (TCE; C$_2$HCl$_3$) (Figure 2) and vinyl chloride (VC; C$_2$H$_3$Cl).

A variety of aliphatic CHCs, including chloroform, have been isolated from marine algae, although the majority are synthetically derived for human use. Historically, chloroform and TCE were used as medical anesthetics; however, they have been subsequently replaced due to their toxicity. Currently, these compounds and other aliphatic CHCs such as DCM are used as industrial solvents for waxes, oils, and grease because they dissolve many organic and inorganic compounds. When these compounds enter the subsurface, they become long-term sources of groundwater contamination. Aliphatic CHCs, THM, and haloacetic acids such as dichloroacetic acid (DCA; CHCl$_2$COOH) (Figure 3) and trichloroacetic acid (CCl$_3$COOH) are formed during chlorination of drinking water that contains dissolved organic matter.

The United States Environmental Protection Agency (USEPA) set MCLs for DBPs in drinking water in the micrograms-per-liter concentration range due to the increased risk of cancer from ingestion of water contaminated with DBPs. Many aliphatic CHCs are amendable to natural or enhanced bioremediation through microbial reductive dechlorination or oxidation pathways.

Aromatic CHCs are hydrocarbons that have chlorine atom(s) bonded to benzene ring(s) made up of six carbon atoms with double bonds between alternate carbon atoms. The aromatic CHCs of current environmental concern are chlorinated benzenes, chlorinated phenols, and polychlorinated biphenyls (PCBs). Chlorinated benzenes are used as solvents and pesticides. Similar to aliphatic CHCs, chlorinated benzenes have been found in abandoned waste sites and in many waste waters and leachates. Chlorobenzene (C$_6$H$_5$Cl) (Figure 4), dichlorobenzene (C$_6$H$_4$Cl$_2$), and trichlorobenzene (C$_6$H$_3$Cl$_3$) are DNAPLs and are therefore long-term sources of groundwater contamination.

Chlorobenzenes (CBs) are amendable to bioremediation through microbial biodegradation. Chlorinated phenols have been used as wood preservatives. For example, pentachlorophenol (PCP; C$_6$HCl$_5$O) (Figure 5) is used as a fungicide and to pressure-treat wood.

PCP has been detected in surface waters and sediments, rainwater, drinking water, aquatic organisms, soil, and food. PCP has been found to be degraded abiotically through photodecomposition and biologically by aerobic and anaerobic microorganisms. Polychlorinated biphenyls (PCBs) are widely used as coolants, plasticizers, solvents, and hydraulic fluids. There are 209 possible congeners of PCBs. The United States banned the manufacturing of PCBs in

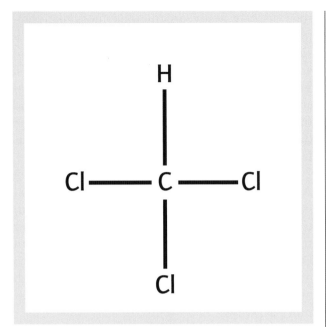

Figure 1 Chloroform

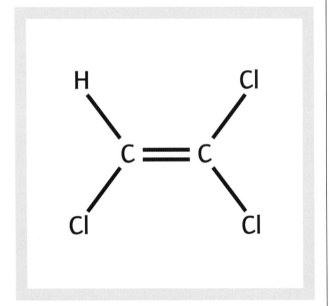

Figure 2 1,2,2-Trichloroethene

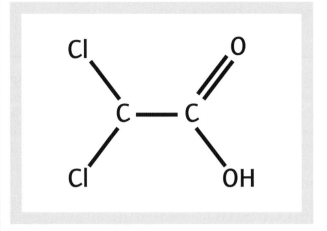

Figure 3 Dichloroacetic acid

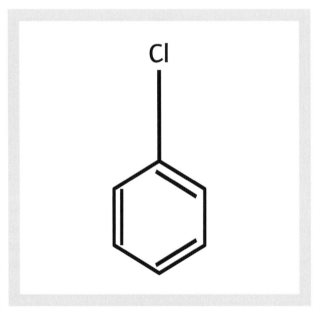

Figure 4 Chlorobenzene

1977. However, PCBs are very stable compounds and are still found in the environment today. While some PCB congeners have been shown to be degraded by microorganisms, these compounds are generally highly resistant to biodegradation. The pesticide dichloro-diphenyl-trichloroethane (DDT; $C_{14}H_9C_{l5}$) (Figure 6) is a chlorinated aromatic compound.

DDT was used to control mosquitoes spreading malaria and lice transmitting typhus,

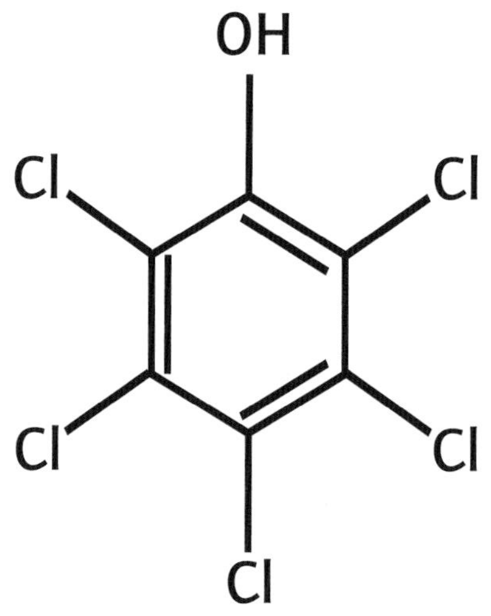

Figure 5   2,3,4,5,6-Pentachlorophenol (PCP)

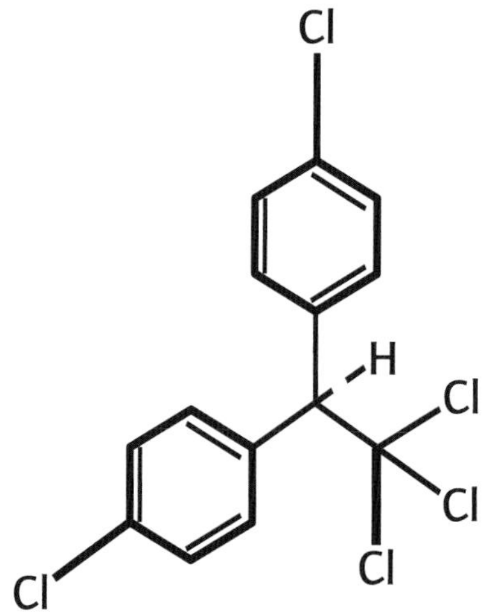

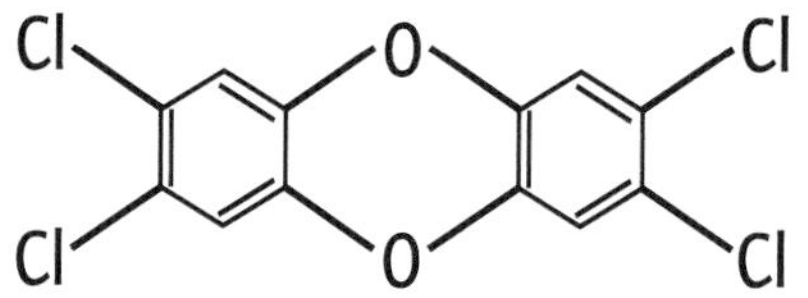

Figure 7   2,3,7,8-Tetrachlorodibenzo-*p*-dioxin (TCDD)

Figure 6   4,4′-(2,2,2-Trichloroethane-1,1-diyl) bis(chlorobenzene)

resulting in dramatic reductions in the incidence of both diseases. However, DDT is a persistent organic pollutant known to cause liver damage, temporary nervous system damage, and reduction in reproductive ability and is a probable human carcinogen. PCBs and DDT are lipophilic (fat loving) and, therefore, tend to bioconcentrate in fatty tissues and fluids, such as breast milk, and can be passed on to fetuses and infants during pregnancy and lactation. Additionally, lipophilic compounds bioaccumulate, with greater concentrations detected at higher levels in the food chain. As a result, PCBs and chlorinated pesticides such as DDT are the major aromatic CHCs in marine fish, mammals, and seabirds.

Heterocyclic CHCs include polychlorinated dibenzodioxins (PCDDs), which consist of two benzene rings fused to one dioxin ring in the middle (Figure 7), and polychlorinated dibenzofurans (PCDFs), which have two benzene rings fused to one furan ring in the middle.

Together, PCDDs and PCDFs are generally known as *dioxins*. Dioxins are produced naturally in forest fires. However, human activity has come to dominate PCDD/F inputs to the environment today. Current exposure levels are a concern because dioxins are toxic, persistent, and bioaccumulate in the environment. Dioxins have been characterized by USEPA as likely human carcinogens and are anticipated to increase the risk of cancer at background levels of exposure.

Penny L. Morrill

See also Environmental Impacts of Manufacturing; Environmental Protection; Groundwater; Water Pollution

Further Readings

Field, J. A., & Sierra-Alvarez, R. (2008). Microbial degradation of chlorinated benzenes. *Biodegradation, 19,* 463–480.

Field, J. A., & Sierra-Alvarez, R. (2008). Microbial degradation of chlorinated phenols. *Review of Environmental Science & Biotechnology, 7,* 211–241.

Muir, D. C. G., Wagemann, R., Hargrave, B. T., Thomas, D. J., Peakall, D. B., & Norstrom, R. J. (1992). Arctic marine ecosystem contamination. *Science of the Total Environment, 122,* 75–134.

U.S. Environmental Protection Agency. (2001). *National primary drinking water standards* (EPA 816-F-01-007). Washington, DC: USEPA, Office of Water (4606).

 # CHLOROFLUOROCARBONS

Chlorofluorocarbons (CFCs) are organic chemical compounds that contain carbon, fluorine, and chlorine atoms. Each CFC is identified by a unique numbering system that describes its structure. The digit in the hundredths place represents the number of carbon atoms in each molecule minus 1, the digit in the tenths place represents the number of hydrogen atoms in each molecule plus 1, and the digit in the units place indicates the number of fluorine atoms. For example, trichlorotrifluoroethane is a CFC that has three fluorine atoms, no hydrogen atoms, and two carbon atoms in each molecule, whence the designation CFC-113, while dichlorodifluoromethane (CFC-12) has one carbon, no hydrogen, and two fluorines.

CFCs are anthropogenic substances and have no natural background level. At room temperature, they are generally volatile, nontoxic, nonflammable, and colorless liquids or gases. These properties make them attractive to industry. Their commercial production, which began in the 1930s and 1940s, was primarily spurred by the need to find safer alternatives to the sulfur dioxide and ammonia refrigerants used at the time. Today, they are used in refrigeration and air conditioning, as cleansing agents for electrical and electronic components, and as foaming agents in shipping-plastics manufacturing. In some parts of the world, they are also used in aerosol propellants. The most common commercial CFCs, registered under the trade name Freon, are trichlorofluoromethane (CFC-11) and dichlorodifluoromethane (CFC-12).

Most CFC gases have both global-warming impacts and ozone-depleting effects. They are powerful greenhouse gases and contribute to global warming and climate change. Studies conducted by various researchers in the 1970s linked CFCs, along with other chlorine- and bromine-containing compounds, to the destruction of atmospheric ozone. Ozone is a trace gas that occurs in high concentrations in the stratosphere, the layer of the atmosphere that is between 10 and 25 km (kilometers) above the earth's surface. Ozone absorbs harmful ultraviolet (UV) radiation in the wavelengths between 280 and 320 nm (nanometers) of the UV-B band, which can cause biological damage in plants and animals. The loss of stratospheric ozone results in more harmful UV-B radiation reaching the Earth's surface. Because CFCs are chemically inert, when released into the atmosphere, they do not react easily and can exist in the troposphere for several decades, entering the stratosphere through the mixing and transport processes of the atmosphere. When CFCs enter the stratosphere, the intense UV radiation breaks them down, releasing chlorine (Cl) atoms, which react with ozone, starting chemical cycles of ozone destruction that deplete the ozone layer.

CFC emissions have rapidly declined since the 1980s, largely due to control of their use under the Montreal Protocol, an international treaty designed to implement the 1985 Vienna Convention for the Protection of the Ozone Layer. The Montreal Protocol was originally signed in 1987, entered into force in 1989, and has been amended seven times. It stipulated that the production and consumption of compounds that deplete stratospheric ozone—CFCs, halons, carbon tetrachloride, and methyl chloroform—were to be phased out at various rates, depending on a country's level of economic development. Some substitute compounds have low ozone depletion potential but high global warming potential (e.g., hydrochlorofluorocarbons, or HCFCs), while others do not deplete ozone but have high global warming potential (e.g., hydrofluorocarbons, or HFCs).

Segun Ogunjemiyo

See also Anthropogenic Climate Change; Stratospheric Ozone Depletion

Further Readings

Albritton, D. L., Watson, R. T., & Aucamp, R. J. (1995). *Scientific assessment of ozone depletion: 1994.* Geneva: World Meteorological Organization.

Cagin, S., & Dray, P. (1993). *Between Earth and sky: How CFCs changed our world and endangered the ozone layer.* New York: Pantheon Press.

Molina, M. J., & Rowland, F. S. (1974). Stratospheric sink for chlorofluoromethanes: Chlorine atom catalyzed destruction of ozone. *Nature, 249,* 810–812.

United Nations Environment Programme. (1997). *Montreal Protocol on substances that deplete the ozone layer.* New York: Author.

 # CHOLERA, GEOGRAPHY OF

Cholera is an acute diarrheal disease with a fecal-oral transmission route and is caused by ingestion of a dose of between 10,000 and 1 million *Vibrio cholerae* bacteria. The disease typically manifests in densely populated areas with high poverty, where water and sanitation remain unimproved. South Asia is the source region of cholera, as its tropical, estuarian environment creates a natural habitat for *V. cholerae* bacteria.

Cholera transmission can be divided into primary and secondary types. Primary cases result from infection by surface water sources where free-living bacteria are part of the natural ecosystem. Thus, primary transmission may infect someone who drinks untreated pond water or eats undercooked shellfish. Those infected by primary transmission may in turn infect others, considered a secondary pathway. Sources of secondary transmission include infection via direct interpersonal contact or through drinking water contaminated with fecal material from an infected person. Environmental factors such as water temperature, salinity, phytoplankton, and algae concentrations control the dynamics of primary transmission, while social structure, economics, and the built environment (e.g., water and sanitation infrastructure) control secondary transmission.

Global History of Cholera

To date, the world has experienced seven cholera pandemics, beginning with the 1817 to 1824 pandemic in the Bay of Bengal, which diffused across the Indian subcontinent, Asia, and the Russian Empire. The second pandemic spread through Europe and the United States, and the third through sixth pandemics affected parts of Asia, Africa, Europe, and South America. All were caused by the classical biotype of cholera. A new biotype of cholera called "El Tor" caused the seventh pandemic, which began on Celebes, Indonesia, in 1961 and spread to Asia in the 1960s and to the Middle East and Africa in the 1970s. By 1991, nearly 140,000 cases of cholera were reported among 19 African nations, and the disease reappeared in Peru after 100 years of absence. A new strain, *V. cholerae* O139, was identified in 1992 during an outbreak on India's eastern coast. Cholera remains endemic in countries throughout the developing world but has been nearly eradicated in industrialized nations.

Cholera was the most frequently mapped disease of the 19th century. Dr. John Snow was an influential figure during this period for his involvement in identifying the source of the 1854 cholera outbreak in London. His map of cholera cases around the heavily trafficked Broad Street Pump helped identify the link between cholera and water. Contrary to miasma theory, which claimed

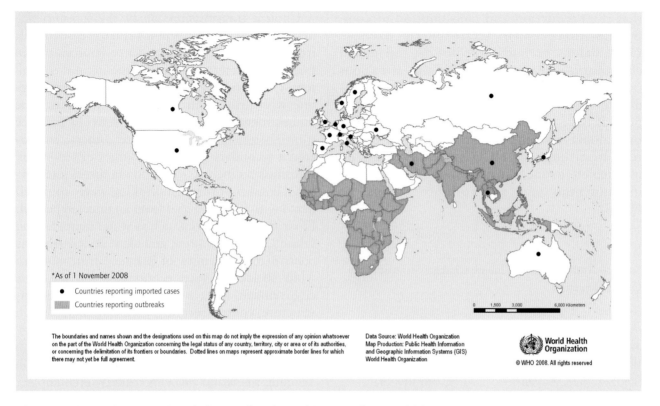

Figure 1 Countries reporting cholera outbreaks and imported cases, 2006–2008

that cholera transmission was airborne, Snow identified the link between the Thames River water from the Broad Street pump and the cholera outbreak in London and had the pump handle removed in 1855.

Cholera in the 21st Century

Cholera outbreaks continue to burden areas of the world lacking safe water and sanitation (see Figure 1). Cholera remains prevalent in South Asia and is common in other resource-poor areas. In August and September 2008, cholera cases were confirmed in Baghdad, Misan, and Babil provinces in Iraq. The reemergence of cholera followed an outbreak in September 2007, with 4,696 cases and 24 deaths. African nations continue to experience cholera outbreaks. In 2008, following floods, Namibia and Angola experienced outbreaks. The Congo, Nigeria, and Guinea Bissau also endured cholera outbreaks. These countries all share a commonality: Political unrest or outright war have disrupted water, sanitation, and health infrastructure, resulting in large cholera epidemics.

Cholera and the Environment

V. cholerae survive in aquatic reservoirs, in which the bacteria consume algae, phytoplankton, copepods, and zooplankton during interepidemic seasons. Increases in local sea surface temperatures influence the growth of phytoplankton, making climate variability an important driver of cholera outbreaks. In areas of South Asia where cholera is endemic, the first outbreak typically occurs in spring, followed by a larger outbreak after the monsoon. Interannual climate variability is influenced by the El Niño Southern Oscillation (ENSO), which affects interannual outbreaks by increasing the water temperature in shallow water bodies and the coastal waters of the Bay of Bengal.

In addition to global climatic events, local environmental factors are important for predicting cholera outbreaks. An increase in ocean chlorophyll concentration, an attachment site for bacteria, and a decrease in freshwater river discharge, which dilutes the number of aquatic bacteria, have been linked to outbreaks in Bangladesh. Cholera outbreaks follow seasonal patterns in countries located at the higher latitudes in both hemispheres. Outbreaks are frequent and constant in countries located within 15° of the equator, where temperatures are generally higher and more constant.

Cholera can be avoided when water, sanitation, and hygiene are good. In the absence of such conditions, however, filtering water through cloth (e.g., South Asian saris) folded a minimum of eight times halves the risk of cholera infection by creating a pore size small enough to remove zooplankton and most phytoplankton containing *V. cholerae*. Surveillance systems and early laboratory identification of cholera cases are critical to interrupting further transmission. The administration of oral rehydration salts (ORS), a mixture of sodium and glucose, is the most common form of rehydration treatment and successfully treats up to 80% of cholera patients. Severely dehydrated patients are treated with intravenous fluids and antibiotics.

Michael Emch and Veronica Escamilla

See also Anthropogenic Climate Change; Disease, Geography of; GIS in Health Care; Health and Health Care, Geography of; Medical Geography

Further Readings

Colwell, R., & Spira, W. (1992). The ecology of Vibrio cholerae. In D. Barua & W. B. Greenough (Eds.), *Cholera* (pp. 107–128). New York: Plenum Medical.

Emch, M., Feldacker, C., Islam, M., & Ali, M. (2008). Seasonality of cholera from 1974 to 2005: A review of global patterns. *International Journal of Health Geographics, 7*(31), 1–33.

Emch, M., Feldacker, C., Yunus, M., Streatfield, P., Thiem, V., Canh, D. G., et al. (2008). Local environmental drivers of cholera in Bangladesh and Vietnam. *American Journal of Tropical Medicine & Hygiene, 78*(5), 823–832.

Koch, T. (2005). *Cartographies of disease: Maps, mapping, and medicine.* Redlands, CA: ESRI Press.

CHORLEY, RICHARD (1927–2002)

Richard John Chorley was a transformative figure in late-20th-century geomorphology. Trained in geography at Oxford University, it was his experience in geology at Columbia University that truly sparked his numerous innovations. At Columbia, he was a student in the Arthur N. Strahler research group that produced so many stellar contributors to modern geomorphology. After brief stints as a junior faculty member in the United States, Chorley returned to Britain and pursued the remainder of his illustrious career at Cambridge University.

Chorley contributed to geography, geomorphology, and environmental sciences in both discrete and overlapping ways that may be seen as occupying at least four spheres. He was a central figure in fundamentally changing geomorphology from the qualitative discipline dominated by the Davisian erosional model. He was a prominent historian of geomorphology and its development. His methodological contributions to geomorphology were so fundamental and his interests so broad that they spilled over into geography at large, as well as into areas such as environmental science. Finally, his interest in teaching, coupled with his historical interests, led to many publications, including very successful books and professional journals (e.g., *Progress in Geography*, which eventually morphed into two journals of similar name).

While Chorley's early publications included direct challenges to Davisian geomorphology and introduction of discrete quantitative techniques (e.g., the lemniscate), he is best known for his comprehensive approach to quantification, which is most clearly seen in his lifelong advocacy of modeling and systems approaches to science. His particular approach to systems—championing the general systems theory of Ludwig von Bertalanffy—has been criticized, sometimes correctly and sometimes incorrectly; however, there is no question that its overall thrust transformed every disciplinary sphere that Chorley sought to influence, especially geomorphology.

By far the most comprehensive history of geomorphology, and one of the most incisive, is the series of four tomes authored by Chorley and other contributors. These volumes span the discipline from its inchoate origins to its present form. Interrupted by Chorley's untimely death, the final volume published in late 2008.

Finally, Richard Chorley contributed mightily to the teaching of the discipline, especially in the early stages of its present modern quantitative form, through innumerable texts, papers, oral presentations, workshops, and conferences.

Colin Edward Thorn

See also Davis, William Morris; Geomorphology; Gilbert, Grove Karl; Penck, Walther; Physical Geography, History of

Further Readings

Chorley, R. J. (1962). *Geomorphology and general systems theory* (USGS Professional Paper 550-B). Washington, DC: U.S. Geological Survey.

Chorley, R. J. (1967). Models in geomorphology. In R. J. Chorley & P. Haggett (Eds.), *Models in geography* (pp. 59–96). London: Methuen.

Chorley, R. J., Dunn, A. J., & Beckinsale, R. P. (1964). *The history of the study of landforms or the development of geomorphology* (Vol. 1). London: Methuen.

Chorley, R. J., & Kennedy, B. A. (1971). *Physical geography: A systems approach.* London: Prentice Hall.

Chorley, R. J., Malm, D. E. G., & Pogorzelski, H. A. (1957). A new standard for estimating drainage basin shape. *American Journal of Science, 255,* 138–141.

 # CHOROLOGY

Chorology is the study of places or regions, usually small ones. The term *chorology* comes from the Greek words for "the science of place," in contrast to *chronology,* the ordering of events in time. Also known as *areal differentiation,* chorology has a long history as a term and concept in geography.

Strabo (64 BC to AD 24), a Greek geographer working for the Romans, advocated a form of chorology in his 17-volume *Geography.* In contrast, Ptolemy (AD 87–150), a Roman geographer and astronomer, maintained that the task of geography is the description of the Earth as a whole; in his eight-volume *Geography,* Ptolemy ridiculed Strabo's emphasis on regions, arguing instead for a holistic view of the Earth. Ptolemy differentiated between *geography* as the study of universals, *topography* as the study of localities, and *chorography* as integrating the two.

The great 17th-century geographer Varens (Varenius, 1622–1650), who wrote the highly influential *Geographia Generalis* in 1650, distinguished between what he called *specific geography* (concerned with the unique character of places) and *general geography,* which was concerned with universal laws. Immanuel Kant (1724–1804), a geographer as well as a philosopher, played an important role in the historical evolution of chorology by arguing that unlike theoretical sciences such as chemistry, geography, like history, was essentially concerned only with the empirical and the unique. His views were hugely influential in subsequent philosophies of space.

Some 19th-century geographers, such as Karl Ritter, emphatically practiced a form of chorology that was instrumental in mapping use values of places in the face of expanding colonial empires. Knowledge of local areas was important to colonial commerce and governance. A notable advocate was Paul Vidal de la Blache (1845–1918), the father of French geography, who studied small French rural areas called *pays* and their associated styles of life, or *genres de vie.* Because the climate of France did not vary much, but lifestyles did, Vidal de la Blache's work was also crucial to the introduction of possibilism to the discipline in the struggle against environmental determinism. His German counterpart, Alfred Hettner (1859–1941), argued in the Kantian tradition that geography was the art of regional synthesis, that is, the pursuit of interrelations in given areas, an aspect that other disciplines ignored. Chorology thus became the basis of geography's disciplinary identity in the early 20th century.

In the 1920s, American geographers adopted chorology in the wake of the demise of environmental determinism. American chorology was personified by Richard Hartshorne (1899–1992), who studied under Hettner, then graduated from the University of Chicago in 1924. In the tradition of Kant, Hartshorne and his fellow chorologists argued that the essence of geography was the regional description of regions, including cultural and physical phenomena. Chorologists advocated getting to know places in great depth, with a healthy regard for cartography and fieldwork. Because large regions are diverse and complex, Hartshorne argued that chorology should focus on small, relatively homogeneous regions. He maintained that regions are essentially mental concepts, that is, subjective tools to find meaning and create order in the landscape. Regions were thus necessarily simplifications and were useful only in as much as the gain in understanding they provided exceeded the loss of detail. Implicit in Hartshornian chorology was the view that location served as a form of explanation, that is, proximity was synonymous with causality, leading to a crude form of spatial determinism reminiscent of Tobler's First Law. Hartshorne maintained that because landscapes exhibit relatively little change in the course of one lifetime, there was no urgent need to study the process of change. In arguing that only by sticking to the facts can we remain objective, Hartshorne's line of thought drew on the philosophical tradition of empiricism, in which facts are simply true without regard for theory.

Chorology collapsed in the 1950s as the reigning model in geography when positivism arose to take its place, although in the popular view, geography and chorology are often erroneously held to be synonymous. The transition began with a famous attack on Hartshorne by Fred Schaefer in 1953. Schaefer claimed that Hartshorne's view of geography as an integrative science concerned only with the unique was simplistic. By refusing to search for explanatory laws, geography condemned itself to being what he called an immature science. Rather than idiographic regions, geographers should seek nomothetic regularities across regions. This critique opened the door to the rise of positivism and the quantitative revolution.

Although traditional chorology died under the positivist onslaught, it did experience something of a resurrection in the 1980s. Some Marxists argued that broad social processes always play out in different ways in different places, as spatial divisions of labor enfolded various places in highly contrasting ways. This perspective led to a renewed respect for the idiographic as more than some prescientific remnant of a theoretically unselfconscious age. What became known as the Localities School approached regions in terms of their historical development as they acquired unique combinations of imprints of different divisions of labor (i.e., investments, labor market practices, cultural forms). In this view, general laws of explanation are only manifested in unique contexts, and localities are not simply unique curiosities. Unlike the earlier tradition of chorology, therefore, this approach eschews empiricism and maintains a central role for theory. Thus, broad processes of globalization cannot be interpreted independently of local chorographic contexts, a notion encapsulated in the term *glocalization*.

Barney Warf

See also Empiricism; Hartshorne, Richard; Hettner, Alfred; Human Geography, History of; Idiographic; Locality; Place; Regions and Regionalism; Varenius; Vidal de la Blache, Paul

Further Readings

Berry, B. (1964). Approaches to regional analysis: A synthesis. *Annals of the Association of American Geographers, 54,* 2–11.

Hart, J. (1982). The highest form of the geographer's art. *Annals of the Association of American Geographers, 72,* 1–29.

Hartshorne, R. (1939). *The nature of geography.* Washington, DC: Association of American Geographers.

Sack, R. (1974). Chorology and spatial analysis. *Annals of the Association of American Geographers, 64,* 439–452.

Schaefer, F. (1953). Exceptionalism in geography: A methodological examination. *Annals of the Association of American Geographers, 43,* 226–229.

Warf, B. (1993). Post-modernism and the localities debate: Ontological questions and epistemological implications. *Tijdschrift voor Economische en Sociale Geografie, 84,* 162–168.

CHOROPLETH MAPS

Choropleth maps can be described as the most commonly created thematic map type, in large part due to the relative ease of their creation using most GIS software packages and also due to the perception that they are easily understood. Choropleth mapping involves aggregating data to an areal unit of measure and then representing these data using color or pattern within the areal unit. Creation of choropleth maps involves normalizing, representing, classifying, and symbolizing the data of interest.

Normalization

Choropleth maps involve mapping data by areal unit, for example by country, state, or county. Because of the disparities in size that exist for all types of areal units, mapping raw totals will often present a skewed representation of the data. To combat this problem, choropleth maps are frequently normalized by a common attribute, such as area or population. Once normalized, the maps are often referred to as *value-by-area* or *value-by-population* maps. By normalizing the raw data, the map is not dominated by the largest unit but rather uses the ratio to highlight the ratios of interest.

Representation

Choropleth maps occur in two distinct types: classed and unclassed. A classed choropleth map presents the data of interest by classifying the values into a distinct number of individual colored classes (Figure 1). Each class is presented as a range of possible values, and the units within each range are represented accordingly.

In contrast, an unclassed choropleth map presents the value of each unit in a distinct color that varies according to the rank of the value with respect to all the other values (Figure 2). Traditionally, both types of choropleth map represent low values with light color and use dark colors to represent greater values.

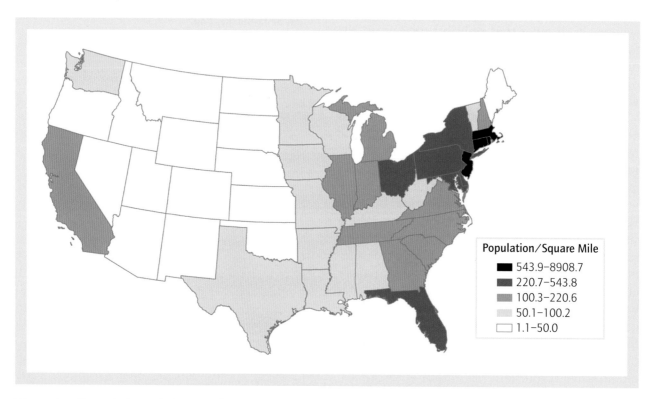

Figure 1   Classed choropleth map of population density for contiguous United States

Source: Adapted by author from U.S. Census Bureau data. Available from www.census.gov/population/www/popdata.html.

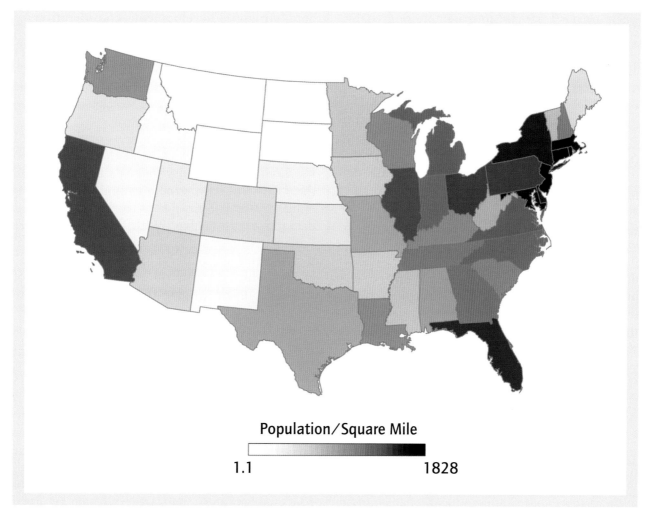

Figure 2 Unclassed choropleth map of population density for contiguous United States

Source: Adapted by author from U.S. Census Bureau data. Available from www.census.gov/population/www/popdata.html.

Data Classification

When a classed choropleth map is created, the classification type must be decided based on the data. One of the most common classification schemes used is the Natural Breaks method. This method involves plotting all data values on a graph and then identifying the largest gaps between the data clusters. Most mapping software will perform this analysis automatically. The other common classification schemes follow more traditional statistical measures, such as quantile classification, defined intervals, or mean/standard deviation classification (most commonly used for diverging data with a meaningful middle). Often, these classification schemes are used as a basis for starting the classification but are modified to accommodate meaningful data values that may fail to be captured by an automated classification method.

Symbolization

Choropleth maps are symbolized in one of three ways, sequentially, diverging, or qualitatively. Pattern is largely limited to qualitative symbolization, though it can be used in combination with color to extend a sequential classification. Black-and-white color tends to restrict the map to a sequential representation—unless combined with pattern. Full color allows for all the options (Figure 3). Sequential symbolization follows a light to

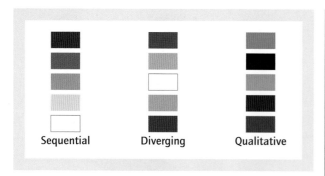

Figure 3 Example of sequential, diverging, and qualitative color schemes

Source: Author.

dark path either through a single hue or from one hue to a second, sometimes through a third. Diverging symbolization places two dark colors at the poles of the symbolization and lightens each toward the middle, where a white or near-white color represents the meaningful middle value. Qualitative colors use differing hues to represent the different values, often with nonnumeric data.

William Buckingham

See also Data Classification Schemes; Isopleth Maps

Further Readings

Brewer, C. A. (2005). *Designing better maps: A guide for GIS users*. Redlands, CA: ESRI Press.
Slocum, T. A., McMaster, R. B., Kessler, F. C., & Howard, H. C. (2005). *Thematic cartography and visualization*. Upper Saddle River, NJ: Prentice Hall.

 ## CHRISMAN, NICHOLAS (1950–)

Nicholas R. Chrisman is a professor at the Université Laval, Canada, and scientific director of GEOIDE, a network that funds research spanning 30 universities across Canada. Over a 30-year period, Chrisman has contributed considerably to the development of GIS and the nascent scientific discussion of its potential and pitfalls, including data structure development, consideration of error, and engagement with sociotechnological dimensions of the use, creation, and consequences of geographic information technologies. Besides his current positions, he has also held academic positions at the University of Wisconsin–Madison and the University of Washington.

Some of his early contributions were as part of the research group working at the Harvard Lab for Computer Graphics and Spatial Analysis in the 1970s on data structures and on programming the first GIS capable of working with vector data in a robust manner. With Tom Peucker (later Poiker), Chrisman coauthored an influential paper titled "Cartographic Data Structures" (1975), which presented a conceptual overview of the issues in developing topological data structures. Connected to programming and application projects while at the Harvard lab, he was part of the group that included James Dougenik, Denis White, Scott Morehouse, Allan Schmidt, and Geoff Dutton, which developed the first GIS capable of overlaying vector data in a robust manner, *ODYSSEY*, a precursor to the Environmental Systems Research Institute's ARC/Info.

Diverse activities at the Harvard Lab and the many interactions with the nascent field through conferences were important for the next phase of Chrisman's professional career, when he returned to academia to work on his PhD dissertation on error in categorical maps, which he completed in 1982 at the University of Bristol. This work on the origins and handling of error also led to publications on the role of time in geographic information and measurement theory.

Parallel to this work, and accompanying a move to his first academic position at the University of Wisconsin–Madison, Chrisman began a long-lasting engagement with the sociotechnological dimensions of the use, creation, and consequences of geographic information technologies. The 1997 book *Exploring Geographic Information Systems* situates the complicated interactions between society and its technologies as nested rings, inextricably linked. The work in this area has become influential in GIScience as well as science and technology studies.

Francis J. Harvey

See also GIScience

Further Readings

Chrisman, N. R. (1989). Modeling error in overlaid categorical maps. In M. Goodchild & S. Gopal (Eds.), *The accuracy of spatial databases* (pp. 21–34). London: Taylor & Francis.

Chrisman, N. R. (1997). *Exploring geographic information systems*. New York: Wiley.

Chrisman, N. R. (1999). A transformational approach to GIS operations. *International Journal of Geographical Information Science, 13*(7), 617–637.

Chrisman, N. R. (2005). Full circle: More than just social implications of GIS. *Cartographica, 40*(4), 23–35.

Chrisman, N. R. (2006). *Charting the unknown. How computer mapping at Harvard became GIS.* Redlands, CA: ESRI Press.

Peucker, T. K., & Chrisman, N. R. (1975). Cartographic data structures. *The American Cartographer, 2*(1), 55–69.

 # CHRISTALLER, WALTER (1893–1969)

Before the German geographer Walter Christaller's central place theory (1933), urban places were viewed in isolation from one another, as unique single entities, differentiated by their position in a hierarchy based on population size. Christaller's dissertation, *Die zentralen Orte in Süddeutschland*, broke from this prevailing view. Instead, Christaller demonstrated the spatial interdependence between places. Interdependence arose because of location relative to other places in the system of places. A place's population size and trade with other places were determined by its location.

Christaller's central place theory concluded that a hierarchy of places would arise as the system of places tended toward spatial equilibrium. Within the hierarchy of places, higher-order places would offer goods and services that required greater numbers of people distributed over larger trade areas for their support. Successively lower-order places would offer goods and services that required fewer people to sustain the activity. Lower-order places with their smaller trade areas are embedded among fewer higher-order places and their larger areas of geographic dominance.

Christaller inductively derived propositions about the population size of places, number of places, and locational distribution of places. He described five sizes of communities:

1. Hamlet
2. Village
3. Town
4. City
5. Regional capital

Christaller was the first to propose a geographically determined hierarchy of places, conforming to specific geometric principles. His hypothetical landscape was composed of trade areas in the shape of hexagons. Hexagons are the most compact of packable shapes. Christaller's reasoning was that depending on the initial assumptions of the model or characteristics of the landscape, development of urban places would occur on the apexes, arcs, or interior of the hexagonal trade area boundaries—the market, transportation, or administration system, respectively. Christaller described the conditions that would bring about the particular geometry of places. In the absence of a transportation network, lower-order places would grow at the location of the apexes, exactly one third of the way between three next-higher-order places. If the landscape included a transportation network that minimized fixed costs and connected the highest-order places, then locational advantage would shift to the arcs of the hexagonal trade areas, with the growth of lower-order places occurring midway between a pair of next-higher-order places. Trade results between the places, and the revenues of a place depend on the geometry that characterizes the system of places. Christaller recognized the importance of internalizing externalities within a hierarchy of administrative zones.

There are a greater number of lower-order places, and they are closer together than higher-order places. Lower-order places depend on higher-order places for the provision of goods and services not available locally, and higher-order places depend on lower-order places to provide demand sufficient for the provision of those goods and services unique to the higher-order places. Because of trade, all places are interdependent in space. The growth opportunities of a place are given by its location relative to other places. Christaller's rigid geometry of place location produces an efficient system for trade, growth, and development.

The German economist August Lösch (1906–1945) generalized Christaller's central place theory in his 1939 book published in German, *The Economics of Location*. Instead of Christaller's distribution of places and their functional order that maximized profits, Lösch's system was based on minimizing the cost of consumers' accessing needed goods and services. Lösch provided a deductive, more formal economic approach to the growth and spatial distribution of urban places.

Christaller's geometry of places has been used to explain and make decisions on settlement growth and contraction. In the post–World War II era, many regions of Eastern Europe adopted highway and development plans that had been created by Christaller during the war. The United States followed suit in the mid 20th century with the development of its Interstate Highway System, based on the German/Christaller model. Christaller's central place theory anticipated which places would benefit from the interstate development and which, being bypassed, would decline.

The policies of Nazi Germany were heavily influenced by ideas of social engineering and the quest for organizational efficiency. Geography, demography, and economics shaped and were in turn shaped by these Nazi policies. Following the completion of Walter Christaller's dissertation on central place theory in 1933, he had a short stint in academia. He joined the Nazi party in 1940 and entered public service with the German government as an applied geographer, performing location analysis within Heinrich Himmler's SS-Planning group. Himmler (1900–1945), one of the most influential people in Nazi Germany, oversaw all police and security forces and was overseer of concentration camps and extermination facilities. Under Himmler's direction, his planning group coordinated the efficient locational distribution of facilities engaged in the killing of millions of people whom the Nazis deemed unworthy. After the fall of the

Nazi regime, Christaller joined the Socialist Party and lived as a practitioner of applied geography until 1969.

August Lösch was strongly opposed to Nazi policies. He refused academic and government posts because he believed them to be too much under the influence of Nazi ideology. Lösch died before the fall of Hitler's Nazi regime.

English translations of Christaller's and Lösch's pioneering works were published in the 1950s. These works propelled economic geography to adopt quantitative methods and formulate rigorous geospatial logic. The publication of these two pioneers' work, and the subsequent body of knowledge that grew out of it, contributed to geography becoming a rigorous science that provides benefit to decision makers in government and private industry.

Grant Thrall

See also Applied Geography; Central Place Theory; Location Theory; Regional Science; Urban Hierarchy

Further Readings

Christaller, W. (1966). *Places in central Southern Germany* (C. W. Baskin, Trans.). Englewood Cliffs, NJ: Prentice Hall. (Original work published 1933)
King, L. (1984). *Central place theory*. Beverly Hills, CA: Sage.

CIRCUITS OF CAPITAL

Developed by Karl Marx in the second volume of *Capital*, the "circuits of capital" approach represents a multifaceted way to understand the cyclical and self-expanding process of capitalist value circulation. Since each circuit is envisioned as a particular moment within a larger process, economic geographers have found it useful in situating local geographies in a wider context.

The circulation process is divided into three main circuits. The first circuit of money-capital (see Figure 1) represents succinctly the goal of any capitalist—starting with money and ending

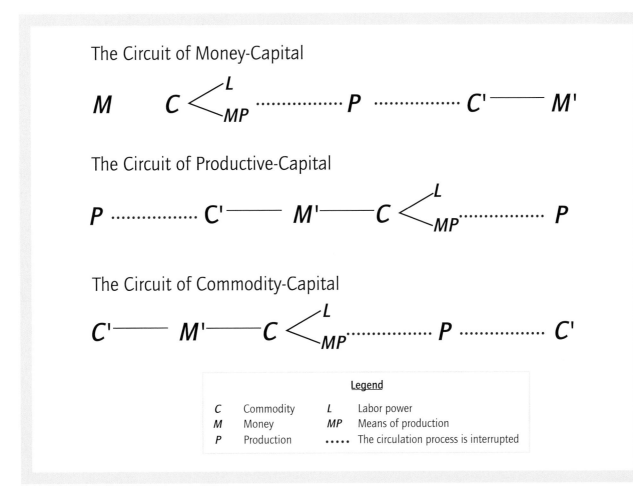

Figure 1   The circuits of capital

Source: Author.

with more. Since most industrial production processes require substantial money-capital to purchase the commodities labor power and means of production ($L + MP$) *before* production takes place, the circuit money-capital can be envisioned as the various types of financial and credit institutions whose overarching goal is to invest money with the aim of gaining more ($M – M'$).

The second circuit of productive-capital represents the periodic renewal of the concrete processes of commodity production (P). Each cycle of production must adapt to produce a specific volume of commodities at a particular speed in accordance with the industry's competitive dynamics. Focusing on this circuit allows a window into the context-specific struggles on the factory floor between capitalists, managers, and workers over the conditions and relations of production.

The third circuit of commodity-capital begins and ends its cycle with a produced commodity (C') that must be sold on the market. While the money obtained through the exchange ultimately will be reinvested in the production process, the process always ends with the renewal of the commodity-capital that must be exchanged. The circuit of commodity-capital can be conceptualized as the various types of distribution, marketing, and retail businesses whose overarching goal is to procure commodities and ensure their exchange.

Finally, the spatial and temporal interruptions of the circulation process—both in bringing commodities on site for production and in transporting them to the market for exchange—are also theorized as potential sources of crisis for the overall system of circulation as a whole.

Geographers have implicitly or explicitly applied the circuits-of-capital approach in a variety of ways. Within the circuit of money-capital, financial centers are studied as particular sites of power in the global market. Within the circuit of productive-capital, Doreen Massey's influential notion of "spatial divisions of labor" examines how localities and production facilities cohere within wider geographical networks of corporate organizational control. Within the circuit of commodity-capital, the commodity chain approach traces the geographical movement of commodities from raw materials to production, processing, retail, and consumption. Finally, following developments from cultural studies, cultural geographers have proposed a fourth "circuit of culture" that suggests that meanings and symbols themselves circulate between and among particular sites and places (e.g., households, advertising agencies, media outlets).

Matthew T. Huber

See also Commodity Chains; Division of Labor; Economic Geography; Harvey, David; Marxism, Geography and; Massey, Doreen; Space of Flows

Further Readings

Harvey, D. (1982). *The limits to capital.* Oxford, UK: Blackwell.

Hughes, A., & Reimer, S. (Eds.). (2004). *Geographies of commodity chains.* London: Routledge.

Marx, K. (1977). *Capital* (Vol. 2). New York: International Publishers. (Original work published 1893)

Massey, D. (1984). *Spatial divisions of labour: Social structures and the geography of production.* London: Routledge.

 # CITIZENSHIP

Citizenship means many things to many people in many different places; thus, ideas of citizenship are unrelentingly geographical. With the help of geographical concepts such as space, place, and scale, one can think about the various dimensions of citizenship by exploring the ways in which geography is so much a part of the political, social, cultural, and economic spheres that include or exclude people as citizens. Hence, geography is fundamental to all the critical issues of citizenship that concern society today. There are many ways in which geographers make connections to various notions and ideas of citizenship. Intersections of geography and citizenship can be traced back to antiquity, when geographers such as Ptolemy and Strabo advocated formalizing citizenship rights to people living within political territories such as the city-states of ancient Greece. Recently, feminist and postcolonial geographers have been concerned with citizenship as a phenomenon that incorporates ideas of inclusion and exclusion or, in other words, questions of who can belong where. This entry first examines the meaning of citizenship as it relates to rights and responsibilities within a given society. It then explores the ways in which geographers have expanded the concept to address issues related to cross-cultural and global citizenship.

Defining Citizenship

Geographical investigations of citizenship concern the relationships between individuals living and working in a particular place (communities, nation-states, etc.) and how these relationships affect everyday lives because of who these individuals are and where they live. As citizens, people are seen to have certain duties and obligations, and in return they can expect certain rights and benefits. Traditionally, at least in democracies, these benefits include various bundles of civil or legal rights (e.g., freedom of speech, assembly, movement, and equality before the law), political rights (e.g., the right to vote and engage in political activity), and economic rights (e.g., rights to social security, welfare, and basic standards of living). Despite the fact that this definition suggests a sense of inclusion and equality, notions of citizenship can also serve to exclude people. Historically, only certain groups of people, namely, property-owning white men, were entitled to these rights of full citizenship. Other groups, such as women and ethnic minorities, have had to fight to have these rights extended to them. Some social groups continue to remain "partial citizens," as they are excluded from particular civil, political, or economic rights; for example, even though members of some minority groups can enjoy legal rights, they may still feel like second-class citizens because of discrimination.

One approach to citizenship and geography is concerned with how the spaces of material human society, represented as administrative political units such as towns, cities, or nation-states, are intertwined with individual social, political, and economic rights and obligations that help define one's membership in a civil society. The way in which political power is organized in particular places is a critical element in determining who is and is not granted certain rights. These geographical connections are usually based on notions of the right to use public space (e.g., parks, streets, plazas), and they are linked to political perceptions of equality, democracy, and liberty. When one thinks of citizenship as a set of rights and responsibilities framed within particular bounded spaces, these spaces suggest an "outside" and an "inside." For example, spaces of citizenship sometimes serve to marginalize women, racial and ethnic minorities, the disabled, and so on. In this sense, geographers are interested in challenging the power structures and the status quo of a place by making claims for social justice and civil rights for those who work or live in a particular place and who may be marginalized by society. Many political geographers are interested in questions of how states, nations, territories, and boundaries play a role in notions of citizenship—for example, what might be at stake when certain people in a place call for political independence from other places, and where boundaries might be placed that mark these new territorial allegiances.

Global Perspectives

Traditional ideas of citizenship as a bundle of social, economic, and political rights that are extended by membership in a nation-state are now being rethought within a global context. Geography has a long tradition of recognizing the importance of global interconnectedness, and geographers are interested in the links between globalization and new forms of citizenship. This reorientation of geographical scale upward involves a range of issues that operate at a global scale, and they require a rethinking of what it means to be a "global citizen." For example, environmental issues such as global warming now require people to create a global sense of responsibility in order to resolve a problem that is transnational in scope. Another issue that geographers consider is how neoliberal global capitalism and regional free trade increasingly contribute to a spatial unevenness of wealth, redefining citizenship in terms of various levels of economic wealth and consumer spending. A geographical perspective is required, therefore, to analyze how local communities and individuals become marginalized through these economic, political, and cultural processes—in other words, how these processes may take the form of geographical isolation and affect one's sense of belonging or citizenship.

Another citizenship issue that goes beyond the borders of a nation-state, and receives much attention from geographers, is the growing cultural and ethnic diversity of many societies, which has led to various debates surrounding multicultural citizenship, especially in the context of immigration and migration. These debates, whether they are about integration of refugees and asylum seekers or the rights of economic migrants, involve examining the geographies of people's movement and settlement and their relationships with new societies. Indeed, these new ideas about citizenship are about how border crossings affect one's sense of cultural and social belonging in different places.

Contemporary human geography provides a spatial lens through which ideas of citizenship are studied across multiple locales, political units, and scales of governance, providing critical insight into the ongoing processes of social, political, and economic restructuring of society. As a result, the study of geography and citizenship makes a key contribution to a better understanding of the rights and responsibilities of citizens and helps make sense of the frequently unequal distribution of opportunities and constraints this creates for some. Geography also contributes greatly to the meanings of citizenship by revealing not only how people living in particular places (especially those living at the bottom rungs of society) are excluded from the bonds of common citizenship but also how those at the top sometimes participate in excluding their fellow citizens. The study of geography and citizenship invites one to continuously revisit and rethink what it is that constitutes today's meanings of citizenship—which is in itself an important step both in making active and participatory citizens and in creating a truly inclusive notion of citizenship. Geographical analysis of citizenship can help one consider the meanings of "belonging" and connectedness with places and spaces; it can also reveal the injustice of who may be excluded from the various rights and benefits that come with being a full citizen. As such, geographies of citizenship can be a valuable way to investigate social and economic justice, a major concern of geographers.

Thomas Chapman

See also Civil Society; Cosmopolitanism; Diaspora; Ethnicity; Globalization; Governance; Identity, Geography and; Immigration; Justice, Geography of; Law, Geography of; Migration; Political Geography; Social Justice; State

Further Readings

Desforges, L., Jones, R., & Woods, M. (2005). New geographies of citizenship. *Citizenship Studies, 9*(5), 439–451.
Mitchell, K. (2003). Educating the national citizen in neoliberal times: From the multicultural self to the strategic cosmopolitan. *Transactions of the British Institute of Geographers, 28,* 387–403.
Parker, G. (2001). *Citizenships, contingency and the countryside: Rights, culture, land and the environment.* London: Routledge.
Valentine, G., & Skelton, T. (2007). The right to be heard: Citizenship and language. *Political Geography, 26,* 121–140.

CIVIL SOCIETY

With the advent of globalization, and the hegemonic dominance of the market-based paradigm of neoliberalism, the term *civil society* has been increasingly popular, particularly in discussions of sustainability and social change in the developing world. Civil society refers to the sector of society existing separate and apart from the state while lying between the state and the individual, or family. Civil society consists of voluntary, or nonmarket, organizations; accordingly, corporations are not part of civil society. The term does not include political parties, which are more closely tied to attempts to influence the state. A variety of organizations, some as seemingly mundane as sport clubs and social associations, constitute civil society; the two most notable and important types are grassroots (or "people's") organizations and nongovernmental organizations (NGOs). The former represent groups of people, such as the urban poor, fisherfolk, and peasant groups, mobilized to pursue a specific cause; the latter are organizations possessing specialized skills, and knowledge, engaging in advocacy at a regional, national, or even international level. NGOs often act on behalf of grassroots organizations. The widespread acceptance of neoliberal policies has led to a profusion of civil society for two reasons. First, many civil society groups provide services for the poor and marginalized (such as soup kitchens and food banks) that were formerly provided by the state; second, many NGOs have come into existence to campaign against development projects (such as large-scale mines or hydroelectric dams) that have the potential to displace the poor and impinge on the environmental resources required for their subsistence. Some components of civil society also engage in advocacy against neoliberalism in general, with its attendant structural adjustment policies, as well as targeting specific projects.

The expansion of low-cost, high-quality telecommunications has greatly assisted the spread of civil society throughout the world. Communications media such as satellite, Internet access, and text messaging have given grassroots organizations in developing regions of the world immediate access to their NGO partners in the developed regions of the world. Such communication allows a prompt diffusion of news affecting people (e.g., environmental accidents) while also making campaign responses to these events easier.

Another event further assisting the global spread of civil society is the World Social Forum. Since its 2001 inception in Porto Alegre, Brazil, the World Social Forum has met at roughly the same time as the World Economic Forum in Davos, Switzerland. Meeting under the mantra of "Another World Is Possible," it has facilitated a veritable "clearing house" of information for civil society. Disparate groups, such as Peruvian *campesinos*, sub-Saharan herders, or Filipino fisherfolk, can meet and coordinate their campaigns, against multinational mining companies or genetically modified organisms, with NGOs from developed countries.

Civil society has its critics. Many allege that unlike corporations (which are accountable to their shareholders) or governments (which are accountable to voters), these organizations act in an accountability vacuum and can do as they wish. It is also alleged that NGOs are dominated by members of elite classes who manipulate their grassroots clientele to serve their own agendas and cater only to their donors, rather than being concerned with the interests of those they "represent." Critics of NGOs argue that as funding causes and recipient countries become more or less fashionable among donor groups, NGOs come and go in pursuit of donor funding; meanwhile, marginalized people who are affected by unfashionable causes and live in unfashionable countries are left

Telecommunications and civil society: A Filipino NGO worker sends a text message with her mobile phone. The satellite dish is used to communicate with a partner NGO in the United Kingdom.

Source: Author.

mired in poverty. Such criticism has been leveled against the world social forum in particular, as some allege that it has become a "festival" for NGOs, where they meet with each other, while the grassroots, whom they represent, are bypassed and excluded.

Notwithstanding these criticisms of civil society, many good things may be said about it. The cause-oriented activism of many groups has stopped undesirable projects from proceeding. In Peru, a highly organized international campaign prevented a Canadian mining company from demolishing half of the town of Tambogrande for the development of an open-pit gold mine. Civil society also provides a peaceful avenue for the poor and marginalized to have a voice that, otherwise, would not be heard; in the absence of these groups, opposition to neoliberalism may become more violent, and the repression of such groups may expand. Many NGOs have become heavily involved with the United Nations and other multilateral agencies and, in doing so, have influenced policy. Multinational corporations and governments, cognizant of the ability of these groups to quickly spread news of their behavior throughout what Marshall McLuhan called the "global village," are more mindful of what they do and are, more often than not, forced to seek a social license for their projects and policies, thus making civil society a major player in global governance.

William N. Holden

See also Governance; Neoliberalism; Nongovernmental Organizations (NGOs); Social Movements; State

Further Readings

Hilhorst, D. (2003). *The real world of NGOs: Discourses, diversity, and development.* London: Zed Books.

Silliman, G., & Noble, L. (Eds.). (1998). *Organizing for democracy: NGOs, civil society, and the Philippine State.* Honolulu: University of Hawaii Press.

Teivainen, T. (2002). The World Social Forum and global democratization: Learning from Porto Alegre. *Third World Quarterly, 23*(4), 621–632.

CLARK, ANDREW (1911–1975)

Andrew Hill Clark was a Canadian-born historical geographer who is widely regarded as the father of the subdiscipline in North America. Clark's career at the University of Wisconsin was cut short when he succumbed to cancer at the age of 64. Clark's development as a historical geographer took a circuitous path but ultimately led to teaching positions at the University of California at Berkeley, Rutgers University, and the University of Wisconsin. It was here that the evolution of his understanding of historical geography became most apparent. More important, perhaps, was his contribution to the discipline through supervision of 19 doctoral dissertations. Many of his students, including David Ward, Cole Harris, James Lemon,

Robert Mitchell, and Arthur Ray, became prominent historical geographers in their own right.

Clark earned a bachelor's degree in 1930 at McMaster University and later enrolled at the University of Toronto, completing his master's degree in 1938. He then received a fellowship to pursue doctoral studies at the University of California, Berkeley, with Carl Sauer. A 2-year teaching position at Canterbury University in New Zealand led to a dissertation on the impact of the introduction of plants and animals by Europeans to South Island, which he completed in 1944. His first academic position was at Rutgers, where he became chair of the geography department in 1949. Two years later, he was appointed to the faculty of the University of Wisconsin, Madison, where he taught until his death in 1975.

Clark's Major Monographs

During his career, Clark wrote three significant monographs and numerous articles on the impacts of European overseas colonization, served as the editor of the Historical Geography of North America series, and was the principal founder of the *Journal of Historical Geography*.

Clark's first book, *The Invasion of New Zealand by People, Plants, and Animals* (1949), provided a synopsis of the environmental changes brought to South Island by British settlers and their plants and animals. The book broke new ground, according to Donald Meinig, by viewing colonization as a process that transformed both the land and the newcomers. In *Three Centuries and the Island: A Historical Geography of Settlement and Agriculture in Prince Edward Island* (1959), Clark used evidence from more than 1,200 maps to document the changing character of the island through time. The uniformity of the island's climate and landforms made possible a comparative analysis of the agricultural practices of different cultural groups.

His final book, *Acadia: The Geography of Early Nova Scotia to 1760*, published in 1968, explores the evolution of Acadia as a region and studies its inhabitants, the Acadians, a French-speaking people who created distinct geographic patterns prior to their dispersal by the British in 1760. It was the Acadians' use of this distinct environment that played an important role in the development and evolution of this region.

Contributions to the Discipline

In 1954, Clark authored the chapter on historical geography in *American Geography: Inventory and Prospect*, which remained the seminal work on the evolution of the subdiscipline for 20 years and provided an assessment of the field in North America at that time. Clark concluded with a plea for more emphasis on the study and teaching of historical geography and training of students in historical geography to ensure the entire discipline's vitality.

Another notable work was his address as Honorary President of the Association of American Geographers, "*Praemia Geographiae*: The Incidental Rewards of a Professional Career," in 1962. One of those rewards was geographers' ability to identify dimensions that permit a holistic understanding of place. Clark noted that field observation was the greatest of these rewards and lamented its widespread decline. In his prologue to the collected essays in honor of Clark, Meinig argued that *Praemia Geographiae* will endure not because it was a philosophical treatise but because it was a very personal reflection on what it means to be a geographer.

Dawn S. Bowen

See also Historical Geography; Human Geography, History of

Further Readings

Clark, A. (1962). *Praemia geographiae:* The incidental rewards of a professional career. *Annals of the Association of American Geographers, 52*(3), 234–235.

Clark, A. (1968). *Acadia: The geography of early Nova Scotia to 1760.* Madison: University of Wisconsin Press.

Meinig, D. (1978). Prologue: Andrew Hill Clark, historical geographer. In J. R. Gibson (Ed.), *European settlement and development in North America: Essays on geographical change in honour and memory of Andrew Hill Clark* (pp. 3–26). Toronto, Ontario, Canada: University of Toronto Press.

CLARK, WILLIAM

Issues of mobility and migration have not only defined William A. V. Clark's research but also initiated his entry into the discipline of geography. After earning his BA and MA degrees from the University of New Zealand in 1960 and 1961, he received a Fulbright fellowship and traveled to the United States, where he received his PhD in geography from the University of Illinois in 1964. Following graduation, he returned to New Zealand, where he lectured in the geography department at the University of Canterbury until 1966. He came back to the United States, first to the geography department at the University of Wisconsin, Madison, in 1966, where he was appointed adjunct associate professor of urban planning in 1968, and then to the geography department at University of California, Los Angeles.

Los Angeles became his home and continues to provide a vantage point for observing the influence of demographic change associated with internal and international population migration flows in urban neighborhoods. Indeed, living in Los Angeles, a place in constant demographic flux, convinced Clark that residential segregation in the urban United States was not simply the outcome of white racism, although it certainly played a role in determining the patterns of black segregation. Rather, it was a much more complex story, including the influence of sometimes rapid demographic change, housing costs and incomes, residential preferences, and ties to the neighborhood. Clark's immense body of work continues to be highly influential and led to his election to the National Academy of Sciences in 2005, which is but one of numerous recognitions of the importance of his research.

Some additional highlights include the Decade of Behaviour Research Award and an Alumni Achievement Award from the College of Liberal Arts and Science at the University of Illinois. He is also an elected fellow of the American Academy of Arts and Sciences and the Royal Society of New Zealand. Clark also holds an honorary doctorate from the University of Utrecht and a DSc from the University of Auckland.

While the important contribution of Clark's considerable body of work has been widely acknowledged, it would be incorrect to assert that it has been received unquestioningly. Clark's work has been challenged and critiqued to varying degrees throughout his career. Specifically, his expert testimony on behalf of school districts in a series of cases on the role school districts played in residential segregation was very polarizing and placed him firmly in the middle of a contentious debate with regard to race relations and the legacy of racial segregation in the United States. Despite the criticism, Clark continued to argue that school districts could not be held responsible for neighborhood-level demographic change that resulted in changes to the racial composition of the school. This argument emerged from rigorous empirical analysis that highlighted the transformative power of demographic transitions on the residential mosaic and a belief that such evidence-based analysis is the foundation on which policy should rest.

In addition to his work examining the effects of demographic change on patterns of residential segregation and separation, Clark has also conducted a comprehensive macrolevel analysis of immigration in California and the United States. In this research, Clark finds the middle ground between two seemingly irreconcilable positions, that is, the view that immigration is a valuable resource and that immigration needs to be drastically curtailed because of the social and economic costs associated with it. In doing so, Clark highlights the geographically uneven and multiscalar nature of immigration, in which both costs and benefits coexist, albeit in different locations and at different scales.

Clark is currently a professor of geography and statistics at the University of California–Los Angeles and continues his work on the impacts of demographic change associated with migration to cities in the United States.

Valerie Ledwith

See also Ethnic Segregation; Migration; Mobility

Further Readings

Clark, W. (1991). Residential preferences and neighborhood racial segregation: A test of the Schelling segregation model. *Demography, 28,* 1–19.

Clark, W. (1992). Residential preferences and residential choices in a multiethnic context. *Demography, 30,* 451–466.

Clark, W. (1998). *The California cauldron: Immigration and the fortunes of local communities.* New York: Guilford Press.

Clark, W. (2003). *Immigrants and the American dream: Remaking the middle class.* New York: Guilford Press.

CLASS, GEOGRAPHY AND

There has been somewhat of a resurgence of interest in class across disciplines, for example, in feminist theory, sociology, and geography. In the United States, there have been several rich, ethnographic accounts of the continued effect of social class, as it intersects with other social positions, all of which challenge the popular myth of North America as a classless society. In the United Kingdom, there has been a stronger tradition of researching class, moving away from initial economist models to more cultural approaches that aim to uncover the interconnected subjective, material, and spatial components of class. The geographical intersection with class emphasizes the ways in which class literally takes place, on and in the landscape. Classed geographies are embodied in individuals' sense of place, in our feelings of belonging, and in our everyday identifications. They are materially enacted in the resources and opportunities that are available or denied to us, and here class division alludes to the stratification within this system. For example, we may see ourselves "located" in terms of our ability or inability to move neighborhoods, to access leisure facilities, to change employment, or to travel abroad. We may position ourselves as alike or dissimilar to our peers, and with an increasingly globalized reference system, different scales of comparison can be mobilized across time and place. Class is often seen as relevant to our uptake of and movements within space, from the everyday level of which neighborhood we live in and on what street, and the restaurants and art galleries we may or may not frequent, to a broader sense of globalized flows in the labor market and the division between the "global rich" and the "global poor." Yet defining class becomes difficult in a climate of supposed classlessness, where class inequalities are thought of as increasingly complex or nonexistent. Rather than moving away from class, many continue to research its fixity and its ability to fix—to deny opportunities in a new global market place.

Defining and Analyzing Class

Changes in the organization of production and consumption, it is often said, make it difficult to describe and analyze social class; occupations that have traditionally defined class have broken down and are now replaced by the service and information industries. Some commentators speak of the "power of flows" taking precedence over the "flows of power," where capital flows are spread throughout interconnected networks, creating a fast-growing electronic economy in which money is increasingly abstract and invisible. In this model, there is no such thing as a global capitalist class as the behavior of capitalists depends on submission to global networks, which is less secure than claims to the ownership of production, foregrounded in Marxist models of class. Here, it also becomes difficult to define who is working class, given the spatial and economic diversity of "workers."

Yet global flows in the location and dislocation of employment may work to sustain class structures rather than undermine them. The irregularity of the flows of globalization is often thought of in terms of increased flexibility or, indeed, renamed as improved "choices" for workers; workers have the "opportunity" to change their places of employment and residence. Such changes and the increased patterns of mobility are often contrasted with the "past" security of traditional working-class jobs, where workers were more spatially fixed, living and working in or near the same locale for their entire lives. Current disputed economic, social, and spatial transformations and continuations serve to redraw disciplinary lines around conceptualizations of class and geography, as the intersection between the two is debated from the macrolevel (such as international globalization and the shape of capitalist restructuring)

to the microlevel (the playing out of global capital on the bodies of specific workers and the changing geographies of inclusion/exclusion).

Some authors have pointed to the displacement of and postindustrial uncertainty about working-class masculinity, as a feminized service sector economy coincides with the production of "feminized" spaces of consumption. Such depictions of loss may at times serve to uphold and valorize a somewhat mythical and romanticized past, where traditional gender and class roles kept people "in their place." A more complex analysis, combining class and gender, highlights the relevance of class to women's employability within service sector transitions and the relegation of, particularly, low-skilled working-class women, to poorly paid, part-time, "pink-collar ghettos." Feminist geographers in particular have challenged the disembodied theories of capitalism, where gender has been frequently sidelined. Yet many have pointed to the role of reproduction in capitalist production; furthermore, processes of reproduction may be as geographically uneven as capital circulation, with domesticity and caring producing a certain class of female *and* male workers.

The disputes surrounding the definition and salience of class point to its social construction, intersecting with the social geographies of place and space, where the landscape is shaped and contested rather than just known. Importantly, the significance of class to geography lies in its attempts to highlight inequalities and ongoing structural patterns, amid economic, social, and spatial shifts. The relationship between class and geography potentially uncovers layers of meanings and sociospatial relations, and Doreen Massey speaks of the dynamics of spatial divisions, which are constantly restructuring themselves, rather than just statically existing. In this framework, such "layers" are conceptualized not just in terms of economics but also as cultural, political, and ideological, with local specificities. Capitalism, for example, is not simply a benign economic system, now responding to a global age; rather, it encapsulates and perpetuates ideas about value and skill, constructing spatialized social divisions, which are then neutralized as natural and in line with dominant political and cultural discourses (such as the "American dream").

Postindustrialism and Class

In postindustrial cities across the world, the transition to leisure-based, service sector economies, during which industrial jobs have all but disappeared, constitutes a major change, economically and culturally. Industrial legacies now appear as outdated histories that cities are often keen to cast aside in their attempts at urban regeneration, producing newly sanitized and upgraded spaces for consumption and leisure (such as shopping and housing complexes). Dualistic properties are often evident in the "regeneration" of urban landscapes, where cities seemingly function as universally accessible, offering generic services and spaces (shops, cafés, select green spaces). Against this apparent sameness, each place may well have its own brand that is carefully managed and shaped to project just the "right" sort of image to the outsider, generating interurban competitiveness on a global scale: What makes city spaces attractive to residents, employees, and tourists? What differentiates such spaces from the suburbs? From the rural?

Such processes and tensions are suggestive of new developments and forms of sociospatial division in postindustrial society, where class is not simply compositional of income and occupation but also a result of vested interests in space and place. As the cultures and traditions of a city are repackaged and reimaged, marketing and advertising make it possible to project a representation of the city that serves to include some of its inhabitants and further exclude other "undesirables," which are classed in specific ways. For example, working-class youth may be specifically excluded from the more prestigious shopping centers and their presence queried elsewhere, assumed to be threatening embodiments of urban disorder. While space and place are shaped, branded, and reimaged to create the "urban dream" (with emphasis on art, culture, consumption, and a café lifestyle), divisions are perhaps reconsolidated, with inhabitants becoming more socially segregated and culturally differentiated perhaps than ever before.

Life Chances and the Sense of Belonging

The impacts of spatial segregation and exclusion can be vividly witnessed in terms of uneven access to crucial social goods, for example, education, health, housing, and employment, where class continues to be a reliable predictor of life chances. The influence of class and geography is apparent when location frequently acts as the basis of access, and inequalities arise from and compound each disadvantage. For example, underachieving schools are frequently situated in deprived areas, while middle-class parents can exercise more choice by moving to a "good" area with a good school (also known as "white flight" in the United States, where white parents sought to move away from black inner-city locations/school). Frequently, working-class areas are represented in metaphors of drowning ("sink" estates) and defecation ("bog standards"), which in neoliberal times, "responsible" parents must avoid, enacting their own good sense rather than relying on welfare intervention. In both the United States and the United Kingdom, social policy frequently invokes notions of "social capital," whereby local networks and communities are felt to add value, embedding individuals in friendship and familial support systems. Viewed more skeptically, this may be seen as an individualist response to structured inequalities, where "the poor" are positioned as failing to generate empowering "social capital," repeatedly situated as failing bodies in failing places.

Location is related to opportunity structures and, more subjectively, to a sense of one's place, as against the place of "others." Areas can be demarcated not only in terms of their facilities and amenities but also in terms of their inhabitants, whose values, lifestyles, and "respectability" are often read through locale. The boundaries of identification (boundaries between "us" and "them") regulate who can fit in these places, with economic, social, and cultural resources embodied in the construction of "professional," "respectable," and "real" identities. The spatialization of class is also apparent in the British terms *Essex girls* and *Scousers* and in the U.S. terms *white trash, hillbillies, rednecks,* and so on, where character is read and assumed from locale (see photo). Furthermore, several researchers have proposed that everyone living in a city has knowledge of "ghetto" estates and the level of danger and exclusion each area presents, where working-class areas are frequently represented as more undesirable and

Hubcap Hurling winner Tater Yarbrough of Jesup, Georgia, throws his last hubcap during the 11th annual Summer Redneck Games, July 8, 2006, in Dublin, Georgia. Started in 1996 as a spoof for the summer Olympics held in Atlanta, the games feature bobbing for pigs' feet, hubcap hurling, and the redneck mud pit belly flop contest for trophies.

Source: Getty Images.

dangerous. New technologies arguably intensify such demarcations, with Web sites and television programs on where to buy and which places to avoid, implicitly aimed at a knowing middle-class consumer. Such spatial segregation fosters new exclusions with the dynamics of capitalist restructuring reproduced at the microlevel, across space and onto the inhabitants who variously occupy it.

The *appearance* of class is written on the landscape and on the bodies of subjects; it is the "fleshy" and "messy" stuff of everyday life. The way one looks is frequently equated with being someone or no one. Such a focus draws on varying theories of identity in space, particularly the ways certain bodies, appearances, and identities are rendered unentitled to occupy a space because of the lack of capitals, bodily or otherwise, to legitimately access that space and to receive interpersonal affirmations and

entitlements within it. Consider, for example, the space of higher education: Who is the typical, entitled university student? What is expected, measured, valued? Pierre Bourdieu makes apparent the classing of spaces and subjectivities, using the concepts of capital and the classed habitus. Bourdieu's model of classed habitus and classed capitals provides a sense of the ways class still affects access to positions, traveling through space. The habitus may be understood as a second skin that is carried around, retaining the "past" as it affects our present; put simply, we carry our history with us, and it affects our objective location and our sense of place in the world. Having the "right" cultural, economic, and social capitals (which when legitimated turn into "symbolic capital," a resource that the working classes "lack") produces opportunities and advantages across various social spheres.

Speaking of trajectories and fields enables an understanding of how bodies have access to different amounts of capitals. This is useful in thinking about how movements through spaces are constituted, facilitated, or impeded. It also points to the entrenched, emotional "value" of spaces, relevant to how social space is viewed and negotiated. Capitals are classed: The distribution of capital runs from those who are best provided with both economic and cultural capital to those who are deprived in both respects. This model of embodied capitals, where resources are attached to classed individuals, challenges straightforward notions of "upward mobility" as gradations of social status continually inform and prescribe movement through space. This contrasts with straightforward, disembodied notions of mobility with its "rises" and "falls," producing a one-dimensional view of social space and ignoring conversions and reproduction strategies. Perhaps the more we have, the easier it is to have more.

Bourdieu's theory of classed capitals and classed habitus is framed solely in terms of class, having little to say about gender or sexuality. The relationship between class and geography, in terms of economic restructuring, opportunity structures, capitals, and identifications, can be queried by an intersectional focus on, for example, class, gender, and sexuality. The geographies of class are complex and inevitably intersectional. Consider, for example, the regeneration of city space previously discussed as also involving the promotion of commercialized lesbian and gay spaces as "niche markets." Borrowing from the success of gay districts in "world cities," such as New York, Sydney, or Paris, "wannabe world cities" have engaged in competitive strategies to attract capital. Such cities have, with varying success, re-created themselves as places of culture and consumption, laying claims to a certain cosmopolitanism that places them as participants in the global economy. Several authors have commented that such regenerations are classed, involving an expulsion of "undesirables" in upgraded leisure space. Here, the connection between the material structuring of space and the everyday movements in and out of space can again be witnessed.

Yvette Taylor

See also Class, Nature and; Circuits of Capital; Consumption, Geographies of; Crisis; Economic Geography; Ethnic Segregation; Everyday Life, Geography and; Feminist Geographies; Gender and Geography; Giddens, Anthony; Harvey, David; Human Geography, History of; Identity, Geography and; Inequality and Geography; Justice, Geography of; Marxism, Geography and; Neoliberalism; Patriarchy, Geography and; Poststructuralism; Race and Empire; Race and Racism; Racial Segregation; Resistance, Geographies of; Segregation and Geography; Smith, Neil; Social Justice; Structuration Theory; Uneven Development; Urban Underclass; Walker, Richard

Further Readings

Adkins, L., & Skeggs, B. (Eds.). (2004). *Feminism after Bourdieu*. Oxford, UK: Blackwell.

Beck, U. (2000). *What is globalization?* Oxford, UK: Blackwell.

Bettie, J. (2002). *Women without class: Girls, race, and identity*. Berkeley: University of California Press.

Bourdieu, P. (1984). *Distinction: A social critique of the judgment of taste*. London: Routledge.

Hennessy, R. (2000). *Profit and pleasure: Sexual identities in late capitalism*. London: Taylor & Francis.

Kefalas, M. (2003). *Working-class heroes: Protecting home, community, and nation in a Chicago neighborhood*. Berkeley: University of California Press.

Lareau, A. (2003). *Unequal childhoods: Class, race and family life*. Berkeley: University of California Press.

Massey, D. (1995). *Spatial divisions of labour*. New York: Routledge.

Reese, E. (2005). *Backlash against welfare mothers: Past and present*. Berkeley: University of California Press.

Taylor, Y. (2007). *Working-class lesbian life: Classed outsiders*. New York: Palgrave Macmillan.

Urry, J. (2000). *Sociology beyond societies: Mobilities for the twenty-first century*. London: Routledge.

Zweig, M. (2000). *The working-class majority: America's best kept secret*. Ithaca, NY: ILR Press.

CLASS, NATURE AND

Taken separately, class and nature constitute two of the most significant categories of analysis within the geographical tradition. As a marker of different forms of social stratification, class has provided a key fulcrum around which political and economic geographers have sought to understand the life paths taken and "life chances" presented to different people. As a complex indicator of both the essential essence of existence and the environmental fabric of being, nature has provided a multidimensional lens for geographers exploring areas as diverse as race relations, the colonial imagination, and late modern forms of environmental change. In these contexts, both class and nature embody key objects of analysis that geographers seek to understand and, at the same time, important conceptual categories for interpreting the operation of various forms of socioeconomic and ecological process. When considered together, the varied relations between class and nature have provided a rich terrain of critical analysis that has, perhaps more than any other arena of study, characterized the contribution of human geographers to the study of the environment.

Marxist Views of Class and Nature

The first systematic, geographical analysis of the relations between class forces and nature was presented in Neil Smith's influential volume *Uneven Development: Nature, Capital and Production of Space*. Smith deploys a Marxist theory of class to reveal the ways in which the social and geographical structures of capitalist society have shaped, and have in turn been conditioned, by certain economic appropriations of nature. Karl Marx's 19th-century reflections on human relations with the natural world were important because they marked a sharp break from the prevailing tendencies at the time to see nature as an object of analysis within the physical sciences but not the fields of economics and politics. In keeping with his broader modes of dialectical thought, Marx asserted that nature was not an eternally unchanging backdrop against which human history is played out but a historically contingent product of prevailing political and economic systems.

It is important to note that the varied social relations associated with a capitalist economy (including systems of market exchange, private property, and regimes of accumulation) have all critically shaped modern relations with the natural world. Smith asserts that it is the particular class formations of capitalism that have determined the specific ways in which modern industrial societies apprehend and transform nature. While Marx himself identified a plethora of socioeconomic classes that were characteristic of capitalism, it is the broad division between the proletariat and the bourgeoisie that Smith focused his analysis on. Following Marx, Smith interpreted the bourgeoisie to be the strata of industrial society that own the means of economic production (viz., agricultural land, factories, mines, etc.) and the proletariat as those whose only property is their own bodies and the associated labor they can sell. While these class positions may be familiar enough, the impacts that they have on socionatural relations are perhaps less easy to discern. It is the various processes of alienation that, in different ways, hold the key to interpreting the ecological significance of such class positions.

Despite working intimately with the things of nature (soils, minerals, timber, livestock, etc.), Marxists argue that the proletariat has historically been alienated from nature to the extent that the construction of their class position has involved the violent expulsion of workers from the collective environmental resources (including pastures, woodlands, and watercourses) that they once shared and used. Where nature was once used for the daily production of sustenance, and respected accordingly, under industrial capitalism, it becomes a thing to be rapidly transformed for wage and exchange. It is not difficult to imagine how the collective management of nature for local use produces a very different ethic of environmental care and understanding from a system that is based on private ownership, waged labor, and the extraction of ecological value for commodity exchange.

In contrast, Marxists claim that the alienation of the bourgeoisie from nature has been predicated not on the demands of waged laboring but on the physical and emotional distance that tends to emerge between propertied classes and the natural world under capitalism. This distance is, in part, based on the geographical separation that patrician classes experience as a consequence of the rapid growth of cities under capitalist social and economic relations. At a less obvious level, this distance is also a product of the elaborate industrial complexes of ecological transformation and supply and the armies of labor associated with capitalism, which work to collectively reduce the need of the bourgeoisie to engage with nature within the processes necessary for their day-to-day reproduction (e.g., eating, drinking, heating their homes).

It is in these contexts that capitalism produces a double motion of alienation from nature. For the working classes, their alienation is more emotional than physical, as they are forced to reevaluate their relations with nature from which they have been legally dispossessed. For the patrician classes, alienation tends to take a much more physical form as they are geographically and economically divorced from contact with the nature on which their existence, and wealth ultimately depend.

External Versus Universal Ideologies

Smith's analysis, however, goes beyond the simple recognition that the class positions associated with capitalism tend to produce differences in the ways in which strata within society engage with nature. Smith claimed that the evolving relations between different classes and nature produce novel ways of understanding the natural world and society's relationship with it. Smith isolated two hegemonic ideologies of nature that have emerged directly from the emergent class positions of capitalism: external ideologies and universal ideologies.

External Ideologies

External ideologies of nature posit the natural world as something that is essentially separate from the human condition. While a belief in the binary distinction between society and nature is not a unique feature of capitalism, it is not difficult to understand how the alienating ecologies of capitalist economies have supported a popular belief in the essential distinction between human and nonhuman worlds. The ideological construction of nature as an external realm, which is emotional and practically disconnected from human experience, has been central to the ways in which complex understanding of nature have been replaced by its much narrower construction as an economic resource. Belief in the essential disconnection between social and natural systems has also been crucial in legitimizing capitalism's domination of the environment in the pursuit of ever-expanding profits.

Universal Ideologies

In contradistinction to external views of the natural world, Smith identified universal ideologies of nature. According to Smith, universal ideologies of nature situate the human species as one among many species in the totality of nature. While universal ideologies of nature also have long precapitalist histories (Smith traced them back to Immanuel Kant, but they structured medieval cosmologies as well), Smith recognized how they were coopted by urban bourgeois elites in their desire to reconnect with the natures they had lost. If humans are just one part of the totality of nature, then reconnecting and respecting nature appear central to the *species-being* of humans. In this context, the rise of capitalist ideologies of universal nature is closely tied to the new discourses of environmental stewardship and care that are associated with the green movement. Notwithstanding this connection, it is also important to note how universal ideologies of nature have been used to justify the class divisions of capitalist society (if humans are a part of nature, then class divisions could be interpreted as being as inevitable as other divisions in the natural world).

Ultimately, the pioneering work of Smith reveals that even though contemporary ideologies of nature predate the class structures associated with capitalism, modern industrial society has witnessed the reconsolidation of certain views of the natural world that both support capitalist economic systems and flow from its contradictions. It is important to note how both external and universal ideologies of nature have been used, despite their oppositional nature, to support and sustain the ecological relations of modern industrial society.

Additional Views of Class and Nature

The broad-ranging reanalysis of class-nature relations first conducted by Smith in the mid 1980s has provided the basis for a rich array of allied analyses of socionatural relations within geography. At one level, questions of class and nature have emerged as a crucial consideration within the work of urban political ecology. Within this broad, also Marxist-inspired, school of analysis, broad questions of class and nature reappear under the rubric of the differential access to resources experienced by different members of the metropolitan community. Consequently, whether it be in terms of access to affordable food, clean water, breathable air, or recreational space, the role of class in determining the differential ability of people to metabolize nature has reemerged as a central theme within urban geography.

Other work, inspired by the politics and philosophies of the environmental justice movement has, however, started to question the centrality of both class and nature as central concerns within the constitution of urban environmental equity. In constructing a distinctively urban-based brand of environmental politics and thought, the environmental justice movement recognizes the importance of class as a factor within the differential environmental conditions of life experienced by metropolitan denizens. Notwithstanding this, one of the key implications of the rise of the environmental justice movement for geography has been a heightened concern with the role of race and gender in the determination of environmental opportunity. Furthermore, the rise of the environmental justice agenda within geography has led to a renewed concern with the role of class, race, and gender within an environmental politics that extends beyond questions of nature (e.g., to include housing quality, workplace environments, and the environmental qualities of public space). To these ends, the environmental justice movement has exposed the ways in which systems of capitalist production and class stratification tend to produce environmental problems that are both ecological and social in their form.

Despite its close association with the work of Marxist geographers, it would be wrong to associate discussions of the relationships between class and nature exclusively with Marxist thought. Work on the form and meaning of landscape within geography has, for example, exposed the intricate braiding of class values and natural landscapes. The work of Denis Cosgrove and Stephen Daniels has revealed the ways in which the landscaping of nature (whether it be in urban parks and gardens or the parklands of wealthy landowners) belies certain bourgeois aesthetics and tastes and is an enduring attempt to mark out class distinctions within the scenery of everyday life. Beyond the deliberate manufacturing of nature within the arts and sciences of landscaping, geographers have also outlined how the selection of zones of nature for conservation (whether in the form of greenbelts, game reserves, national parks, or areas of outstanding natural beauty) reflects the projection of certain elite class values into the countryside and the associated subjection of working environments to bourgeois control.

Mark Whitehead

See also Class, Geography and; Critical Studies of Nature; Distribution of Resource Access; Ecological Imaginaries; Environmental Justice; Ethnicity and Nature; Gender and Nature; Marxism, Geography and; Nature; Nature-Society Theory; Political Ecology; Race and Nature; Smith, Neil; Social Construction of Nature; Uneven Development

Further Readings

Burkett, P. (1999). *Marx and nature: A red and green perspective.* London: Macmillan.
Cosgrove, D., & Daniels, S. (Eds.). (1989). *The iconography of landscape: Essays on the symbolic representation, design and use of past environments.* Cambridge, UK: Cambridge University Press.
Gottlieb, R. (2005). *Forcing the spring: The transformation of the American environmental movement* (Rev. ed.). Washington, DC: Island Press.
Harvey, D. (1997). *Justice, nature and the geography of difference.* Oxford, UK: Blackwell.
Heynen, N., Perkins, H. A., & Roy, P. (2007). Failing to grow "their" own justice? The co-production of racial/gendered labour and Milwaukee's urban forest. *Urban Geography, 28,* 732–754.
Keil, R. (2003). Progress report: Urban political ecology. *Urban Geography, 24,* 723–738.
Pulido, L. (1994). Restructuring and the contraction and expansion of environmental rights in the United States. *Environment & Planning A, 26,* 915–936.
Smith, N. (2008). *Uneven development: Nature, capital, and production of space* (3rd ed.). Athens: University of Georgia Press.

CLIENT-SERVER ARCHITECTURE

In the client-server architecture, clients and servers communicate with one another via a network connection. This architecture is composed of a distributed computing system with software on both the client and the server. The client may initiate a session while the server sits waiting for requests from clients. A server can be used by one or many clients depending on its intent and capabilities. A client can also use one or many servers depending on its intent and capabilities.

Common Examples

Common examples of client-server architectures include e-mail exchange, database access, and Web access. The most obvious example is Web access. In this case, the Web browser (e.g., Microsoft Internet Explorer or Mozilla Firefox) is the client. When one types a Web address into the Web browser, the browser (the client) initiates a connection to the Web address (the server). The server responds based on the request sending back information to be displayed in the Web browser. In this example, the response is typically a Web page.

Geospatial Examples

Geospatial examples include Google Maps and Microsoft Virtual Earth. Both of these examples use a Web browser as the client to connect to servers hosting large amounts of geospatial data that are returned, in the form of maps, and displayed by the server. Another common geospatial example is virtual globes, such as NASA World Wind and Google Earth. In this case, the client no longer uses a Web browser; rather, a custom application is installed. Both NASA World Wind and Google Earth have highly customized servers that are designed to provide spatial information to a specific client; this creates the base data for the virtual globe. These clients also have the ability to connect to any number of other servers that adhere to certain Web standards. Additionally, desktop geographic information system (GIS) software, such as ArcGIS Desktop (from the Environmental Systems Research Institute), ENVI (from ITT), and Quantum GIS (open source), can be considered clients in client-server architecture. All these products have the ability to connect to both local and remote sources of data. A user can access data in a remote database or via geospatial Web standards, and in doing so, these software products are acting as clients contacting a server. In these cases, the data being returned are often very different from what is returned to a Web browser as these clients can deal with many complex remote sensing and GIS data formats. However, the underlying architecture remains the same—the client makes a request to a server that is awaiting such requests, and the server processes the request and returns a response.

Open Geospatial Consortium Standards

The client-server architecture relies on standards for communication for clients and servers to easily communicate with one another. In the case of general Web communications, this standard is the Hypertext Transfer Protocol (HTTP). Additional standards can be used in addition to HTTP for specific types of communications. The Open Geospatial Consortium (OGC) has developed and maintains a number of standards for requesting and receiving geospatial data via client-server architecture. While these standards are too numerous and complex to cover them all in detail here, a brief overview of some of the more common standards is warranted. The Web Map Service (WMS) is used for returning maps as images. It is important to note that the returned map is not data; it is just a picture of data. The Web Feature Service (WFS) is used to request/return feature data such as points, lines, and polygons. The Web Coverage Service (WCS) is used to request/return raster data. Unlike WMS, both WFS and WCS return actual data, not just images. These services among others are commonly used to allow clients to request data from servers via the Web.

Thin Clients Versus Thick Clients

There are two readily accepted types of clients, often referred to as "thin clients" and "thick clients." A Web browser would be an example of a thin client. In this case, the server does all the processing, while the Web browser (the thin client) is only concerned with sending input to the server and displaying the response. A desktop GIS product is an example of a thick client. In this case, the client can do a great deal of processing on its own and is not reliant on a server connection. A Web browser is highly reliant on the server for content to display, while a desktop GIS product has a great deal of functionality that is not dependent on the server. The key difference is that the thick client uses much more of the local computer's resources, while a thin client relies heavily on the server to do the majority of the processing. The capabilities of the thick clients vary greatly. In the case of desktop GIS products, a user could carry out entire projects without ever connecting to a server as a client. However, virtual globes such as Google Earth (also a thick client) rely heavily on the server for content but use much more of the local computer resources for processing, displaying, and interacting with the data than a common Web browser. Unlike a desktop GIS product though, virtual globes are not as useful without a connection to a server for content.

Stateless Versus Stateful

The connections established between client and server can be stateless or stateful. In a stateless transaction, the client sends a request, and the server responds. After sending its response, the transaction is complete, and future transactions in no way depend on the previous transaction. In a stateful transaction, the server remembers the client between requests (e.g., by using cookies). An example of this might be an individual visiting the Web site of a data portal and logging in to his or her account. The server creates a session with the client, and it knows certain information (in this example perhaps name, affiliation, data type preferences, previous data ordered, etc.) about the user. In this case, one request may depend on a previous request. For example, the client might request the addition of a certain data file to a shopping cart. The server could respond with a list of items in the shopping cart. In the next request, the client could continue browsing for more data or choose to request the data. The server could carry out this request by preparing

the data for download and responding with a location to download the requested data files.

Synchronous Versus Asynchronous

Connections between clients and servers can be synchronous or asynchronous. A Web browser makes a good example. In the case of a synchronous connection, when the client requests new information from the server, the displayed page is replaced by a new page returned by the server. This new page could be completely different, or it might only change a small amount of the content on the page. In the case of an asynchronous connection, the client can send a request to the server in the background without changing the display or behavior of the current page. On receipt of the server's response, the client can update a small part or the entire page depending on the request and response. A good example of this would be an interactive mapping site. A map may be embedded in a Web page. Using a synchronous connection, zooming in on the map would require reloading the whole page with new information. Using an asynchronous connection, the user could specify an area to zoom in to, and the client could request an update to the map from the server. In this case, the server would respond with an updated map, but the rest of the Web page would be unchanged.

Distributed GIS

Recent advances in network infrastructure have allowed the rise of technologies such as virtual globes, which require intensive network communications between the client and the server. Many have attributed the rise of virtual globes with changing the way people (geographers and non-geographers alike) access and visualize data. These advances also allow a move toward distributed GIS, which is also inherently based on a client-server architecture. In this case, a client could connect to a server to request actual GIS processing tasks as opposed to requesting only data. For a simple example, a GIS client might make a request to a server that included a set of points that the user wanted to perform a spatial interpolation on. Rather than having complex software with this capability, the user could submit the data and the request to the server, the server would carry out the interpolation and return an interpolated surface as its response. However, moving around data sets such as this can require a great deal of network bandwidth and server processing power. However, it seems that current advances in infrastructure, servers, and software are bringing us closer to wide-ranging distributed GIS services.

Sharolyn Anderson and Benjamin T. Tuttle

See also Database Management System; Distributed Computing; Geospatial Semantic Web; GIS Web Services; Google Earth; Internet GIS; Neogeography; Open Geospatial Consortium (OGC); Virtual Globes

Further Readings

Bambacus, M., Yang, P., Evans, J., Cole, M., Alameh, N., & Marley, S. (2007). An interoperable portal supporting prototyping geospatial applications. *URISA Journal*, 19(2), 33–39.

The Open Geospatial Consortium, Inc.: www.opengeospatial.org

Peng, Z., & Tsou, M. (2003). *Internet GIS*. Hoboken, NJ: Wiley.

Scharl, A., & Tochtermann, K. (Eds.). (2007). *The geospatial Web: How geobrowsers, social software and the Web 2.0 are shaping the network society*. London: Springer.

Tuttle, B. T., Anderson, S., & Huff, R. (2008). Virtual globes: An overview of their history, uses, and future challenges. *Geography Compass*, 3, 1478–1505.

CLIMATE: DRY

Although the scarcity of water resources in dry climates makes them inimical in modern times for habitation and settlement, dry climates were more favorable for early civilizations. Dry climatic conditions provided a suitable niche for the establishment of Neolithic and Chalcolithic settlements as debilitating waterborne diseases such as malaria and schistosomiasis, which inflict the more humid areas, were less common in arid and semiarid zones.

Flaming mountain, Xinjiang, in northwestern China. Cut off from marine climatic influences and surrounded by high mountains, the region has a dry climate and is extremely hot in summer.

Source: Xiaoping Liang/iStockphoto.

Dry climates have unique qualities, both majestic and fearsome, that have intrigued travelers and investigators alike (see photo). Some hyperarid climates have almost Martian-like landscapes with distinct aeolian features. These climates pose a potential danger for those attempting to travel there because of harsh and extreme weather elements such as high temperature, solar radiation, and lack of humidity. People living in this climate regime were forced to adapt to the very meager, sporadic, and uncertain land productivity and had developed a nomadic lifestyle where grazing was, and in some areas is still, the only economic activity.

Definition

A dry climate is defined simply as an area with perennial water deficit, little biological productivity, little carrying capacity, and restricted human

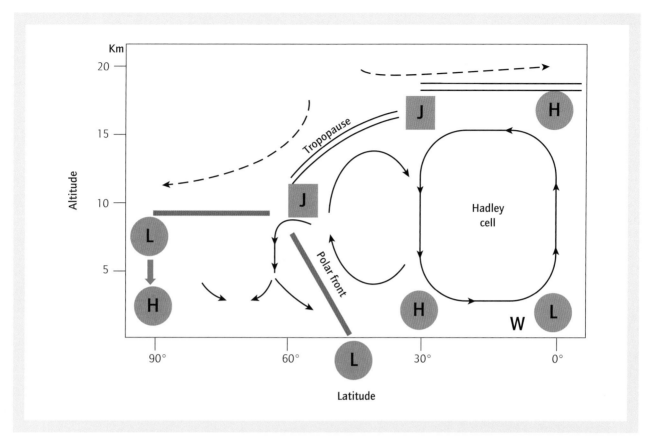

Figure 1 A schematic diagram showing the general circulation patterns along with latitudinal pressure distribution

Source: Author.

activities. According to the Köppen classification, there are five major climate regimes: tropical rainy, dry, humid temperate, humid cool, and polar. Each regime has subdivisions based on precipitation distribution and/or temperature regimes. A dry climate is simply defined as that which has a substantial negative water balance (precipitation is much less than potential evaporation), either permanently or for the larger part of the year.

There is a clear difference between aridity and drought. The first term refers to a condition in which long and sustained conditions of low and highly variable precipitation are the norm in a given area. For instance, the Sahara in North Africa has been typified by very low precipitation for the past several thousand years, and thus, the climate of the Sahara is either arid or hyperarid. On the other hand, drought is a temporary weather condition in which precipitation ceases for a long enough time to cause agricultural, hydrological, and/or meteorological drought. Drought occurs

almost everywhere, but its occurrence is more frequent and severe in marginal climates, where precipitation tends to fail more often (e.g., marginal areas separating two contrasting climates—the Great Plains and the Sahel region in Africa).

Location of Dry Climates

There are distinctive geographic zones where dry climates prevail. The major belt of dry climates is located in the subtropical zones between about 20° and 30° north and south of the equator (Figure 1). In the Northern Hemisphere, this includes the Great Sahara in North Africa, the Arabian Peninsula, the Gobi desert, the Sonora desert in North Mexico, and the southwest of the United States, and in the Southern Hemisphere, it includes the Atacama and Patagonia deserts in South America, the Namibian and Kalahari deserts in Southern Africa, and the Australian desert. The reason for the formation of these deserts in these

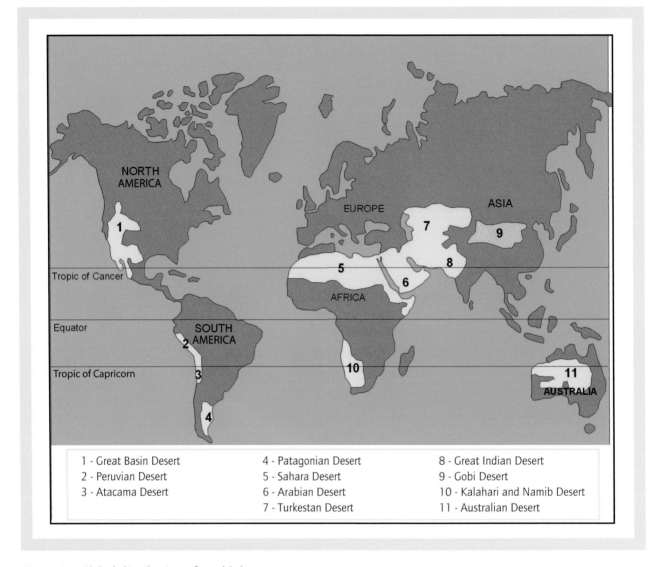

Figure 2 Global distribution of world deserts

Source: Author.

zones is directly connected to air subsidence manifesting the general circulation (see Figure 2). Irregular land-ocean distribution, however, breaks up this high-pressure zone into cells that allow some areas to receive ample amount of rainfall (Hare, 1983). When an air parcel subsides, it heats up adiabatically, and thus, its relative humidity decreases, leading to diminishing chances for cloud formation. These zones are typified by a high atmospheric pressure and become sources of air masses.

In addition to the general circulation, there are three other factors that contribute to the formation of dry climates, albeit at a smaller scale. These include the following: (1) sheer remoteness from moisture sources (e.g., Northern Asia, the Central Sahara, and Central Australia), (2) location in the lee wind (e.g., the Jordan Valley; Nevada and Idaho, East Oregon, and East Washington, which are located behind the Sierra Nevada and the Cascades), and (3) the presence of cool oceanic currents (compare the precipitation regime over Florida on the one hand and southwestern California and Baja California on the other, southeastern South America and its western part: the Chilean coast,

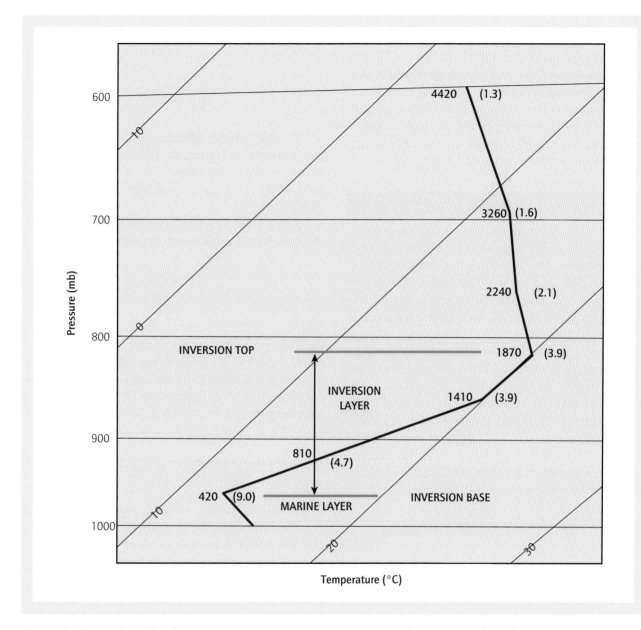

Figure 3   Vertical profile of air temperature and moisture within the lower troposphere for a marine system within the arid zone

Source: Adapted from Krishnamurti, T. N. (1979). *Compendium of meteorology* (Vol. 2, Pt. 4, Tropical Meteorology, WMO No. 364). Geneva: World Meteorological Organization.

Zimbabwe and Angola, the southwestern parts of Australia and its southeastern parts). Although these paired areas are approximately located at the same latitudes, the eastern parts enjoy plenty of precipitation due to the warm oceanic currents, which encourage deep convective clouds, whereas the cool oceanic currents in the western parts cause a persistent inversion layer deep enough to suppress any vertical development, leading to dry conditions for the most part of the year. Figure 3 shows the vertical temperature and moisture profile within the lower troposphere over areas experiencing cold oceanic currents along with air subsidence. The figure shows that air temperature increases from about 12 °C at 420 m (meters) at the inversion base to about 20 °C at 1,870 m at the inversion top. Conversely, atmospheric moisture decreases from 9 g/kg (grams per kilogram) at 420 m to 3.9 g/kg at the inversion top. This sharp moisture reduction indicates a significant detachment between the base of the inversion and the rest of the atmosphere due to strong mixing suppression resulting from negative buoyancy. This typical atmospheric profile ($\partial\theta/\partial z > 0$) with its deep inversion thickness suppresses any vertical movement, which in turn results in the predominance of aridity.

Dry Climate Indices

Climate classification serves many objectives, including the following: (a) to establish a distinct spatial pattern, (b) to reduce the number of elements needed, and (c) to characterize association between/among phenomena (e.g., flora and climate). Many investigators have attempted to classify dry climates (e.g., Köppen, Meigs, Thornthwaite, Budyko). Wladimir Köppen classified the world climates into five major classes, and each of them was given a letter designation code: tropical humid (A), dry (B), warm temperate (C), cool temperate (D), and polar (E). Dry climates are empirically determined based on precipitation amount, its annual distribution, and the average annual temperature. Köppen proposed the following criteria for identifying dry climates:

$$\frac{P}{10} < (2T + 28), \text{ 70\% of } P \text{ in the warm season,} \quad (1)$$

$$\frac{P}{10} < 2T, \text{ 70\% } P \text{ in the cool/cold season,} \quad (2)$$

$$\frac{P}{10} < (2T + 14), P \text{ distributed throughout the year,} \quad (3)$$

where P and T are the average annual precipitation (in millimeters [mm]) and air temperature (degrees Celsius [°C]), respectively. This climate is further subdivided into two subclasses based on precipitation, dry (Bw) or semidry (Bs). A climate is semidry steppe (BS) if precipitation exceeds one half the temperature criteria mentioned in Equations 1 to 3:

$$\frac{P}{10} > (T + 14), \text{ 70\% of } P \text{ in the warm season,} \quad (4)$$

$$\frac{P}{10} > (T), \quad (5)$$

$$\frac{P}{10} > (T + 7). \quad (6)$$

Dry and semidry climates are subdivided either into dry/semidry tropical (BW$_h$ or BS$_h$) if the average annual temperature ≥18 °C or dry/semidry cold if the average annual temperature <18 °C (BW$_k$ or BS$_k$).

C. W. Thornthwaite attempted to classify dry climates based on a moisture index,

$$I_m = \frac{100s - 60d}{PE}, \quad (7)$$

where I_m is the moisture index (dimensionless), s the water surplus (mm), d the water deficiency (mm), and PE the annual potential evaporation, determined primarily by air temperature. The calculations of s and d are made on a monthly basis for the entire year.

$$s = \sum_{i=1}^{12} (P_i - P_{E_i}), \quad d = \sum_{i=1}^{12} (P_i - P_{E_i}). \quad (8)$$

Semiarid climates, termed *humidity provinces* by Thornthwaite, have a short growing season that sustains the growth of short grass, the dominant natural vegetation. For instance, in semidry areas within the Mediterranean region, short perennial grass grows during the period from February through the middle/end of March. Dry climates have, on average, moisture deficiency year-round, $\sum s_i = 0$. Annual potential evaporation there is usually high and frequently exceeds 2,000 to 2,500 mm in most tropical desert areas. Thus, the moisture index is highly negative and could exceed −50. The above classifications are empirically based.

M. I. Budyko provided a generic classification based solely on precipitation and net radiation as a surrogate for potential evaporation. Budyko defined an aridity index (Φ),

$$\Phi = \frac{R_n}{\lambda_E P}, \quad (9)$$

where λ_E is the latent heat of evaporation (joules per kilogram) and P is the average annual precipitation. R_n is the net radiation (watts per square meter), defined by

$$R_n = S(1 - \alpha) + \varepsilon_a \sigma T_a{}^4 - \varepsilon_s \sigma T_a{}^4 \qquad (10)$$

where S is the solar radiation, ε_a and ε_s the atmospheric and surface emissivities, respectively, σ the Steffan Boltzmann constant, and T_a the air temperature. For very humid areas, $\Phi > 1$; for intermediate climate, $\Phi \sim 1$; and for dry areas, $\Phi < 1$. In very dry areas, Φ could simply reach 40. When Φ exceeds 10, the area potential becomes extremely small, and as such, natural vegetation growth is restricted to wady beds and alluvial fans.

Dry areas are classified into subcategories based on the degree of aridity. The following, from the United Nations Environment Programme's *World Atlas of Desertification*, provides the aridity index along with the percentage of global land area covered by each category:

Category	Aridity Index (Φ)	% Global Land Area
Hyperarid	>20	7.5
Arid	$20 < \Phi < 5$	12.1
Semiarid	$5 < \Phi < 2$	17.7

The main shortcoming of the aridity index proposed by Budyko is that precipitation seasonality, or more specifically its effectiveness, is not taken into account. Precipitation effectiveness, defined loosely as P-PE for a given month, is higher for areas receiving their bulk precipitation in winter than in summer.

There is a clear association between climate and natural vegetation. Native vegetation is a mirror image of the entire spectrum of all climatic elements and represents the best climate indicator as it integrates all climate and soil elements. Natural plants growing in dry climates may be classified into two types—drought evaders and drought resistant. The first type is composed of seasonal plants that emerge under favorable conditions of moisture and soil temperature. These plants emerge and end their growth cycle within a short period of time, leaving behind a bank of seeds that may lay dormant for several years or more,

awaiting the right conditions to restart their cycle again. Drought-resistant plants withstand the harsh conditions (very low soil moisture, high solar radiation, and excessive heat) via physiological adaptation. This includes root distribution (shallow or deep), morphological characteristics (barrel-like plants such as cacti), waxy/few or no leaves, leaf shedding, large root bulbs, and dormancy. These physiological adaptations are meant to harvest as much soil moisture as possible and/or to conserve soil moisture in order to avoid lengthy drought conditions and excessive heat (e.g., Leopold, 1961; Miller, 1985).

Climatic Elements

Dry climates are characterized by plenty of sunshine, large diurnal and annual temperature amplitudes, and very little precipitation, and as such little soil moisture.

Solar Radiation

Although extraterrestrial solar radiation reaches its maximum (~13.1×10^3 MJ m^{-2} yr.$^{-1}$ [megajoules per square meter per year]) at the equator, the corresponding maximum insolation reaching the ground surface is recorded in the subtropical high-pressure zones (~8.3×10^3 MJ m^{-2} yr.$^{-1}$), where deserts are situated. This gives an average annual atmospheric transmission of about 0.7 compared with the equatorial zones, where atmospheric transmission is about 0.35 to 0.4. The relatively low zenith angle of the sun and little cloud cover are responsible for this significant amount of global radiation reaching the ground surface in the arid zones. Although many of the economic woes of deserts are partially attributed to the high levels of solar radiation there, this element is a promising economic asset that can be harnessed for the sake of development of these lands (e.g., heating and cooling, desalination of brackish water, electricity generation).

Temperature

Air temperature represents a mirror image of the partitioning of net radiation at the surface-atmosphere boundary. The highest air temperature (57.6 °C) was recorded in Azizia, Libya, in the Northern Sahara. Still higher air temperatures

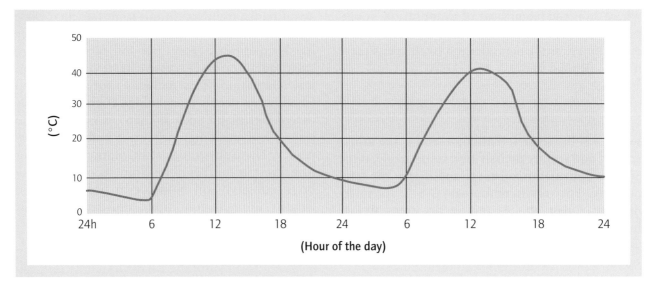

Figure 4   Diurnal surface temperature for a sand dune in Algeria, in the Northern Sahara desert during the beginning of April

Source: Adapted from Krishnamurti, T. N. (1979). *Compendium of meteorology* (Vol. 2, Pt. 4, Tropical Meteorology, WMO No. 364). Geneva: World Meteorological Organization.

were probably experienced and had gone unobserved in these hyperarid zones. An interesting question would be the midday surface temperature of a desert landscape. The energy balance at the surface-atmosphere boundary devoid of soil moisture may be expressed in the following form (e.g., Oroud, 1998):

$$S(1 - \alpha) + \varepsilon_a \sigma T_a^4 - \varepsilon_s \sigma T_s^4$$
$$- K_h(T_s - T_a) - K_\lambda(T_s - T_1) = 0. \quad (11)$$

The first, second, third, fourth, and fifth terms are absorbed solar radiation, atmospheric radiation, emitted radiation, sensible heat flux, and subsurface heat flux, respectively. If surface temperature is linearized ($\varepsilon_s \sigma T_s^4 \sim \varepsilon_s \sigma T_a^4 + 4\varepsilon_s \sigma T_s^3 (T_s - T_a)$), then Equation 10 can be written for the temperature difference between the surface and that of the contiguous atmosphere within the surface layer (ΔT) as follows:

$$\Delta T = \frac{S(1 - \alpha) - \sigma T_a^4 \delta \varepsilon}{Q_r + K_h + K_\lambda} \quad (12)$$

where $Q_r = 4\varepsilon_s \sigma T_a^3$ (W m^{-2} K^{-1} [watts per square meter per Kelvin]) and K_h and K_λ are the sensible and conduction heat transfer coefficients, respectively. For a typical subtropical desert, the

midday solar radiation reaching the surface in summer is about 1,000 W/m^2, and with typical values of $\alpha \sim 0.35$, $Q_r \sim 7$ W m^{-2} K^{-1}, $K_h \sim 13$ to 15 W m^{-2} K^{-1}, and $K_\lambda \sim 3$ W m^{-2} K^{-1}, the above expression yields a ΔT of about 20 to 25 °C. Thus, the midday surface temperature of a typical desert surface would be about 65 to 70 °C. For darker surfaces, such as an asphalt pavement or a basalt outcrop, or surfaces with very small roughness lengths, such as playas, ΔT is expected to be larger than what was obtained above, and T_s could reach about 75 to 80 °C. With such strong vertical temperature gradients, frequent dust devils are a common daytime phenomenon. Additionally, optical illusions (mirage) are monotonous features over shimmering surfaces such as paved surfaces and desert playas.

Due to clear skies, little/or nonexistent soil moisture, and low heat capacity of desert soils, surface temperatures there have significant diurnal swings. A desert surface heats rapidly during the early morning hours till it reaches its maximum 1 or 2 hours after noon. The reverse, although less dramatic, is true during the nighttime hours, with a temperature drop of several degrees Celsius near sunset. Near dawn, surface temperatures are usually 35 to 40 °C lower than they were at midday. Figure 4, taken over a sand dune in early April in the Sahara, shows this diurnal temperature trend clearly. Surface frost is also a common feature of desert surfaces during the winter months. Dew or hoar frost typically forms over these surfaces when adequate atmospheric moisture content and calm winds prevail. Dew has been reported to be a major moisture source for natural vegetation growing in these dry realms.

Ibrahim M. Oroud

See also Anthropogenic Climate Change; Arid Topography; Biome: Desert; Climate Types; Desertification; Drought Risk and Hazard; Dunes; Environmental Management: Drylands; Köppen-Geiger Climate Classification; Playa; Xeriscaping

Further Readings

Budyko, M. I. (1958). *The heat balance of the Earth's surface*. Washington, DC: Office of Technical Services, Department of Commerce.

Hare, K. (1983). *Climate and desertification: A revised analysis* (Report WCP-44). Geneva: World Climate Applications Programme, World Meteorological Organization and United Nations Environment Programme.

Leopold, A. S., & the Editors of Life. (1961). *The desert*. New York: Time.

Meigs, P. (1953). World distribution of arid and semiarid homoclimates. In *Reviews of Research on Arid Zone Hydrology* (pp. 203–210). Paris: UNESCO, Arid Zone Research.

Miller, E. W. (1985). *Physical geography*. London: Miller.

Oliver, J. (1987). Climatic classification. In J. Oliver & R. W. Fairbridge (Eds.), *The encyclopedia of climatology* (pp. 221–237). New York: Van Nostrand Reinhold.

Oroud, I. M. (1998). The influence of heat conduction on evaporation from sunken pans in a hot, dry environment. *Journal of Hydrology, 210*, 1–10.

Steila, D. (1988). Drought. In J. E. Oliver & R. W. Fairbridge (Eds.), *The encyclopedia of climatology* (pp. 388–395). New York: Van Nostrand Reinhold.

Thornthwaite, C. W. (1948). An approach to a rational classification of climate. *Geographical Review, 38*, 55–94.

United Nations Environment Programme. (1992). *World atlas of desertification*. Washington, DC: Author.

CLIMATE: MIDLATITUDE, MILD

The midlatitude mild climate zone describes a spatially expansive land area of the Earth that stretches from approximately 30° to 55° north and south latitudes. This climate zone features less extreme climate conditions throughout the year than neighboring climate zones such as the colder, snowier midlatitude severe climate zone located generally to the north of this region. Although the weather and climate of this zone is characteristically mild year-round, several areas situated within these latitudinal boundaries are not classified as having a mild midlatitude climate. More extreme climate conditions prevail in these reaches, such as areas of high elevation and areas in arid and semiarid regions. Considering these exceptions, most of the spatial land area of this climate zone is located on the North American and European continents. The midlatitude mild climate zone has moderate temperatures and precipitation totals throughout the year and large-scale climate controls such as the presence of subpolar low-pressure centers.

Nevertheless, there is much diversity within this climate zone. Climate can vary considerably with situation in the higher- or lower-latitude locations of the zone as well as with situation in the marine (coastal) or continental areas. For instance, in the lowest latitudes, precipitation in the form of rain is abundant with higher annual temperatures, which allows these areas to retain a yearly vegetative ground cover and green foliage. Higher latitudes of the climate zone have more variable temperatures and precipitation throughout the year, and vegetative ground cover and foliage can be sparse to nonexistent in the fall and winter months. An important feature of this climate zone is the impact of large-scale weather and climate phenomena, such as the El Niño/Southern Oscillation, drought, floods, hurricanes, and tornadoes, which can have significant positive and negative influences on the people and environments of this region. Moving into the future, there is also much uncertainty as to the location and extent of climate change impacts in the midlatitude mild climate zone.

Latitudinal Climate Variability

The thermal and moisture conditions of the midlatitude mild climate zone are frequently warm and dry in summer months and cool and wet in winter months. There is considerable climate variability within this zone (particularly in the winter season), which can be distinguished by latitude. In the highest latitudes of this climate zone, temperatures are generally cooler, averaging nearly 0 °C in winter and 23 °C in summer. Temperatures can reach well below freezing during the winter season, especially in areas situated above and below 43° north and south latitudes, respectively. For areas located in latitudes below and above 35° north and south, respectively, temperatures are much higher in winter and are rarely below freezing for extensive periods of time.

Rainfall is common in the summer and spring months in all latitudes, though in the fall and winter months, troughs of low pressure can develop over the region, leading to snowfalls in higher-latitude locations. A common winter feature along the North American east coast are nor'easter storm systems, which can bring below-freezing temperatures, heavy snows, freezing rain, sleet, and strong winds. For most of this climate zone, the close vicinity of the Pacific and Atlantic Oceans provides abundant moisture sources for precipitation. In summer, the lowest latitudes of the midlatitude mild climate zone experience the highest temperatures, though all locations can have warm to hot summers as dry, tropical air masses move across the climate zone. Occasionally, humid summer weather conditions that are more common in the lower latitudes are experienced in the high-latitude locations of this climate zone, leading to health threats and discomfort for the populations, who are less accustomed to this summer weather. Frequently, these conditions correspond to the presence of moist, tropical air masses.

Marine and Continental Locations

In the midlatitude mild climate zone, the climate conditions of a given location can also vary with situation near a water body or within the continental interior. In marine areas (considered the coastal locations), the close proximity to ocean waters moderates thermal and moisture conditions. Here, temperatures are less variable between the summer and winter seasons than at locations further inland, typically with an annual range of only 20° to 25 °C. In this climate zone, these conditions are often experienced on the western sides of continents. With oceans serving as an abundant moisture source, weather throughout the year can also be more humid and wet. However, marine locations of this climate zone typically receive the most precipitation in the fall and winter seasons. Maritime air mass movements contribute to the moist weather. Winter rains and cloudy skies are common in these areas. In addition, advection fogs are a regular feature of summer mornings in the marine locations of the midlatitude mild climate zone. In the lower latitudes of the climate zone, the presence of subtropical high-pressure centers offshore can direct moist weather systems away from the region, resulting in warm to hot and relatively dry summers. Warm and cool offshore currents and sea breezes may also influence the weather of marine areas in the climate zone.

For locations within the continental interiors, which are only found in this climate zone in the Northern Hemisphere, more variable annual weather and climate conditions are experienced. The weather and climate conditions experienced here are more variable on a regular basis throughout a season than in any other place on Earth. Midlatitude cyclones move across the continent and control the climate variability in these areas. These systems direct tropical and polar air masses, and the associated fronts and weather patterns, across the region. Polar air masses dominate the continental interiors in winter, especially in the higher latitudes. Tropical air masses are more often directed into the continental interior in summer. Regardless of the season, when tropical and polar air masses meet, milder weather conditions prevail. This often occurs as two air masses mix and modify after storms develop and pass across the area. The moisture conditions associated with these air masses come from the Pacific and Atlantic Oceans as well as the Gulf of Mexico and affect the continental interior precipitation totals and intensity as well as cloud cover. The instability associated with convection in this region also can lead to frequent thunderstorms, especially in southern locations and in the summer season.

Vegetative Cover and Human Adaptations

The vegetation of the midlatitude mild climate zone reflects the moderate temperatures and relative wealth of moisture across the region (see photo). Forests and grasslands form an abundant vegetative cover across the midlatitude mild climate zone. Broadleaf and deciduous forests are located across the zone. Hardwood (e.g., oak, maple, and cherry) and softwood (e.g., pine, spruce, and fir) forests are plentiful, though varieties that shed their leaves in winter are more common in the higher latitudes. Green foliage is common in the fall and winter months in the lowest latitudes of this climate zone.

For centuries, the forests of this climate zone have been considered an important multiple-use resource for humans occupying these lands. Areas with grasslands native to the climate zone are extensive, though many of these locales have been physically transformed into economically productive agricultural regions. For instance, many of the prairies native to the U.S. Great Plains are now farmlands in the form of pasture, croplands, and rangelands. In the marine locations of the climate zone, particularly on the western coasts, vegetative cover can become more patchy and sparse. As an example, in Mediterranean biomes located within this climate zone, chaparral vegetation is abundant. The warmer, drier climate of these areas has resulted in larger human populations than those found in the grasslands. Wetlands (e.g., swamps and marshes) are unique environments that are also common across the midlatitude mild climate zone. In recent years, conservation and preservation efforts are increasing to protect these areas from modifications of the land surface associated with increased human occupation and growing populations.

Large-Scale Weather and Climate Features

Variations from typical climate and weather conditions can prevail over the midlatitude mild climate zone with the influence of large-scale forcings and periods of anomalous weather. The El Niño/Southern Oscillation is an example of a large-scale forcing that can dramatically affect the climate of this zone and one that climatologists continue to investigate to better develop and understand these relationships. Other forcings include phenomena such as the North Atlantic Oscillation and volcanic eruptions. El Niño/Southern Oscillation impacts in this climate zone can influence widespread periods of below-average or above-average temperatures and precipitation. Marine locations, particularly on the North American west coast, can experience more significant impacts such as flooding, droughts, landslides, and severe storms. In higher latitudes, significant snowfalls can occur with El Niño and La Niña regimes. The impact on the economic conditions in this climate zone can also be severe, occasionally resulting in billions of dollars spent on long-term recovery.

Periods of anomalous weather that may or may not arise from large-scale forcings, such as hurricanes, tornadoes, floods, and droughts. can have a great impact on the climate and weather of the midlatitude climate zone. Hurricanes generating in the lower latitudes between June and November migrate into these regions, often growing in strength before they hit land. High winds and storm surges come with intense rainfall and can produce flooding in coastal areas as well as widespread damage, as prominently seen with Hurricane Katrina in 2005. Tornadoes, which are unique to this climate zone, develop in minutes but also leave long-term impacts on the areas they move across. Typically, tornadoes and severe thunderstorms develop over continental interior locations in the summer months. However, it is not uncommon to see tornadoes develop around warmer coastal locations that experience rapidly moving air masses, leading to abrupt changes in atmospheric stability. Though floods are common with hurricanes, they can also affect areas of the midlatitude mild climate zone wherever heavy rains are abundant, particularly for long periods of time. During the spring season, flooding can

The vegetation of the midlatitude mild climate zone reflects the moderate temperatures and relative wealth of moisture across the region. Pictured here is Orange County Park in Warwick, New York.

Source: Morguefile.

occur in higher-latitude locales as winter snow packs melt, rainfall becomes more prevalent, stream flow peaks, and the ground remains saturated for lengthy intervals. Droughts can occur anywhere within this climate zone but are often more of a threat to locations that remain drier throughout the year. Summertime is usually when the onset of droughts is experienced. These periods of anomalous climate can last from months to years and cause dangerous forest fires, depleted water supplies, enhanced erosion, and desertification.

Climate Change

Future climate changes in the midlatitude climate zone are uncertain, though several areas are considered to be more at risk, with the threat of a warmer climate and changes to the normal moisture regimes of the climate zone. Coastal areas appear to be the most vulnerable to a changing climate, particularly with regard to rising sea level. In addition, the most productive agricultural areas of this climate zone may change. It is hypothesized that higher-latitude locations that experience warming may have longer growing seasons and may be able to grow crops that cannot thrive at present with the lower temperatures than the future predictions. This means that these areas might look forward to heightened total agricultural production. Variations in precipitation amounts, and possibly intensity, could lead to a change in the vegetative cover and duration of green foliage throughout the climate zone. In addition, storm track movements could vary somewhat from normal if polar and tropical air masses became more prominent in the higher latitudes.

Melissa L. Malin

See also Biome: Midlatitude Deciduous Forest; Climate: Midlatitude, Severe; Climate Types; Climatology

Further Readings

Barry, R. G., & Chorley, R. J. (2003). *Atmosphere, weather and climate* (8th ed.). New York: Routledge.

Botkin, D. B., & Keller, E. A. (2007). *Environmental science: Earth as a living planet* (6th ed.). Hoboken, NJ: Wiley.

Christopherson, R. W. (2006). *Geosystems* (7th ed., pp. 276–281). Upper Saddle River, NJ: Pearson Prentice Hall.

Gill, A. E. (1982). *Atmosphere: Ocean dynamics*. San Diego, CA: Academic Press.

Lutgens, F. K., & Tarbuck, E. J. (2004). *The atmosphere* (9th ed., pp. 415–455). Upper Saddle River, NJ: Pearson Prentice Hall.

Lydolph, P. (1985). *The climate of the Earth*. Totowa, NJ: Rowman & Allanheld.

CLIMATE: MIDLATITUDE, SEVERE

Severe climates of the midlatitudes include humid continental and subarctic climates. These climate types are typified by harsh winters and large annual temperature ranges due to continentality. Because the climate types are constrained to latitudes between about 40° and 70° within large land areas, they are not found in the Southern Hemisphere, where there is very little land area at these latitudes. They are restricted to Europe, Asia, and North America and include a significant proportion of the global population.

Both the subarctic and the humid continental climate subtypes have no true dry season, receiving precipitation throughout the year. Many places receive more precipitation in the summer than in the winter, usually as a result of local convection or midlatitude cyclones. Winter precipitation arrives mostly as snow and is predominantly the result of cyclonic activity. Humid continental and subarctic climates correspond to the Köppen climate type known as "microthermal," with the letter designation "D" (Figure 1).

Humid Continental

Regions with a humid continental climate have a large annual temperature range and four distinct seasons with mild to hot summers and long, severe winters. Humid continental is more temperate than the subarctic type due to its lower latitudes, around 40° N to 55° N. Mean annual precipitation typically ranges from 50 to 100 cm (centimeters), or 20 to 40 in. (inches), but this average usually decreases with increasing latitude. Proximity to the shoreline also influences precipitation, so in the United States and Southern Canada, for instance, precipitation decreases with increasing distance from the Atlantic shoreline, demonstrating reduced moisture content in the atmosphere.

Humid continental climates have the Köppen letter codes of *Dfa*, *Dfb*, *Dwa*, and *Dwb*. *Dfa* and *Dwa* are characterized by hot summers, while *Dfb* and *Dwb* have mild summers. The "f" and "w" refer to precipitation, with "f" meaning consistent monthly precipitation (≥6 cm) and "w" meaning a relatively dry winter season.

In the hot summer zones (*Dfa* and *Dwa*), January temperatures typically average below 0 °C (32 °F) and July temperatures average between 18 °C (65 °F) and 24 °C (75 °F). Weather in the summer is often humid, and severe thunderstorms that produce hail and/or tornadoes can occur. North of these zones are areas with mild summers (*Dfb* and *Dwb*). Here, the warmest months do not exceed 22 °C (72 °F) on average, and winters are more severe, with greater amounts of snowfall.

A large proportion of the population of the United States, Canada, Eastern Europe, and Asia live in humid continental climates, mostly along the eastern part of these continents. Humid continental regions with hot summers include the Eastern and Midwestern United States, from the Atlantic coast to approximately the

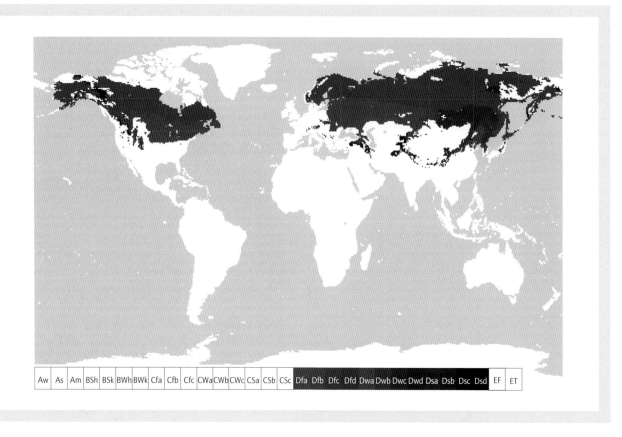

| Aw | As | Am | BSh | BSk | BWh | BWk | Cfa | Cfb | Cfc | CWa | CWb | CWc | CSa | CSb | CSc | Dfa | Dfb | Dfc | Dfd | Dwa | Dwb | Dwc | Dwd | Dsa | Dsb | Dsc | Dsd | EF | ET |

Figure 1 World Köppen map of microthermal "D" climates. Humid continental and subarctic climates are only found in the Northern Hemisphere. Subarctic climates are the more severe of the two since they are north of humid continental climates.

Source: From *Koppen classification worldmap* D. (2006). Retrieved January 10, 2010, from http://commons.wikimedia.org/wiki/File:Koppen_classification_worldmap_D.png. Reprinted with permission.

100th meridian; East Central Europe; Northern China; and Northern Korea. Regions with mild summers include the New England and Great Lakes regions of the United States, South and Central Canada, Scandinavia, Eastern Europe, and Russia.

The humid continental climate supports a diversity of ecosystems present in certain biomes. Humid continental biomes include midlatitude deciduous forests and midlatitude grasslands. Where midlatitude deciduous forests are present, climate conditions afford enough moisture to support tree growth. The seasonality present in this climate corresponds to the seasonal cycle of deciduous trees, in which the leaves change color and fall in the autumn and regrow in the spring. In the summer months, a dense canopy of broadleaf trees forms an almost continuous canopy. In

the Eastern United States, dominant tree species found in deciduous forests include maple, oak, hickory, sugar maple, aspen, poplar, cottonwood, and elm, among others.

Midlatitude grasslands are found in the drier humid continental regions (*Dfa*, *Dwa*) or where there is a local factor that reduces available moisture. For example, the Great Plains of the United States are located in the rainshadow of the Rocky Mountains, so the climate favors grass over trees due to decreased precipitation. In general, grasslands found in this climate type are usually in the interior of continents, far from moisture sources. These grasslands provide an almost continuous cover of grass and flowering plants, which lie dormant in the winter and sprout in the spring. Fire is common in the grassland biome, and the ecosystem is adapted to it. Before European settlers began suppressing fire, it kept trees and woody shrubs from growing, maintaining an ideal habitat for grazing animals such as bison. Now that fire has been suppressed for many decades, trees and shrubs are expanding into the periphery of grasslands.

Animals living in these biomes usually hibernate in winter and live off the land in the other seasons. Many animals are camouflaged to blend into the biome they are adapted to. Examples of midlatitude or temperate forest animals found in North America are brown bears, beavers, deer, raccoons, and opossums. Smaller animals and insects include earthworms (also abundant in grasslands), snakes, scorpions, newts, and frogs, to name a few. The forest is an inviting habitat for many bird species such as eagles, hawks, owls, cardinals, woodpeckers, and sparrows. Grassland animals are adapted to windy, dry conditions. They are grazers, burrowers, and predators. Grasslands support a diversity of animals, including invertebrates such as grasshoppers and beetles and mammals such as gophers, antelopes, and coyotes.

Climate and vegetation determine the dominant soils of a location, and in humid continental regions, alfisols and mollisols are the dominant soil types. Alfisols correspond to midlatitude forests and mollisols to midlatitude grasslands. Both soil types (especially mollisols) are appealing for agriculture, because they have a high clay content and can retain nutrients and moisture. Because of the combination of a humid climate and rich soils, many grasslands and forests found in this climate have been cleared and plowed for farmland. The Corn Belt of the Midwest is located within this climate region (mostly with hot summers), and soybeans, alfalfa, and hay are familiar crops in addition to corn. Humid continental locations with mild summers support crops such as potatoes, wheat, and peas.

Subarctic

Subarctic climates are located north of humid continental climates at about 50° N to 70° N and occupy the northernmost extent of midlatitude severe climates. More than half of Canada and Alaska and much of Russia have a subarctic climate. Summer temperatures are quite cool, cooler than in humid continental regions. A more drastic difference is the extremely cold winter temperatures, where in some locations they can remain below freezing for up to 7 mos. (months). The annual temperature range is greatest in this climate than in any other due to its high latitudinal location and vast changes in insolation throughout the year. In the summer months, daylight hours increase when the North Pole tilts toward the sun, with an average day lasting 17 to 22 hrs. (hours). In the winter, when the North Pole tilts away from the sun, nights last 18 hrs. or more.

Even though there are four seasons in this climate, it is characterized by a mild summer, a long, harsh winter, and very brief periods of spring and autumn. Summer is very brief, lasting from only 1 to 3 mos. Summer temperatures average around 10 °C (50 °F). Winter arrives as early as October, and some areas experience average monthly winter temperatures of less than −15 °C (5 °F) for as long as 3 to 4 mos.

Precipitation is also lower than in humid continental regions, usually ranging from about 12 to 50 cm (5–20 in.) annually. Summers are wetter than winters, as almost all the annual precipitation falls as rain during this season. The source of summer precipitation is primarily due to poleward migration of midlatitude cyclone tracks. The other months are quite dry, as the area is dominated by cold, dry air masses caused by a polar high-pressure system, resulting in little precipitation. In some years, snow may accumulate to depths of up to 1 foot or more, persisting because of the bitterly cold winters.

Köppen symbols for the subarctic regions include *Dfc*, *Dfd*, *Dwc*, and *Dwd*. The *Dfc* and *Dfd* Köppen subtypes describe a subarctic climate with a severe winter, cool summer, and no dry season. In contrast, the *Dwc* and *Dwd* both have dry winters, with *Dwd* and *Dfd* having the coldest winters. *Dwc* is found only in far Eastern Russia and is only distinguished from *Dfc* due to winter drought. *Dwd* and *Dfd* are only found in extreme northeastern Siberia, with the coldest months averaging below −38 °C (−36.4 °F). Thus, *Dfc* is by far the most common subcategory of subarctic climates. It is found in a wide belt across North America and Eurasia. Here, there is year-round precipitation and temperatures above 10 °C (50 °F) for 1 to 4 mos. of the year.

A common characteristic of subarctic climates is permafrost, permanently frozen subsoil. During brief summers, the upper layers of soil thaw to depths of 1 to 4 m (meters), or 3 to 12 ft. (feet). The portion that remains frozen year-round is permafrost. Since water cannot penetrate frozen soil, the soil becomes waterlogged, forming bogs and swamps during the summer months. Massive populations of black flies and mosquitoes explode at this time, providing food for numerous bird species. Plants that have been dormant through the cold months explode with growth, accessing the meltwater during the short growing season.

Soils found in the subarctic are inceptisols, gelisols, and spodosols. Inceptisols—young, undeveloped soils—form chiefly through mechanical processes such as frost action and glacial grinding. Spodosols correspond to areas with conifer trees. The plant litter from the trees is very acidic, which causes nutrients to be leached out of the soil. Usually found north of spodosols, gelisols are associated with cold climates with permafrost within 2 m of the surface. Because these soils develop very slowly due to cold temperatures, gelisols contain vast amounts of organic matter.

One biome associated with subarctic climates is the boreal forest, or taiga. It is characterized by coniferous trees with needle-shaped leaves with primarily mosses and lichens on the forest floor. Evergreen tree species found in the taiga include the jack pine, balsam fir, white and black spruce, larch, fir, and pine as well as deciduous species such as birch, aspen, willow, and rowan. Most of the region's trees are xerophytes, or plants adapted to dry conditions. Where soils are too saturated to sustain tree growth, the landscape is covered by peat, grasses, and sedges. Some tundra is found at the periphery of subarctic climates, but it is most common in Köppen's E climate (polar). Tundra is usually found north of the taiga, where tree growth is restricted by cold temperatures and a short growing season.

Animal life in the tundra is less abundant than in the midlatitude forests and grasslands further south. Subarctic animal habitants include caribou, fox, otter, mink, ermine, squirrel, lynx, sable, moose, bears, reindeer, and the wolf. Few people live in this region, and those who do may rely on trapping, as animals in the region grow heavy fur pelts for warmth.

Subarctic climates are not advantageous for human settlement since the growing season is short and the soils are poor for agriculture. The scarce population of the region makes a living by trapping, mining, logging, and fishing. Further difficulty arises when attempting to build atop permafrost. Heat loss from buildings must be mitigated to avoid melting the surrounding subsoil. Melting permafrost under buildings causes their collapse. Therefore, building in subarctic regions is very expensive and challenging, so there are few cities found here, and they are generally small with few roads and no railways connecting them. Because of this, transportation is limited to helicopters, airplanes, and, in the summer, boats. The largest city in this climate is Murmansk, Russia, with a population of more than 300,000. This extreme northwestern seaport of Russia remains icefree year-round due to the North Atlantic Ocean current. Many settlements in the subarctic region occur where the climate is modified by an ocean current, making the climate milder than it would be in a continental location.

Trish Jackson

See also Biome: Boreal Forest; Biome: Midlatitude Deciduous Forest; Biome: Midlatitude Grassland; Biome: Tundra; Climate: Midlatitude, Mild; Köppen-Geiger Climate Classification; Permafrost; Soils; Wetlands

Further Readings

Aguado, E., & Burt, J. E. (1999). *Understanding weather and climate*. Upper Saddle River, NJ: Prentice Hall.

Arbogast, A. F. (2007). *Discovering physical geography*. Hoboken, NJ: Wiley.

Blue Planet Biomes. (2009). *World climates*. Retrieved June 3, 2009, from www.blueplanet biomes.org/climate.htm

Enchanted Learning. (2009). *Biomes*. Retrieved June 3, 2009, from www.enchantedlearning.com/biomes

McKnight, T. L., & Hess, D. (2000). Climate zones and types: The Köppen system. In *Physical geography: A landscape appreciation* (pp. 200–201). Upper Saddle River, NJ: Prentice Hall.

Peel, M. C., Finlayson, B. L., & McMahon, T. A. (2007). Updated world map of the Köppen-Geiger climate classification. *Hydrology and Earth Systems Science, 11*, 1633–1644.

CLIMATE: MOUNTAIN

Climate is the fundamental factor driving the natural environment, setting the stage on which all physical, chemical, and biological processes operate. The influence of climate on environmental processes is exaggerated in mountains. Immense environmental gradients occur over short distances as a result of the diverse topography and highly variable nature of the energy and moisture fluxes in mountains. Nevertheless, predictable patterns and characteristics are found within these heterogeneous systems; for example, temperatures normally decrease with elevation, while cloudiness and precipitation increase; it is usually windier in the mountains, the air is thinner and clearer, and solar radiation is more intense.

Mountains themselves, by acting as a barrier and elevated surface, affect regional climate and modify passing storms. Their influence may be felt for hundreds of kilometers, making the surrounding areas warmer or colder, or wetter or drier than they would be in their absence. The exact effect of the mountains depends on their location, size, and orientation with respect to the moisture source and direction of the prevailing winds.

Mountain climates occur within the framework of the regional climate and are controlled by the same factors, including latitude, prevailing winds, altitude, continentality, and the character of regional storms.

Latitude

The distance poleward from the equator governs the angle at which the sun's rays strike the Earth, the length of the day, and thus the amount of solar radiation arriving at the surface, which dictates temperatures. In the tropics, the sun is always high overhead at midday, and the days and nights are of nearly equal length throughout the year. With increasing latitude, however, the height of the sun changes during the course of the year, and days and nights become longer or shorter depending on the season.

The distribution of mountains in the global circulation system, which has a strong latitude distribution, dictates their basic climate. Mountains near the equator are under the influence of the equatorial lows and receive daily precipitation on their east-facing windward slopes. In contrast, mountains located around 30° latitude experience considerable aridity. Further poleward, mountains receive heavy winter precipitation on westward slopes facing the prevailing westerlies. Polar mountains are cold and dry year-round.

Continentality

The relationship between land and water has a strong influence on the climate of a region. Water heats and cools more slowly than land, so the temperature ranges between day and night and between winter and summer are smaller in marine areas than in continental areas. Generally, the more water dominated an area is, the more moderate its climate. An extreme example is a small oceanic island, on which the climate is essentially that of the surrounding sea: a marine climate. The other extreme is a central location on a large land mass such as Eurasia, far removed from the sea: a continental climate with hot summers and cold winters.

Altitude

Essential to mountain climatology are the changes that occur in the atmosphere with increasing altitude, principally the decrease in temperature, air density, water vapor, carbon dioxide, and aerosols and the increase in cloudiness, precipitation, and solar energy received during clear-sky conditions. The sun is the ultimate source of energy, but little heating of the atmosphere takes place directly. Rather, solar radiation passes through the atmosphere and is absorbed by the Earth's surface, which then reradiates infrared radiation, which heats the atmosphere from the bottom up (i.e., the greenhouse effect). Mountains are part of the Earth, too, but they present a smaller land area at higher altitudes in the atmosphere, so they contribute less to heating the atmosphere above them. A mountain peak is analogous to an oceanic island in the continentality effect. The smaller the island and the farther it is from large land masses, the more its climate will be like that of the surrounding sea. In contrast, the larger the island or mountain area, the more it modifies its own climate. This mountain mass effect can be a major factor in the regional climate.

Looking up Fisher Creek at Fisher Peak, located in North Cascades National Park, Washington. This landscape demonstrates the variety of different slope angles and aspects, rock types, active geomorphic surfaces, and different vegetation communities in a few square kilometers of area in a typical mountain environment. Noticeable aspect influence is seen on slope processes and vegetation. The south-facing slope (*left*) has more vegetation cover due to warmer temperatures and exposure to sun. The north-facing slope (*right*) has less vegetation and more active talus slopes, due to more vigorous freeze-thaw activity.

Source: Author.

The density and composition of air controls its ability to absorb and hold heat. The density of the air at sea level (standard atmospheric pressure) is generally expressed as 1,013 mb (millibars), or 760 mm (millimeters) or 29.92 in. (inches) of mercury. Near sea level, pressure decreases at the rate of approximately 1 mb/10 m (meters), or 30 mm/300 m or 1 in./1,000 ft. (feet) of altitude. Above 5,000 m (20,000 ft.), atmospheric pressure begins to fall off exponentially.

The ability of air to hold heat is a function of its density. At higher altitudes, air molecules are spaced farther apart, so there are fewer molecules in a given parcel of air to receive and hold heat. Similarly, the composition of the air changes rapidly with altitude, losing water vapor, carbon dioxide, and suspended particulate matter. These constituents, important in determining the ability of the air to absorb heat, are all concentrated in the lowest reaches of the atmosphere. Water vapor is the chief heat-absorbing constituent, and half of the water vapor in the air occurs below an altitude of 1,800 m (6,000 ft.). It diminishes rapidly above this point and is barely detectable at altitudes above 12,000 m (40,000 ft.).

Solar Radiation

The influence of the sun on temperatures becomes more exaggerated and distinct with elevation. The time lag, in terms of energy flow, between inputs and temperature change is greatly compressed in mountains. The alpine environment has perhaps the most extreme and variable radiation climate

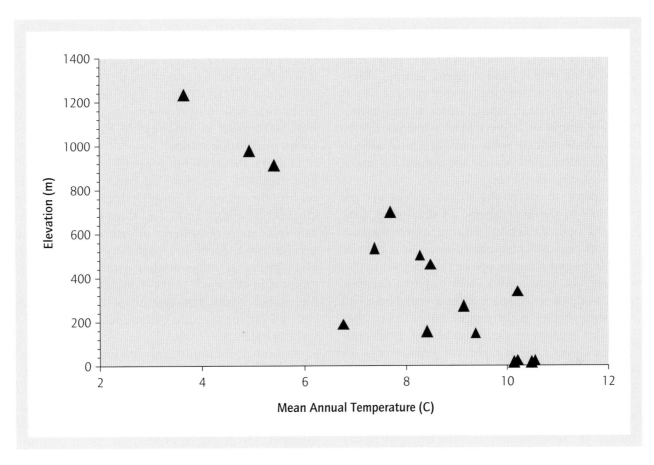

Figure 1   Mean annual temperatures from a number of stations throughout the North Cascades range of Washington

Source: Created by author with data from the National Climate Data Center, 1971–2000.

on Earth. The thin clean air allows very high solar intensities, and the topographically complex landscape provides surfaces with a range of different exposures, reflections, and shadowing from the nearby peaks. Although the air next to the ground may heat up very rapidly under the direct rays of the sun, it may cool just as rapidly when the sun is blocked. The most striking aspect of the vertical distribution of solar radiation in the atmosphere is the rapid depletion of short-wavelength energy at lower elevations. This attenuation results from the greater density of the atmosphere in the lower troposphere. Thus, the alpine environment receives considerably more ultraviolet radiation (UV) than the lower elevations.

The closer to perpendicular the sun's rays strike a surface, the greater their intensity. The longer the sun shines on a surface, the greater the heating that takes place. In mountains, every slope has a different potential for receiving solar radiation. Mountains are composed of a wide range of surface types: snow, ice, water, grasslands, extensive forests of various types and coverages, desert shrub, and bare bedrock and soil (see Fisher Peak photo). This wide-ranging variety of surface characteristics affect the receipt of incoming solar radiation. Dark-colored features, including vegetation, absorb rather than reflect radiation, receiving increased amounts of energy. Snowfields, glaciers, and light-colored rocks have a high reflectivity (albedo), so that much of the incoming shortwave energy is reflected back into the atmosphere.

Temperature

The decrease of temperature with elevation is one of the most visceral and fundamental features of mountain climate (Figure 1). Change of temperature with elevation is termed the *environmental lapse rate*, averaging 6.4 °C/1,000 m (3.5 °F/1,000 ft.), although values vary with season, time of day, and regional setting. While the general trend of declining temperatures is commonly observed, temperatures within mountains are found to be very heterogeneous due to the aforementioned solar radiation and albedo changes over small scales, the mountain mass, and the inversion effects.

Temperature inversions within valleys are the exception to the general rule of decreasing temperature with elevation. During a temperature inversion, the lowest temperatures occur in the valley bottom and increase up the slopes. Eventually, however, the temperatures will begin to decrease again, so that a warmer zone, the thermal belt, will experience higher night temperatures than either the valley bottom or the upper slopes. Inversions form as slopes cool at night, the heavier cold air begins to slide down the slope, flowing underneath and displacing the warm air in the valley. Temperature inversions are best developed under calm, clear skies, where there is no wind to mix and equalize the temperatures, and the transparent sky allows the surface heat to be rapidly radiated and lost to space. If the valley is blocked, a pool of relatively stagnant colder air may collect, but if the valley is open, there may be a continuous movement of air down the watershed.

Precipitation and Clouds

The increase of precipitation and clouds with elevation is well documented. This phenomenon is part of a complex of processes known as the orographic effect. As air is lifted over the mountains, it is cooled by expansion and mixing with cooler air at higher elevations (i.e., adiabatic processes) until condensation occurs and clouds form. This forced ascent of air is most effective when mountains are oriented perpendicular to the prevailing winds and when the slopes are steeper. Mountains also act to block/dam storms, slowing their progress and allowing longer-duration precipitation events.

While heavy precipitation occurs on the windward side of mountains, the leeward side typically receives considerably less precipitation because the air is descending with much of the moisture already removed. This creates the rainshadow effect—an arid area on the downwind side of mountains. As the air is forced to descend, it will be heated by compression (adiabatic heating) and will lead to clear, dry conditions. This is a characteristic phenomenon on the lee side of mountains and is responsible for the famous foehn or chinook winds.

Despite these useful generalities, many local and regional variations in precipitation occur within mountains. Complex local topography creates funneling effects (i.e., through passes) that can increase atmospheric moisture content and precipitation, even downwind from the funnel. High peaks or ridges within a range can create "mini-rain-shadow" zones even in the center of a high range. Significant quantities of precipitation can fall on the lee side of mountains because of spillover effects.

A number of cloud forms are unique to mountain environments. All of them are stationary clouds, which continually form on the upwind edge and dissipate on the lee edge of the cloud, thus appearing to remain in the same location for long periods. A cap or crest cloud forms over the top of an isolated peak or ridge. They resemble a cumulous cloud, although they may be streamlined or have streamers of cirrus forms. Banner clouds are cap clouds that extend downwind from the peak like a flag waving in the wind. Lenticular clouds are lens-shaped clouds formed in regularly spaced bands parallel to the lee side of a mountain barrier (see Mt. St. Helens photo). These streamlined cloud features are formed by the interaction of high-velocity winds with the mountain barriers.

Winds

Mountains are among the windiest places on Earth. They protrude relatively high into the atmosphere, where there is less friction to retard the air movement. The wind is usually greatest on mountains oriented perpendicular to the prevailing wind, on the windward rather than the leeward side, and on isolated, unobstructed peaks rather than those surrounded by other peaks. The reverse situation may exist in valleys, since those oriented perpendicular to the prevailing winds are protected, while those oriented parallel to the wind

Looking southeast at Mt. St. Helens in October 2004 at sunrise. The clouds are lenticular clouds forming as winds blow right to left over the mountain.

Source: Author.

may experience even greater velocities than the peaks, owing to funneling and intensification.

Winds that blow upslope and up the valley during the day and downslope and down the valley at night are common. The driving force for these winds is differential heating and cooling, which produces air density differences between slopes and valleys and between mountains and adjacent lowlands. During the day, slopes are warmed more than the air at the same elevation in the center of the valley; the warm air, being less dense, moves upward along the slopes. Similarly, mountain valleys are warmed more than the air at the same elevation over adjacent lowlands, so that the air begins to move up the valley. At night, when the air cools and becomes dense, it moves downslope and down the valley under the influence of gravity. This flow is responsible for the development of temperature inversions.

Slope winds consist of thin layers of air, usually less than 100 m (330 ft.) thick, and can be as thin as a few centimeters. In general, the upslope movement of warm air during the day is termed *anabatic flow*, and the downslope movement of cold air during the night is referred to as *katabatic flow*, or a gravity or drainage wind. Katabatic winds, in the strict sense, are local downslope gravity flows caused by nocturnal radiative cooling near the surface under calm, clear-sky conditions or by the cooling of air over a cold surface such as a lake or glacier. Since slope winds are entirely thermally induced, they are better developed in clear weather than in cloudy weather, on sun-exposed rather than on shaded slopes, and in the absence of overwhelming synoptic winds. Local topography is important in directing these winds; greater wind speeds will generally be experienced in ravines and gullies than on broad slopes.

Foehn Wind

Of all the ephemeral climatic phenomena of mountains, the foehn wind (pronounced "fern" and sometimes spelled *föhn*) is the most intriguing.

Similarly formed winds have local names worldwide, including "chinook" in North America. The foehn, known in the Alps for centuries, is formed as synoptic winds blow over a mountain crest and down the lee side, where it is heated adiabatically. Many legends, folklore, and misconceptions have arisen about this warm, dry wind that descends with great suddenness from mountains. They are, however, responsible for rapidly melting the snow (occasionally resulting in flooding) and increasing the incidence of fires.

Microclimates

In addition to the climatic characteristics reviewed above, it should be emphasized that there are substantial variations in climates over very short distances within mountains. Mountain environments are exceedingly spatially complex in terms of vegetation types and structures, disturbances, geology, hydrology, soils, and topography (see Fisher Peak photo). All vary in composition (i.e., species, canopy characteristics, or rock types), and variations occur across a range of slopes and aspects. The climate over each of these surfaces, or microclimates, can differ significantly because of variations in net radiation, soil and air temperature, humidity, precipitation accumulation (amount and form), soil moisture, and winds. The thin atmosphere at high elevation means that surfaces facing the sun on a clear day can warm dramatically, while shaded surfaces remain cold. Other effects may arise according to valley orientation with respect to the mountain range, valley cross-profile, and the effect of winds and cold air drainage. These steep environmental gradients (both elevational and spatial) result in the juxtaposition of diverse vegetation assemblages with narrow ecotones.

The resolution of most weather station networks in mountains is far too coarse to capture the spatial variability of microclimates in mountains. Maps of climatic variables are often interpolated from existing meager data sets, using assumed or empirical relationships with elevation. These models are unable to demonstrate local deviations in trends, and when combined with map scale, microclimates are typically eliminated from most maps of mountains.

Climate Change and Variability

Temporal variability of climate is an important natural component of the Earth's climate system. Climatic variability (i.e., occurrence of certain climatic events) is different from climatic change, which is a permanent change in climatic conditions. However, changes in variability are one likely consequence of climatic change. All temporal climate records demonstrate some degree of interannual variability. Every mountain location has its record high and low temperatures, snowfall, rain event, drought, and wind speed. Temperature, precipitation, and the resulting runoff are often related to distant forcing mechanisms, such as the El Niño/Southern Oscillation. Several other periodic, yet chaotic, perturbations to the climate system have been linked to increased climatic variability.

Mountain and glacier environments are especially sensitive to climate changes and variability. Many climate changes have been detected in mountain records, as illustrated by the glacial and interglacial climates of the Pleistocene, by the Holocene glacial fluctuations, and over the period of instrumental records. The current general scientific consensus is that global climate is in the process of warming due to anthropogenic inputs of greenhouse gases into the atmosphere. Different magnitudes of warming, and even cooling, are predicted for different mountainous regions of the world. Precipitation, in particular, is predicted to both increase and decrease in different regions because of changes in general circulation. In mountains, higher temperatures will cause a higher percentage of annual precipitation to fall as rain (i.e., higher snowlines) as well as accelerate summer ablation. Longer snowfree periods will increase the evaporative demands and lower the soil moisture, increasing the dominance of drought-tolerant species. Climate models in mountainous regions, however, tend to be rather poor because of coarse resolution, topographic smoothing, and local effects not captured by the models.

Warming temperatures will lead to reduced snow pack, glacial recession, and altered downstream runoff characteristics (i.e., seasonality and magnitude) over the next several decades. If glaciers entirely disappear from mountains, then the melt season, especially the late melt-season discharge, will decrease substantially. Changes to mountain hydrology will have significant consequences not only in the mountains themselves but also in the populated lowlands that depend on the runoff as a water resource.

In addition to external (i.e., global) climate change, mountains are subject to a variety of internal disturbances that influence climate. Among the major natural disturbances are fire, landslides, floods, and insect infestations. The spatial and temporal patterns of these disturbances have been altered by human activities, as well as the added impacts of settlement, timber harvest, exotic species introductions, and the damming of rivers. Small-scale disturbances increase the landscape and climate heterogeneity, while large-scale disturbances, such as those caused by humans (especially agriculture), act to homogenize mountain environments.

Climate change is considered a major threat to mountain ecosystems. Since many organisms living in mountains survive near their tolerance range for climatic conditions, even minor climatic changes could have a significant impact on alpine ecosystems. Vegetation zones will migrate altitudinally in response to warming temperatures, possibly eliminating some biomes, although the adaptations will likely be more complex. Because mountain tops are smaller than their bases, biomes shifting upslope will occupy smaller and more fragmented areas, reducing populations and increasing genetic and environmental pressures.

Andrew J. Bach

See also Adiabatic Temperature Changes; Atmospheric Circulation; Biota and Topography; Chinooks/ Foehns; Climate Change; Climate Types; Glaciers: Mountain; Wind

Further Readings

Barry, R. G. (2008). *Mountain weather and climate* (3rd ed.). Cambridge, UK: Cambridge University Press.
Whiteman, C. D. (2000). *Mountain meteorology: Fundamentals and applications.* Oxford, UK: Oxford University Press.

CLIMATE: POLAR

Polar climates are the result of low total solar energy receipt but large seasonal variations in solar energy receipt at the Earth's North and South poles. In summer, incoming solar radiation greatly exceeds outgoing terrestrial radiation during the 24-hr. (hour) polar day, while in winter, the reverse occurs during the 24-hr. polar night. In addition, the high albedo coefficient (the parameter characterizing the diffused reflection of sunlight from the surface of an object) of snow and ice covers in polar areas results in even less solar energy receipt. As a result, solar radiation receipt above the 60th degree of latitude is two to ten times smaller than in the tropical climatic zone.

The regions of high latitudes vary geographically. They are represented by extensive land areas, such as parts of the Asian and North American continents, the huge ice sheets of Greenland and Antarctica, as well as the different oceanic masses, frequently covered by sea ice, occurring around the Antarctic continent or in the vicinity of the North Pole. Thus, the climate of different parts of the polar provinces depends on its location and the type of the substratum. Although there is no question about the astronomical delimitation of polar areas due to their location above the polar circles (66°33′39″, on both the Northern and the Southern Hemispheres), the climatic and environmental conditions are much more diversified, moving the border of the region either to the north or to the south.

The Arctic

The limits of the Arctic are marked by the 100 °C isotherm of July (the warmest month of the year in the Northern Hemisphere). Another criterion is the boundary of 15 kcal cm^{-2} yr.$^{-1}$ (kilocalories per square centimeter per year), or 62.7 kJ cm^{-2} yr.$^{-1}$ (kilojoules per square centimeter per year). This zone corresponds to the transition between the taiga and tundra ecosystems, and its position fluctuates between 50 °N and 700 °N. In such a range, the influences of harsh climatic conditions extend over an area of almost 27 million km^2 (square kilometers). The inner part of the region, delimited with the 50 °C isotherm of

The glacial mountains of Spitsbergen, Norway's most northern territory

Source: Morguefile.

July, is called the high Arctic, with the landscape dominated by herbaceous and bare-ground tundra (with a majority of lichens and mosses) and polar desert.

Polar climates can be identified as either maritime or continental. In the Northern Hemisphere, the warm Norwegian Current of the Northern Atlantic provides significant energy fluxes into the adjacent polar land masses. The current flows along the Scandinavian coast and then continues far into the north. Mild maritime influences are observed up to the 80° latitude, and sea ice does not accumulate along most of the Norwegian and Svalbard coasts. Typically (e.g., the Bear Island on the Northern Atlantic), monthly average temperatures oscillate between –100 °C in February and 50 °C in July. In the coastal areas of the inner part of the Arctic Ocean, temperatures in winter drop down to about –300 °C due to the presence of sea ice, but in summer, the temperatures are still under the control of the open sea and reach 50 °C (e.g., Barrow in Alaska and Tiksi in the northeastern Siberian coast).

The most continental climates, with the highest yearly temperature ranges, appear far inland within the Siberian region, where monthly average temperatures in January reach almost –500 °C and in July can reach 150 °C. Furthermore, the lowest registered air temperatures in the Northern Hemisphere (officially –69.80 °C), caused by cold air stagnation in the valleys (temperature inversions) as in Vierkhoyansk and Oymyakon, occur in polar climates. In parts of the Arctic, air temperatures rarely rise above 0 °C. Over the high Greenland ice sheet, they range between –600 °C in winter and –120 °C in summer, and in the Arctic interior, temperatures might rise in the latter season to up to 0 °C because heat exchange between the ocean and the atmosphere through the permanent but partly loosened sea ice cover provides an avenue for energy transfer.

Precipitation in the Arctic is usually low. In most regions, the annual precipitation total does not exceed 500 mm (millimeters); however, in southeastern Greenland, precipitation totals reach at least 1,200 mm/yr. (year) and can go up to 2,000 mm/yr. In other parts of the Arctic, precipitation totals are determined by distance from a coast and the advection of humid air masses across the Arctic. As a result, the driest parts (polar desert), with yearly precipitation less than 200 mm, are located in the central area of the Arctic interior, the Northern Canadian region, and continental Siberia. At the sea ice edge and along coastal areas, precipitation rises to 400 mm/yr. and even more in higher-located mountain massifs such as the Alaska or the Scandinavian range. Seasonally, winter precipitation dominates, with more than 90% in the form of snow, while in the summer period, up to 60% might be liquid. In coastal areas, fog is typical, whereas visibility is typically better in the dry, non-polluted inland areas.

Thermal and precipitation conditions in the arctic and subarctic regions depend on penetration into these regions of extratropical cyclones (rare) and the formation and path of polar cyclones and anticyclones. The semipermanent high-air-pressure systems located over the polar sea ice and the polar continental areas typically inhibit the penetration of warm air masses into the polar regions. However, beyond these regions, a dynamic zone of cyclone formation develops, especially in winter. Large thermal contrasts between the ice-covered areas and icefree oceans generate large air pressure gradients and strong winds. The maximum wind velocities that are observed in the Atlantic region may exceed 150 km/hr. (kilometers per hour), while in other areas they are up to 100 km/hr., with averages during nonstorm conditions in the range between 15 and 25 km/hr.

Antarctica

Difficulties in establishing and supplying permanent antarctic stations have resulted in few long-term studies of the antarctic climate. Delimitation of the area that experiences polar conditions in the Southern Hemisphere is based on circumantarctic oceanic-water individuality and biodiversity. In the Antarctic Convergence zone, between 48° S and 610° S, the cold antarctic shelf waters come into contact with the warmer waters of the three oceans—the Atlantic, the Pacific, and the Indian.

The whole of Antarctica—an area of about 52.5 million km²—includes the Antarctic continent and the so-called subantarctic islands, such as Falklands, South Georgia, and South Sandwich Islands, as well as those located further to the south, for example, South Shetlands and the Balleny Islands. Average monthly air temperatures fluctuate between 100 °C in January and 40 °C in July in the most outer maritime areas, at the latitudes south of the American continent. In the austral summer, in the coastal zone of the Antarctic Peninsula, temperatures reach 0 °C. Typically, the continental coastal regions lie within the isotherm −40 °C in January and −240 °C in July. However, in the central part of the Antarctic Plateau (Eastern Antarctica), temperatures below −280 °C in summer and −600 °C in winter are typical. The lowest temperature recorded in Antarctica—actually the lowest ever recorded on the surface of the Earth—was −89.20 °C on July 21, 1983, at Vostok Station—78°28′00″ S, 106°48′00″ E. In contrast, the highest temperature recorded on the Antarctic continent was 14.60 °C in Hope Bay (63°23′42″ S, 56°59′46″ W) and Vanda Station (77°31′00″ S, 161°40′00″ E) on January 5, 1974.

The circumpolar Antarctic belt of cyclones that migrates from west to east, a result of the very dry cold air masses that form over the glaciated Antarctic continental interior, descend to the coast and mix with the warmer, humid oceanic air masses on the periphery of the Antarctic, bringing midwinter thaws and precipitation to the coast. The very cold dry air can hold very little moisture, and consequently, precipitation totals of less than 50 mm/yr. are normal. Strangely then, most of the snowfall occurs in the coastal zones and in the surrounding archipelagos, where the total precipitation can reach 600 mm/yr. As a result, most of the snow accumulation in the center of Antarctica occurs as drifting snow blowing in from the coast, and thus, most of the continent is classified as a polar desert. The cold katabatic winds draining off the interior of Antarctica have been recorded at speeds of 327 km/hr. on July 1972 at the station Dumont d'Urville (66°39′46″ S, 140°00′05″ E). High-velocity winds (exceeding on average 50 km/hr.) are especially characteristic of the ice sheet surfaces sloping down to the coastal areas.

Climate Changes in Polar Regions

Contemporary weather data collected during the past century have shown enhanced warming in the polar regions. It is hypothesized that much of this warming is due to (1) changes in global CO_2 (carbon dioxide) levels; (2) changes in the albedo of ice and snow; (3) faster heat transfer in cold and dry air masses, and (4) decreasing sea ice cover. In the 20th century in the Arctic, the fastest temperature rises were observed in the 1950s and over the past three decades, when in some regions (such as Central Siberia, Alaska), the temperature rise was estimated at 0.4 to 0.80° C per decade; however, the western coast of Greenland and Baffin Bay region experienced a corresponding cooling. In Antarctica, the warming is especially noticeable in winter and in spring. The yearly average temperature values are at the level of 0.2 to 0.40° C warmer per decade.

This warming has resulted in widespread glacial retreat, a decrease in the area extent of sea ice and in sea ice thickness in the Arctic Ocean basin, ice shelf collapse, the rise of the snowmelt line, and a shift in the northern tree limit. These processes have been accompanied by an increase in rainfall but not snowfall. Climate modeling predicts continued warming of polar areas. In some forecasts, the temperature in the Arctic will increase between 3 and 50° C, with more warming in winter than in summer. Arctic sea ice cover is predicted to totally disappear in summer by the second half of the 21st century. The trends of climate changes in Antarctica, conditioned on the existence of a huge ice sheet, are not explicitly predictable.

Grzegorz Rachlewicz

See also Albedo; Atmospheric Circulation; Atmospheric Energy Transfer; Climate Change; Ice; Poles, North and South

Further Readings

King, J. C., & Turner, J. (2007). *Antarctic meteorology and climatology.* New York: Cambridge University Press.

Przybylak, R. (2003). *The climate of the Arctic.* Norwell, MA: Kluwer Academic.

Serreze, M. C., & Barry, R. G. (2005). *The Arctic climate system.* New York: Cambridge University Press.

CLIMATE: TROPICAL HUMID

Tropical humid climate is one of several climate categories found on Earth. To understand it better, it is important to define the term *climate*. The climate of an area is the average characteristic condition of the atmosphere near the Earth's surface for at least 30 years. This includes the region's general pattern of weather conditions, seasons, and weather extremes such as hurricanes, droughts, or rainy periods. Various scientists classify climate differently based on different criteria, and the whole classification process is still inconclusive. Robert Christopherson classifies climate into six climate categories—namely, tropical, desert, mesothermal, microthermal, polar, and highland. Tropical climates are truly winterless. Understanding the dynamics of tropical humid climate is extremely important because it is Earth's most extensive climate category, occupying about 36% of the Earth's surface, including both ocean and land areas. This climate category comprises Earth's largest tropical rain forests and other biomes, which are very important sinks for atmospheric carbon, without which global warming will be enhanced.

Causes of Tropical Humid Climate

Tropical humid climates experience warm temperatures throughout the year because of the high angles of incoming solar radiation and nearly continuous penetration of moist air masses over these regions. Tropical climate regions are broadly constrained by the Tropic of Cancer to the north and the Tropic of Capricorn to the south. Persistent high angles of incident solar radiation within the tropics cause consistently high temperatures throughout the year as the solar radiation received per square meter is high. In the higher latitudes, the persistent lower angles of incident solar radiation result in less radiation received per square meter. While the humid air masses release heat to the environment through the condensation of water vapor, the main causal factors for tropical climate are the following:

1. Consistent day length of about 12 hours throughout the year

2. Persistent high angles of incident solar radiation

3. The passage of the intertropical convergence zone (ITCZ), which brings rain as it shifts seasonally following the location of the overhead sun

4. Unstable maritime air masses in the region (Although air masses of polar origin occasionally invade the tropical and equatorial zones, the tropical humid climate is almost wholly dominated by the tropical and equatorial air masses.)

Tropical humid regions are characterized by high rainfall totals in most months of the year. For rain to occur, there must be some mechanism for vertically lifting the humid air, referred to as "atmospheric lifting mechanisms," so that the air is cooled and condensation occurs, forming clouds and rainfall. There are four types of atmospheric lifting mechanisms, the convergent, convectional, orographic, and air mass frontal processes. The convergent atmospheric lifting process occurs as air blown by southeast and northeast trade winds converges along the equatorial region, forming the ITCZ. As the rising air cools, clouds and rain develop. Areas under ITCZ have extensive air uplift, causing towering cumulonimbus cloud development and high annual precipitation. The convectional atmospheric lifting process occurs when air mass from a maritime source region passes over a warmer continental region. Such moist air is heated, becomes less dense and rises, condenses, and form clouds and precipitation. The orographic atmospheric lifting process occurs when the physical presence of a mountain acts as a topographic barrier to the migrating air mass or the incoming maritime air, which forces the air to rise upslope. Again, as air rises, it cools and condenses into clouds and precipitation. Frontal atmospheric lifting is not very dominant in the tropics but does occur in some areas. In the frontal atmospheric lifting process, the cold front (the leading edge of an advancing air mass) conflicts with the warm air mass front, forcing the warm, less dense air mass to rise and resulting in clouds and precipitation, similar to the orographic atmospheric lifting process.

Characteristics and Classification of Tropical Humid Climatic Regions

Generally, the mean rainfall in the humid tropics is about three times the world average. The trade winds and the ITCZ are responsible for much of these totals. Tropical climates stretch northward to the tip of Florida and South to Central Mexico, Central India, and southeast Asia. The only interruptions of tropical climates across the equatorial region are the Andes of Central and South America, the volcanoes of the rift valley of East Africa, and the Himalayas of South Asia. These higher elevations experience lower temperatures. For instance, Mount Kilimanjaro is less that 4° south of the equator, but it is 5,895 m (meters), or 19,340 ft. (feet), above sea level and has permanent glacial ice on its summit. Such mountainous sites fall within the highland climate category. Rainfall amounts and temperature are used to classify the tropical humid climatic regions. Tropical humid climates can be classified into three distinct regimes or subcategories: (1) tropical rain forest (ITCZ present year-round), (2) tropical monsoon region/forest (ITCZ present for 6 to 12 mos. [months]), and (3) tropical savanna (ITCZ present for less than 6 mos.).

Tropical Rain Forest Climate

Tropical rain forest climate is hot and muggy year-round. It supports dense tropical rain forests. Rainfall is heavy and occurs as frequent showers and thunderstorms throughout the year. Average annual rainfall varies from about 70 to 100 in. (inches), or 175 to 250 cm (centimeters). Temperatures are high, and they change little during the year. The coolest month has an average temperature no lower than 18 °C (64 °F). The temperature difference between day and night is greater than the temperature difference between the hottest and colder months. Frost and freezing temperatures do not occur. Plants grow all year-round.

Tropical rain forest regions are dominated by maritime equatorial (mE) air masses. The wet equatorial belt with heavy rainfall occurs around the equator in the Amazon River basin, South America; in the Congo River basin, equatorial Africa; in much of the African coast from West Nigeria to Guinea; and in southeastern Asia. The

Tropical forest at Sachatamia, Equador, May 2005

Source: Courtesy of Mark Welford.

prevailing warm temperatures and high moisture content of the mE air mass favor abundant convectional rainfall triggered by local heating and trade wind convergence. The thunderstorms peak each day from midafternoon to late evening inland and earlier in the day where marine influence is strong along the coastlines. Rainfall follows the migrating ITCZ, which shifts northward and southward with the position of the overhead sun throughout the year. Water surpluses are enormous, creating the world's greatest stream discharges in the Amazon and Congo rivers. High rainfall sustains lush ever-green broadleaf tree growth, producing the Earth's equatorial and tropical rain forests. The leaf canopy is so dense that little light diffuses to the forest floor, leaving the ground surface dim and sparse in plant cover.

Tropical Monsoon Climates

Tropical monsoon climatic regions are located between 5° and 25° north and south latitudes, which constitute a narrow coastal zone of the larger continents of America, Africa, and Asia. These regions receive more than 150 cm (60 in.) of annual precipitation. Rainfall brought by the ITCZ falls for 6 to 12 mos. of the year. The dry season occurs when the ITCZ moves away from the area. The coastal areas within the tropical rain forest belong to the monsoon climate category and usually experience seasonal variation of wind and precipitation. Mountains prevent the cold air masses from Central Asia from getting into the monsoon areas, resulting in its high average annual temperatures. Besides the Indian subcontinent, tropical monsoon climate is experienced in northeastern Australia, Madagascar, and Central America, where trade wind coasts are found with moist maritime tropical (mT) air mass and heavy orographic rainfall. These climatic regions experience cool, dry and warm, wet weather.

Tropical monsoon forests comprise three, less distinct layers. With trees up to 30 m tall, the top canopy layer supports lianas, epiphytes, and climbing plants. The middle woody shrub layer reaches a height of 15 m. The ground layer of shrubs and undergrowth receives ample sunlight to support the growth of shrubs, ferns, and palms. The availability of sunlight allows the species in the forest floor to grow densely.

Tropical Savanna Climates

Tropical savanna climate occurs at latitudes 5° to 20° north and south globally, but in Asia, it occurs mostly between 10° N and 30° N, just south of the Himalayan ranges. Temperatures vary more in tropical savanna climates than in tropical rain forest regions. The tropical savanna regime can have two temperature maximums during the year because the sun moves between the equator and the tropics. The ITCZ reaches the tropical savanna climate regions for about 6 mos. or less of the year as it migrates with the overhead sun. Summers are wetter than winters because convectional rains accompany the shifting ITCZ when it is overhead. This produces a notable dry condition when the ITCZ is farthest away and high pressure dominates. Seasonal alternation occurs between dominance of moist mT and mE air masses along the ITCZ and dry continental tropical (cT) air masses of the subtropical high-pressure belt. As a result, there is a very wet season at the time of high sun and a very dry season at the time of low sun. Cooler temperatures accompany the dry season but give way to a very hot period before the rains begin. This climate occurs mostly in India, Indochina, West Africa, East Africa, Southern Africa, South America, both north and south of the Amazon lowland, and the north coast of Australia. Marked dryness from June to October defines the changing dominant pressure systems rather than annual changes in temperature.

There are actually two very different seasons in a savanna: a very long dry season (winter) and a very wet season (summer). In the dry season, only an average of about 4 in. of rainfall occurs. Between December and February, no rain will fall at all. In the summer, there is a lot of rain. In Africa, the monsoon rains begin in May. An average of 15 to 25 in. of rainfall occurs during this time. It gets hot and very humid during the rainy season. Every day, the hot, humid air rises off the ground and collides with the cooler air above and turns into rain. In the afternoons in the summer savanna, the rain pours for hours.

Danielson R. Kisanga

See also Atmospheric Moisture; Biome: Tropical Rain Forest; Climate Types; Monsoons

Further Readings

Fleischfresser, L. (1998). *Tropical climate modeling: Basic concepts and methods.* Berlin: Verlag.

McGregor, G., & Nieuwolt, S. (1998). *Tropical climatology: An introduction to the climates of the low latitudes.* New York: Wiley.

Michael, G. (2001). *Tropical rainforest.* Retrieved December 29, 2008, from www.blueplanetbiomes.org/rainforest.htm

National Aeronautics and Space Administration. (2008). *Intertropical convergence zone.* Retrieved December 29, 2008, from http://earthobservatory.nasa.gov/IOTD/view.php?id=703

 # CLIMATE CHANGE

This entry reviews climate change records on all timescales—instrumental, historical, and geological—and their primary causes. The characteristics of glacial phases are detailed, followed by a discussion of the climate of the Holocene and the past millennium.

Changes of climate operate over all timescales. For convenience, we may recognize variations as decadal to century, millennial, 10^4 to 10^6 yrs. (years), and $\geq 10^8$ yrs., each with its dominant forcing factors. Solar variability, volcanism, and changes in atmospheric composition occur on all timescales. Astronomical changes in Earth's orbital characteristics occur for more than 20,000 to 400,000 yrs. with distinct periodicities; these are about 23,000 yrs. (precession of the equinox), 41,000 yrs. (axial tilt of Earth), and 100,000 and 400,000 yrs. (orbital eccentricity of Earth). Geological processes of continental drift and tectonism (mountain building) occur on the longest timescales. In addition, there are many feedback effects in the climate system that may amplify or dampen an initial forcing.

The temporal characteristics of climate changes may involve step changes in mean values, gradual trends in means or variability, and oscillations of varying length. For instrumental records spanning approximately the past 150 yrs., anomalies are commonly calculated with reference to a 30-yr. base period (e.g., 1971–2000), which is considered to represent a "mean" or reference climatic state.

Instrumental Records

Instrumental records began generally in the 17th to 18th centuries, but early instruments were not well calibrated or were poorly exposed. Standard instruments and careful exposure became widespread in the mid 19th century, along with organized station networks and standard observing times. However, various temperature scales—Fahrenheit, Celsius, Reamur—were in use. Precipitation records remain one of the least consistent records due to the effects of airflow over the gauge, especially for solid precipitation, gauge characteristics (dimensions, rim design, and height above ground), shielding devices, and gauge siting. The frequency and time of gauge reading also introduces inhomogeneities. There are more than 50 different gauge types in use in national weather services, creating inhomogeneity in cross-border amounts. Sea surface temperatures are another problem area. Originally, readings were made with a thermometer inserted into a canvas bucket of seawater. Later, engine room intake temperatures were used. However, the changeover was gradual, and as ships increased in size, the depth of the water sampled by the engine room intake got deeper. Estimates of wind speed at sea, prior to the invention of anemometers, involved the use of the Beaufort wind scale (developed in 1805 by Admiral Beaufort), where categories 0 (calm) to 12 (hurricane) were based on the wind effects on wave state. In spite of the inhomogeneity problems, careful selection and adjustment of the records has enabled a consistent picture of climate variations over the period of instrumental records to be obtained. In the case of temperature, this reveals that temperatures were higher at the end of the 20th century than at the end of the 19th century in almost all locations on Earth.

The longest satellite climate record is of the Northern Hemisphere snow cover since 1966. Consistent global sea ice data began with passive microwave satellite measurements in 1979.

Historical Records

Various historical records, notably in Europe, China, and Japan, document extremes of weather—floods, droughts, heat waves, severe winters, lakes and rivers freezing, snowfalls, storms, and so on—as well as indirect records of agricultural production and harvests. These include written documents such as weather diaries, monastic or manorial rolls and ships' logs, iconographic records, landscape paintings of glaciers, and so on. Each of them must be examined critically, however, as they are subjective observations that are for the most part qualitative. The records span much of the past millennium and become more frequent from the 17th century onward. There is, for example, a documentary history of climate in Iceland from AD 1500 to AD 1800.

Proxy Records

The determination of climatic conditions for most of Earth's history depends on the use of proxy indicators of temperature, precipitation, winds, humidity, and so on. Proxies can be grouped into three general categories: geophysical variables, biological variables, and chemical variables. Geophysical variables include accumulation layers in ice cores extracted from polar ice sheets and high-altitude ice caps not subject to summer melt; electrical conductivity in ice cores, indicating the presence of sulfates associated with explosive volcanic eruptions; ice-rafted detritus layers in ocean sediments, reflecting iceberg transport;

temperatures in ground boreholes; and pedological and geomorphological evidence of past cold/hot and dry/wet conditions.

Biological variables are diverse and span a range of timescales. They include annual growth rings in trees whose width may reflect growing-season temperatures or moisture availability, pollen grains accumulating in lakes and bogs that reflect the local and regional vegetation, plant macrofossils in packrat middens that reflect local vegetation, and plant and animal fossils in the geological record. Microorganisms in ocean sediments also provide paleoclimatic information, based on the isotopic composition of carbonate in the test (shell) of the organisms and the distribution of organisms geographically and over time.

Chemical variables include many records from ice cores—sediments (from lakes and the oceans) and stalagmites in caves. In ice cores, these include anions and cations, indicating air chemistry; $\delta^{18}O$ and δD (deuterium), which indicate air temperature; and gases trapped in air bubbles, indicating atmospheric composition. Other sources include $\delta^{18}O$ in tropical corals, indicating seawater temperature and $\delta^{13}C$ in trees, which indicates carbon dioxide concentrations in the atmosphere.

The Geological Record

There have been at least four major Ice Ages in Earth's history. The Neoproterozoic (~850–630 mya [million years ago]) was a time when land ice and sea ice covered most of Earth—a state known as "Snowball Earth." There were also major glaciations in the Ordovician and Silurian (460–430 mya), the Permo-Carboniferous (350–260 mya), and the late Cenozoic. The latter began with mountain glaciation around 40 to 35 mya and full glaciation in East Antarctica around 34 mya and around 26 mya in West Antarctic. In Central Arctic, the ice-rafted detritus found on the Lomonosov Ridge dates from about 46 mya (i.e., sea ice and glacial ice), with perennial arctic ice by 14 mya. Until recently, it was thought that arctic glaciation began on Greenland, Iceland, and Alaska much later, between 11 and 7 mya. As a consequence of this history, it appears that the Earth would have had a state of unipolar glaciation in the Southern Hemisphere during the period ~32 to 14 mya, but this is a picture that is the subject of much discussion at present.

During the last phase of glaciations, there were 41,000-yr. oscillations from ~2.8 mya, changing to ~100,000-yr. oscillations around 1.2 mya. The former represent the effects of the axial tilt of Earth and the latter the eccentricity of Earth's orbit around the sun. There is also a 23,000-yr. precession cycle that may amplify or dampen the seasonal contrasts according to the timing of the perihelion of Earth's orbit about the sun. These oscillations were first fully documented by Milutin Milankovitch, a Serbian astronomer.

The last four glacial/interglacial cycles each span around 100,000 to 120,000 yrs., with only 10% of that time as warm as today (Figure 1). Marine isotopic stage (MIS) 11 was a much longer interglacial cycle that occurred between 425,000 and 395,000 yrs. ago; identified in the Dome-C ice core from East Antarctica, it was unusual in that the Milankovitch variables were identical to the present day and the CO_2 (carbon dioxide) level was like that of the preindustrial period. Overall, there were high sea surface temperatures in high latitudes, a strong thermohaline circulation, and a higher sea level than at present.

Sea level fell by about 135 m during the glacial maxima, with major ice sheets in North America and Fennoscandinavia, and rose 5 to 6 m above the present levels during the last interglacial (the Eemian), around 130,000 to 125,000 yrs. ago, when parts of the Greenland and West Antarctic ice sheets melted. Temperatures were slightly higher than now, even though the atmospheric CO_2 levels were lower than at present. The last glacial phase began around 115,000 yrs. ago and reached maxima about 75,000 and 20,000 yrs. ago. The temperature record, the extent of the glacial ice, and the sea level during these cycles mimic a sawtooth pattern over time.

The last glacial phase is marked by episodes of very pronounced climatic shifts. These are thought to be associated with changes in the North Atlantic thermohaline circulation (THC). There is a strong, ~1,500-yr. cycle known as Dansgaard-Oeschger (D-O) Oscillations identified in Greenland ice cores. These are clustered in packets within longer-term Heinrich events (H6 ~ 68–60

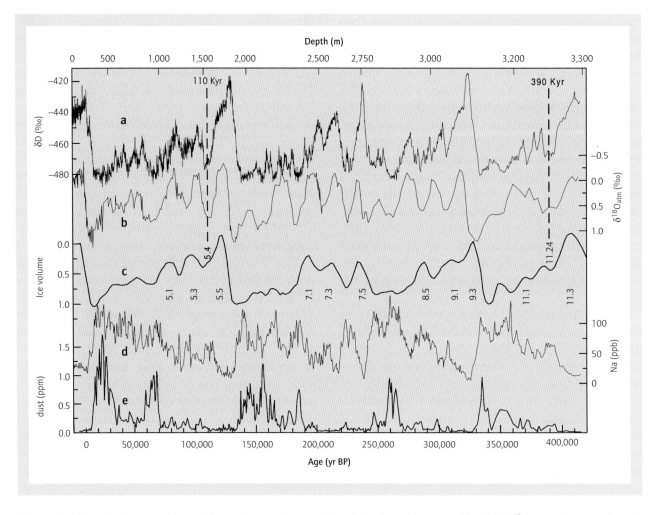

Figure 1 Vostok time series and ice volume: time series of (a) deuterium profile; (b) δ ^{18}O atm (atmosphere) profile obtained combining published data and 81 new measurements performed below 2,760 m; (c) seawater δ ^{18}O (ice volume proxy) and marine isotope stages; (d) sodium profile obtained by a combination of published and new measurements with a mean sampling interval of 3–4 m (concentrations are expressed in nanograms per gram or parts per billion); and (e) dust profile combining published data and extended below 2,760 m, every 4 m on average (concentrations are expressed in milligram per gram or parts per million)

Source: Adapted by Michael Mann from Petit, J. R., Jouzel, J., Raynaud, D., Barkov, N. I., Barnola, J. M., Basile, I., et al. (1999). Climate and atmospheric history of the past 420,000 years from the Vostok ice core, Antarctica. *Nature*, 399, 429–436 (specifically, temperature and CO$_2$ from Figure 3, p. 431, and dust from Figure 2, p. 430).

kya [kiloyears ago] to H1 ~ 14–7 kya), at a spacing of about 7.2 kya, where each successive D-O event has a smaller amplitude than its predecessor. Heinrich events involve the deposition of ice-rafted detritus from icebergs that calve off the eastern margin of the Laurentide ice sheet and East Greenland and melt in the northern North Atlantic. The injection of freshwater greatly reduces the THC in the North Atlantic. The D-O oscillations appear to represent instabilities in the North Atlantic circulation that may be driven by anomalies of salinity arising from precipitation minus evaporation anomalies. The massive iceberg discharge (Heinrich) events appear to occur with every fourth or fifth D-O oscillation, perhaps because the ice sheet needs time to build up again. A "binge-purge" oscillation in the eastern margins of the North American Laurentide ice sheet has been proposed to account for the Heinrich events.

The Northern Hemisphere ice sheets over Northern North America and Fennoscandinavia retreated rapidly between 17,000 and 1,400 yrs. ago, and then, there was a brief return to very cold conditions—the Younger Dryas—between approximately 12,700 and 11,500 yrs. ago BP. This episode was mainly concentrated in the North Atlantic sector and is thought to be the result of a significant reduction or shutdown of the North Atlantic thermohaline circulation in response to the sudden influx of freshwater from Lake Agassiz and/or the Arctic Ocean and deglaciation in North America. Following this event, the climate warmed rapidly, and this led to total deglaciation of the northern continents, with the last Laurentide ice disappearing in Eastern Canada around 6,500 yrs. ago. The postglacial period, known as the Holocene, is defined as starting around 10,000 ^{14}C yrs. BP (11,700 yrs. ago).

During the last ice age, vegetation zones were displaced equatorward, and there was a large and varied megafauna that became extinct around 12,000 yrs. ago. The respective roles of warming climate and human hunting and the use of fire in this process remain controversial.

In the tropics and subtropics, there were wetter and drier intervals with higher/lower lake levels across Africa and elsewhere. In Africa, there was a cold, arid phase about 70,000 yrs. ago, followed by a slight climatic amelioration and then a second aridity maximum around 22,000 to 13,000 yrs. ago. Conditions then quickly became warmer and moister, leading up to the Holocene "optimum" of greater rain forest extent and vegetation covering the Sahara, with high lake levels. This allowed the expansion of people and associated animal husbandry into areas of North Africa that are hyperarid today. Conditions then turned much more arid and similar to the present.

The Holocene

The Holocene is subdivided into five intervals based on climatic fluctuations: Preboreal (10–9 kya), Boreal (9–8 kya), Atlantic (8–5 kya), Subboreal (5–2.5 kya), and Subatlantic (2.5 kya to the present). Conditions were warmest between 9,000 and 5,000 yrs. BP, and then, cooling

occurred around 150 yrs. ago. The early-Holocene warm period is sometimes referred to as the Hypsithermal or Climatic Optimum, but the latter terminology raises the question of "optimal" for what? It is attributed to increased solar radiation receipts resulting from the occurrence of perihelion in July rather than in January as at present. This made boreal summers warmer than now and led to tree species advancing northward in Eurasia and North America. It also led to enhanced monsoonal circulations as a result of the warmer landmasses. Conditions during most of the Holocene were much less variable than during the glacial periods. However, there was a sudden cool and dry phase in many areas about 8,200 yrs. ago that lasted a few centuries. This is considered to be the result of an outburst of freshwater from proglacial lakes around the margin of the rapidly waning Laurentide Ice Sheet.

The Past Millennium

Temperatures over the past millennium have fluctuated about 1 to 2 °C (Figure 2). A Medieval Warm Period, dated about AD 900 to AD 1250, affected Europe and the North Atlantic sector. This interval saw the Viking settlement of Western Greenland, which subsequently died out as conditions deteriorated. This warm epoch was followed by the Little Ice Age (AD 1350–1850), during which period there were major advances of mountain glaciers in many mountain areas of the Northern Hemisphere. Temperatures declined between about 0.5 and 1.0 °C but fluctuated considerably on decadal scales.

The cause of these variations is attributed primarily to fluctuations in solar output and volcanic activity. Solar output (across all wavelengths) varies about 0.1% over the ~11-year solar cycle. Solar output is increased during sunspot episodes due to the increased emission from the brighter solar faculae, which outweighs the decrease from the darker (and cooler) sunspots. During the Maunder Minimum (about AD 1645–1715), there was a prolonged absence of sunspot. However, it is not known how much total solar irradiance decreased during that time; estimates vary from ~0.3% to as little as 0.1%.

Explosive volcanic eruptions inject particles and sulfur dioxide into the stratosphere, and this

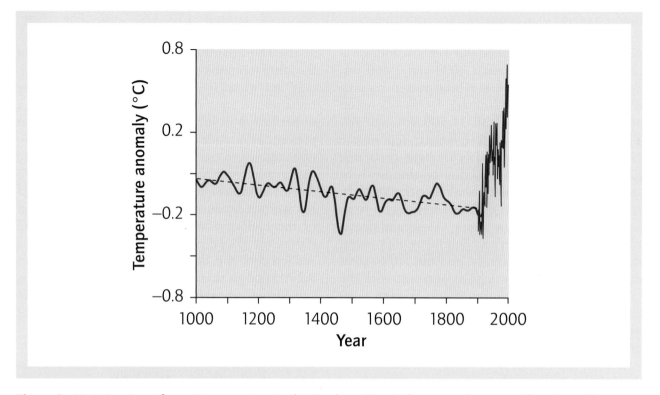

Figure 2   Variation in surface air temperature in the Northern Hemisphere over the past millennium. The reconstructed 40-year smoothed values are plotted for AD 1000–1880, together with the linear trend for AD 1000–1850 and observed temperatures for AD 1902–1998. The reconstruction is based on estimates from ice cores, tree rings, and historical records and has two standard error limits of about ± 0.5 °C during AD 1000–1600. The values are plotted as anomalies relative to AD 1961–1990.

Source: Adapted from Mann, M. A., Bradley, R. S., & Hughes, M. K. (1999). Northern Hemisphere temperatures during the past millennium: Inferences, uncertainties, and limitations. *Geophysical Research Letters, 26,* 759–762. Courtesy of M. E. Mann, University of Virginia.

reduces surface temperatures by 1 to 3 °C for up to 3 yrs. after the event. The cataclysmic eruption of Mount Tambora in April 1815 lowered global temperatures by as much as 3° C. Most of the Northern Hemisphere experienced sharply lower temperatures during the succeeding summer months. In parts of Europe and Eastern North America, 1816 was known as "the year without a summer."

Temperatures in the 20th century rose to an initial peak in the 1930s and 1940s, then declined or leveled off, and have risen continuously since the mid 1980s. Eleven of the 12 years, from 1995 to 2006, ranked as the warmest years in the instrumental record since AD 1850. The early-mid-century warming is potentially attributable to natural variability (solar output and volcanic eruptions), while higher sulfate and aerosol levels are believed to have caused the subsequent decline.

Over the period 1960 to 1990, there was a "global dimming," with a reduction in direct solar radiation by 4% to 6% attributed to this effect. With air pollution controls and industrial decline in Eastern Europe and Russia, a "global brightening" succeeded this. Increasing concentrations of greenhouse gases (CO_2, CH_4 [methane] and N_2O [nitrous oxide]) have led to an accelerated warming in the last two decades of the 20th century. Carbon dioxide levels were 35% to 40% higher than the preindustrial levels (AD 1850) in the first decade of the 21st century as a result of the combustion of fossil fuels and deforestation. It should be noted that there is a close correlation between the greenhouse trace gas concentrations and global temperature fluctuations over the past 700,000 yrs., as shown by the ice core records from Antarctica.

Rising temperatures have been due in part to positive feedbacks that are playing an increasingly important role, especially in the northern high latitudes. A record minimum sea ice extent occurred in late summer 2007, which means increased absorption of solar radiation by the Arctic Ocean and a positive feedback. The rate of decline in Arctic sea ice since the late 1990s is faster than what the climate model simulations have predicted. The Greenland ice sheet is undergoing accelerated melting and increased ice loss due to dynamical processes. In the tundra, the seasonal active layer that thaws annually is getting deeper in the northern continental areas, and there is warming and thawing of permafrost. This process is also observed on the Tibetan Plateau.

Roger G. Barry

See also Anthropogenic Climate Change; Atmospheric Circulation; Carbon Cycle; Climate: Polar; Climate Policy; Geologic Timesale; Global Climate Change; Ice

Further Readings

Barker, P. F., Diekmann, B., & Escutia, C. (2007). Onset of Antarctic glaciation: Deep-sea research, Part II. *Topical Studies in Oceanography, 54*(21/22), 2293–2307.

Bradley, R. S., Briffa, K. R., Cole, J., Hughes, M. K., & Osborn, T. J. (2003). The climate of the last millennium. In K. D. Alverson, R. S. Bradley, & T. F. Pedersen (Eds.), *Paleoclimate, global change and the future* (pp. 105–141). Berlin, Germany: Springer.

Darby, D. A. (2008). Arctic perennial ice cover over the last 14 million years. *Paleoceanography, 23,* PA1S07.

Edgar, K. M., Wilson, P. A., Sexton, P. F., & Suganuma, Y. (2007). No extreme bipolar glaciation during the main Eocene calcite compensation shift. *Nature, 448,* 908–911.

Hansen, J. E., Sato, M., Ruedy, R., Lo, K., Lea, D. W., & Medina-Elizade, M. (2006). Global temperature change. *Proceedings of the National Academy of Science, 103*(14), 288–293.

Lamb, H. H. (1977). *Climate: Present, past and future: Vol. 2. Climatic history and the future.* London: Methuen.

MacAyeal, D. (1993). Binge/purge oscillations of the Laurentide ice sheet as a cause of the North Atlantic's Heinrich events. *Paleoceanography, 8,* 775–784.

Mann, M. E., & Jones, P. D. (2003). Global surface temperatures over the past two millennia. *Geophysical Research Letters, 30*(15), 1820.

Moberg, A., Sonechkin, D. M., Holmgren, K., Datsenko, N. M., & Karlen, W. (2005). Highly variable Northern Hemisphere temperatures reconstructed from low- and high-resolution proxy data. *Nature, 433,* 611–617.

Ogilvie, A. E. J. (1992). Documentary evidence for changes in the climate of Iceland, A.D.1500 to 1800. In R. S. Bradley & P. D. Jones (Eds.), *Climate since A.D. 1500* (pp. 92–117). London: Routledge.

Peltier, W. R., & Sakai, K. (2001). Dansgaard-Oeschger oscillations: A hydrodynamic theory. In P. Schaefer, W. Ritzrau, M. Schlüter, & J. Thiede (Eds.), *The northern North Atlantic: A changing environment* (pp. 423–439). Berlin, Germany: Springer.

Sarnthein, M., Dreger, D., Erlenkeuser, H., Grootes, P., Haupt, B., Jung, S., et al. (2001). Fundamental modes and abrupt changes in North Atlantic circulation and climate over the last 60 ky: Concepts, reconstruction and numerical modeling. In P. Schaefer, W. Ritzrau, M. Schlüter, & J. Thiede (Eds.), *The northern North Atlantic: A changing environment* (pp. 365–410). Berlin, Germany: Springer.

Solomon, S., Qin, D., Manning, M., Chen, Z., Marquis, M., & Averyt, K. B. (Eds.). (2007). *Climate change 2007: The physical science basis* (Contribution of Working Group I to the Fourth Assessment Report of the Intergovernmental Panel on Climate Change). Cambridge, UK: Cambridge University Press.

Street-Perrot, A. F., & Harrison, S. P. (1984). Temporal variations in lake levels since 30,000 yr BP: An index of the global hydrological cycle. *American Geophysical Union, 29,* 118–129.

Stroeve, J., Holland, M. M., Meier, W., Scambos, T., & Serreze, M. (2007). Arctic sea ice decline: Faster than forecast. *Geophysical Research Letters, 34,* L09501.

CLIMATE POLICY

Climate policy refers to government or private actions designed to lower anthropogenic greenhouse gas (GHG) emissions or adapt to climate change. Emissions are rising worldwide and could result in substantial warming, sea-level rise, changes in precipitation patterns, and climate instability if the present trends continue. While knowledge of the greenhouse effect from increasing atmospheric levels of carbon dioxide (CO_2) due to fossil fuel combustion dates back to an 1896 paper by the Swedish physical chemist Svante Arrhenius, interest in climate policy began only in the early 1980s. At that time, the international climate community was preoccupied with the newly discovered hole in the stratospheric ozone layer. An international convention, or treaty, designed to protect it was approved in Vienna in 1985, and specific protocols were agreed to in Montreal in 1987. These treaties have proven to be effective mechanisms for international cooperation in research, monitoring, and action. The 1987 Montreal Protocol began the process of phasing out chlorofluorocarbons, a class of chemicals that are the main ozone-depleting substances. International treaties on the more complicated issue of climate change were approved in 1992 and 1997, but they have had limited impact and compliance thus far.

The United Nations Environment Programme (UNEP), the initiator of the 1985 Vienna Convention for the Protection of Stratospheric Ozone, also called for a convention on climate change. UNEP saw this as a mechanism to gain international agreement to reduce CO_2 and other trace gas emissions, based on the Vienna experience. Increasingly sophisticated climate modeling projected a doubling of atmospheric CO_2 levels above preindustrial levels by 2030. UNEP's efforts triggered the creation of the Intergovernmental Panel on Climate Change (IPCC) in 1988 under the auspices of the World Meteorological Organization and UNEP. The IPCC is an international body that works with thousands of scientists around the world to study the greenhouse phenomenon and its regional effects and to suggest mitigation strategies. This organization went on to jointly share the 2007 Nobel Peace Prize with former U.S. Vice President Al Gore for its cumulative work in climate change science and policy.

Framework Convention of 1992

The IPCC issued its first scientific assessment of climate change in 1990. Underscoring the potential serious consequences of climate change on the global environment and people, it presented its report to the Second World Climate Conference in Geneva in October 1990. The United Nations General Assembly subsequently created an Intergovernmental Negotiating Committee charged with developing a climate convention. The climate treaty was finished at the 1992 United Nations Conference on Environment and Development held in Rio de Janeiro.

The main objective of the Framework Convention on Climate Change (1992) was to achieve "stabilization of GHG concentrations in the atmosphere at a level that would prevent dangerous anthropogenic interference with the climate system" (p. 4). Short-term goals included establishment of national GHG inventories and returning GHG emissions (other than those covered by the Montreal Protocol) to 1990 levels by the end of the 20th century. A sticking point between then U.S. President George H. W. Bush and European leaders was targets and timetables. Consequently, emissions reduction was made voluntary. However, this goal did not cover developing nations such as China, India, and Brazil, and thus, it is called the *common but differentiated responsibility* principle. This issue became a more vexing problem 5 years later in Kyoto, Japan. Nonetheless, as a first step, the climate convention was eventually ratified by 192 nations, though compliance was essentially ignored as attention turned to the Kyoto Protocol.

Kyoto Protocol of 1997

After 2 weeks of intense negotiations, the Kyoto Protocol was adopted on December 11, 1997. Six GHGs are covered: (1) CO_2, (2) methane, (3) nitrous oxides, (4) sulfur hexafluorides, (5) hydrofluorocarbons, and (6) perfluorocarbons. This agreement requires 37 Annex I countries (industrialized countries that have accepted obligations to cut GHG emissions and submit annual inventories) and the European Union to reduce their emissions by a collective average of 5.2% below the 1990 levels by 2008 to 2012. These emission targets are termed *assigned amount units*. However, various countries negotiated different obligations, from the strictest (an 8% cut for most European countries) to the lenient (an 8% increase for Australia and a 10% increase for Iceland). Russia and Ukraine agreed to only freeze their emissions, which, given their economic collapse and associated emissions decline in the first part of the 1990s, some have criticized as creating "hot air" credits that could be traded elsewhere. To underscore this point, Russia's GHG emissions fell by 32% from 1990 to 2004, at which point it signed the Protocol.

The Kyoto Protocol has other major provisions. Non-Annex I countries, including China and India, are not required to cut emissions, since they successfully argued that because they contributed very few of the GHGs during the pre-treaty industrialization period, to do otherwise would be inequitable. Indeed, Annex I parties are also obligated to provide technical and financial assistance to other countries for climate studies and projects. The Protocol set up several flexibility mechanisms, such as emissions trading between Protocol parties that were designed to increase the cost-effectiveness of implementation. These provisions remain controversial since they could be abused. They include a Clean Development Mechanism (CDM) and Joint Implementation (JI). The CDM allows for the generation of new carbon credits through emission reduction projects in Non-Annex I countries, while JI allows any Annex I party (including the private sector) to earn an Emission Reduction Unit from an emission reduction or emission removal project in another Annex I country. An important requirement in all cases is the establishment of a reliable baseline of future emissions, so credit is not given for cuts that would have happened anyway. Finally, Article 3 of the Protocol permits parties to make adjustments in their GHG inventory through verifiable changes resulting from human-induced land use, land use change, and forestry (LULUCF) activities that have occurred since 1990. These projects include afforestation, reforestation, and improved forest management—all positive adjustments generating Removal Unit (RU) credits—and deforestation (a negative adjustment). Certified emission reduction credits in the form of CDMs, JI, and RU credits can be traded, subject to some geographic restrictions.

Unlike the Framework Convention, the Kyoto Protocol has binding enforcement provisions. These require any Annex I country not in compliance with its emissions obligation to make up the difference plus an additional 30% by the completion of the second commitment period in 2016. Furthermore, that country would not be allowed to trade emissions.

The Kyoto Protocol took effect on February 16, 2005, following ratification by Russia. As of late 2009, 186 countries plus the European Union (EU) have ratified it. The United States signed the Protocol but did not ratify it, and this country has been the only one on record to have proclaimed that it had no intention of signing it.

National Policies and Compliance

By 2009, few countries were on track to meet their Kyoto emission targets. Based on available emissions reports, only the United Kingdom, France, Germany, Sweden, and Norway seemed likely to meet their emissions obligations for 2012, with the United Kingdom and Norway being significantly assisted by CDM and LULUCF projects. Germany's progress in cutting GHGs was aided by economic restructuring in Eastern Germany, with Germany agreeing to a more stringent EU emissions target for 2012.

Overall EU emissions, when counting LULUCF activities, are down 5% below baseline 1990 levels, but this pattern is driven by the actions of the few nations noted earlier. Some countries have had great difficulty in cutting emissions, such as Spain, which was about 50% above its 1990 level in 2004. Even Japan, host of the Kyoto meeting and known for its high levels of energy efficiency, has seen its GHG emissions rise 5% since 1990. Thus, while the Annex I countries may meet the overall emissions target under Kyoto, much of the progress can be attributed to a few countries, as well as the precipitous decline in emissions in Eastern Europe after the fall of communism in the region. As a result, even before the first commitment period (2008–2012) was reached, the parties to the treaty began to discuss a successor to the Kyoto Protocol, which includes active participation by the United States and developing countries.

While the United States has not ratified the Kyoto Protocol, it released plans under President Bill Clinton and President George W. Bush to lower its emissions. Both plans relied on voluntary measures and close cooperation with industry, with no new legislation passed as of late 2009. In the meantime, U.S. GHG emissions rose 14.7% from 1990 to 2006. Given the slow federal response, numerous U.S. states and cities have developed their own GHG reduction initiatives. These include most of the northeastern states and California (which will return its GHG emissions to the 1990 level by 2020). At the municipal level, more than 1,000 cities have signed a Climate Protection Agreement to meet or exceed the original 1990 Kyoto target in their cities, a nationwide program that was started by the mayor of Seattle.

Australia, like the United States, had not ratified the Kyoto Protocol, citing similar economic concerns as the rationale. However, Prime Minister Kevin Rudd signed the Protocol in December 2007, soon after he was elected. It is possible that Australia may meet its Kyoto obligation when the effects of LULUCF activities are included.

Developing countries have been very active in international negotiations on climate change. Although not yet obligated to reduce their rapidly growing GHG emissions under the Kyoto Protocol, several have climate change policies. These include China (now the world's leading GHG emitter), India, Brazil, Costa Rica, Mexico, and South Africa. Climate change projects in these countries are financed through the Global Environment Facility, the CDM, and an Adaptation Fund established by Kyoto.

The Post-Kyoto Period

With the Kyoto Protocol set to expire in 2012, there have been attempts to strengthen and expand the reach of any subsequent global climate change treaty. The most significant came at a meeting in Bali in December 2007, where issues such as the need to address climate change adaptation, foster technological innovation, and promote wider participation were identified as requiring attention in the "roadmap" to a new or revised climate treaty. This process was significantly advanced through an agreement reached at the December 2009 Copenhagen international climate change summit and will continue in 2010.

Conclusion

While a broad consensus exists among the international scientific community as to the presence of, and anthropogenic contribution to, global climate change, agreement on what the next steps should be post-Kyoto is fluid. With the United States, China, and India responsible for half of global GHG emissions, their leadership in the next round of climate treaties is essential, as is continued progress by Annex I nations to hold the line and prevent backsliding. Participation by non–Annex I nations will likely entail technological transfer, continued investment in GHG offset credits by Annex I nations, and assistance with climate change adaptation. Many of the nations most vulnerable to climate change, particularly in Africa and South Asia, would require such incentives to address the equity of past emissions during industrialization by Annex I nations as well as unequal responsibility for existing emissions. Popular support for more forceful action in the United States is rising, as is evident by the array of state and local initiatives. Moreover, the rising global price of carbon-based energy has contributed to a change in consumer behavior, in turn broadening support for national action.

Future climate negotiations will be challenging in a global context where the need for consensus among almost 200 nations weakens the resulting agreement. Thus, more progress might emerge among smaller multiparty agreements similar to regional trading blocs, as within the EU or between China, the United States, and other major emitters, with low-emission developing nations joining as external incentives and internal conditions permit.

Barry D. Solomon and Michael K. Heiman

See also Anthropogenic Climate Change; Carbon Trading and Carbon Offsets; Greenhouse Gases; United Nations Environment Programme (UNEP)

Further Readings

Aldy, J., & Stavins, R. (2008). Climate policy architectures for the post-Kyoto world. *Environment, 50*(3), 7–17.

Bailey, I. (2007). Climate policy implementation: Geographical perspectives. *Area, 39,* 415–417.

Bailey, I. (2007). Market environmentalism, new environmental policy instruments and climate policy in the United Kingdom and Germany. *Annals of the Association of American Geographers, 97,* 530–550.

Brody, S. D., Zahran, S., Grover, H., & Vedlitz, A. (2008). A spatial analysis of local climate change policy in the United States. *Landscape and Urban Planning, 87,* 33–41.

Hecht, A. D., & Tripak, D. (1995). Framework agreement on climate change: A scientific and policy history. *Climatic Change, 29,* 371–402.

Solomon, B. D. (1995). Global CO_2 emissions trading: Early lessons from the US Acid Rain Program. *Climatic Change, 30,* 75–96.

United Nations. (1992). *United Nations Framework Convention on Climate Change.* Retrieved January 25, 2009, from http://unfccc.int/resource/docs/convkp/conveng.pdf

Victor, D. G. (2004). *Climate change: Debating America's policy options.* New York: Council on Foreign Relations Press.

 # CLIMATE TYPES

Climate is generally referred to as the average (or mean) weather, but it deals with much more, including variations from average values, extreme events, classification, and long-term changes. Climate is much better defined as the statistical summary of atmospheric elements, such as temperature and precipitation, at a location or over a region over a long period. The values of the statistics vary from one location to another based on the control factors. By using the statistics, knowledge of the control factors, and a classification system, it is possible to separate or group locations into climate types. Sometimes, two locations that are near one another have very different climates, while two other locations may be far apart and have very similar climates. There are many classification systems available, but one of the most widely used is the Köppen system, which determines climate type based on temperature and precipitation data.

Climate Data and Statistics

Many observations are necessary to calculate the statistical properties. Thousands of locations around the world take hourly measurements of temperature, precipitation, wind speed and direction, pressure, and many other atmospheric elements. These measurements are the sources for much of the climate data that scientists use. By focusing on one location over a long period, such as 30 yrs. (years), it is possible to calculate average values for the elements and also examine the variation of these elements over time. For example, the average temperature for November 1 in Boston, Massachusetts, is 13.8 °C (57 °F). If this value is compared with the temperature for individual years, it becomes obvious that some years are cooler, some years are warmer, and, in rare cases, some years may be the same as the average.

Using the same data, it is possible to examine seasonal patterns. When are the cool months, and when are the warm months? Is precipitation evenly distributed throughout the year, or are there any dry and wet periods? If there are wet and dry periods, during which seasons do they occur? To answer these questions, and others, it is necessary to understand some of the statistics that are calculated using atmospheric measurements. The following values are calculated over a long period, such as 30 yrs. The average daily temperature is obtained by adding the minimum and maximum temperatures for a day and dividing by two. Adding all the average daily temperatures during a month and dividing by the number of days results in the average monthly temperature. The average annual temperature is the average of the 12 average monthly temperatures during a year. The annual temperature range is the difference between the warmest average monthly temperature and the coolest average monthly temperature.

Similar values are determined for precipitation. It is important to note that precipitation refers to liquid amounts; any solid precipitation (snow, sleet, freezing rain, and hail) is melted to give a liquid equivalent. Average monthly precipitation is the total precipitation during a month, obtained by adding daily average precipitation amounts. Average annual precipitation is the total precipitation from January 1 to December 31. It is the sum of the 12 average monthly precipitation values.

Once these values are calculated for a location, it is possible to understand the climate at that location. It would be extremely difficult, if not impossible, to find two stations that have exactly the same climate conditions. For example, the climate of Boston, Massachusetts, is quite different from the climates of Honolulu, Hawaii; Seattle, Washington; and Beijing, China. Although it is possible for different climate conditions to exist over short horizontal distances, locations within a region tend to have similar climates, because they are influenced by the same climate controls or factors that affect long-term atmospheric conditions.

Climate Controls

Climate controls include latitude, proximity to water, geographic position and prevailing winds, elevation, ocean currents, and pressure and wind systems. General patterns exist for each of these controls. Temperature tends to decrease as latitude increases. Water has a moderating effect on temperatures, in that locations near water usually have smaller temperature ranges. Proximity to water also affects precipitation amount; precipitation typically decreases as distance from large water bodies increases. If the prevailing wind over a continent is from west to east, then a location on the western coast will have a different climate from one on the eastern coast. The western location will be affected by air coming from the water, but the eastern location will be affected by air that has been over land for a long period. Therefore, the western location tends to be cooler during summer and warmer during winter, while the eastern location is warmer during summer and cooler during winter. As elevation increases, temperature tends to decrease, and precipitation tends to increase. Locations near cold ocean currents are typically cooler than locations near warm ocean currents. Finally, areas under the influence of high pressure year-round are usually dry; areas under the influence of low pressure for much of the year tend to be wet; and areas where the pressure pattern alternates between high and low pressure tend to have wet and dry seasons. It is important to remember that these descriptions are generalizations. In addition, multiple climate controls usually affect a given region, and the interaction of the controls complicates the climate patterns.

Locations that have similar controls have similar climates. When these locations are near one another, such that the controls result in a homogeneous set of conditions, the locations are in a climate region. It is also possible for locations to be thousands of miles apart yet have similar climates. For example, the climate in coastal Southern California is very similar to the climate throughout the Mediterranean region. A climate similar to Southern New England could be found in parts of Northern Europe or Eastern Asia. Conditions in Tucson, Arizona, are similar to those in parts of Australia.

Climate Classification

Because similar climates are affected by similar controls, it is possible to organize the climates into classes or types. This process is called climate classification. Classification helps bring order to the large amount of climate data and facilitates communication of the data. It is essentially a shorthand to help recognize variations and similarities in climate conditions and characteristics. There are many classification systems, ranging from those that may be applied worldwide to those that are very specific and may only be useful in a certain region.

There are three main approaches to determining climate types—empirical, genetic, and applied. Empirical classification is based on observed elements, such as average monthly and annual temperatures and precipitation values. This approach focuses on the numbers. The genetic approach is based on causes. It attempts to show relationships between an atmospheric element, such as temperature or precipitation, and the factors that affect the element. Applied classification is based on the effect of climate. This approach may consider how climate affects vegetation or human comfort.

The approaches are the basis for the many classification systems. Several systems are introduced here before turning to the most widely used system. Grosswetter, a genetic approach, investigates large-scale atmospheric circulation patterns and how they steer weather systems. The Strahler classification system, developed by Arthur N. Strahler in 1951 and later modified by John E. Oliver, focuses on the influence of atmospheric circulation on air mass type. It is a combination of the empirical and genetic approaches. Charles W. Thornthwaite developed an applied system that he later modified in 1948. It focuses on measurements of temperature and potential evapotranspiration (how much water could leave plants given an endless supply) and vegetation type. The most widely taught and therefore the most popular system is the Köppen system. It involves an empirical approach.

Köppen Climate Classification System

Wladimir Köppen developed the Köppen classification system in 1918. It was revised several times by him and modified by others since its inception; however, the basic structure of the system has remained intact. Köppen was trained in climatology and botany, and he used his knowledge of both fields to construct a system that attempted to explain vegetation distribution based on temperature and precipitation data. That is, the type of vegetation that grows in a region is related to the climate conditions in that region. The main climate data are average monthly temperatures, average monthly precipitation, and average annual temperature.

The Köppen classification system divides the world into four major climate types that are based on temperature and a fifth that is based on both temperature and precipitation. Each type is assigned a capital letter. Those types based on temperature include tropical and humid (A-type climates), midlatitude mild (C-type climates), midlatitude severe (D-type climates), and polar (E-type climates). The remaining climate, which considers temperature and precipitation, is designated as dry (B-type climates). Each of these major types is subdivided using specific information, such as precipitation timing and amount and temperature extremes. The subdivisions, designated by one or

two additional letters, are discussed in more detail below. By using a letter system, Köppen simplified the process of labeling climate regions.

A-Type Climates

The A-type climates extend from the equator (0°) to approximately 15° to 25° north and south of the equator. These climates are very warm year-round, with the average temperature of every month higher than 18 °C (64.4 °F). They also tend to have a small annual temperature range (several degrees) and much annual precipitation (generally greater than 150 cm [centimeters], or 59 in. [inches]). Because temperature does not vary greatly, the timing of the precipitation, which falls almost entirely as rain, usually defines the seasons. The vegetation that grows in these climates is known as megatherms, which are large leafy plants.

Subdivisions of the A-type climates include tropical wet or tropical rain forest (Af), tropical monsoon (Am), and tropical wet and dry or tropical savanna (Aw). Tropical wet climates, as their name implies, have no dry season. They receive at least 6 cm (2.4 in.) of precipitation every month. Tropical monsoon climates have a short dry season, which may last several months. During the driest month(s), precipitation must be less than 6 cm. The tropical wet and dry climates also receive less than 6 cm of precipitation during the driest month, but they have a very well-defined dry season during the winter. These areas tend to have a slightly larger temperature range than the other A-type climates, with their warmest temperatures occurring just prior to the wet season.

B-Type Climates

The B-type climates cover large areas. They are found predominately between 20° and 35° north and south of the equator, but they can be closer to the equator or extend farther north and south of 35°. If they are outside the general latitude range, they tend to occur on expansive landmasses and/or regions surrounded by, or on the leeward side of, mountain chains. Increasing distance from water bodies limits the amount of available moisture, and mountain ranges can block the flow of moisture. The main characteristic, therefore, is a moisture deficit, such that potential evaporation and transpiration exceed the precipitation for the year. Precipitation varies from one year to the next. The vegetation in these climates is known as xerophytes. These plants are generally short, and many of them have thorns, such as cacti.

The subdivisions of the B-type climates—arid (BW) and semiarid (BS)—are based on annual temperature and precipitation. Arid climates are true deserts. Semiarid climates are known as steppes or semideserts; they receive more precipitation and have less evaporation than the true deserts. They separate the true deserts from more humid areas. The subdivisions are further differentiated using temperature. If the average annual temperature exceeds 18 °C (64.4 °F), an *h* is added as a third letter. If the average annual temperature is lower than 18 °C, a *k* is added as a third letter. Each subdivision has a name: BWh climates are known as subtropical, or low-latitude, deserts; BWk climates are midlatitude deserts; BSh climates are subtropical, or low-latitude, steppes; and BSk climates are midlatitude steppes.

C-Type Climates

The C-type climates have a variety of names, including midlatitude mild, humid mesothermal, and midlatitude moist subtropical. They cover a large range of latitudes, generally beginning at the northern and southern limits of the A-type climates—approximately 15° to 25°—and extending to as far as 70° to 75° north and south. They are also typically found on the eastern and western sides of continents. The temperature range is larger than in the A-type climates. Winter temperatures remain fairly mild, with temperatures during the coolest month lower than 18 °C but higher than –3 °C (26.6 °F). Summers are warm and humid, with temperatures during the warmest month higher than 10 °C (50 °F). Precipitation type and amount vary, but it tends to fall mostly as rain, although snow is not uncommon in the higher latitudes. In general, midlatitude cyclones cause much of the winter precipitation, and showers and convective thunderstorms dominate during the summer months. While vegetation appearance varies, it fits under the category known as mesotherms, which includes types of conifers, deciduous trees, and some scrubland.

There are many subdivisions of the C-type climates. The Cs-type climates are known as Mediterranean or dry summer subtropical. At least one winter month must have three times more precipitation than the driest summer month, and 1 mo. (month) must have less than 3 cm (1.2 in.) of precipitation. Cw-type climates are called subtropical monsoon or midlatitude wet-and-dry, mild winter. These areas have dry winters, with at least 10 times as much precipitation in a summer month as in the driest winter month. The Cs- and Cw-type climates are further separated by adding an *a* or a *b* as a third letter. An *a* indicates that the average temperature in the warmest month is higher than 22 °C (71.6 °F), while a *b* indicates that it is lower than 22 °C, and at least 4 mos. have temperatures higher than 10 °C. The third subdivision is Cf, which is known as the midlatitude rainy, mild winter. It is separated into Cfa (humid subtropical), Cfb, and Cfc, known as marine west coast climate. A *c* as the third letter indicates that only for 1 to 3 mos. is the temperature higher than 10 °C.

D-Type Climates

The D-type climates also have several names, including midlatitude severe, humid microthermal, and midlatitude moist continental. These climates begin at the poleward extent of the C-type climates—approximately 45° to 70°—and extend to 80° in the Northern Hemisphere only. They are not found in the Southern Hemisphere, because there are few large landmasses in the mid to high latitudes. D-type climates have large temperature ranges due in part to the great variation in hours of daylight and in the solar angle through the year and the heating and cooling properties of land. Summers are warm to cool, with at least 1 mo. having temperatures higher than 10 °C, and winters are cold or severe, with temperatures in the coldest month falling below –3 °C. Precipitation falls mainly as rain during the summer; winter precipitation falls as snow or rain. Vegetation is known as microtherms, which includes types of conifers and deciduous trees.

There are two secondary divisions of the D-type climates—Df (midlatitude rainy, cold winter) and Dw (midlatitude wet-and-dry, cold winter). The meaning of the second letter is the same as described above. A third letter reveals details of the temperature, including *a*, *b*, and *c*, which are described above, and *d*, which is used when the coldest month has temperatures lower than −38 °C (36.4 °F). Dfa-, Dfb-, Dwa-, and Dwb-type climates are also known as humid continental, while Dfc-, Dfd-, Dwc-, and Dwd-type climates are also known as subarctic.

E-Type Climates

The final major climate type in the Köppen system is the E-type climates, which are known as polar. These climates are found poleward of the D-type climates in Asia, Europe, North America, Greenland, and Antarctica. Temperatures are cold year-round, with temperatures in the warmest month lower than 10 °C, due primarily to low sun angles. Many of these locations experience 24 hours of darkness for some portion of the year. Precipitation falls mainly as snow, although rain can also occur. Precipitation is limited, but it remains on the ground for long periods due to the cold temperatures. Vegetation is limited or nonexistent. Where plants do grow, they have a short growing season. The vegetation is known as hekistotherms, which includes grasses, some wildflowers, and lichens.

There are only two subdivisions of the E-type climates—ET and EF. ET-type climates are known as tundra. These areas have an average temperature in the warmest month between 0 °C (32 °F) and 10 °C. Vegetation can grow here. EF-type climates are known as ice caps. Every month, the temperature is below 0 °C so that most of the land is either frozen or continually covered by snow and/or ice.

H-Type Climates

A sixth climate type has been added to Köppen's original system—H-type climates. These are known as mountain or highland climates. These climates are shown on maps when mountainous terrain leads to rapid changes in climate type over short horizontal distances. In extreme cases, climate types can transition from A or B to C to D and possibly to E from the base of a mountain or mountain range to the peak.

Richard R. Brandt

See also Atmospheric Circulation; Atmospheric Variations in Energy; Climate: Dry; Climate: Midlatitude, Mild; Climate: Midlatitude, Severe; Climate: Mountain; Climate: Polar; Climate: Tropical Humid; Climate Change; Climatology; Köppen-Geiger Climate Classification

Further Readings

Aguado, E., & Burt, J. E. (2007). *Understanding weather and climate*. Upper Saddle River, NJ: Prentice Hall.

Barry, R. G., & Carlton, A. M. (2001). *Synoptic and dynamic climatology*. New York: Routledge.

Barry, R. G., & Chorley, R. J. (2003). *Atmosphere, weather, and climate*. New York: Routledge.

Bridgman, H. A., & Oliver, J. E. (2006). *The global climate system: Patterns, processes, and teleconnections*. New York: Cambridge University Press.

Critchfield, H. J. (1998). *General climatology*. Upper Saddle River, NJ: Prentice Hall.

Oliver, J. E., & Hidore, J. J. (2002). *Climatology: An atmospheric science*. Upper Saddle River, NJ: Prentice Hall.

Rohli, R. V., & Vega, A. J. (2007). *Climatology*. Sudbury, MA: Jones & Bartlett.

Yarnal, B. M. (1994). *Synoptic climatology in environmental analysis: A primer*. Hoboken, NJ: Wiley.

CLIMATIC RELICT

Sometimes referred to as *glacial relicts*, *climatic relicts* are taxa that were more widely distributed in the past when climatic conditions were more conducive to a greater range. During times of optimum climatic conditions, taxa may expand their ranges into new regions. As global climates have changed, climatic relict taxa have adapted by reducing their ranges to smaller refugia, where residual communities remain isolated (or disjunct) from their ancestral community.

Taxa can be relict due to shifts in entire climate zones or due to changes in single climatic variables, such as temperature or precipitation. In particular, a number of climatic relict taxa remain due to climate shifts during the early Holocene, when conditions began to warm following the cooler glacial-interglacial conditions of the Pleistocene. While this phenomenon is observable in both floral and faunal taxa, it is expected that the number of climatic relict taxa will increase over the next century due to anthropogenic climate change.

Floral Climatic Relics

While the number of climatic relict taxa is unknown (estimated to be several hundred species in Eurasia alone), Norwegian mugwort (*Artemisia norvegia*) is a commonly cited example. Reconstructions of past glacial environments show that the alpine plant was widely found throughout Central Europe during the last glacial maximum. Its extent decreased when the climate warmed and is now restricted to the less forested coastal mountains of Norway, the Ural Mountains, and two isolated communities in the Scottish Highlands where cooler conditions persist.

Climatic relics do not have to occur exclusively due to climatic warming. For example, the European strawberry tree (*Arbutus unedo*) is currently found throughout the Iberian Peninsula and around the south and west coastlines of the Mediterranean Sea. However, its range extended farther north during the early-to-mid Holocene, when the climate was warmer than at present. Today, two disjunct populations of the tree remain in Western Ireland, where the North Atlantic Gulf Stream buffers the effects of the colder Irish climate.

Faunal Climatic Relics

Faunal climatic relics typically require less time than floral taxa to expand their ranges during periods of optimum climatic conditions. Species such as the musk ox (*Ovibos moschatus*) are capable of traveling large distances but are limited by the extent of their habitat. Fossil evidence suggests that during phases of Pleistocene glaciation, the musk ox expanded its range into Europe north of the Pyrenees and Alps beginning around 130,000 yrs. (years) ago. This range receded with the glaciers to its current-day relict position around the arctic tundra of Greenland and Canada.

Changes in climate can also affect marine taxa, resulting in climatic relict species such as the staghorn coral (*Acropora cervicornis*). The coral is typically found in the warm waters of the Caribbean coinciding with the 18 °C monthly minimum seawater isotherm. However, the range of the staghorn coral was wider during the early-to-mid Holocene, when the Western Atlantic waters were warmer. The cold-sensitive staghorn coral can now only be found, in addition to the Bahamas, to the east of Florida, where the Florida Current helps maintain a relict population of the species.

Future Climatic Relics

Climatic relics, particularly glacial relics, can occasionally represent the last of any remnants of a species before becoming extinct. A number of species have been identified by the Intergovernmental Panel on Climate Change (IPCC) and the International Union for Conservation of Nature and Natural Resources as changing their ranges in response to current-day climate change. One such species is the Joshua tree (*Yucca brevifolia*), which is losing range from the south due to increasing temperatures and a reduced ability to migrate due to the extinction of the giant sloth (which would spread seeds in their dung). It is expected that the Joshua tree will either become extinct or its range will slowly migrate north from its current position in the Mojave Desert region. Whether this species will become a climatic relict by finding a refuge from the warming Mojave climate remains in question.

Conversely, the American pika (*Ochotona princeps*) has the ability to move to a cooler location but is running out of new habitats to migrate to. Donald Grayson found that the average elevation of the pika around 40,000 to 10,000 yrs. ago was 1,750 m (meters). However, since the gradual warming of the Holocene, the species has been forced to migrate upslope to where average temperatures are cooler, at an average elevation of about 2,320 m. This has resulted in a fragmented climatic relict distribution whereby many members of the species are isolated on mountaintops, which limits their ability to migrate when climatic conditions become unsuitable.

It is expected that if the current-day climate continues to warm, many populations of American pika that are currently found in the Great Basin region will begin to run out of cooler elevations they can retreat to.

Maria Caffrey

See also Biogeography; Biota and Climate; Climate Change; Extinctions; Glaciers: Continental; Glaciers: Mountain; Global Climate Change; Island Biogeography

Further Readings

Cox, C. B., & Moore, P. D. (2005). *Biogeography: An ecological and evolutionary approach.* Malden, MA: Blackwell Science.

Grayson, D. (2005). A brief history of Great Basin pikas. *Journal of Biogeography, 32,* 2103–2111.

Lomolino, M. V., Riddle, B. R., & Brown, J. H. (2006). *Biogeography* (3rd ed.). Sunderland, MA: Sinauer.

MacDonald, G. M. (2003). *Biogeography: Introduction to space, time and life.* New York: Wiley.

Parry, M. L., Canziani, O. F., Palutikof, J. P., van der Linden, P. J., & Hanson, C. E. (2007). *Climate change 2007: Impacts, adaptation and vulnerability* (Contribution of Working Group II to the Fourth Assessment Report of the Intergovernmental Panel on Climate Change). Cambridge, UK: Cambridge University Press.

Shogren, E. (2008, February). *Outlook bleak for Joshua trees.* Retrieved January 16, 2009, from the National Public Radio Web site: www.npr.org/templates/story/story.php?storyId=17628032

CLIMATOLOGY

Climatology is one of the primary subdisciplines of physical geography, along with biogeography and geomorphology. Climatology is the scientific study of the world's vast and ever-changing climate systems through time. The spatial nature of the discipline makes it unique to geography, though it is very much an interdisciplinary field of study that often integrates information from related sciences (e.g., meteorology, oceanography, ecology, anthropology, and geology) to understand climate variability and its impacts on Earth and human systems. Within the field, climate is investigated on a range of timescales, generally from decades to centuries. This is why it is often regarded as the long-term study of weather in which statistical patterns of weather phenomena can be detected. In many ways, the study of climate is more expansive and diverse than this definition proposes. This point is demonstrated by the variety of subdisciplinary studies within climatology, the wide array of data, resources, and methodologies used within the field, and the emerging relevance of climatology in today's society, where we are experiencing evident climate change with far-reaching impacts extending to energy, economics, and policy issues. There are six major subdisciplines within climatology: (1) paleoclimatology, (2) microclimatology, (3) physical climatology, (4) synoptic and dynamic climatology, (5) hydroclimatology, and (6) health/medical climatology. The use of geospatial techniques, from geographic information systems to remote sensing, is frequently incorporated in all areas of climatological investigations, as are geostatistics and climate models. Data sources for climatological examination range from instrumental to proxy and reconstructed records.

Climatology Studies

Paleoclimatology

Paleoclimatology is an important subdiscipline of climatology that focuses on the examination of past climates on Earth. Roughly, the temporal period for studying past climates is considered to span beyond the intervals that instrumental data records provide, ranging from approximately 50 years ago to several million years ago and earlier. This type of climatological study is aimed at uncovering the extent to which climate has changed throughout geologic time while revealing the internal and external influences (of natural and anthropogenic origin) on past climate conditions and climate change. Examples of some of these influences include variations in Earth's orbital parameters, plate tectonics, volcanic eruptions, and the extensive logging and clear-cutting practices of the early North American settlers. In addition, paleoclimatologists often evaluate how climate variability in the distant past may have assisted in the formation, arrangement, and deformation of the structure of Earth's present and past landforms and water bodies as well as atmospheric composition. For instance, extensive periods of wetter than average conditions in the Pleistocene led to the creation of many paleolakes, which are mostly dried up at present. To assist many paleoclimatological investigations, proxy data records are used from sources such as tree rings, glacial ice cores, pollen, lake and ocean sediments, and oxygen isotopes. These data are critical in providing long-term evidence by which climate conditions of the distant past can be evaluated. Often, climate reconstructions of meteorological variables such as air temperature, pressure, precipitation, and stream flow variability can be created from such sources.

Microclimatology

The localized study of unique climate conditions at or near the ground surface is another subdiscipline of climatology, known as *microclimatology*. This includes the assessment of vertical variations in climatological and meteorological conditions within 1 to 2 meters of Earth's surface. In addition, the horizontal variability of climate between site-specific locations is also examined in microclimatology. Important influences to the microclimate of a given area can include longwave and shortwave radiation flux, soil characteristics, topography, evapotranspiration rates, and the type and extent of vegetative cover. These elements are often highly variable between locations and between atmospheric layers located further from the surface layer. Microclimatology is closely connected to the study of larger-scale climate studies, as factors such as cloud cover and the movement of air masses and weather patterns may also affect the microclimate of an area. Of course, the nature of interactions between humans, animals, insects, and technology with the ground surface can also lead to unique and variable microclimate conditions. This field within climatology uses a host of monitoring systems, instrumentation (i.e., anemometers, hygrometers, thermometers, rain gauges, staff gauges, evaporimeters, and lysimeters), and sensors for observational measurements of climatological data.

Physical Climatology

Physical climatology is a subdisciplinary study of climatology that investigates the principles, processes, and interactions of the present air-sea-land interface in relation to the past, so that future changes to the climate system may be understood and predicted. This field incorporates working understandings of atmospheric physics and geophysical fluid dynamics, particularly the ways in which energy is distributed across the planet, so that areas of sensitivity may be identified and assessed. Physical climatologists play a critical role in modeling the state of the Earth's climate. The models produced by physical climatologists allow for the assessment of large and small hypothetical changes to controlled components of the climate system, so that their impacts on other areas of the system may be examined. This is currently the predominant way in which various anthropogenic climate change scenario impacts are forecast. Outputs generally display some level of variation between models, as they are highly dependent on the algorithms developed by physical climatologists. Therefore, researchers in this realm of climatology must present an accurate depiction of real-world climatic factors, from the amount of transpiration produced by a single leaf on a given day to the decadal patterns of global oceanic circulation.

Synoptic and Dynamic Climatology

Synoptic and dynamic climatology describes the subdisciplinary area of climatology that explains the spatial and temporal variability of atmospheric circulation patterns while examining subsequent impacts on the state of the climate system. Large-scale circulation influences the strength and persistence of pressure centers, air mass advections, and the movement of storm tracks. Therefore, to understand weather experienced at the surface, synoptic and dynamic climatological research aims at characterizing the nature of atmospheric circulation variability. This can include the study of jet streams, planetary waves, oscillatory modes of climate variability (e.g., the El Niño/Southern Oscillation and North Atlantic Oscillation), circulation cells, blocking phenomena, and synoptic weather types. Additional research within the field focuses on how atmospheric circulation patterns influence the environment and humans. Often, this involves the assimilation of many different climate and weather components, from physical observations to climate model outputs, to understand the conditions affecting a unique area.

Hydroclimatology

The study of the world's climate system in relation to the hydrosphere is known as *hydroclimatology*. Hydroclimatologists examine the spatial and temporal variability of moisture conditions at and beneath the surface as well as within the atmosphere. This can include investigations of precipitation and snowfall, snow cover/extent, snow pack depth and stability, stream flow, atmospheric and soil moisture, evapotranspiration, and water resource management and distribution. Periods of significant, anomalous hydroclimatic variability—for instance, during times of drought or flood—are a focus of this field. A multitude of changes in atmospheric conditions and land surface processes leading to hydroclimatic variability are also frequently examined within this subdiscipline of climatology. Hydroclimatology research stems from data collected across a dense network of stations around the world that monitor various elements of the hydrosphere as well as climate model outputs. However, it is the synthesis of this information with spatial analyses that truly separates this discipline from traditional hydrology and hydrometeorology investigations.

Health and Medical Climatology

The subdisciplinary study of climatology known as *health and medical climatology* focuses on how climate variability influences the quality of human health and general human biology and physiology. This field of study focuses on how the human body responds to natural climate flux as well as anthropogenic alterations to the climate system. Research conducted within this subdiscipline can include examinations of summer climate conditions and how the frequency and duration of human exposure to heat and humidity can lead to various levels of stress and even death for different age groups. Alternatively, a health/medical climatology investigation of winter climate might assess how the human body adjusts to below-average temperatures in winter with various levels of exposure (e.g., from the time it takes to shovel a driveway to the duration of an outdoor hockey game). A very different type of analysis in this field might consider how climate variability influences the prevalence and spread of diseases such as malaria and shistosomiasis or how migratory populations are affected by changing climate conditions. A great deal of national and global cooperation between climatologists, government, and health/medical professionals and scientists is required (especially with data sharing) to conduct investigations in this field.

Climatology Resources

Geospatial Techniques

As a spatial discipline, climatology frequently incorporates the use of geospatial techniques in research analyses. Some of these resources include geographic information systems (GIS), satellites, remote sensing, and global positioning systems (GPS). The use of GIS as a computing tool for data-mapping and display purposes is widespread in the climatological community. GIS is also considered a science that can aid climatologists with the storage, organization and arrangement, modification and manipulation, and synthesis and analysis of data sets. This is particularly useful when working with large or cumbersome data sets, which are often incorporated in climatological investigations. For example, statistical operations can be performed within a GIS that allow for data gathered at point locations across a region, continent, or hemisphere to be displayed on a map with various spatial interpolations. Climatologists can select and change very specific criteria within a GIS while working with and displaying data, such as the colors used, map projection type, the resolution size, or the spatial area that the data will be confined to. The ease with which GIS allows a researcher to amend or alter the methodologies of an examination is a highly favorable feature that has led to its popularity among climatologists. Of further interest, the temporal component of data sets can also be incorporated into a GIS for climatological assessment. For example, there are tracking features within a GIS that allow for the examination of tropical storm and hurricane paths to be conveyed and compared through time. Recently, GIS specialists and scientists have been working to further increase the capabilities of GIS for evaluating temporal data.

Many data used for climatological analyses are regularly collected by geospatial techniques with satellites and remote sensing. Over the past three decades, these techniques have become increasingly sophisticated and are widely employed to gather climate data. Radar is a well-known variety of one of these techniques that is extensively used within the field of climatology. Currently, there are far more climate data available from these sources than there are research investigations to use them. Both satellites and remote sensing allow for components of the climate system (ranging from atmospheric water vapor content to sea surface temperatures to the color of vegetative cover) to be monitored and collected by a host of sensors placed at various distances from Earth's surface. These techniques are advantageous to climatology because of the precision with which data can be measured as well as the continuous intervals at which data can be temporally gathered. In addition, data can be transferred rapidly to a computer for processing and examination, and bias from human error, instrumentation, and ground conditions is generally avoided with these techniques. A variety of satellite data receivers, or GPS units, are available to further aid climatologists in research investigations. These are particularly useful in conducting outdoor field experiments such as those performed in hydroclimatology, microclimatology, and paleoclimatology.

Geostatistics

As with geospatial techniques, climatological assessments also rely heavily on the use of geostatistics. Geostatistics encompass a wide range of statistical methods and statistical models, as well as the development of statistical algorithms that are used for the purpose of environmental data analysis. Geostatistics are important in all aspects of climatology studies. They can be used in the preliminary steps of a research investigation, for instance, to determine how stations should best be distributed across a study area based on the kind of climatological analysis being conducted. Geostatistics may also be the primary way in which large data sets are examined in a research investigation, once collected. Examples of these kinds of evaluations include regression analyses, cluster analyses, and principal components analyses. In addition, geostatistics assist climatologists in proving the validity and relevance of their results with tests for data significance, variance, and correlation. Climatologists often employ geostatistical procedures within statistical computing and programming packages as well as within a GIS.

Climate Models

Climate simulation models are important for climatology research to investigate the sensitivity of Earth's climate to various external and internal forcings. Climate models are useful because the collective algorithms run within an individual computer model serve as numerical representations of the physical and chemical state of the climate system at a given time and are useful for short- and long-range climate predictions. There are many different kinds of climate simulation models in operation around the world that examine climate on numerous spatial scales. Some climate models are designed as three-dimensional replications of the coupled atmosphere-ocean system, such as general circulation models, while others simulate a specific component of the climate system (e.g., modeling the global carbon cycle). All are useful for projecting complex climatic responses to change, especially since many different hypothetical scenarios can be considered within one climate model. For example, a climate model is considered a highly beneficial resource for projecting temperature variability responses to increasing fossil fuel emissions. However, the impact on other areas of the climate system, such as the likelihood of drought and flood or changes to sea ice extent and permafrost cover, may also be examined.

Climate Data

To understand present and past climate conditions and make accurate forecasts of the future climate, climatologists make use of a variety of climate data resources. Instrumental data records collected at individual sites along the surface are one type of data used for research. These data are collected all over the world, from places such as first-order weather station networks, cooperative station networks, ocean buoys, and even ship observations. Measurements may be taken of variables such as temperature, precipitation, cloud cover, wind speed, and wind direction with time. In addition, real-world observational data are collected from remote sensing with ground sensors, aircrafts, and satellites. Composited data are also used based on surface observations and observations of climate conditions throughout the atmosphere, hydrosphere, and lithosphere (e.g., sea surface temperatures for territorial regions of the Pacific Ocean). Approximations of climate from natural sources, ranging from tree rings to ocean sediments, further provide important information that climatologists draw from, particularly in paleoclimate investigations. From a combination of these sources, climatologists can statistically reconstruct the state of the climate system at various time intervals. This can make available a number of reconstructed data records that may be used for other climatological assessments. Another type of data that are available to climatologists stem from approximations of climate based on computer-generated model simulations of actual, real-world conditions. Climate data are used on a range of timescales, as hourly, daily, twice daily, weekly, monthly, yearly, or longer periods may all be necessary temporal intervals for a particular investigation. Furthermore, climate data are assessed in terms of averages, maximums, minimums, totals, and extents.

Climatology and Climate Change

Over the past few decades, global climate change has become a powerful topic for world discussions and debate, with climate programs, committees, and policies directed at coping with the rate at which human activities may be affecting natural climate flux. Climatology is at the center of the issue, as the current and future directions of these initiatives rely on evidence provided by trained climate researchers. Primarily, it is through climatology that changes occurring within the climate system through time are understood. Climatology studies also help educate the public and develop a comprehension of how climate change may be beneficial and detrimental to society and the environment. For instance, rising temperatures that result in sea-level rise may have a host of negative environmental and economic impacts on coastal environments as well as on polar and subpolar climate zones. However, some higher-latitude and higher-elevation regions may find that the growing seasons of certain crops are extended, and therefore more productive, with increasing temperatures, which could improve the economy of these areas.

Government, industry, and individuals must make decisions on how to prepare for all possible scenarios of climate change. The justification for enacting and enforcing change (e.g., with policy, incentives, restrictions, and mandates) aimed at influencing human activity and interactions with nature stems in great part from physical evidence uncovered by climatology research. In some instances, decision makers seek direct consultation from climatologists, and in others, climatologists make recommendations to decision makers and the general public. Climatology reports are also frequently issued from climate groups and agencies on matters concerning climate conditions and climate change. This has been especially prevalent in today's society where energy consumption and conservation are concerned, for example, with such groups or agencies promoting the use of alternative energy sources.

Melissa L. Malin

See also Anthropogenic Climate Change; Climate Change; Climate Policy; Climate Types; Geostatistics; GIS in Environmental Management; Models and Modeling

Further Readings

Barry, R. G., & Carleton, A. M. (2001). *Synoptic and dynamic climatology*. New York: Routledge.

Barry, R. G., & Chorley, R. J. (2003). *Atmosphere, weather and climate* (8th ed.). New York: Routledge.

Christopherson, R. W. (2006). *Geosystems* (7th ed., pp. 174–317). Upper Saddle River, NJ: Pearson Prentice Hall.

Davis, J. C. (2002). *Statistics and data analysis in geology* (3rd ed.). New York: Wiley.

Geiger, R., Aron, R. H., & Todhunter, P. (2003). *The climate near the ground* (6th ed.). Lanham, MD: Rowman & Littlefield.

Longley, P. A., Godchild, M. F., Maguire, D. J., & Rhind, D. W. (2005). *Geographical information systems and science* (2nd ed.). Chichester, UK: Wiley.

Lutgens, F. K., & Tarbuck, E. J. (2006). *The atmosphere* (10th ed., pp. 196–246). Upper Saddle River, NJ: Pearson Prentice Hall.

Lydolph, P. (1985). *The climate of the Earth*. Totowa, NJ: Rowman & Allanheld.

Salby, M. L. (1996). *Fundamentals of atmospheric physics*. San Diego, CA: Academic Press.

CLOUDS

A cloud is any visible mass of liquid drops or ice crystals suspended above the surface of the Earth. In addition to their obvious importance in producing precipitation, clouds play an important role in large-scale circulation by cycling water and energy through the atmosphere. Because of their generally high albedo but widely varying temperature, clouds constitute an important feedback in the global radiation balance and are a significant source of the uncertainty in estimates of global climate change. A simple overview of clouds can proceed from an introduction to the traditional cloud classification scheme and a discussion of rudimentary cloud microphysical processes. The development of global cloud climatologies and methods of assessing the influence of clouds on climate are included as being particularly relevant to the discipline of geography. It is helpful first to distinguish between clouds and precipitation.

Clouds and Precipitation

Cloud particles, whether composed of liquid drops or ice, are referred to generally as *hydrometeors*, a term that also includes precipitation-size particles. Although also composed of liquid drops or ice-phase particles, and inextricably linked with cloud processes, *precipitation* is distinguished from *cloud* in that precipitation-sized particles have a nonnegligible fall velocity. Cloud droplets, on the other hand, are generally characterized by a terminal fall velocity much smaller than the cloud updraft magnitude (i.e., $v_t \ll w$). This distinction leads to definitions and descriptions that refer to clouds as being *suspended* (as mentioned above) or *floating* in the atmosphere.

Cloud Classification

The cloud taxonomy in current use was originally formulated by the pharmacist and amateur meteorologist Luke Howard in 1803 and adopted formally in 1956 by the World Meteorological Organization (WMO) on publication of the *International Cloud Atlas*. Clouds are classified based on 10 mutually exclusive categories derived from their structural characteristics. These 10 cloud *genera* are cirrus (Ci), cirrocumulus (Cc), cirrostratus (Cs), altocumulus (Ac), altostratus (As), nimbostratus (Ns), cumulus (Cu), stratocumulus (Sc), stratus (St), and cumulonimbus (Cb).

Additional classifications include *species*, which ascribe to clouds details of shape and structure, and *varieties*, which represent particulars of cloud arrangement and optical properties. An exhaustive cloud classification also includes a description of supplemental features and *accessory* clouds. Last, a description of cloud origin, termed *mother cloud*, can be applied when one cloud forms from another. These five categories permit an extraordinary number of cloud types. Typical meteorological practice, however, employs only the genera and occasionally a species, for example, *cumulus humilis* (fair weather cumulus).

The 10 cloud genera are consistent with a cloud classification by altitude. Examples of low cloud types are cumulus, stratocumulus, and stratus. Altocumulus, altostratus, and nimbostratus are midlevel clouds. Upper-level clouds include cirrus, cirrocumulus, and cirrostratus. Cumulonimbus, which frequently span the entire depth of the troposphere, are typically included in the low-cloud category.

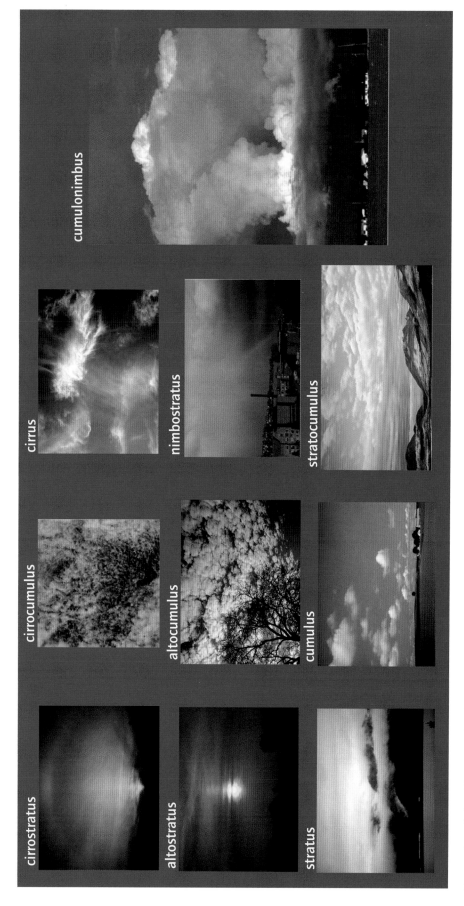

Examples of the 10 WMO cloud genera

Source: Nimbostratus cloud photo taken by Dr. M. A. LeMone, www.windows.ucar.edu/tour/link=/earth/Atmosphere/images/nimbostratus1_big_jpg_image.html. All other cloud photos, copyright © John Day, 2006, www.cloudman.com.

Although nearly all national meteorological services employ the WMO cloud classification, other useful cloud taxonomies exist. A cloud classification based on hydrometeor composition (liquid, ice, or mixed phase) reflects the importance of cloud formation or growth mechanisms and is important when retrieving cloud properties (e.g., liquid water content or mean drop size) from remote sensing platforms such as satellites or radar. The International Satellite Cloud Climatology Project (ISCCP), begun in 1982, developed a nine-category cloud classification based on cloud top pressure and cloud optical depth obtained from satellite observations. These nine ISCCP categories have much in common with the 10 WMO cloud genera.

Cloud Formation and Development

Clouds originate from the condensation or deposition of water vapor. Cloud formation requires the presence of water vapor and some mechanism to achieve supersaturation conditions, defined as a relative humidity over 100%. Most commonly, a supersaturated water vapor field comes about via cooling of air as it flows over topography, ascends in the vicinity of frontal surfaces, or rises in buoyant convective updrafts. As the rising air expands and cools, relative humidity increases above 100%, and embryonic cloud droplets form on soluble atmospheric aerosol called *cloud condensation nuclei* (CCN). Without the presence of these naturally occurring CCN, a relative humidity of approximately 400% would be required to nucleate cloud droplets, a consequence of the free energy barrier associated with the creation of the curved droplet surface. The role of soluble CCN is to reduce the equilibrium vapor pressure over the droplet surface, a mechanism commonly referred to as the *solute effect* (Raoult's law). The interplay between the effects of curvature and dissolved solute is described using Köhler theory.

Once a critical size or relative humidity is reached, the droplet is said to be *activated* and subsequently grows by condensation. Condensational growth of small droplets is rapid and can explain the existence of typical cloud drops of radius ~10 μm (micrometers). Condensation is not, however, sufficient to explain the growth of precipitation-sized droplets. Once a small number of 20-μm droplets are present, rapid droplet growth of precipitation-sized droplets can occur through the process of collision-coalescence, where larger droplets fall and collect smaller cloud droplets, with the large drops growing in the process. The large drops ultimately fall below the cloud base as precipitation.

The requirements for the nucleation of ice-phase hydrometeors are somewhat different. Ice particles can form either directly from the vapor phase or via the freezing of supercooled liquid drops. Homogeneous nucleation—the freezing of liquid drops without the need for any nuclei—occurs near −40 °C. Nucleation at warmer temperatures requires the presence of ice nuclei, broadly categorized as either deposition nuclei or freezing nuclei. Suitable ice nuclei are typically insoluble, with a crystalline structure similar to that of ice. Kaolinite, a clay mineral, is one of the naturally occurring ice nuclei, while insoluble inorganics such as silver iodide have been employed to nucleate ice particles in cloud-seeding experiments. Ice-phase nucleation is highly temperature dependent, with appreciable ice nucleation only beginning at −10 °C. Often, clouds −10 °C or warmer contain very little ice.

When cloud physicists began flying instrumented aircraft through cold clouds, it was noted that the concentration of ice-phase particles was much higher (often by orders of magnitude) than the concentration of ice nuclei. This discrepancy necessitated a secondary ice production mechanism or ice multiplication process. The most commonly accepted explanation of ice multiplication is the Hallet-Mossop process (rime splintering), which occurs over a narrow temperature range when graupel interacts with supercooled liquid droplets, ultimately resulting in the production of a large number of small ice splinters.

Small, just-nucleated ice-phase particles can grow very efficiently by vapor deposition, which is the ice-phase analog to condensational growth for liquid droplets. The presence of pristine ice crystals evinces depositional growth, with the actual temperature and humidity characteristics determining the particular crystal structure, or *habit*, that forms. Possible crystal habits are dendrites, columns, plates, and needles.

Depositional growth in mixed-phase clouds, composed of both ice crystals and supercooled liquid water, is governed by the Bergeron-Findeisen mechanism. Because the equilibrium vapor pressure over liquid water is much greater than that over ice, the interior of a cloud that is saturated with respect to liquid water (i.e., a relative humidity of 100%) is greatly supersaturated with respect to a surface of ice. This difference promotes the depositional growth of the ice-phase hydrometeors and evaporation of the liquid water to maintain 100% relative humidity. Rapid depositional growth of the ice particles will continue as long as sufficient supercooled liquid water is present.

Ice-phase particles can also grow by accretion of supercooled liquid water in a process called *riming*, the mechanism responsible for hail growth, and by aggregation with other ice particles.

Developing Global Cloud Climatologies

Surface meteorological observations often contain information about clouds, usually in the form of sky condition (a measure of the fractional sky coverage by clouds) and cloud base height. Cloud observations may, where applicable, contain descriptions of multiple cloud layers, though of course not all layers may be visible to an observer at the surface. Stephen Warren and Carole Hahn developed and digitized a global climatology of clouds from surface-based observations over the land and ocean. The land-based observations are taken from regular surface observations at fixed locations, while the ocean-based observations are based on cloud observations made by ships at sea.

The advent of polar orbiting and geostationary satellites over the past few decades has made possible a comprehensive cloud climatology spanning the entire globe. Satellites have the advantage of consistent spatial and temporal coverage, but they are nevertheless limited by the fact that conventional passive techniques cannot fully sample multiple cloud layers. This limitation also applies to surface-based remote sensing of clouds. The most significant consequence of this limitation is that climatologies of single-layer clouds are better represented relative to regimes where multiple cloud layers dominate. It is expected that this problem will be ameliorated somewhat in the coming years with the advent of profiling cloud radars, both land and satellite based, optimized to collect cloud information relevant for climate.

Influence of Clouds on Global Climate

Because of their brightness relative to the underlying Earth surface, clouds exert a profound impact on the global radiation balance. Whether specific clouds constitute a cooling or a warming effect (in the *global mean* sense) depends on cloud albedo and cloud top temperature. In the shortwave, cloud feedbacks are always negative; that is, clouds always exert a radiative cooling effect. In the long-wave, however, cloud top temperature ultimately determines the magnitude of warming. A low cloud whose top is perhaps only a few degrees cooler than the surface results in nearly the same outgoing long-wave flux as if the cloud were not there. A cloud high in the atmosphere, on the other hand, because of its cold temperature, yields a much smaller upward flux and a significant warming effect relative to the warm surface below. Whether the shortwave *albedo effect* (always cooling) or the long-wave *greenhouse effect* (either warming or near neutral) dominates depends on the cloud thickness and cloud top height. In general, low clouds such as marine stratocumulus cool the climate system. Upper-level clouds, particularly high, thin cirrus, warm the system.

Cloud radiative properties are strongly modulated by the character of the aerosol on which the cloud droplets form, a dependence that can lead to a series of complicated aerosol-cloud-precipitation interactions. These cloud-mediated radiative feedbacks are termed *aerosol indirect effects*. For example, clouds forming in an environment with higher CCN concentrations will be brighter, reflecting more solar radiation back to space (known as the *first indirect* or *Twomey* effect). Additionally, a higher concentration of CCN tends to suppress the formation of precipitation in warm clouds, in principle leading to longer cloud lifetimes, since liquid water remains in the cloud rather than raining out (the *second indirect*, *cloud lifetime*, or *Albrecht* effect). However, the second aerosol indirect effect is poorly constrained in this simple relationship between a reduction of precipitation and greater cloud lifetime, since it neglects important cloud dynamical feedbacks.

These aerosol-cloud-precipitation feedbacks must be understood so that they may be adequately represented in global climate models.

The Intergovernmental Panel on Climate Change (IPCC) in the Fourth Assessment report stated that cloud feedbacks remain the largest uncertainty in assessments of global climate change. Understanding cloud feedback mechanisms is crucial, since any change (natural or anthropogenic) in Earth's climate would most likely result in a change of cloud properties, with important consequent feedbacks.

David B. Mechem

See also Albedo; Atmospheric Moisture; Precipitation Formation; Weather and Climate Controls

Further Readings

Houze, R. A., Jr. (1993). *Cloud dynamcis*. San Diego, CA: Academic Press.

Lohman, U., & Feichter, J. (2005). Global indirect aerosol effects: A review. *Atmospheric Chemistry and Physics, 5,* 717–737.

Rogers, R. R., & Yau, M. K. (1989). *A short course in cloud physics*. Boston: Butterworth Heinemann.

Stephens, G. L. (2005). Cloud feedbacks in the climate system: A critical review. *Journal of Climate, 18,* 237–273.

Warren, S. G., & Hahn, C. J. (2006, February). *Climatic atlas of clouds over land and ocean.* Retrieved December 31, 2008, from www.atmos.washington.edu/CloudMap

World Meteorological Organization. (1956). *International cloud atlas* (Vol. 1). Geneva: Author.

World Meteorological Organization. (1956). *International cloud atlas* (Vol. 2). Geneva: Author.

 # CLUSTERS

Clusters have long been a subject of much debate within human geography. While historically informed by location theory and regional science, the cluster has evolved as a highly interdisciplinary concept. While the concept of the cluster is particularly poignant in human and, more specifically, economic geography, it is an important concept across the social sciences. In simplest terms, a cluster can be understood as a group of similar or linked firms within a defined geographical area, although in addition to its importance as an empirical project, the concept provides a basis for considering the relationship between actors and space.

The term *cluster* was first coined and subsequently popularized by the regional scientists Stan Czamanski and Luiz Augusto de Q. Ablas in 1979; however, it was the seminal work of Alfred Marshall in the late 19th century that provided the foundation for these debates. Without explicit reference to clusters, Marshall's study of the British textile industries in Lancashire and Yorkshire, potteries in Stoke on Trent, and metallurgical industries in Sheffield identified the existence of "industrial districts." Marshall observed that the high capital costs associated with such Fordist industries meant that they tended to remain in a place for a long time once located, a characteristic fundamental to the establishment and sustenance of the business and the sociocultural relationships of industrial districts.

Through cooperation and competition among firms, Marshall found, industrial districts offered advantages not available to those outside the locale. Indeed, so great were the advantages of the industrial district that it became a significant factor in the location decisions of firms. More specifically, Marshall identified four aspects of external economies whereby industries sought to extend and refine social and economic relationships as well as physical infrastructure through the scale of production: (1) the reduction of transportation costs, (2) the creation of a common specialized labor force, (3) the specialization of input products, and (4) an intangible dimension that Marshall (1890) referred to as how "the mysteries of trade become no mysteries, but are as it were in the air" (p. 271).

Marshall's work on industrial districts, agglomeration economies, and external economies provided an important focus for much of the cluster literature and the basis for more contemporary theorizations of the cluster. While the neologisms for the term *cluster* have become ever more numerous with the changing nature of the

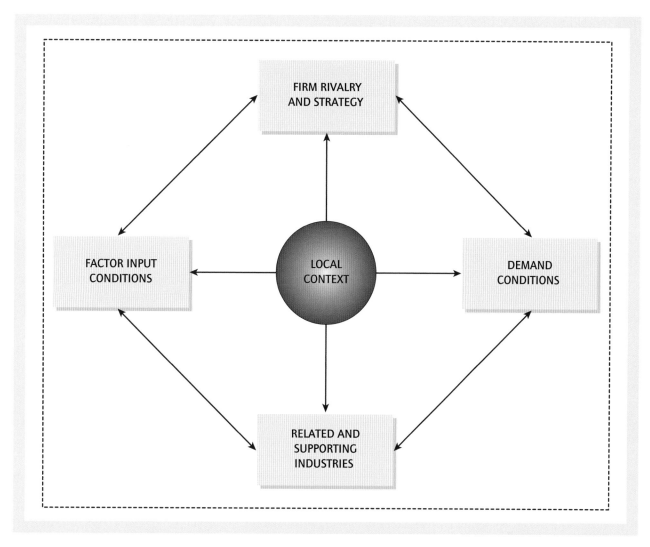

Figure 1   Porter's competitive diamond. Geographical clustering intensifies interactions within the competitive diamond.

Source: Vorley, T. (2008). The geographic cluster: A historical review. *Geography Compass, 2*(3), 803.

political-economic landscape, the essence of cluster theory continues to be based on the principles of economic localization and increasing returns. The theoretical evolution of cluster debates has not been orderly but rather comprises a diverse range of interrelated perspectives, and economic geography has embraced this intellectual sophistication in the study of clusters.

One of the most prominent recent contributions to the literature has been the "rebranding" of the cluster concept by the economist Michael Porter. Despite reiterating the characteristics of the Marshallian industrial district, Porter's refashioned concept draws on Joseph Schumpeter's notion of creative destruction. Thus, it represents more than a rediscovery or reinvention of the cluster concept. The Porterian model identified competition as the catalyst to clustering, which, based on Porter's *competitive diamond*, helps understand the interaction of factors that affect competitiveness and productivity. The model shown in Figure 1 finds the intensity of relations among the four spheres of the competitive diamond enhanced through clustering and reinforces what Porter termed the *importance of local context*.

Porter has had a profound impact on the contemporary clusters debate in both academic and policy-making communities. However, his work has also sustained some vehement critiques, from his use of aggregated data rather than empirical studies to his overly flexible interpretations of geographical scale and the use of competitiveness as the theoretical basis for his conceptualization of the cluster. Despite the impact Porter has had on the contemporary cluster debate, the literature has continued to diversify, both as a heuristic and as a technical concept, as clusters have been positioned as part of the institutional matrix at regional, national, and supranational scales.

The transition from industrial to knowledge-based economy has added a new dimension to the cluster literature, whereby the emergence of knowledge as a core factor of production has resulted in organizational changes with significant geographical implications. This newfound focus on colocation and proximity incorporates numerous topics, ranging from regional innovation systems to technopoles that embody the innovative, high-technology, knowledge-intensive nature of what may be classified as *new industrial districts*. Further to local, place-based conceptualizations of the cluster, the focus on knowledge raises the issue of relational proximity. This is considered in the work on *buzz* and *pipelines*, which seeks to explain both the local and extralocal dimensions of clusters as collectively integral to the innovative development of clusters themselves. Another dimension of the cluster debate is the temporality of clusters. While Marshall recognized traditional industrial districts to be long-standing agglomerations, and although this continues to be the case with contemporary conceptualizations of clusters, there has been a focus on exhibitions and conferences as examples of temporary clusters.

In addition to the work on clusters as an empirical project, more conceptual debates have sought to address what appears as a highly confused literature with multiple and conflicting claims about the nature, meaning, and form of clusters. In acknowledging the complexities of the clusters literature, Paul Benneworth and Nick Henry contend that debates of colocation and proximity have become too big to be encompassed by any single theorization; instead, they advance the multiperspectival approach as a more nuanced way for understanding clusters. Recognizing the scope of the cluster concept, the multiperspectival approach permits the coexistence of multiple theoretical perspectives, thereby emphasizing the dynamism of the cluster concept.

The concept of the cluster has generated significant interest within and beyond economic geography since Marshall's seminal work on industrial districts. While theoretical and empirical accounts continue to present alternative directions for cluster theory, the cluster remains an important yet elusive concept and as such can be seen to represent a work in progress.

Tim Vorley

See also Agglomeration Economies; Industrial Districts; Innovation, Geography of; Knowledge Spillovers; Learning Regions

Further Readings

Asheim, B., Cooke, P., & Martin, R. (Eds.). (2006). *Clusters and regional development: Critical reflections and explorations.* London: Routledge.

Bathelt, H., Malmberg, A., & Maskell, P. (2004). Clusters and knowledge: Local buzz, global pipelines and the process of knowledge creation. *Progress in Human Geography, 28,* 31–56.

Benneworth, P., & Henry, N. (2004). Where is the value added in the cluster approach? Hermeneutic theorising, economic geography, and clusters as a multiperspectival approach. *Urban Studies, 41*(5/6), 1011–1023.

Czamanski, S., & Ablas, L. (1979). Identification of industrial clusters and complexes: A comparison of methods and findings. *Urban Studies, 16,* 61–80.

Karlsson, C., Johansson, B., & Stough, R. (Eds.). (2004). *Industrial clusters and inter-firm networks.* Cheltenham, UK: Edward Elgar.

Marshall, A. (1890). *Principles of economics.* London: Macmillan.

Porter, M. (1990). *The competitive advantage of nations.* New York: Free Press.

Porter, M. (1998). *On competition.* New York: Free Press.

Vorley, T. (2008). The geographic cluster: A historical review. *Geography Compass, 2*(3), 790–813.

COAL

Coal is the most abundant fossil fuel, though its global distribution is highly uneven, and it powered the Industrial Revolution. Readily combustible and solid, it is formed in ecosystems where plant matter has been preserved by water and mud from oxidation and biodegradation. The fuel is black, brownish-black, or brown. Coal reserves are classified into categories or ranks based on the degree of *coalification*, or progressive alteration, that has occurred over time—this refers to changes in its energy and carbon content and environmental constituents. Ranks include anthracite (most desirable), bituminous (most abundant), subbituminous, and lignite (least desirable). Coal use often has been limited because of high emissions of sulfur dioxides, nitrogen oxides, particulates, and, most recently, carbon dioxide (it has the highest carbon emission rate among the fossil fuels). Other properties affect the desirability of coal—for example, heating value/volatile matter, moisture content, ash content and ash characteristics, and hardness. This entry reviews the geography of coal in countries with the largest proved reserves, primarily the United States and China, and briefly discusses its future in a carbon-constrained world.

United States

U.S. proved coal reserves account for 28.6% of the world total, more than 50% higher than the next largest country (Russia). Coal has provided the largest single source of domestically produced energy in the United States since 1984, and it has been one of the three most commonly consumed energy resources since the mid 19th century. The vast majority of coal is used for electricity generation, with the remainder dedicated for coking coal in the steel industry and other industrial applications, with a small quantity used in heating homes or offices or exported.

Two main U.S. coal regions exist: one each east and west of the Mississippi River. The eastern market developed first, with its coal older and of higher rank. The main mining areas are in the Appalachian states of West Virginia, Kentucky, Ohio, and Virginia, as well as Pennsylvania (Figure 1). In general, eastern coal has higher heating value and more sulfur and is mined by both underground and surface (strip) mining methods. There are also significant coal reserves and mining in the Illinois Basin, which includes Illinois, Western Indiana, and Western Kentucky. Eastern mines are known for high production costs, being more labor intensive, and having higher unionization rates. However, United Mine Workers of America membership has steadily fallen for decades.

The Western state of Wyoming is by far the dominant coal producer in the nation, and it has been so since 1988. Almost all its output comes from 20 large surface mines, with thick seams, high capital requirements, and low labor and production costs. All subbituminous and lignite coal in the United States is in the west. Several other Western states provide much smaller coal output, including Texas, whose ample lignite supply is usually consumed close to the mines because of low energy content and high production costs.

China

China's coal sector is more dominant than any other country's. While it is only third in reserves (after the United States and Russia), China is the world's largest coal producer and user, and the fuel accounts for 69% of its total primary energy consumption. In contrast to the United States, coal use in China is much more concentrated in industrial applications and residential heating, though electricity generation is still the main application.

Northern China, Shanxi Province in particular, has most of the country's readily accessible coal and almost all the large state-owned mines (Figure 2). In contrast, coal areas of Southern China are generally higher in sulfur and ash, and more restrictions have been placed on their development. Following a decline in China's coal use from 1997 to 2000, due to restructuring of the industry and the closure of many small mines, output has been rapidly increasing ever since.

The process of restructuring China's coal sector has continued as production expands. Traditionally, the industry was spread out among large and local state-owned mines and thousands of village and town mines. There were 28,000

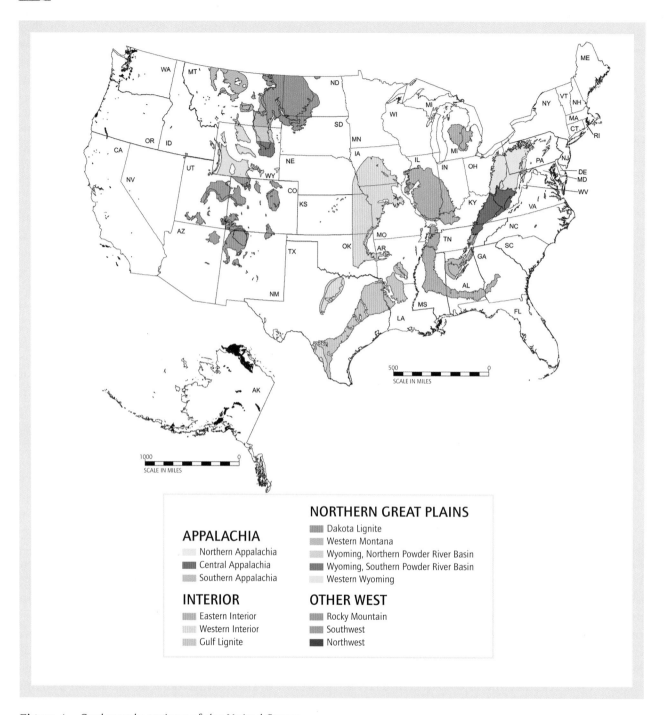

Figure 1 Coal supply regions of the United States

Source: Energy Information Administration, U.S. Department of Energy, Washington, DC.

coal mines in China at the start of 2006, compared with just 1,400 in the United States. The National Development and Reform Commission released a plan in 2006 to consolidate the industry into five or six giant conglomerates in the main coal-producing provinces, attract more foreign investment, and close down all small mines by 2015. The smaller mines are considered inefficiently managed, with inadequate capital and poor safety records.

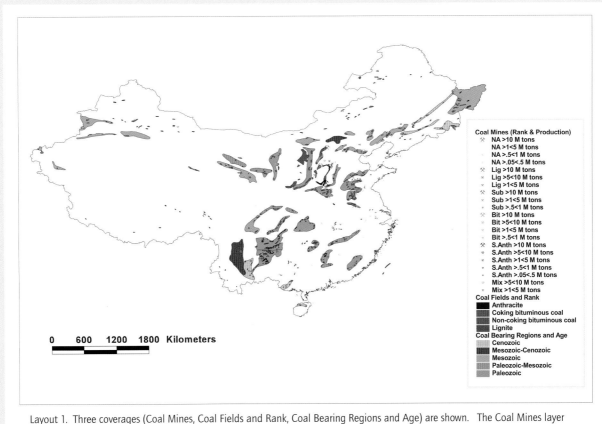

Layout 1. Three coverages (Coal Mines, Coal Fields and Rank, Coal Bearing Regions and Age) are shown. The Coal Mines layer shows the locations of major coal mines, type of coal (NA=not available, Lig=Lignite, Sub=Sub-bituminous, Bit=Bituminous, S. Anth=Sub-Anthracite, Mix=mix of coal ranks), and approximate annual yields of coal in millions of metric tons (M). The Coal Fields and Rank coverage displays locations of individual coal fields and associated coal rankings. The Coal Bearing Regions and Age layer shows potential coal-bearing areas of common geologic age.

Note: some coalfields are to small to be shown on map.

Data sources: U.S. Environmental Protection Agency Report ERP 430-R-96-005, USGS Open-File Report 00-047.

Figure 2 Coal mines, fields, and regions of China

Source: Courtesy of the U.S. Geological Survey from Open File Report OF 01-318, 2004, Reston, Virginia.

Other Countries

While Russia has the second largest coal reserves in the world, it is much less reliant on coal than is China. It has, however, seen a revival of its coal industry in the past decade. Among other former Soviet Republics, Kazakhstan and Ukraine are also significant producers. Besides China, the next most important coal regions in Asia are India and Indonesia, both of which have rapidly expanded output in the past decade. Other significant players in the global coal industry and trade include Australia (with the fourth largest reserves) and South Africa (by far the dominant market in Africa).

The Future

Given the world's large coal reserve base, the fuel cannot be ignored, especially as oil and gas scarcity becomes more apparent in the 21st century. However, the need to control carbon dioxide emissions has given pause to further coal development in many countries. Consequently, only a few countries seem intent on rapidly expanding coal production (e.g., China, Indonesia, Australia, and India), and even here, technologies to remove or sequester the carbon from coal combustion or conversion have become increasingly attractive.

Barry D. Solomon

See also Anthropogenic Climate Change; Climate Policy; Energy Policy; Petroleum

Further Readings

Elmes, G. A., & Harris, T. M. (1996). Industrial restructuring and the United States coal-energy system, 1972–1990: Regulatory change, technological fixes, and corporate control. *Annals of the Association of American Geographers, 86,* 507–529.

Hudson, R. (2002). The changing geography of the British coal industry. *Transactions of the Institution of Mining and Metallurgy (Section A), 111,* 180–186.

Thomson, E. (2002). *The Chinese coal industry: An economic history.* Oxford, UK: Routledge.

Coastal Dead Zones

As river drainage basins collect the runoff from the continents, they carry the flush of agricultural runoff, fertilizers, manure, sewage, and other waste. This discharge load acts as nutrient enrichment that forces *eutrophication* (from the Greek *eutrophos,* meaning "well nourished"). Huge spring blooms result as phytoplankton in the upper *photic zone* of coastal waters flourish. Anthropogenic (human-forced) pollution is damaging some 400 estuarian and coastal marine systems worldwide. These damaged ecosystems are Earth's *coastal dead zones.*

As algae die, they drift to the seabed and accumulate in the bottom sediments, feeding bacterial action in the *benthic zone.* The bacteria consume oxygen as they process the organic debris. As a function of this bacterial respiration, dissolved oxygen (DO) levels decline. The biological oxygen demand (BOD) of bacteria feeding on the

decay exceeds the dissolved oxygen, and *hypoxia* develops, killing any fish that venture into the area. Benthic-dwelling animals (flounder, crab, lobsters) and immobile animals (worms, clams, oysters) are killed. These low-oxygen conditions act as a *limiting factor* on marine life.

Conditions are worsened by physical factors and can vary seasonally. Stratification of waters by temperature produces a warmer surface layer and cold deep layers following spring blooms and the approach of summer warmth. Some recovery is possible from seasonal depletions, but hypoxic conditions can spread spatially with the increasing influx of nutrients over time. Thus, the dead zone, a region of oxygen-depleted (hypoxic) water off the coast of Louisiana in the Gulf of Mexico has expanded to more than 22,000 km² (square kilometers), or 8,500 mi.² (square miles), each summer (first photo). The Mississippi drainage system handles the runoff for 41% of the continental United States, delivering the nutrient load to this region of the Gulf. In fact, all along the Texas coast, an extensive dead zone prevails, stretching in a discontinuous concentration some 32 km (20 mi.) from the coastline, fortified by nutrient-laden runoff and coastal septic systems.

Various events can worsen conditions, such as heavy precipitation from storms or hurricanes tracking inland, and produce periodic episodes of eutrophication. Dead zones occur in episodes such as the one in North Carolina in 1999. In short succession during September and October, Hurricanes Dennis, Floyd, and Irene delivered several feet of precipitation to the state, each storm falling on already saturated ground. In North Carolina, more than 10 million hogs, each producing 2 tons of waste per year, were located in about 3,000 agricultural factories. These operations collected almost 20 million tons of manure into hundreds of open lagoons, many set on river floodplains. The three-hurricane downpour flushed out waste lagoons into wetlands and streams and on to Pamlico Sound and the ocean, producing a spreading dead zone and an environmental catastrophe.

When the 1993 Midwest floods hit, the Mississippi river's nutrient discharge was such that the size of the dead zone in the Gulf of Mexico doubled. Following Hurricane Katrina in 2005, the dead-zone region was overwhelmed by the

The large region of low-oxygen water often referred to as the *Dead Zone* spreads across the Gulf of Mexico in what appears to be an annual event. NASA satellites monitor the health of the oceans and spot the conditions that lead to a dead zone. This image shows how ocean color changes from winter to summer in the Gulf of Mexico. Aqua shows highly turbid waters that may include large blooms of phytoplankton extending from the mouth of the Mississippi River all the way to the Texas coast. When these blooms die and sink to the bottom, bacterial decomposition strips oxygen from the surrounding water, creating an environment in which marine life find it difficult to survive. Red and orange represent high concentrations of phytoplankton and river sediment.

Source: NASA/Goddard Space Flight Center Scientific Visualization Studio.

discharge carrying all the sewage, chemical toxins, oil spills, household pesticides, animal wastes and carcasses, and nutrients. The agricultural, feedlot, and fertilizer industries dispute the connection between their nutrient input and the dead zone.

In other parts of the world, the connection between nutrient inflow and eutrophication leading to depleted dissolved oxygen and hypoxia is well established. In Sweden and Denmark, a concerted effort to reduce nutrient flows into their rivers reversed hypoxic conditions in the Kattegat Strait (between the Baltic Sea and the North Sea). The Black Sea is chronically plagued by dead-zone conditions. However, after regional political changes and the decline in state farms, fertilizer use is down more than 50% since 1990, and for the first time in recent decades, the Black Sea is getting several months' break from its year-round hypoxia. As oxygen levels improved, so did ecosystem recovery, although still not to original biological conditions. In water bodies such as the Thames and Mersey river estuaries in England and the Hudson and East rivers in New York, reducing nutrient inflows has healed dead zones.

As with most environmental situations, the cost of mitigation is cheaper than the cost of continued damage to marine ecosystems. For the Mississippi river drainage, a 30% cut in nutrient inflow upstream is estimated to increase dissolved oxygen levels by more than 50% in the dead-zone region of the Gulf.

One mitigating strategy might be to mandate application of only the levels of fertilizer needed. In addition to reducing nutrient outflows from farmland, this action would lead to a savings in overhead costs for agricultural interests. Also, action is needed to begin dealing with animal wastes in many of the world's drainage basins, including the Mississippi.

The strategy is simply to keep fertilizers and manure and other nutrients on the land and not in the water. Inorganic fertilizers are chemically produced through artificial nitrogen fixation at factories using the Haber-Bosch process, developed in 1913. The annual production of synthetic fertilizers now is doubling every 8 years; worldwide, some 1.82 million tons (i.e., almost 2 million tons) are produced per week. Humans fix more nitrogen per year than all terrestrial sources combined, overwhelming natural denitrification systems. Anthropogenic sources of fixed nitrogen first exceeded the normal range of naturally fixed nitrogen in 1970.

Robert J. Diaz, an important scientist researching dead zones, considers hypoxia and anoxia as evidence of the widespread deleterious anthropogenic influence on coastal systems. He ranks this as a major global environmental problem. The ongoing decline in dissolved oxygen and the rate of that decline are unprecedented over such a short time span. Understanding and mitigating this problem requires the spatial analysis tools of geographic science.

Robert W. Christopherson

See also Agricultural Land Use; Agriculture, Industrialized; Biogeochemical Cycles; Coastal Zone and Marine Pollution; Environmental Impacts of Agriculture; Nutrient Cycles

Further Readings

Diaz, J. D., & Rosenberg, R. (2008). Spreading dead zones and consequences for marine ecosystems. *Science, 321*, 926–929.

Goddard Earth Sciences Data and Information Services Center, NASA. (2008, May). *Science focus: Dead zones*. Retrieved January 11, 2010, from http://disc.gsfc.nasa.gov/oceancolor/scifocus/oceanColor/dead_zones.shtml

Rabalais, N. N., & Turner, R. E. (Eds.). (2007). *Coastal hypoxia: Consequences for living resources and ecosystems*. Washington, DC: American Geophysical Union.

COASTAL EROSION AND DEPOSITION

Coastal erosion is the term used for the removal of land or sediment from a coastline, whereas *coastal deposition* is the addition of sediment to a coastline. Factors influencing erosion and deposition include the rate of sediment supply to the coast, the shape of the underlying geology, the rate at which coastal processes transport sediment, and the rate and direction of sea-level change. These can ultimately be attributed to tectonics, climate, geology, and eustasy, all of which vary spatially and temporally. It is important to understand these processes, given the significance of the coast for habitat, recreation, and resources.

Coastal morphodynamics, a systematic approach to integrating the processes and boundary conditions that influence coastal erosion and deposition, was introduced by Don Wright and Bruce Thom in 1977. This integrates the effects of instantaneous processes (e.g., wave and swash processes including incident waves, rip currents, bed return flows, and infragravity waves), long-term processes such as sea-level change, and geological processes (e.g., tectonic changes). Using this approach, the coasts can be divided into three categories: prograding, eroding, and stationary (where net sediment movement is zero) coasts.

The Twelve Apostles, on the southern coast of Australia, formed by bedrock eroding at a variable rate

Source: Natalie Townsend.

Coastal Erosion

Eroding coasts, often termed *retrograding*, *encroaching*, or *receding coasts*, result from a net negative sediment budget. This can occur due to factors including wave energy increase, wave direction change, currents, relative sea-level rise, and losses to the coastline by tidal and aeolian transport. The less aggregated and finer grained the beach, the easier it is to transport and remove the sediment from the coast.

On rocky coasts, constant wave pounding, chemical and mechanical weathering, abrasion, and hydraulic action play a major role in degrading the parent rock into finer particles for deposition elsewhere. The ease with which the bedrock is eroded depends on the characteristics of the parent material. This may result in the formation of rock stacks such as the Twelve Apostles on the southern coast of Australia (see photo).

Other coastal features formed by coastal erosion include beach and dune scarps on sandy coasts. This often leads to infrastructure being compromised or stranded on the beach or, ultimately, the loss of human life. An example of eroded dunes due to a net loss of sediment is on the northern French coast (see photo), where stationary World War II concrete bunkers are gradually being exposed.

Coastal Deposition

Coastal deposition, often referred to as *progradation* or *transgression*, occurs when there is a net

The erosion along the northern French coast has left relict World War II bunkers exposed along the coast.

Source: Denis Marin.

sediment input. Deposition frequently results in the development of large alluvial deltas (e.g., Mississippi and Rhone deltas), beach ridge plains (e.g., Normandy coast, France), Chenier plains (e.g., Princess Charlotte Bay, Northern Australia; see photo), or mud flats. The nature of each shoreline depends on the origin of the sediment supply (i.e., whether it is dominated by river sediments or reworked beach sediment) and on wave conditions.

Factors Affecting Coastal Erosion and Deposition

In the following sections the most common factors influencing coastal erosion and deposition are reviewed.

Wave Climate

Waves and their induced sediment disturbance, infragravity waves, bed return flows, and rip currents are mechanisms for sediment transport. The amount of sediment that is transported depends on the incident wave climate, which ultimately depends on geographically variable wind climate. An increase in wind speed leads to an increase in wave energy. Greater wave energy and wave length also enable larger grain sizes to be transported and larger volumes of sediment to be disturbed and displaced due to the increased depth of the wave base (D):

$$D = 1/2L,$$

where L is the wave length.

Storm surge and increased rip intensity on wave-dominated coasts often accompany elevated wave conditions. Storm surge accelerates erosion by exposing higher land elevations than normal to these more energetic wave conditions and by raising the water table and saturation of the soil. Increased rip current intensity also transports greater volumes of sediment and larger grain sizes offshore, often beyond a depth from which they may be returned onshore by normal wave action or above the swash zone as a berm deposit that may also include debris. This may lead to permanent or transient alterations to the shoreface during storms and hurricanes.

On a sandy coast, the movement of sediment offshore is a consequence of the beach adjusting to a flatter offshore profile that tends to dissipate more of the wave energy prior to reaching the shoreface. With a lowering of wave energy, beach water tables are lowered, and sediment is deposited on the upper shoreface. This is because the

An example of a chenier ridge in Princess Charlotte Bay, Northern Australia. Both chenier plains and beach ridge plains can be observed as parallel rows of sediment ridges and vegetation on the shore.

Source: Dylan Horne.

sediment-laden swash drains downward as it runs up the beach, leaving the entrained sediment behind. This results in a steeper beach profile. Thus, small waves tend to form steep profiles, and larger waves tend to form flatter profiles. This is because steep beaches tend to reflect incoming waves, and flatter beaches tend to dissipate wave energy in the nearshore zone. The beach may change continuously from steeper to flatter according to changes in wave energy.

Longshore currents are generated when waves approach the beach at an angle. For example, a general northward trend in sediment transport occurred in Eastern Australia at lower sea levels (before approximately 6,000 years BP), when the coast consisted of long straight beaches aligned at an angle to the dominant southeast swell generated in the Tasman Sea. This led to the development of several sand islands located at the end of

the transport path, off the coast of Southern Queensland—for example, Fraser Island. Locally, a change in wave direction can lead to beach rotation, where the beach appears to rotate within an embayment due to erosion and deposition at its opposite ends. Beach rotation occurs as a consequence of beach profile readjustment, so that the beach angle remains perpendicular to waves. These shifts in wave direction may be seasonal or aseasonal in nature, for example, El Niño/La Niña or winter/summer cut/fill cycles.

Longshore transport may initiate exchange of sediment between adjacent sandy beaches through headland bypassing, where sediment accumulation occurs on the downdrift side of a headland due to the currents. This sediment then moves subaqueously around the headland as a downdrift spit and migrates shoreward. Some of the sediment will result in net accretion as a sand wave;

Mangroves aid accretion by increasing wave attenuation and entraining sediment in the pneumatophores. Also seen in this picture is the effect of erosion by the running stream as the tides ebb and flow or as rain water is added to the estuary.

Source: John Rae, Oyster Bay, Australia.

however, most remains subaqueous. The sand wave then migrates and merges with the down-drift beach, with a topographically controlled rip moving in advance, causing localized erosion.

Dunes

Wind and washover transfer sand into and out of the nearshore zone and dunes. Wind transports fine-grain sand into aeolian dunes, which is then trapped by pioneering sand plants, such as *Spinifex* spp., and accumulates into ridges. These ridges may be eroded or transported landward into mobile transgressive dunes during storms or if the vegetation is disturbed.

Tidal Currents

Tidal currents are important in areas with tidal ranges greater than 2 m (meters; macro- and mesotidal beaches) or in channels where flow velocities are fast. These currents move sediment with the direction of flow or incise and erode into adjacent deposits. Tides also alter the water table level, causing small-scale erosion and deposition cycles.

Reefs, Tombolos, and Biota

Reefs and islands are natural offshore features resulting in reduced wave energy in their lee. This can result in the formation of salient features or the development of a tombolo. If a tombolo

The undermining and collapse of a sea wall structure on the Northern French coast due to high erosion rates and inappropriately engineered structures

Source: Author, Wissant, France.

feature is formed, adjacent beaches are protected from incident wave energy, leading to changes in beach morphodynamics and patterns in deposition and erosion. Sea grass, kelp beds, and mangroves (see Oyster Bay photo) also aid shoreline accretion by reducing incident wave energy through wave attenuation.

Fragments of coral may be eroded from the reefs and deposited onshore, creating coral beaches. Reef islands themselves tend to vary in behavior. High wave energy, coral shingle, and boulder motu islands prograde during storms and redistribute the sediment during fair-weather conditions, whereas less exposed sandy cays erode during storms and recover during fair-weather conditions.

Chemical Effects

Apart from chemical weathering of parent rock due to the slight acidity of sea water, in temperate to tropical regions, beaches may become cemented into beach rock by an excess of calcite crystals. During erosive conditions, the presence of beach rock leads to accelerated erosion due to a decrease in beach permeability. Eventually, it may be completely eroded, leaving behind a newly formed reef.

Human Interference

Sea walls are shore-parallel features built on the subareal beach between the shoreline and foredune with the aim of protecting adjacent property and stopping shoreline retreat. If the sea wall is built impermeable and within the swash zone, the water table rises, and increased wave reflection and turbulence are initiated against the wall. This ultimately exacerbates erosion. Ultimately, this can result in erosion of the adjacent beach or undermining and failure of the structure (see Wissant photo).

Breakwaters are also shore-parallel structures, built away from the shoreline with the aim of decreasing wave energy in their lee and protecting the coast from erosion. The presence of these features may have no impact on the system if the erosion is due to longshore transport. Alternatively, they can cause deposition in their lee, building up a salient feature or, possibly, tombolo.

Groynes, training walls, and jetties are shore-perpendicular structures extending into the surf zone and put in place to minimize erosion. However, they often exacerbate erosion, particularly on coastlines dominated by drift transport. The presence of these structures causes beach realignment due to up-drift accumulation of sediment and down-drift erosion. Wave refraction around the groyne and induced rip currents increase the potential for offshore transport of sediment and beach erosion.

Changes to beach sediment may occur due to beach nourishment, sand mining for soil aggregates or heavy minerals, dredging and spoil dumping, and increased runoff and soil erosion due to land development, clearing, and degradation. This not only has a direct impact on the removal and deposition of sediment but also alters the beach slope and wave dissipation conditions, altering its response to wave climate changes.

Sea-Level Change

The sea level may be altered through processes including an increase in water temperature and changes to the volume of land ice (*eustasy*), tectonic movement and land subsidence (*isostasy*; e.g., the North American Pacific Coast), and deglacial uplift (*glacio-isostasy*; e.g., Sweden).

Erosion or deposition may occur with a change in sea level, depending on the rate and direction of the sea-level change and the rate of sediment supply. If the rate of sediment supply exceeds the rate of sea-level rise or sediment accumulation occurs during sea-level fall, then deposition occurs. If the rate of sea-level rise exceeds the rate of sediment supply, then erosion occurs. If the rate of sediment supply is equal to the rate of sea-level change, then the shoreline remains stationary. If sea-level rise is a consequence of climatic changes, there may be exacerbated consequences of erosion through increased wave energy.

The most widely accepted model for sea-level change was initially developed by Per Bruun, termed the *Bruun rule*. In his papers, Bruun formulated how shorelines retreat to rising sea levels and undergo deposition under falling sea levels to maintain an equilibrium profile. The equilibrium profile is based on the assumption that the shoreface maintains a constant profile slope for any given depth, which balances offshore movement by gravity and water currents and onshore movement by wave motion. This shape is maintained through sea-level changes, so that the amount of sediment eroded from the upper shoreface is equal to the amount deposited on the lower shoreface. The resultant landward recession distance due to sea-level rise according to the Bruun rule is

$$R = S(L/B + h) = (S)1/\tan\phi$$

where R is the recession distance, S is the amount of sea-level rise, L is the length of the profile, B is the berm height, h is the wave base closure depth, and ϕ is the profile slope angle.

However, the validity and applicability of the Bruun rule in real-life situations has been the subject of debate. It has been pointed out that the assumptions made in the model are unrealistic, that the nearshore system is not closed and sediment is frequently exchanged with dunes or seaward of the shoreface, and that the equilibrium shoreface is unlikely to exist or be present at every coast. Thus, various modifications were made to the Bruun rule, which resulted in the development of three main coastal responses: erosion, barrier rollover, and overstepping (Figure 1). The erosional response is a typical example of the Bruun

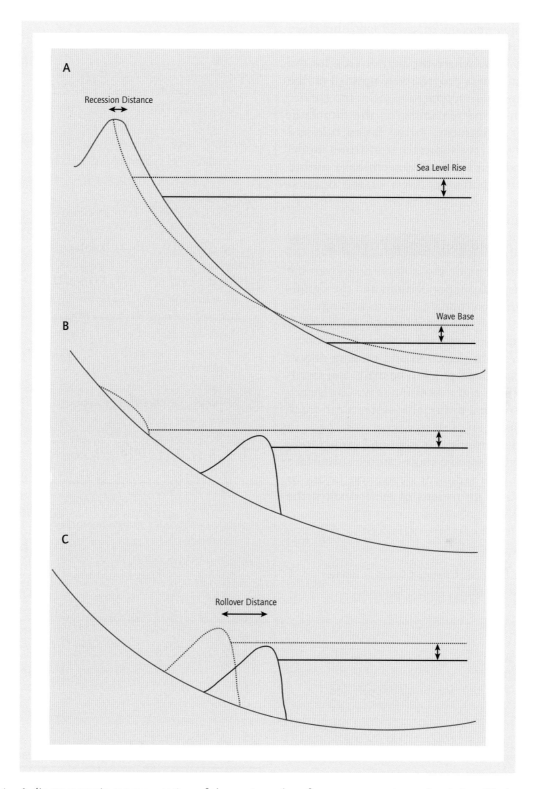

Figure 1 A diagrammatic representation of the various shoreface responses to sea-level rise: (A) the general Bruun rule of erosion, where offshore dispersal of sediment occurs and seabed rise equals sea-level rise; (B) overstepping, where the barrier is drowned and the sea level rises above it; and (C) barrier rollover, where the barrier moves landward, recycling the sediment

Source: Adapted from Carter, R. W. G. (1989). *Coastal environments: An introduction to the physical, ecological, and cultural systems of coastlines*. London: Academic Press.

Rule, whereby erosion of the upper shoreface occurs and the sediment is then transported offshore. *Rollover* is where the barrier recycles the sediment and moves progressively upward on the shoreface; and *overstepping* identifies the barrier being drowned and remaining on the shoreface. More recent adjustments to the Bruun rule and new models (e.g., the shoreface translation model [STM] introduced by Peter Cowell) have further minimized the potential flaws of the original Bruun model.

Shoreline Monitoring and Predictions

Shoreline retreat modeling is important given the globally dense coastal populations and the significance of coastal resources. New technology and models are now used to further understand shoreface behavior and aid in such predictions. For example, light detection and ranging allows precise beach topographies to be made and monitored for erosion and deposition (e.g., Cat Island, Mississippi); global positioning systems can track changes in shoreline features on a daily basis in 3D (three dimensions); and ground penetrating radar and other seismic tools can see subsurface profiles to show sediment deposition layers or erosion surfaces up to tens of meters below the land surface.

Elspeth Rae

See also Anthropogenic Climate Change; Climate Change; Coastal Hazards; Dunes; Tsunami

Further Readings

Anthony, E. J. (2000). Marine sand supply and Holocene coastal sedimentation in Northern France between the Somme estuary and Belgium. *Geological Society (London), Special Publications, 175,* 87–97.

Bruun, P. (1988). The Bruun Rule of erosion by sea level rise: A discussion of large scale two- and three-dimensional usages. *Journal of Coastal Research, 4,* 627–648.

Carter, R. W. G. (1989). *Coastal environments: An introduction to the physical, ecological, and cultural systems of coastlines.* London: Academic Press.

Cooper, J. A. G., & Pilkey, O. H. (2004). Sea-level rise and shoreline retreat: Time to abandon the Bruun Rule. *Global and Planetary Change, 43,* 157–171.

Cowell, P. J., Roy, P. S., & Jones, R. A. (1995). Simulation of large-scale coastal change using a morphological behaviour model. *Marine Geology, 126,* 45–61.

Curray, J. R. (1964). Transgressions and regressions. In R. C. Miller (Ed.), *Papers in marine geology* (pp. 175–203). New York: Macmillan.

Houghton, J. T., Ding, Y., Griggs, D. J., Noguer, M., van der Linden, P. J., & Xiaosu, D. (Eds.). (2001). *Climate change 2001: The scientific basis* (Contribution of Working Group I to the Third Assessment Report of the Intergovernmental Panel on Climate Change [IPCC]). Cambridge, UK: Cambridge University Press.

Ranasinghe, R., McLoughlin, R., Short, A., & Symonds, G. (2004). The Southern Oscillation Index, wave climate, and beach rotation. *Marine geology, 204,* 273–287.

Thom, B. G. (Ed.). (1984). *Coastal geomorphology in Australia.* Sydney, New South Wales, Australia: Academic Press.

Woodroffe, C. D. (Ed.). (1994). *Coastal evolution: Late Quaternary shoreline morphodynamics.* Cambridge, UK: Cambridge University Press.

Wright, L. D., & Short, A. D. (1984). Morphodynamic variability of surf zones and beaches: A synthesis. *Marine Geology, 56,* 93–118.

Wright, L. D., & Thom, B. G. (1977). Coastal depositional landforms: A morphodynamic approach. *Progress in Physical Geography, 1,* 412–459.

 # COASTAL HAZARDS

A *coastal hazard* is a process occurring at the coastline that can cause damage to people living, working, or visiting the coastline. The damage can be physical, in terms of injury, loss of lives, and ruined properties, or economic, social, or cultural, in terms of interrupting the use of the

coastline (e.g., navigation, fisheries, and recreational use). It is commonly studied under the broad topic of geography because there are spatial variations to both the processes that cause the hazard and the behavior of the people who are affected by the hazard. Coastal hazards are mainly grouped into flooding (or inundation) hazards and erosion hazards but can also include swimming and navigation hazards.

People living near the coast depend on the sea for food, trade, and transport. In modern Western societies, where the wealthier economies allow the average person more time for recreation and relaxation, coastal properties have become an increasingly valuable asset. This has fueled a greater interest in using science to protect these properties from coastal hazards. However, not all interest in coastal hazards has been inspired by wealth protection. Advances in our understanding of coastal processes and our ability to provide increasingly accurate predictive computer models have been driven both from the top, by international efforts to avoid repeats of devastating coastal disasters (e.g., the Indian Ocean tsunami on December 26, 2004), and from the bottom up by local issues (e.g., such as surf life-saving organizational activities to decrease incidents of drowning).

Flooding Hazards

Coastal flooding is a hazard not only because of the damage that saltwater can inflict on property but also because the associated deep water, strong currents, and turbulence can cause drowning. Moreover, floating objects torn free by turbulent waters can injure and exacerbate inundation damage.

The water level at the coastlines at a single instant in time is made up of (1) the mean sea level, (2) storm surge, (3) the tide, and (4) run-up. The first influence on mean sea level to consider is that the ocean is intimately connected to the atmosphere. Currents and water levels respond to changing wind, pressure, and air temperature patterns. These properties change dramatically from season to season, from year to year because of climatic effects such as El Niño/La Niña, and over much longer timescales because of influences such as the arrival and passing of ice ages. The sea

level can also experience relative changes, as the landmasses can sink or rise slowly due to effects such as isostatic rebound (the lift in land masses that follows retreat of continental glaciers). It is therefore difficult to measure mean sea level at a site, let alone determine if it is changing in any permanent sense. Thus, the definition of mean sea level depends on the time interval over which the mean is calculated. For the most part, this is a yearly average and so varies from year to year and depending on location.

Storm surge is the rise in water level caused by the atmospheric changes that accompany storm events. The reduced air pressure lifts the surface of the water up in an effect called the *inverse barometer effect*. Strong winds that circulate around low-pressure events can cause seawater to pile up against the coastline. The relative importance of the pressure and wind effects depends on the location and intensity of the storm, with the latter more important in tropical cyclones and hurricanes. The lifting of the water surface (by wind or pressure) can initiate a large-scale wave or a long wave, which propagates away from the storm at a speed that is controlled by the depth of the coastal waters. If the storm travels at the same speed and direction as the wave, there is ample time for energy transfer, and the wave can get large. Some semi-enclosed areas such as bays and basins can resonate with the same period as the long wave, resulting in an extra large storm surge. This is the case of storm surges in the North Sea and the reason why the Netherlands has needed to invest so much money in storm surge defense structures.

The tide is the wave that is excited by the gravitational attraction of the sun and the moon. High tide corresponds to the crest of the wave and can occur either twice daily or once daily, depending on the location. The axial tilt of the Earth relative to the sun and moon cause, in part, this spatial variation. This large-scale wave propagates around the world at a speed that depends on the depth of the water. It reflects at the boundaries created by landmasses to create standing waves, which rotate around basins and in some cases cause resonance in these basins and bays. Thus, the size of the tide depends on the characteristics of basins and so varies with location, in addition to the variations caused by the Earth's axial tilt. Temporal changes depend on the proximity and orientation of the sun and the moon, which vary with latitude, season, and the phases of the moon. The largest and most hazardous high tides occur when the gravitational effect of the sun and the moon combine (spring tide) and the sun and the moon are closest to Earth in their orbits (perigee and perihelion).

Run-up is the added elevation of the water level caused by waves. In deep water, the currents inside waves form circles or orbits, which are almost closed. The consequence is that the water does not move significantly inside waves. When these waves break, these orbits change dramatically and the water rushes on average forward near the surface. The result is water piling up at the shoreline in an effect that is called *set-up*. The magnitude of set-up depends on how quickly the waves lose energy across the surf zone, which depends on the seabed topography. On top of the set-up is *swash*, or the movement of the water line up and down with each incoming wave. These two effects combine to make run-up. In the event of a tsunami, the hazard consists of the associated run-up (only on a much larger scale).

Calculating the flooding hazard for a site is simply accomplished by adding together estimates of the size of these four effects. Of these, the run-up is probably the least well predicted by existing models. Obviously, the flooding hazard depends on the probability of these four flooding hazards occurring at the same time.

Coastal Erosion Hazards

The waves, currents, and elevated water levels that occur during storms can cause erosion of the coastline. This is a hazard because it can cause properties and roads to fall into the sea (see photo). When waves break and the currents rush forward on average near the surface of the water, continuity dictates that there must be a return flow or an undertow near the bottom. This return flow will quite often converge along the shore to make offshore streams of water called *rip currents*. Elevated water levels, waves, and turbulence near the coastline caused by breaking mean that it is easy for sand grains to get suspended or entrained from the seabed. This sand is then transported seaward by the undertow and rip currents, where it is deposited outside the zone of

breaking waves where the currents die out. The net result is a movement of sand from the beach face seaward into sandbars during storm events.

Shorelines that have steep slopes cause waves to reflect, and reflected waves have strong currents just seaward of the shoreline. This is why harbor walls are often designed with sloping seaward sides. It is also why building seawalls can actually exacerbate erosion by stirring and eroding the sand in front of the seawall. Predicting the extent of erosion during a storm is very difficult because not only is a water-level prediction needed, it is also necessary to predict the behavior of waves and currents very close to the shoreline as well as the mechanisms for entraining sand (which depend on the composition and shape of sand grains).

Coastal Hazard Zone

A coastal hazard zone is a tool for managing coastal development. It can be defined as a distance from the shoreline that is likely to experience significant erosion and flooding in a time frame that covers the lifetime of the building (e.g., 50 or 100 years). Coastal hazard zones are being continually updated as scientific advances and better records of storm events allow more realistic predictions to be incorporated. The increasing value of coastal land has put pressure on governments to minimize hazard zones. In some places, it is difficult to get insurance coverage for homes located in a coastal hazard zone, or there are stronger building restrictions, such as requiring houses to be relocatable or on stilts.

Defenses Against Coastal Hazards

Defense against coastal hazards most commonly involves a "hard" solution such as a protective barrier. For example, in the case of flooding hazard, defenses consist of floodgates that can be shut or walls or dykes to stem the flow of water. Temporary barriers are often constructed out of sandbags. Defending against coastal erosion can be accomplished by protecting the shoreline using a seawall, piles or bags of rocks, tires, concrete structures, or giant sandbags. The piles of rocks/ structures are also aimed at removing energy from the waves by increasing the action of friction.

Erosion caused by a northeaster on the Outer Banks in North Carolina, 1987

Source: National Oceanic and Atmospheric Administration.

Coastal erosion can also be minimized by stopping the wave energy from reaching the shoreline. This can be accomplished by building an offshore structure that causes waves to break seaward of rather than at the shoreline. If currents tend to cause sand to move alongshore, walls projecting from the shoreline (*groins*) are built to trap the sand. The negative side of hard solutions is that they are costly, look out of place, and often cause a hazard in themselves. For example, rocks can come loose during storms, and injury can occur on decaying, poorly maintained seawalls where rusting steel reinforcements have become exposed.

More recently, there has been a focus on finding ways to protect the shoreline that also enhance or maintain its natural character and amenity value. The dunes that commonly back natural beaches are great reservoirs of sand that are tapped into during storm events. Keeping these reservoirs stocked for emergencies is a vital part of a natural beach's ability to cope with storm events. Therefore, "soft" solutions are increasingly used, which are to enhance the dune by planting, removing nonnatural species, installing fences to trap sand, and limiting or channeling pedestrian access. An added benefit is that the natural beauty of the coastline is enhanced, increasing the potential for tourism and restoring increasingly rare natural habitats.

Many problems that have been associated with coastal protection works can be avoided by education about the inherent variability of natural systems. In many cases, the most economical approach is retreat, recognizing that the coastal hazard zone was never optimal for human habitation.

Coastal Hazards and Climate Change

Predictions of climate change have required revaluation of coastal hazard zones. Not only is the sea level predicted to change, but so too is the frequency and severity of storm events. The vast complexity of the ocean-atmosphere system makes it nearly impossible to make predictions of potential changes, let alone provide an indication of the risk associated with the scenarios. The Intergovernmental Panel on Climate Change reports are our best guess, and most governments base their precautionary approach on these predictions.

Karin Bryan

See also Anthropogenic Climate Change; Coastal Erosion and Deposition; Coastal Zone and Marine Pollution; El Niño; Global Sea-Level Rise; Hurricanes, Risk and Hazard; La Niña; Natural Hazards and Risk Analysis; Oceans; Tsunami; Tsunami of 2004, Indian Ocean

Further Readings

Beatley, T., Brower, D. J., & Schwab, A. K. (2002). *An introduction to coastal zone management* (2nd ed.). Washington, DC: Island Press.

Garrison, T. S. (2007). *Oceanography* (6th ed.). Florence, KY: Brooks/Cole.

The H. John Heinz III Center for Science, Economics and the Environment. (1999). *The hidden costs of coastal hazards*. Washington, DC: Island Press.

Komar, P. D. (1997). *Beach processes and sedimentation* (2nd ed.). Upper Saddle River, NJ: Prentice Hall.

Schwartz, M. L. (Ed.). (2005). *Encyclopedia of coastal science*. New York: Springer.

 # COASTAL ZONE AND MARINE POLLUTION

The coastal zone is defined as the interface between the land and sea where the land and sea influence each other. Its landward and seaward boundaries are not fixed. If the water depth is extended to 200 meters (m), the waters cover an area of about 7% (26×10^6 square kilometers [km^2]) of the surface of the global ocean. This area accounts for at least 15% of oceanic primary production, 80% of organic matter burial, 90% of sedimentary mineralization, 75% to 90% of the oceanic sink of suspended river load, and about 50% of the deposition of calcium carbonate. In addition, 90% of the world fish catch and its overall economic value and at least 40% of the value of the world's ecosystem services and natural capital are in this area. Fisheries provide livelihoods to about 40 million people worldwide, and 80% of them are in Asia; the three top fisheries producers are China, Peru, and the United States.

About 40% of the world's people live within 100 km of the coastline, or 7.6% of Earth's total land area. Each year, roughly 50 million more people move into this narrow fringe of coastal land worldwide. By 2050, with the world's population projected to be around 9 billion, some 60% will be living within 60 km of the sea. From another perspective, the contiguous coastal area less than 10 m above sea level covers 2% of the world's land area but contains 10% of the world's population and 13% of the world's urban population. This low-lying zone has 21% of the world's cities, with populations over 5 million; in Asia, 32% of the population lives in such coastal cities.

Population growth has a direct bearing on the extent and magnitude of physical alterations of coastal areas and their consequent destruction. The most damaging actions are changes in land use, including draining wetlands and mangroves for agriculture or settlements; building structures such as ports, seawalls, dams, aquaculture installations, and tourist facilities; and overuse of resources, including overfishing, water use, and sand and gravel extraction.

Several examples of the types of physical alterations currently occurring in coastal environments give an idea of the magnitude of the problem. Of the world's wetlands, 50% have been lost over the past century, and the percentage is higher for mangroves. Already 34% of the world's coral reefs are lost or seriously damaged, and another 20% are threatened in the next 20 to 40 years. The destruction of habitats varies from region to region, but all show deterioration of some kind.

Climate-induced changes in sea level are likely to increase the risk of inundation in many low-lying parts of the coastal zone. Hotspots of coastal vulnerability are the populated deltas, especially the megadeltas of Asia; low-lying urban areas; and atolls, where the stresses on natural systems coincide with low human adaptive capacity and high exposure. Regionally, South Asia, southeast and East Asia, Africa, and small islands are the most vulnerable. Of growing concern is the impact of increasing atmospheric carbon dioxide concentrations and the resultant acidification of the sea. Acidification is expected to disrupt calcium carbonate formation, affect the oxygen metabolism of animals, and influence the availability of nutrients.

Despite its relatively small area, the coastal zone plays a significant role in the global biogeochemical cycles because virtually all land-derived materials (water, sediments, dissolved and particulate nutrients, etc.) enter this region in surface runoff or groundwater flow. The recent increases in coastal urban population and changes in land use practices have led to rapid and big changes in sediment supplies and increases in nutrient, pollutant, and pathogen loadings to coastal waters.

Marine Pollution

Marine pollution occurs due to the harmful effects of the entry into the ocean of various pollutants (chemicals, particles, and industrial, agricultural, and residential wastes) and the spread of invasive species. The pollutants come from three main sources: (1) land-based sources (44%), (2) atmospheric inputs (33%), and (3) maritime transport (12%).

Sewage

The discharge of untreated sewage is a major source of pollution of coastal waters. Sewage generally contains organic carbon, nutrients and human pathogens, as well as industrial chemicals, oils, and greases. The problem occurs mainly in the developing countries, where only a small portion of the domestic wastewater is collected and treated by sewage treatment plants, the majority of which do not work efficiently or reliably.

With the rising coastal populations, nearly 90% of the sewage entering the coastal zones in many developing countries is estimated to be raw and untreated. Untreated wastes entering the seas

vary from 10% to 20% in Western Europe to more than 50% in the Mediterranean Sea and more than 80% in the Caribbean and Latin America, East Asia, South Asia, West Africa, and the southeast Pacific. While the effects of discharging raw sewage into water bodies are generally local, some transboundary effects occur in certain areas. Overall, the control of pollution from sewage continues to be a major problem, particularly in the developing countries, where the constraints are financial and not technical.

Persistent Organic Pollutants

Persistent organic pollutants (POPs) are highly toxic and stable organic chemical substances that accumulate in organisms and persist for years and even decades in the marine environment. They consist of pesticides, industrial chemicals, and associated by-products. POPs pose risks to human health, principally through the ingestion of marine food. The Stockholm Convention on Persistent Organic Pollutants of 2001 is an international agreement to address the problem of POPs. It identified an initial set of 12 chemicals because they have the common hazardous characteristics of toxicity, persistence, and bioaccumulation and are capable of traveling vast distances via water and air. Of these 12, 9 are pesticides.

Since international controls were instituted, the situation has improved considerably, although the problem still persists in developing regions dependent on agriculture. It tends to be worse at high latitudes. For example, the Arctic has a relatively uniform distribution of POP levels in various sites, and the indigenous people are particularly exposed because of their diet.

Heavy Metals

Mercury, lead, and cadmium are metals of concern because of their high toxicity in certain forms and their ability to be transported over large distances in the atmosphere. Their effects are most pronounced near factories and in industrialized estuaries. A well-known example is Minamata disease, caused by severe mercury poisoning arising from the discharge of methyl mercury from a Chisso Corporation chemical factory in the late 1950s, which bioaccumulated in shellfish and fish eaten by the local population around Minamata Bay, Japan.

Other metals of concern are arsenic, copper, nickel, selenium, tin, and zinc. Tributyltin, used in antifouling paints for ships and boats, and its derivatives disrupt the body's normal hormonal functions and have proven to be much more persistent in the environment than expected.

While most developed countries have taken the necessary steps to address the problem of heavy metals, this global progress is offset by new sources of pollution in emerging economies. An increasingly serious problem globally is "electronic waste," particularly the disposal of computers and mobile phones in China and southeast Asia, which contain thousands of different materials, many of them toxic.

Excessive Nutrients

Land-based activities such as agricultural runoff (fertilizer), atmospheric releases from fossil fuel combustion, and sewage and industrial discharges are the main contributors to nutrient overenrichment of the oceans. The global average for nitrogen loading in the coastal areas has doubled in the past century, making the coastal areas the most highly chemically altered in the world. Excessive nutrient content encourages the rapid growth of phytoplankton and results in algal blooms. Red tides or harmful algal blooms are caused by some algae species, producing toxins that cause shellfish poisoning and present serious health hazards to human consumers.

Worldwide, "coastal dead zones," or hypoxic areas of oxygen deprivation and devoid of life, numbered nearly 150, having doubled every decade since 1960. While many coastal dead zones are small coastal bays and estuaries, areas in marginal seas of up to 70,000 km^2 are also affected. The primary cause of these dead zones varies. For example, the very large dead zone in the Gulf of Mexico is caused primarily by nitrogen from agricultural runoff. In the Baltic Sea, northern Adriatic Sea, Gulf of Thailand, Yellow Sea, and Chesapeake Bay, the dead zones result from a combination of agricultural runoff, nitrogen compounds from fossil fuel burning being deposited from the air, and discharges of human wastes. Control varies from region to region as the amount of nutrients entering the oceans tends to vary significantly over time and from region to region.

Marine Litter

Marine litter consists mainly of slowly degradable or nondegradable substances found in coastal areas and oceans across the world. Dispersed easily by marine currents and wind, it is found even in remote places far from human population centers. Of the litter entering the oceans, 70% lands on the seabed, 15% is deposited on the beaches, and 15% remains floating on the surface. Marine litter has a wide range of ecological, environmental, and socioeconomic impacts, including ingestion by and entanglement with the marine life, fouling of coastlines, and interference with navigation. For example, it is estimated that plastic litter kills more than 1 million birds and 100,000 marine mammals and sea turtles each year.

Marine litter originates from sea and land. The major sea-based sources include shipping and fishing activities; offshore mining and extraction; legal and illegal dumping at sea; abandoned, lost, or otherwise discarded fishing gear; and natural disasters. The major land-based sources include wastes from dump sites, rivers, and floodwaters; industrial outfalls; storm water discharge; untreated municipal sewerage; littering of beaches and coastal recreation areas; coastal tourism and recreation; fishing industry activities; ship-breaking yards; and natural storm-related events.

The problem of marine litter has steadily grown worse, despite national and international efforts to control it. The conventions or treaties that are most relevant to marine litter control are the MARPOL 73/78 (International Convention for the Prevention of Marine Pollution From Ships) and its Annex V (which prohibits the at-sea disposal of plastics and garbage from ships), the London Convention (Convention for the Prevention of Marine Pollution by Dumping of Wastes and Other Matter) and the Basel Convention (Convention on the Transboundary Movements of Hazardous Wastes and Their Disposal). While information, education and public awareness programs are proving useful, there is still room for improvement.

A pair of goats and a pig forage for food among the rubbish and debris on this "beach" in Mumbai, India.

Source: Gordon Dixon/iStockphoto.

Oil Pollution

A lot of the petroleum that pollutes the ocean comes from natural sources, such as natural seeps in the seabed, and cannot be stopped. By comparison, the largest portion of the annual input of oil into the coastal and marine environment worldwide comes from land-based sources as a result of ordinary, everyday activities on land—for example, atmospheric dispersion of volatile fractions, storm sewers, and sewage treatment plants. Oil spills, which epitomize the problem of oil pollution, account for 10% to 15% of all the oil that enters the ocean every year. Oil spills damage natural habitats, smother aquatic communities and are generally toxic to aquatic life, taint seafood, contaminate water supplies and generally affect human health, and foul coastlines and beaches. Although oil pollution from marine transportation has reduced significantly, land-based runoff is expected to increase.

Radioactive Substances

Radioactive substances enter the coastal and marine environment due to a variety of activities, including nuclear power generation and the use of radioactive materials in medicine, industry, research, and space exploration as well as military operations. All contemporary uses of large amounts of radionuclides should conform to the International Safety Standards for Protection Against Ionizing Radiation and the Safety of Radiation Sources. Although accidents can occur, the impacts on human health and the environment are generally minor. The situation in the marine environment in this respect is stable, and controls on routine discharges are generally stringent.

Future Issues and Challenges

Four global issues have emerged as requiring priority in the marine environment: (1) sewage and the management of municipal wastewater, (2) nutrient overenrichment, (3) marine litter and physical alteration, and (4) habitat destruction. Other emerging challenges include the coastal dead zones; new chemicals, numbered by the thousands, released into the environment; and sea-level rise, which is expected to cause salinization as well as physical alteration, posing a major challenge for coastal management.

Poh Poh Wong

See also Agrochemical Pollution; Chemical Spills, Environment, and Society; Coastal Dead Zones; Global Sea-Level Rise; Heavy Metals as Pollutants; Nonpoint Sources of Pollution; Oceans; Oil Spills; Point Sources of Pollution; United Nations Environment Programme (UNEP); Water Pollution

Further Readings

United Nations Environment Programme. (2006). *The state of the marine environment: Trends and processes.* The Hague, Netherlands: Coordination Office of the Global Programme of Action for the Protection of the Marine Environment from Land-Based Activities of the United Nations Environment Programme.
United Nations Environment Programme, Global Programme of Action for the Protection of the Marine Environment from Land-Based Activities: www.gpa.unep.org

 # COLD WAR, GEOGRAPHY OF

The Cold War was a geopolitical division between the Western allies and the Soviet Union that emerged in the late 1940s, shortly after the end of World War II, and continued through the collapse of the Soviet Union in 1991. During World War II, the United States was allied with the United Kingdom and the Soviet Union against Germany, Italy, and Japan. However, within several years of the war's conclusion, this global division of power changed dramati-

cally to one where the United States and its allies in Western Europe and Japan were aligned against the Soviet Union and its satellite states in Eastern Europe.

From the U.S. perspective, the Cold War geopolitical model of the world saw the United States and its allies (the "First World") competing with the Soviet Union and its allies (the "Second World") for control of those places not directly aligned with either (the "Third World"). After the 1949 communist takeover of Eastern Europe, the Soviet Union also sought to extend its influence over parts of the Third World. For U.S. foreign policymakers in the late 1940s and early 1950s, the Soviet Union was an inherently evil and expansionist state bent on world domination that must be stopped, in the parlance of the day, "wherever it reared its ugly head." On the other hand, Soviet foreign policymakers in the late 1940s saw the United States as an aggressive, expansionist state determined to destroy the Soviet Union.

Thus, the Cold War saw skirmishes between the two superpowers and their surrogates in Latin America and the Caribbean, Africa, the Middle East, and South and southeast Asia. In this Cold War, local events, movements, and revolutions in the Third World were not understood in local terms (e.g., as the overthrow of a local despotic ruler or as part of a local uprising trying to overthrow the government to bring greater social welfare to the poor) but rather were interpreted as possible gains or losses for the First World or Second World based on how they fit into the grand global "battle" between "democracy" and "communism." Hence, in the name of stopping potential communist threats, the United States helped overthrow governments or intervened directly in Iran and Guatemala in the 1950s, Vietnam in the 1960s, Chile, Laos, and Cambodia in the 1970s, and Nicaragua, El Salvador, and Grenada in the 1980s.

The Soviets were certainly not immune to such meddling, as seen in their interventions in Hungary in 1956 and Czechoslovakia in 1968 and in their 10-year battle to control Afghanistan throughout the 1980s.

Guiding U.S. actions under this Cold War global geopolitical model were two crude "geographical" principles, containment and the domino theory, which engage with geography as being primarily about location.

Under the concept of containment, the United States established a policy that the forces of the Soviet Union and China should be bottled up and not allowed to expand beyond their spheres of influence, whether beyond China itself or in terms of the Soviet Union beyond that which they had established in the immediate aftermath of World War II in Eastern Europe. Hence, the United States set up a series of military alliances with countries surrounding the perimeter of China and the Soviets' sphere, arguing that if any of these countries were attacked, the United States would step in and protect them. These alliances ranged from the North Atlantic Treaty Organization (NATO), set up in 1949 between the United States and the non-Soviet-dominated countries of Western Europe, to ANZUS, an alliance between the United States, Australia, and New Zealand. Any attack on these allied countries would be treated by the United States as if it were an attack on itself.

However, it might be tough to sell the American public on the necessity of fighting the forces of communism anywhere in the world, especially if that was half a world away (e.g., in Vietnam). So in the 1950s, the concept of the "domino theory" was introduced, which suggested that the forces of communism were cunning and that one victory in some far-flung part of the world would lead to a series of victories in the neighboring countries, which would fall to the communists like a row of falling dominoes. Through this metaphor, distant events were able to be linked back home. Thus, intervention in southeast Asia was sold to the American public as necessary because once the countries in southeast Asia "fell," then, like a row of falling dominoes, communist forces would ultimately continue until they reached American soil.

The domino theory continued to be used well into the 1980s to sell the American public on the importance of containing communism. For example, Ronald Reagan sold his administration's support of the *contra* rebels fighting the socialist Sandinista government in Nicaragua to the American public in part by arguing that the United States had to defeat the Sandinistas because it was only a 2-day drive from Nicaragua to South Texas. Despite the fact that the domino theory has fallen into disfavor, it is still occasionally invoked today.

One of the appeals and dangers of conceptualizing global geopolitical divisions of power in broad terms such as the Cold War's tripartite division is its ability to take a large variety of complex unique places with different histories, cultures, and languages and homogenize them under broad, sweeping generalizations and categorizations. Such conceptualizations represent grand oversimplifications of the world, as when we refer to the Cold War as being a battle between the First and Second Worlds for control of the Third World. Recognize that lumped together in a Third World that the First or Second World felt compelled to respond to potential threats for fear of their going "communist" or "Western" were places as different as Angola, Chile, Cuba, Grenada, Iran, Nicaragua, and Vietnam. Rather than recognizing the vast differences and complexities among these places, they were seen in simple uniform terms as part of an undifferentiated Third World.

Jonathan Leib

See also Domino Theory; Geopolitics; Political Geography

Further Readings

Agnew, J. (2003). *Geopolitics: Re-visioning world politics* (2nd ed.). London: Routledge.

Nijman, J. (1992). The limits of superpower: The United States and the Soviet Union since World War II. *Annals of the Association of American Geographers, 82,* 681–695.

O'Sullivan, P. (1982). Antidomino. *Political Geography Quarterly, 1,* 57–64.

Ó Tuathail, G., & Agnew, J. (1992). Geopolitics and discourse: Practical geopolitical reasoning in American foreign policy. *Political Geography, 11,* 190–204.

Ó Tuathail, G., Dalby, S., & Routledge, P. (Eds.). (1998). *The geopolitical reader.* London: Routledge.

COLLABORATIVE GIS

Local and global problems such as land use and land cover change, poverty and deprivation, health and diseases, and global warming are all directly linked to the intricate biophysical and sociopolitical networks operating within geographic space. Resolving the complex issues embedded in these problems requires the involvement of multiple interest groups, reliable baseline geographic data, and relevant decision support technologies. A collaborative geographic information system (GIS) provides the integrated structure in which to embed the interest groups, geographic data, and relevant technologies for planning and decision outcomes.

Collaborative GIS is defined as a combination of interrelated theories, tools, and technologies focusing on, but not limited to, the structuring of human interactions in group spatial planning and decision processes. For planning tasks, the collaborative GIS facilitates the collective development of the stages needed to achieve a desired outcome. For decision tasks, the collaborative GIS facilitates the generation of decision options and selection among alternatives. To implement the overall integrated structure, the collaborative GIS design uses GIS technology as a platform on which to combine the geographic data, interest group interactions, and structured processes that would lead to agreeable planning, problem-solving, and decision-making outcomes. GIS technology has been in constant development since the 1960s. Today, a GIS is considered to be an organized arrangement of computer hardware, software, and trained personnel to manage, transform, and analyze spatially referenced data in order to find solutions for unstructured real-world problems.

Components of a Collaborative GIS

The main components of a collaborative GIS are shown in Figure 1. A meaningful problem initiates the establishment of a collaborative GIS. Choices about whether to implement the collaborative process in a same-place or same-time configuration as compared with a different-place or different-time configuration is influenced by the extent of the problem being addressed and the wishes of any organization that is directly

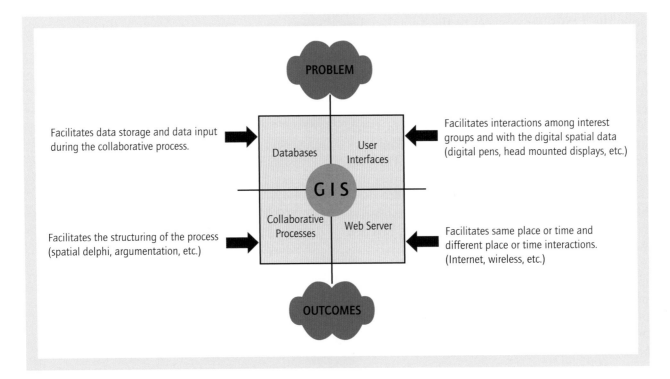

Figure 1 Components of a collaborative GIS

Source: Authors.

Application Areas	Summary Description
Natural resources management	Collaborative GIS is used primarily to develop scenarios and to forecast eventual outcomes. It has also been used to generate agreement on conflicting issues among multiple interest groups. In addition, collaborative GIS processes have been used to generate model-based tools for natural resources management tools.
Disaster management	Collaborative GIS is used primarily to integrate multiple personnel and data resources in a command and control setting. It has also been used to collect field data from mobile devices to monitor evolving disaster situations.
Health and disease management	Collaborative GIS is used primarily to integrate field data from the public for use in identifying disease hotspots in remote areas.

Table 1 Application areas of collaborative GIS

Source: Authors.

sponsoring the decision process. The core of the collaborative GIS design contains a GIS that combines the geospatial databases, user interfaces, Web servers, and collaborative process components. There are two key roles for the geospatial databases. First, they will contain the baseline data that the multiple interest groups will use to develop a well-defined problem to be addressed within a specified time frame. Second, they will store the new information generated from the collaborative GIS process as the interest groups work toward an agreeable solution. The user interfaces allow the multiple interest groups to interact with each other and with the existing geospatial data. Standard user interfaces such as the computer mouse and keyboard are usually not sufficient for dealing with digital map data on the computer screen. User interfaces for collaborative GIS provide the end user with a richer experience and include digital pens for sketch mapping, head-mounted displays for simulated real-world experiences, and multitouch screen displays for rapid database browsing and query. The Web server allows the collaborative GIS group members to make use of the Internet to interact with an extended public audience, such as citizens, experts, or government representatives. This is useful as it removes the barriers of location and time for those interested in contributing to the decision process. The collaborative process component is a set of well-established methods that can be used by facilitators to structure the process toward agreeable solutions to the problem. Structuring approaches such as shared workspace, argumentation mapping, spatial Delphi, real-time conferencing, and sketch maps are widely used. The collaborative GIS process generates solutions in the form of digital maps, visualization products, and data attributes in response to the problem. Implementing the solutions is done within an organizational and technical framework to ensure continued development of the collaborative GIS process.

Linking Collaborative GIS With GIScience

Geographic information science (GIScience) is the assumptions, theories, and methods that inform the practical design and application of GIS. As a discipline, GIScience is now well established, with a research agenda containing 12 well-defined themes and multiple research challenges. Collaborative GIS is positioned as a subarea of the "GIS and Society" research theme, together with other implementations of spatial decision support systems such as planning support systems, spatial understanding support systems, and geocollaboration systems.

Applications of Collaborative GIS

Collaborative GIS designs have been applied widely in multiple contexts. Table 1 summarizes some of the more popular areas of application.

Suzana Dragicevic and Shivanand Balram

See also GIScience; Neogeography; Public Participation GIS

Further Readings

Balram, S., & Dragicevic, S. (2006). Collaborative geographic information systems: Origins, boundaries, and structures. In S. Balram & S. Dragicevic (Eds.), *Collaborative geographic information systems* (pp. 1–22). Hershey, PA: Idea Group.

Jankowski, P., & Nyerges, T. (2001). *Geographic information systems for group decision making: Towards a participatory geographic information science.* New York: Taylor & Francis.

MacEachren, A. M., & Brewer, I. (2004). Developing a conceptual framework for visually-enabled geocollaboration. *International Journal of Geographical Information Science, 18*(1), 1–34.

Wright, D. J., Goodchild, M. F., & Proctor, J. D. (1997). Demystifying the persistent ambiguity of GIS as "tool" versus "science." *Annals of the Association of American Geographers, 87*(2), 346–362.

COLONIALISM

Intimately associated with the development of capitalism was Europe's conquest of the rest of the globe. This process, often euphemistically called the "Age of Exploration," can be viewed as the expansion of capitalism on a global scale. If the geographies of capitalism are typified by uneven spatial development, then colonialism involved the construction of uneven development on a global scale, with Europe at the center and its colonies on the world periphery. This theme is found in many theories of world development, such as world-systems theory.

Colonialism was simultaneously an economic, political, and cultural project. It was also an act of conquest, by which a small group of European powers came to dominate a very large group of non-European countries. Culturally, as Edward Said points out, this process involved the emergence of the distinction between the "West" and the "Rest." In conquering the "Orient," Europeans came to discover themselves as Westerners, often in contrast to other people, whom they represented in highly erroneous terms.

Everywhere, colonized people fought back against colonial rule. Examples include the Inca rebellions against the Spanish, the Zulu attacks on the Dutch Boers, the great Indian Sepoy uprising of 1857, and the Boxer Rebellion in China in 1899–1901. Yet Western powers, armed with guns, ships, and cannons, effectively dominated the entire planet. While a few countries were nominally independent, such as Thailand, the only one to escape colonialism substantively was Japan, which, under the Tokugawa Shogunate, closed itself off from the world until 1868.

Colonialism had profound implications for both the colonizers and the colonized, which is why a sophisticated understanding of economic geography must include some understanding of this process. Globally, colonialism produced the division between the world's developed and less developed countries. Colonialism changed Europe too, increasing the formation of capitalist social relations and markets as well as nation-states in Western Europe. Prior to colonialism, Europe was a relatively poor and powerless part of the world, compared with the Muslim world, India, or China; afterward, Europe became the most powerful collection of societies on the planet.

The Temporal and Spatial Unevenness of Colonialism

It is important to note that colonialism did not occur in the same way in different historical moments and different geographic places. Temporally, there were two major waves of colonialism, one associated with the preindustrial mercantile era and the other with the Industrial Revolution. From the 16th century, when colonialism began, until the early 19th century, Western economic thought was characterized by mercantilism, in which state protection of private interests was justified as necessary for the national well-being. During this period, the primary colonial powers were Spain and Portugal, and their primary colonies were in the New World and parts of Africa.

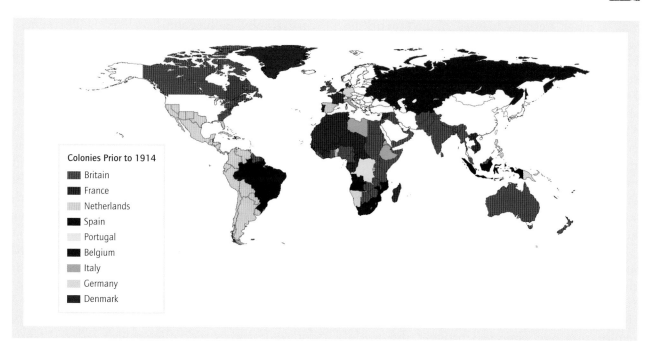

Figure 1   Colonial empires prior to 1914. Different European powers conquered various regions of Earth at different moments in time.

Source: Author.

Following the Napoleonic wars, which ended in the Treaty of Vienna in 1815, the European powers were relatively weak. This opening provided an opportunity for nationalists in Latin America, led by Simon Bolivar, to break away and become independent countries. Thus, the number of colonies declined sharply in the early 19th century. However, during the subsequent phase of industrialization, characterized economically by the ideology of free trade, the number of colonies grew again. This time, Britain emerged as the world's premier power and, along with France, colonized large parts of Africa and Asia. Finally, the number of colonies declined rapidly after World War II, in the era of decolonization.

Spatially, colonialism was also uneven. Different colonial empires had widely varying geographies, as illustrated by the distribution of empires at their peak in 1914, the eve of World War I (Figure 1). The British Empire, which encompassed one quarter of the world's land surface, stretched across every continent of the globe, including Canada, much of the Middle East, large parts of Eastern and Southern Africa, the Indian subcontinent, and parts of southeast Asia. The French ruled in parts of Western Africa and in Indochina.

The Portuguese had Brazil; Goa in India; Angola, Mozambique, and Guinea Bissau in Africa; Timor; and Macau. The Belgians possessed Congo, the private domain of Leopold II. Italy was in Ethiopia and Libya. Even Germany, late to unify and to industrialize, controlled parts of Africa (Togo, Namibia, Tanganyika) and New Guinea.

How Did the West Do It?

What allowed Europe to conquer the rest of the planet? The answers to this question are not simple. Obviously, they do not lie in any innate superiority of the Europeans. Indeed, for much of history, Europe was relatively weaker and poorer than the countries it conquered. By the 16th century, however, Europe did come to possess several technological and military advantages over its rivals.

Jared Diamond, in his famous book *Guns, Germs, and Steel*, maintains that Western societies enjoyed a long series of advantages by virtue of geographical accident. Agriculture in the West, centered on wheat, was productive and could sustain large, dense populations. Old World societies were often stretched across vast East-West axes, or regions with common growing seasons. There was a long Western history of metalworking, which not only increased economic productivity but also allowed for the manufacture of weapons such as guns and cannons. By the Renaissance, Europeans had become highly skilled at building ships and navigating the oceans. And by the late 18th century, the West had discovered inanimate energy, which offered numerous economic and military advantages. In addition, the Europeans unleashed diseases (if unintentionally), particularly smallpox and measles, on the New World, which not only provided them with an unintended advantage but also led to labor shortages.

Central to this process were the oceans. The new European relationship to the seas was forged through the navigational revolution of the 15th century, which allowed open-ocean sailing rather than coastal voyages. The ensuing voyages of exploration and conquest brought the world's maritime regions within the expanding circuits of European capital circulation and accumulation. Colonialism transformed the oceans from barriers into a means of accessing the lands beyond, even if those were unknown or only dimly perceived. Often this process was born out of dire necessity: Portuguese and Spanish entries into the Atlantic Ocean were in many respects an attempt to circumvent the Turkish domination of the Middle East (including the seizure of Constantinople in 1453), a force that motivated Europeans to find new ways to the lucrative Asian trade. The eventual result was a series of colonial empires that bound together vast regions separated by enormous maritime spaces. Because sea voyages did not allow penetration of continental interiors, ship-based empires were confined to coasts, littorals, archipelagos, and islands. Oceanic discoveries and conquests also propelled a gradual restructuring of European spatiality, with a decline in the Italian city-states and the Hanseatic League and a rise of the Atlantic sea powers, shifting the primary locus of trade and balance of power from the Mediterranean and Baltic to the Atlantic, in which not one but several empires operated simultaneously. Under Henry the Navigator, Lisbon became the new global center of navigational and cartographic expertise. Henry's headquarters at Sagres pioneered the European exploration of the Atlantic and West Africa and Portugal's entry into Asia, and it revolutionized

cartography. Sailing ships such as the Portuguese caravel, the only ship capable of sailing anywhere on the planet, and its successor, the galleon, were able to sail into the wind and in shallow waters in all weathers, cut weeks from time at sea, and allowed sailors ("men of the sail") to return home more quickly and reliably, initiating a feedback loop in which geographic knowledge became cumulative and its growth self-reinforcing and continuous. Bartholomeu Dias returned from rounding Southern Africa in 1488, which rendered Christopher Columbus's original intentions superfluous and encouraged him to think about a westward voyage. Portuguese expansion replaced the older land-and-sea trade networks with an all-ocean one, greatly lowering the costs of bringing goods from Asia to Europe. Around 1600, the Dutch began using a new vessel, the *fluytschip*, for trade across the Baltic, which reduced labor costs by half in waters that had been cleared of pirates. The fluytschip, produced in large numbers, had as much an impact on European trade as did the steamship centuries later. Using this technology to dominate European ocean shipping transformed Amsterdam from a backwater of the Hapsburg Empire into one of the first great commercial and banking centers of mercantile capitalism. Similarly, the rise of Britain was enabled by the "race-built" ships, which were the first to consciously emulate the streamlined design of a fish. Unlike the Venetians, Portuguese, or Spanish, who compelled all traders to travel in government-organized convoys, the Dutch and English used privately owned but state-charted monopoly corporations such as the Hudson Bay Company, the British East India Company, and the Dutch Vereenigde Oost-Indische Compagnie (VOC) or Dutch East India Company.

Others scholars maintain that the West's advantages were not just technological but also political. The Western "rational" legal and economic system stressed secular laws and the importance of property rights. Yet others point out that unlike the Arabs, Mughal India, or China, Europe was never united politically. Indeed, every time one European power attempted to conquer the others, it was defeated, as exemplified by France in the early 19th century and Germany twice in the 20th century. Europe's very political separation into competing powers—in contrast to the

Muslim world, India, or China—fostered intense rivalries and arms races that generated a self-reinforcing series of changes. In the view of world-systems theory, colonialism witnessed numerous hegemons dominate the global economy at successive historical moments, including Portugal, Spain, the Netherlands, Britain, and, later, the United States, a pattern that may be indicative of long-wave cycles of capital accumulation in the capitalist economy. The lack of centralized political authority created a climate in which dissent and critical scholarship were tolerated. For example, the French Huguenots, Protestants in a predominantly Catholic country, could flee persecution by moving to Switzerland, where they started the Jura district watch industry. Similarly, when Columbus failed to obtain financing for his voyages from the Italians and Portuguese, he could switch to Spain, whose king ultimately consented.

A Historiography of Conquest

To appreciate colonialism, it is necessary to briefly delve into its specifics in different times and places. This short review is intended to demonstrate that colonialism meant quite different things under different contexts (i.e., it was historically and geographically specific).

Latin America

Home to wealthy and sophisticated civilizations such as the Inca, Mayan, and Aztec cultures, Latin America was one of the first regions to be taken over by the Europeans. Two years after Columbus arrived, Spain and Portugal struggled over who owned the New World—a contest settled by the Pope through the Treaty of Tordesillas in 1494. The conquistadors who spread out over Mexico and Peru annihilated the Aztec and Incan states, respectively. In large part, this was accomplished through the introduction of smallpox, which killed 50 million to 80 million people within a century of Columbus's arrival, the largest act of genocide (whether intentional or not) in human history.

Subsequent to the extermination of tens of millions of Native Americans, the New World was methodically transformed into a global periphery

that aided greatly in the development of Europe, allowing that continent to escape domestic ecological constraints by tapping into new supplies of fish, lumber, minerals, and other resources. Of course, chronic labor shortages in the Western Hemisphere were alleviated largely through the forced importation of 10 million to 20 million African slaves, a phenomenon that reflects the repeated misery visited on that highly marginalized continent by successive waves of capitalism. In keeping with early modernity's maritime emphasis, Spanish colonialism generated a series of coastal entrepots, such as Havana, Veracruz, and Cartegena. Sugar plantations, which became prototype factories of a sort, were among the first to subject their labor force to the discipline of commodity production.

Under the philosophy of mercantilism, which stressed bullion, or precious metals, as the key to national wealth, the Spanish took home large quantities of silver from the New World. The silver mines in Central Mexico were among the largest in the world, and in the enormous Potosi mines in Bolivia, two million Aymara Indians perished digging silver. Argentina takes its name from *argentium*, the Latin word for silver, and is home to the Rio Plata. Most of this metal was taken back to Spain in galleons and provided an enormous base of capital that financed economic activities throughout Europe. By far the most important contribution of the New World to the Old was silver; 10 times more of it than gold was shipped back to Spain in galleons (financing luxuries such as the palace at El Escorial). Silver became the primary medium for integrating the world's spaces in the 16th and 17th centuries and the force that produced a series of distinctively novel geographies (Figure 2). New World silver became the primary specie integrating the mercantile world economy, and it essentially allowed Europe to "buy itself a ticket on the Asian train" (Frank, 1998, p. xxv). American silver was so ubiquitous that merchants from Boston to Havana, Seville to Antwerp, Murmansk to Alexandria, Constantinople to Coromandel, Macao to Canton, and Nagasaki to Manila all used the Spanish peso or piece of eight (*real*) as the standard medium of exchange. Given Europe's large trade deficits with Asia, two thirds of European payments for Asian imports consisted of precious metals. From Spain, silver flowed to England, France, and the Netherlands to purchase manufactured goods, from where much of it went to Russia to finance imports of furs. Russian silver, in turn, flowed along the Volga River to Persia. Spanish silver also found its way directly to the Ottomans and thence to the Indian Ocean. Much of the Mughal Empire in India was financed by silver, a considerable portion of which went to purchase horses from Central Asia. Indian silver also flowed through Malacca to China, or at times overland via the Silk Road routes. At the easternmost end of this network, China, in which the Ming dynasty was remonetizing the economy from paper to silver, was the "sink" for half to three fourths of the world's silver flows between 1500 and 1800. Indeed, the 17th century witnessed a broad financial crisis across eastern and Southern Asia associated with this restructuring. When Spain established its Manila colony in the Philippines in 1571, it became possible, for the first time, for New World silver to travel back to Europe in two directions, that is, in galleons across both the Atlantic and the Pacific Oceans.

Spain also introduced the *encomienda* land grant system into the New World, giving large tracts of lands to potential rivals to the Spanish throne. This set of practices was an extension of the ancient latifundia system practiced in Roman Iberia. As a small landed aristocracy consolidated its hold, the distribution of farmland became highly uneven, with a few wealthy landowners and large numbers of landless *campesinos*. This pattern continues in the present, indicating how colonialism has shaped the geographies of the contemporary world.

The Spanish Empire in the New World was largely ended by the independence movements of the 1820s that followed the Napoleonic Wars (although it took the Spanish-American War of 1898 to finish it off). After gaining independence, the erstwhile Spanish Empire broke up into a series of independent countries stretching from Mexico to Argentina. The Portuguese Empire in Brazil did not fragment in the same way, leaving that nation as the giant of Latin America.

Finally, the Caribbean exhibited a complex mosaic of colonial powers. The Spanish dominated Cuba, Eastern Hispaniola, and Puerto Rico, exterminating the native Taino Indians. The British colonized Jamaica, the Bahamas, and many smaller islands in the Windward and Leeward Islands chain. The French took Haiti and small islands such as Martinique and Guadeloupe, trading others periodically with the British. The Dutch also settled in Curacao and Aruba.

North America

The colonialization of North America proceeded along somewhat different lines. Here, the Spanish were active in Florida and in the Southwest. A century before the Pilgrims arrived at Plymouth Rock in 1620, the Spanish had control of what is now Texas and California. The French took over Quebec and the St. Lawrence River valley, only to lose it to Britain in the 18th century, and the Mississippi River valley, with the key port of New Orleans, only to sell it to the United States in the Louisiana Purchase of 1803. The Russians crossed the Bering Straits and seized Alaska but sold it to the United States in 1867. The Dutch established colonies in New Amsterdam, including Haarlem and Brueklyn, but the British captured the area in 1664 and renamed it New York. Britain emerged as the dominant power in North America, controlling New England and the Piedmont states along the eastern seaboard. Canada was colonized largely through the famous Hudson Bay Company, then the world's largest, which controlled much of the region's fur and fish trade.

British colonialism in North America began with a series of port cities on the east, typically at the mouths of rivers (e.g., Boston, New York, Philadelphia). After the independence of these colonies in the Revolutionary War, the settlers moved west, across the Appalachians in the early 19th century and across the Great Plains and Rocky Mountains somewhat later. Railroads opened up this region to the east. As in Latin America, this process involved the wholesale eradication of Native American peoples, theft of their land, and commodification of territory.

Africa

Colonialism in Africa was unique, as it was everywhere. This process included slavery, the kidnapping of roughly 20 million people and exporting them to the New World, where they

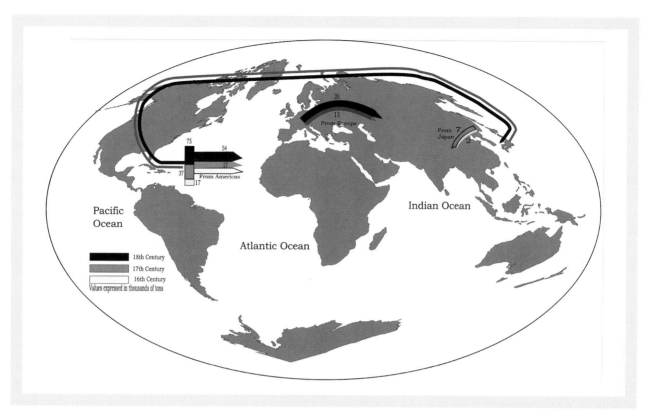

Figure 2   Global silver flows, 16th to 18th centuries. Silver from the New World largely financed Europe's entry into the global economy.

Source: Adapted from Frank, A. (1998). *ReOrient: Global economy in the Asian age.* Berkeley: University of California Press.

were used to compensate for the labor shortages brought on by the decimation of the Native Americans. The capture of slaves robbed these societies of young adults in their prime working years and sometimes occurred with the assistance of local kings who sought to profit from the trade, using the revenues to buy British textiles and whiskey. Slaves were generally taken from Western Africa, and the largest numbers were brought to work in the sugar plantations in Brazil and the Caribbean. Others were brought to work in the cotton and tobacco plantations of the American South. In Eastern Africa, there long existed a smaller system of slavery operated by the Arabs.

Africa is rich in minerals, and the colonialists were quick to seize on that fact. In the 19th century, when the European powers penetrated from the coastal areas into the interiors, mines for copper, gold, and diamonds were opened, often using slave labor. These remain the basis of many African economies today.

Perhaps the most important colonial impact on Africa was the political geography that Europe constructed. Following the famous Berlin Conference of 1884, when the European powers drew maps demarcating their respective areas of influence, the colonies of Africa bore no resemblance whatsoever to the distribution of indigenous peoples. Roughly 1,000 tribes were collapsed into about 50 states. Some groups were separated by colonial boundaries; many others, with widely different cultures and economic bases, were lumped together. Not surprisingly, on attaining independence in the 1950s and 1960s, many African states, including Angola, Congo, Rwanda, Liberia, and Sudan, have been wracked by numerous civil wars and tribal conflicts, in which millions have perished.

The Arab World

The Arab world, which was one of the most powerful and sophisticated centers of world culture from the 7th through the 15th centuries, had been colonized long before the Europeans arrived. The expansion of the Ottoman Empire over several centuries saw one group of Muslims, the Turks, dominate another group of Muslims, the Arab peoples of the Middle East and North Africa. Gradually, starting in the late 18th century, European powers encroached on this vast domain. In 1798, the French seized Egypt from the Ottomans, only to lose it to Britain 4 yrs. (years) later. The British used Egypt as a source of cotton, growing large plantations along the Nile River and building the Suez Canal in 1869.

After World War I, the Ottoman Empire collapsed, and the French and British seized its Arab colonies. The French took over Morocco, Algeria, and Tunisia, as well as Syria and Lebanon. The British assumed control over Palestine, Jordan, much of which became Israel in 1948, as well as Iraq and the sheikdoms of the Persian Gulf. Many Arabs, who initially welcomed the Europeans as liberators, quickly learned that the new boss was very similar to the old one in its suppression of indigenous liberties.

South Asia

As the first of the European powers to infiltrate Asia and set up foreign colonies, Portugal, the world-system's first hegemon, played a uniquely important role in the construction of the colonial world system. Portugal's invasion of the Indian Ocean in the early 16th century pulled the plug on the centuries-old Arab-Asian trade and set the stage for the Ottoman conquest of the Arab world. Portugal's seizure of Malacca in 1511 was followed in short order by Timor, Macau, and Formosa. Given how far such places were from Europe—a round-trip voyage to Asia could take 3 yrs.—direct control over colonial entrepots was often loose at best.

The Indian subcontinent later became the jewel in the crown of the British Empire. Starting in the 17th century, the British East India Company established footholds in this domain, founding the city of Calcutta (now Kolkata) in 1603. India, which is predominantly Hindu, had long been controlled by Muslim rulers, chiefly the Mughals, whose power was gradually usurped by the foreigners. A vast land that stretched from Muslim Afghanistan in the east through Hindu India to Buddhist Burma, the Indian colony was the largest colonial possession on Earth.

Britain had enormous economic impacts on this land. In the 19th century, British textile imports flooded India, destroying the Mughal textile industry, a classic example of colonial deindustrialization. This event later became symbolically important in the independence movement after World War II. Indian laborers were exported throughout the British Empire, including the Caribbean, Eastern Africa, and parts of the Pacific, such as Fiji. Tea plantations were established in Bengal and Ceylon (now Sri Lanka). Britain also built railroads to facilitate the extraction of Indian wealth.

In 1857, a mass uprising against the British took place known as the Sepoy Rebellion. Encouraged by the local raj, or the Mughal ruler of Calcutta, the revolt claimed the lives of tens of thousands of Indians when it was crushed. Although it failed, the action forced the British crown to assume direct control over this land rather than administer it through the East India Company.

East Asia

East Asia, comprising China, Korea, and Japan, also had a unique colonial trajectory. Japan, as noted earlier, was never colonized. When it emerged from a long period of isolation that began in the early 17th century and lasted until 1868, Japan rapidly Westernized and industrialized and became the only non-Western power to challenge the West on its own terms, gradually expanding its power through Northeast Asia. Korea, which opened up under the threat of force in the 1870s, was taken over by Japan in 1895 and annexed in 1905, and Japan held it until the end of World War II. A similar situation held in Taiwan. Later, Japan extended its colonial hold over Manchuria, parts of China, and, briefly, most of southeast Asia.

China, however, was a different story. Under the rule of the Manchus, foreigners from the north (Manchuria) who ruled the Chi'n dynasty from 1644 until the revolution of 1911, the Chinese government was weak and corrupt. Except for a few cities along the coast, China was never formally colonized; rather, European control operated through a pliant and cooperative government. Chinese coolie labor was exported to British colonies in southeast Asia and to the United States, where Chinese labor built railroads in the American West. British, French, German, and American trade interests purchased vast amounts of Chinese tea, silks, spices, and porcelains. In fact, Britain ran a negative balance of trade with China in the 18th and early 19th centuries, which it rectified by introducing large amounts of opium. By the 1830s, opium addiction was widespread in China, creating severe social disruptions in Chinese society. Nonetheless, profits were more important, and the British trade balance was restored.

In a rare moment of defiance, the Manchu government resisted the opium imports, and Britain and China fought two short, nasty conflicts, the Opium Wars of the 1840s. The British won easily. As compensation, the British seized "treaty ports," coastal cities such as Hong Kong and Shanghai, where Western, not Chinese, law applied. Britain held Hong Kong until 1997.

Chinese resentment against the Manchus culminated in the Taiping Rebellion, a huge uprising in the southern part of the country led by the Chinese Christians. Lasting from 1851 to 1864, this rebellion led to the deaths of more than 20 million people but was ultimately crushed by the Chinese government with Western backing. The shorter Boxer Rebellion of 1899–1901 was more explicitly anti-Western. These revolts set the stage for the successful nationalist revolution of 1911, which ended Manchu rule and initiated the Republic of China.

Southeast Asia

The peninsula of Indochina and the islands of southeast Asia, long home to a rich and diverse series of peoples and civilizations, were conquered by different European powers. The Philippines was Spain's only Asian colony and served as the western terminus of the trans-Pacific galleon trade. Administered as part of Mexico, the Philippines was heavily shaped by Spanish rule, which affected land use patterns (including sugar plantations) and the language and made it the only predominantly Catholic country in Asia. Spanish rule ended in 1898, when the United States took over, and it became independent after World War II.

The French controlled much of Indochina, including Vietnam, Laos, and Cambodia. French rule shaped the design and architecture of cities such as Saigon, and large numbers of Vietnamese converted to Catholicism. French domination did not end until 1954, with the defeat at Dienbienphu, an event that laid the foundations for American military involvement there.

Britain was also a major power in southeast Asia, controlling Burma and, only informally, the economy of Thailand. These were made into rice exporters for other parts of the British Empire. The British controlled the colony of Malaya (later Malaysia), including the strategically critical Malacca Straits. Sir Edmund Raffles founded the city of Singapore as a naval station and commercial center to exercise British control over this region. Malaya, like other colonies in the area, became a major producer of rubber products, as well as timber and tin.

Indonesia, now the fourth most populous country in the world, was dominated by the Dutch for several hundred years. Dutch rule, starting with the founding of the Batavia colony on Java (now Jakarta) in the 18th century, gradually expanded to include the other islands. The primary institution involved was the Dutch East Indies Company, a chartered crown monopoly similar to the British East Indies and British West Indies Companies. Indonesia became a significant source of spices, tropical hardwoods, rubber, cotton, and palm oils. In all of southeast Asia, Chinese immigrants came to play major roles in the economy as bankers and shopkeepers.

Oceania

Australia, New Zealand, and the Pacific Ocean islands, which are commonly grouped as the region of Oceania, formed yet another domain of colonialism. Australia was inhabited for thousands of years by indigenous aborigines, most of whom were eradicated by the British as they exerted control. The native Tasmanian population was completely exterminated. The continent served originally as a penal colony for criminals. In the 19th century, it became a significant exporter of wheat and beef.

New Zealand, also a British colony, had a much larger native population, the Maori, who continue to form a significant presence there. The introduction of refrigerated shipping in the late 19th century turned this island nation into a major producer of lamb and dairy products.

Finally, the countless islands of the Pacific Ocean, home to varying groups of Polynesians, Micronesians, and Melanesians, were conquered by the British and the French. Captain Cook sailed through in the 17th century, paving the way for followers. Fishing and whaling interests used these islands to refuel in the 19th century. Following World War II, when the United States drove Japan out of the Pacific, America became the leading political force in the area.

The Effects of Colonialism

By now, it is abundantly evident that colonialism had enormous consequences on colonized places, the effects of which are often still felt today. Above and beyond simple, brutal annihilation of entire societies, colonialism entailed a series of massive economic, political, cultural, and spatial transformations.

Often, colonialism involved traumatic consequences for the people who were conquered. At times, this involved open genocide, such as in Australia, in which more than 90% of the aboriginal population was exterminated. In the New World, disease led to the deaths of tens of millions. The African slave trade devastated tribal societies on that continent. While this is the bluntest expression of colonial control, it serves as a reminder that the European conquest of the world was often violent.

The incorporation of colonies into a worldwide division of labor led above all to the development of a primary economic sector in each of them. Primary economic activities are those concerned with the extraction of raw materials from the Earth, including logging, fishing, mining, and agriculture. In Latin America, Africa, and southeast Asia, cash crops such as sugar, cotton, fruits, rubber, and tobacco were grown in the plantation system for sale abroad. Silver, tin, gold, and other metallic ores were mined using slave labor or peasants working in slavelike conditions. Mercantilist trade policies worked to suppress

industrial growth in the colonies. This process is largely responsible for the fact that many developing countries today export low-valued goods and must import high-valued ones.

Colonialism brought with it great inequality in the colonized societies. Often, colonial powers used a small, native elite to assist them in governing the colonies, typically drawn from an ethnic minority. For example, the French used the Alawites in Syria, a sect of Islam neither Sunni nor Shiite; the Germans and the Belgians favored the lighter-skinned Tutsi over the darker-skinned Hutu in Rwanda and Burundi; and the British relied on the Muslim Mughal rulers to govern Hindu India.

For the bulk of the population, colonialism entailed declining economic opportunities, a theme central to dependency theory. Traditional patterns of agriculture were disrupted, often with disastrous effects. Land use patterns favored the colonialists, while indigenous peasants had to pay for the costs of their own exploitation with taxes. People living in dry climates in Western Africa, for example, coped well with drought until the British forced them into a system of cash cropping. Huge famines shook India, Egypt, and China in the 19th century.

As the colonial societies became polarized, so too did the spaces they occupied. Ports, which were central to European maritime trade and control, became important centers of commerce, often to the detriment of traditional capitals further inland. For example, in Peru, Lima displaced the Incan city of Cuzco; in Western Africa, the famous trade city of Timbuktu declined as new maritime routes flourished; in India, coastal cities such as Calcutta (now Kolkata), Bombay (now Mumbai), and Madras (now Chennai) displaced the Mughal capital of Delhi; in Burma (now Myanmar), Mandalay fell far behind coastal Rangoon; in Vietnam, the imperial capital Hue declined in the face of Saigon; and in Indonesia, the traditional center of Jogjakarta was marginalized by the Dutch port city of Batavia (later Jakarta). As cities grew and offered more opportunities, millions of people left the poorer rural areas in waves of rural-to-urban migration. From the coasts, railroads extended colonial control into the interior, often reaching into mineral-rich regions or plantations. A long-term consequence of this design is that the road and rail networks of developing countries often bear little resemblance to the distribution of the population that lives there and their needs; rather, they were constructed to facilitate the export of raw materials to the colonizing country and, today, to the global economy.

The nation-state, as we observed earlier, was fundamentally a European creation. Nonetheless, it was widely dispersed around the world as colonies were made into states. In Africa, where this process was the most notorious, it led to the formation of unstable states with highly artificial borders. Similarly, Burma and Afghanistan were creations of the British. Even India, which is more culturally diverse than all of Europe, was stitched into one country; surprisingly, it was partitioned only into two upon independence, including Pakistan. (Later, in 1971, Bangladesh split from Pakistan.) Unlike Europe, where states were centered on some degree of ethnic similarity, the states of much of the developing world were too diverse to be understood in the same terms as in Europe. In such societies, where local religious, ethnic, and tribal loyalties supersede nationalism, political conflicts can impair economic growth and development. Of course, the United States is also very diverse ethnically and culturally, but it emerged under very different historical circumstances from the former European colonies in the developing world. Above all, the United States was primarily a colonizer rather than a colony, able to exert military and economic power over other countries and ultimately rising to become the world's premier superpower, and that makes all the difference.

Colonialism is not simply an economic or political process but also a cultural and ideological one. Western economic and political control was accompanied by the imposition of Western culture. Missionaries, for example, sought to spread Christianity throughout the colonial world, sometimes successfully (e.g., Latin America, the Philippines) and sometimes not (e.g., China); in Africa, Christianity jostled for converts alongside Islam. Missionaries both legitimized colonial rule as God's work and at times established schools and hospitals. School systems in colonies, generally set up to benefit the ruling elite, offered extensive instruction in the history and culture of the colonial country but little about the society in which the students lived. More broadly, colonialism may be seen as one chapter in the broader process by which global capitalism homogenizes lifestyles, values, and role models around the world, turning disparate sorts of peoples into ready consumers.

The End of Colonialism

The European empires were long-lived, lasting almost half a millennium. Yet ultimately, they collapsed. In Latin America, this process began relatively early, following the Napoleonic wars, which weakened the core sufficiently to allow the periphery to break away. In Africa and Asia, the end of colonialism came much later, following World Wars I and II, which, similar to the global geopolitical situation of the early 19th century, saw the European powers self-destruct.

The international environment following World War II provided an ideal opening for various nationalist and independence movements in the Arab world, Africa, India, and southeast Asia. Sometimes the communists were involved. The Japanese occupation of Indochina, the Philippines, and Indonesia destroyed the myth of European invincibility. Often, independence movements were led by intellectuals educated in the West, such as Ghana's Kwame Nkrumah, Vietnam's Ho Chi Minh, and India's Mohandas Gandhi. Moreover, the Cold War rivalry between the United States and the Soviet Union allowed political leaders in the developing world to play the superpowers off each other.

Independence movements succeeded, sometimes peacefully, as in India, and often violently, as exemplified by the Vietnamese and the Algerian defeat of the French in their respective countries. As a result of this process of decolonization, the number of independent states multiplied rapidly in the 1950s, 1960s, and 1970s. Today, there are very few official colonies remaining, although as the theory of neocolonialism suggests, colonialism survives in practice if not in name.

Barney Warf

See also Columbus, Christopher; Decolonization; Dependency Theory; Diamond, Jared; Exploration; Gama, Vasco da; Imperialism; Magellan, Ferdinand; Orientalism; World-Systems Theory

Further Readings

Blaut, J. (1993). *The colonizer's model of the world: Geographical diffusionism and Eurocentric history*. New York: Guilford Press.

Boorstin, D. (1983). *The discoverers*. New York: Random House.

Chirot, D. (1985). The rise of the West. *American Sociological Review, 50,* 181–195.

Diamond, J. (1999). *Guns, germs, and steel*. New York: W. W. Norton.

Frank, A. (1998). *ReOrient: Global economy in the Asian age*. Berkeley: University of California Press.

Hugill, P. (1993). *World trade since 1431: Geography, technology and capitalism*. Baltimore: Johns Hopkins University Press.

Marks, R. (2007). *The origins of the modern world: A global and ecological narrative from the fifteenth to the twenty-first century* (2nd ed.). Lanham, MD: Rowman & Littlefield.

Mignolo, W. (1995). *The darker side of the Renaissance: Literacy, territoriality, and colonization*. Ann Arbor: University of Michigan Press.

Modelski, G., & Thompson, W. (1988). *Seapower in global politics, 1494–1993*. Seattle: University of Washington Press.

Pomeranz, K., & Topik, S. (1999). *The world that trade created: Society, culture, and the world economy, 1400–the present*. Armonk, NY: M. E. Sharpe.

Said, E. (1978). *Orientalism*. New York: Vintage Books.

Wallerstein, I. (1979). *The capitalist world-economy*. Cambridge, UK: Cambridge University Press.

 # COLOR IN MAP DESIGN

One of the principal methods of communicating geographic information is visually on a map. For both two-dimensional paper maps and digital maps displayed on a computer screen, color is often an essential component. The use of color in map design is a method of symbolization where hues, sometimes with varying amounts of saturation and intensity, are employed to symbolize features and differentiate between values for quantitative and qualitative spatial data. A hue is a distinct wavelength of visible light. The purest hues include blue, green, and red. Saturation refers to the amount of color, or hue, displayed on a surface area, and the intensity of color is associated with the brightness of hues. When applied properly, the use of color in maps enables more effective communication of spatial information. Used improperly, color will not only decrease the effectiveness of communication but can also lead to false interpretations of the information. An awareness of the intricacies of color use in map design and the incorporation of common cartographic conventions developed to address some of those intricacies will promote more effective communication of spatial information in maps.

Color can be used in both raster and vector data to represent a range of values within cells or to give more meaning to point, line, and area features. There are many considerations when using color in map design that collectively make the process of choosing appropriate colors intricate. For example, color provides an additional amount of contrast to map symbols that can add weight to the meaning and/or values of features on a map. In addition, certain colors have connotative and associative meanings. Moreover, interpretation of the meaning may vary for different map users. Thus, a brief discussion of these and other considerations is in order. Some considerations include color connotation and psychological reactions, color association, color interactions, and light source.

Different colors and variations in color intensity may have a variety of connotations. For example, when viewing a map of voting results by county within California for a proposition, counties could be symbolized with color to represent those that voted in favor of the proposition versus those that voted against. If we chose red to symbolize those in favor and blue to symbolize those that voted against, both colors are the same level of purity and the same intensity and saturation. Consequently, one color does not stand out over another. Thus, these colors provide equal weight and importance to each result. In addition, the colors are clearly distinct. A connotation that might result from the use of red and blue is in the relation of these colors to the Republican and Democratic parties, respectively. The use of blue for those counties that voted in favor might imply that the decision is also supported by the Democratic Party. If,

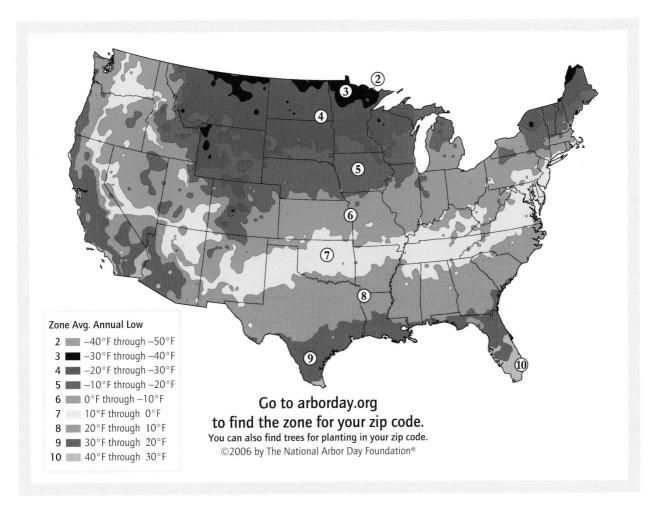

Figure 1 Map of hardiness zones in the United States

Source: © 2006 The National Arbor Day Foundation.

however, we chose red for those in favor and yellow, a less pure color resulting from the combination of red and green, for those against, the red counties would stand out more on the map and potentially harbor a connotation that the decision to vote in favor of the proposition is more important than the decision to vote against. When a hue is used with different saturation levels and or intensities to create a range of color for symbolizing features on a map, lower saturation and intensity levels might be interpreted as having less value than those categories that are represented with higher saturation and intensity levels. This might provide clarity when representing quantitative data in a map of precipitation. However, when used with qualitative data, such as a map of dominant ethnicity, the use of those same colors might imply that there is an order to the data that does not exist.

The use of color can evoke psychological reactions by map users. Blue and green are often described as being cool or mellow colors, while red and yellow are warm colors. Feelings are linked with color as well. For example, red can be associated with aggression. Consequently, when choosing color in map design, consideration should be given to potential psychological reactions to color.

Through the observation of color in the real world, humans develop expectations about colors in everyday life and how those same colors are used to represent certain features on a map. For example, the color blue is associated with water and green is associated with vegetation. Using magenta to symbolize an ocean can create confusion for the map reader. Consideration should be given to associations when choosing colors. In Figure 1, blue shades are used for colder climates and reds for warmer ones. A temperature map that symbolizes warmer areas with blues and cooler areas with reds would be confusing to the map user and require more scrutiny of the map legend.

The interactions between colors are important considerations as well. For example, the appearance of one color may change when displayed on different backgrounds. As a result, a single color defined in a map legend may actually appear as two different colors within the map when displayed with different background colors. In addition, the ability of a map user to differentiate between similar colors improves when the colors are displayed side by side but decreases when the colors are spread out throughout a map and are no longer adjacent.

The intended use of the map and the subsequent light source for the map user should also be considered when choosing colors. The appearance of colors will vary significantly when viewed on different computer monitors, in direct sunlight, in incandescent light, and in fluorescent light.

Common cartographic conventions have been established regarding the use of color in map design. In general, hue is used to symbolize different categories of data, while saturation and intensity are employed to represent numerical values in data. In addition, the use of certain colors to represent common map features has been established. For example, contour lines are depicted in brown, water features are cyan, roads are red, human-made features such as buildings and borders are black, and vegetation is represented with green. Maintaining a balance of color on a map is also important. When symbols vary significantly in areal size on a map, larger areas are typically symbolized with lighter colors, while smaller areas are symbolized with darker colors to achieve more visual balance.

Judd Michael Curran

See also Cartography; Map Design; Map Visualization

Further Readings

Brewer, C. A. (2005). *Designing better maps: A guide for GIS users.* Redlands, CA: ESRI Press.

Garo, L. A. B. (1998). *Color theory.* Retrieved March 14, 2009, from http://personal.uncc.edu/lagaro/cwg/color/index.html

Krygier, J., & Wood, D. (2005). *Making maps: A visual guide to map design for GIS.* New York: Guilford Press.

Robinson, A. H., Morrison, J. L., Muehrcke, P. C., Kimerling, A. J., & Guptill, S. C. (1995). *Elements of cartography* (6th ed.). Hoboken, NJ: Wiley.

COLUMBUS, CHRISTOPHER (CA. 1451–1506)

Christopher Columbus (or Colón in Spanish) is widely held to be the first European to "discover" the New World of the Americas, which was a major turning point in world history. In 1484, while still in Portugal, Columbus's challenge was not about the shape of the Earth but about the distance needed to travel to the Indies and the Spice Islands. Columbus's miscalculations came from his reading of Ptolemy, who declared each degree of the Earth to be 50 mi. (miles) and the total circumference to be 18,000 mi. In addition to classical sources, Columbus read and annotated *The Travels of Marco Polo* (1298), *The Travels of Sir John Mandeville* (1371), Pius II's *Historia Rerum Ubique Gestarum* (1477), Pierre d'Ailly's *Imago Mundi* (1480–1483), and the Bible, among many other books preserved in the Biblioteca Colombina by Ferdinand Columbus. These texts confirmed the existence of the Antipodes, suggested the possibility of arriving in the East by traveling west, and provided him with myths and fables as he explored unknown geographies.

Genoa, with its flourishing maritime environment, is considered Columbus's birthplace and the port city where he spent his first 21 years. In this context, it is easy to understand his profound interest in cartography and navigation. Genoa was at the time a maritime commercial center where some of the best mapmakers of portolano charts of the Mediterranean resided. Columbus arrived at the Portuguese coast in 1576 and joined his younger brother Bartholomew in Lisbon. There they made and sold navigation charts and met and interviewed mariners who had new information to update old maps and charts. These were the circumstances that supported the story of the *piloto desconocido*, the unknown pilot who arrived from an unknown hemisphere and returned sick to Lisbon, dying at Columbus's home. This story was later used as evidence in the important legal battles between Columbus's heirs and the Spanish Crown and undermined Columbus's own claims of conquest and promises contracted under the Capitulaciones de Santa Fe (April 17, 1492).

As with many of Columbus's actions, his intellectual contributions to physical and human geography are the subject of many ethical, legal, and historical debates that were initiated after Europe recognized the significance of his journeys and explorations. The same questions still permeate the historiography on Columbus: Was he erudite enough? Were his actions the product of his own genius, or did he simply use the knowledge and resources provided by others? And for those who opposed the Spanish treatment of the Amerindian populations, what role did he play in the expansion of slavery and colonialism? A whole series of early colonial chroniclers writing about Columbus came to the rescue of his historical legacy; among them were his own son, Ferdinand Columbus; the Dominican advocate of Amerindian rights, Bartolomé de las Casas; and other well-known historians such as Pietro Martire d'Anghiera, Gonzalo Fernández de Oviedo, and Hernán Pérez de Oliva. Their accounts reveal their own ideological positions on Spanish geopolitical aims of domination and provided evidence of Columbus's life and writings that still help readers understand the political, religious, and cultural milieu of the Columbian experience and the first years of European colonization.

Both Bartolomé de las Casas and Ferdinand Columbus protested the 1507 map by Martin Waldseemüller, the first portrait of the lands explored by Columbus that depicted a continent separated from Asia; the new land was named "America" after Amerigo Vespuci. These biographers and many others who favored Columbus's heirs in the legal claims they had against the Spanish court emphasized the extensive reconnaissance undertaken by Columbus at the service of Spain. The Caribbean, South America, and parts of Central America (e.g., Honduras, Costa Rica, Panama) were all recognized in his journeys without knowing that these regions belonged to a separate continent.

A sketch of the northwest coast of Hispaniola is the first illustration of that island attributed to Columbus that registered the Americas in European consciousness. It was drawn in 1492 as he surveyed the region believed to be Cipango (Japan); Columbus, experienced in cartography, used his drawings as instruments to account for his experience and name regions he encountered on behalf of the Spanish Crown. In his texts, the need to map the regions explored is crucial; not only did he reassure the Crown that his letters would be accompanied by his own maps and charts, but he also included textual geographical descriptions.

With Christian cosmology and medieval *mappamundis* in mind, the speculation that Columbus believed that he had found Paradise is crucial to understanding how his sense of geography was informed by religion. After encountering volumes of freshwater gushing into the ocean and the different currents of the Orinoco River, in his account of the third journey, Columbus placed himself in the middle of the four rivers of Paradise: the Nile, Euphrates, Tigris, and Ganges. The shape of the world is depicted using the metaphor of a woman's breast, like a pear, with Paradise located on the nipple. The search for Paradise, the conquest of Jerusalem, and the conversion of indigenous populations to Christianity became some of Columbus's last pleas to the imperial court. The lands, free labor, gold, and wealth of the New World had become secondary to an enterprise that more importantly represented the triumph and supremacy of Catholic Spain over the rest of Europe.

Santa Arias

See also Cartography, History of; Exploration; Human Geography, History of; Magellan, Ferdinand

Further Readings

Columbus, F. (1992). *The life of the Admiral Christopher Columbus by his son Ferdinand* (B. Keen, Ed.). New Brunswick, NJ: Rutgers University Press.

Gómez, N. (2008). *The tropics of empire: Why Columbus sailed south to the Indies*. Cambridge: MIT Press.

Harley, J. (1990). *Maps and the Columbian encounter*. Milwaukee: University of Wisconsin Golda Meir Library.

Morison, S. (1942). *Admiral of the ocean sea: On the life of Christopher Columbus* (2 vols.). Boston: Little, Brown.

Zamora, M. (1993). *Reading Columbus*. Berkeley: University of California Press.

COMMODITY CHAINS

A commodity chain encompasses a set of interrelated activities associated with the production of one good or service such as manufacturing, consumption, design, retailing, marketing, and advertising. In commodity chain analyses, the economy is conceptualized as an assemblage of individual chains, each possessing a unique logic, organization, and spatiality. Using the lens of the commodity chain, geographers have explored the dynamics of production and consumption in a range of commodities, including fashion, furniture, footwear, gold, diamonds, fruits, vegetables, coffee, chocolate, and cut flowers. The focus is on the vertical relationships between buyers and suppliers of a particular product, tracing the flow of material resources, value, finance, and knowledge—as well as signs and symbols—between sites along a chain. Processes of coordination and competition among actors operating at the same node or function are given less weight. This tradition thus marks a break with horizontal approaches to the economy, which focus on general activities such as retailing or on one stage in the life of a commodity such as manufacturing.

The notion of a commodity chain is related to a number of similar concepts that also involve tracing connections between interrelated sites, such as filieres, global value chains, systems of provision, circuits, and actor-networks. The growing attractiveness of this methodology relates to a broader "cultural turn" in human geography, which aims to bring together economy and culture, production and consumption, and the material and the symbolic. The interest in commodity chains can also be traced to a growing political concern among consumers with the origins, quality, and ethics of products and with the vast distances and inequalities that often separate producers and consumers. A key problem is that the images and meanings of consumer goods often contain few traces of the production processes and environments that created them. The aim of a commodity chain analysis is to uncover these connections.

The Global Commodity Chain

The work of Gary Gereffi is a key reference point in the commodity chain literature. Gereffi uses the term *global commodity chain* (GCC) to describe the complexity of interfirm relationships within globally dispersed production networks. According to Gereffi, commodity chains have three main dimensions. The first is an input-output structure, encompassing a group of products and services linked together in a sequence of value-adding activities. Second, individual chains possess a distinct territoriality, a pattern of geographical diffusion or concentration of raw materials, production, export, and marketing activities. Finally, commodity chains are characterized by a particular governance structure.

Governance Structures

Of these attributes, governance structures have received the greatest attention in the literature. Governance structures refer to relationships of power and authority that determine how financial, material, and human resources, as well as economic surpluses, are allocated within the commodity chain. Governance defines the functional division of labor along a chain, as well as delineating the terms of chain membership and exclusion.

As one example of governance, the state is a key actor in mediating the dynamics of commodity chains. Through labor legislation, training, education, and foreign investment programs, governments play a critical role in regulating chain relationships. In addition to the state, a range of other institutions govern chains such as trade organizations and unions.

While external actors are important, the literature on commodity chains has tended to foreground internal governance mechanisms and, in particular, the nature of interfirm linkages along a chain. Two types of governance are commonly associated with commodity chains. The first includes producer-driven chains that are typical of capital-intensive industries such as aerospace, computers, and automobiles. In these chains, transnational producers, typically in core countries, play a central role in coordinating production networks, including backward and forward linkages.

The second type of governance structure is represented by buyer-driven commodity chains, where large retailers and brand-name marketers play the dominant role in establishing decentralized production networks in a variety of exporting countries. These dynamics are common in labor-intensive consumer goods industries such as garments, footwear, toys, and electronics. Production is generally executed by tiered networks of contractors that make finished goods according to the design specifications of retailers or marketers. In buyer-driven chains, profits derive from innovations in design, sales, and branding rather than from advantages in scale, volume, or technology.

Many commentators chart a shift in recent years from producer-driven to buyer-driven chains in key sectors of the economy. One example is found in the case of fresh fruit and vegetable chains, where supermarkets exercise heightened control in determining what products will be offered to consumers and when, as well as dictating the quality, appearance, and packaging of produce. To achieve this control, large grocery chains have been shifting from sourcing products through wholesale markets to establishing their own tightly managed supply chains.

While these ideal types constitute a useful heuristic device, it is also important to recognize that different types of governance may coexist within a sector and that governance structures change over time and across space. Chains may not be exclusively producer or buyer driven, and different forms and degrees of power may accrue at different sites in the chain, not just those assigned "key agency."

Revised Typology of Governance

Empirical research has revealed tremendous variation and complexity in models of governance. Moving beyond this binary formulation, Gereffi and his colleagues have recently elaborated a more nuanced portrait of commodity chain governance based on three factors: (1) the complexity of information involved in a transaction, (2) the extent to which knowledge can be codified (communicated or traded), and (3) the capabilities of suppliers with regard to the requirements of the transaction. Based on these criteria, they offer a revised typology identifying five types of governance.

The first form of governance is markets. This situation applies to transactions where the complexity of information exchanged is low and knowledge is easily codified. Product specifications are typically simple, requiring little input from buyers. Under these conditions, transactions can be governed with little coordination. The second variation in governance is represented by modular value chains. Here, technical standards simplify the process of interaction by creating uniform component, product, and process specifications. Suppliers often have the ability to supply full modules, which internalizes knowledge and lowers the need for direct monitoring of suppliers. As with simple market exchange, codification enhances the ability to exchange information with little coordination, and the cost of switching partners is low. The third model of commodity chain governance is found in relational value chains. In these chains, transactions are more complex, and product specifications and knowledge cannot be codified. Buyers and suppliers exchange tacit knowledge (knowledge based on shared norms and conventions, which cannot be easily communicated). Because suppliers have high competencies, relations of mutual dependence develop, which are regulated through reputation, as well as social and spatial proximity. The cost of shifting to new partners is high. A fourth form of governance is associated with captive value chains where the complexity of product specification is high, as is the ability to codify information. However, supplier capabilities are low, leading to a high degree of intervention and control by the lead firm. Suppliers are often engaged in a narrow range of tasks, such as basic assembly. Under these conditions, suppliers face considerable switching costs and are therefore "captive." A final type of governance is hierarchy. This form occurs when products are complex and specifications cannot be codified. If capable suppliers cannot be found, then lead firms will tend to manufacture products in-house.

Each type of governance involves weighing the costs and benefits of outsourcing, and entails different levels of coordination and power asymmetry between buyers and suppliers, ranging from low in the case of markets to high in the case of hierarchy.

A prominent area of enquiry is how chain governance affects the possibilities for industrial upgrading among firms in developing countries. Upgrading can take place in a number of capacities, including possibilities for upgrading processes through reorganizing production systems or implementing new technology, upgrading products by shifting into the production of more sophisticated lines, or upgrading functions by expanding into new areas such as research, design, or marketing. A central issue surrounds the role of buyers in transferring knowledge to their suppliers. Researchers have found that chains organized according to more hierarchical relations may facilitate product and process upgrading but are unlikely to enable firms to expand into design or marketing. These patterns elucidate how chain governance influences whether and how firms in developing countries can realize gains from their participation in global commodity chains.

Alternatives to the GCC Approach

While the GCC approach has proven useful in understanding the unique power dynamics and inequalities associated with individual commodity chains, authors such as Ian Cooke, Philip Crang, and Peter Jackson call for a conception of commodities using a less linear logic of "circuits" and "networks." Rather than searching for a singular or determining logic of a chain and tracing relationships of causality, these accounts unsettle the prioritization of any one moment in the commodity's circulation. The aim is to furnish a more

contextual analysis of the multiple meanings attached to commodities at different times, places, and phases. These accounts consider how goods are displaced from one site to another and the way knowledges animate and are animated by the process of displacement. The notion of circuits allows for dense webs of interdependence between sites.

Extending these ideas, recent accounts also draw on actor-network theory to explore the multistranded relationships between firms, states, and organizations in individual commodity networks. In actor-network theory, power is theorized as a composite of the actions; agency is an effect generated by a network of heterogeneous, interacting materials, including not only actors located at various sites in the chain but also technologies, natures, and the commodity itself. Agency is granted to human and nonhuman actors in the network. Nodes do not exist as concrete objects per se but are constituted in the networks of which they are a part.

Approaches informed by the literature on circuits of culture and actor-network theory extend commodity chain analysis in fruitful directions, furnishing a dynamic and fluid understanding of power along a commodity chain. These approaches also incorporate processes of circulation and feedback between a wide variety of actors.

Applications of Commodity Chain Analysis

Regardless of the metaphors and concepts used to understand the linkages between sites, a commodity chain analysis provides a helpful framework for understanding some of the central characteristics and geographies of contemporary capitalism. Labor organizations are increasingly using the tool of commodity chain analysis to map the connections between workers around the world as well as between workers and consumers at a variety of sites and scales. Unions have begun organizing across commodity chains and developing transnationally networked forms of mobilization (such as alliances with other unions and ethical trade initiatives). The commodity chain approach not only holds analytical appeal, but it also affords the political potential to challenge relations of exploitation, inequality, and injustice across space. This is not to deny that there are problems associated with ethical consumption initiatives and other strategies designed to mobilize actors along the chain. Acting politically through commodities is often associated with contradictions and ambivalences. However, commodity chain analysis provides us with an interesting approach to understanding not only the dystopias of modern commodity systems but also the twists and turns of ethical consumption practices that may further processes of marginalization and uneven development. In following commodity narratives and engaging with these political productivities, there is hope though for questioning and ultimately transforming the nature of commodity production and consumption.

Deborah Leslie

See also Actor-Network Theory; Consumption, Geographies of; Cultural Turn; Economic Geography; Fair Trade and Environmental Certification; Food, Geography of

Further Readings

Cook, I., & Crang, P. (1996). The world on a plate: Culinary culture, displacement and geographical knowledges. *Journal of Material Culture, 1,* 131–153.

Fine, B. (2002). *The world of consumption: The material and cultural revisited.* London: Routledge.

Gereffi, G., Humphrey, J., & Sturgeon, T. (2005). The governance of global value chains. *Review of International Political Economy, 12,* 78–104.

Gereffi, G., Korzeniewicz, M., & Korzeniewicz, R. (Eds.). (1994). *Commodity chains and global capitalism.* Westport, CT: Greenwood Press.

Hartwick, E. (1998). Geographies of consumption: A commodity-chain approach. *Environment and Planning D: Society and Space, 16,* 423–437.

Hughes, A., & Reimer, S. (Eds.). (2004). *Geographies of commodity chains.* London: Routledge.

Jackson, P. (2002). Commercial cultures: Transcending the cultural and the economic. *Progress in Human Geography, 26,* 3–18.

Leslie, D., & Reimer, S. (1999). Spatializing commodity chains. *Progress in Human Geography, 23,* 410–420.

COMMON POOL RESOURCES

Common pool resources (CPRs) are resources for which the exclusion of prospective users is difficult and whose utilization by one person affects the availability or quality of the resource for others (a property known in the literature as subtractability). Examples of CPRs include pastures, fishing grounds, irrigation systems, groundwater basins, forests, atmosphere, wildlife, and, more recently, biodiversity and the "digital commons." CPRs contrast with private goods, where exclusion is feasible, and public goods, which do not exhibit subtractability. The management challenge of CPRs derives from the resultant user incentives. The benefits of exploitation tend to accrue to individual users, but the costs of use, in terms of a reduced resource base or higher extraction costs, are spread among all users. At the same time, the cost of investment in the resource (e.g., groundwater recharge structures) is with the individual investors, but the benefits can often be garnered by all users, creating a "free rider" problem. The dual result is often assumed to be an overexploited and degraded system.

This assumption led to the formulation of Garrett Hardin's famous but now somewhat discredited "Tragedy of the Commons" thesis. Hardin put forward the view that rational users of the commons would continue making demands on the resource as long as the expected benefits of their action equaled or exceeded the expected costs. Moreover, since the self-interested individual user ignores the costs imposed on others, rational individual decisions would, in the long run, lead to destruction of the commons. However, later work, in particular by Elinor Ostrom and her associates, showed that such pessimistic outcomes were not universal. The large consensus that emerged over the past two decades is that the tragedy of the commons that Hardin referred to was indeed the tragedy of open-access resources (*res nullius,* or no property) and not that of common property resources (*res communes*). The confusion, it became apparent, was that CPRs could be either ungoverned under an open-access regime—this is what Hardin referred to as tragedy of the commons—or managed as a private or government property, or as common property.

Geography and Common Pool Resources

At a basic level, the use of CPRs revolves around humans, their environment, and the spatial interaction between the two—the core subject matter of human geography. Geographers' work in the field has tended to focus on applied issues of nature-society relations within common property or open-access resource regimes rather than on theoretical aspects of the commons per se. For example, Gordon Matzke and Nontokozo Nabane examined the outcomes of communally owned forests and wildlife in Zimbabwe. Toward the other scalar extreme, Itay Fischhendler and Eran Feitelson studied the spatial dimensions of transboundary rivers in explaining water management regimes between the United States and Mexico. A notable exception is Paul Robbins, who used empirical studies of cattle grazing in Western India to challenge assumptions about commons "ownership" (e.g., private, state, and open access) and resource outcomes. Another exception is Mark Giordano, who attempted to develop a spatially explicit framework for considering differential commons outcomes and management responses. More recently, Lisa Campbell used this framework to understand sea turtle conservation policies in Costa Rica.

Almost all scholars working on the commons have acknowledged the role of scale and space in managing CPRs. For example, nested institutional arrangements that operate at different sociopolitical scales have been often cited as one of the preconditions of successful governance of the commons. The issue of scale is especially useful because in recent times many governance issues of the commons are at a global scale (e.g., climate change, loss of biodiversity), while much of the accrued knowledge on governance of the commons stems from small-scale systems and institutions (e.g., traditional irrigation systems, local fishing grounds). Hence, an appreciation of scale is important for transposing lessons from local cases to global ones. From a geographical perspective, equally, if not more, important is the understanding of spatial issues as they relate to the use of CPRs.

Yet, curiously, geography as a discipline has not deployed geographical concepts of scale and space in developing a theoretical understanding of CPRs, despite the considerable empirical work referred to above.

A Geographical Typology of Common Pool Resources

Some geographers have characterized the commons as a spatial relationship between the resource and the resource users. Based on spatial interaction, Giordano has developed an explicitly geographical typology of CPRs by classifying them into one of three categories—open access, fugitive, and migratory. Figure 1 illustrates the spatial aspects of private, open-access, fugitive, and migratory resources.

An open-access resource, defined from a geographic perspective, is one in which resource boundaries and the right of several users to use the resource intersect. In such a case, while the parties involved in resource extraction enjoy the full benefits of resource use, they share only a part of the cost. In the absence of any property rights regime (be it private, public, or common), such spatial interlinkage between the resource and the user leads to overexploitation of the resource. Overgrazing in a pasture that is totally unowned is an example of an open-access resource problem.

A fugitive resource is one where although the rights domain of two users, say A and B, are clearly demarcated (e.g., say Countries A and B) but the resource itself travels in a unidirectional way (say a river that flows from Country A to Country B). Here too, the threat of overexploitation exists, since at least one of the users can gain the full benefit of resource use, without paying the full cost of resource exploitation. An example of a fugitive resource problem is that of pollution in a transboundary river. An upstream country can pollute the river and get benefits from it, but it does not have to share the cost of pollution that the downstream country faces in consequence.

A migratory resource is similar to fugitive resource in that user right domains are separate, but it is different from fugitive resource in that the movement of the resource is bidirectional. An example is that of wildlife, birds, and fish that migrate from one domain to the other and back. Here again, like open-access resources, but unlike

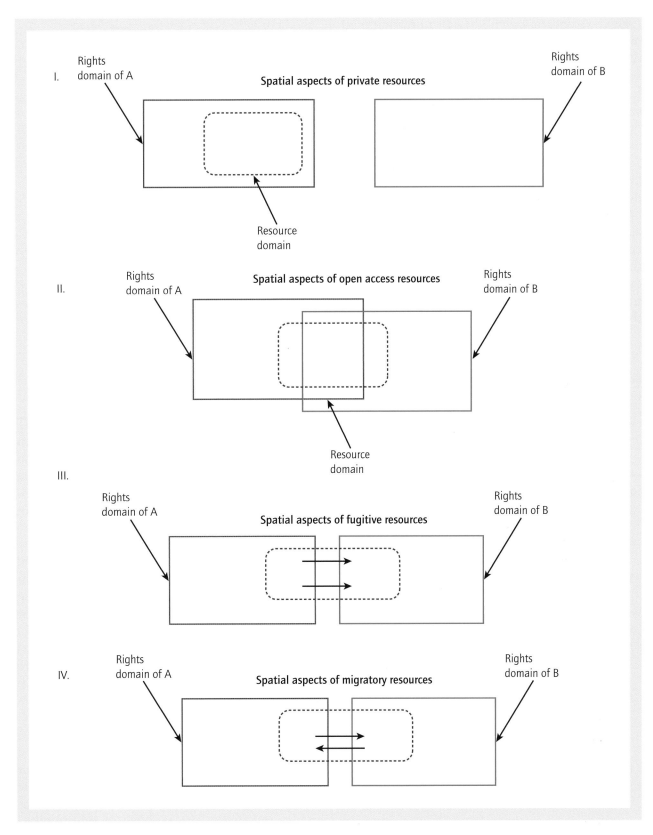

Figure 1   Spatial adjustment as a mechanism for resolving river basin conflicts: the U.S. versus Mexico case

Source: Giordano, M. (2003). The geography of the commons: The role of scale and space. *Annals of Association of American Geographers, 93*(2), 320.

fugitive resources, the full benefit of exploitation may accrue to the party using the resource, but the costs of resource exploitation have to be shared by both the parties. However, resource users' perception regarding optimal rates of resource use might vary significantly between open-access and migratory resources, leading to very distinct outcomes.

This threefold classification of CPRs from a geographical perspective of space emphasizes the point that resource governance regimes ought to be different depending on the spatial interlinkages between the resource users and the resource endowments. Therefore, the spatial interaction of the resource and the resource users has to be explicitly understood within a geographical framework of space and scale in order to effectively govern the commons.

Aditi Mukherji and Mark Giordano

See also Common Property Resource Management; Commons, Tragedy of the; Community-Based Conservation; Externalities

Further Readings

Campbell, L. M. (2007). Local conservation practice and global discourse: A political ecology of sea turtle conservation. *Annals of the Association of American Geographers, 97,* 313–334.

Dietz, T., Ostrom, E., & Stern, P. (2003). The struggle to govern the commons. *Science, 302,* 1907–1912.

Fischhendler, I., & Feitelson, E. (2003). Spatial adjustment as a mechanism for resolving river basin conflicts: The US-Mexico case. *Political Geography, 22,* 557–583.

Giordano, M. (2003). The geography of the commons: The role of space and scale. *Annals of Association of American Geographers, 93,* 365–375.

Hardin, G. J. (1968). The tragedy of the commons. *Science, 162,* 124–142.

Matzke, G. E., & Nabane, N. (1996). Outcomes of a community controlled wildlife utilization program in a Zambezi Valley community. *Human Ecology, 24,* 65–85.

Ostrom, E. (1990). *Governing the commons: The evolution of institutions for collective action.* Cambridge, UK: Cambridge University Press.

Ostrom, E., Burger, J., Field, C. B., Norgaard, R. B., & Policansky, D. (1999). Revisiting the commons: Local lessons and global challenges. *Science, 284,* 278–282.

Ostrom, E., Gardner, R., & Walker, J. (1994). *Rules, games and common pool resources.* Ann Arbor: Michigan University Press.

Robbins, P. (1998). Authority and environment: Institutional landscapes in Rajasthan, India. *Annals of the Association of American Geographers, 88,* 410–435.

COMMON PROPERTY RESOURCE MANAGEMENT

Common property resource management regimes are defined as a set of institutional arrangements that regulate the use of common pool resources (CPRs, also called the commons). The institutional arrangement or property rights governing CPRs could be no property (open access), private property, government property, and common or communal property. Owing to the historical development in the field of knowledge on the commons, there is much confusion between the nature or characteristics of a CPR and the management mechanisms forged to address the problems of the commons. CPRs are those natural and human-constructed resources where it is difficult to exclude potential users and the benefits derived from the resource are subtractable. These are called the principles of difficulty of exclusion and subtractability. Examples of CPRs are irrigation systems, groundwater basins, fishing grounds, pastures, forests, the atmosphere, wildlife, biodiversity, digital commons, and so on. The problem of subtractability makes CPRs particularly vulnerable to congestion, overuse, and degradation, while the difficulty of exclusion creates a "free rider" problem (where some users consume more than their fair share of a resource), which in turn threatens the long-term sustainable use of the resources.

In his hugely influential, but now controversial, 1968 thesis of "The Tragedy of the Commons," Garrett Hardin argued that because of the unique characteristic of the commons, the users of the commons were trapped in a behavioral situation that ultimately led to the destruction of the very resources they depended on. To avoid the tragedy from playing out, Hardin suggested that the commons should be either privatized or managed by the government in which right to entry and use could be controlled. Hardin's thesis proved to be one of the most influential works in the environmental field and has had a huge sway over policy making in the field of natural resources management. However, later scholars such as Elinor Ostrom (among others) found that Hardin's simple policy prescription that the tragedy of the commons could only be avoided by privatizing the resource or putting it under government control was erroneous. This was because it disregarded a range of cases where CPRs were governed effectively under common or communal property regimes and where the predicted tragedy did not occur. The large consensus that emerged over the past two decades is that the tragedy of the commons that Hardin referred to was indeed the tragedy of open access (res *nullius*, or no property) and not that of common property (res *communes*). The confusion, it became apparent, was that CPRs could either remain ungoverned under an open-access regime—this is what Hardin referred to as "Tragedy of the Commons," or be managed as private property, government property, or common property and that the absence of a property regime was often confused with a common property regime.

Types of Property Rights Systems for Governing CPRs

Commons or CPRs may be held under four categories of property rights: open-access, private, government, and common/communal property. It should be noted that these are idealized analytical categories. In reality, CPRs may be governed under overlapping combinations of these four kinds of property rights regimes.

Open access is defined as the absence of property rights, where there is no restriction on the use of natural resources by anyone. Under such a case, access to the resource is open to and free for all, and it is the classic case where Hardin's tragedy of the commons plays out most often. Examples of open-access resources are the global atmosphere and climate, noncoastal ocean fisheries, and a common pasture.

Under private property, the rights to use the commons are exclusively transferred to a private entity (usually an individual or a corporation), which means that the commons are no longer subject to difficulty of exclusion. Such private property rights are usually held up in the state in the form of legal rights. Examples include privately held forests or inland fishing grounds leased to private entrepreneurs.

Under state or government ownership, the rights to the commons are exclusively vested in the state, which in turn can create the rules of the games vis-à-vis use, allocation, and management of the resource. Examples include pastures and forestlands that are vested in or controlled by the state. Unlike the other two property regimes (private and common property), which depend on the state to enforce their property rights, in case of state ownership, the state also has the coercive powers of enforcement.

Under common or communal property rights regime, the resource is held jointly by a well-defined group or community of people, so that from an outsider's perspective the common property regime has the appearance of a private good (since outsiders are excluded), while for the insider it resembles a CPR (since the problems of difficulty of exclusion and subtractability are faced by the members of the group who own the resource). Examples include indigenous irrigation systems managed by communities all over the world. Recent scholarship has found that many of our natural resources are indeed managed under a common property regime, and hence, the tragedy of the commons has not played out as frequently as one might have expected if these resources were left ungoverned. It should be noted that the most successful cases of common property regimes relate to the local commons (e.g., irrigation systems, pastures, inland fisheries) rather than the global commons (the atmosphere, air pollution, ocean fisheries, etc.).

These four property rights regimes have different implications for the two aspects of the CPRs—that is, ease of excluding potential users (exclusion) and possibility of regulating use and users (subtractability). In the case of open-access or no property rights, both exclusion of potential users and regulation of use are difficult, if not virtually impossible. In the case of private property, excluding potential users is easier; however, there are cases, such as the right to exploration of oil or groundwater, where exclusion becomes difficult even when private property rights are invoked. Under private property regimes, the possibility of controlling use and users is also much easier, while it is not so under state or government ownership due to the bureaucratic nature of such control. A wealth of literature in the past three decades has shown that under common property resource management regimes, outsiders can be effectively excluded and the behavior of insiders can be influenced by rules and regulations, so that the resources held under such regimes can be effectively governed with long-term positive feedbacks. The conditions under which common property regimes succeed have become a major area of research in recent times.

Conditions for the Success of Common Property Regimes

Since the early 1980s, a large number of social scientists, including anthropologists, economists, political scientists, human geographers, and sociologists, have contributed to the burgeoning study of common property resource management. They have found that defying the doomsday prediction of the tragedy of the commons, CPR users around the world often forge institutional arrangements and management regimes that help them use CPRs equitably for a long duration of time with no negligible equity losses. This observation has led to the next analytical question, that is, under what conditions have such successful institutional arrangements for managing the commons been forged? Much of what is known about the conditions under which common property regimes can succeed stems from local commons managed by small user groups and communities; knowledge about global commons is much more limited.

Arun Agrawal has come up with and added to a comprehensive list of critical factors that ensure sustainable use of CPRs or the commons. These enabling or critical conditions can be divided into four main categories:

1. Resource system characteristics

2. Group characteristics

3. Institutional arrangements

4. External environment

Furthermore, interaction between (1) and (2), that is, the relationship between resource and resource users' characteristics, and interaction between (1) and (3), that is, the relationship between resource characteristics and institutional arrangements, make for the fifth and sixth critical conditions for the success of common property regimes (Table 1). Within each of these broad conditions, there are several factors, and the presence or absence of each factor influences the governance of the commons, and often these interactions are poorly understood.

Conclusion

An important conclusion that can be drawn from the list of factors shown in Table 1 is that there are numerous factors that encourage or impede the successful management and sustainability of common property resources. Almost all these factors emerge from case studies of local commons managed by small communities. This set of factors, as can be seen, turns out to be quite large and, thus, is a potential barrier to the formulation of an overarching theory of common property resources and their management. Also, the lack of a general theory of common property resource management makes the governance of global commons (the atmosphere, oceans, biodiversity, etc.) much more difficult than that of local

A. Resource system characteristics
1. Small size
2. Well-defined boundaries
3. Low levels of mobility
4. Possibilities of storage of the benefits of the resource
5. Predictability

B. Group characteristics
1. Small size
2. Clearly defined boundaries
3. Shared norms
4. Past successful experiences and social capital
5. Appropriate leadership
6. Interdependence among members
7. Heterogeneity of endowments, homogeneity of identities and interests
8. Low levels of poverty

C. Institutional arrangements
1. Rules are simple and easy to understand
2. Locally devised access and management rules
3. Ease in rule enforcement
4. Graduated sanctions
5. Availability of low-cost adjudication
6. Accountability of managers to the users

D. External environment
1. Technology: low-cost exclusion technology and ease of adoption
2. Low levels of interaction with outside markets and gradual change in interaction with outside markets
3. State: governments not to undermine local authority, supportive external sanction mechanisms, appropriate external aid if needed and nested governance at different sociopolitical levels

E. Relationship between resource and group characteristics
1. Overlap between resource domain and rights domain
2. High dependence on the resource
3. Perceived fairness in allocation of benefits
4. Low levels of users demand
5. Gradual change in levels of demand

F. Relationship between resource system and institutional arrangements
1. Match extractible quantity with its regeneration capacity

Table 1 Critical factors determining success of common property management regimes

Sources: Adapted from Agrawal, A. (2001). Common property institutions and sustainable governance of resources. *World Development, 29,* 1649–1672; Wade, R. (1988). *Village republics: Economic conditions for collective action: Lessons from cooperative action in South India.* Oakland, CA: ICS Press; Ostrom, E. (1990). *Governing the commons: The evolution of institutions for collective action.* Cambridge, MA: Cambridge University Press; Baland, J. M., & Platteau, J. P. (1996). *Halting degradation of natural resources: Is there a role for rural communities?* Oxford, UK: Clarendon Press.

commons, because knowledge of global commons is more limited.

Aditi Mukherji

See also Biodiversity; Common Pool Resources; Commons, Tragedy of the; Oceans; Sustainable Fisheries

Further Readings

Agrawal, A. (2001). Common property institutions and sustainable governance of resources. *World Development, 29,* 1649–1672.

Baland, J. M., & Platteau, J. P. (2000). *Halting degradation of natural resources: Is there a role for rural communities?* New York: Oxford University Press.
Feeney, D., Berkes, F., McKay, B. J., & Acheson, J. M. (1990). The tragedy of the commons: Twenty-two years later. *Human Ecology, 18,* 1–19.
Hardin, G. (1968). The tragedy of the commons. *Science, 162,* 1243–1248.
Ostrom, E. (1990). *Governing the commons: The evolution of institutions for collective action.* Cambridge, UK: Cambridge University Press.
Ostrom, E. (1999). Coping with tragedies of the commons. *Annual Review of Political Science, 2,* 493–535.

Ostrom, E., Burger, J., Field, C. B., Norgaard, R. B., & Policansky, D. (1999). Revisiting the commons: Local lessons, global challenges. *Science, 284,* 278–282.

Wade, R. (1994). *Village republics: Economic conditions for collective action in South India.* Oakland, CA: ICS Press.

COMMONS, TRAGEDY OF THE

The tragedy of the commons is the idea that in an open natural system, the depletion of commonly held resources is inevitable. Competing individual and group interests place resource users in a trap that results in overexploitation and eventually decimation of common resources. While the idea of a "commons" was noted by many scholars, dating as far back as Aristotle, it is largely attributed to Garrett Hardin and his seminal 1968 article of the same name. His article focuses on the issue of overpopulation; however, its central impact has been its theories related to natural resource management. Given its influence on understandings of common resources and the architecture of natural resource policies, the tragedy of the commons is an important concept for geographers, particularly those working in the subfield of political ecology who explore issues surrounding natural resource use.

The Commons

The commons refers to a set of resources that are used by many people but privately owned by no one. Examples include fisheries, forests, water resources, the atmosphere, and pastures. Usage of the term *commons* dates to Medieval Europe, where it referred to the meadows that peasants were allowed to live on under the manorial system. The commons or common pool resources possess two important characteristics: (1) excludability—the resource is such that excluding users from access is either extremely costly or physically impossible—and (2) subtractability—exploitation of the resource by one user has negative consequences for the rest of the users by reducing the quantity or quality of the resource.

The Tragedy

In his article, Hardin used the analogy of herdsmen to explain how conditions of the commons will lead to the tragedy of resource decimation. A herdsman grazing livestock in open pasture will seek to maximize his gain, so he will add an additional animal to graze on the field. The animal in turn will deplete the quality and quantity of the pasture resource. The herdsman will receive all the profit from the sale of that piece of livestock. However, the cost of the depletion of the pasture will be shared by all the users, so it will only cost him a small fraction of the profits. There are no incentives for individual users to be protective of the common resource, and "rational" actors will continue to deplete resources for their own gain, ultimately leading to ruin for all.

Hardin emphasized the futility of an appeal to conscience in the management of common resources. He suggested that telling resource users to voluntarily limit their use of resources would place them in a double bind. First, they will be condemned for taking more than their share of resources. However, if they don't take their share, users will still feel condemned for standing by while other users openly exploit the commons. Resource users cannot be expected to limit use on their own because they will always be worried that "free riders" will continue to exploit resources while they refrain from taking their share.

Examples

There are many examples of the tragedy of the commons—situations in which the open access to a common resource has led to its decline. This concept can be presented as the root of many of the major environmental problems that humanity faces. Some of the most prominent examples are in the arena of natural resources.

Fisheries are a kind of commons problem, and the decline of fisheries worldwide as well as the collapse of stocks, such as the Canadian cod stock in 1992, are largely attributed to this tragedy of the commons. Forests and pastures or grazing lands are also considered common pool resources, and large-scale deforestation and overgrazing, particularly in developing countries, are other examples of the tragedy of the

Fire in the Amazon rain forest. A significant percentage of deforestation in the Amazon results from cattle ranches and soybean cultivation, while the rest mostly results from small-scale subsistence agriculture. Although each individual farmer benefits from cutting the forest, collectively their behavior destroys a commonly held resource.

Source: iStockphoto.

commons (see photo). The decline of harvested animals, including the world's species of whales or all the large land birds of New Zealand, is a historical example of this phenomenon.

Hardin also suggested that both overpopulation and pollution are examples of the tragedy of the commons. Individuals contribute to population while minimally feeling the effects of the added resource user on the system. In the case of pollution, individuals and corporations have little financial incentive to limit pollution, as they only minimally, if at all, feel the economic consequences of their pollution. This behavior leads to continued pollution of the commons. Global warming is considered one of the most recent examples of the tragedy of the commons, representing degradation of the global atmospheric commons.

The concept of the commons is not limited to the environment or natural resources. Some unconventional examples of the commons, which might be subject to Hardin's theories, include highway systems or traffic levels, government funds or subsidies such as welfare, and issues affecting public health, such as the proper use of antibiotics.

Solutions

Given the grim prospects for common resources, Hardin envisioned only two possible solutions to reverse the trend toward degradation: socialism or privatization. By socialism, he meant government ownership and control of resource use through the development of legal restrictions and incentives. With privatization, users are given ownership over a restricted part of the resource, converting it from common property to private property. To Hardin, this will lead to increased stewardship of natural resources as any cost of depletion or degradation will only be subsumed by the private users.

Influence

The metaphor of the tragedy of the commons has captured the imagination of the modern world. The concept has underlain the development of many contemporary natural resource regulatory structures. Its conceptions of common property are frequently taught in the university education system in economic, political, and environmental contexts. Hardin's article spawned a surge of research into real-world cases of common pool resources to examine some of his theories from a more empirical basis. This work by scholars from several different disciplines, including geography, ecology, political science, economics, conservation biology, and natural resource management, has led to findings that complicate some of the ideas outlined in his arguments.

Critiques

Researchers have critiqued Hardin's pessimistic understanding of human nature and the commons predicament. The tragedy of the commons suggests that the destruction of commonly held resources is inevitable; however, empirical research has identified a multitude of historical and present-day cases in which this has not happened. Hardin stated that "rational" human actors would always choose their own benefit over that of the group, leading to the inexorable decline of common resources. Commons scholars, of whom the political scientist Elinor Ostrom is the most prominent, have found many cases in which groups of users have worked together to limit exploitation. Instead of watching the decimation of important resources before their eyes, she finds that "rational" actors will work to develop systems that protect their decline. Throughout the world, there is evidence of groups developing varied and innovative solutions to successfully sustain common resources such as fisheries, forests, grazing lands, and water

resources, even without the external force of a governmental institution.

Another common critique of the tragedy of the commons is the overly narrow set of solutions that Hardin offered to counteract the tragedy. Hardin suggested that the only possible means of evading resource demise is through privatization or centralized government control. Some researchers argue that Hardin overlooked an entire category of successful commons governance—communal property. Communal property is a type of commons systems in which a community of users has excluded outsiders from access to a particular resource and developed its own extragovernmental set of norms or policies regarding use of the resource. These systems are often local and culturally based. Successful examples of communal property systems include management of hunting and fishing lands by the James Bay Amerindian community in Subarctic Canada, water-sharing plans in the irrigation regimes of Northern New Mexico, management of lobster fisheries by local fishermen in Maine, and modern fishing cooperatives in Japan. Some scholars believe that Hardin's limited understanding of commons solutions may have prevented lawmakers from considering other possibilities, including communal property, in natural resource policy.

Additionally, critics argue that Hardin's proposed solutions are not always effective at preventing resource decline. In some cases, privatization has not only prevented the conservation of resources, but it has also increased the rate of their decline. Some scholars argue that, unchecked, private users have no incentive to maintain resources if they can make a large profit from the decimation of the resources in the present. Private users can always be tempted to exploit the resource to turn it into capital, which, once invested, may grow at a faster rate than money gained by the sustainable extraction of resources ever could.

These scholars argue that socialism or large-scale governmental control of resources has not proven immune to failure either. Top-down governance of resources can degrade the authority of local, community-based forms of communal property management, which have worked to protect resources throughout history. When these local forms of resource governance lose authority, chaos and degradation of previously protected resources can result. In addition, state-controlled resources are subject to corruption, which can ultimately lead to resource decline. Corruption amid government-controlled resources has been a widespread problem, particularly in less developed countries. For example, poor state control has been attributed to the decimation of India's forests and the dire state of Russia's fisheries.

Critics have also attacked Hardin's ideas from a social justice perspective. To Hardin, the inevitable future decline of resources has created an emergency situation that supersedes questions of social justice or acceptability. Critics suggest that his ideas have justified large-scale government seizure and control of resources as well as a broadscale movement of natural resource policy toward privatization, both of which have contributed to social problems. Top-down government control of natural resources has in many cases lacked input from the very users who are most influenced by changes in resource policy. Some scholars argue that while sustaining natural resources, this type of governance may ignore factors important for the economic and cultural sustainability of communities who use the resources.

Resource privatization schemes include the distribution of public lands (e.g., grazing and forested lands) to private users and quota systems in fisheries that award a certain percentage of fish stocks to particular users—once awarded, these quotas can be bought and sold on the market. These privatization efforts have led to questions of equity because a chosen few users are awarded access to resources, gaining a windfall of economic benefit, while others remain unfairly excluded. In addition, privatization schemes create generational inequity where the first generation of users is freely granted access to a resource while subsequent generations must pay prohibitively high amounts to access the resource. Critics argue that privatization schemes have often favored large industries, which have significant capital, leading to the consolidation of resources with large corporations, to the exclusion of local, community-based users.

Current Research

Since Hardin's article appeared in 1968, thinking about the commons has been significantly expanded and refined. Recent scholars argue that instead of Hardin's prescriptive solutions of privatization or socialism, the keys to successful commons governance are more contingent and context dependent. However, most commons scholars agree that without some kind of system or intervention, common pool resources are subject to large-scale degradation.

Geographers have contributed to commons research in many ways, examining the biological, social, political, and economic aspects of various common pool resource management scenarios. Current commons research focuses on examining both successful and failed cases of common pool resource management to learn what types of institutions promote the sustainability of common resources. Scholars continue to develop theories regarding the qualities that are necessary for sustaining common pool resources in a variety of contexts.

Laurie S. Richmond

See also Common Pool Resources; Common Property Resource Management; Environmental Management; Political Ecology

Further Readings

Agrawal, A. (2003). Sustainable governance of common-pool resources: Context, methods, and politics. *Annual Review of Anthropology, 32,* 243–262.

Baden, J. A., & Noonan, D. S. (Eds.). (1998). *Managing the common* (2nd ed.). Bloomington: Indiana University Press.

Feeny, D., Berkes, F., McCay, B. J., & Acheson, J. M. (1990). The tragedy of the commons twenty-two years later. *Human Ecology, 18,* 1–19.

Giordano, M. (2003). The geography of the commons: The role of scale and space. *Annals of the Association of American Geographers, 93,* 365–375.

Hardin, G. (1968). Tragedy of the commons. *Science, 62,* 1243–1248.

Ostrom, E. (1990). *Governing the commons.* Cambridge, UK: Cambridge University Press.

Ostrom, E., Burger, J., Field, C. B., Norgaard, R. B., & Policansky, D. (1999). Revisiting the commons: Local lessons, global challenges. *Science, 284,* 278–282.